AF794291

Applied Nanoindentation in Advanced Materials

Applied Nanoindentation in Advanced Materials

Edited by

Atul Tiwari
University of Hawaii at Manoa
Honolulu, USA
and
Flora Coatings, Center for Entrepreneurial Innovations,
Phoenix, AZ, USA

Sridhar Natarajan
Chief Medical Examiner/Director, Lubbock County Medical
Examiner's Office, Lubbock, Texas, USA
and
Colonel, Medical Corp, United States Army Reserves
(Retired) and former United Stated Navy Nuclear
Submarine Officer Gold Dolphin Insignia

WILEY

This edition first published 2017

Registered Offices
John Wiley & Sons, Inc., 111 River Street, Hoboken, NJ 07030, USA
John Wiley & Sons Ltd, The Atrium, Southern Gate, Chichester, West Sussex, PO19 8SQ, UK

Editorial Office
9600 Garsington Road, Oxford, OX4 2DQ, UK

For details of our global editorial offices, customer services, and more information about Wiley products visit us at www.wiley.com.

Wiley also publishes its books in a variety of electronic formats and by print-on-demand. Some content that appears in standard print versions of this book may not be available in other formats.

Library of Congress Cataloging-in-Publication Data

Names: Tiwari, Atul, editor, Natarajan, Sridhar, editor.
Title: Applied nanoindentation in advanced materials / edited by Atul Tiwari, Sridhar Natarajan
Description: Chichester, UK ; Hoboken, NJ : John Wiley & Sons, 2017. |
Includes bibliographical references and index.
Identifiers: LCCN 2016054038 (print) | LCCN 2017006539 (ebook) | ISBN 9781119084495 (cloth) | ISBN 9781119084525 (pdf) | ISBN 9781119084518 (epub)
Subjects: LCSH: Materials–Testing. | Nanotechnology. | Nanostructured materials–Formability.
Classification: LCC TA410 .A675 2017 (print) | LCC TA410 (ebook) | DDC 620.1/153–dc23
LC record available at https://lccn.loc.gov/2016054038

Cover Design: Wiley
Cover Images: (inset images) © Georgy Shafeev /Shutterstock;
(inset middle image) courtesy of Atul Tiwari;
(background) © Ozerina Anna / Shutterstock

Set in 10/12pt WarnockPro by SPi Global, Chennai, India

10 9 8 7 6 5 4 3 2 1

Contents

List of Contributors

James B. Adams
President's Professor
Materials Science and Engineering Program
School for Engineering of Matter Transport and Energy
Arizona State University
Tempe, AZ 85287, USA

W. Aperador
Department of Engineering, Universidad Militar Nueva Granada
Bogotá, Colombia

M.R. Ayatollahi
Fatigue and Fracture Laboratory
Center of Excellence in Experimental Solid Mechanics and Dynamics
School of Mechanical Engineering
Iran University of Science and Technology
Narmak
Tehran, 16846, Iran

B.B. Aydelotte
Lethal Mechanisms
RDRL-WML-H
US ARL, APG
MD 21005-5066
USA

Valeri Barsegov
Department of Chemistry
University of Massachusetts
Lowell, MA 01854
USA

Ben D. Beake
Micro Materials Ltd.
Willow House, Ellice Way
Yale Business Village
Wrexham
LL13 7YL, UK

R. Becker
Impact Physics
RDRL-WMP-C, US ARL
APG, MD 21005-5066
USA

E. M. Bringa
Facultad de Ciencias Exactas y Naturales
Univ. Nac. de Cuyo - CONICET
Mendoza 5500
Argentina

Silvio Francisco Brunatto
Plasma Assisted Manufacturing Technology & Powder Metallurgy Group
Department of Mechanical Engineering
Universidade Federal do Paraná, Curitiba
Paraná, Brazil
e-mail: brunatto@ufpr.br

H.H. Caicedo
Department of Anatomy and Cell Biology, University of Illinois at Chicago USA and National Biotechnology and Pharmaceutical Association, Chicago USA

J.C. Caicedo
Tribology, Powder Metallurgy and Processing of Solid Recycled Research Group
Universidad del Valle, Cali, Colombia

Zhangwei Chen
Department of Earth Science and Engineering
Royal School of Mines Building
Imperial College London
South Kensington, London
SW7 2BP, UK

Zhaoyu Chen
Applied Mechanics
Saarland University
Geb. A4.2, 66123 Saarbrücken
Germany
e-mail: zh.chen@mx.uni-saarland.de

J.D. Clayton
Impact Physics
RDRL-WMP-C, US ARL
APG, MD 21005-5066
USA

Murat Demiral
Department of Mechanical Engineering
Çankaya University
Ankara 06790
Turkey

Stefan Diebels
Applied Mechanics
Saarland University
Geb. A4.2, 66123 Saarbrücken
Germany
e-mail: s.diebels@mx.uni-saarland.de

Daniel Esqué-de los Ojos
Doctor, Departament de Física
Universitat Autònoma de Barcelona
Facultat de Ciències
E-08193 Bellaterra, Spain

Christian Ganser
Institute of Physics
Montanuniversitaet Leoben
8700 Leoben, Austria

Y. Gao
Physics Department and Research Center OPTIMAS
University Kaiserslautern
Kaiserslautern, 67663
Germany

Marc J. Anglada Gomila
Universitat Politècnica de Catalunya
CIEFMA, Campus Diagonal Besòs - Edif. DBI, Av. d'Eduard Maristany
10-14, 08019 Barcelona
Spain

and

Universitat Politècnica de Catalunya
Research Center in Multiscale Science and Engineering
Campus Diagonal Besòs - Edif. DBC
Av. d'Eduard Maristany
10-14, 08019 Barcelona
Spain

S.K. Gullapalli
Department of Mechanical Engineering
University of Texas at El Paso
El Paso, Texas 79968, USA

Louis G. Hector, Jr.
Senior Research Scientist
Materials and Processes Laboratory
General Motor R&D Center
Warren
Michigan 48090-9055, USA

C.D. Hilton
Oak Ridge Institute for Science and Education
US ARL, APG, MD 21005-5069
USA

S. Huth
Dr.-Ing., Hilti Corporation
9494 Schaan
Liechtenstein

Anne Jung
Applied Mechanics
Saarland University
Geb. A4.2, 66123 Saarbrücken
Germany
e-mail: anne.jung@mx.uni-saarland.de

A. Karimzadeh
Fatigue and Fracture Laboratory
Center of Excellence in Experimental Solid Mechanics and Dynamics
School of Mechanical Engineering
Iran University of Science and Technology
Narmak
Tehran, 16846, Iran

J. Knap
Computational Sciences
RDRL-CIH-C, US ARL
APG, MD 21005-5066
USA

Olga Kononova
Department of Chemistry
University of Massachusetts
Lowell, MA 01854
USA

and

Division of Applied Mathematics
Moscow Institute of Physics and Technology
Moscow region, 141700
Russia

P.-L. Larsson
Department of Solid Mechanics
Royal Institute of Technology
Teknikringen 8 D
SE-10044, Stockholm
Sweden
e-mail: plla@kth.se

Carlos Maurício Lepienski
Department of Physics
Universidade Federal do Paraná, Curitiba
Paraná, Brazil
e-mail: lepiensm@física.ufpr.br

Tomasz W. Liskiewicz
Institute of Functional Surfaces
School of Mechanical Engineering
University of Leeds
Woodhouse Lane, Leeds
LS2 9JT, UK

Qiang Liu
Wolfson School of Mechanical, Electrical and Manufacturing Engineering
Loughborough University
LE11 3TU, UK

G. Martinez
Department of Mechanical Engineering
University of Texas at El Paso
El Paso, Texas 79968, USA

Kenneth A. Marx
Department of Chemistry
University of Massachusetts
Lowell, MA 01854
USA

A. Mina
Tribology, Powder Metallurgy and Processing of Solid Recycled Research Group
Universidad del Valle, Cali, Colombia

Shojiro Miyake
Dr, Nippon Institute of Technology
Miyashiro-machi
Saitama 345-8501
Japan

Mandhakini Mohandas
Centre for Nanoscience and Technology
Anna University
Chennai 25, India

M. Mozafari
Bioengineering Research Group
Nanotechnology and Advanced Materials Department, Materials and Energy Research Center (MERC), Tehran, Iran

Alagar Muthukaruppan
Polymer Composite Lab Departnment of Chemical Engineering
Anna University
Chennai 25, India

Vahid Nekouie
Wolfson School of Mechanical, Electrical and Manufacturing Engineering
Loughborough University
Leicestershire, UK

M. Noor-A-Alam
Department of Mechanical Engineering
University of Texas at El Paso
El Paso, Texas 79968, USA

Jan Perne
RWTH Aachen University
Templergraben 55, 52056 Aachen
Germany

Emilio Jiménez Piqué
Universitat Politècnica de Catalunya
CIEFMA, Campus Diagonal Besòs - Edif. DBI, Av. d'Eduard Maristany
10-14, 08019 Barcelona
Spain

and

Universitat Politècnica de Catalunya
Research Center in Multiscale Science and Engineering, Campus Diagonal Besòs - Edif. DBC, Av. d'Eduard Maristany
10-14, 08019 Barcelona
Spain

F. Pöhl
Dr.-Ing., Ruhr-Universität Bochum
Universitätsstr.150 44801 Bochum
Germany

A. Rahimi
Fatigue and Fracture Laboratory
Center of Excellence in Experimental Solid Mechanics and Dynamics
School of Mechanical Engineering
Iran University of Science and Technology
Narmak
Tehran, 16846, Iran

Rezwanur Rahman
Department of Petroleum and Geosystems Engineering
The University of Texas at Austin
TX 78705, USA

C.V. Ramana
Department of Mechanical Engineering
University of Texas at El Paso
El Paso, Texas 79968, USA

Oscar Rodríguez de la Fuente
Departamento de Física de Materiales
Universidad Complutense de Madrid
Madrid 28040, Spain

Joan Josep Roa Rovira
Universitat Politècnica de Catalunya
CIEFMA, Campus Diagonal Besòs - Edif. DBI, Av. d'Eduard Maristany
10-14, 08019 Barcelona
Spain

and

Universitat Politècnica de Catalunya
Research Center in Multiscale Science and Engineering
Campus Diagonal Besòs - Edif. DBC
Av. d'Eduard Maristany
10-14, 08019 Barcelona
Spain

Anish Roy
Wolfson School of Mechanical, Electrical and Manufacturing Engineering
Loughborough University
Leicestershire, UK

E.J. Rubio
Department of Mechanical Engineering
University of Texas at El Paso
El Paso, Texas 79968, USA

C.J. Ruestes
Facultad de Ciencias Exactas y Naturales
Univ. Nac. de Cuyo - CONICET
Mendoza 5500
Argentina

Norbert Schwarzer
Saxonian Institute of Surface Mechanics
SIO
Tankow 2
18569 Ummanz / Rügen
Germany
www.siomec.de
e-mail: n.schwarzer@siomec.de

Donald J. Siegel
Associate Professor
Department of Mechanical Engineering
University of Michigan
Ann Arbor
MI 48109-2133, USA

Vadim V. Silberschmidt
Wolfson School of Mechanical Electrical and Manufacturing Engineering
Loughborough University
Leicestershire, UK

Francisco J. G. Silva
Auxiliar Professor
Department of Mechanical Engineering
ISEP – School of Engineering
Polytechnic of Porto
Rua Dr. Antònio Bernardino de Almeida
431, 4200-072 Porto
Portugal

Jordi Sort
Professor, Institució Catalana de Recerca i Estudis Avançats (ICREA) &
Departament de Física
Universitat Autònoma de Barcelona
Facultat de Ciències
E-08193 Bellaterra, Spain

Yi Sun
Professor, Harbin Institute of Technology
Harbin
People's Republic of China
e-mail: sunyi@hit.edu.cn

Christian Teichert
Institute of Physics
Montanuniversitaet Leoben
8700 Leoben
Austria

W. Theisen
Prof. Dr.-Ing., Ruhr-Universität Bochum
Universitätsstr.150 44801 Bochum
Germany

H.M. Urbassek
Physics Department and Research Center
OPTIMAS
University Kaiserslautern
Kaiserslautern, 67663
Germany

Mei Wang
Dr, OSG Corporation
Honnogahara 1-15
Toyokawa 442-8544
Japan

Fanlin Zeng
Professor
Harbin Institute of Technology
Harbin, People's Republic of China
e-mail: zengfanlin@hit.edu.cn

Jun Zhong
Professor
School of Materials Engineering
North China Institute of Aerospace
Engineering
Langfang 065000
P.R. China

Preface

Natural materials consist of self-aligned nanoscopic domains with a well-defined pattern and unique surface characteristics. The properties of such materials can therefore be predicted and evaluated using conventional analytical techniques. The adjustment and alignment of atomic or molecular compounds in natural materials is generally consistent and identical until the natural source is changed. The physical and chemical properties however changes if materials are modified using man made techniques. It is critical to revaluate the properties before such transformed materials can be used for further applications. It is apparent that properties of synthetic or man-made materials could vary significantly from source to source due to the variable processing conditions. The service life and performance of products from such materials therefore largely depends on the mechanical properties inherited within the nanoscopic domains.

The mechanical properties must be checked using various analytical techniques available to the material scientists. It is worth mentioning that the failure in an industrial product appears at the macroscopic level at the end of the fatigue cycle; the origin of failure however occurs much earlier in the nanoscopic domain. The conventional mechanical analyses such as those achieved out of universal testing techniques could give an indication of material's strength at the macroscopic domain, but the inception of failure can't be recognized in such cases as these techniques fail to uncover in nanoscopic regimes. Such analyses could turn out even more complicated in complex synthetic materials with anisotropy.

The inventions of nanoindentation techniques have further allowed scientists and engineers to evaluate natural and synthetic materials and design them per the product requirement. The techniques have given even greater control over understanding and tracing of failures in materials. New methods and accessories are under continuous development/improvement that can be affectively used with nanoindenter instruments to draw useful information that was not available earlier. Although, the technique is extremely valuable and can be applied to study a wide range of materials, not enough books are available to educate new students and scholars. A book that teaches the technique with the help of applied examples is therefore immediately required.

This book contains twenty six chapters written carefully to cover the fundamentals and experimental methodologies associated with the use of nanoindentation techniques. The first section of this book discusses the methods and experimentation applied to advanced materials. The second section of the book is meant for the advanced learners and discusses various modeling and simulations implemented on the theoretically designed materials. A complete in-depth understanding on nanoindentation analysis can be achieved after going through the two sections of this book.

The first thirteen chapters in the applied experimental section will help readers in learning the advanced concepts needed to understand the use of nanoindentation. The subsequent thirteen chapters in the second section are dedicated to the studies on modeling and simulations in nanoindentation. Residual stress plays a significant role in the physical properties of coatings and is therefore dealt in the beginning of the book followed by the characterization of diamond like carbon films. The thermal barrier coatings, thin films, extremely hard coatings, macro-porous materials, plasma nitrided parts, defective surfaces, hybrid polymeric nanocomposites, hybrid foams and cellulosic fibers are studied in proceeding chapters. Finally, the properties of stainless steel and metallic glasses are described at the end of this section. Molecular dynamics modeling and continuum modeling is discussed in the beginning part of the simulation and modeling section followed by the treatment of the subject with finite element analysis and atomistic simulation. The simulations of nanoindentation on advanced ceramics, thin hard coatings and single crystals have been contributed in the following chapters. The atomistic simulations at nanoscale and multiscale modeling in polymer-polymer nanocompoiste have been detailed in the final chapters of the second section.

We believe that this book could be a key reference for the students and scholars from diverse science and engineering background such as those from chemical or mechanical engineering, biotechnology related to biomaterials, paints and coatings, composites and nanocomposites, geosciences and many more. We are confident that readers will appreciate the efforts rendered in publishing this book of significant technical importance. The editors are thankful to the organizers of Technical Corrosion Collaboration (TCC), particularly Daniel J. Dunmire, Director, and Richard Hays, Deputy Director, Corrosion Policy and Oversight Office in the Office of the Under Secretary of Defense (Acquisition Technology & Logistics), USA for providing a platform to work on this book. Suggestions on the modifications and possible new inclusions are always welcomed.

USA

Atul Tiwari, PhD, FRSC
Sridhar Natarajan, M.D., M.S.

Part I

1

Determination of Residual Stresses by Nanoindentation

P-L. Larsson

Department of Solid Mechanics, Royal Institute of Technology, Stockholm, Sweden

1.1 Introduction

Residual stresses and strains in a material can be determined by using various experimental measuring techniques. Examples of such techniques include for example indentation crack techniques [1], fracture-surface analysis, neutron and X-ray tilt techniques [2], beam bending, hole drilling [3], and layer removal [4]. These methods can, however, be both complicated and expensive and therefore, sharp indentation testing, being the method of interest in this chapter, can be a very attractive alternative. It goes almost without saying that this can be of substantial practical importance as the effects of residual stress and strain fields in materials can be considerable with respect to, for example, fatigue, fracture, corrosion, wear, and friction.

Until approximately 20–30 years ago, the influence of residual stresses and residual strains on the results given by a sharp indentation test, in comparison with the corresponding results for a material without residual stresses or residual strains present, i.e. a virgin material, has been studied only occasionally, and then mainly experimentally. This is in contrast to sharp indentation or hardness testing of virgin materials which is a well-known experimental method used for determination of the constitutive properties of conventional materials such as metals and alloys. The method has of course benefitted substantially due to the development of new experimental devices like the nanoindenter (Pethica *et al.* [5]), enabling an experimentalist to determine the material properties from extremely small samples of the material. Indentation testing is for example a very convenient tool for determining the material properties of thin films in ready-to-use engineering devices.

Returning now to the case when residual fields are present, it should be mentioned that already in 1932 Kokubo [6] studied several materials subjected to applied tensile and compressive uniaxial stress. The Vickers hardness was measured and some very small influence from sign and size of the applied stress was found. However, the observed effect of stress on the hardness value was so small that no decisive conclusions could be drawn from these investigations. These results were confirmed somewhat later by Sines and Carlsson [7].

More recently, starting with the study by Doerner and Nix [8], several interesting experimental investigations dealing with this issue were presented, cf. also [9, 10]. The

Applied Nanoindentation in Advanced Materials, First Edition. Edited by Atul Tiwari and Sridhar Natarajan.

basic features of the problem were completely understood, however, until Tsui *et al.* [11] and Bolshakov *et al.* [12] investigated, by using nanoindentation as well as numerical methods, the influence of applied stress on hardness, contact area and apparent elastic modulus at indentation of aluminum alloy 8009, an almost elastic-ideally plastic material. Qualitative results of interest were presented as it was shown that the hardness was not significantly affected by applied (residual) stresses while the amount of piling-up of material at the contact contour proved to be sensitive to stress (piling-up increased when the applied stresses were compressive and decreased at tensile stresses).

Based on the results in [11, 12], further studies have been presented, cf. e. g. [13–20], more directed towards the mechanical behavior of the problem. Perhaps being the first to address this issue, Suresh and Giannakopoulos [13] derived, by making certain assumptions on the local stress and deformation fields in the contact region, a relation between the contact area at indentation of a material with elastic residual stresses (and plastic residual strains) and the corresponding contact area at indentation of a material with no stresses present. The analysis in [13] was restricted to equi-biaxial residual stress and strain fields but it should be mentioned that, for the forthcoming discussion, that these authors clearly distinguished between tensile and compressive residual stresses. The relation was, however, approximated with close to linear functions.

The physical understanding of the problem was further developed by Carlsson and Larsson [14, 15], in a combined theoretical, numerical and experimental investigation. A more detailed discussion of the results achieved in [14, 15] will be presented in forthcoming sections below but in short, these authors showed that good correlation between predictions and numerical/experimental results could be achieved if the material yield stress, in relevant indentation parameters, was appropriately replaced by a combination of yield stress and residual stress. Most of the results presented by Carlsson and Larsson [14, 15] were related to equi-biaxial residual stress states but in [15] the derived relations were extended to apply also for more general residual stress fields. In the latter case though, high accuracy results could not be achieved. Furthermore, the accuracy was worse for compressive residual stresses as shown by Larsson [21].

The latter issue was addressed by Rydin and Larsson [20] and very accurate relations linking both compressive and tensile residual stresses to the size of the contact area were presented. Based on the achievements in [20], Larsson [22] attacked the problem pertinent to general residual stresses and presented relations yielding predictions of high accuracy also when neither uniaxial nor equi-biaxial stress state could be assumed.

Below then in the next section, the results presented in [14, 15, 20 and 22] will be explained in detail and it is demonstrated how these findings can be used for determining the residual stresses on the surface of a body. Furthermore, possible improvements of the approach using previous findings concerning the size of the contact area, see e.g. Larsson and Blanchard [23], is discussed as well as the appropriate choice of indenter geometry.

For obvious reasons the presentation here is very much focused on the approach taken by the author. It should be clearly stated though that there are many other research groups, some of them have already been mentioned above, that have suggested alternative approaches yielding promising results. For one thing, another possible approach to the determination of residual stresses by indentation methods is to apply inverse modelling. This has been attempted in a number of studies where perhaps the most general one was presented by Bocciarelli and Maier [19]. These authors used, together with the standard global indentation properties, the shape of the residual imprint at indentation as a parameter in order to arrive at a unique inverse solution.

Further progress regarding the understanding of the problem concerning residual stresses and indentation was achieved by Huber and Heerens [24] and Heerens *et al.* [25] as these authors analyzed the corresponding problem of residual stress determination using spherical indentation testing. This is a more involved problem (as compared to sharp indentation testing) due to the existence of a characteristic length. Indeed, when elastic and plastic effects are of similar importance self-similarity of the problem is lost and a correlation between the indentation contact pressure and the residual stress state as attempted by Huber and Heerens [24] and Heerens *et al.* [25] becomes very much involved. Despite of this though, also other investigators, see e.g. Swadener *et al.* [26], have suggested that spherical indentation is an attractive approach for residual stress determination. The main reason behind this is that indentation variables are more sensitive to residual stresses in this case (as compared to sharp indentation testing).

Despite the discussion right above, presently sharp indentation is adhered to due to the fact that hardness and relative contact area are independent of indentation depth (due to the fact that the problem is mathematically self-similar with no characteristic length) and this is a particular advantage at interpretation of the results. Furthermore, the emphasis on nanoindentation testing also suggests that sharp indentation is the feature of most interest presently.

1.2 Theoretical Background

The basic foundation of the analysis by Carlsson and Larsson [14, 15], as confirmed by finite element calculations, is that a residual stress field will alter the magnitude but not the principal shape of the field variables involved. This immediately suggests that classical indentation analysis still applies but have to be corrected based on the residual stress. In short, it was shown by Carlsson and Larsson [14, 15] that it is possible to correlate the magnitude of the residual stress field with the well-known Johnson [27, 28] parameter:

$$\Lambda = E \tan \beta / (\sigma_y(1 - \nu^2)) \tag{1.1}$$

In Equation (1), E is the Young's modulus, ν the Poisson's ratio, σ_y the flow stress and β is the angle between the sharp indenter and the undeformed surface of the material, see the (cone) indenter geometry schematically shown in Figure 1.1. Furthermore, in Equation (1.1) elastic-ideally plastic material behavior is assumed.

Johnson [27, 28] suggested that the outcome of a sharp indentation test on an elastic-ideally plastic material falls into one out of three levels, see Figure 1.2, characterized by the parameter Λ in Equation (1.1). In Figure 1.2, H is the material hardness here and in the sequel defined as the average contact pressure. The three levels are schematically shown in Figure 1.2 where in level I, $\Lambda \leq 3$, very little plastic deformation occurs during the indentation test and an elastic analysis of the problem will be sufficient. In level II, $3 < \Lambda < 30$, plastic deformation spreads over the contact area. Finally, in level III, $\Lambda > 30$, pertinent to most engineering metals and alloys, rigid plastic conditions dominate as plastic deformation is present over the entire contact area and elasticity no longer has any effect on the hardness.

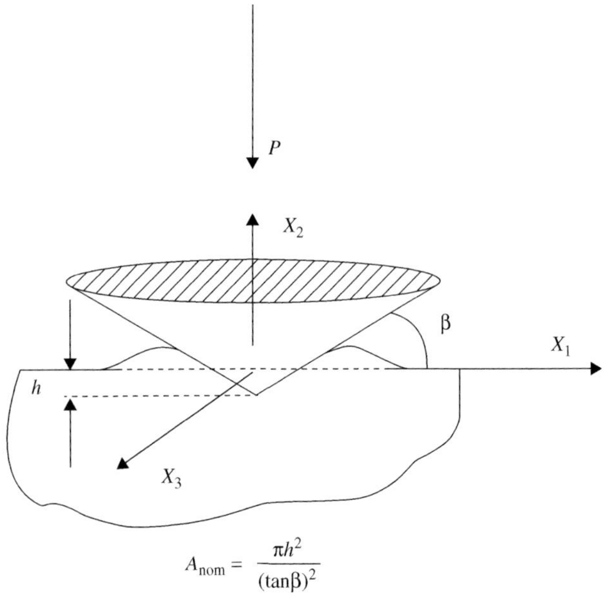

Figure 1.1 Schematic of the geometry of the cone indentation test.

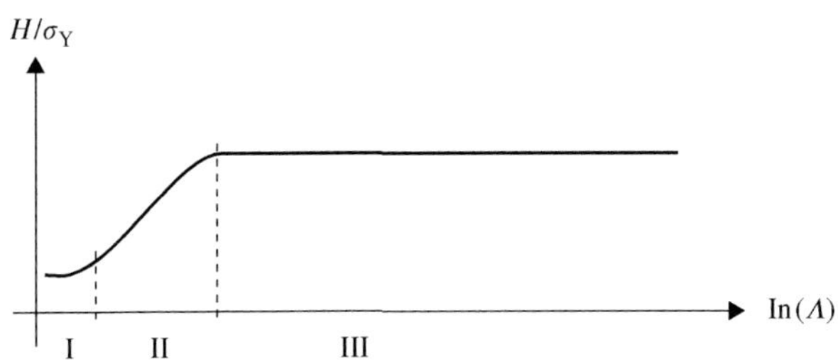

Figure 1.2 Normalized hardness, H/σ_y as a function of In Λ, Λ defined according to Equation (1.1). Schematic of the correlation of sharp indentation testing of elastic-ideally plastic materials as suggested by Johnson [27, 28]. The three levels of indentation responses, I, II and III, are also indicated.

From theoretical, numerical and experimental results [11, 12, 14, 15] it is, as mentioned above, a well-established fact that the material hardness is not noticeably influenced by stresses at sharp indentation testing. The relative contact area, however, here and throughout this chapter defined as:

$$c^2 = A / A_{nom} \tag{1.2}$$

A being the projected true contact area and A_{nom} the nominal contact area as defined in Figure 1.1 for cone indentation, can be directly related to the material state (it should be noted in passing that if $c^2 < 1$ (sinking-in) the resulting contact area is smaller than what could be expected from purely geometrical considerations and the other way around if $c^2 > 1$ (piling-up)). This finding is of fundamental importance when indentation testing is used to determine residual fields and, subsequently, it was shown by Carlsson and Larsson [14, 15] that when the residual (or applied) stress field is equi-biaxial the relation:

$$c^2 = c^2(\varepsilon_{res}, \sigma_{res} = 0) - 0.32\ln(1 + (\sigma_{res}/\sigma_y(\varepsilon_{res}))) \tag{1.3}$$

can be expected to give results of high accuracy at tensile stresses but worse at compressive stresses [21]. In Equation (1.3), c^2 is the relative contact area for a material with a (equi-biaxial) residual stress field σ_{res} present (and possibly a (von Mises) effective residual strain field ε_{res}), $c^2(\varepsilon_{res}, \sigma_{res} = 0)$ is the corresponding relative contact area for a material with no residual stress and $\sigma_y(\varepsilon_{res})$ is the material flow stress when the effective plastic strain equal ε_{res}.

In case of ideally-plastic behavior, initially assumed here for simplicity but not necessity, Equation (1.3) reduces to:

$$c^2 = c^2(\sigma_{res} = 0) - 0.32\ln[1 + (\sigma_{res}/\sigma_y)] \tag{1.4}$$

as then the yield stress of the material is independent of the residual strain field.

Equations (1.3) and (1.4) were derived by Carlsson and Larsson [14, 15] based on the fact that the stress state in the contact region closely resembles the stresses arising at indentation of a virgin material with an initial material yield stress $\sigma_y + \sigma_{res}$. This was shown by careful and comprehensive numerical investigations of the behavior of the indentation induced stress fields as well as deformation fields close to the contact boundary for materials with and without residual stresses.

With this as a background it is then possible to correlate the experimentally determined c^2-value with the residual stress state based on the universal curve schematically shown in Figure 1.3 by introducing an apparent yield stress:

$$\sigma_{y,\text{apparent}} = \sigma_y + \sigma_{res} \tag{1.5}$$

in Λ in Equation (1.1) according to:

$$\Lambda = E\tan\beta/(\sigma_{y,\text{apparent}}(1 - \nu^2)) \tag{1.6}$$

The usefulness of this feature rests on the fact that elastic effects are more pronounced for c^2, than for the material hardness, as also shown in Figure 1.3, and as a result, level II is the dominating region for this parameter.

As mentioned above Equation (1.4) is accurate when a tensile residual stress is at issue but not so at compressive fields. The reason for this is that a compressive residual

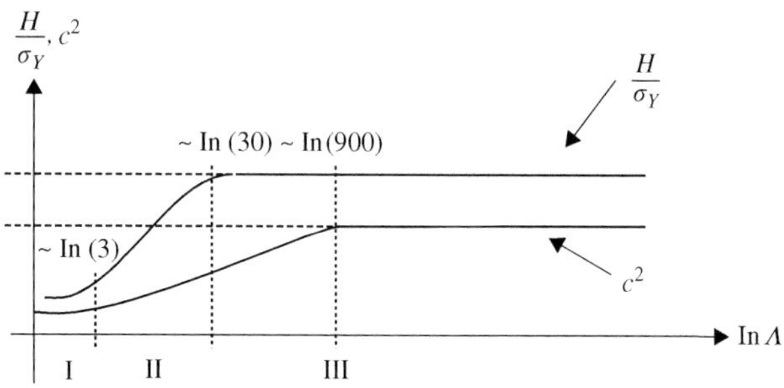

Figure 1.3 Normalized hardness, H/σ_y, and area ratio, c^2, as functions of ln Λ, Λ defined according to Equation (1.1). Schematic of the correlation of sharp indentation testing of elastic-ideally plastic materials. The three levels of indentation responses, I, II and III, are also indicated.

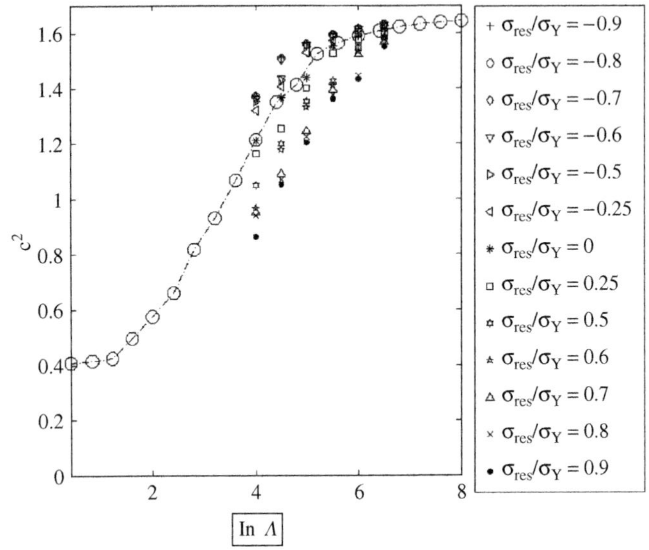

Figure 1.4 The area ratio, c^2, as function of ln Λ, Λ defined according to Equation (1.1). Cone indentation of elastic-ideally plastic materials is considered. *Source*: Rydin 2012 [20]. Reproduced with permission of Elsevier.

stress state will, cf. Equation (1.5), reduce the apparent yield stress $\sigma_{y,\text{apparent}}$ leading to a stronger influence from level III indentation effects. This problem was accounted for by Rydin and Larsson [20] (Figures 1.4, 1.5) and in this work it was found, from studying the yield surface at particular points around the contact boundary, that replacing Equation (1.5) with the expression:

$$\sigma_{y,\text{apparent}} = \sigma_y + F\sigma_{res}, \tag{1.7}$$

where

$$\begin{aligned} F &= 0.52, \quad \sigma_{res} < 0 \\ F &= 1.77, \quad \sigma_{res} > 0, \end{aligned} \tag{1.8}$$

gave results of very high accuracy both in tension and compression. Explicitly, Rydin and Larsson [20] suggested that the relation:

$$c^2 = c^2(\sigma_{res} = 0) - 0.35\ln(1 + (F\sigma_{res}/\sigma_y)) \tag{1.9}$$

should replace Equation (1.4) above. It was shown by Rydin and Larsson that Equation (1.9) improved very much on the situation as compared with the results from Equation (1.4). High accuracy predictions in both tension and compression were achieved as depicted in Figures 1.4–1.6 where in particular the excellent agreement in Figure 1.6, pertinent to results based on Equation (1.9), should be noted.

The model by Carlsson and Larsson [14, 15] is based on the fact that the indentation induced in-plane stresses at the contact boundary are compressive and approximately equi-biaxial also when general residual stress states are considered (as shown by extensive finite element calculations). Following the discussion above about the equi-biaxial case, a direct extension would be, as also suggested by Carlsson and Larsson [15], to

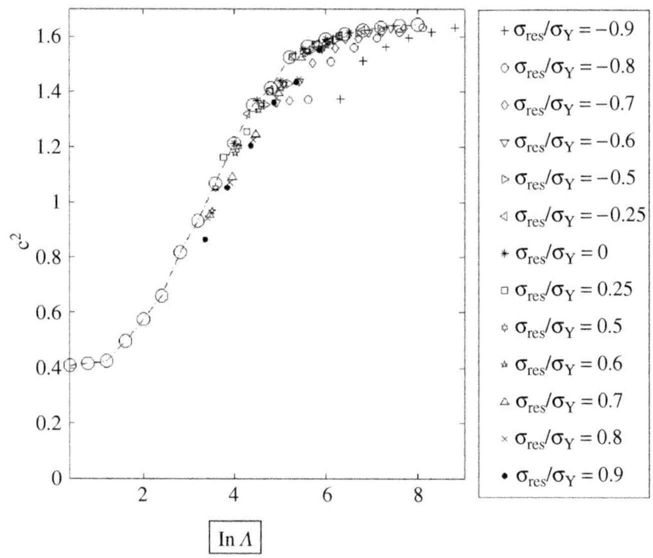

Figure 1.5 The area ratio, c^2, as function of ln Λ, Λ defined according to Equation (1.6) with the yield stress σ_y replaced by the apparent yield stress $\sigma_{y,\text{apparent}}$ in Equation (1.5). Cone indentation of elastic-ideally plastic materials is considered. *Source*: Rydin 2012 [20]. Reproduced with permission of Elsevier.

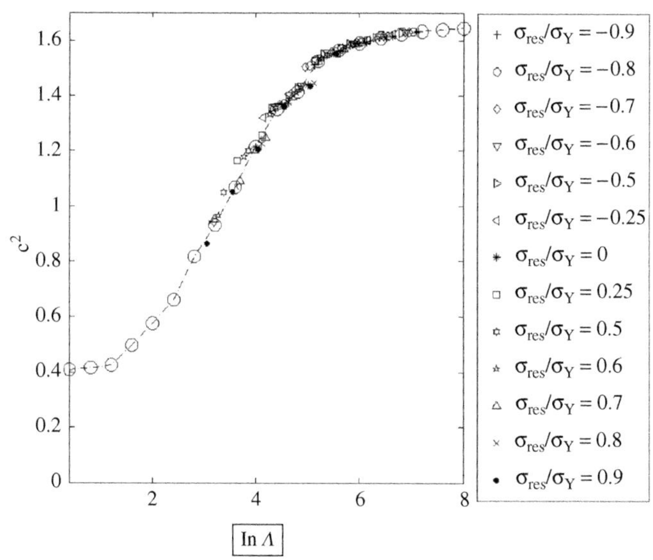

Figure 1.6 The area ratio, c^2, as function of ln Λ, Λ defined according to Equation (1.6) with the yield stress σ_y replaced by the apparent yield stress $\sigma_{y,apparent}$ in Equation (1.7). Cone indentation of elastic-ideally plastic materials is considered. *Source*: Rydin 2012 [20]. Reproduced with permission of Elsevier.

determine the apparent yield stress when an indentation induced compressive and equi-biaxial stress field σ_{ind} is superposed over the surface residual stress field in the material.

The von Mises yield criterion then becomes:

$$\sigma_y{}^2 = (1/2)((\sigma_1 - \sigma_{yind})^2 + (\sigma_2 - \sigma_{yind})^2 + (\sigma_1 - \sigma_2)^2) \tag{1.10}$$

where σ_{yind} is the apparent yield stress at indentation, $\sigma_{yind} > 0$, while σ_1 and σ_2 are the principal stresses representing the surface residual stress field in the material. The principal stresses are indicated in Figure 1.7 where also the resulting elliptic contact area (at general residual stresses) is shown as defined by the semi-axes a_1 and a_2.

In the equi-biaxial case the quantity σ_{res} in Equation (1.5) represents the change of the apparent yield stress at indentation. Consequently, it was suggested by Carlsson and Larsson [15] that this quantity could represent also a general residual stress field when determined from the expression:

$$\sigma_{res} = \sigma_{yind} - \sigma_y. \tag{1.11}$$

In Equation (1.11), σ_{yind} is determined from Equation (1.10) and it goes almost without saying that ideally plastic material behavior is assumed.

As already mentioned above, and as also pointed out by Carlsson and Larsson [15], the predictive capability of Equation (1.5), and thereby also Equations (1.10, 1.11), deteriorates substantially at compressive residual stresses. In the equi-biaxial case this was, as also mentioned above, corrected by the results derived by Rydin and Larsson [20] and the basic results in [20] were used by Larsson [22] in order to determine prediction also in a general case. Explicitly then in [22], the relation between the relative contact area c^2 and the residual stress state σ_{res}, determined from Equations (1.10) and (1.11), were

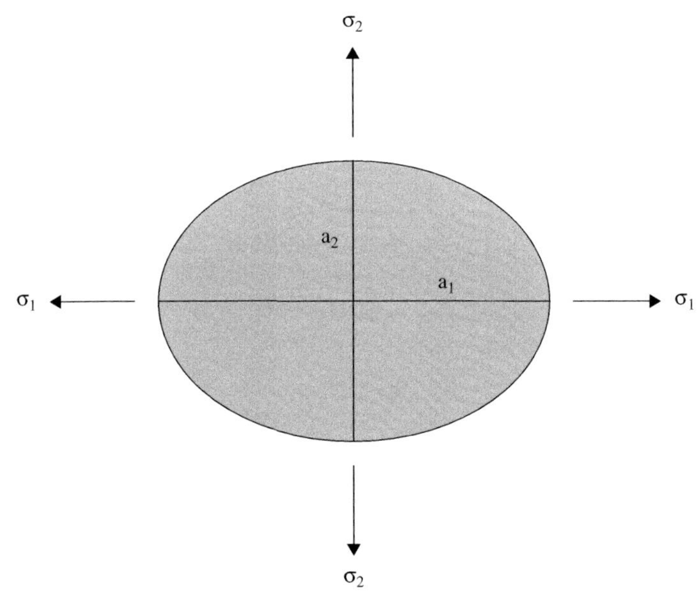

Figure 1.7 Schematic of the contact area (shaded) at indentation. The principal residual stresses and the corresponding semi-axes of the elliptical contact area are also indicated.

expressed by Equation (1.9) also generally. In short, Larsson [22] reported high accuracy predictions based on this approach. In this context it should be clearly stated that the nature of the stress state, based on the ratio σ_1/σ_2, enters the analysis by Larsson [22] through σ_{res}. However, it is only possible to derive the magnitude of the residual stresses involved based on such approach and in order to also determine explicit values on the ratio σ_1/σ_2 additional experimental information is needed. Such information can, as suggested in [22], be given from the elliptic shape of the contact area, see Figure 1.7, i.e. the value on the ratio a_1/a_2.

It was shown, however, by Larsson and Blanchard [23] that even though such an approach is possible (the influence from the ratio σ_1/σ_2 on a_1/a_2 can be proven) such influence is very weak in case of cone indentation and will not be of practical use in an experimental situation. A more complex indenter geometry would then be advantageable and possibly a Knoop indenter should be relied upon for this purpose remembering the rhombic shape of this indenter. This issue, however, remains to be investigated.

The relations presented are, as already stated above, pertinent to elastic-ideally plastic behavior. In order to extend the validity of the present approach to strain-hardening materials, it is possible to draw upon results from a previous study by Larsson [29]. In this study it is assumed that the indented material is well described by a power law material with a uniaxial stress–strain relation according to:

$$\sigma(\varepsilon_p) = \sigma_y + \sigma_0(\varepsilon_p)^{1/n}, \tag{1.12}$$

where σ_0 and n are material constants and ε_p is the accumulated effective plastic strain. It was then shown that at level II cone indentation the nominal contact area could always be expressed as:

$$c^2 = C_1(n) + C_2(n)\ln(\Lambda_h) \tag{1.13}$$

with

$$C_1(n) = -0.07 + 0.75(1/n) - 0.29(1/n)^2 \tag{1.14}$$

and

$$C_2(n) = 0.30 - 0.29(1/n) + 0.09(1/n)^2 \tag{1.15}$$

Derived from curve-fitting based on the results by Larsson [29]. In Equation (1.13) Λ_h is the Johnson's parameter [27, 28] in Equation (1.1) also accounting for strain-hardening according to:

$$\Lambda_h = E\tan\beta/[\sigma_r(1-\nu^2)] \tag{1.16}$$

where σ_r is a stress measure representing in an average sense the plastic strain-hardening of the indented material. Traditionally, the suggestion by Tabor [30] (where σ_r is the flow stress at $\varepsilon_p = 0.08$) is used but it was shown by Larsson [31] that the choice:

$$\sigma_r = 0.392\sigma(\varepsilon_p = 0.02) + 0.608\sigma(\varepsilon_p = 0.35) \tag{1.17}$$

yields better accuracy in a general situation. It should be noted in passing that the actual values on the constants in Equations (1.13, 1.14, 1.15 and 1.17) are pertinent to cone indentation with an angle $\beta = 22°$, see Figure 1.1. These constants will change in case of

Vickers and Berkovic indentation, the latter being more pertinent to nanoindentation, as discussed in detail in [31].

Returning to Equation (1.13) and recalling that the constant $C_2(n)$ determines the slope of the Johnson-curve [27, 28] in a situation where strain-hardening effects are present. It is then straightforward, when strictly following the reasoning above leading to Equation (1.3) and (1.9), to derive the relation:

$$c^2 = c^2(\sigma_{res} = 0) - C_2(n)\ln(1 + (F\sigma_{res}/\sigma_r)) \tag{1.18}$$

In (16) it has been tentatively assumed that the apparent representative stress is changed due to a residual stress state σ_{res} based on Equation (1.10) according to:

$$\sigma_{r,apparent} = \sigma_r + F\sigma_{res}, \tag{1.19}$$

see Equation (1.7), and that any (von Mises) effective residual strain ε_{res} can be neglected. In this context it should be immediately emphasized that it remains to determine the validity of Equations (1.18) and (1.19) as, for one thing, the variation of F at plastic strain-hardening is not known.

In summary then, from the discussion above it is hopefully clear that the basic theoretical foundation exists for accurate determination of residual stress field by nanoindentation. In particular when it comes to equi-biaxial residual stresses in low hardening materials, a full theory is available. This is also the case to be discussed in detail below in the context of practical applications. However, concerning the effects from plastic strain-hardening and general biaxiality a complete theory is not yet, as also discussed above, available even though the relations (1.10, 1.11, 1.18 and 1.19) are of direct relevance for at least qualitative predictions. These issues will also be discussed further.

1.3 Determination of Residual Stresses

As just mentioned above, in this section a solution strategy for the determination of residual stresses by indentation will be discussed and outlined in the context of the theory presented above. The solution strategy will mainly concentrate on equi-biaxial residual stresses in low hardening materials but also a general approach is discussed. In most cases, no distinction is made between standard indentation and nanoindentation but when so required, this will be specified.

1.3.1 Low Hardening Materials and Equi-biaxial Stresses

What is considered then first is a low hardening material accurately described by classical Mises plasticity. It is assumed that the material constants at issue are known from experiments on a virgin material (a material with no residual stresses or strains present). Furthermore, it is also assumed that the hardness, $H(\sigma_{res} = 0)$, and relative contact area, $c^2(\sigma_{res} = 0)$, of the virgin material is known from previous experiments.

Accordingly, the first step in the procedure concerns the determination of the virgin properties$H(\sigma_{res} = 0)$, and $c^2(\sigma_{res} = 0)$. It should be immediately emphasized that these properties are independent of any residual strain fields present due to the fact that only ideal plasticity (or close to ideal plasticity) is considered. This, however, will be discussed in some more detail below.

Furthermore in this context and in the context of particular issues related to nanoindentation, it is important to emphasize that when determining indentation properties the contact area should always be determined from optical measurements. As a standard procedure at nanoindentation, the contact area is determined from the indentation load–indentation depth (P-h) relation according to the procedure suggested by Oliver and Pharr [32]. However, such an approach can give results of low accuracy leading to erroneous conclusions as shown by Bolshakov *et al.* [12].

In the next step the surface of the material with residual stresses is indented and the hardness, $H(\sigma_{res})$, and the relative contact area, $c^2(\sigma_{res})$, are determined. The reason for recording also the hardness values, $H(\sigma_{res} = 0)$, and $H(\sigma_{res})$, is as mentioned above to check the invariance of hardness at ideal plasticity (or close to ideal plasticity). The two simple steps so described, nanoindentation of the material in a virgin and in a stressed state, constitute (together with the material characterization of the virgin material) the experimental part of the procedure aiming at residual stress determination.

The third and final step in this procedure concerns explicit determination of the equi-biaxial residual surface stress σ_{res} from Equation (1.9). Remembering that both $c^2(\sigma_{res} = 0)$ and $c^2(\sigma_{res} = 0)$, as well as the material yield stress σ_y, are known from the introductory experiments it is then a straightforward task to calculate σ_{res} (as σ_{res} is the only unknown in Equation (1.9)). The only consideration that has to be made is the explicit value on the constant F in Equation (1.9). In short, if $c^2(\sigma_{res}) > c^2(\sigma_{res} = 0)$ this implies that the residual stress state is compressive and, consequently, if $c^2(\sigma_{res}) < c^2(\sigma_{res} = 0)$ tensile residual stresses are present on the surface of the material. Based on the sign of σ_{res}, the value on F can be determined in a straightforward manner according to Equation (1.8).

Clearly, in the case of low hardening materials and equi-biaxial stresses it is at least in theory a rather straightforward task using the present approach to determine the relevant variables describing the residual stress field. However, if these restrictions do not apply, the situation becomes much more difficult and indeed, for some particular cases additional research is needed as outlined above.

1.3.2 General Residual Stresses

One of the complicating issues concerns the case when equi-biaxiality cannot be assumed. However, as mentioned above a solution approach for this case has been suggested by Larsson [22], based on Equations (1.10 and 1.11) together with Equation (1.9), yielding high accuracy predictions in a situation when the explicit value on the ratio between the residual stresses, σ_1 / σ_2, is known, for example in a uniaxial situation, see Figure 1.8 pertinent to an ideally-plastic material. If this ratio is not known, however, further information is needed for a complete determination of the residual stresses in the material.

An obvious candidate to provide such additional information would of course possibly be given from the elliptic shape of the contact area, see Figure 1.7, i. e. the value on the ratio a_1 / a_2. However, as mentioned above, it was shown by Larsson and Blanchard [23] that even though such an approach is possible this influence from the ratio σ_1 / σ_2 on a_1 / a_2 is very weak in case of cone indentation, cf. e. g. results by Larsson and Blanchard [23, 33] as shown in Figures 1.9 and 1.10 for two different values on the Johnson [27, 28] parameter Λ, and most likely also for other highly symmetric indenters such as the

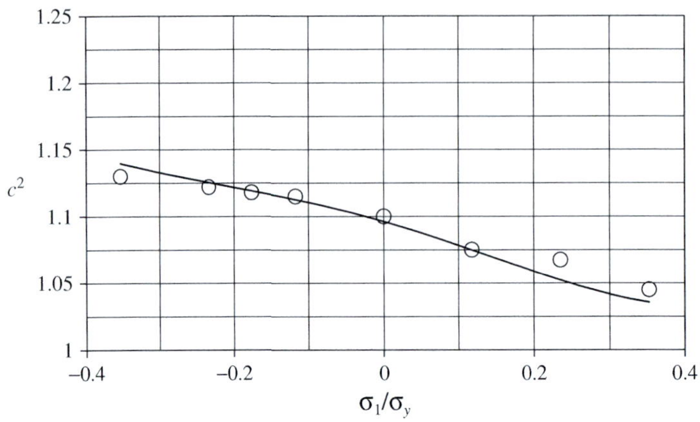

Figure 1.8 Berkovich indentation of an aluminum alloy 8009 ($E = 82.1$ GPa, $\nu = 0.31$, $\sigma_y = 425.6$ MPa (this is the peak stress after a small amount of initial work-hardening), $c^2(\sigma_{res} = 0) = 1.10$). The area ratio c^2 is shown as function of an applied uniaxial stress (ratio) σ_1 / σ_y. (O), experimental results by Tsui *et al.* [11]. (—), theoretical predictions by Larsson [22]. *Source*: Larsson 2014. Reproduced with permission of Springer.

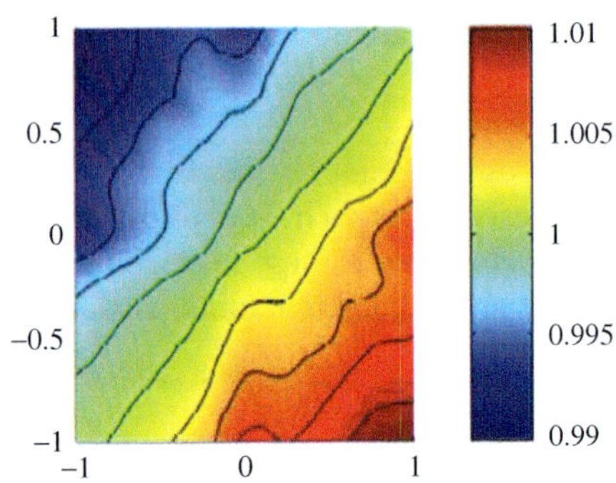

Figure 1.9 Semi-axes ratio a_1 / a_2, see Figure 1.7, as function of the principal stress ratios σ_1 / σ_y (horizontal axis) and σ_2 / σ_y (vertical axis). Explicit values on a_1 / a_2 are determined by the colors on the right hand side of the figure. The value on the Johnson [27, 28] parameter is $\Lambda = 100$. *Source*: Larsson 2012 [23]. Reproduced with permission of Elsevier.

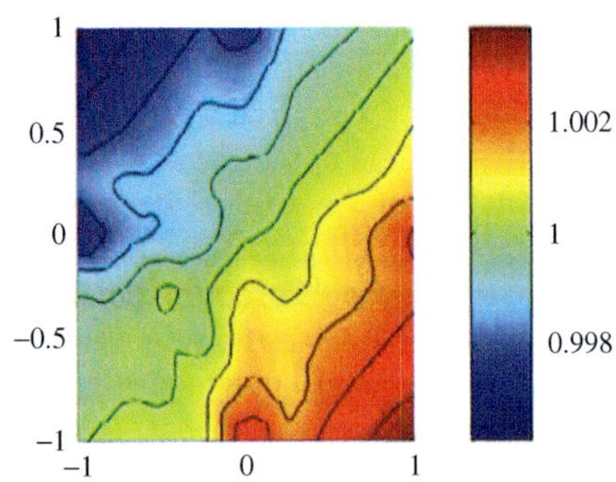

Figure 1.10 Semi-axes ratio a_1 / a_2, see Figure 1.7, as function of the principal stress ratios σ_1 / σ_y (horizontal axis) and σ_2 / σ_y (vertical axis). Explicit values on a_1 / a_2 are determined by the colors on the right hand side of the figure. The value on the Johnson [27, 28] parameter is $\Lambda = 300$. *Source*: Larsson 2012 [23]. Reproduced with permission of Elsevier.

Vickers and the Berkovic indenters. As mentioned previously the Knoop indenter could then be a possible choice of indenter in order to improve the influence on a_1 / a_2 from the stress ratio σ_1 / σ_2. This is definitely an important future research direction which, at least when numeric analysis is concerned, would not introduce any fundamental difficulties as finite element analyses of sharp indentation is now very much a standard task, cf. [34–38] for early efforts. Indeed, such analysis have been conducted previously, [39–41], but not in the context of determination of residual stresses.

1.3.3 Strain-hardening Effects

Further complications related to the theory outlined above concerns plastic strain-hardening effects. In this context, Carlsson and Larsson [14, 15] used Equation (1.3) for this feature primarily in the equi-biaxial case. It should then be noticed that the information needed in this case are not only the quantities $H(\varepsilon_{res}, \sigma_{res})$ $c^2(\varepsilon_{res}, \sigma_{res})$, for the stressed material, but also the stress-free quantities $H(\varepsilon_{res}, \sigma_{res} = 0)$ and $c^2(\varepsilon_{res}, \sigma_{res} = 0)$ where ε_{res} represents, as indicated above, the influence from residual plastic deformation due to plastic strain-hardening on the global indentation properties.

It is to be expected that the hardness values are independent of any residual stresses, cf. [12] and [14], but this quantity should be used to determine ε_{res} from the original uniaxial stress–strain curve via an appropriate relation between hardness and plastic strain hardening, cf. Tabor [30] and Larsson [31].

If equi-biaxiality can be assumed it is then a straightforward matter, based on the experimental information achieved, to determine the ressidual stress σ_{res} via Equations (1.18 and 1.19) where of course also (if necessary) a residual field ε_{res} can be accounted for in Equation (1.18) according to:

$$c^2(\varepsilon_{res},\ \sigma_{res}) = c^2(\varepsilon_{res},\ \sigma_{res} = 0) - C_2(n)\ln(1 + (F\sigma_{res} / \sigma_r)) \tag{1.20}$$

as discussed just above. As stated previously, it remains, however, to determine the validity of Equations (1.18 and 1.19), and of course Equation (1.20) as, for one thing, the variation of F at plastic strain-hardening is not known.

Finally, in this context, it should also be mentioned that when equi-biaxiality is lost at strain-hardening plasticity the situation becomes even more involved. In theory, Equations (1.10 and 1.11), together with Equation (1.20), could be applied as in the corresponding ideally-plastic case but again, the validity of such an approach needs to be investigated in more detail. Clearly, again as in the ideally-plastic case, further information is needed for a complete determination of the residual stresses.

1.3.4 Conclusions and Remarks

It should be emphasized that the discussion above is essentially restricted to cone indentation. However, as shown by Carlsson and Larsson [15], basically the same solution strategy could be applied to pyramid indenter geometries such as the Vickers and Berkovich indenters, the latter being more pertinent to nanoindentation, even though the details might be different. This issue concerns for example the definition of the representative stress in Equation (1.17), as discussed in detail by Larsson [31], where for pyramid indenters this relation yields:

$$\sigma_r = 0.5(\sigma(\varepsilon_p = 0.02) + \sigma(\varepsilon_p = 0.35)). \tag{1.21}$$

Finally, it is worth mentioning that the present approach could very well be applied to other types of contact problems as well. One of these problems could be scratching and scratch testing where correlation of material and contact properties, in the spirit of Johnson [27, 28], have been discussed for some time now, cf. [42–51]. It remains, however, to undertake an analysis that incorporates also residual stresses in this type of correlation.

References

1 Marshall DB, Lawn BR. Indentation of brittle materials. ASTM Special Technical Publications 889, 26–46; 1985.
2 Hehn L, Zheng C, Mecholsky JJ, Hubbard CR. Measurements of residual-stresses in Al2O3/Ni laminated composites using an X-ray-diffraction technique. Journal of Materials Science 30, 1277–1282; 1995.
3 Rendler N.J, Vigness I. Hole-drilling strain-gauge method of measuring residual stresses. Experimental Mechanics 13, 45–48; 1973.
4 Flavenot JF, Nikulari A. Measures des contraintes residuelles, method de la fleche. Les Memories Technique du Cetim 31, 6–42; 1977.
5 Pethica JB, Hutchings R, Oliver WC. Hardness measurements at penetration depths as small as 20 nm. Philosophical Magazine A48, 593–606; 1983.
6 Kokubo, S. On the change in hardness of a plate caused by bending. Science Reports of the Tohuko Imperial University ser.1, 21, 256–267; 1932.
7 Sines G, Carlson R. Hardness measurements for determination of residual stresses. ASTM Bulletin 180, 357; 1952.
8 Doerner MF, Gardner DS, Nix WD. Plastic properties of thin films on substrates as measured by submicron indentation hardness and substrate curvature techniques. Journal of Materials Research 1, 845–851; 1986.
9 LaFontaine WR, Yost B, Li CY. Effect of residual stress and adhesion on the hardness of copper films deposited on silicon. Journal of Materials Research 5, 776–783 (1990).
10 LaFontaine WR, Paszkiet CA, Korhonen MA, Li CY. Residual stress measurements of thin aluminum metallizations by continuous indentation and X-ray stress measurement techniques. Journal of Materials Research 6, 2084–2090; 1991.
11 Tsui TY, Oliver WC, Pharr GM. Influences of stress on the measurement of mechanical properties using nanoindentation. Part I. Experimental studies in an aluminum alloy. Journal of Materials Research 11, 752–759; 1996.
12 Bolshakov A, Oliver WC, Pharr GM. Influences of stress on the measurement of mechanical properties using nanoindentation. Part II. Finite element simulations. Journal of Materials Research 11, 760–768; 1996.
13 Suresh S, Giannakopoulos AE. A new method for estimating residual stresses by instrumented sharp indentation. Acta Materialia 46, 5755–5767; 1998.
14 Carlsson S, Larsson PL. On the determination of residual stress and strain fields by sharp indentation testing. Part I. Theoretical and numerical analysis. Acta Materialia 49, 2179–2191; 2001.

15 Carlsson S, Larsson PL. On the determination of residual stress and strain fields by sharp indentation testing. Part II. Experimental investigation. Acta Materialia 49, 2193–2203; 2001.
16 Lee YH, Kwon D. Measurement of residual-stress effect by nanoindentation on elastically strained (100)W. Scripta Materialia 49, 459–465; 2003.
17 Lee YH, Kwon D. Stress measurement of SS400 steel beam using the continuous indentation technique. Experimental Mechanics 44, 55–61; 2004.
18 Lee YH, Kwon D. Estimation of biaxial surface stress by instrumented indentation with sharp indenters. Acta Materialia 52, 1555–1563; 2004.
19 Bocciarelli M, Maier G. Indentation and imprint mapping method for identification of residual stresses. Computational Materials Science 39, 381–392; 2007.
20 Rydin A, Larsson PL. On the correlation between residual stresses and global indentation quantities: Equi-biaxial stress field. Tribology Letters 47, 31–42; 2012.
21 Larsson PL. On the mechanical behavior at sharp indentation of materials with compressive residual stresses. Mater Design 32, 1427–1434; 2011.
22 Larsson PL. On the determination of biaxial residual stress fields from global indentation quantities. Tribol Lett 54, 89–97; 2014.
23 Larsson PL, Blanchard P. On the correlation between residual stresses and global indentation quantities: Numerical results for general biaxial stress fields. Materials Design 37, 435–442; 2012.
24 Huber N, Heerens J. On the effect of a general residual stress state on indentation and hardness testing. Acta Materialia 56, 6205–6213; 2008.
25 Heerens J, Mubarok F, Huber N. Influence of specimen preparation, microstructure anisotropy, and residual stresses on stress-strain curves of rolled Al2024 T351 as derived from spherical indentation tests. Journal of Materials Research 24, 907–917; 2009.
26 Swadener JG, Taljat B, Pharr GM. Measurement of residual stress by load and depth sensing indentation with spherical indenters. Journal of Materials Research 16, 2091–2102; 2001.
27 Johnson, K.L.: The correlation of indentation experiments. Journal of the Mechanics and Physics of Solids 18, 115–126; 1970.
28 Johnson KL. Contact Mechanics. Cambridge University Press, Cambridge; 1985.
29 Larsson PL. On the mechanical behavior of global parameters in material characterization by sharp indentation testing. Journal of Testing and Evaluation 32, 310–321; 2004.
30 Tabor D.: Hardness of Metals. Cambridge: Cambridge University Press, Cambridge; 1951.
31 Larsson PL. Investigation of sharp contact at rigid plastic conditions. International Journal of Mechanical Science 43, 895–920; 2001.
32 Oliver WC, Pharr GM. An improved technique for determining hardness and elastic modulus using load and displacement sensing indentation experiments. Journal of Materials Research 7, 1564–1583; 1992.
33 Larsson PL, Blanchard P. On the invariance of hardness at sharp indention of materials with general biaxial residual stress fields. Materials Design 52, 602–608; 2013.
34 Bhattacharya AK, Nix WD. Finite-element simulation of indentation experiments. International Journal of Solids and Structures 24, 881–891; 1988.

35 Bhattacharya AK, Nix WD. Analysis of elastic and plastic deformation associated with indentation testing of thin films on substrates. International Journal of Solids and Structures 24, 1287–1298; 1988.
36 Laursen TA, Simo JC. A study of the mechanics of microindentation using finite-elements. Journal of Materials Research 7, 618–626; 1992.
37 Giannakopoulos AE, Larsson PL, Vestergaard R. Analysis of Vickers indentation. International Journal of Solids and Structures 31, 2679–2708; 1994.
38 Larsson PL, Söderlund E, Giannakopoulos AE, *et al.* Analysis of Berkovich indentation. International Journal of Solids and Structures 33, 221–248; 1996.
39 Giannakopoulos AE, Zisis T. Analysis of Knoop indentation. International Journal of Solids and Structures 48, 175–190; 2011.
40 Zisis T, Giannakopoulos AE. Analysis of Knoop indentation strain hardening effects. International Journal of Solids and Structures, 48, 3217–3231; 2011.
41 Giannakopoulos AE, Zisis T. Analysis of Knoop indentation of cohesive frictional materials. Mechanics of Materials 57, 53–74; 2013.
42 Bucaille JL, Felder E, Hochstetter G. Mechanical analysis of the scratch test on elastic and perfectly plastic materials with three-dimensional finite element modeling. Wear 249, 422–432; 2001.
43 Bucaille JL, Felder E, Hochstetter G. Experimental and three-dimensional finite element study of scratch test of polymers at large deformations. Journal of Tribology 126, 372–379; 2004.
44 Felder E, Bucaille JL. Mechanical analysis of the scratching of metals and polymers with conical indenters at moderate and large strains. Tribology International 39, 70–87; 2006.
45 Bellemare S, Dao M, Suresh S. The frictional sliding response of elasto-plastic materials in contact with a conical indenter. International Journal of Solids and Structures 44, 1970–1989; 2007.
46 Wredenberg F, Larsson PL. On the numerics and correlation of scratch testing. Journal of the Mechanics of Materials and Structures 2, 573–594; 2007.
47 Ben Tkaya M, Zidi M, Mezlini S, *et al.* Influence of the attack angle on the scratch testing of an aluminium alloy by cones: Experimental and numerical studies. Materials Design 29, 98–104; 2008.
48 Wredenberg F, Larsson PL. Scratch testing of metals and polymers—experiments and numerics. Wear 266, 76–83; 2009.
49 Aleksy N, Kermouche G, Vautrin A, Bergheau JM. Numerical study of scratch velocity effect on recovery of viscoelastic-viscoplastic solids. International Journal of Mechanical Sciences 52, 455–463; 2010.
50 Bellemare S, Dao M, Suresh S. A new method for evaluating the plastic properties of materials through instrumented frictional sliding tests. Acta Materialia 58, 6385–6392; 2010.
51 Larsson PL. On the correlation of scratch testing using separated elastoplastic and rigid plastic descriptions of the representative stress. Materials Design 43, 153–160; 2013.

2

Nanomechanical Characterization of Carbon Films

Ben D. Beake[1] *and Tomasz W. Liskiewicz*[2]

[1] *Micro Materials Ltd., Willow House, Ellice Way, Yale Business Village, Wrexham, UK*
[2] *Institute of Functional Surfaces, School of Mechanical Engineering, University of Leeds, Woodhouse Lane, Leeds, UK*

2.1 Introduction

The term 'thin film technologies' relates to coatings manufactured typically by Chemical Vapour Deposition (CVD) or Physical Vapour Deposition (PVD) processes. Both methods produce functional thin films, typically less than 10 μm thick, with superior mechanical properties. Carbon films are thin film coatings which consist predominantly of the chemical element carbon, while Diamond-like Carbon (DLC) is a generic term used to describe a range of amorphous carbon films. These include hydrogen free DLC (a-C), hydrogenated DLC (a-C:H), tetrahedral amorphous carbon (ta-C), hydrogenated tetrahedral amorphous carbon (ta-C:H), and those containing silicon or metal dopants, such as Si DLC and Me DLC. The first report of successful deposition of DLC coating dates back to 1971, when Aisenberg and Chabot processed hydrogen free diamond-like carbon films using carbon ions [1].

2.1.1 Types of DLC Coatings and their Mechanical Properties

Diamond-like carbon is a metastable form of amorphous carbon and has a mixture of sp^3 and sp^2 bonding and the mechanical properties of DLC films vary with sp^2/sp^3 ratio [2]. Sp^3 hybridized diamond type bond has configuration resulting in strong C-C bonds, while graphite has a three-fold coordinated sp^2 hybridized bond configuration forming weak bonding between the atomic planes. Coating deposition techniques use typically hydrocarbon gases as a source of carbon and thus DLC coatings contain a certain amount of hydrogen. Hence, depending on the hydrogen content and ratio of sp^2/sp^3 bonds, DLC coatings are divided in different groups, as shown in Figure 2.1, and the ratio of sp^3/sp^2 bonds and the hydrogen content in the coating determine the properties of DLC films. The hydrogen content can vary from less than 1% in non-hydrogenated DLC films to about 60% in hydrogenated DLC films. With a wide range of structural and mechanical properties, DLC coatings can be tailored to various applications where high hardness, low friction, good wear and corrosion resistance and chemical inertness are required properties. Mechanical properties of DLC coatings are compared with their sp^3 ratio and hydrogen content in Table 2.1.

Applied Nanoindentation in Advanced Materials, First Edition. Edited by Atul Tiwari and Sridhar Natarajan.

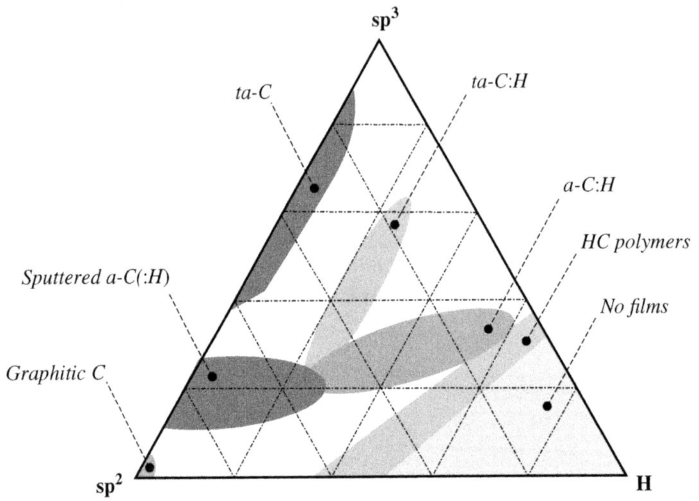

Figure 2.1 Ternary phase diagram of DLC films. *Source:* Robertson 1999 [3]. Reproduced with permision of Cambridge University Press.

Table 2.1 Properties of various types of carbon.

	sp^3 (%)	H (%)	Density (g cm^{-3})	Hardness (GPa)
Diamond	100	0	3.5	100
Graphite	0	0	2.3	
Glassy C	0	0	1.3–1.5	3
Evaporated C	0	0	1.9	3
Sputtered C	5	0	2.2	
ta-C	80–88	0	3.1	80[a)]
ta-C:H	70	30	2.4	50
a-C:H hard	40	30–40	1.6–2.2	10–20
a-C:H soft	60	40–50	1.2–1.6	<10

a) Note. With the advent of more advanced methods of calibration and experiment design, more recent values are somewhat lower than this.
Source: Robertson 1992 [4]. Reproduced with permission of Elsevier.

2.1.2 Carbon Films Processing Methods

DLC coatings are metastable materials deposited by non-equilibrium processes where energetic ions interact with the surface, namely CVD and PVD techniques. CVD involves chemical volatile precursors, which react with the substrate surface to produce the desired deposit [5], while PVD coating is produced by the condensation of a vaporized form of the material onto various surfaces [6].

During CVD process gaseous reactants can be deposited onto a substrate by employing endothermic chemical reactions and providing additional energy for initiation. In conventional CVD processes developed by Arkel and Boer [7], thermal energy is employed for necessary activation. The required temperature is typically in the range of 800–1100 °C, which often leads to dimensional distortion of metallic components and the necessity of an additional heat treatment operation after CVD deposition. One of the ways to decrease CVD process temperature is to replace the thermal energy component with another source of energy. This has been achieved in Plasma-Enhanced Chemical Vapor Deposition (PECVD) process, where electrical energy is used to generate a glow discharge (plasma) in which the energy is transferred into a gas mixture, allowing for a decrease in the deposition process temperature to 300–350°C. Employment of plasma allows the transformation of the gas mixture into reactive radicals (ions, neutral atoms, molecules and other highly excited species) by collision in the gas phase, which allows the maintenance of a low temperature of the substrate material. DLC coatings deposited by means of PECVD method are typically processed in a chamber where acetylene (C_2H_2) is cracked in the plasma and carbon molecules become available for coating growth on the substrate.

According to Bunshah [8], all coatings deposition techniques under reduced pressure in which thermal evaporation, sublimation and spraying phenomena are employed, can be classified as PVD methods. Currently, dozens of variations and modifications of PVD techniques are available [6], all of them however have the following two attributes in common:

1) Based on physical processes taking place under reduced pressure in the range of $10–10^{-5}$ Pa;
2) Conducted on cold substrates or heated ones below 500°C.

Coatings deposited by PVD methods are usually formed by beams of atoms and/or ions. For evaporation or sublimation of the particles, electron beam or arc discharge is used, while for magnetron sputtering usually Ar+ ions accelerated in the electric field are employed. The above techniques are reactive in nature and atoms are moved into a vapor state through sublimation, metal evaporating from the surface of the molten pool, or by sputtering the solid magnetron target.

Different ways of DLC processing result in different mechanical properties of coatings. Table 2.2 shows a comparison of mechanical properties and an indication of industrial use of DLC coatings deposited using PECVD and PVD methods.

2.1.3 Residual Stresses in Carbon Films

DLC films are usually deposited with high levels of intrinsic compressive stress [2, 4, 10], due to ion bombardment during deposition process and competition between implantation and relaxation. The residual stress varies with sp^2/sp^3 bonding ratio and the coating thickness. Since the high level of residual stress is detrimental to the coating performance due to adhesion weakening, several strategies have been employed to minimize the level of residual stress. These include duplex, multilayering, metal doping and bias grading to reduce stress and avoid stress concentrations and eliminate interfacial cracking.

Indentation fracture behavior and toughness of ta-C thin films deposited using pulsed laser deposition (PLD) were examined by Jungk *et al.* [11]. This deposition process produced films with a compressive stress of 7 GPa as measured using wafer curvature technique. By comparing the indentation behaviour of free-standing DLC films and films on

Table 2.2 PECVD and PVD types of DLC coatings (after [9]).

	Metal-doped DLC	Amorphous hydrogenated DLC	Silicon-doped DLC	Sputtered DLC	Hydrogen-free DLC
Type	WC-C:H	a-C:H	a-C:H-Si	a-C	ta-C
Method	PVD/PECVD	PECVD	PECVD	PVD	PVD
Hardness ($HV_{0.05}$)	800–2200	1500–3500	1500–2500	2000–4000	3000–7000
Coefficient of friction	0.1–0.2	0.05–0.15	0.05–0.1	0.05–0.1	0.02–0.1
Internal stress (GPa/μm)	0.1–1.5	1–3	1–3	2–6	1–3
Thickness (μm)	1–10	1–10	1–10	1–3	1–3
Industrial use	yes	yes	yes	yes	yes
Mass production	+ + +	+ + +	++	+ + +	++

Table 2.3 Residual stress in DLC coatings resulting from different coating thickness and deposition technique.

Coating/substrate	Deposition process	Residual stress (GPa)	Reference
110 nm ta-C on Si	PLD	−7	11
600 nm DLC on Si	CVD	−3.8	12
0.1–1.9 μm DLC on Si	Sputtered	−1 to −1.5	13
0.9–1.1 μm DLC on Si	PECVD	−0.1 to −0.7	14
0.9–1.1 μm DLC on Steel	PECVD	−0.03 to −0.24	14

Silicon substrate, Lee *et al.* [12] determined the residual stress of a 600 nm CVD DLC on Si with a reduced elastic modulus of 493 GPa as −3.8 GPa. Fu *et al.* [13] deposited DLC films with different thicknesses (from 100 nm to 1.9 μm) using sputtering of graphite. They found an increase in film stress with film thickness increased from 100 to 300 nm, followed by gradual stress decreases above 300 nm film thickness and the change in sp^2/sp^3 ratio in the films was pointed out as a reason for film stress change. Residual stress measurements of TiN, DLC and MoS2 coated surfaces have been summarised by Holmberg *et al.* and related to their tribological fracture behaviour [14] (Table 2.3).

2.1.4 Friction Properties of Carbon Films

DLC coatings provide low friction and superior wear resistance. They provide surface protection of components designed to operate in fully lubricated regime during strap-up when mixed or boundary regime exists. In gasoline engines, application of DLC coating reduced friction by 25% resulting in an overall 4% improvement of fuel efficiency [15]. The role of abrasive particle size on the wear of DLC coatings was investigated in

[16]. The abrasive particle size was varied under the mixed lubrication regime and no delamination or any major damage to the coating was observed in the case of lubricating fluid having no sand. However, three-body abrasive wear leading to internal cracking, spallation and sequential removal of coating was observed with sand added in the lubricating fluid as an abrasive medium. The three-body wear mechanism was attributed to the entrainment of abrasive particles into the tribological contact.

2.1.5 Multilayering Strategies

Hard tetrahedral amorphous carbon coatings are also characterized by a high Elastic Modulus, which is typically 6–10 times higher than their hardness [17]. This leads to high internal compressive stresses within the coating, limits its thickness and is a challenge in terms of adhesion to the coating/substrate interface. This problem is typically overcome by designing multi-layered coating architectures composed of a top DLC coating and adhesive interlayers, e.g., Ti or Cr. Properties of DLC coatings can be further modified by the addition of dopants, i.e., elements such as nitrogen, silicon, oxygen, fluorine or metals (Ti, Nb, Ta, Cr, Mo, W, Ru, Fe, Co, Ni, Al, Cu and Ag). Such modifications are made mainly to lower internal stress within the coating and improve adhesion but also to functionalize coating by lowering friction coefficient, reduce surface energy or modify electrical properties [18–20].

Corrosion performance of DLC coatings and role of interlayers on 316 steel were studied in [21]. It has been found that a-SiNx interlayer provided the best adhesion and significantly improved the corrosion resistance of the DLC. A two-order of magnitude improvement in the corrosion current density was recorded for that interlayer comparing to the DLC with the nitrided interlayer. Also, as a result of the formation of a passive silicon oxide film at the electrode/electrolyte interface, Si-doped DLC coatings exhibited improved corrosion barrier properties. Corrosion protection of DLC coatings doped with silicon deposited on magnesium alloy was studied in [22]. Incorporation of silicon in the DLC coating improved its corrosion protection and the corrosion performance improved with increasing silicon content in the coating. Further surface analysis carried out in this study indicated a presence of a thick oxide layer on the Si-DLC coating surfaces. It was observed that insulating properties of this oxide layer results in high anti-corrosion properties of the Si-DLC films. It was concluded that very low internal stress in the Si-DLC coatings is one of the reasons for the high anti-corrosion performance of this system.

Depner-Miller *et al.* investigated a hybrid coating architecture, where the DLC layer was deposited on top of top of a multilayer PVD coating system [23]. Residual compressive stresses and hardness of the coating system were measured and the abrasive degradation of the investigated coatings was found to depend mainly on coating hardness under combination of abrasive and fatigue loading with the hardest DLC coating showing least abrasive degradation. In contrast, the same coating showed the most severe fatigue damage when tested under cyclic loading. A sandwich structure with a polymeric amorphous DLC layer was proposed by Neuville as a novel approach for erosion-resistant coatings [24]. The composite architecture of the coating, composed of a soft polymeric amorphous carbon, an adhesion layer and a hard elastic anti-wear PECVD layer, showed promising results and its performance was theoretically analyzed.

2.1.6 Applications of Carbon Films

A wide range of available amorphous carbon coating architectures and a number of possible deposition methods allow tailoring of the coating functionality to many applications. DLC coating properties include high hardness, low friction, electrical insulation, anti-corrosion, chemical inertness, optical transparency, biological compatibility, ability to absorb photons selectively, smoothness and resistance to wear. Hence, DLC coatings have found many practical applications across industrial sectors, including razor blades, computer hard drives, silicon solar cells, MEMS applications, orthopedic implants, optical lenses, and finally cutting tools and internal combustion engine components.

2.1.7 Optimization/testing Challenges

With a variety of carbon film architectures and properties come optimization and testing challenges as, very often, focus on combination of properties is needed. Therefore, a multi-technique approach to optimize DLC behavior has been postulated by the authors. The approach combines overload tests with fatigue and durability experiments and it includes a suite of nano-mechanical testing methods, namely nano-indentation, nano-scratch, nano-impact and nano-fretting. Table 2.4 matches the DLC coating systems evaluated by the authors over the last decade with the nano-mechanical techniques described in more detail in this chapter.

2.2 Factors Influencing Reliable and Comparable Hardness and Elastic Modulus Determination

2.2.1 The International Standard for Depth-sensing Indentation: EN ISO 14577–4 : 2007

To optimize DLC films it is important that their mechanical properties such as hardness and elastic modulus be reliably determined. Progress from the EU INDICOAT project in particular [37] resulted in the development of the ISO standard (14577) [38]. The standard identifies the four key parameters – force, displacement, instrument compliance and indenter shape – that influence the quality of the results and provides the necessary methodology to accurately calibrate these. Refinements to the analysis procedures have been proposed recently [39]. Fused silica is commonly used as reference material as it is isotropic, though recently sapphire and tungsten have been considered as they are more sensitive to the instrument stiffness. The most recent section of this standard, ISO 14577-4, 'Test method for metallic and non-metallic coatings' concentrates on determination of these properties on coatings independently of any substrate contribution. In particular it emphasizes that the well-known '1/10 rule' - indentation to less than 1/10 of the total film thickness to determine coating-only properties independently of substrate influence - should only be applied to the determination of hardness and does not work for elastic modulus due to the elastic stress field extending much further into the substrate than the plastic stress field. Improved strategies for elastic modulus determination are based on performing indentations to a range of indentation depths and extrapolating the results to zero penetration depth to obtain a result free of substrate influence.

Table 2.4 DLC coating systems evaluated in this chapter.

DLC system[a)]	Nano-mechanical techniques	Reference
FCVA[a)] 5–80 nm ta-C on Si	Nanoindentation and nano-scratch	25
	Nano-impact	26
	Nanoindentation and nano-scratch	27
	Nanoindentation	28
	Nanoindentation, nano-scratch and nano-fretting	29
FCVA 80 nm ta-C on Si	Nano-fretting	30
CFUBMS[b)] 70 nm a-C:H on Si	Nanoindentation and nano-scratch	27
	Nano-fretting	30
CFUBMS 200 nm–1.1 μm a-C on Si	Nanoindentation, nano-scratch and nano-wear	31
CFUBMS 1 μm a-C on Si	Nano-fretting	30
PECVD[c)] 140–610 nm a-C:H and Si:a-C:H on glass	Nanoindentation, nano-scratch	32
CFUBMS 3.5 μm a-C:H(Ti) on steel	Nanoindentation and nano-scratch (25 to −30 °C)	33
CFUBMS 2.5–2.8 μm a-C:H and a-C on steel	Nanoindentation and nano-impact	34
PACVD[d)] 2.3–4.5 μm a-C:H on steel	Nanoindentation and nano-impact	34
CVD[e)] 570–620 nm a-C:H on Si	Nano-impact	35
PECVD[e)] 2.8–2.9 μm a-C:H and Si:a-C:H on steel[e)]	Nanoindentation and nano-fretting	36
Balinit[e)] C Star 2 μm a-C:H:W[e)]	Nanoindentation and nano-fretting	36

a) Ultra-thin filtered cathodic vacuum arc
b) Closed field unbalanced magnetron sputtering deposition process
c) Plasma-enhanced chemical vapour deposition
d) Plasma-assisted chemical vapour deposition
e) Where coatings are multi-layered, the total thickness is given but only the composition of the outer DLC is described for simplicity.

For hard coatings on soft substrates the hardness of the coating can only be determined with a sharp indenter that causes sufficient yielding within the coating before the substrate yields (i.e. the maximum shear stress occurs within the coating causing plastic deformation whilst the stress in the substrate is lower than the substrate yield stress). A plateau in the measured hardness vs. h_c/t_c, (h_c = contact depth, t_c = film thickness, h_c/t_c is commonly referred to as the relative indentation depth or RID, although other terms such as contact depth ratio [40] have also been proposed) or the value taken at RID = 0.1 is usually taken as the coating hardness. However, strictly this should be considered as a lower bound for the hardness since it may not be possible to reach the true hardness of the coating before plastic yield of the substrate occurs. An example from the INDICOAT project is included in the ISO standard, where for DLC coatings on M2 steel with nominally the same composition but different thickness (0.46, 1.5 and 2.5 μm) the measured hardness of the thinnest coating was lower, at 15 GPa, than that measured on both the thicker coatings (18 GPa) due to the onset of substrate yield before fully developed plasticity could be reached. Chudoba has investigated the minimum depth (in relation to the tip radius) where constant hardness could be expected for a conical indenter with a spherical end cap [41]. More recently the same author has noted that the plateau corresponding to the coating hardness is typically from a depth of 0.2 R (where R is the indenter radius) to RID = 0.1 [42]. For hardness measurements on coatings of different thicknesses to be directly comparable it follows that the minimum thickness be at least $2R$. Since the tip radius of a typical Berkovich indenter is 100–200 nm it follows that the minimum thickness is of the order of 200 nm. Care must be taken in comparing hardness measurements for coatings thinner than this. For ductile materials like metals Tabor determined an empirical relationship between hardness and yield stress of $H/Y \approx 3$ [43]. However, lower values are found for materials with higher H/E. For a 2.1 μm DLC on Si, Richter and co-workers reported $H/Y = 1.57$ [44, 45]. Liskiewicz and co-workers reported $H/Y \approx 1$ for a-C:H and Si-doped a-C:H multilayer coatings deposited by PECVD on M2 steel.

Since the measured hardness at $h_c/t_c = 0.1$ can provide a reliable measure of the coating hardness the corresponding elastic modulus at the same depth has been taken as the coating modulus. However, this is incorrect since the elastic stress field extends much further than the plastic stress field so that the value at $h_c/t_c = 0.1$ will contain appreciable contribution from the substrate. To circumvent this ISO 14577–4 recommends a linear extrapolation of measurements at different depths to zero depth. Although FEA has shown that in general the relationship is non-linear, nevertheless the approach can provide a more robust measurement of modulus than using a value at a depth of $h_c/t_c = 0.1$. An accurate value of the coating thickness is not required to obtain a reliable intercept value. For non-spherical indenters such as the Berkovich the effective contact radius (a) is approximated by a circle with the same area as the projected area of contact of the indenter (i.e. $a = \sqrt{(A_p/\pi)}$). For hard/brittle coatings the data should be in the range $a/t_c < 2$. For 1–3 μm TiN coatings on stainless steel Jennett and co-workers have shown that measurement of E for different thickness coatings were almost indistinguishable when plotted the same values of a/t_c [46]. Data were extrapolated to $a/t = 0$ to obtain the coating-only modulus which varied by only 1% between the films. Although there is no 'safe depth range' at which the elastic response of a coated system to indentation is completely free of substrate influence. However, for relatively thick films it is possible to obtain a reasonably estimate by restricting the indentation depth to around 5% of the coating thickness (i.e. RID = 0.05) if coating and substrate have similar moduli [26]. ISO 14577-4 provides flowcharts to aid the reliable determination of coating-only hardness and elastic modulus. In the case of multi-layered coatings it is not generally possible to apply the ISO extrapolation approach although it may provide a reasonable estimate of the modulus of the outer layer if the sub-layers are sufficiently thin and broadly similar in stiffness to the substrate, as can be the case for bond layers. For multi-layered coatings more complex analytical treatments such as Film Doctor [47] can offer more reliable determination of the stiffness of the individual layers, provided the properties of the substrate are well known. Spherical indenters can be used to minimise plasticity and cracking and determine elastic modulus more precisely. Experimental determinations of the variation in indentation modulus with depth for a 2 μm DLC on M2 tool steel by 9.5 μm end radius diamond were found to be in good agreement with elastic estimates in the range $0 \leq t_c/a \leq 5$ [37, 48].

2.2.2 Challenges in Ultra-thin Films

As shown in Table 2.3, the residual stress in DLC films can reach 10 GPa, limiting the maximum thickness that can be deposited without delamination. For example, ta-C are limited to 200 nm (PLD) or 180 nm (FCVA) and sputtered a-C are limited to ~1 μm when deposited at high substrate bias voltage [49]. Multilayering is a common strategy when higher thickness is required. For reliable hardness measurements it is necessary to have a fully developed plastic zone. On ultra-thin hard carbon films (approximately less than ~100 nm) the hardness measured by nanoindentation may be less than the true film hardness, either at very shallow penetration depths where the indenter is more spherical in shape and the contact is essentially elastic (so that the mean pressure is less than the hardness) or at greater depths where plastic yield in the substrate occurs before the plastic zone is fully developed within the coating. For an ultra-thin DLC film analysed with a Berkovich indenter it is necessary to consider the combined influence of the substrate and indenter blunting on the measured data. A typical example is shown in Figure 2.2 [28]. It can be seen that the measured hardness first increases then decreases with increasing contact depth.

The most common approach is to choose the intermediate depth region as a measure of the film hardness since this is less influenced by tip geometry and substrate behaviour [50]. To extract the film-only hardness from data including some substrate contribution the system can be treated as a two-layer composite system where the overall hardness is the sum of volume law of mixtures contributions as suggested by Burnett and Rickerby [51]. Alternatively, Bhattacharya and Nix [52] used FEM to develop an analytical equation for a hard film on a softer substrate:

$$H_{\text{measured}} = H_s + (H_f - H_s)\exp(-k_1 * \text{RID}) \quad (2.1)$$

This can provide a good fit to experimental data beyond the tip blunting region but assumes a perfectly sharp tip and is not applicable to data at low depth. Lemoine and

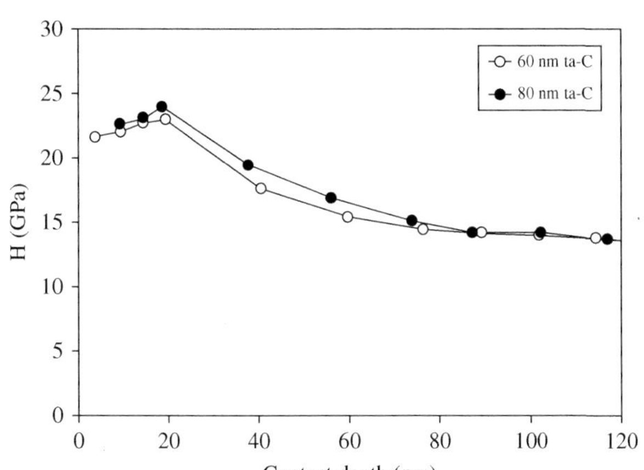

Figure 2.2 Contact depth dependence of measured hardness on 60 and 80 nm ta-C films on Si. *Source:* Beake 2010 [28]. Reproduced with permission of Taylor & Francis.

co-workers have modified this analytical treatment to accommodate the effect of tip blunting [40].

$$H_{\text{measured}} = H_s(h_c) + [H_f(h_c) - H_s(h_c)] \exp(-k_1 * \text{RID}) \tag{2.2}$$

Using this treatment they reported best-fit values of 29 and 38 GPa for 10 nm and 50 nm ta-C on Si respectively, which were larger than those determined by the application of Equation (2.1). Although ta-C films have been reported to have hardness in excess of 80 GPa at very low penetration depths the accuracy of the indenter area function and the method used to determine it may also affect this. In other cases of coatings with reported super-hardness (over 80 GPa), notably the TiSiN nanocomposites reported by Veprek and co-workers it has later been proved that the results were significantly in error due to the incorrect area function being used [53].

2.2.3 Indenter Geometry

Sharper indenters, and indenters with smaller apical angles such as the cube corner have been used to measure coating-only hardness since they promote plastic deformation in the coating before substrate yield. Nevertheless, cracking is more likely. Jungk and co-workers reported that although the indentation of 110 nm ta-C films with a cube corner indenter was completely elastic below 1.5 mN with multiple discrete displacement jumps due to cracking being observed above this threshold load [11]. Martinez and Esteve have shown that by applying Hertz's theory to the indentation of totally elastic materials that hardness is dependent on indenter angle [54]. It is therefore not possible to directly compare hardness results obtained with cube corner indenters to with those obtained with more commonly used Berkovich indenters. For Berkovich indenters, at shallow depths the hardness is a function of indentation depth due to the non-ideal shape of the indenter [55]. For example, Lemoine and co-workers reported that $H(h) \propto h^{0.38}$ for the first 20 nm [50]. Hertzian contact mechanics shows that the measured hardness increases with the depth in an asymptotic manner $H(h) \propto h^{0.5}$ until it reaches the value which would be measured with an ideally sharp indenter. Deviations from ideal geometry are also commonly reported for spherical diamond indenters, which are typically sharper than their nominal radii at low penetration depths. For example Herrmann *et al.* reported that when measured by AFM the actual tip radius of a nominal 10 μm indenter was 8.3 μm at 100 nm increasing 9.0 μm at 300 nm and was close to the nominal value at 1.0 μm depth [56]. Similarly, Ni *et al.* reported that the effective radius of a spherical indenter used to check on robustness of FEA-supported relationships between PI and H/E_r varied from ~8 μm at 180 nm to ~14 μm at 1850 nm indentation depth with indentation data on copper being fitted to a power-law dependency using a constant elastic modulus for the sample throughout the displacement range [57].

2.2.4 Surface Roughness

The international standard for depth-sensing indentation, ISO14577, states that the surface roughness should be less than 5% of the maximum displacement to eliminate the influence of roughness on nanoindentation data. In practice this may be difficult to achieve, particularly for relatively thin films deposited on non-mirror polished substrates. High surface roughness therefore is a challenge for nanoindentation instrumentation. Instruments must be able to apply high load to mitigate against it and have high rigidity in the plane normal to the indentation. Depending on the lateral stiffness of the loading head in the instrument, surface roughness may have a significant influence on scratch and indentation behaviour. If the lateral stiffness is sufficiently high the indentation loading head does not move laterally on contact with the surface. In this situation the high surface roughness results in data scatter since the indenter will contact on asperity peaks and in the valleys between them. In principle, both the root mean square (or arithmetic) roughness and its wavelength may influence the resultant measured values of hardness and modulus:-

1) Where the wavelength is large relative to the shape of the indenter the effect of roughness is increase the area of contact and the increase the hardness measured assuming an initially flat surface.
2) When the wavelength of the roughness is small compared to the indenter size (asperity contact) the effect is opposite and hardness is reduced.

Bouzakis and co-workers have reported that for TiAlN coatings with R_a roughness 0.05–0.35 μm, the effect of increasing roughness was to widen the distribution of results whilst the mean values remained unchanged [58], suggesting the roughness should not too significantly influence the mean values of hardness and elastic modulus, i.e. the mean values of H and E are not necessarily adversely affected and statistical reliability simply be achieved by making more measurements. For multilayered CVD coatings with high sample roughness (R_a = 0.19–0.46 μm) typical of tool coatings it has been reported that mean values are acceptable provided sufficiently large numbers of indentations are performed [59].

However, if the lateral rigidity of the loading head is not high enough the indenter can slide down the sides of an asperity contact and settle in a dip resulting in a systematic under-estimation of the real area of contact. Since the hardness is determined assuming indentation into a flat surface, a systematic over-estimate of the hardness of the sample (roughness induced indentation size effect) ensues. High surface roughness can be a significant factor when the mechanical properties are measured by oscillatory techniques (CSM method). Using the CSM method Zhang and Huan reported high (> 60 GPa) hardness on 500 nm DLC when deposited on bearing steel or alloy steel but the standard deviation in hardness in both cases was of the order of ± 20 GPa when the indentation depth was under 100 nm which may question the reliability of their results [60].

With the exception of arc-deposited coatings, deposition of DLC coatings typically does not significantly increase the surface roughness of the substrate. For characterisation, DLC coatings are often deposited on mirror polished Si. Ta-C coatings on Si have been reported to have surface roughness around 3 nm [25]. Results when deposited on Si are directly relevant to MEMS applications but the indentation behaviour can be complex, with phase transformations of the Si substrate, cracking of coating and/or substrate and delamination. Peng and co-workers have reported even lower values [61]. For tribological applications it is more relevant to determine behaviour on the substrate of interest, such as high-speed steel, stainless steel or titanium. The surface roughness of DLC coatings has found to influence their tribological behaviour [62–65]. Smoother coatings can minimise contact stress concentrations and thus reduce the risk of cracking and/or plastic deformation [63]

2.3 Deformation in Indentation Contact

2.3.1 The Relationship Between *H*/*E* and Plastic and Elastic Work in Nanoindentation

DLC coatings are hard and elastic, and when deposited on steel substrates they are typically harder but of similar or lower modulus so have considerably higher H/E and can provide improved load support. The ratio of the hardness to modulus (H/E) has been found to correlate with tribological behaviour more closely than hardness alone. In sliding/abrasion several authors have shown that a high H/E is generally beneficial [66–68]. Cheng and Cheng have used finite element analysis to show that H/E_r is related to a dimensionless plasticity index (PI) which is the plastic or irreversible work done during indentation (W_p) divided by the total elastic (W_e) and plastic work done during the indentation, over a wide range of H/E values, with the relationship taking the apparent form [69, 70]:

$$PI = W_p/(W_p + W_e) = 1 - x(H/E_r) \tag{2.3}$$

where x is a constant and E_r is the reduced indentation modulus. The FE results suggested x should be around 5. Although E_r, the reduced indentation modulus (sometimes referred to as E'), appears in the equation directly, the plane strain modulus $E^* = E/(1 - \nu^2)$ is sometimes quoted (e.g. in ISO 14577) and it is important in presenting nanoindentation data to specify clearly which modulus is being reported (E_r, E^* or E). Several authors have tested this equation experimentally. To eliminate the possibility of results being affected by a non-standard indenter geometry Beake *et al.* used several different Berkovich indenters and found values are higher on most materials, reporting [71] x = 6–7 for copper, aluminium, tungsten, bearing steel, stainless steel and WC-Co, whilst only fused silica and K7 glass show lower x values very close to those predicted by Cheng and Cheng [70]. Choi and co-workers have also re-investigated the Cheng and Cheng relationship and found that these two regimes exist [72], for their data x ~ 5 for materials which sink-in (glasses) whilst they found x ~ 7 for materials which exhibit pile-up (metals). More recently, its applicability to high H/E coated systems has been studied. When Berkovich indenters are used values on DLC are around 6, consistent with Hard TiN-based nanocomposites (~6.4) [73] and TiAlCrN/TiAlCrSiYN multilayers (5.7) [74]. When comparing values for different coatings it should be noted that it can additionally depend on (i) the peak indentation load (higher loads result in greater substrate contribution to elastic modulus and may result in cracking) (ii) whether the coating exhibits full plasticity at this load (otherwise the measured hardness will be lower (iii) elastic accommodation of deformed material under the indent (iv) indenter rounding at the low indentation depths required to measure coating properties. Milman [75] and Cheng and Cheng [69, 70] have also noted that another dimensionless index easily obtained from nanoindentation, the ratio of the residual depth (h_r) to the maximum indentation depth (h_m) is approximately equivalent to the plasticity index (Eqn. [1.4]).

$$h_r/h_m \sim W_p/W_{total} \tag{2.4}$$

Beake and co-workers investigated the relationship between the plasticity index and the ratio of the residual to maximum indentation depth for a range of metallic and non-metallic bulk samples, finding good correspondence (to within 0.01) for all the metallic samples but on non-metals (fused silica and K7 glass) with higher H/E the *PI* is slightly lower than h_r/h_m [71]. The above discussion shows that for a Berkovich indenter in principle *PI*, h_r/h_m and H/E can be used almost interchangeably, though the *PI* has the potential advantages of (i) negating the requirement for the indenter area function (since H and E are not being determined) (ii) using the entire indentation curve rather than only the unloading segment (iii) easier to determine the areas accurately than the exact residual displacement (iv) it is a direct measurement of the energies involved in the indentation process.

Studies into the interrelationships between *PI*, h_r/h_m and H/E have not been restricted to Berkovich indenters. Using dimensional analysis and finite element calculations Ni and co-workers studied spherical indentation in elastic–plastic solids with work hardening [57]. By analysing FEA generated indentation curves they determined from a least squares fit that $h_f/h_{max} = W_p/W_t$ and suggested that this relationship is universal and does not depend on the mechanical properties of the material, the indenter radius or the indentation depth but it should be noted that experimental data for the only high H/E material studied, fused silica, was found to be off the best fit-line. Ni and co-workers [57] applied their FEA-supported approach to determine an approximately linear relationship between H/E_r and *PI*, which for a fixed h_{max}/R could be expressed as

$$PI = \mathrm{B}\,(H/E_r) + 1 \tag{2.5}$$

Least squares fitting to their FEA-generated data enabled them to propose that the ratio of hardness to reduced modulus can be obtained by determining h_{max}/R and the plasticity index without the requirement to analyse the unloading curve according to

$$H/E_r = 0.5928(h_{max}/R)0.62^{*}PI \tag{2.6}$$

2.3.2 Variation in *H/E* and Plasticity Index for Different DLC Films

Depending on their structure, DLC films cover very wide range of hardness, from sp^2-rich films which can have hardness as low as 5 GPa, to FCVA ta-C with ~85 % sp^3 which have been reported to have hardness of 70–80 GPa, approaching that of diamond. Chartidies surveyed the nanomechanical literature and has determined how hardness and elastic modulus for different types of DLC vary with density, sp^3 content and coordination [76]. This study clearly shows that for different types of DLC there is an approximately linear relationship between the mechanical properties (H, E) and the structural nature (co-ordination number, sp^3 content) of the film. Different types of films were reported to have different slopes in plots of H vs. sp^3 content and when extrapolated to 100 sp^3 the hardness was found to vary from 46–81 GPa [76]. The reason for such a wide discrepancy is not immediately obvious. However, several of the films were under 100 nm which leads to significant difficulties in determining the true thin film hardness from nanoindentation due to the influence of substrate yield and tip blunting on the measured hardness data. It may be that the view of the differences in thickness for the different types of films that the hardness measurements may not be directly comparable between the various studies.

Although DLC films cover a wide range of hardness, the variation in H/E is much smaller, with values commonly around $H/E = 0.1$. Gogotsi and co-workers [77]

Table 2.5 Nanoindentation data for amorphous carbon coatings on M42 tool steel.

Coating	Hardness (GPa)	Reduced modulus (GPa)	H/E_r	H^3/E_r^2 (GPa)
a-C	13.9 ± 0.5	161.6 ± 3.9	0.086	0.102
a-C:H	17.0 ± 0.4	128.8 ± 2.5	0.132	0.296

noted that other carbon films, such as carbide-derived carbon films, glassy carbon and pyrolytic carbon also have H/E close to 0.1. Michler and co-workers noted that-although DLC films were commonly prepared with $E/H = 9$–10, deposition by bipolar pulsed DC PACVD could produce films with $E/H \sim 6 - 7$ [78, 79]. Deposition of a-C under high bias typically results in high compressive residual stress. Zhang *et al.* reported [80] a value of 4.1 GPa for a-C deposited at −150 V. Increasing the sp^2 content in the DLC can provide stress relief. Zhang and co-workers have been able to produce amorphous carbon films with much lower stress by several different approaches: (i) doping with Al (ii) preparing a nanocomposite nc-TiC/a-C(Al) by co-deposition of Ti and Al (iii) bias-grading during deposition [49, 80–82]. Doping with the Al significantly lowered the residual stress to 0.4 GPa due to the combined influence of increasing sp^2 fraction and lower hardness from the incorporation of the soft metal itself. The reduction in stress was associated with an increase in plasticity. Similar behaviour has been reported on CFUBMS coatings. In Table 2.5 the mechanical properties of a 2.8 μm a-C:H (Dymon-IC from TEER Coatings) on M42 tool steel are contrasted with a 2.5 μm sp^2-rich a-C (Graphit-IC from TEER Coatings) on the same substrate [34]. Due to the high surface roughness on the M42 steel it was necessary to perform a large number of indentations. The mean values of H and E with their standard errors were from over 60 indentations to 200 nm. The differences in mechanical properties influence the tribological behaviour as discussed in following sections.

2.3.3 Cracking and Delamination

The occurrence of cracking and/or coating-substrate debonding during contact loading depends on the test conditions (such as indenter geometry, applied load, loading rate) and properties of the film, substrate and the interface between them. The low yield stress of metallic substrates such as Ti or stainless steel limits the ability to accurately determine the hardness of DLC coatings when deposited on them, particularly for films with low thickness. Although the yield stress of metals and alloys shows a similar indentation size effect to that hardness at the nano-scale [83] nevertheless, values remain much lower than most DLC coatings. At the onset of plastic flow in the substrate the load is essentially transferred to the substrate. As the coating cannot accommodate the resulting substrate deformation tensile stresses develop parallel to the surface within the coating and cracking can occur [84]. Zhang and Huan [60] observed steps in curves during indentation of 500 nm DLC on 9Cr18 and 40CrNiMo steel substrates and interpreted these as sub-surface cracks due to the large differences in hardness between the DLC layer and the substrates. Dramatic pop-ins have been observed on DLC films. These pop-ins can be associated with the onset of through-thickness cracking and can be very similar in magnitude to the film thickness. The actual (on-load) indenter displacement at the onset of the film failure can also be similar to the film thickness but the correspondence is somewhat fortuitous in that the relationship between the indentation depth during loading and the contact depth is dependent on the indenter geometry. Chen and co-workers have proposed a method for estimating coating toughness when no excursions in the load-displacement occur as has been found for thin coatings where picture-frame cracks occur [85]. The method relies on separating the fracture energy from other contributions to the irreversible work done in indentation and for oxide coatings similar values of fracture toughness were obtained to those determined by other methods. Galvan and co-workers differentiated the load-displacement curves of nc-TiC/a-C:H nanocomposite coatings on stainless steel, associating the depth at which the curvature begins to decrease with the onset of plastic flow in the much softer stainless steel substrate [84]. A coating with high H^3/E^2 and lower elastic modulus than the stainless steel spreads the load over a larger volume. This delays the onset of plastic deformation in the substrate and subsequent cracking and chipping of the coating due to the development of tensile stresses on the coating side close to the substrate-coating interface that result in cracks. A similar approach has been proposed by Klemberg-Sapieha and co-workers in multilayer design for erosion protection [86, 87]. The occurrence of a noticeable change in the indentation curve and the extent of cracking can be correlated to differences in mechanical properties between film and substrate. When the properties of the film are similar to the substrate (e.g. a-C on Si), then slope changes and pop-in events are less likely. For this reason, substrate pre-treatment (e.g. nitriding) can be especially useful in minimising so-called egg-shell failures of hard coatings on much softer substrates. The choice of indenter geometry can influence the type of cracking. Faisal and co-workers performed multiple-cycle indentation to constant load to assess susceptibility of 100 and 500 nm DLC on Si to cracking and delamination [88]. Using a Berkovich indenter appreciable film and substrate cracking and film removal resulted in an increase in indentation depth with each cycle. They found that when a conical indenter with 10 μm end radius was used the load-displacement curve often showed a backward depth deviation (i.e. apparent indentation depth reducing with each reloading cycle) and subsequent AFM imaging of the worn area revealed blistering of the coating due to delamination. FE modelling suggests that location of the maximum stress near the coating-substrate interface was responsible.

2.3.4 Coatings on Si: Si Phase Transformation

To more fully understand the indentation response of DLC films on Silicon it is first necessary to obtain data on uncoated Silicon wafers with the same probes as the DLC films are to be tested with. Phase transformations and lateral cracking have been observed in Silicon in indentation and brittle fracture in a wide variety of mechanical contacts. It is now well established that a phase transformation from semiconductor to metallic behaviour takes place beneath the contact site and that the characteristic 'pop-out' event occurring during unloading is a consequence of that phase transformation and its accompanying volumetric expansion.

Beake *et al.* used a R = 4.6 μm probe to study the behaviour of ta-C on Si and uncoated Si(100) [89]. With this probe the critical load for pop-in was 40 mN on the uncoated Si(100). Pop-outs during unloading were observed after loading to peaks loads in excess of ~60 mN. The load required for pop-outs unloading from 100 or 200 mN

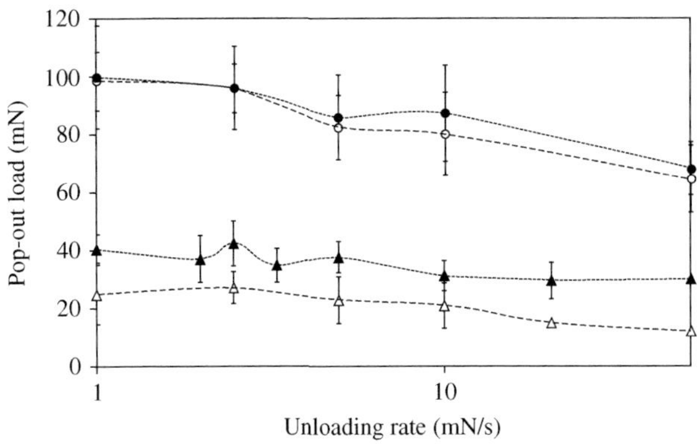

Figure 2.3 Variation in critical load for pop-out with unloading rate on Si (filled symbols) and 80 nm ta-C (open symbols). The peak load was 200 mN for the circles and 100 mN for the triangles.

peak load decreases with loading rate from 1–50 mN/s (Figure 2.3). Cracking around the indentation site occurs at ≥200 mN with a radial-lateral crack system developing at >300 mN. The presence of the DLC film affects the phase transformation behaviour of the Si altering the critical loads for pop-ins and pop-outs. Thicker coatings have enhanced load-carrying ability in comparison with thinner coatings and consequently coating thickness influences the stress fields generated in the substrate and alters the loads/depths at which stress-induced phase transformations. As well as reducing the load transferred to the substrate, hard coatings also distribute the applied load over a wider area consequently resulting in phase transformation and plastic deformation in the Si over a wider area. Beake and co-workers noted that for taC on Si pop-ins are bigger on than on uncoated Si [29].

2.4 Nano-scratch Testing

In the design of coatings for improved tribological performance, a key issue to be resolved is the relationship between the coating mechanical properties and its adhesion to the substrate. It has been found that the mechanical response of a coated component is underpinned by the film/substrate system deformation behaviour rather than the adhesion strength alone [90–100]. Both scratch resistance and interfacial toughness/adhesion under light and heavy loaded sliding contact can be assessed through nano-scratch experiments. In a progressive load nano-scratch test film performance throughout the load range can be assessed in a single test. Nano-scratch testing using nanomechanical test instruments such as the NanoTest overcomes the limitations of atomic force microscopy (AFM) in terms of measurement stability and excessive tip wear on hard carbon films and is an important tool in simulation of single asperity contact in tribological experiments. It is a flexible method, which can be easily tailored to represent different application scenarios by using various types of tip materials and geometries as well as different test kinetics with a constant or ramped applied load in a single or repetitive scratch mode.

Nano-scratch test data on a range of DLC coatings have been critically analysed to determine the sensitivity to intrinsic and extrinsic factors, impact of scan speed and loading rate, influence of probe radius and geometry, estimation of tip contact pressure, film stress and thickness, and the role of ploughing on the load dependence of friction [101]. The nano-scratch results are from experiments performed using NanoTest systems fitted with tangential force sensors and spherical indenters as scratch probes are presented and discussed. The nano-scratch tests described include (i) constant load nano-scratches (ii) ramped load nano-scratch tests and (iii) multi-pass repetitive unidirectional constant load nano-scratch tests (nano-wear). The progressive load nano-scratch tests were performed as a topography-progressive load scratch-topography multi-pass procedure. The progressive load '3-scan' scratch tests, constant load single scratches and repetitive scratch tests (nano-wear) were all performed using NanoTest systems (manufactured by Micro Materials, Wrexham, UK) fitted with tangential force sensors and spherical indenters as scratch probes. The end radii of the scratch probes were calibrated by nanoindentation. The design of the NanoTest combines high frictional sensitivity with sufficiently high lateral stiffness for accurate friction determination. Data were corrected in the NanoTest software to allow estimation of mean contact pressure by an in situ method described below without recourse to post-test imaging of scratch tracks.

The critical load in the macro-scale scratch test can be influenced by a range of extrinsic and intrinsic factors with intrinsic factors including scratching speed, loading rate, tip radius and extrinsic factors such as substrate hardness, and the roughness, thickness and friction of the coating 5. Additionally, the mechanical properties of the coating also play a critical role in nano-scratch testing, with even quite small changes in H/E capable of altering the locus of film failure relative to the moving probe [99, 100].

Coatings for mechanical property evaluation by nanoindentation have often been deposited on Silicon as a readily available and smooth choice of substrate. Unsurprisingly, to date many reports of nano-scratch testing have also involved thin coatings deposited on Si substrate. This has the advantage that coatings are smoother but adds complexity since the combined response of the coating-substrate system can be influenced by the contact-induced phase transformation of the Si substrate. The nanomechanical and nanotribological characterisation of thin films on Si has involved nanoindentation with Berkovich indenters in combination with nano-scratch testing with spherical indenters [27, 29, 31, 99, 100]. A key motivation was to understand the interplay between film thickness and interfacial toughness for MEMS and other protective thin film applications. The reliability of Si-based MEMS devices is limited by stiction forces when contact occurs. Applying very thin, low surface energy and low stress coatings can alleviate this problem but it is critical that they are deposited optimally. To fully understand the influence of the Si on the behaviour of DLC thin films on Si, the behaviour of Si in spherical indentation and scratching has been studied in depth.

2.4.1 Scan Speed and Loading Rate

The influence of loading rate and scan speed on the critical load in the nano-scratch test has been investigated for DLC and Si-doped DLC coatings on glass, DLC on Si and ta-C on Si [29, 32]. In these studies either (i) scratch speed and load rate were both varied to probe a wide range of dL/dx or (ii) the loading rate was varied with a concurrent increase in scan speed so that the tests were done at the same dL/dx but covering a wide range of loading rate. These parametric investigations were performed since in macro-scale scratch testing a strong dependence has been reported.

There was considerably less sensitivity to the scan parameters in the nano-scratch testing of DLC and Si-doped DLC coatings on glass than was observed in macro-scale scratch testing. No marked dependence on critical load was found on (1) scratching speed, (2) loading rate, or (3) (dL/dx) when dL/dx is much less than 1 N/mm, whilst critical loads only slightly increase at higher dL/dx (1–5 N/mm). When testing a 80 nm ta-C film on Si minimal rate dependence was found over a wider range of dL/dx, with the critical load for film failure being (113 ± 15) mN over a 100-fold variation in dL/dx from 0.1−100 mN/μm. Nano-scratch testing of 450 nm and 962 nm ECR-CVD DLC films on Si with a 6.5 μm end radius probe was performed with loading rate 0.167-16.7 mN/s and concurrent increase in scan speed from 0.5−50 μm/s so that the tests were done at the same dL/dx [102]. The 100-fold increase in loading rate with concurrent increase in scan speed resulted in only a 10% increase in the L_{c2} critical loads. These results from these three studies suggest that nano-scratch tests under significantly different loading conditions can be compared directly.

2.4.2 Influence of Probe Radius

As behaviour in the nano-scratch test is critically dependent on the radius of the test probe, to compare the behaviour of different DLC films it is preferable that the same or very similar indenter radii are used [101]. Sphero-conical probes with end radius typically 1−10 μm are commonly used in nano-scratch testing with sharper probes being used to test thinner films. The choice of probe size enables the entire force range of the instrument (500 mN) to be used and to reliably separate out the various transition points without recourse to post-test SEM or AFM imaging. Testable DLC film thickness is in the range 5 nm to 1500 nm. An additional benefit of using probes with larger radii is that they are less susceptible to tip wear. When testing thicker (e.g. ≥2 μm) DLCs micro-scale or conventional macro-scale scratching (load range 0–100 N; probe radius 200 μm) can be performed. Micro-scratch testing (load range 0–30 N; typical probe radius 25 μm) has the advantage that the maximum stress may be positioned close to the interface so that the results become more sensitive to real variations in interfacial strength whilst in the conventional scratch test with the larger radius probe the maximum stress is far into the substrate at film failure. For bulk materials transitions occur when critical pressures are reached, as in nanoindentation. Assuming constant mean pressure (for a bulk material displaying no indentation size effects) leads to power law dependence with exponent of two for the variation of the critical load with probe radius. For coatings the behaviour may be more complex when failures occur not solely at a critical pressure but are also associated with exceeding a critical bending strain in the coating that will be dependent on the radius of the probe.

2.4.3 Contact Pressure

The 3-scan procedure (topography scan–progressive scratch–topography scan) enables identification of failure mechanisms, such as the role of stress, in more detail. The first reported multi-pass test of this type was described in detail by Wu and co-workers from IBM in 1989 [103, 104]. In 2006 Beake and co-workers showed that by removal of the instrument compliance contribution to the measured deformation the true nano-scratch and nano-wear depth data could be displayed after levelling [99]. It is also important that the test instrumentation have sufficiently high lateral stiffness so that the scratch track is not significantly affected by surface roughness. By performing three-scan progressive load nano-scratch tests it is possible to determine the critical load for the onset of non-elastic deformation since this is the load at which the residual scratch depth is no longer zero.

The progressive load multi-pass scratch technique has been improved by a simple analytical treatment of the scratch depth data [27]. The method enables the yield stresses and the pressure required for the failure of the film to be estimated from contact mechanics, assuming the geometry of indentation, provided spherical indenters are used. The contact depth (h_p) in a spherical indentation contact is given by

$$h_p = (h_t + h_r)/2 \tag{2.7}$$

where h_p is the contact depth, h_t is the on-load scratch depth and h_r is the residual depth from the final scan. The contact radius (a) is determined from Equation (2.3), where R is the indenter radius.

$$a = \sqrt{(2Rh_p - h_p{}^2)} \tag{2.8}$$

$$P_m = L/\pi a^2 \tag{2.9}$$

The contact pressure P_m at any point along the scratch track is given by Equation (2.9), where L is the applied load. To apply this approach to the nano-scratch data it is necessary to assume that: (i) the presence of a tangential load does not influence the pressure distribution too greatly so that the measured friction coefficient is well below 0.3 (ii) the radius of the indenter is constant (iii) the sliding speed is sufficiently slow and contact sufficiently close to elastic that the load is supported on the rear of the indenter (iv) the indenter can reach the bottom of the scratch track in the final topographic scan. These conditions can be met in the nano-scratch test when spherical probes are used, although the approximation becomes less accurate as the contact geometry moves away from Hertzian conditions due to increasing friction or plasticity. The methodology has been validated for thin films on Silicon wafers with good agreement between (i) scratch hardness independently determined from optical measurements of scratch widths (ii) contact pressures for film and substrate yield events. The approach described above can also provide an estimate of the contact pressure during the progressive load nano-scratch test, at least at low contact forces where the friction coefficient is sufficiently low. It can provide an estimate of the mean contact pressure required to produce plastic deformation. Once failure and delamination occurs it is not generally possible to apply this treatment.

The Hertzian treatment is well suited to the nano-scratch testing of DLC films with spherical probes of ~5 μm end radius due to their intrinsic low friction and high H/E so that contact remains elastic or close to it over a wide load range. As an example, Figure 2.4 shows loading curves from indentation and scratch tests on a 462 nm ECR-CVD DLC film deposited on Si(100). The coefficient of friction was 0.06 at yield (20 mN) gradually increasing with load to 0.09 at 100 mN. The increase is accompanied by a small and gradual reduction in scratch recovery to 88 % at 100 mN.

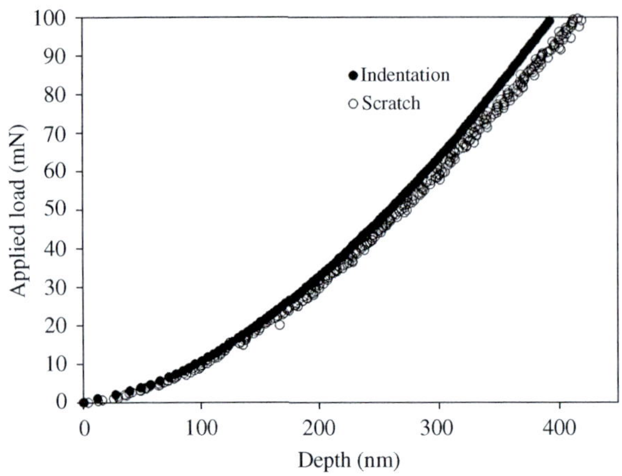

Figure 2.4 Similarity of loading curves from indentation and scratch tests on a 462 nm ECR-CVD DLC film deposited on Si(100). R = 6.5 microns.

2.4.4 Role of the Si Substrate in Nano-scratch Testing

Although Si is well characterised by nanoindentation, less is understood about its behaviour under the type of more complex loading geometries that can occur in practical tribological situations, particularly at the nano- and micro-scale. An increased understanding of the scratch resistance of Si is important for (1) the design of thin coatings on Si for MEMS devices and (2) simulation of ductile-regime machining of silicon in Single Point Diamond Turning (SPDT) [105, 106]. A systematic study of contact damage occurring on the surface of Si(100) has been carried out, using the same R = 4.6 μm spherical probe for nanoindentation, nano-scratch and nano-fretting experiments (with the latter described in a later section) (Figure 2.5).

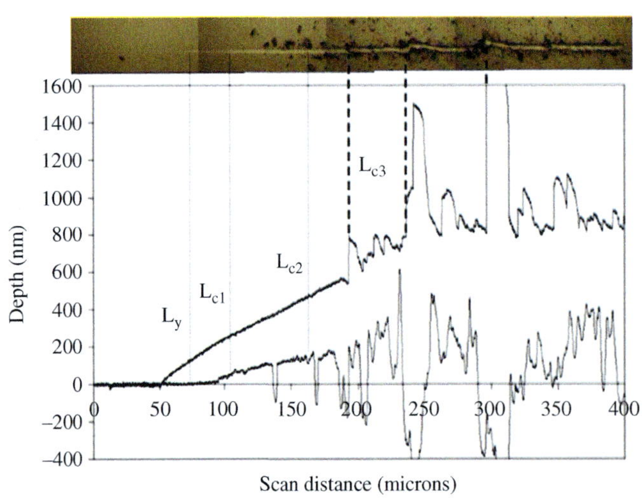

Figure 2.5 On-load and residual depth data on Si(100). The linear load ramp started after 50 microns.

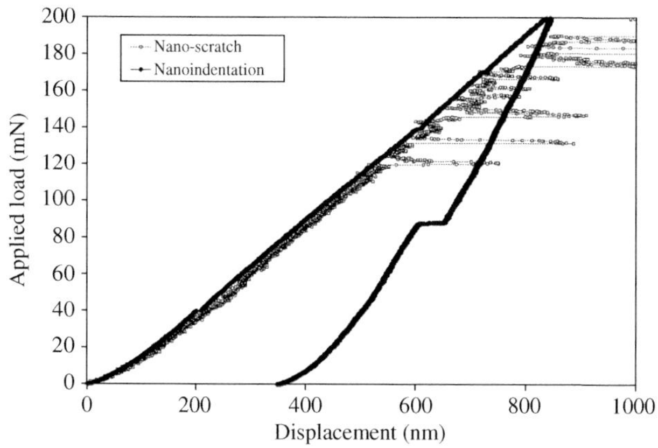

Figure 2.6 Nanoindentation and scratch loading curve on Si(100).

Table 2.6 Critical loads in nanoindentation and nano-scratch testing of Si(100).

	Critical load (mN)	
	Nanoindentation	**Nanoscratch**
L_y	40 ± 5	37 ± 5
L_{c1}	~80–95	56 ± 4
L_{c2}	>200	106 ± 7
L_{c3}	>300	155 ± 16

The main transitions in the response of Si(100) when scratched by spherical indenters are:

L_y yield : onset of non-elastic deformation

L_{c1} pop-in is observed in the residual depth data and accompanied by a colour change in the optical micrograph of the scratch track. Debris (ductile chipping) forms at the side of the scratch track above this.

L_{c2} marks the onset of more pronounced chipping at the sides of the track and edge cracking.

L_{c3} is lateral cracking (brittle chipping) and is marked by a more jagged scratch track beyond this point due to probe twisting due to the highly uneven load support of the cracked surface. The scratch tracks are completely straight up to this point due to the high lateral stiffness of the instrumentation.

As was shown for DLC on Si in Figure 2.4, the indentation and scratch response of Si is very similar at low load. Figure 2.6 contrasts typical loading curves. As the load increases the deformation in the nano-scratch test becomes more marked. The critical loads corresponding to the above transitions are summarised in Table 2.6.

The first pop-in marks the onset of non-elastic behaviour in indentation and scratch testing. In both tests the behaviour just after the pop-in is almost completely elastic since the pop-in is usually only a few nm and the residual scratch depth is only a few nm. Analysis of the critical pressure required for the first yield/pop-in to occur in the nano-scratch test gives 12–13 GPa, very similar to the hardness determined from analysis of nanoindentation unloading curves. The friction coefficient at yield is sufficiently low (~0.07) that the stress distributions remain similar. Zok and Miserez have shown that in comparison with indentation, tangential loading with friction promotes yield according to $P_y/P_y^0 = 1 - 2.9\mu^2$ here P_y^0 is the yield load in the absence of friction, P_y is the yield load with friction, and μ is the friction coefficient [107]. Experimentally, the Friction coefficient at failure is ~ around 0.2. The radial-lateral crack system develops at >300 mN in indentation [108], with more extensive lateral cracking requiring >400 mN and a penetration depth of >2 μm. In contrast, in nano-scratch testing, lateral cracking (L_{c3}) requires only ~150–170 mN, due to the different stress distribution.

2.4.5 Failure Behaviour of ta-C on Si

Wear and stiction forces limit the reliability of Silicon-based micro-systems. To improve durability in mechanical contact it is desirable to alter the mechanical and interfacial behaviour of the silicon surfaces by the application of very thin, low surface energy and low stress coatings. Ultra-thin filtered cathodic vacuum arc (FCVA) ta-C films are being considered as protective overcoats for Si-based MEMS devices. Fretting, nano-scratch and nanoindentation of different thickness (5, 20 and 80 nm) ta-C films deposited on Si(100) has been performed using spherical indenters to investigate the role of film thickness, probe radius, tangential loading, contact pressure and deformation mechanism in the different contact situations. The influence of the mechanical properties and phase transformation behaviour of the silicon substrate in determining the tribological performance (critical loads, damage mechanism) of the ta-C film coated samples has been evaluated by comparison with data on uncoated Silicon described in the previous section.

Nano-scratch and nanowear resistance of FCVA ta-C films has been studied with spherical probes of 100–1200 nm resulting in critical loads in the range 0.1–7.5 mN [25, 109–111]. Examination of the scratch track widths revealed that the effective radius of the indenter used in the study by Beake and Lau [25] was only ~1.1 μm at the shallow depths required to fracture the ta-C films. The nano-scratch data using the 1.1 μm probe on 20 and 60 nm ta-C is in good agreement with data by Lemoine and co-workers using a ~1200 nm indenter on 10 and 50 nm ta-C films also produced by FCVA [111] and 20 nm FCVA films tested by Li and Bhushan with a ~1 μm probe [110]. It is of interest to investigate the variation in measured critical load with the radius of the spherical probe for these thin films. Very small radius probes (100 nm) have been shown to get progressively worn away on repeated testing of hard films [110]. In the nano-scratch testing of a range of 5–150 nm thin films deposited on silicon by FCVA and CFUBMS precision polished probes with end radii of 3.1 and 9.1 μm were used. These probes were less likely to blunt during the testing, had well defined radius and produced film failures at higher forces easily observed by optical microscopy without recourse to post-test SEM imaging. Ta-C films deposited by FCVA have high hardness, but the films can be highly stressed [112, 113]. In contrast the closed field unbalanced magnetron sputtering deposition process (CFUBMS) is capable of producing dense, low stress coatings, which are

Table 2.7 Influence of ta-C film thickness on critical loads.

ta-C thickness (nm)	L_y (mN)	L_{c1} (mN)	L_{c2} (mN)
5	12.5 ± 1.2	23.7 ± 0	41.2 ± 1.2
20	17.9 ± 1.9	28.3 ± 2.9	54.5 ± 1.9
80	13.2 ± 0.7	35.7 ± 8.5	91.3 ± 3.8

$R = 3.1$ μm

typically somewhat softer than the films deposited by FCVA [114, 115]. For brevity, only the results on the ta-C films are discussed below. Increasing the ta-C thickness from 5 to 80 nm was found to increase the critical load for film delamination by a factor of two (Table 2.7). Above the critical load for edge cracking, L_{c1}, but prior to the total film failure some isolated small delamination events could occasionally be seen in the scratch track. SEM images show periodic cracking across the scratch track between L_{c1} and L_{c2} in ta-C [29, 111]. At film failure the abrupt change in depth is very close to the film thickness. Nano-scratch testing of ta-C films has shown that they are sufficiently thin to not show large area delamination and scratch resistance increases with effective H/E.

With a 3.1 μm end radius probe a contact pressure of about 12 GPa was required for yield on the 5, 20 and 80 nm films. The yield stress of Si is known to be very close to this (around 11.3 GPa) [44]. The Hertzian method of analysing the nano-scratch data suggests the onset of non-elastic deformation in the nano-scratch test is due to substrate yield rather than film deformation. The contact radius calculated from Equation (2.3) is in the range 300–700 nm at the onset of yield. Maximum stresses are generated at a distance of $\sim 0.5a$ under the surface in a Hertzian indentation contact, which for these very thin films is deep into the Si substrate, consistent with substrate yield occurring initially as described above. Although the presence of a tangential load would produce a less uniform pressure distribution and result in the maximum stress shifting towards the surface [116, 117], for the 5–80 nm hard films, with measured friction coefficients of around 0.1–0.15 at yield, yield would be expected to begin in the silicon.

Nano-scratch tests have been also been performed on these ta-C films with a 4.6 μm radius probe. Figure 2.7 shows the variation in the critical load for the total film failure of an 80 nm FCVA ta-C film on Si wafer with the radius of the test probe. The critical load follows a power law dependence on the probe radius of the type shown in Equation (2.10) where a is a measure of the effective adhesion strength in the nano-scratch test. The best fit curve to the data in Figure 2.7 is given by a curve of $y = 8.4x^{1.76}$ [101].

$$L_c = aR^m \tag{2.10}$$

Hertzian analysis of contact pressure at failure using the 4.6 μm probe gives a contact pressure of (15 ± 1) GPa for the 80 nm film at L_{c2}, which is close to the value of 14.5 GPa previously determined with the 3.1 μm probe. By extending the force range it was possible to study the lateral cracking (L_{c3} failure) that occurs on Si at higher load and investigate whether the ta-C films could protect the substrate or not. The variation in the critical load for lateral cracking with film thickness is shown in Table 2.8.

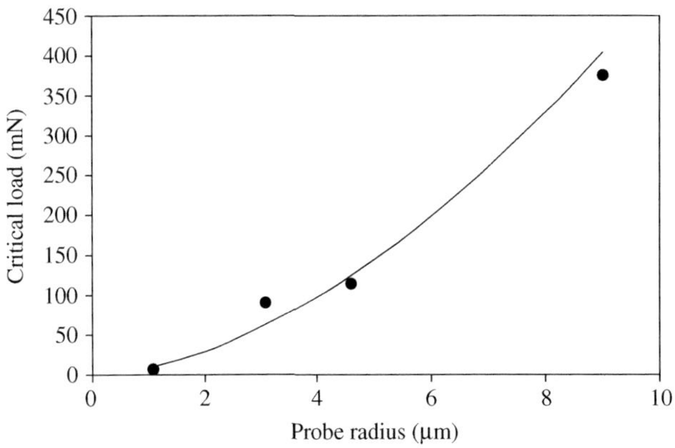

Figure 2.7 Influence of probe radius on the critical load for total film failure of 80 nm ta-C film on Si.

Table 2.8 Critical load for lateral cracking in nano-scratch tests.

	Si	5 nm ta-C	20 nm ta-C	80 nm ta-C
L_{c3} (mN)	130 ± 10	152 ± 35	163 ± 45	276 ± 20

The critical load for the L_{c3} failure increased dramatically on the 80 nm ta-C and the lateral cracking was less pronounced. For the uncoated, 5 and 20 nm ta-C film sample the scratch track is more jagged after L_{c3} due to the more extensive lateral cracking. The nano-scratch test data show that tangential loading promotes the formation of large lateral cracks on all the ta-C film samples at much lower load than in nanoindentation as was found for uncoated Si(100) [89]. The occurrence of large pop-ins at high load when Silicon is indented with spherical indenters has been shown to be highly stochastic in nature, with Oliver and co-workers reporting a threshold load of (350 ± 100) mN with a 4.3 μm probe and larger pop-ins (>1 μm) more likely at >500 mN [108]. In high load nanoindentation tests the thin hard ta-C films showed no obvious protection against lateral cracking, which occurred only occasionally, irrespective of the ta-C film thickness. However, in the nano-scratch test the tangential loading promoted dramatic film failure at much lower load (e.g. ~115 mN for 80 nm ta-C) with extensive lateral cracking occurring below 300 mN. Marked differences emerged for the 80 nm film where the critical load for this L_{c3} failure increased dramatically and the lateral cracking was less pronounced. These thin films do not show the large-area delamination outside the scratch track commonly observed on failure of thicker hard and elastic (high H/E) coatings discussed in more detail below. It appears that when the film fails in the scratch track, the intact ta-C film outside the scratch track is capable of maintaining a measure of load support and providing some protection. The deposition of <100 nm amorphous carbon films on Si could be a promising strategy for improving reliability of Si-based MEMS devices as none of the very thin films tested underwent stress-related delamination failures that occur behind the indenter during the nano-scratch testing of thicker amorphous carbon films.

2.4.6 Film Stress and Thickness

The film thickness is a key parameter influencing the critical load in the nano-scratch test. Varying the thickness of a hard carbon coating can change the deformation and fracture behaviour so it is reasonable to expect that the variation in critical load and load carrying capability in mechanical contact will be a complex function of coating thickness. This has been investigated in the nano-scratch test as a function of substrate bias during the deposition of DLC films on Si. The stress in the film may vary with thickness, as is expected for graded films and has also been reported for 100–2000 nm sputtered DLC on Si, where maximum compressive stresses of ~1.5 GPa were reported for films with 300–700 nm thickness [13]. Even if composition, mechanical properties and intrinsic stress do not vary through the film thickness altering it can have opposing effects: 1) thicker films that are harder than the underlying substrate can provide more load support and so delay the onset of the substrate deformation that is often the precursor of film failure (higher critical load) 2) thicker films can be more highly stressed and more easily through-thickness crack and delaminate when deformed (lower critical load) since the driving force for spallation to reduce stored elastic energy is greater. It is also important to consider the test probe geometry as this will have an effect on the stress distribution and the significance of substrate yield prior to film failure. It is not always the case that harder and stiffer films can offer protection to the underlying substrate and the degree of enhancement to the load carrying capacity is probe radius dependent. Chudoba and co-workers noted that for 2 μm TiN on steel loaded with very large radii the higher modulus of the film results in a stress concentration at the interface and consequently premature failure of the system in comparison to an uncoated sample [44].

For well-adhered coatings with relatively low stress, scratch and wear tests often show the same trends with H/E as the behaviour is largely dominated by load bearing capability. For many well-bonded coatings the scratch test is not really an adhesion test since differences in mechanical properties and coating thickness can affect the critical loads more greatly than differences in coating-substrate bonding strength. Larsson *et al.* [90], and Ichimura and Ishii [91] reported a linear relationship between critical load and substrate hardness for TiN coatings on steel and recently Wang *et al.* reported that critical loads for 1.4 μm DLC and Zr-DLC on Ti were strongly correlated with substrate hardness [118].

It was shown in the previous section that in the nano-scratch testing of 5–80 nm FCVA ta-C films on Si with spherical probes the critical load for film failure increases with their thickness. At the critical load the maximum von Mises stress is in the substrate so that film failure is driven by the deformation of the Si substrate due to plasticity and phase transformation. The increasing critical load for thicker films is due to their ability to better protect the Si substrate. Nano-scratch studies on a-C films deposited on Si by CFUBMS and Si:a-C:H films deposited on glass by plasma enhanced chemical vapour deposited (PECVD) have also shown that the critical load for total film failure in the nano-scratch test can be strongly correlated with film thickness reflecting enhanced load support and substrate protection. Provided the films are not too stressed the ratio L_c/t_f can be approximately constant in the nano-scratch testing of DLC films. The critical load normalised by film thickness for 141 nm DLC, 366 nm and 614 nm Si-doped DLC films on glass was found to be approximately constant (0.04–0.06 mN/nm) [101].

However, thin PVD films can be highly stressed and these high internal stresses coupled with shear stresses in the scratch test can lead to poor adhesion and dramatic coating spallation. In nano-scratch testing of 450 nm and 962 nm ECR-CVD DLC films deposited on Si without interlayers it was found that the thicker film failed at around 60% of the critical load of the thinner film to due high stress. Substrate bias during deposition can significantly alter film stress [119]. Shi and co-workers reported nano-scratch data for 200 nm and 1000 nm a-C films deposited with varying substrate bias voltage (−20 to −140V) by CFUBMS [31]. At 200 nm thickness the highest H/E films performed best since the intrinsic stress in the film was relatively low. For ~1 μm a-C films they observed a general relationship between H/E and the scratch test critical loads. When H/E_r is ~0.09–0.1 there was a tendency of the a-C films to delaminate behind the moving probe at low critical load. This behavior is not confined to DLC films, A similar correlation was found for ~0.8 μm Ti-Si-N nanocomposites films on Si using spherical indenters and hard 0.9–1.5 μm nitride-based films on Si using Berkovich indenters [99, 100]. For the TiSiN films higher H/E led to higher critical loads for elastic–plastic transition and also for the total film failure occurring in front of the probe. However, when H/E_r was >0.09 film failure was always by tensile failure behind the probe (also described as unloading failure). The high tensile stress behind the probe in the nanoscale scratch test can lead to complete failure for hard films on silicon. Similarly, failure of harder TiFeN and TiFeMoN films with $H/E_r > 0.11$ was accompanied by delamination outside the scratch track. Hard films with $H/E_r \leq 0.11$ possess a more optimum combination of hardness and toughness for applications where they will be exposed to high shearing forces and strain in the film in this case is more readily relieved by intergrain cracking. The results of all these studies show the same trends. For optimum durability in highly loaded sliding contact, when films are deposited to around 1 μm thickness they should not be designed to be as hard and have as high a ratio H/E_r as possible. For high H/E films energy dissipation by mechanisms such as plastic deformation or localised intergranular fracture is less likely during the highly loaded sliding of these hard and elastic films, and ultimately the stored elastic energy is instead relieved by dramatic fracture resulting in a larger delaminated area.

2.4.7 Repetitive Nano-wear by Multi-pass Nano-scratch Tests

A ramped scratch to failure can be considered as an 'overload' test. More information can be provided by repetitive multi-pass scratch (nano-wear) tests at sub-critical load. Constant load, unidirectional multi-pass scratch testing was first described by Bull and Rickerby in 1989 [120] and has been shown to be an effective low cycle fatigue test. The same approach can be applied to micro- and nano-scratch testing. Constant load nano-wear tests are often used to determine rates of sliding/abrasive wear and investigate the role of fatigue. The low-cycle nano-wear experiments can be more informative regarding the influence of thin film stress leading to poor adhesion than single scratch tests. When compared to progressive load nano-scratch testing, nano-wear testing has the advantage that the load can be varied to tune the maximum von Mises stress to be close to the coating-substrate interface and minimize (and potentially eliminate in some cases) substrate deformation as a precursor of coating debonding. Single constant load scratches can also be informative if the load is chosen to position

Table 2.9 Evolution of scratch recovery and mean pressure during nano-wear test on a-C films at 50 mN using a 4 μm probe.

Substrate bias during deposition	H/E_r	% scratch recovery in wear cycle 1[a)]	% scratch recovery in wear cycle 10	Mean pressure in wear cycle 1 (GPa)[b)]	Mean pressure in wear cycle 10 (GPa)
−20 V	0.076	89	77	9.0	7.9
−120 V	0.089	93	82	13.4	10.6

a) Scratch recovery is defined as (100% x [on-load depth - residual depth]/on-load depth).
b) Estimated from Hertzian analysis after correction of raw depth data for the influence of sample slope and topography, and instrument compliance.

the maximum stress is close to the interface. A NanoTest fitted with a cold stage been used to investigate the nanotribological and nanomechanical properties of a functionally graded a-C:H(Ti)/TiCN/TiN/Ti coating on M2 steel from 25 °C to −30 °C [33]. This multilayered DLC coating failed by cracking and spallation in the room temperature nano-scratch tests. Analysis of fracture sections showed these failures originated from, or close to, the interface between the top a-C:H(Ti) layer and the TiCN layer. Decreasing the test temperature improved its properties with testing at 0 °C resulting in increased H, H/E_r and H^3/E_r^2, consistent with improved crack resistance and wear resistance, with further modest improvements when the test temperature was reduced to −30 °C.

Shi and co-workers performed sub-critical load nanowear testing at two load levels (50 and 150 mN) to investigate the performance of 1 μm a-C films described above that fail in the progressive load nano-scratch test at a critical load of ~200 mN [31]. The load chosen for the low load test ensured that the maximum stresses were within the film. In the lower load wear test the contact is almost completely elastic with residual wear depths under 100 nm. Table 2.9 illustrates how scratch recovery and contact pressure vary between the first and last cycles for films deposited under high (−120 V) or low (−20 V) substrate bias. The films deposited at high bias showed lower residual depth and higher scratch recovery consistent with their higher H/E. An increase in residual wear depth and decrease in scratch recovery with each wear cycle confirm that the low load multi-pass wear test is a fatigue process.

In the low load test the location of maximum stress is within the coating so coating-properties dominate and interfacial strength is less important. By increasing the wear load the test can be tuned to place the max shear close to the interface so that with increasing wear cycles delamination occurs readily. At 150 mN Hertzian analysis shows that the maximum von Mises stress is very close to the interface when using a 4 micron probe. Under these conditions plasticity and micro-fracture dominate and the harder films deposited under high bias perform poorly. The behaviour of 1 μm a-C films with a very high ratio of H/E, deposited under high substrate bias, was found to be very strongly dependent on the test conditions, with the films performing well at low load wear but very poorly in more highly loaded contact. They exhibit low critical loads in progressive load scratch tests and when they fail in the nano-wear test there is extensive delamination outside of the scratch track. The combination of nano-wear tests at different loading levels provides information regarding the suitability of the a-C films for contact applications.

2.4.8 Load Dependence of Friction

The friction force in the nano-scratch test can be deconvoluted into its interfacial and ploughing components so that the interfacial friction can be reported (Eqn. 2.11):-

$$\mu_{\text{total}} = \mu_{\text{interfacial}} + \mu_{\text{ploughing}} \tag{2.11}$$

The interfacial friction component can be determined by different approaches: (1) performing constant load friction test at very low force where contact is completely elastic and the ploughing contribution is zero (2) performing repetitive scratches to eliminate the ploughing contribution (3) performing progressive load scratch and extrapolating the low load friction data to zero load. Typically the friction coefficient at yield is in the region of 0.05. The observed frictional behaviour on repetitive sliding is explained by changes to the ploughing component of friction and smoothing of asperities, both of which alter the contact area and therefore can influence the friction [121]. When friction and surface roughness are low the loading curves in indentation and scratch tests can be surprisingly similar as illustrated in Figures 2.4 and 2.6. In nano-scratching Si with a R = 4.6 μm diamond probe, the friction coefficient at pop-in is low (~0.07), similar to the value of (0.08 ± 0.01) that marks the onset of ploughing in nano-scratching with a 1 μm tip [122]. Although the critical loads are very different, there is fair agreement when comparing friction coefficients at the same representative deformation level, in this case, the onset of non-elastic behaviour. Charitidis has reviewed the load dependence of the friction coefficient in progressive load nano-scratch testing of various DLC film systems reporting [76] that the friction tended to increase non-linearly although the probe radius was not constant in these studies and so the state of deformation at a given applied load will vary. With a R = 4.6 μm diamond probe the friction coefficients during nano-scratch testing of 80 nm ta-C were ~0.09 at yield, ~0.14–0.15 at L_{c2} film failure and ~0.21–0.24 at the onset of lateral cracking. The friction coefficients at yield and the onset of lateral cracking are close to those on the uncoated Si when tested with the same probe. The friction coefficient increases with load, altering the stress distribution so that the contact becomes more asymmetric with a pronounced tensile stress maximum at the trailing edge, so that film failure can be by exceeding a critical tensile stress.

2.5 Impact and Fatigue Resistance of DLC Films Using Nano-impact Testing

DLC films are hard and elastic films so that almost complete recovery occurs in indentation and plasticity is minimised. They typically have high H/E of around 0.1 and can have very high H^3/E^2. Crack opening is commonly driven by tensile stresses. As DLC films are typically deposited with high compressive stress due to ion bombardment during the deposition this may benefit toughness since extra energy is required to overcome that stress before the crack can be driven in tension. Increasing $H^3/E_r^{\,2}$ in the film has been associated with high resistance to cracking and hence high fracture toughness. For example, Musil and Jirout [123] have shown that, for indentation load up to 0.5 and 1 N with a Vickers indenter, resistance to cracking is correlated with the film resistance to plastic deformation, H^3/E^2, so that films with less plasticity would have higher resistance to cracking.

However, it is well known that DLC coatings commonly display poor durability under impact or dynamic loading conditions. To optimise the performance of DLC coatings operating in demanding contact loading conditions it is useful to assess their mechanical behaviour under more relevant loading conditions than solely in nanoindentation. Films that show higher dynamic toughness should combine (1) a high load threshold for the initiation of cracking (2) mechanisms to minimise and retard crack propagation, i.e. it is necessary to display a certain degree of damage tolerance. When coatings are developed for highly loaded contact applications higher plasticity in indentation (i.e. lower H/E) has been correlated with higher durability under service conditions. Toughness is not the same as plasticity but the two properties can be very well correlated since plasticity is an effective mechanism for stress relief. Under highly loaded conditions cracks can propagate to failure where dissipative mechanisms such as plasticity do not exist.

In view of this apparent contradiction regarding the optimum coating mechanical properties for toughness, a repetitive nanomechanical test (the nano-impact test) has been used to take a more direct approach to the assessment to film toughness and damage tolerance under dynamic loading. The test measures the degradation of surface engineered materials from repeated localized stresses. The repetitive contacts in the nano-impact test are actual impact events where the probe leaves the surface between each subsequent impact which occurs at the same location on the sample surface.

Although nanoindentation can be used to assess toughness as a measure of resistance to crack initiation and overload failure it is not possible to monitor crack propagation under repetitive, oscillating loading conditions in the conventional quasi-static nanoindentation test and for this the dynamic nano-impact test is preferred. Nano-impact testing is a nanomechanical test technique involving high strain rate contact. The test extends the range of nanomechanical test techniques to much higher strain rates (typically $\sim 10^2-10^3\ s^{-1}$ [124]) than are possible in standard quasi-static nanoindentation (commonly $\sim 10^{-2}-10^{-4}\ s^{-1}$). This high strain rate contact allows much closer simulation of the performance of coatings systems under highly loaded intermittent contact and the evolution of wear under these conditions. The progression of surface damage is followed using the depth-sensing capability of the nanotest.

Originally conceived as a test method to assess the degradation of surface engineered materials to repeated localised stresses, with single and multiple impact configurations and different indenter geometries available, a much wider range of applications are now being developed (illustrated in Table 2.10). A single impact is essentially a high strain rate indentation test where the 'dynamic hardness' (or impulse hardness) of the material, i.e. its effective hardness at a given strain rate, can be determined from analysis of impact data using the approach outlined by Schuh, van Vliet and others [124–126, 136]. Of more importance to DLC coatings is the repetitive nano-impact test. When repetitive or multiple impacts are performed at the same location on the sample surface the resultant behaviour depends strongly on the ductility or brittleness of the sample. On a ductile material there is a gradual increase in probe penetration depth, which decreases with continued impacts, particularly for strongly work hardening materials. In contrast, for a brittle material there are often several abrupt increases in probe depth during the tests that are associated with cohesive and/or interfacial failures. Jennett and Nunn have shown that there is a clear increase in the energy absorbed for impacts causing immediate fracture [156]. By investigating individual impact depth vs. time curves in detail

Table 2.10 Applications of nano-impact testing.

Type of nano-impact test	Property investigated	Material system	Reference
Single impulse	Strain rate sensitivity	Polymers	125
	Dynamic hardness	Al, Mg alloys, NiW	124, 126
	Dynamic H/E	Al-12Si UFG cladding	127
	Energy absorption	Tissues and tissue surrogates	128
Multiple impact	Fracture resistance	NiAl coatings deposited by HVOF, APS and CS	129
		PEO coatings on Ti6Al4V	130
		Nanocomposite films on polymers	131
		TiAlN coatings	132, 71
		CrAl(Si)N coatings	133, 134
	Fracture toughness	Intermetallic $Al_{12}(Fe, Cr)_3Si_2$ coatings on stainless steel	135
	Evolution of dynamic hardness and debonding	Sol–gel on stainless steel	136
	Debonding	DLC on Si	35, 88
		250 nm Pt-Ir coatings on WC-Co	137
	Erosive wear	EB-PVD Thermal barrier coatings	138

it is possible to separate film behaviour into distinct regimes as the test progresses: (i) load support (pre-impact) (ii) initial impact and (iii) subsequent impacts. This procedure enables the role of film properties on load-support, initial resistance to crack nucleation and subsequent crack propagation to be determined. The nano-impact test does not immediately provide a value of the quasi-static fracture toughness K_{1c} but rather an assessment of the sample's resistance to fatigue fracture, or effective dynamic toughness, under dynamic and repetitive loading that can be more representative of actual contact conditions in applications. Experimental parameters such as probe geometry, acceleration distance, coil force, impact angle, number of cycles and test frequency can be controlled to alter the severity of the test and its duration. A cube corner diamond indenter whose geometry induces high contact strain is often chosen as the test probe, as this high contact strain is beneficial in inducing fracture within a quite short test time. Impact testing is a fatigue process, so repeat tests are necessary to obtain a statistical measure of the coating's resistance to fracture. In contrast to macro-scale impact testing where this requires different specimens and is very time consuming, testing at the nano-scale enables many instrumented repeat tests to failure to be performed on the same specimen in a very short time. The time-to-fracture and final impact depth typically show a stochastic response. Impact resistance of different coatings can be compared by the time required for failure (fracture) to occur in 50% of the tests, enabling comparison between similar materials. Fracture probability can be estimated by ranking the time-to-failure events in order of increasing fatigue resistance and then assigning a probability of failure (Equation (2.12))

$$P(\mathrm{f}) = n/(N+1) \tag{2.12}$$

to the nth ranked failure event in a total sample size of N, analogously to the treatment of distributions of failure stresses in Weibull statistics. The probability of failure with a given time (usually the end of the test) is a convenient way to compare different samples. By combining fracture probability data at different loads it is possible to produce a graph of the time or number of impacts required for fracture to occur in 50% of tests vs. impact force. The energy supplied for coating failure can be calculated from the time-to-failure data using Equations (2.13) and (2.14).

$$\text{Energy per impact} = \text{force} \times \text{accelerating distance} \tag{2.13}$$

$$\text{Total energy supplied} = \text{number of impacts} \times \text{energy per impact} \tag{2.14}$$

For typical test conditions of 1 impact every 4 seconds the number of impacts to failure = 0.25 × time-to-failure in seconds.

The technique has been used to test the impact resistance of > 2 μm DLC on steel. The coating systems studied were (a) compositionally graded a-C:H and a-C coatings deposited on M42 tool steel by closed field unbalanced magnetron sputter ion plating (CFUBMSIP) (b) a-C:H/Cr coating on steel and (c) PACVD a-C:H coatings deposited on M2 tool steel [34]. Typical impact test parameters for testing a-C:H coatings on steel are: (i) probe cube corner diamond (ii) impact angle 90° (iii) applied load 1–15 mN (iv) accelerating distance 15 μm (v) impact frequency 0.25 Hz (vi) test duration 300 s (75 impacts in total) (vii) ten repetitions at different locations of the coating surface for each load chosen. The sharp cube corner indenter has been replaced by Berkovich, 10 μm or 25 μm diamond probes when testing thinner DLC films on Si [26, 35, 88]. The fracture properties of 5–80 nm tetrahedral amorphous carbon (ta-C) thin films deposited on silicon by FCVA have also been studied. When testing these much thinner ta-C a well-worn Berkovich diamond probe impacting with an applied load of 0.120.3 mN was used. The results of these studies are discussed below.

2.5.1 Compositionally Graded a-C and a-C:H Coatings on M42 Tool Steel

In loaded contact coating-substrate systems can fail by: (i) delamination of the coating from the substrate, including interfacial cracking (adhesive failure) (ii) fracture of the coating (cohesive failure) (iii) failure of the substrate (fracture or plastic deformation) (iv) a combination of the above. Increasing the impact load changes the stress distribution and consequently both the mode and location of failure. Beake *et al.* studied the load dependence of the impact response on compositionally graded a-C and a-C:H coatings on M42 tool steel. The hydrogen-free amorphous carbon coatings produced by the CFUBMSIP process are reported to have a graphitic structure, and that this predominantly sp^2 bonding imparts low stress and hence good adhesion [139]. Hardness and elastic modulus data on the a-C and a-C:H coatings are shown in Table 2.5 [34, 73]. The a-C coating is lower in hardness but stiffer and consequently has lower H/E and H^3/E^2 than the a-C:H coatings. It has been reported that the hardness of a-C coatings is strongly dependent on deposition conditions; it is possible to produce much harder hydrogen-free graded coatings by the CFUBMSIP process. However, it is possible

Table 2.11 Number of impacts required for fracture in 50% of tests vs. impact force for amorphous carbon coatings on M42 tool steel.

Coating	Number of impacts required for fracture in 50 % of the tests		
	At 1 mN	At 5 mN	At 15 mN
a-C	206	284	283
a-C:H	7	3	2

Table 2.12 Load dependence of final probe depth in nano-impact testing amorphous carbon coatings on M42 tool steel.

Coating	Penetration depth after impact test (μm)		
	At 1 mN	At 5 mN	At 15 mN
a-C	0.6 ± 0.2	1.5 ± 0.2	2.5 ± 0.2
a-C:H	1.0 ± 0.4	2.1 ± 0.2	3.9 ± 0.3

that the lower hardness of the amorphous carbon sample tested is beneficial for impact resistance. It has been reported that a-C coatings with higher hardness have lower adhesion (i.e. low critical load in scratch tests), the authors citing the influence of residual stress on hardness and adhesion. In contrast the higher hardness and sp^3 bonding in a-C:H result in lower plasticity and reduced durability to repetitive impact. The a-C:H coating always showed a shorter time to fracture, larger change in depth on fracturing and a larger final depth at the end of the impact test. At 5 mN fracture on a-C:H involved ring cracking. At 1 mN the fracture remains within the coating (cohesive). In contrast, the a-C coating with greater plasticity showed only minor cohesive cracking after ~280 impacts at the same (or higher) impact forces (Tables 2.11 and 2.12).

The a-C coating system exhibits enhanced impact resistance. This is consistent with other studies showing improved wear resistance for a-C coatings developed using the CFUBMS process [80, 81, 139], although the nano-impact testing shows the difference more dramatically. Amorphous carbon coatings produced by the CFUBMSIP process are reported to have a graphitic structure, and this sp^2 bonding imparts low stress and hence good adhesion. Good wear resistance, particularly under dynamic loading situations, has been reported for coatings with high H/E [140], since this is related to the elastic strain-to-break. Instead toughness (relative toughness can be estimated from the extent of elastic recovery exhibited on unloading in indentation, and the E/H ratio) is more critical under these conditions. Although they have similar hardness, the large difference in E/H between the a-C and a-C:H coatings is consistent with this explanation. Matthews *et a.l* have suggested a parameter such as $(E_c - E_s)/E_s$ may be useful as measure of the stress transfer to substrate, and reported for systems of similar H/E ratio, a correlation between resistance to dynamic loading and this parameter [140]. The

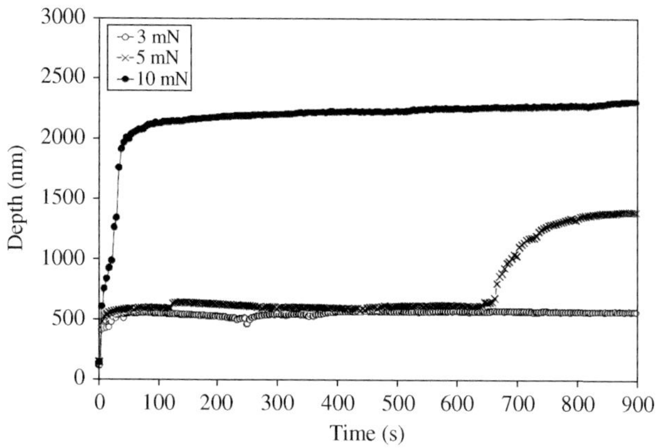

Figure 2.8 Influence impact load on behaviour of DLC/Cr on steel.

a-C coating is closer in modulus to the steel substrate and also both coatings are more compliant than the steel and hence elastic strain-to-break arguments may not apply in the same way as for much stiffer ceramic coatings.

2.5.2 DLC/Cr Coating on Steel

A strong load dependence was found in the nano-impact response of a 2.3 μm DLC/Cr Coating on steel which had a hardness of 21 GPa and reduced modulus of 210 GPa so that $H^3/E_r^{\ 2} = 0.21$ GPa. Figure 2.8 shows how the applied load influences the time-to-fracture. It is clear that the impact load has a dramatic effect on the coating lifetime. When the impact load is low the coating is durable and when fracture occurs it is cohesive, but above a threshold load the through-thickness cracking occurs readily and leads to removal of the coating (final depth > coating thickness).

Bousser and co-workers have shown that for 8–13 μm CrN and CrSiN films on stainless steel the ratio of indentation depth to film thickness (d/h) at which circular cracking in Vickers indentation occurs was correlated with plasticity and inversely correlated with H^3/E^2, suggesting that, under these highly loaded conditions, the inability of the coating to accommodate the substrate deformation drives the cracking process [87]. At lower impact load, the ratio of impact depth to film thickness is low, so higher H^3/E^2 in the film is beneficial by preventing cracks from forming in the first place. At higher impact load, effective durability and damage tolerance requires resistance to both formation and propagation of cracks under dynamic loading and films with high H^3/E^2 may not perform well. The optimum mechanical properties for improved dynamic toughness therefore depend on the severity of the impact conditions. Suppressing dislocation activity to produce yet harder films (very high H^3/E^2) is beneficial for low load contacts but can be detrimental to their durability under highly loaded contacts as suitable dissipative mechanisms are not available.

2.5.3 PACVD a-C:H Coatings on M2 Steel

Nanoindentation data showed that the hardness and modulus of the three coatings tested with compositions detailed in Table 2.13 was rather similar, with $H \sim 22$ GPa

Table 2.13 Time-to-fracture, probe depth-at-fracture and final depth data from nano-impact of PACVD DLC coatings on M2 steel.

Coating system	Total thickness (μm)	Time-to-50% fracture (s)	Depth at first fracture (μm)	Final depth at 1800 s (μm)
a-C:H	2.3	91	0.88 ± 0.02	1.8 ± 0.2
a-C:H/TiN interlayer	3.0	113	0.89 ± 0.06	1.5 ± 0.2
Multilayered a-C:H	4.5	81	0.84 ± 0.02	1.3 ± 0.1

and $E_r \sim 190$ GPa. The time-to-50%-fracture data provided an estimate of the relative impact performance under these conditions, although the DLC/TiN coating showed a different distribution in time-to-fracture. The thicker multilayered coating has better initial load support under these impact conditions, but this is associated with poor durability (shorter time to fracture). The fracture on this coating was less extreme with a lower change in depth on failure and low erosive wear rate both leading to very low final depth. Final depth data show large differences between the samples due to (i) differences in depth change at fracture(s) and (ii) the rate of continued impact damage after this.

The time-to-fracture data for the experimental PACVD materials were rather similar. Resistance to impact-induced cohesive fracture under these conditions can be ranked in the order DLC/TiN ≥ DLC ≥ multilayered DLC. Although time-to-fracture and final impact depth can be highly stochastic, for these coatings the depth at which the first (cohesive) failure occurred was remarkably constant, presumably reflecting the very similar mechanical properties of the outer DLC layer [34]. The main differences in behaviour between the PACVD coatings occur after the initial impact-induced cohesive fracture. After fracturing, the enhanced wear rate observed for the thinnest a-C:H coating is due to reduced ability to support the impact load after fracture since the steel substrate is softer. Wear rates are lower on the other samples since, being thicker; they are more able to continue to protect the substrate in their damaged state. This is consistent with the composite (as opposed to coating-only) hardness or H/E ratio of the coating/substrate system controlling the subsequent coating removal rate after the initial impact-induced cohesive fracture. In general, film thickness has been found to influence nano-impact behaviour to a much smaller extent than nano-scratch testing. For example, critical loads for film failure in nano-scratching ta-C films on Si and a-C:H-based films on glass scale with film thickness but the number of impacts required for failure can show an opposite trend. In nano-impact tests at low load (5 mN, cube corner indenter) a 140 nm a-C:H with $H^3/E_r^2 = 0.21$ required almost twice as many impacts to fail as a 600 nm Si-doped a-C:H with $H^3/E_r^2 = 0.16$.

2.5.4 DLC Films on Si-film Thickness, Probe Geometry, Impact Force and Interfacial Toughness

Film failure by fracture with high contact strain (e.g. cube corner) indenters is associated with abrupt increases in probe depth during the test. In contrast when spherical probes are used at low impact forces a backward depth evolution can be observed (i.e. the on-load probe depth decreases with continued impacts) [88, 136]. This behaviour has been confirmed by post-test AFM imaging of impact scars showing volume uplift

Table 2.14 Impact behaviour of DLC films on Si wafer.

RF Deposition Power/W	film thickness / μm	Maximum probe depth / μm	Mean time-to-first-failure/s	% cohesive failure	Energy to failure / μJ
105	0.56	0.57	1032	12.5	7.3
115	0.53	0.47	855	37.5	6.0
125	0.62	0.44	813	62.5	5.6

and has been interpreted as due to interfacial delamination without (or before) the accompanying fracture that results in the increase in depth. Under these conditions stresses are relatively low and impact-induced plasticity is minimised. This raises the intriguing possibility that the impact test may be used to assess adhesion and interfacial toughness. Fasial and co-workers studied the nanoscale fatigue failure of 100 nm sputtered DLC on Si using different indenters and test methodologies. The hardness, modulus and compressive stress in the film were 12.5, 153 and 0.9 GPa respectively. They reported that in the repetitive impact testing this film the geometry of the impact probe influenced the deformation [88]. Using a 10 μm end radius spheroconical indenter resulted in delamination and uplift at low load. FE analysis of the position of the maximum von Mises stress suggested the blistering and delamination occurred when the maximum stresses were located in the vicinity of the film-substrate interface. A similar backward depth evolution with this probe was also observed at low forces in multiple-load cycling nanoindentation tests and confirmed by AFM images showing volume uplift (>100 nm). When testing ~500 nm CVD DLC films on Si with a 25 μm end radius spherical indenter at 1 mN, Beake and co-workers reported the probability of interfacial and cohesive failures varied with deposition power (Table 2.14) [35]. Films deposited at higher RF power failed more rapidly but films did not delaminate as readily as films deposited at lower RF power suggesting improved adhesion strength. With this less sharp probe the energy required to drive failure was quite large.

When a standard Berkovich probe was used Fasial and co-workers found that the 100 nm DLC film failed after a few impacts at loads as low as 100 μN resulting in increases in probe depth. Post-test AFM confirmed film failure with blistering and pile-up outside the impact crater and delamination and substrate failure in the centre of the impact. To reduce the severity of the test they also performed fatigue tests where the probe did not leave the sample between tests, linearly loading to peak load in 1 s before unloading to 90% of peak load in 1 s (cycle frequency ~0.5 Hz) [141]. Under these conditions they could also cause failure of the stressed thin film but due to the lower strain rate the failures occurred after longer much fatigue cycles. The high strain rate in the nano-impact test, where each cycle is a true impact event, is more efficient at promoting film failure.

The results can be compared with those in an earlier study of the nano-impact behavior of 5, 20, 60 and 80 nm ta-C thin films deposited on silicon by FCVA using a well-worn Berkovich indenter at sub-mN forces [26]. It that work it was found that the 60 and 80 nm ta-C films also failed after only a few impacts even though the probe was blunter. Final depths (Table 2.15) were lower due to the shorter test duration and blunter geometry

Table 2.15 Variation in the final wear depth after nano-impact testing of ta-C films on Si at 100 μN.

ta-C film thickness (nm)	Mean final on-load depth (nm)[a]
80	212 ± 13
60	209 ± 10
20	171 ± 7
5	156 ± 8

a) ± represents the standard error in the mean from 20 measurements. The on-load depth is the sum of the elastic, plastic and fracture deformation.

and it some cases a slight backwards depth evolution was observed immediately prior to the increase of depth resulting in total film failure over a few impacts. The resistance to impact-induced fracture decreased as the film thickness increased.

Multiple factors are likely to contribute to this including (i) differences in their bending strength (ii) location of peak stresses relative to the interface (iii) relaxation of compressive stresses. The thickness alters the relative ability of the films to deflect (elastically or plastically) with the Si substrate under load, with the lower elastic strain to yield of the thicker films resulting in lower resistance to fracture [10]. For some applications, multilayer stacks of alternating hard and soft thin layers have been used with improved performance compared to single thicker coatings. These multilayers are designed so the individual hard and stiff layers in the stack do not exceed their bending stress in compression and the shear takes place in the softer layers. The maximum stresses are further away from the coating-substrate interface for the thinner ta-C films, which being softer than the thicker ta-C are likely to have lower compressive stress. The 60 and 80 nm ta-C films appear to be less resistant to impact induced damage than the underlying Si under these conditions. Under these conditions the impact-induced stresses are not high enough to cause phase transformation or lateral cracking in the Si substrate but the fatigue process causes film failure.

2.6 Wear Resistance of Amorphous Carbon Films Using Nano-fretting Testing

Fretting wear is a specific surface destruction process, where the reciprocating sliding motion with relatively small amplitude is responsible for the debris formation and loss of material by interfacial shear work. Currently there is no formal nano-fretting definition available in the literature, however for the purpose of simplicity the nano/micro-wear type of experiments carried out in this work are called nano-fretting. Performing the tests at this scale with well-defined spherical probes as model single asperity contacts simplifies the contact geometry and provides a route for deeper understanding of the fundamental phenomena.

2.6.1 Nano-fretting: State-of-the-art

Nano-fretting data have been reported by several research groups carrying out experiments on various, commercially available and purpose-built equipment. Varenberg *et al.* studied partial and gross slip fretting behaviour of 3.1 μm diameter scanning probe microscopy probes tested against Si wafers [142]. A substantial increase of the coefficient of friction was noted at the transition between partial and gross slip fretting regimes by testing a range of sliding amplitudes from 5 to 500 nm. In collaboration with the same research group, Yoon *et al.* investigated impact of surface roughness and materials combination on fretting wear in a micro-spherical contact using 200 μm diameter spherical tip under 20 mN load and 6.5 μm displacement amplitude [143]. It was found that the materials combination has a greater effect on fretting wear behaviour than surface roughness in this small scale experiments.

Nano-fretting behaviour of monocrystalline silicon for potential application in MEMS devices operating in vacuum conditions was studied by Yu *et al.* using AFM tips [144, 145]. The energy ratio related to the transition from partial to gross slip regime was measured and compared to the same energy ratio observed in classic macroscale fretting. The authors looked also at 2 nm thick DLC coatings deposited on Si(100) and carried out tests against SiO_2 microspheres under vacuum and air conditions [146]. They found that DLC coating reduced significantly adhesion and friction force in air conditions comparing to Si(100) substrate, however wear behaviour was quite similar to that of Si. Wilson *et al.* focused on C and Cr doped amorphous C films and carried out small scale fretting experiments using a modified nanoindenter (NanoTest) with 300 μm diameter ruby tip under 10–200 mN applied load and 2–14 μm displacement amplitude [147–149]. The authors identified two distinct fretting wear regimes, with classic W-shaped wear scar under low oscillation amplitude and full U-shaped wear scar at larger amplitudes. Finally, micro-wear behaviour of DLC and TiN coatings using microtribometer setup under reciprocating sliding was investigated by Achanta et al., where the wear mechanism and topographical changes within the wear tracks were investigated using AFM analysis [150, 151].

Nano-fretting results from experiments performed using NanoTest systems are presented and discussed in this section. The instrument's hardware was modified to include an additional oscillating stage unit that employs a multilayer piezo stack to generate sample motion. Loading and sample movement are controlled independently in the instrument software so that the hardware modification to enable sample oscillation does not require significant software alteration and the loading and contact routines of the commercial nanoindentation module can be used. The sinusoidal piezo driving voltage is produced by a signal generator + amplifier + transformer combination.

The type of experiments described include (i) nano-fretting of thin DLC films on Si, and (ii) nano-fretting of thicker DLC films on steel. The results are discussed in terms of friction evolution, wear behaviour, occurrence of critical failure events, influence of probe radius and geometry, and finally role of ploughing on friction evolution.

2.6.2 Nano-fretting of Thin DLC Films on Si

Comparative behaviour for nano-fretting of various sub-200 nm DLC coatings deposited on Si with a 25 μm spheroconical probe is shown at 10 mN in Figure 2.9a [30]. During the duration of the 10 000 cycle test no abrupt film failure events were

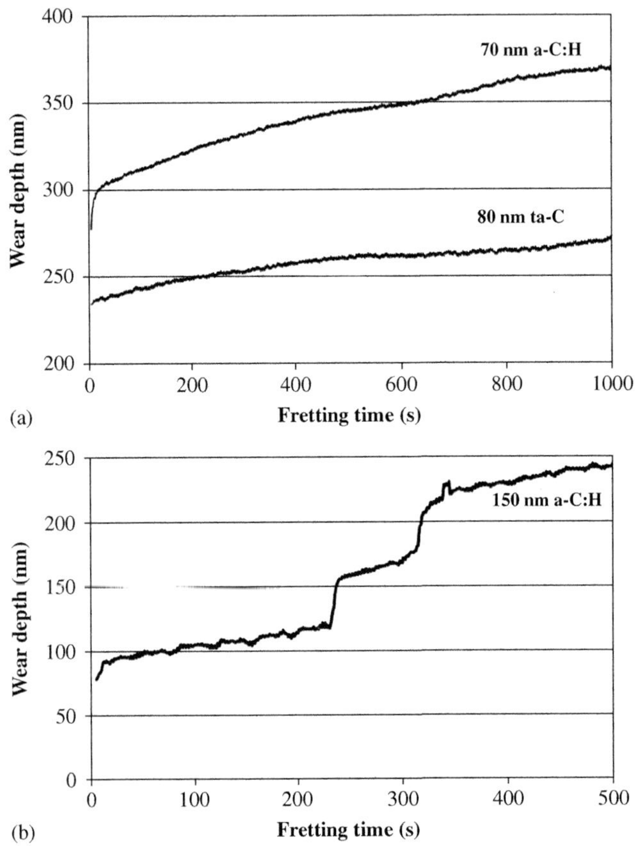

Figure 2.9 Nano-fretting behaviour of thin DLC films on Si: a) 70 and 80 nm a-C:H films, applied load 10 mN, 25 μm spheroconical test probe.; b) 150 nm a-C:H film, applied load 1 mN, Berkovich pyramidal test probe.

observed but it is evident that the evolution of the probe depth under load of the thin coatings varies considerably, with the 80 nm ta-C being the most resistant. To induce film failure rapidly at low load a sharper pyramidal Berkovich test probe is required. Micro-wear of a 150 nm a-C:H at 1 mN with this probe is shown in Figure 2.9b. The abrupt changes in probe depth after 240 and 320 s are consistent with multiple failure events, as previously observed in nano-scratch tests [27].

The wear behaviour during the accelerated micro-wear test was a strong function of probe geometry. In ramped load scratch tests the on-load depth before failure can be much greater than the film thickness when testing very thin films with large radius probes. This can occur without plasticity in the film, as under loading the thin films can accommodate deformation of the substrate, bending into the hole if substrate plasticity occurs [152]. This method can be used to obtain an estimate of the micro-wear kinetics of materials by comparing evolution of the probe depth under load from a constant load test without sample oscillation applied to the same test under oscillated sample conditions.

In [29] small scale fretting of 5, 20 and 80 nm thickness ta-C films deposited on Si(100) was investigated using spherical indenters. It was found that fretting wear occurred at significantly lower contact pressure than is required for plastic deformation and phase transformation in indentation and scratch experiments. EDX analysis performed across fretting wear scars revealed a reduction in the coating thickness and oxygen incorporation in the worn region. The absence of any abrupt changes in depth or friction showed that deformation proceeded by a fatigue mechanism with a gradual wearing away of the film. SEM analysis also suggested distinctive gross-slip type of damage with scratches generated during sliding present within the fretting wear scar area. The geometry of the wear scars correlated with the loads applied: reduction in wear track length and wider wear tracks were observed with increased load.

2.6.3 Nano-fretting of DLC Coatings on Steel

Thicker DLC coatings with varied composition and mechanical properties, as shown in Table 2.16, were studied in [36]. Coatings were deposited on M2 grade steel substrates by means of PECVD method.

Observed differences in resistance to nano-fretting wear were correlated in this work with differences in the coatings Y/E and H/E_r ratio. This is consistent with observations at the macro-scale by Leyland and Matthews, that wear resistance often correlates better with H/E than H alone [66]. The observed frictional behaviour on repetitive sliding is explained by changes to the ploughing component of friction and smoothing of asperities, both of which alter the contact area and therefore can influence the friction. In a repetitive nano-scratch test the initial friction force typically decreases due to smoothing out surface asperities. Such initial decreases are occasionally seen in the nano-fretting friction (Figure 2.10). As the fretting test progresses, the increasing friction is observed due to increasing contact area resulting from wear and decreasing contact pressure. The increase in friction through the test is much greater on coating C, reflecting the much greater wear on this sample.

In general, higher wear rate was associated with increasing friction, and a transition to lower wear rate with more steady friction. Studies of the relative importance of yield stress and microstructure on the evolution of friction and wear of metallic materials during micro-scale repetitive low-pass sliding, have concluded that yield stress plays the dominant role on the evolution of friction as the friction was almost independent of the grain size but decreased with increasing hardness [153–155]. The lower initial friction on coating A is consistent with its higher hardness and reduced ploughing contribution.

Table 2.16 Composition, thickness and mechanical properties of DLC coatings.

Coating	Coating type	Thickness (μm)	H (GPa)	E_r (GPa)
Coating A	Cr + W-C : H + DLC	2.9	23.4 ± 1.1	194.9 ± 8.9
Coating B	Cr + CrW + W-C : H + Si-DLC	2.8	16.2 ± 0.6	143.0 ± 3.1
Coating C	CrN + a-C : H : W	2.0	11.5 ± 0.9	146.1 ± 5.6

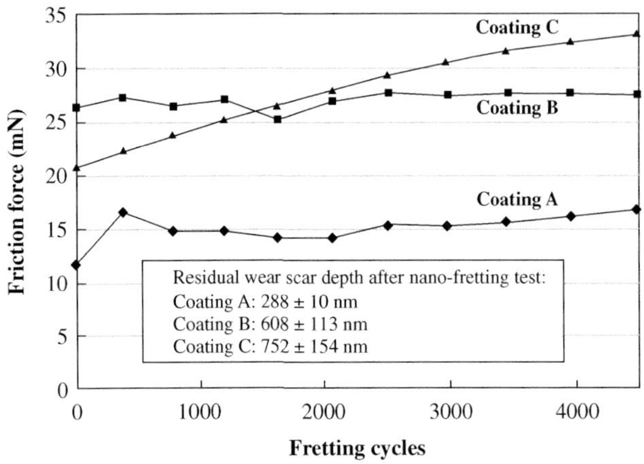

Figure 2.10 Evolution of friction force as a function of fretting cycles for the three DLC coatings.

2.7 Conclusion

Depending on their composition DLC films can have a very wide range of mechanical properties. Accurate nanomechanical characterisation is a critical step in their efficient optimisation for demanding applications. In this chapter we have highlighted the key factors that are required for accurate hardness and elastic modulus determination and the further complications that arise when measuring on ultra-thin DLC films. In addition to hardness and modulus nanoindentation can provide information about film plasticity and its correlation to H/E and, depending on film composition and indenter geometry, cracking may also occur during loading and the role of substrate deformation is discussed.

For more complete characterisation the results of the nanoindentation tests are combined with additional nanomechanical and nano-tribological test capability available in the NanoTest system, namely nano-scratch, nano-impact and nano-fretting. These tests provide complementary information about the durability of DLC films in these more complex mechanical loading situations. The nano-scratch test is currently being standardised in Europe and the factors influencing the critical loads and friction-load dependence of DLC films are discussed. It is shown that substrate yield can play a key role in the nano-tribological behaviour, particularly for ultra-thin films on Silicon. By performing repetitive nano-scratch tests at a sub-critical load it is possible to tune the maximum stress close to the coating-substrate interface so that the test is more sensitive to high stress in the film and at the interface. DLC films have an optimum H/E for enhanced durability in the nano-scratch test.

DLC coatings show poor durability under dynamic loading conditions; hence the nano-impact test has been used to take a more direct approach to the assessment to film toughness and damage tolerance under dynamic loading. The test measures the degradation of coatings from repeated localized stresses. Nano-impact testing has shown that a-C:H films are susceptible to fracture and delamination and show a strong load dependence in their behaviour under repetitive contact, however hydrogen-free amorphous coatings with optimum H/E can show improved durability. Therefore, for higher dynamic toughness, films should combine high load threshold for the initiation of cracking and a mechanisms to minimise and retard crack propagation.

The micro-wear kinetics of DLC coatings can be obtained by comparing evolution of the probe depth under the same loading conditions in nanoindentation (no oscillation) and nano-fretting (with sample oscillation). The nano-fretting technique is complementary to ramped nano-scratch and repetitive nano-scratch methods by allowing higher cycle test, extended duration and significantly lower contact pressure. The presence of the ta-C film on Si substrate affects the phase transformation behaviour of the Si by providing load support, reducing the effective load reaching the substrate and spreading the deformation out over a wider area. Nano-fretting wear occurs in that case at significantly lower contact pressure than is required for plastic deformation and Si phase transformation in nano-indentation and nano-scratch testing. The nano-fretting frictional behaviour of DLC coatings deposited on steel was explained by changes to the ploughing component of friction and smoothing of asperities, both of which alter the contact area and therefore can influence the friction. As the test progresses increasing friction is observed due to increasing contact area resulting from wear. In general, when the severity of nano-fretting tests is controlled by changing the probe geometry, it has been found that under more severe conditions the DLC films show poorer performance potentially limiting their application in such conditions.

By combining the results of all the nanomechanical tests performed with the NanoTest system it is clear that (i) DLC films show strong load dependence in their properties with extensive fracture at high load (ii) there is an optimum H/E which is related to the contact conditions in the actual application (iii) a multi-technique nanomechanical characterisation provides more useful data for design and coating system optimisation than nano-indentation alone.

Analytical tools for nano-scale materials testing have rapidly developed over the last decades. Improved resolution and efficiency of DLC coating testing techniques is needed for development of new applications. Moreover, integration of measurement data obtained from nano-mechanical testing provides also reliable inputs to improved predictive coating wear models.

References

1 S Aisenberg, R Chabot. Ion beam deposition of thin films of diamond-like carbon, Journal of Applied Physics 1971; 42: 2953–2958.

2 J Robertson. Diamond-like amorphous carbon. Materials Science and Engineering R 2002; 37 : 129–281.

3 J Robertson. Deposition and properties of diamond-like carbon. Material Research Society Symposium Proceedings 1999; 555:12.

4 J Robertson. Mechanical properties and structure of diamond-like carbon. Diamond and Related Materials, 1992;1: 397–406.

5 J Geiser. Models and Simulation of Deposition Processes with CVD Apparatus: Theory and Applications. Nova Science Publishers; 2009.

6 DM Mattox. Handbook of Physical Vapor Deposition (PVD) Processing. William Andrew; 2010.

7 AE van Arkel, JH de Boer. Darstellung von reinem Titanium-, Zirkonium-, Hafnium- und Thoriummetall. Zeitschrift fur Anorganische und Allgemeine Chemiel, 1925; 48: 345.
8 RF Bunshah. Deposition Technologies for Films and Coatings. New Jersey, Noyes Publications; 1982.
9 IHI Hauzer Techno Coating BV. DLC Coating: http://www.hauzertechnocoating.com/en/plasma-coating-explained/dlc-coating/ (Accessed: 5 December, 2014).
10 XL Peng, TW Clyne. Residual stress and debonding of DLC films on metallic substrates. Diamond and Related Materials 1998; 7: 944–950.
11 J.M. Jungk, B.L. Boyce, T.E. Buchheit *et al.* Indentation fracture toughness and acoustic energy release in tetrahedral amorphous carbon diamond-like thin films. Acta Materialia 2006; 54(15): 4043–4052.
12 Y-H. Lee, K. Takashima, D. Kwon. Micromechanical analysis on residual stress-induced nanoindentation depth shifts in DLC films. Scripta Materialia 2004; 50(9), 1193–1198.
13 Y.Q. Fu, J.K. Luo, A.J. Flewitt, *et al.* Laser micromachining of sputtered DLC films. Applied Surface Science 2006; 252(13): 4914–4918.
14 K. Holmberg, H. Ronkainen, A. Laukkanen, *et al.* Residual stresses in TiN, DLC and MoS_2 coated surfaces with regard to their tribological fracture behaviour. Wear 2009; 267(12): 2142–2156.
15 H. Maruyama. The Potential of the CDM and CO_2 Reduction from Road Transportation. Institute for Transport Policy Studies, Japan; 2007.
16 T. Haque, D. Ertas, A. Ozekcin, H.W. Jin, R. Srinivasan. The role of abrasive particle size on the wear of diamond-like carbon coatings. Wear 2013; 302: 882–889.
17 A. Grill. Diamond-like carbon: State of the art. Diamond and Related Materials 1999; 8: 428–434.
18 M. Grishke, A. Hieke, F. Morgenweck, H. Dimigen. Variation of the wettability of DLC-coatings by network modification using silicon and oxygen. Diamond and Related Materials 1998; 7, 454–458.
19 C. Donnet, J. Fontaine, A. Grill, *et al.* Wear-resistant fluorinated diamondlike carbon films. Surface and Coatings Technology 1997; 94–95: 531–536.
20 A. Grill, V. Patel, C. Jahnes. Novel low k dielectrics based on diamond like carbon materials. Journal of the Electrochemical Society 1998;145: 1649–1653.
21 M. Azzi, P. Amirault, M. Paquette, *et al.* Corrosion performance and mechanical stability of 316L/DLC coating system: Role of interlayers. Surface and Coatings Technology 2010; 24: 3986–3994.
22 J. Choi, S. Nakao, J. Kim, *et al.* Corrosion protection of DLC coatings on magnesium alloy. Diamond and Related Materials 2007; 16: 1361–1364.
23 U. Depner-Miller, J. Ellermeier, H. Scheerer, *et al.* Influence of application technology on the erosion resistance of DLC coatings. Surface and Coatings Technology 2013; 237: 284–291.
24 S. Neuville. An approach for erosion-resistant coatings by the use of sandwich a-C:H/ta-CNx. Surface and Coatings Technology 1998; 106: 277–281.
25 B.D. Beake, S.P. Lau. Nanotribological and nanomechanical properties of 5–80 nm tetrahedral amorphous carbon films on silicon. Diamond and Related Materials 2005; 14: 1535.

26 B.D. Beake, S.P. Lau and J.F. Smith. Evaluating the fracture properties and fatigue wear of tetrahedral amorphous carbon films on silicon by nano-impact testing. Surface and Coatings Technology 2004; 177–178: 611.
27 B.D. Beake, S.R. Goodes, B. Shi. Nanomechanical and nanotribological testing of ultra-thin carbon-based and MoST films for increased MEMS durability. Journal of Physics D: Applied Physics 2009; 42: 065301.
28 B.D. Beake, G.A. Bell, S.R. Goodes *et al.* Improved nanomechanical test techniques for surface engineered materials. Surface Engineering 2010; 26, 37.
29 B.D. Beake, M.I. Davies, T.W. Liskiewicz *et al.* Nano-scratch, nanoindentation and fretting tests of 5–80 nm ta-C films on Si(100). Wear 2013; 301: 575.
30 T.W. Liskiewicz, B.D. Beake, J.F. Smith. In situ accelerated micro-wear: A new technique to fill the measurement gap. Surface and Coatings Technology 2010; 205, 1455.
31 B. Shi, J.L. Sullivan, B.D. Beake. An investigation into which factors control the nanotribological behaviour of thin sputtered carbon films. Journal of Physics D: Applied Physics 2008; 41: 045303.
32 B.D. Beake, A.A. Ogwu, T. Wagner. Influence of experimental factors and film thickness on the measured critical load in the nanoscratch test. Materials Science and Engineering A, 2006, 423; 70.
33 J. Chen, G.A. Bell, B.D. Beake, H. Dong. Low temperature nano-tribological study on a functionally graded tribological coating using nanoscratch tests. Tribology Letters 2011; 43: 351.
34 B.D. Beake and J.F. Smith. Nano-impact testing: an effective tool for assessing the resistance of advanced wear-resistant coatings to fatigue failure and delamination. Surface and Coatings Technology 2004; 188–189C: 594.
35 B.D. Beake, S. R. Goodes, J.F. Smith, *et al.* Investigating the fracture resistance and adhesion of DLC films with micro-impact testing. Diamond and Related Matererials 2002; 11: 1606–1609.
36 T.W. Liskiewicz, B.D. Beake, N. Schwarzer and M.I. Davies, Short note on improved integration of mechanical testing in predictive wear models. Surface and Coatings Technology 2013; 237: 212.
37 N.M. Jennett *et al.* Determination of the hardness and modulus of thin films and coatings (INDICOAT) EU project contract SMT4-CT98-2249, NPL Report MATC(A) 24 May 2001, ISSN 1473–2734.
38 ISO14577, Metallic materials: Instrumented indentation test for hardness and materials parameters, Parts 1–4. International Standards Organization, Geneva; 2002.
39 T. Chudoba and N.M. Jennett. Higher accuracy analysis of instrumented indentation data obtained with pointed indenters. Journal of Physics D: Applied Physics 2008; 41: 215407.
40 P. Lemoine, J.P. Quinn, P.D. Maguire, *et al.* Intrinsic mechanical properties of ultra-thin amorphous carbon layers. Applied Surface Science 2007; 253: 6165–6175.
41 T. Chudoba, (2006) Nanostructured coatings. In: A. Cavaleiro and J.Th.M. De Hosson, eds. Measurement of hardness and Young's modulus of coatings by nanoindentation. Springer, New York, pp. 216–260.
42 T. Chudoba. presentation at 41st ICMCTF, San Diego, USA (2014).
43 D Tabor. The Hardness of Metals. Clarendon Press, Oxford, UK, 1951.

44 T. Chudoba, N. Schwarzer and F. Richter. Steps towards a mechanical modelling of layered systems. Surface and Coatings Technology 2002; 154: 140.
45 T. Chudoba, M. Griepentrog, A. Dück, *et al.* Young's modulus measurements on ultra-thin coatings. Journal of Materials Research 2004; 19(1): 301.
46 N.M. Jennett, G. Aldrich-Smith and A.S. Maxwell. Validated measurement of Young's modulus, Poisson ratio and thickness for thin coatings by combining instrumented nanoindentation an acoustical measurements. Journal of Materials Research 2004; 19: 143–147.
47 Film Doctor, SIOMEC, http://www.siomec.de/FilmDoctor
48 R.A. Hunt, L.N. McCartney. Use of models to predict deformation during the elastic nano-indentation of coatings, NPL Report MATC (A) 45, July 2001. National Physical Laboratory, Teddington, UK.
49 S. Zhang, X.L. Bui, X.T. Zeng, X. Li. Towards high adherent and tough a-C coatings. Thin Solid Films 2005; 482: 138–144.
50 P. Lemoine, J.F. Zhao, J.P. Quinn, *et al.* Hardness measurements at shallow depths on ultra-thin amorphous carbon films deposited onto silicon and Al_2O_3-TiC substrates. Thin Solid Films 2000; 379: 166–172.
51 P.J. Burnett, D.S. Rickerby. The mechanical properties of wear resistant coatings 1. Modelling of hardness behaviour. Thin Solid Films 1987; 148: 403–416.
52 A. Bhattacharya, W. Nix. Analysis of elastic and plastic deformation associated with indentation testing of thin films on substrates. International Journal of Solids and Structures 1988; 24(12): 1287.
53 A.C. Fisher-Cripps, S.J. Bull, N. Schwarzer. Critical review of claims for ultra-hardness in nanocomposite coatings. Philosophical Magazine 2012; 92: 1601–1630.
54 E. Martinez and J. Esteve. Nanoindentation hardness measurements using real-shape indenters: application to extremely hard and elastic materials. Applied Physics A, 2001; 72: 319–324.
55 FM Borodich, LM Keer, CS Korach. Analytical study of fundamental nanoindentation test relations for indenters of non-ideal shapes. Nanotechnology 2003;14: 803–808.
56 K Herrmann, NM Jennett, W Wegener, *et al.* Progress in determination of the area function of indenters used for nanoindentation. Thin Solid Films 2000; 377: 394–400.
57 W. Ni, Y.-T. Cheng, C.-M. Cheng, D.S. Grummon. An energy-based method for analysing instrumented spherical indentation experiments. Journal of Materials Research 2004; 19: 149–157.
58 K.-D. Bouzakis, N. Michailidis, A. Siganos. Research activities of the laboratory for machine tools and manufacturing engineering (EEDM) during the last three years in the fields of manufacturing engineering, emphasizing on cutting tools' coatings. Proceedings of T.H.E. Conference; 2005.
59 B.D. Beake and N. Ranganathan. An investigation of the nanoindentation and nano/micro-tribological behaviour of monolayer, bilayer and trilayer coatings on cemented carbide. Materials Science and Engineering A 2006; 423: 46.
60 Th. Zhang, Y. Huan. Nanoindentation and nanoscratch behaviours of DLC coatings on different steel substrates. Composites Science and Technology 2005; 65: 1409–1413.

61 X.L. Peng, Z.H. Barber, T.W. Clyne. Surface roughness of diamond-like carbon films prepared using various techniques. Surface and Coatings Technology 2001; 138: 23–32.
62 J. Jiang, R.D. Arnell. The effect of substrate surface roughness on the wear of DLC coatings. Wear 2000; 239: 1–9.
63 R. Gahlin, M. Larsson, P. Hedenqvist. ME-C:H coatings in motor vehicles. 2001; Wear 249: 302–309.
64 P. Harlin, P Carlsson, U. Bexell, M. Olsson. Influence of surface roughness of PVD coatings on tribological performance in sliding contacts. Surface and Coatings Technology 2006; 201: 4253–4259.
65 P. Harlin, U. Bexell, M. Olsson. Influence of surface topography of arc-deposited TiN and sputter-deposited WC/C coatings on the initial material transfer tendency and friction characteristics under dry sliding contact conditions. Surface and Coatings Technology 2009; 203: 1748–1755.
66 A. Leyland, A. Matthews. On the significance of the H/E ratio in wear control: a nanocomposite coating approach to optimised tribological behaviour. Wear 2000; 246: 1.
67 A. Leyland, A. Matthews. Design criteria for wear-resistant nanostructured and glassy-metal coatings. Surface and Coatings Technology 2004; 177–178: 317–324.
68 G.S. Fox-Rabinovich, S.C. Veldhuis, V.N. Scvorstov, *et al.* Elastic and plastic work of indentation as a characteristic of wear behavior for cutting tools with nitride PVD coatings. Thin Solid Films 2004; 505: 469–470.
69 Y.-T. Cheng, C.-M. Cheng. Relationships between hardness, elastic modulus, and the work of indentation. Applied Physics Letters 1998; 73: 614.
70 Y.-T. Cheng, C.-M. Cheng. Scaling, dimensional analysis, and indentation measurements. Materials Science and Engineering R 2004; 44, 91–149.
71 B.D. Beake, G.S. Fox-Rabinovich, S.C. Veldhuis, S.R. Goodes. Coating optimisation for high-speed machining with advanced nanomechanical test methods. Surface and Coatings Technology 2009; 203: 1919.
72 Y. Choi, H.-S. Lee and D. Kwon. Analysis of sharp-tip indentation load-depth curve for contact area determination taking into account pile-up and sink-in effects. Journal of Materials Research 19 (2004) 3307.
73 B.D. Beake, S.R. Goodes, J.F. Smith, *et al.* Using nanomechanics to optimise coatings for cutting tools. In: (ed. S Zhang) Handbook of Nanostructured Thin Films and Coatings, Mechanical Properties pp. 205–244, CRC Press; 2010.
74 B.D. Beake, G.S. Fox-Rabinovich, Y. Losset, K. Yamamoto *et al.* Why can TiAlCrSiYN-based adaptive coatings deliver exceptional performance under extreme frictional conditions? Faraday Discussions 2012; 156: 267–278.
75 Yu V Milman. Plasticity characteristic obtained by indentation. Journal of Physics D: Applied Physics 2008; 41: 074013.
76 C.A. Chartidis. Nanomechanical and nanotribological properties of carbon-based thin films: A review. International Journal of Refractory Metals and Hard Materials 2010; 28: 51–70.
77 Y. Gogotsi, S. Welz, D.A. Ersoy and M.J. McNallan. Conversion of silicon carbide to crystalline diamond-structured carbon at ambient pressure. Nature 2001; 411: 283–287.

78 T. Michler, M. Grischke, I. Traus, *et al.* Mechanical properties of DLC films prepared by bipolar pulsed DC PACVD, Diam Relat Mater 7 (1998) 1333–1337.
79 T. Michler, M. Grischke, I. Traus, K. Bewilogua and H. Dimigen, DLC films deposited by bipolar pulsed DC PACVD. Diamond Related Materials 1998; 7: 459–462.
80 S. Zhang, X.L. Bui, Y. Fu, H. Du. Development of carbon-based coating on extremely high toughness with good hardness. International Journal of Nanoscience 2004; 3: 571–578.
81 S. Zhang, X.L. Bui, Y. Fu, *et al.* Bias-graded deposition of diamond-like carbon for tribological applications. Diamond Related Materials 2004; 13: 867–871.
82 Y.X. Wang, S. Zhang. Toward hard yet tough ceramic coatings. Surface and Coatings Technology 2014; 258: 1–16.
83 B.D. Beake, T.W. Liskiewicz. Comparison of nano-fretting and nano-scratch tests on biomedical materials. Tribology International 2013; 63: 123.
84 D. Galvan, Y.T. Pei, J.Th.M. De Hosson. Deformation and failure mechanism of nano-composite coatings under nanoindentation. Surface and Coatings Technology 2006; 200: 6718–6726.
85 J. Chen. Indentation-based methods to assess fracture toughness for thin coatings., Journal of Physics D: Applied Physics 2012; 45: 203001.
86 S. Hassani, M. Bielawski, W. Beres, *et al.* Surface and Coatings Technology 2008; 203: 204–210.
87 E. Bousser, M. Benkahoul, L. Martinu, JE Klemberg-Sapieha. Effect of microstructure on the erosion resistance of Cr-Si-N coatings. Surface and Coatings Technology 2008; 203 776.
88 N.H. Faisal, R. Ahmed, S. Goel *et al.* Influence of test methodology and probe geometry on nanoscale fatigue failure of diamond-like carbon film. Surface and Coatings Technology 2014; 242: 42–53.
89 B.D. Beake, T.W. Liskiewicz , J.F. Smith. Deformation of Si(100) in spherical contacts: Comparison of nano-fretting and nano-scratch tests with nano-indentation. Surface and Coatings Technology 2011; 206: 1921–1926.
90 M. Larsson, M. Olsson, P. Hedenqvist, S. Hogmark. Mechanisms of coating failure as demonstrated by scratch and indentation testing of TiN coated HSS: On the influence of coating thickness, substrate hardness and surface topography. Surface Engineering 2000: 16: 436–444.
91 H. Ichimura and Y. Ishii. Effects of indenter radius on the critical load in scratch testing. Surface and Coatings Technology 2003; 165: 1–7
92 K. Holmberg, H. Ronkainen, A. Laukkanen ,K. Wallin. Friction and wear of coated surfaces: scales, modelling and tribomechanisms. Surface and Coatings Technology 2007; 202: 1034–1049.
93 K. Holmberg, H. Ronkainen, A. Laukkanen, *et al.* Tribological analysis of TiN and DLC coated contacts by 3D FEM modelling and stress simulation. Wear 2008; 264: 877–884.
94 P.A. Steinmann, Y. Tardy, H.E. Hintermann. Adhesion testing by the scratch test method: influence of intrinsic and extrinsic parameters on the critical load. Thin Solid Films 1987; 154: 333–349.
95 H. Ollendorf, D. Schneider. A comparative study of adhesion test methods for hard coatings. Surface and Coatings Technology 1999; 113: 86–102.

96 S.J. Bull and D.S. Rickerby. New developments in the modelling of the hardness and scratch adhesion of thin films. Surface and Coatings Technology 1990; 42: 149–164.
97 P.J. Burnett, D.S. Rickerby. The relationship between hardness and scratch adhesion. Thin Solid Films 1987: 154: 403–416.
98 S.J. Bull. Failure modes in scratch adhesion testing. Surface and Coatings Technology 1991; 50: 25–32.
99 B.D. Beake, V.M. Vishnyakov, R. Valizadeh, J.S. Colligon. Influence of mechanical properties on the nanoscratch behaviour of hard nanocomposite TiN/Si3N4 coatings on Si. Journal of Physics D: Applied Physics 2006; 39: 1392–1397.
100 B.D. Beake, V.M. Vishnyakov, A.J. Harris. Relationship between mechanical properties of thin nitride-based films and their behaviour in nano-scratch tests. Tribology International 2011; 44: 468.
101 BD Beake, AJ Harris, TW Liskiewicz. Review of recent progress in nanoscratch testing. Tribology 2013; 7: 87.
102 NANOINDENT-PLUS project. Standardising the nano-scratch test, FP7/2007-2013. grant agreement NMP-2012-CSA-6-319208.
103 T.W. Wu, R.A. Burn, M.M. Chen, P.S. Alexopoulos. Micro-indentation and micro-scratch tests on sub-micron carbon films. Materials Research Society Symposium Proceedings 1989; 130: 117–122.
104 T.W. Wu. Microscratch and load relaxation tests for ultra-thin films. Journal of Materials Research 1991; 6: 407.
105 S. Goel, X. Luo, R.L. Reuben. Wear mechanism of diamond tools against single crystal silicon in single point diamond turning process. Tribology International 2013; 57: 272–281.
106 S. Goel, X. Luo, A. Agrawal, R.L. Reuben. Diamond machining of silicon: A review of advances in molecular dynamics simulation. International Journal of Machine Tools and Manufacture 2015; 88: 131–164.
107 F.W. Zok, A. Miserez. Property maps for abrasion resistance of materials. Acta Materialia 2007; 55: 6365.
108 D.J. Oliver, B.R. Lawn, R.F. Cook, *et al.* Giant pop-ins in nanoindented silicon and germanium caused by lateral cracking. Journal of Materials Research 2008; 23: 297.
109 X. Li, B. Bhushan. Micro/nanomechanical and tribological characterization of ultra-thin amorphous carbon coatings. Journal of Materials Research 1999; 14: 2328.
110 S. Sundararajan, B. Bhushan. Micro/nanotribology of ultra-thin hard amorphous carbon coatings using atomic force/friction force microscopy. Wear 1999; 225–229: 678.
111 P. Lemoine, J.P. Quinn, P. Maguire, J.A. McLaughlin. Comparing hardness and wear data for tetrahedral amorphous carbon and hydrogenated amorphous carbon thin films. Wear 2004; 257: 509.
112 D. Sheeja, B.K. Tay, K.W. Leong, C.H. Lee. Effect of film thickness on the stress and adhesion of diamond-like carbon coatings. Diamond and Related Materials 2002; 11: 1643.
113 D. Sheeja, B.K. Tay, S.P. Lau, K.W. Leong, C.H. Lee. An empirical relation for critical load of DLC coatings prepared on silicon substrates. International Journal of Modern Physics B 2002; 16: 958.

114 S. Yang, D. Camino, A.H.S. Jones, D.G. Teer. Deposition and tribological behaviour of sputtered carbon hard coatings. Surface and Coatings Technology 124 (2000) 110.
115 S. Yang, S., X. Li, N.M. Renevier and D.G. Teer. Tribological properties and wear mechanism of sputtered C/Cr coating. Surface and Coatings Technology 2001); 142–144: 85–93.
116 R.D. Arnell. The mechanics of the tribology of thin-film systems. Surface and Coatings Technology 1990; 43–44: 674.
117 H. Djabella, R.D. Arnell. Finite element analysis of the contact stresses in elastic coating/substrate under normal and tangential load. Thin Solid Films 1993; 223: 87.
118 C.T. Wang, A. Escudeiro, T. Polcar, *et al.* Indentation and scratch testing of DLC-Zr coatings on ultrafine-grained titanium processed by high-pressure torsion. Wear 2013; 306: 304–310.
119 X.T. Zeng, S. Zhang, X.Z. Ding, D.G. Teer, Comparison of three types of carbon composite coatings with exceptional load-bearing capacity and high wear resistance. Thin Solid Films 2002; 420–421: 366–370.
120 S.J. Bull, D.S. Rickerby. Multi-pass scratch testing as a model for abrasive wear. Thin Solid Films 1989; 181(1–2): 545–553.
121 S. Achanta, T.W. Liskiewicz, D. Drees, J.-P. Celis. Friction mechanisms at the micro-scale. Tribology International 2009, 42, 1792–1799.
122 B. Bhushan, B.K. Gupta, M.H. Azarian. Nanoindentation, microscratch, friction and wear studies of coatings for contact recording applications. Wear 1995; 181–183(2): 743–758.
123 J. Musil, M. Jirout. Toughness of hard nanostructured ceramic thin films. Surface and Coatings Technology 2007; 201(9–11): 5148–5152.
124 J.R. Trelewicz, C.A. Schuh. The Hall-Petch breakdown at high strain rates: Optimizing nanocrystalline grain size for impact applications. Applied Physics Letters 2008; 93(17): 171916.
125 G. Constantinides, C. Tweedie, D. Holbrook *et al.* Quantifying deformation and energy dissipation of polymeric surfaces under localized impact. Materials Science and Engineering A: Structural Materials Properties Microstructure and Processing 2008; 489(1–2): 403–412.
126 G. Constantinides, C. Tweedie, N. Savva *et al.* Quantitative impact testing of energy dissipation at surfaces. Experimental Mechanics 2009; 49(4): 511–522.
127 J. Arreguin-Zavala, J. Milligan, J. Davies *et al.* Characterization of nanostructured and ultrafine-grain aluminum-silicon claddings using the nanoimpact indentation technique. Journal of Metals 2013; 65(6): 763–768.
128 Z.I. Kalcioglu, M. Qu, K.E. Strawbecker *et al.* Dynamic impact indentation of hydrated biological tissues and tissue surrogate gels. Philosophical Magazine 2011; 91(7–9): 1339–1355.
129 W.B. Choi, Y. Wu, S. Sampath *et al.* Modified Indentation techniques to probe inelasticity in Ni-5%Al coatings from different processes. Journal of Thermal Spray Technology 2009; 18(1): 65–74.
130 J.M. Wheeler, C.A. Collier, J.M. Paillard *et al.* Evaluation of micromechanical behaviour of plasma electrolytic oxidation (PEO) coatings on Ti-6Al-4V. Surface and Coatings Technology 2010; 204(21–22): 3399–3409.

131 J. Moghal, J. Bird, A. Harris *et al.* Nanomechanical study of thin film nanocomposite and PVD thin films on polymer substrates for optical applications. Journal of Physics D-Applied Physics, 2013; 46(48): 485303.
132 GS Fox-Rabinovich, BD Beake, SC Veldhuis, *et al.* Impact of mechanical properties measured at room and elevated temperatures on wear resistance of cutting tools with TiAlN and AlCrN coatings. Surface and Coatings Technology 2006; 200: 5738.
133 A. Mosquera, *et al.* Advantages of Nanoimpact Fracture Testing in Studying the Mechanical Behavior of CrAl(Si)N Coatings. Nanoscience and Nanotechnology Letters 2010; 2(4): 352–356.
134 A. Mosquera, L. Mera, G. S. Fox-Rabinovich, R. *et al.* Statistical analysis of nanoimpact testing of hard CrAl(Si)N coatings. Materials Research Society Symposium Proceedings 2011; 1339.
135 E. Frutos, J.L. Gonzalez-Carrasco. A method to assess the fracture toughness of intermetallic coatings by ultramicroindentation techniques: Applicability to coated medical stainless steel. Acta Materialia 2013; 61(6): 1886–1894.
136 J.M. Wheeler, A.G. Gunner. Analysis of failure modes under nano-impact fatigue of coatings via high-speed sampling. Surface and Coatings Technology 2013; 232: 264–268.
137 F. Klocke, K-D. Bouzakis, K. Georgiadis *et al.* Adhesive interlayers effect on the entire structure strength of glass molding tools Pt-Ir coatings by nano-tests determined. Surface and Coatings Technology 2011; 206(7): 1867–1872.
138 J. Chen, B.D. Beake, R.G. Wellman *et al.* An investigation into the correlation between nano-impact resistance and erosion performance of EB-PVD thermal barrier coatings on thermal ageing. Surface and Coatings Technology 2012; 206: 4992–4998.
139 D. Camino, A.H.S. Jones, D. Mercs, D.G. Teer. High performance sputtered carbon coatings for wear resistant applications. Vacuum 1999; 52(1–2): 125–131.
140 J.C.A. Batista, C. Godoy, G. Pintaude, *et al.* An approach to elucidate the different response of PVD coatings in different tribological tests. Surface and Coatings Technology 2003;174–175: 891–898.
141 R Ahmed, YQ Fu, NH Fasial. Fatigue at nanoscale: an integrated stiffness and depth sensing approach to investigate the mechanisms of failure in diamondlike carbon. Journal of Tribology 2012; 134: 012001.
142 M. Varenberg, I. Etsion, G. Halperin. Nanoscale fretting wear study by scanning probe microscopy. Tribology Letters 2005; 18(4): 493–498.
143 Y. Yoon, I. Etsion, F.E. Talke. The evolution of fretting wear in a micro-spherical contact. Wear 2011; 270: 567–575.
144 J.X. Yu, L.M. Qian, B.J. Yu, Z.R. Zhou. Nanofretting behaviours of monocrystalline silicon (100) against diamond tips in atmosphere and vacuum. Wear 2009; 267: 322–329.
145 J. Yu, L. Qian, B. Yu, Z. Zhou. Nanofretting behaviour of monocrystalline silicon (100) against SiO_2 microsphere in vacuum./ Tribology Letters 2009; 34: 31–40.
146 L. Chen, M. Yang, J. Yu, *et al.* Nanofretting behaviours of ultrathin DLC coating on Si(100) substrate, Wear 2011; 271: 1980–1986.
147 G.M. Wilson, J.F. Smith, J.L. Sullivan. A nanotribological study of thin amorphous C and Cr doped amorphous C coatings. Wear 2008; 265: 1633–1641.

148 G.M. Wilson, J.F. Smith, J.L. Sullivan. A DOE nano-tribological study of thin amorphous carbon-based films. Tribology International 2009; 42: 220–228.

149 G.M. Wilson, J.L. Sullivan. An investigation into the effect of film thickness on nanowear with amorphous carbon-based coatings. Wear 2009; 266: 1039–1043.

150 S. Achanta, D. Drees, J.-P. Celis. Friction and nanowear of hard coatings in reciprocating sliding at milli-Newton loads. Wear 2005; 259: 719–729.

151 S. Achanta, D. Drees, J.-P. Celis. Investigation of friction on hard homogeneous coatings during reciprocatings tests at micro-Newton normal forces. Wear 2007; 263: 1390–1396.

152 S.J. Bull, A.M. Korsunsky. Mechanical properties of thin carbon overcoats. Tribology International 1998; 31: 547–551.

153 C.T. Wang, N. Gao, M.G. Gee, *et al.* Effect of grain size on the micro-tribological behavior of pure titanium processed by high-pressure torsion. Wear 2012; 280–281: 28–35.

154 T. Hanlon, A.H. Chokshi, S. Manoharan, S. Suresh. Effects of grain refinement and strength on friction and damage evolution under repeated sliding contact in nanostructured metals. International Journal of Fatigue 2005; 27: 1159–1163.

155 A. Singh, M. Dao, L. Lu, S. Suresh. Deformation, structural changes and damage evolution in nanotwinned copper under repeated frictional contact sliding. Acta Materialia 2011; 59: 7311–7324.

156 N.M. Jennett, J. Nunn. High resolution measurement of dynamic (nano) indentation impact energy: a step towards the determination of indentation fracture resistance. Philosophical Magazine 2011; 91: 1200–1220.

3

Mechanical Evaluation of Nanocoatings under Extreme Environments for Application in Energy Systems

E.J. Rubio, G. Martinez, S.K. Gullapalli, M. Noor-A-Alam and C.V. Ramana

Department of Mechanical Engineering, University of Texas at El Paso , USA

3.1 Introduction

Energy and aerospace technologies, which often encounter the physical, chemical, thermal and mechanical extremes, are currently placing stringent 'requirements and demands' on the choice of materials and their performance. The fundamental understanding of the structure-property relationship and the interaction of extreme environments with materials can be used to design, develop and implement strategies to improve the functionality and durability of the materials for such applications. However, accomplishing this challenging task requires utilization of selected, well-designated characterization tools that enable the understanding and evaluation of materials under extreme environmental conditions. One such analytical tool is the nanoindetation method, which provides information on the mechanical and themo-mechanical behavior various types of materials.

Application of indentation techniques has extensively contributed to the better understanding of the mechanical characteristics of bulk and/or macroscale materials. Such a detailed understanding of the mechanics by utilization of indentation measurements has provided crucial information leading to substantial advancement and optimization of materials for application in energy and aerospace technologies. However, the current and emerging technologies involving miniaturization of materials poses a stringent requirement on the mechanical and thermal durability, which requires the extensive utilization of nanoindentation to evaluate the mechanical properties of the materials at the micro-to-nanometer scale. Standard nanoindentation technique can be used to determine information related to elastic modulus and hardness values for thin films and coatings; however, more advanced methods must be applied to obtain data related to surfaces and interfaces. Residual stress in growing thin layers, scratch surface resistance and film-substrate adhesion information are some examples to mention [1].

Nanoindentation technique can be beneficial and provide highly useful information, which will help selectively choose the given material for practical applications. This is quite important especially when the given application involves extreme environments, such as high pressures and high temperatures, where thermal stresses, coefficient of thermal expansion, among others have strong influence on the materials' properties, phenomena and performance. Furthermore, in addition to the operating extremes, the

Applied Nanoindentation in Advanced Materials, First Edition. Edited by Atul Tiwari and Sridhar Natarajan.

other secondary factors due to processing can modify the structural and intrinsic properties of materials which demands a deeper understanding of the mechanical characteristics in order to optimize the performance of component materials in the overall energy system.

The focus of this chapter is towards micro and nanostructured coatings operating under extreme environments of power generation systems. Specific emphasis is made on the mechanical testing and evaluation of the coatings. However, note that the number of energy systems that require the understanding of the mechanical properties of their functional materials is constantly increasing as the research dedicated to improving the efficiency of energy conversion systems and power plants has been advancing. Therefore, presentation and discussion of every material related to power generation will be beyond the scope of this book chapter; nevertheless, we will focus in this chapter on materials and issues related to two specific energy technologies. The first set of materials considered are the thermal barrier coatings (TBCs), which are most specific to turbines for energy harvesting and combustion-related processes, where extremely high temperatures, pressures and highly corrosive environment are encountered. The coal-based power plants require extensive utilization of TBCs. Turbine blades are mainly composed out of Ni-based superalloys, but their high-temperature capability is improved due to the incorporation of a multilayer micro-size coating system called Thermal Barrier Coatings (TBCs). Their incorporation allow a temperature gradient 100–300°C from the operational temperature to the turbine component [2]; however, the multilayer design provides issues, particularly in the mechanical loads that each component needs to handle, and where nanoindentation provides valuable information for the failure, improvement, and operation of the system. The second set of materials discussed are typical to the nuclear power plants for energy production, where radiation and high energy particles affect the overall mechanical characteristics of the functional materials inside the nuclear reactor. Recent approaches have been demonstrated that nanocrystalline materials show promising performance for radiation tolerance, where bulk mechanical testing fail to evaluate, nanoindentation techniques allow in-depth analysis of the superficial effect of radiation [3]. The two sets of coatings materials and their performance evaluation using nanoindetation is discussed in the subsequent sections presented below.

3.2 Thermal Barrier Coatings

Thermal barrier coatings (TBC) constitute a complex system, where the complexity is attributed to the fact that it is a multilayer, multi-component material system. The main purpose and challenging function of TBC is to provide protection to the components inside combustion turbines from high temperatures. During system operation, temperature can rise up to ~1300°C, and the incorporation of TBC allows the turbine to function properly due to the temperature gradient of ~100 to 300°C from the surface of the superalloy to the high-temperature gases, this reduction in temperature makes the system to operate with a higher efficiency [2]. As shown in Figure 3.1, TBCs are composed of several layers that increase the complexity of the system, especially at high temperatures, The TBC component parts and their specific functions are:

- Substrate: It is composed of a nickel (Ni) or cobalt (Co) based superalloys, and air-cooled from the inside of the component which allows a difference in temperature

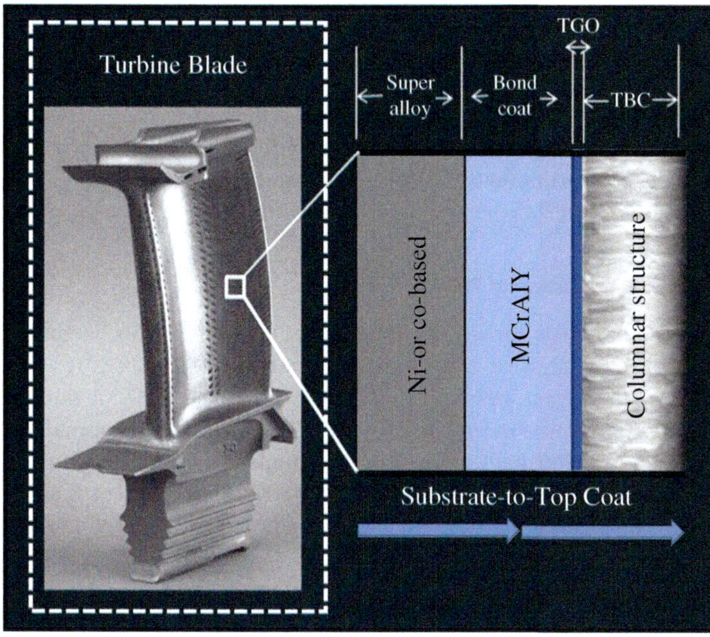

Figure 3.1 Schematic diagram of the thermal barrier coatings (TBCs). The individual component layers are as shown in the diagram.

from the inside of the wall. The superalloys are manufactured with up to 12 different metallic components, in order to enhance certain capabilities of the material, such as ductility, oxidation resistance, high-temperature tolerance, and hot-corrosion resistance [2].

- Bond coat: The purpose of bond coat is to provide oxidation resistance. It is a metallic layer of ~75 to 150 μm. It is typically composed from MCrAlY coatings where M can stand for Ni and/or Co, and deposited via plasma sprayed, high velocity oxyfuels, low pressure plasma sprayed or EB-PVD. Usually the composition of the bond coat is the following: 15–25%Cr, 10–15%Al and 0.2–0.5%Y (in wt-%), and consists in a β-NiAl phase in a γ-Ni-based matrix [4]. During operation, temperature in the bond coat can reach ~700°C and higher, which causes this layer to oxidize and for the next layer of the TBC system. Bond coats are engineered such that aluminum is the preferred oxidizing element, and the formation of uniform, defect free $\alpha\text{-}Al_2O_3$ is created, at the same time, composition of the bond coat maximizes adhesion of the oxide layer [2].
- Thermally grown oxide (TGO): TGO is $\alpha\text{-}Al_2O_3$ layer created by the oxidation of the aluminum content inside the bond coat material. This layer is ~1-10 μm in thickness. It is formed on top of the metallic bond coat and below the ceramic top coat, and it provides the bonding between the top and the bottom layer. Once this layer is formed, it becomes a diffusion barrier that retards the oxidation of the bond coat. TGO and interface between, TGO/bond coat and TGO/top coat are the regions where most of the failure occurs [2].
- Top coat: This ceramic based layer provides the thermal insulation for the system, typically it is composed out of Y_2O_3 stabilized ZrO_2 (YSZ). Top coat has to offer a

low value of thermal conductivity at high temperature, and in the particular case of YSZ it shows values of ~2.3 W/mK at 1000°C, mostly due to the phonon scattering caused by the oxygen vacancies created by the incorporation of Yttrium. Furthermore, this material should have high thermal expansion coefficient (CTE) to reduce the stresses arising by the thermal expansion mismatch with the other components. The commonly employed deposition methods for the ceramic top coats are EB-PVD and plasma spraying techniques [2].

3.2.1 Nanoindentation Characterization of TBCs

The complexity of the TBC system allows different mechanical failure mechanisms; however, the final failure of the system occurs when spallation of the top coat occurs. The most common mechanism of fracture is caused at the location of the TGO, due to the cycling processes inside a combustion turbine, stresses arise by the CTE mismatch, which drive final fracture due to thermal fatigue. The mechanism of failure is similar between deposition methods; however, common crack propagation and failure is allocated near TGO region or the interfacing with the other layers, but the overall failure mechanism has not been completely understood so far.

Optimizing the life of TBC systems requires the maximum understanding of the failure mechanism, which is dependent on the mechanical properties of the components, such as Young's modulus (E), fracture toughness (Kc), and hardness (H), among many other properties that nanoindentation allows. Not only that, but the computational work required to fabricate an accurate model capable of helping to determine when failure may occur, and optimize material selection; this predefined model will require the understanding of the different mechanical properties that nanoindentation is capable of providing [5].

Mechanical properties of top coats has been extensively studied, due to the importance of this particular section and the stresses that it will suffer during operation, starting by analysis of different deposition techniques, comparing APS vs EB-PVD. Each deposition technique provides different crystallization conditions, and modify important intrinsic values for the materials such thermal conductivity; in the case of the mechanical properties, the analysis starts with the Young modulus (E) of YSZ because this value relates directly to the material tolerance to thermal fatigue, and can be calculated using the gradient of load-displacement curve at the unloading section, S, using the following equations:

$$S = \frac{dP}{dh} = \sqrt{\frac{4A}{\mu}}E_r \qquad (3.1)$$

$$\frac{1}{E_r} = \frac{1\text{-}\nu_s^2}{E_s} + \frac{1\text{-}\nu_i^2}{E_i} \qquad (3.2)$$

where P is the applied load, h is the indentation depth, A is the contact area, ν is the Poisson's ratio and the subscripts r, s and i correspond to reduce modulus, sample modulus and indenter material modulus, respectively. At the same time hardness in the top coat can be calculated with the following equation:

$$H = \frac{P_{max}}{A} \qquad (3.3)$$

where P_{max} is the maximum load and A is the projected area. Once *E* and *H* are calculated, nanoindentation technique allows us to calculate fracture toughness as the critical stress factor K_c for the top coat layer; using the following equation proposed by Lawn *et al.* [6] and modified by Laugier [7]:

$$K_c = 1.073k\left(\frac{E}{H}\right)^{1/2}\frac{P}{c^{3/2}} \tag{3.4}$$

where k is a calibration constant equal to 0.016, *c* is the crack length measured from the center of contact to the end of the corner radial crack. The value of *c* cannot be calculated using only nanoindentation , but methods such as Scanning Electron Microscope (SEM) had been used in the past, by measuring the crack length at the indentation crack region. These methodologies can also be employed to understand the mechanical characteristics of the bond coat material by indenting the layer before the top coat layer is deposited, and avoiding any oxide layer formation [8]. Also, this evaluation techniques can helps us determine the materials aging, and degradation due to the exposure to high temperature, by testing the TBCs mechanical properties after heat treatments and high heat flux environments exposure [9].

In recent years, nanoindentation technique has also been employed to more advanced understanding. Zhu *et. al.* [10] calculated the residual stress of the TBC systems using nanoindentation technique by applying the Suresh and Giannakopoulos [11] theoretical model using a Berkovich indenter, to determine the equi-biaxial stresses, which is based on:

$$\sigma_H = \left(\frac{A_0}{A} - 1\right) p_{\text{ave}} \tag{3.5}$$

where σ_H is the stress component parallel to the indentation axis induced by the residual stress applied on the indenter, p_{ave} is the average contact pressure (hardness), which can be calculated with Equation 3.3, A is the contact area of the stressed sample, and A_0 is the contact area of the virgin or unstressed material. Figure 3.2 shows the schematic representation of the influence of residual stress in the material.

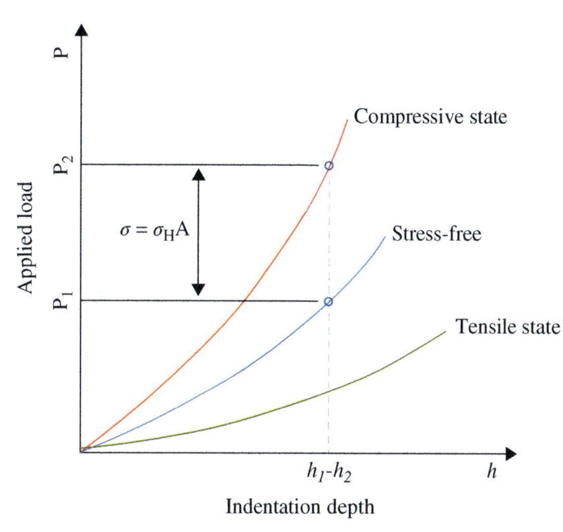

Figure 3.2 Schematic representation of influence of residual stress in the load vs depth plot presented by Zhu *et al.* *Source*: Zhu 2012 [10]. Reproduced with permission of Springer.

In this case hardness is assumed to be equal in the case of stress and unstressed materials. Unstressed samples were achieved by removing metallic substrates by submerging the TBC system into hydrochloric acid until the substrate is removed. In the case of Berkovich tip, the contact area can be calculated as $A=24.56h_c^2$, where h_c is the contact depth and is calculated as:

$$h_c = h_{max} - \varepsilon \frac{P_{max}}{S} \tag{3.6}$$

where h_{max} is the maximum indentation depth, and ε is a constant for indentation geometry; for Berkovich is $\varepsilon=(2^*(\pi-2)/\pi)$. The importance of calculating the residual stress for thermal barrier coatings is due to the high impact that the tensile or compressive stresses affect during operation. Tensile residual stress may cause crack formation and propagation of the coating if the stress's magnitude exceeds the tensile strength of the coating. Compressive stress tend to close surface cracks, and improves fatigue properties, but extensive compressive stress can cause spallation on the coating. Further in this chapter we will present some examples of these already mentioned techniques to characterize new top coat candidates.

3.2.2 Mechanical Properties of Hafnium-based TBCs

Thermal barrier coatings (TBC) are ceramic materials used for protecting the metal components from high temperatures in gas turbine engines. The traditionally used materials for TBCs are made of Zirconium oxide (ZrO_2) doped with 7wt% Yttrium oxide (Y_2O_3) which has an operation limit of 1200°C. Increasing need for pushing the operation temperature has focused the search for new materials to replace the existing materials. Hafnium oxide based coatings were found out to be a viable replacement material. Sputtering targets made from hafnium oxide doped with 7.5 wt% yttrium oxide (YSH), different ratios of hafnium oxide and zirconium oxide doped with yttrium oxide (YSHZ) were used to deposit coatings to study their properties for application as TBCS. The coatings were deposited at a sputtering power of 100 Watts with substrate temperatures varying from room temperature to 500°C. These coatings should have an elastic modulus less than 250 GPa and a hardness value more than 6 GPa to qualify for a TBC for gas turbine application. Nanoindentation technique has been used to measure and compare the mechanical properties of the YSH and YSHZ films. These properties were also used to evaluate the effect of different dopant concentrations in the YSHZ films.

Figure 3.3 shows the hardness (H), modulus of elasticity (E_s), reduced modulus of elasticity (E_r) and stiffness (S) for YSH and various compositions of YSHZ coatings. It is evident (Figure 3.3) that YSH shows the maximum value in all mechanical strength when compared to all other compositions. The optical images show the damage on the sample surface after nanoindentation (Figure 3.4).

One of the most important microstructural parameters that influence the mechanical properties is grain size. There is a critical grain size at which the mechanical properties start decreasing drastically. This critical grain size in pure hafnia is ~2 μm. The grain size of YSH and YSHZ coatings employed for mechanical property evaluation are in the range of ~5–20 nm, which is quite low compared to the critical grain size where the mechanical properties degrade in pure hafnia. No drastic change in mechanical properties with the variation of composition in YSHZ may actually be due to the fact the grain

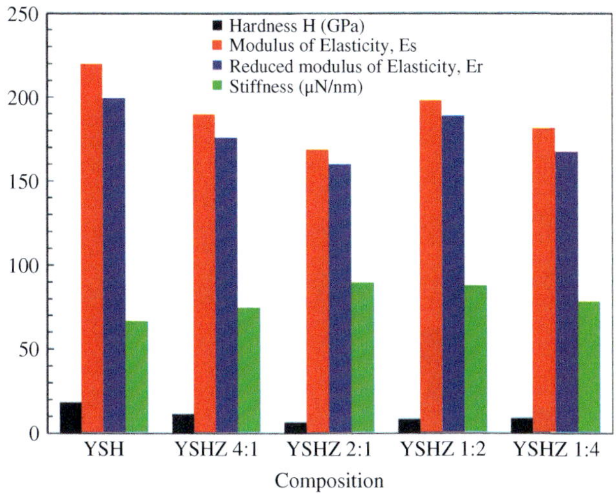

Figure 3.3 Mechanical properties of YSH and YSHZ coatings measured using nano-indentation.

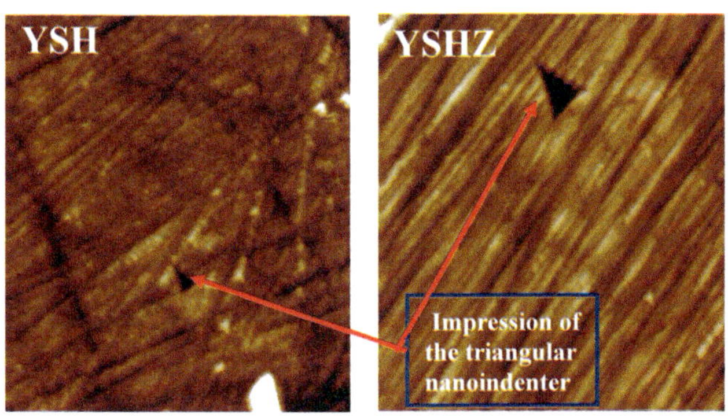

Figure 3.4 Optical images of YSH and YSHZ coatings after nano-indentation.

size is extremely low to induce changes in mechanical characteristics. The interpretation of the mechanical properties could be based on the residual strain energy stemmed from the anisotropy of thermal expansion. If the grains are big enough, micro-cracks are initiated which is responsible for the reduction in mechanical strength. This is not the case in these films as no phase change apart from cubic structure was evident in the operating range of temperature (up to 1300°C) which could be the origin of significant thermal expansion anisotropy. Since there is no transformation toughening in either YSH or YSHZ coatings, the hardness, Young's modulus and stiffness values are the reflection of the combination of individual material properties of hafnia, zirconia and yttria. These properties in fully or partially stabilized hafnia or hafnia-zirconia mixed compositions are strongly influenced by the grain size. The high value of hardness and Young's modulus is, perhaps, due to the dense columnar structure which is clearly seen in both YSH and YSHZ. As was reported, the hardness value of fully stabilized hafnia decreases with the content of alloying oxides. This observation is also visible here when zirconia is incorporated in YSH structure; YSHZ exhibits lower hardness values than

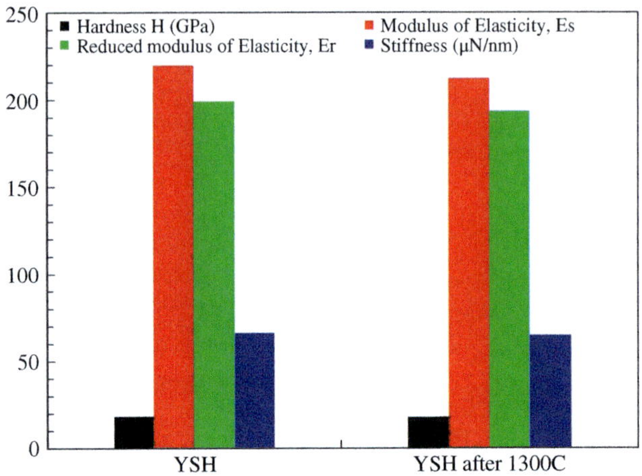

Figure 3.5 Mechanical properties of YSH coatings before and after exposure to 1300°C.

YSH. Although the hardness and Young's modulus are very high and close to YSH, there is slight randomness among all the compositions of YSHZ. The randomness is due to the experimental uncertainty resulted due to the columnar structure of both YSH and YSHZ. Since the nanoindentation was performed on the surface of the samples, it is really hard to identify from the top of the surface exactly where the columns are located. If the nano-indenter is placed on the column it will show higher hardness value compared to that obtained if the indenter is placed in between the columns. Figure 3.5 shows the mechanical properties before and after exposure to 1300°C used for high temperature XRD. YSH coatings show the similar mechanical properties before after exposure to 1300°C indicating the high mechanical stability.

3.3 Nanoindentation Evaluation of Coatings for Nuclear Power Generation Applications

Nuclear reactors can expose their component materials to extreme conditions such as high temperatures (~1000°C), high fluence radiation, and highly corrosion environment. High fluence radiation is most critical and is capable of causing formation of point defects in the materials (interstitial and substitutional). It has been demonstrated that free surfaces and interfaces such as grain boundaries and inter-phase interfaces can act as freely-migrating point defect sinks [12]. In the past, materials have been exposed to reactor radiation, and their mechanical properties were evaluated after exposure; however, this type of tests are costly, and the time needed to perform them has increased [13]. Low energy proton beam, and ion-induced damage are techniques capable of simulating point defect induced damage, but their penetration depth is limited, which enables the nanoindentation technique to calculate the mechanical properties and to assess the level of damage caused by the irradiation.

Simulating point defects and flaws encountered inside nuclear reactor by ion irradiation and the evaluation of the change in mechanical properties via nanoindentation has been studied since 1986, when Zinkler and Oliver evaluated the mechanical properties of ion-irradiated copper and Cu-Zr alloys. These authors employed nanoindentation to quantify the damage levels caused by heavy ion-irradiation. This methodology was required due to the difficulties encountered to reduce the testing area to only the irradiated region [14].

The range of materials investigated to interact with the harvesting of material via nuclear reaction is very broad, and depends on the type of reactor that is going to be employed. For example, the use of ferritic/martensitic Cr-steels is a common material employed inside nuclear reactor and they are candidates for the next generation IV fission and future fusion reactors [15]; $Ti_3(Si,Al)C_2$ has demonstrated promising results in terms of forming part of Gas Fast nuclear Reactors (GFR) [16]. Tungsten is also been considered as one candidate for generation of IV fission and future fusion reactors [12].

One important tool that is necessary to mention, is the use of Stopping and Range Ions in Matter (SRIM) which allows identifying the dose profile of the irradiated samples. It is well known that low and heavy ion irradiation does not produce homogeneous 'flat' profiles on the materials surfaces. Heavy ions have smaller penetration which causes less homogenous profile, compared to high energy ions (light ions). Hosemann *et. al.* [13] discussed specific issues that must be necessarily taken into account when nanoindentation technique is used to evaluate ion-irradiation damage on materials, based on their finding in the irradiation profile as shown in Figure 3.6. Some of the issues mentioned are directly related to calculation of hardness of the material, which is strongly affected by ion-irradiation, and as the irradiation areas are not homogenous, their values can be affected depending on the zone where the indentation is made because of the defect density on each area; for that reason they suggest the use of depth control indentation instead of load controlled which can provide different hardness reading [13].

3.3.1 Evaluation of W-based Materials for Nuclear Application

Tungsten (W) has been considered for the nuclear reactor applications because of its uniqueness of its properties including high thermal strength, good thermal conductivity, high melting point, low sputtering rates and good neutron irradiation. Nanocrystalline W and W-Y coatings with a thickness in the range of 1.1 μm were deposited onto silicon (100) and sapphire (C-plane) using RF sputtering technique. Nanoindentation technique

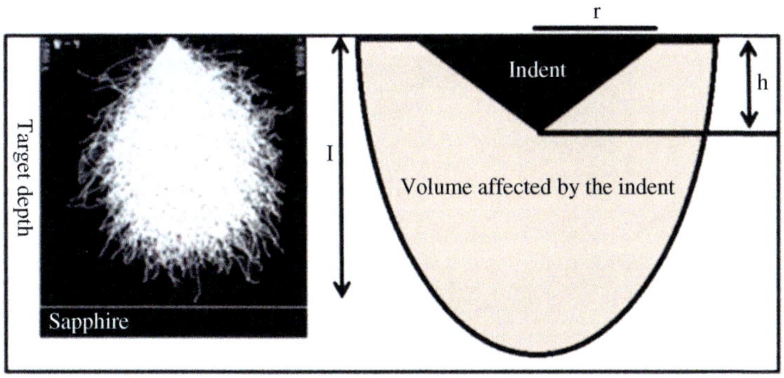

Figure 3.6 The changing dose profile (blue), the indentation size effects (red) and effects caused by "contamination" are shown as well as the affected volume of the indentation test presented by Hosemann *et. al. Source*: Adapted from Hosemann 2009 [13].

was utilized to characterize mechanical properties of the films as deposited and after subjected to varied doses of heavy ion irradiation. Nanoindentation results support the claim that inclusion of Yttrium (Y) into W matrix helping stabilize the grain boundaries and interfaces upon thermal effects and nuclear radiation irradiation exposure. Specifically, enhanced irradiation tolerance and phase stability of W-nanocrystalline W-Y films compared to pure, nanocrystalline W films is demonstrated. Pure Tungsten, Tungsten-Yttrium 95-5 wt.% and 90-10 wt.% with purity of 99.999% were utilized to carry out the sample fabrication to obtain pure W films and W-Y films with increasing Y content. A sputtering power of 120 watts was used for all depositions with variable deposition time to obtain the desired 1.1 μm sample thickness, allowing an indentation depth of 100 nm to characterize mechanical properties. The deposition was carried out under six Ar pressures in the range of $P_{Ar} = 0.43 \times 10^{-2}$–$3.4 \times 10^{-2}$ mbar, Ar flow was controlled using as MKS mass flow meters. Before each deposition, the W-Y target was pre-sputtered for 10 min with the sample shutter closed in order to obtain stable plasma for deposition and rotated 2–3 rpm for lateral isotropy of the film. The deposition times were calculated accordingly to the sputtering yield of the experiment to produce films with the mentioned thickness. The samples were deposited at a constant substrate temperature (T_s) of 200°C.

Irradiation-induced damage, microstructure evolution, grain growth, and phase transformation are all possible failure mechanisms for materials subjected to extreme environments. Nanoindentation was utilized to characterize property degradation of materials induced by ion and neutron irradiation. The samples were then irradiated at room temperature by Au^{3+} ions with a mass of 196 amu and accelerated using an energy beam of 5 MeV in an angle of incidence of 7°. The irradiated area was set for 2 mm by 2 mm enough to allow for mechanical nanoindentation characterization after the experiment. Using the stopping and range of ions in Matter (SRIM 2013) simulation package the beam current (80 nA) and irradiation times (28 sec, 4 min 34 sec and 45 min 42 sec) were calculated to produce a final distribution of ions in the target no deeper than 700 nm into the film and a fluence damage (ions/cm^2) of 1.0×10^{14}, 1×10^{15} and 1×10^{16} ions/cm^2, respectively. SEM cross-sectional images of pure W films before and after irradiation with a fluence of 1 E16 ions/cm^2 are shown in Figure 3.7. It is

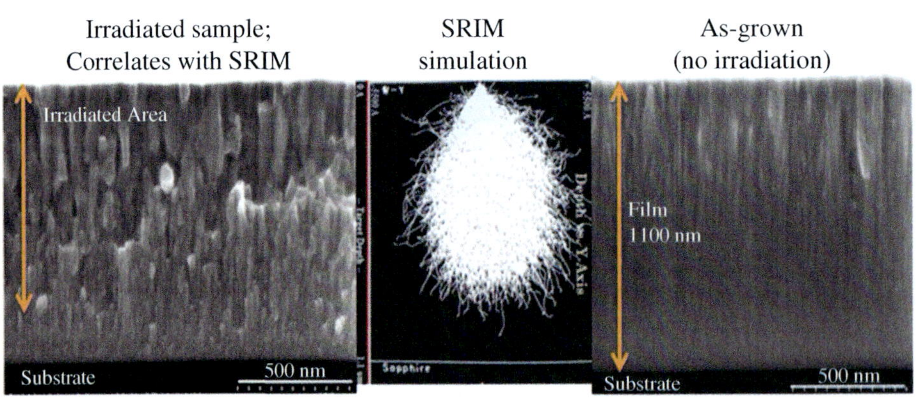

Figure 3.7 Simulation of Au^{+3} (5.0MeV) ion distributions and scanning electron microscopy for irradiated and non-irradiated pure W samples.

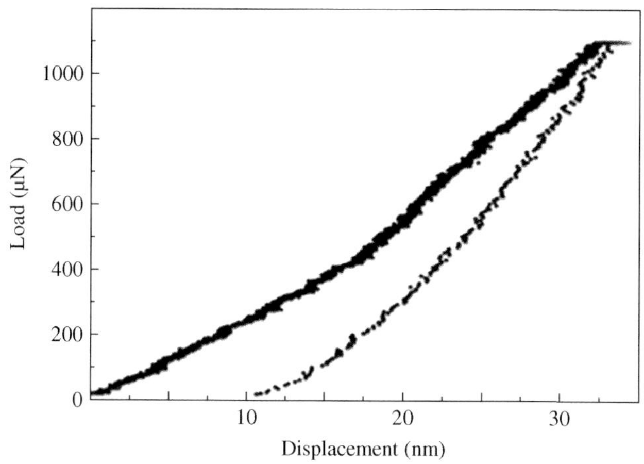

Figure 3.8 Load vs displacement plot of nanoindentation curve of W-Y sample.

demonstrated the comparison between the simulated cross section and the actual sample cross sections before and after irradiation. Figure 3.7 clearly demonstrates the accuracy of the simulation in terms of the penetration and ion density of the Au^{+3}.

Optimizing thin film properties requires a well-understood relationship between processing parameters, modeling, and model confirmation through novel testing techniques. The calculation of nano-mechanical properties was performed using a Hysitron TI 750 TriboIndenter where indentation and nano scratch testing were used to quantify degradation of materials. Depth-controlled indentation tests were performed on each sample to determine the maximum load that could be applied in the film without penetrating more than 10% of the total film thickness.

Tests were performed using a diamond Berkovich tip with a radius of curvature 396 nm to calculate hardness and reduced modulus of each film. Sixteen indents were performed and averaged on each sample, with peak indentation loads in the range of 13–5 μN according to the thickness of each sample. The hardness was calculated from the indentation data according to the Oliver–Pharr method (Figure 3.8) [17].

This technique serves as a measure of the loss of the stiffness of an elastic isotropic material as a result of being damaged by heavy ions such as Au^{3+} ions. The Young's modulus results are plotted in Figure 3.9. The plot is separated to depict the Young's modulus versus the fluence damage. The plot also separates the results as three increasing sputtering (0.63, 1.2 and 3.4 $mbar^{-2}$) pressures and increasing yttrium content of the film. The as deposited Young's modulus values vary from 140 to GPa. Overall the samples grown with higher percentages of yttrium meaning W-Y10%, show the highest values of Young's modulus. Let us first analyze the sputtering pressure phenomena with respect to mechanical performance. For pure tungsten films, the lower sputtering pressures yield the highest values of Young's modulus at 106 MPa. For the 5% yttrium content low sputtering pressure will yield 122 MPa but pressure 1.2 $mbar^{-2}$ will yield a little higher at 124 MPa. In the same manner the sample with 10% wt. of yttrium will yield the highest modulus values at even higher pressures yielding 140 MPa at PAR = 3.4 $mbar^{-2}$. Refereeing to the modulus values with fluence damage it can be observed that the higher values as deposited and after damage are of those of higher yttrium content.

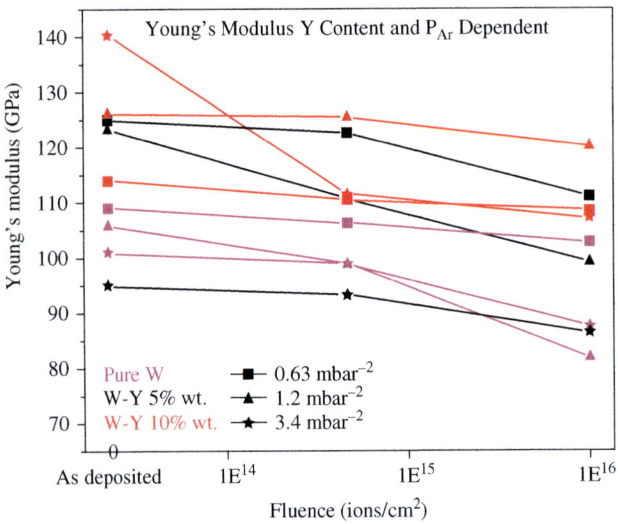

Figure 3.9 Simulation of Au^{+3} (5.0 MeV) ion distributions and scanning electron microscopy for irradiated $1e^{15}$ ions/cm^2 and non-irradiated pure W sputtered at 0.43 mbar^{-2}.

Using nanoindentation technique to characterize mechanical properties nanomaterials have been shown to have excellent radiation resistance in some cases, exhibiting very low swelling and retaining mechanical properties despite being irradiated. The addition of yttrium atoms into the tungsten matrix has shown to improve the irradiation bombardment of heavy Au 3+ when compared to pure tungsten alone. It is believed that the addition of yttrium creates grain boundaries that are capturing interstitial ions and recombining to eliminate vacancies. It is also demonstrated that this recombination 'self-healing' method is also possible at low temperatures. However, the driving mechanism for this phenomenon is not completely understood and more research needs to be conducted in this system.

3.4 Conclusions and Outlook

The mechanical and thermo-mechanical properties of materials play key role in several of the technologies involving extreme environmental conditions. Nanoindetation testing and analysis of the materials and mechanics is extremely useful in such applications. The nanoindetation testing provides information on the elastic modulus, hardness, strain-hardening, cracking, phase transformations, creep, fracture toughness, and energy absorption. In this chapter, we have considered two representative cases of thermal barrier coatings for coal-based power generation systems and structural alloys for nuclear energy applications to demonstrate the application of nanoindetation technique to derive an understanding of the materials' mechanical properties. The hardness (H), modulus of elasticity (E_s), reduced modulus of elasticity (E_r) and stiffness (S) of Yttria-stabilized Hafnia (YSH) and various compositions of yttria-stabilized hafnia-zirconia derived from nanoindentation method are discussed. Mechanical stability and durability up to a temperature of 1300°C is demonstrated for YSH coatings. For nuclear applications, intrinsic tungsten (W) and radiation-irradiated W coatings evaluated based on nanoindetation demonstrated the irradiation tolerance of W- and W-Y alloy coatings. While the *in-situ* indentation under extreme conditions of thermodynamic variables are critical, the capabilities are limited due to various reasons. However, the future applications of nanoindenation seems wide and open. The scale of deformation being very small, the technique can be applicable to materials at the reduced dimensionality, which include nanomaterials, thin films, surface and interfaces, where the link between the microstructure and mechanics of the materials is fundamental and critical to the technology development.

Acknowledgments

The first technical focus of the manuscript (TBCs) is based upon work supported by the Department of Energy under Award Number DE-FE0000765. The second technical focus of the manuscript (Structural Materials for Nuclear Applications) is based upon work supported by the Department of Energy National Nuclear Security Administration (NNSA) under Award Number DE-FNA0000979.

References

1 Fischer-Cripps AC. Nanoindentation, 2nd edn. New York: Springer; 2004.
2 Padture NP, Gell M, Jordan EH. Thermal barrier coatings for gas-turbine engine applications. Science 296, 280–284; 2002.
3 Yu KY, Liu Y, Sun C *et al.* Radiation damage in helium ion irradiated nanocrystalline Fe. Journal of Nuclear Materials 425, 140–146; 2012.4. Darolia, R Thermal barrier coatings technology: critical review, progress update, remaining challenges and prospects. International Materials Review 58, 315–348, 2013.
4 Bouzakis KD, Lontos A, Michailidis N *et al.* A Determination of mechanical properties of electron beam-physical vapor deposition-thermal barrier coatings (EB-PVD-TBCs) by means of nanoindentation and impact testing. Surface and Coatings Technology 163–164, 75–80; 2003.
5 Lawn BR, Evans AG, Marshall DB. Elastic/plastic indentation damage in ceramics: The median/radial crack system. Journal of the American Ceramic Society 63(9–10), 574–581; 1980.
6 Laugier MT. New Formula for indentation toughness in ceramics. Journal of Materials Science Letters 6, 355–356; 1987.
7 Jang HJ, Park DH, Jung YG *et al.* Mechanical characterization and thermal behavior of HVOF-sprayed bond coat in thermal barrier coatings (TBCs). Surface and Coatings Technology 200, 4355–4362; 2006.
8 Shang FL, Zhang X, Guo XC *et al.* Determination of high temperature mechanical properties of thermal barrier coatings by nanoindentation . Surface Engineering 30, 283–289; 2014. .
9 Zhu J, Xie H, Hu Z, Chen P, Zhang Q. Cross-sectional residual stresses in thermal spray coatings measured by Moiré interferometry and nanoindentation technique. Journal of Thermal Spray Technology 21, 810–817; 2012.

10 Suresh S, Giannakopoulos AE. A new method for estimating residual stresses by instrumented sharp indentation. Acta Materialia 46, 5755–5767; 1998.
11 Wang H, Gao Y, Fu E *et al.* Effect of high fluence Au ion irradiation on nanocrystalline tungsten film. Journal of Nuclear Materials 44, 189–194; 2013.
12 Hosemann P, Vieh C, Greco RR *et al.* Nanoindentation on ion irradiated steels. Journal of Nuclear Materials 389, 239–247; 2009.
13 Zinkler SJ, Oliver WC Mechanical property measurements on ion-irradiated copper and Cu-Zr. Journal of Nuclear Materials 141–143, 548–552; 1986.
14 Heintze C, Bergner F, Hernandez-Mayoral M. Ion-irradiation-induced damage in Fe–Cr alloys characterized by nanoindentation. Journal of Nuclear Materials 417, 980–983; 2011.
15 Liu XM, Le Flem M, Bechade JL, Monnet I. Nanoindentation investigation of heavy ion irradiated $Ti_3(Si,Al)C_2$. Journal of Nuclear Materials 401, 149–153; 2010.
16 Oliver WC, Pharr GM. An improved technique for determining hardness and elastic modulus using load and displacement sensing indentation experiments. Journal of Materials Research 7, 1564–1583; 1992.

4

Evaluation of the Nanotribological Properties of Thin Films

Shojiro Miyake[1] *and Mei Wang*[2]

[1] *Dr, Nippon Institute of Technology, Miyashiro-machi, Saitama, Japan*
[2] *Dr, OSG Corporation, Toyokawa, Japan*

4.1 Introduction

Recently, with the high performance and miniaturization of various machines and devices in the field of advanced industries, variations in the micro amounts of wear and friction have become problematic. Therefore, in relation to the tribological characteristics, the importance of extreme surfaces is increasing. For example, because the load in the head-medium interface of a magnetic disk is small, the friction phenomenon, called nanotribology, which the extreme surface generates, affects the surface of the extreme surface. Therefore, lubricating thin films has been used [1, 2]. Furthermore, micro- and nano-systems have been developed in the fields of micro- and nano-technology. Surface-forming technologies and surface evaluation techniques have become important [3–5]. Nanotribology has become an important agenda in the field of advanced technologies.

Here, the topic related to the evaluation of the nanotribological properties in thin films is reviewed. Our recent results on the evaluation of the nanoindentation hardness, nanometer wear and force-modulation using an atomic force microscope (AFM) [1–5] are referred to. In this chapter, the evaluation methods of various thin films are addressed. The first method outlines the nanoindentation of thin films. The second method presents the evaluation of thin films undergoing nano friction and wear tests. The third method outlines the evaluation of the force modulation using an AFM. The last part introduces several examples for the complex evaluation of the other properties of thin films.

4.2 Evaluation Methods of Nanotribology

The science and technology that microscopically deals with various phenomena at the vicinity of a surface, related to friction, wear and lubrication on the atomic scale is called nanotribology. Friction is the interaction between atoms on opposing surfaces and wear is the intermolecular binding destruction caused by friction (interaction). For example, for a number of friction times for an information machine and device, a wear depth of one atomic layer or less per pass is required. As an atom is the smallest unit, wear is only

Applied Nanoindentation in Advanced Materials, First Edition. Edited by Atul Tiwari and Sridhar Natarajan.

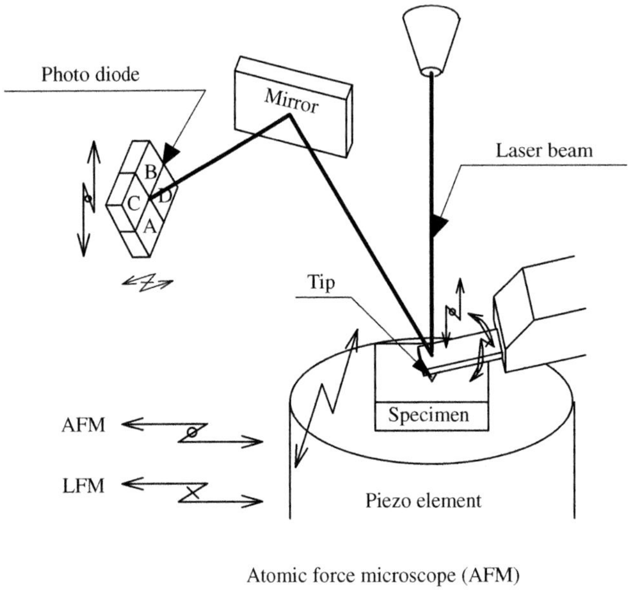

Figure 4.1 Processing method using an atomic force microscope (AFM).

on the per-unit basis. Rather than using the conventional concept of a continuous wear volume, the discontinuous wear of an atomic unit occurred.

To evaluate the tribological properties of these nano-scale films, a scanning probe microscope (SPM) was used to observe the surfaces on an atomic scale [4, 5]. Using the probe shown in Figure 4.1, the nanoscale mechanical properties, such as the friction and wear characteristics, could be evaluated using an AFM. The dynamic evaluation could be realized using a force modulation method by applying vibrations.

4.3 Nanotribology Evaluation Methods and Examples

4.3.1 Nanoindentation Evaluation

To accurately evaluate the surface mechanical properties, hardness measurements were carried out to give an indication of the surface intensity; these are relatively easy to perform. There are three hardness measurement methods that have been used for films [2, 4]: (1) the hardness can be expressed in terms of the deformations formed by pressing an indenter on a superhard material; (2) the surface can be dynamically deformed or indented to obtain the rebound or dynamic hardness. In this method, a diamond-tipped hammer is dropped onto the test surface from a fixed height and the hardness is expressed in terms of the energy of impact and the size of the resulting indentation; and (3) the hardness can be measured by scratching the film with a material that has a high hardness. In general, the hardness measured using method (1) is called the indentation hardness. When evaluating the surface of a film, the results obtained are affected by the underlying substrate because the surface of the film may be penetrated when the indentation load is large. For this reason, so-called nanoindentation tests have been applied, as hardness cannot be evaluated using conventional microhardness tests.

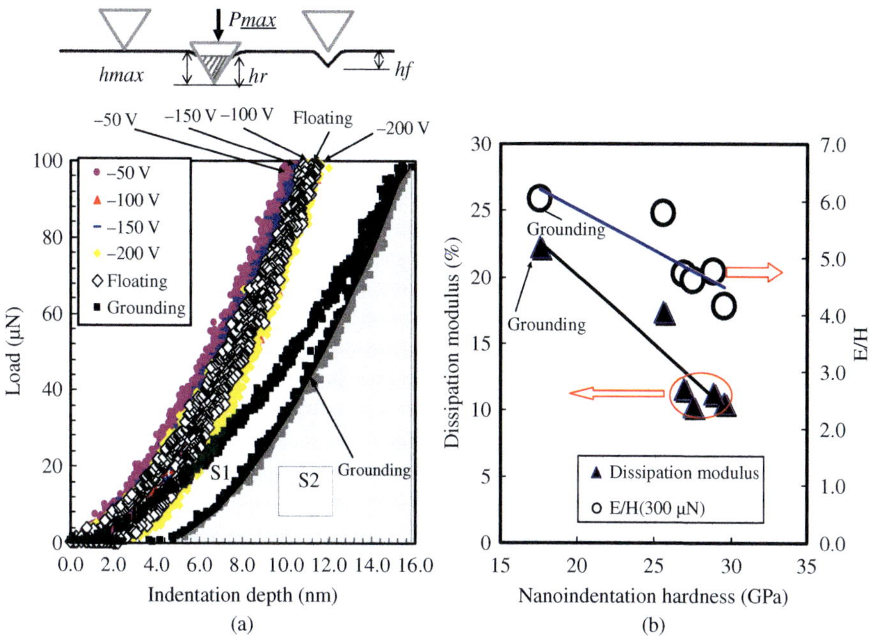

Figure 4.2 (a) Nanoindentation curve and (b) the dependences of the dissipation modulus on the substrate bias voltage (b) for FCVA-DLC films.

In a nanoindentation test, which uses a piezoelectric element, the indentation depth (h) of a test sample is measured with a displacement sensor during the application of an ultra-small load (with a magnitude ranging from μN to mN) by a diamond indenter. From this, the mechanical properties of the surface layer, with a monolayer thickness, of a solid thin-film can be evaluated. The mechanical properties of diamond-like carbon (DLC) films deposited by a bend type filtered cathodic vacuum arc (FCVA) technique have been investigated. Figure 4.2a shows the curves obtained from FCVA-DLC nanoindentation measurements [6]. The indentation hardness (H) was determined by the maximum load (W) divided by the residual indentation area (Ar): H = W/Ar. The Young's modulus (E) was calculated from the slope of the stress-strain unloading curve [1, 3]. The plasticity index, which indicates the plastic deformability of a material (E/H), was also evaluated [2, 6]. Deformation energy analysis is effective to evaluate the mechanical properties of thin films [6, 7]. The total deformation energy (S1+S2) caused by nanoindentation is the integral of the loading curve. The storage energy (S2) is the integral of the unloading curve and the dissipated energy (S1) is the total energy minus the stored energy. The dissipation modulus is the dissipated energy divided by the total deformation energy. Figure 4.2b shows the dissipation modulus curve for the E/H of FCVA-DLC as a function of the hardness. The dissipation modulus and E/H decreased with increasing hardness, indicating that plastic deformation barely occurred. The E/H for most materials ranges from 14 to 20 [2]. In contrast, the DLC films prepared by the FCVA technique under suitable conditions have an E/H = 7.0, which is significantly low. This value indicates that FCVA-DLC films are highly resistant to plastic deformation.

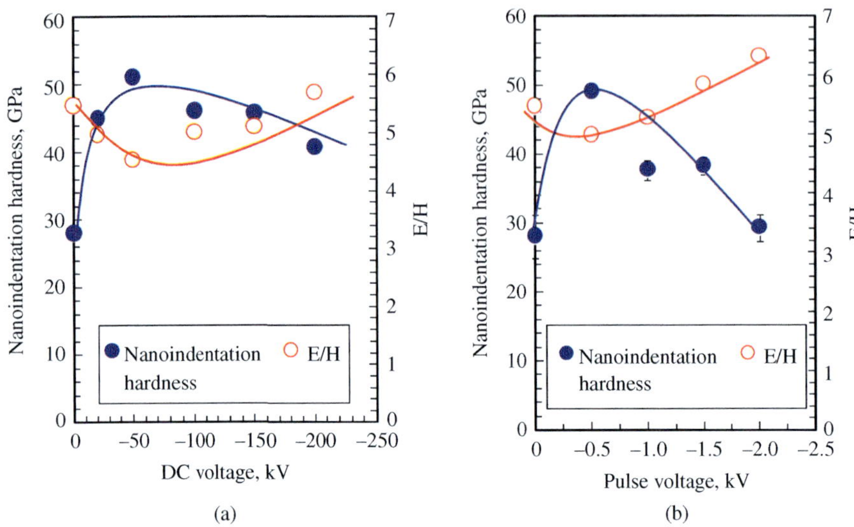

Figure 4.3 Effect of (a) a DC bias voltage and (b) a pulse bias voltage on the nanoindentation hardness (H).

DLC films were deposited by FCVA plating with a bent filter, a graphite target and argon gas [8]. The substrates were subjected to direct current (DC) and pulse bias voltages during film deposition. To deposit the films, a bias voltage, either a DC bias voltage from −50 to −200 V or a pulse bias voltage from −0.5 to −2.0 kV was applied to the substrate. The DLC films were also deposited under a floating voltage or they were grounded, without a bias voltage. They were then compared with the biased DLC films. Miyake et al. reported that the structure and nano- and macrotribologial properties of the DLC films deposited by FCVA are greatly affected by bias voltage [8].

Nanoindentation tests were carried out to evaluate the mechanical properties of the FCVA-DLC films with an AFM (with a Hysitron digital instrument). A Berkovich diamond indenter, which is a three-sided pyramid, was used with a load of 300 μN and a tip radius of approximately 200 nm. During the experiments, both the loading and unloading times were 5 s. Figure 4.3 [8] shows that the H values of the DLC films with either a floating voltage or grounded were 45 and 29 GPa, respectively. The DLC films deposited with a DC bias voltage had higher H values than those of the grounded DLC films. The DLC films exhibited a maximum H of approximately 50 GPa when the DC bias voltage was −50 V. The hardness of these films decreased with an increasing DC bias voltage. With a pulse bias voltage, the DLC films had a maximum hardness when the peak voltage was −0.5 kV. The hardness decreased with an increasing peak voltage. The proper bias voltages for high hardness were obtained under both DC and pulse bias voltages. The same tests were carried out on the DLC films deposited by electron beam ion plating and sputtering. The H values of both films were approximately 20 GPa. The H of the silicon substrate was 13 GPa.

A scatter diagram showing the modulus of dissipation and the E/H dependence on H is presented in Figure 4.4. A high positive correlation was observed between the modulus of the dissipation, E/H and H, because their correlation coefficient (R) and coefficient of determination (R2) were as high as 0.82 and 0.65, respectively. The relationships

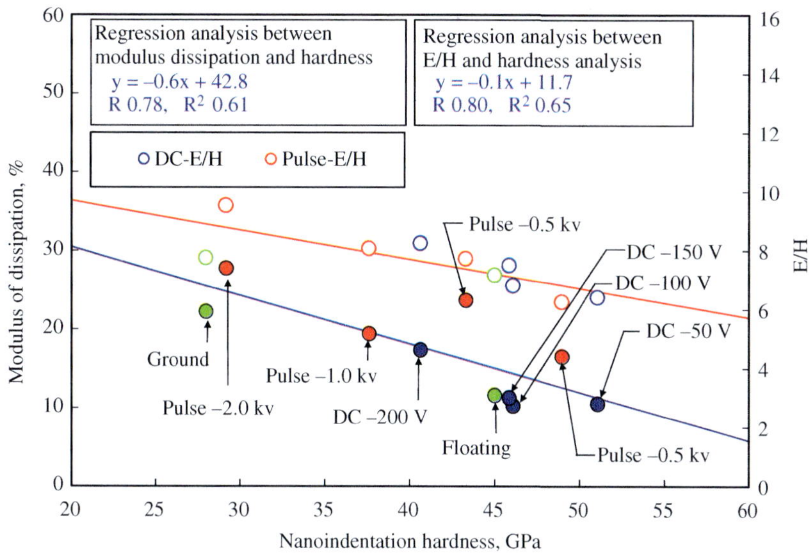

Figure 4.4 Scatter diagram of the modulus of dissipation (E/H) and H for DLC films deposited with 10 different bias voltages.

between the moduli of the dissipations in the nanoindentation tests are shown in Figure 4.4 [8].

All of the tested DLC films deposited while grounded, with a floating or a DC pulse bias voltage could be expressed in one master curve. The hardest – V DC biased DLC films showed low modulus of dissipation and E/H values. Therefore, these FCVA-DLC films had excellent resistances to plastic deformation. Moreover, when the DC bias voltage was reduced from −50 V to −150 V, both the modulus of dissipation and E/H increased. The modulus of dissipation and E/H values of the DLC films deposited while grounded were larger than those of the other films. The modulus of dissipation values of the DLC films deposited while grounded or with a −2. kV pulse peak voltage were approximately 25% and 28%, respectively. Under the same test conditions, the modulus of dissipation for the DLC films deposited by electron beam ion plating were approximately 25%, which is similar to those of the DLC films deposited while grounded.

For very hard materials, it is difficult to induce plastic deformation. The E/H value ranges from 14 to 20 for conventional materials such as metals. For the DLC films deposited by the FCVA method with a bias voltage, the E/H value was very small, about 10.0. From the nanoindentation tests, it could be concluded that the DLC films deposited by the FCVA method exhibited superior resistance to plastic deformation [8].

Another example was used to reveal the effects of the extremely thin nitrogen-containing carbon films deposited by reactive ion plating on the indentation hardness and micro scratch hardness [9]. Thin carbon layers containing nitrogen were deposited using a reactive ion-plating apparatus. Carbon was evaporated by an electron beam and then reacted with an ionized nitrogen gas, generated by a hot cathode discharge [9]. The Knoop hardness was measured at a load of 0.1 N and then the microindentation hardness was measured using a Fischer hardness tester H100 at 4 mN with a Vickers hardness indenter [9], as shown in Figure 4.5.

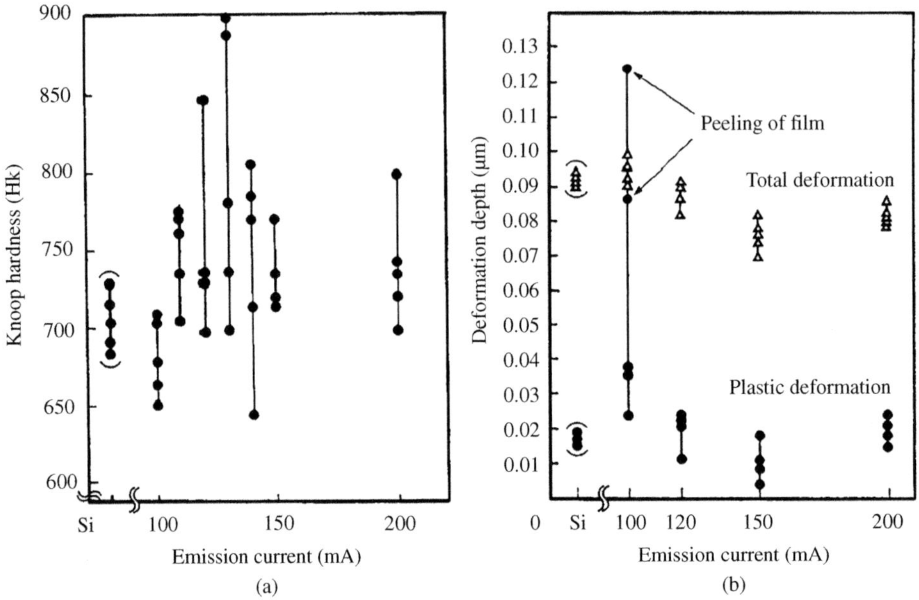

Figure 4.5 Micro-indentation hardness of a nitrogen-containing carbon film (a) Knoop hardness and (b) Indentation depth.

4.3.2 Nanowear and Friction Evaluation

To obtain a material structure that reduces the friction force, a method to form flexible thin films with a low shear resistance is widely known for hard materials. It is possible to realize an extremely-low friction by using a super-hard material as a friction constituent material, supplying an appropriate lubricant in a surface material configuration and reducing the shear resistance of the contact regions on a surface.

Atomic-scale wear is a problem in nanotribology. Reducing the wear to several atomic layers or less during use, that is, achieving zero-wear, is needed from a practical viewpoint. A surface material design to reduce the atomic-scale wear has been proposed [10]. For ultra-hard films with excellent friction characteristics and durability in their material composition, it is expected that low-energy surfaces could be obtained by modifying the surface.

In the hardness measurements on ultrathin films, the film hardness was measured under ultralow loads to avoid the effects from the substrate. During the measurements, errors occurred, owing to variations in the conditions on the film surfaces. Therefore, nanowear tests over square regions on the surface were introduced to reduce the error by averaging over several regions [6, 9]. For instance, the nanowear properties of the nitrogen-containing carbon (CN) films were evaluated by scanning in the same way as that with an AFM [9]. As the surface roughnesses of the thin films changed with the film deposition conditions, this led to changes in the nanoindentation hardness of the films. Although reliable evaluation was difficult, the effect of the nitrogen content on the properties of CN films was clarified using nanowear tests over a square region. The wear depth of the CN films was about one-tenth of that of the C (carbon) films. Thus, a clear difference between the CN and C films was detected [9]. The atomic-scale wear could also be evaluated by selecting appropriate loading conditions for the nanowear

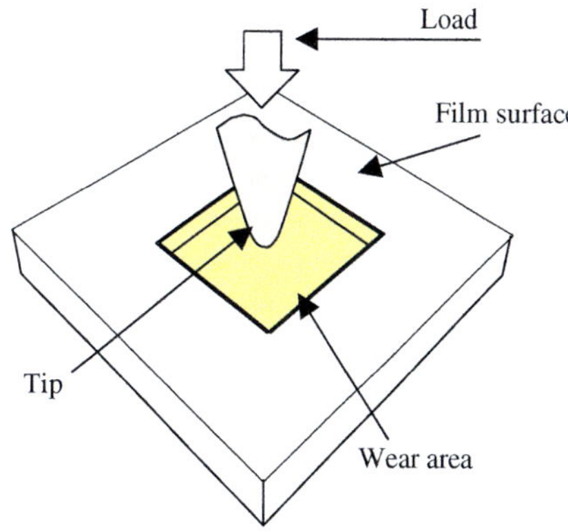

Figure 4.6 Schematic of the nanowear process.

tests. The nanowear shape was dependent on the bias voltage for the FCVA-DLC films [6, 9]. Differences in the wear depths on the order of approximately 1 nm, equivalent to an atomic radius, could be evaluated from the profile of the friction surfaces, obtained by applying various bias voltages. Therefore, appropriate conditions for film deposition were also clarified [6, 11].

4.3.2.1 Nanowear Properties

Nanowear tests were performed to evaluate the wear resistance on the nanometer scale using an AFM [9, 10, 12], as shown in Figure 4.6. The tip was supported by a parallel-leaf spring unit with a small spring constant. The diamond tip was slid along the surface of the sample by a PZT (lead zirconate titanate) scanner, which moved the sample during contact, loading and scanning. The samples coated with DLC films were scanned with this scanner. The test conditions were as follows: the radius of the Berkovich diamond indenter tip was nearly 200 nm, the load was 30 μN, the scan range was 500 × 500 nm^2 and the friction speed was 4.0 μm/s. After the wear tests, the nanowear volumes were measured with an AFM. The changes in the surface profile were measured with a load of less than 1 μN. Then, the average wear depth was evaluated.

The nanowear tests were carried out with an AFM to evaluate the dependence of the nanowear resistance properties of the FCVA-DLC films on the bias voltage [8]. Figure 4.7a shows the nanowear profiles of the DLC films deposited with DC = −50 V (I) and a −1.5 kV pulse bias voltage (II) compared with ground (III). Quantificational analysis was carried out to evaluate the wear depth on the atomic scale [8]. Histograms of the profiles near the wear areas were calculated. For each DLC film, two peaks corresponding to the top surface and the surface of the wear groove were obtained, as shown in Figure 4.7 [8]. The wear depth was estimated from the distance between the central values of the two peaks. The atomic radius of carbon is approximately 0.07 nm. The atomic scale of the wear depth could be evaluated using this method. For the DLC films deposited while grounded, the peak of the top surface was broad, caused by the large surface roughness and the peak of the bottom surface was separated from the top surface because of the deep wear depth. In contrast, on the DLC films prepared with DC = −50 V, the wear groove was flat and the surface roughness was low. A difference of nearly 0.6 nm between the two peaks of these DLC films was observed. For the DLC films deposited with a pulse bias voltage of −1.5 kV, the wear depth was 1.8 nm, whereas for those deposited while grounded, the wear depth was 2.5 nm.

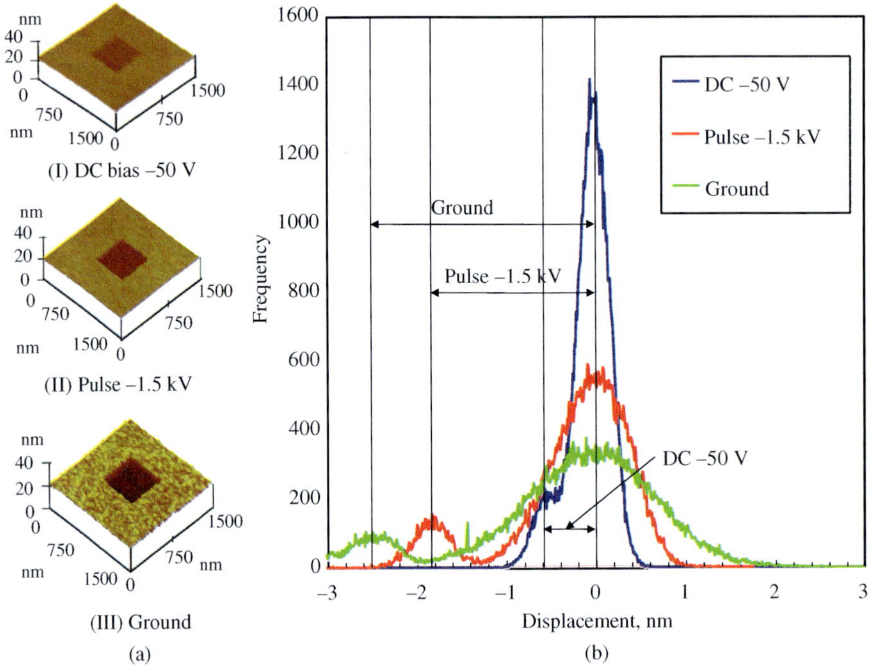

Figure 4.7 Nanowear profile and histogram of a wear surface (a) Nanowear profile and (b) Histogram of wear surfaces.

Figure 4.8 shows a scatter diagram of the wear depth and wear volume for the films deposited with 10 different bias voltages [8]. A high positive correlation was found between the wear depth and the wear volume and R = 0.91. The wear depths of the films deposited with a pulse bias voltage were slightly greater than those of the films deposited with a DC bias voltage. The wear depths of the DLC films deposited with either a pulse bias voltage of −2.0 kV or while grounded exhibited similar large values.

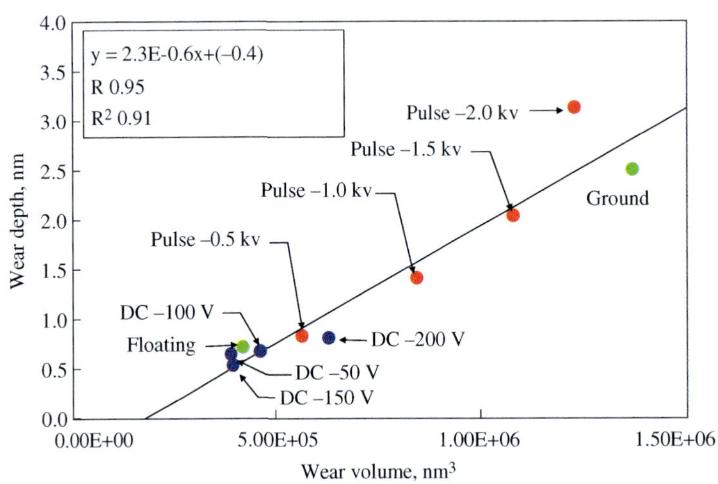

Figure 4.8 Scatter diagram of the wear depth and wear volume for DLC films deposited with 10 different bias voltages.

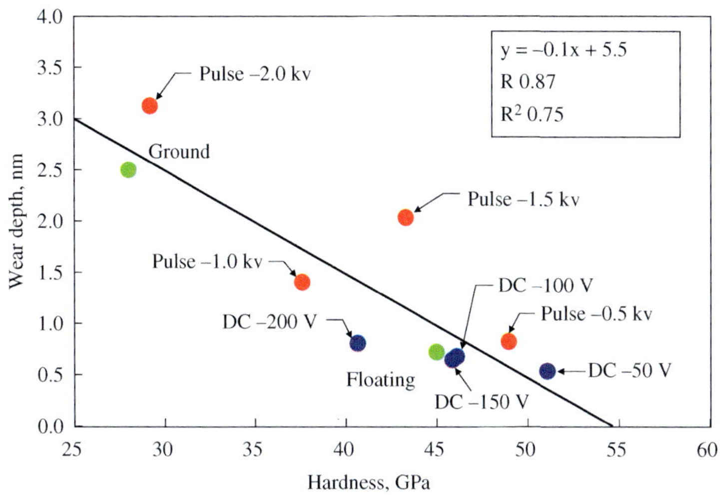

Figure 4.9 Scatter diagram of the wear depth and nanoindentation hardness for DLC films deposited with 10 different bias voltages.

Figure 4.9 shows a scatter diagram of the wear depth and H for the films deposited with 10 different bias voltages [8]. A strong negative correlation was found between the wear depth and H, and R = 0.87. The extremely thin wear depths, which were less than 4 nm, revealed that the clear differences in the nanotribological properties of the films deposited under different conditions could be evaluated.

4.3.2.2 Frictional Properties with Different Lubricants

Ball-on-disk type friction tests were carried out to evaluate the tribological properties of the FCVA-DLC films, as shown in Figure 4.10 [8]. By rotating the samples under the

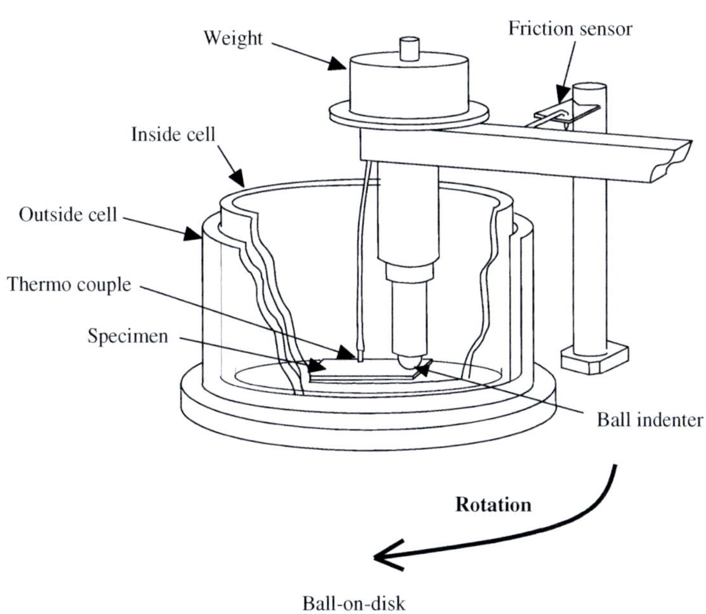

Figure 4.10 Schematic of the ball-on-disk test.

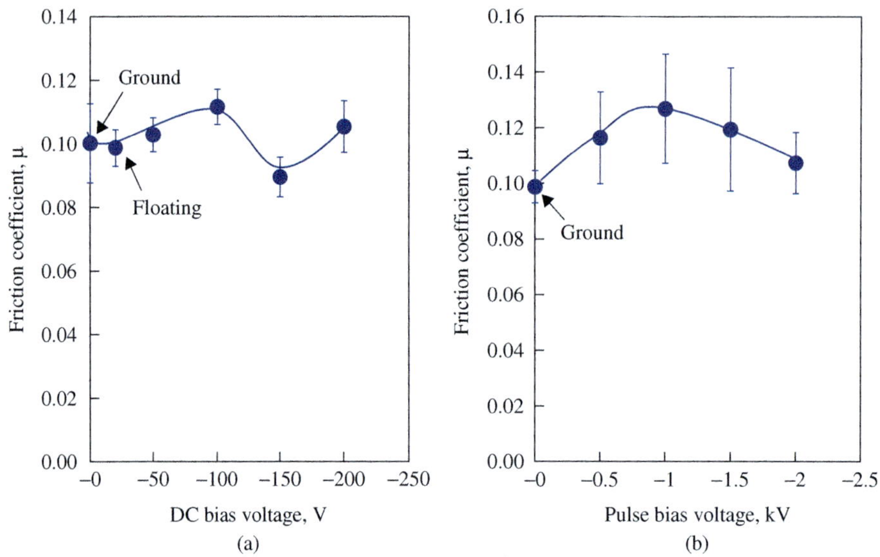

Figure 4.11 Average friction coefficients of DLC films deposited with (a) a DC bias and (b) a pulse bias voltage, lubricated using Z-20 with MoDTC.

application of a specific load on a 6 mm-diameter SUS440C ball indenter, the friction force was measured using a strain gauge-type friction sensor. The tribological properties of these DLC films under dry and boundary lubrication conditions were investigated. The dependence of the frictional properties on the rotation cycles was evaluated under a load of 5.0 N, a rotation speed of 100 rpm, a rotation radius of 4 mm and a total number of rotating cycles of 6000. The velocity was approximately 31.4 mm/s. The testing temperature was maintained near 293 K. After the friction tests, the wear traces of the samples were observed using an optical microscope and a three-dimensional profile meter. Moreover, to investigate the boundary lubrication properties of the DLC films, Z-20 lubricants of molybdenum dithiocarbamate (MoDTC) and water were used.

The evaluation of the boundary lubrication properties was carried out using Z-20 with an additional modifier (MoDTC) lubricant. The average friction coefficients of the DLC films are shown in Figure 4.11. The DLC films deposited with all of the DC bias voltages tested and those grounded exhibited low friction coefficients near 0.1 [8]. Under boundary lubrication with refined water, the friction reduction effect was less than that obtained using Z-20 with the MoDTC lubricant, as shown in Figure 4.12. The changes in the friction properties compared with those obtained under dry conditions were different with refined water [8].

The DLC films were damaged during the friction tests [8]. Under dry conditions, no visible damage to most of the FCVA-DLC film-coated samples was observed. Figure 4.13 shows an example of the wear tracks formed with the use of Z-20 with the MoDTC lubricant [8]. The damage to the DLC films deposited with a DC bias voltage and a floating voltage was difficult to observe with an optical microscope. In contrast, clear damage to the DLC films deposited with a pulse bias voltage and while grounded was observed. Under water boundary lubrication, less damage was observed

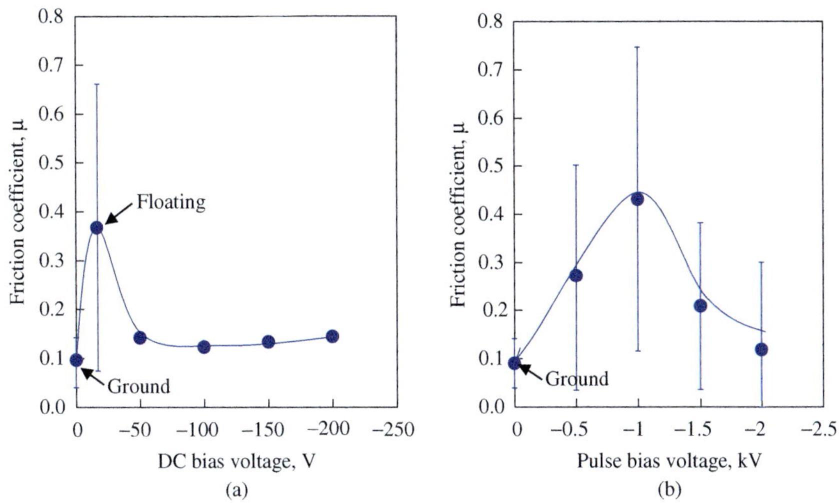

Figure 4.12 Average friction coefficient of DLC films deposited with (a) a DC bias and (b) a pulse bias voltage, lubricated with refined water.

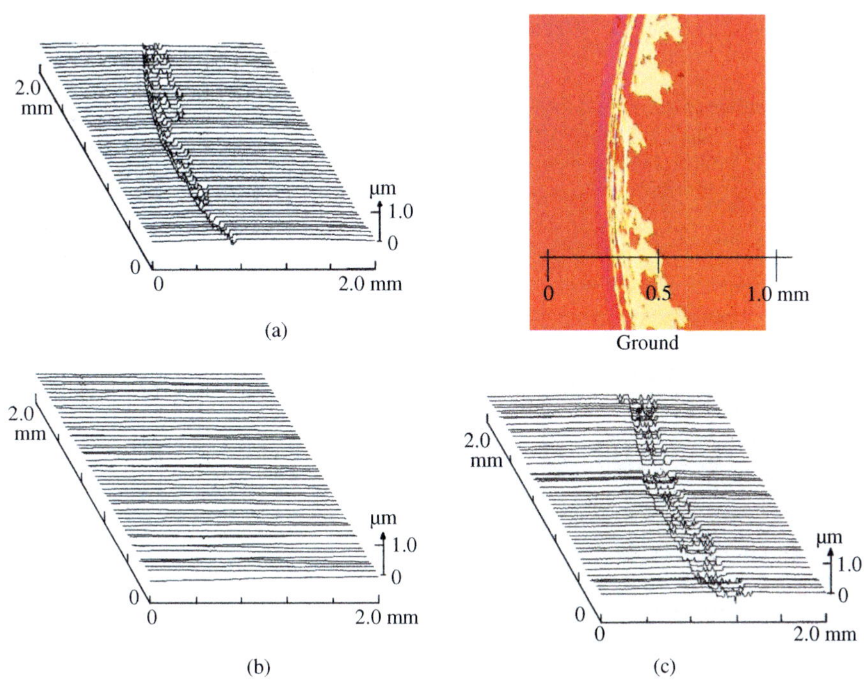

Figure 4.13 Wear profiles of DLC films deposited (a) while grounded, (b) with a DC of −100 V and (c) with a pulse bias of −1.0 kV, lubricated using Z-20 with MoDTC.

on the DLC films deposited with a DC bias voltage than on the others, as shown in Figure 4.14 [8]. Conversely, the DLC films deposited with a pulse bias voltage, a floating voltage or while grounded showed considerable damage, especially those deposited while grounded, which showed the most damage.

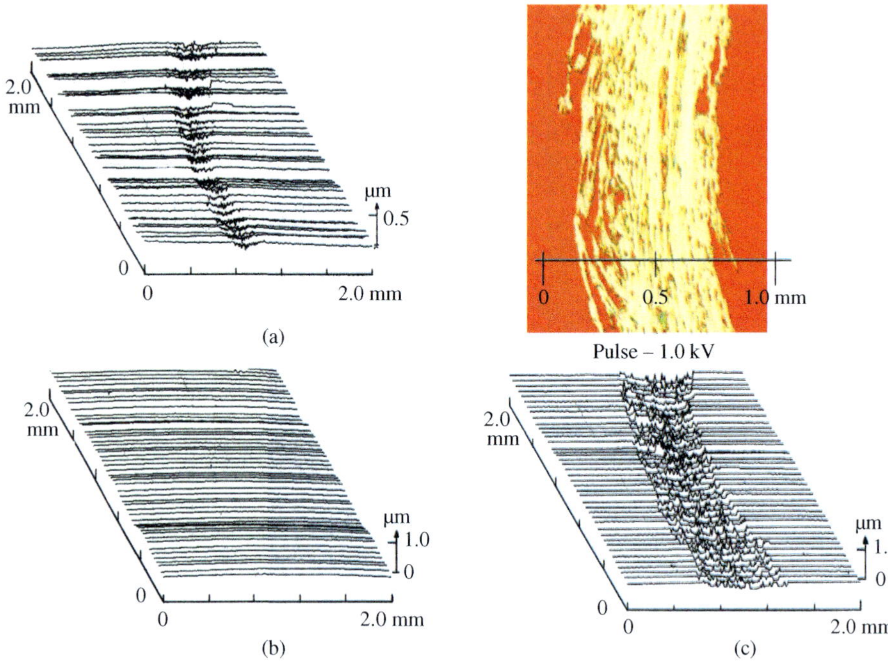

Figure 4.14 Wear profiles of DLC films deposited (a) while grounded, (b) with a DC of −100 V and (c) with a pulse bias of −1.0 kV, lubricated with water.

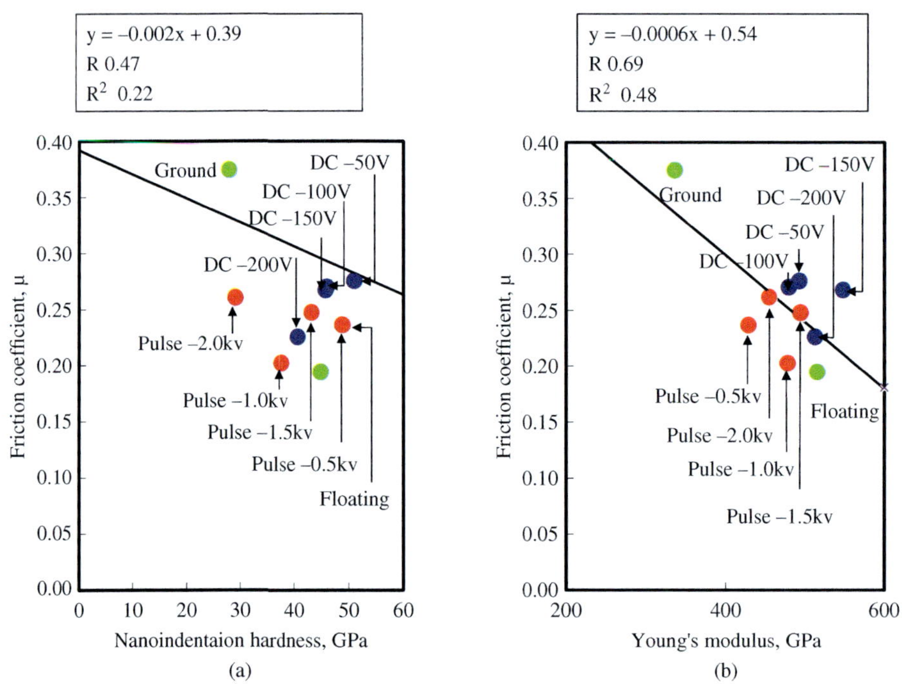

Figure 4.15 Scatter diagram of the friction coefficient, nanoindentation hardness and Young's modulus (a) Friction coefficient dependence on nanoindentation hardness in dry condition and (b) Friction coefficient dependence on Young's modulus in dry condition.

Generally, the friction coefficient depends on the hardness and Young's modulus of a material. The scatter diagram of the friction coefficients shows a dependence on both the hardness and Young's modulus for the 10 types of DLC films deposited with a bias voltage under dry conditions. A negative correlation between the friction coefficient and the hardness is shown in Figure 4.15 [8]. The R values of the friction coefficient with both hardness and E were as large as 0.47 and 0.69, respectively, as shown in Figure 4.9. This decrease in the friction coefficient was attributed to the decrease in the contact area with an increasing hardness and Young's modulus [8].

4.3.2.3 Nanowear and Frictional Properties, Evaluated with and without Vibrations

Extremely thin protective DLC films with thicknesses of 1, 2 and 100 nm were deposited on silicon wafers (Si [100] surfaces) using the FCVA ta-C and electron cyclotron resonance plasma chemical vapor deposition (ECR-CVD) methods [8, 13–16]. Nanowear tests with and without vibrations were carried out on the extremely thin DLC films deposited using the FCVA and ECR-CVD methods [13].

The nanowear tests were performed by sliding a tip with a load and scanning over a 1000 nm × 1000 nm area. The test conditions included 500–4500 nN loads and a 1.8 kHz scan frequency, with and without vibrations, in the Z- (vertical) or X-direction (horizontal) relative to the sample, as shown in Figure 4.16. The vibrations in the Z- or X-directions were added simultaneously during the scan using the force modulation mode of the AFM [13]. Figures 4.17 and 4.18 show the surface profiles of the FCVA-DLC and ECR-CVD-DLC films after the vibration nanowear tests at 200°C in a vacuum with a 4500 nN load [13]. For the FCVA-DLC films, owing to the vibrations, the removal of the hard abrasive wear debris produced from the FCVA-DLC films resulted in a

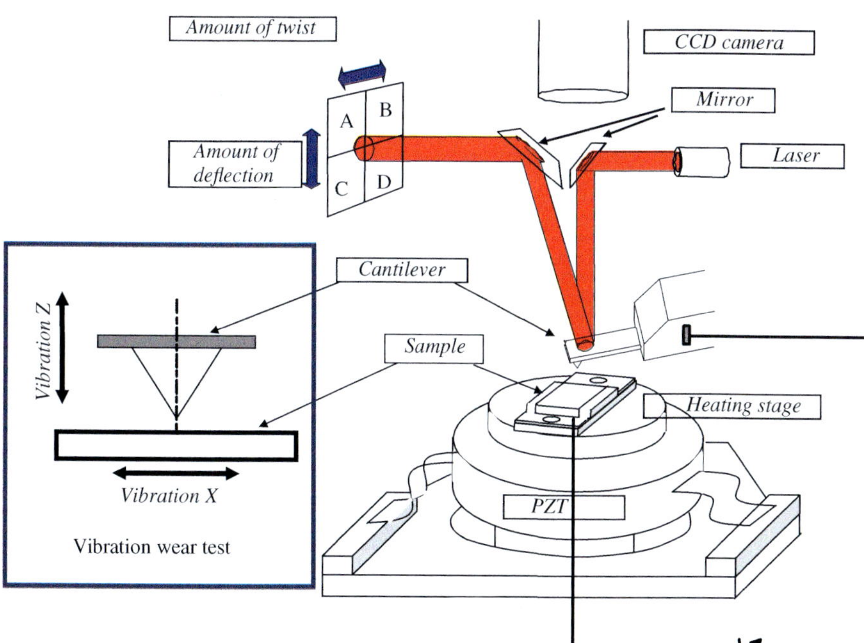

Figure 4.16 Vibration nanowear test using an environmentally controlled AFM.

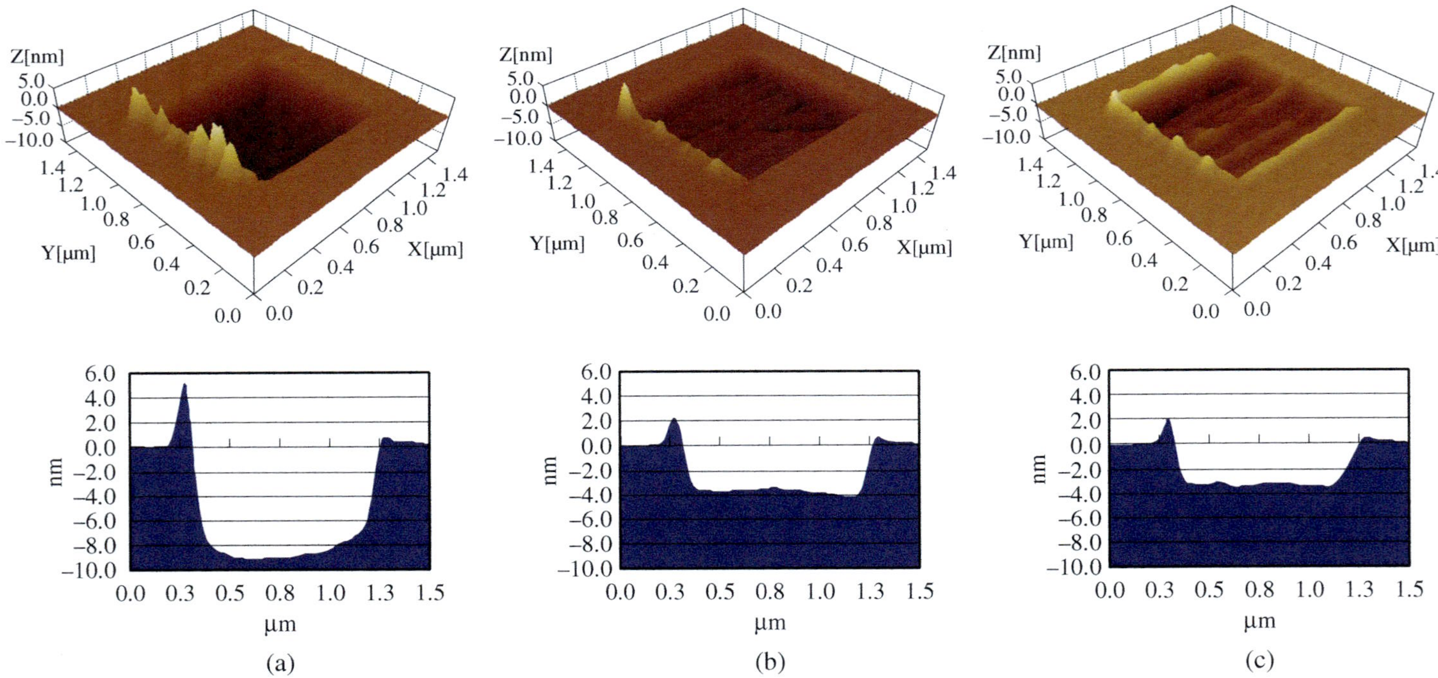

Figure 4.17 Surface profiles after nanowear tests at 200°C under a vacuum (FCVA, load: 4500 nN) (a) Without vibration, (b) Vibration Z and (c) Vibration X.

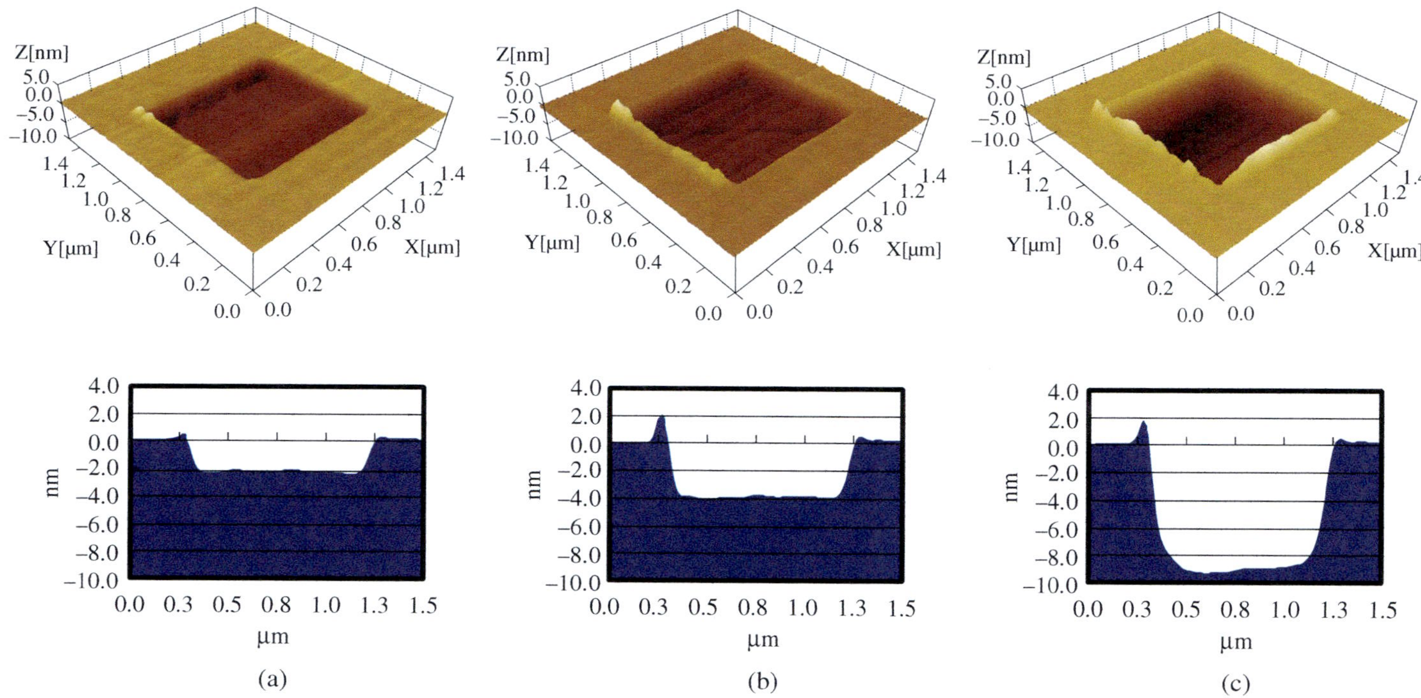

Figure 4.18 Surface profiles after the nanowear tests at 200°C under a vacuum (ECR, load: 4500 nN) (a) Without vibration, (b) Vibration Z and (c) Vibration X.

decrease in the wear. However, for the ECR-CVD-DLC films, the wear depth increased with the vibrations. The lubricous tribochemical products were removed from the ECR-CVD-DLC films by the vibrations, resulting in an increase in the wear [13].

4.3.3 Evaluation of the Force Modulation

SPMs make the detection of the interactions between the tip and sample possible, with a high spatial resolution. An AFM was used to profile the topographies of the samples by scanning a probe over the surfaces of the samples. Furthermore, the AFM could also be used to probe certain mechanical properties of the samples by measuring the response of the sample with an increased or decreased load by the tip [17, 18]. With the further development of AFMs, a lateral modulation friction force microscope (LM-FFM) can be used to evaluate rough surfaces, such as hard disks, with a high resolution [19]. This method involves applying a small periodic lateral movement between the tip and the sample and using a lock-in technique to measure the torsional displacement of the cantilever that results from scanning the tip over the sample to determine the lateral force. Using force modulation, a LM-FFM can form images by mapping the quantities related to the friction and viscoelasticity properties of the samples [20, 21]. The amplitude and phase responses of a cantilever can be measured using the lock-in technique [22]. As a result, the amplitude and phase images provide a complementing contrast mechanism for the AFM imaging process. Furthermore, the amplitude and phase images might be more sensitive than the topography and friction in resolving microscopic features [23].

By using SPM with a force modulation method, it is possible to perform vibration processing to evaluate the surface friction force, viscoelasticity and dynamic properties. Force modulation involves the application of vertical [24] and horizontal [21] vibrations while modulating the force. To investigate the effect of the boundaries in multilayer films on the defect suppression, evaluation of the boundary is necessary.

Moreover, to ensure the reliability of magnetic disk devices, improving the nanotribological properties is important. Therefore, it is necessary to clarify the performance of the lubricating and protective films on the disk surface when subjecting it to friction. Thus, the effect of a lubricant on the protective films, such as the effect of a 1-nm-thick lubricating layer on preventing surface fractures [25] and the effect of applying a lubricant [26], have been evaluated with lateral vibration friction tests.

To evaluate the nanoprocessing properties of a nanoperiod multiplayer film, nanoprocessing was performed using the lateral vibration processing test technique with a LM-FFM (Seiko Instrument Inc.) [21, 23]. In a LM-FFM, an AFM with a cantilever equipped with an extremely sharp electric conductive diamond tip with a radius of approximately 45 nm was used to evaluate the surface profile and microprocessing properties. The spring constant of the cantilever (nanosensor; CDT-NCHR) was about 30 N/m.

Figure 4.19a shows the processing method using a LM-FFM. Processing tests were conducted by sliding the tip along each sample with a load, both with and without applying lateral vibrations with an amplitude of 10 nm and a frequency of 1 kHz. In the LM-FFM profiles, the topographies of the samples were obtained by scanning the tip over the surfaces of the samples while probing the mechanical properties of the samples by measuring the amplitude and phase responses of the samples, where a small load of 100 nN was applied by the tip [17–19, 21, 23].

Nanoprocessing was performed during the processing tests by applying lateral vibrations to the samples with an amplitude of 10 nm and a frequency of 1 kHz. The typical processing marks formed at loading forces of 2100 and 2850 nN are shown in

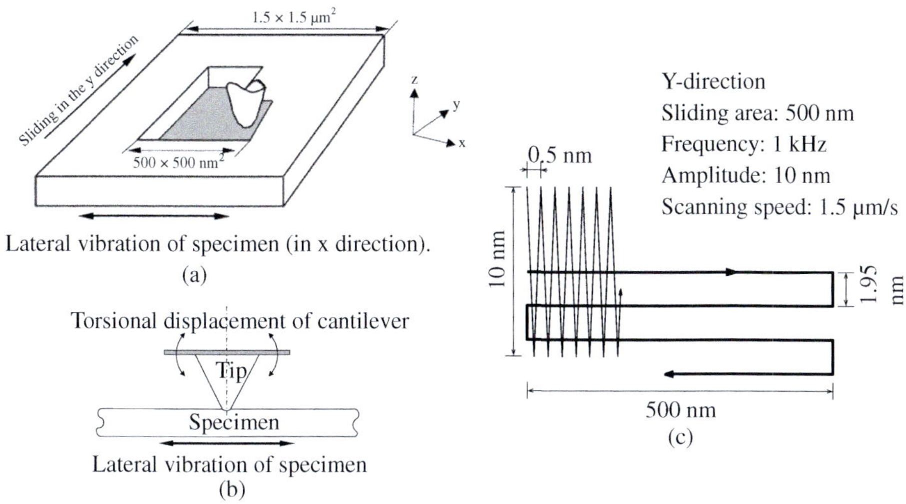

Figure 4.19 Torsional displacement of (a) a tip, (b) the locus of the processing tip and (c) Schematic of AFM tip trajectory.

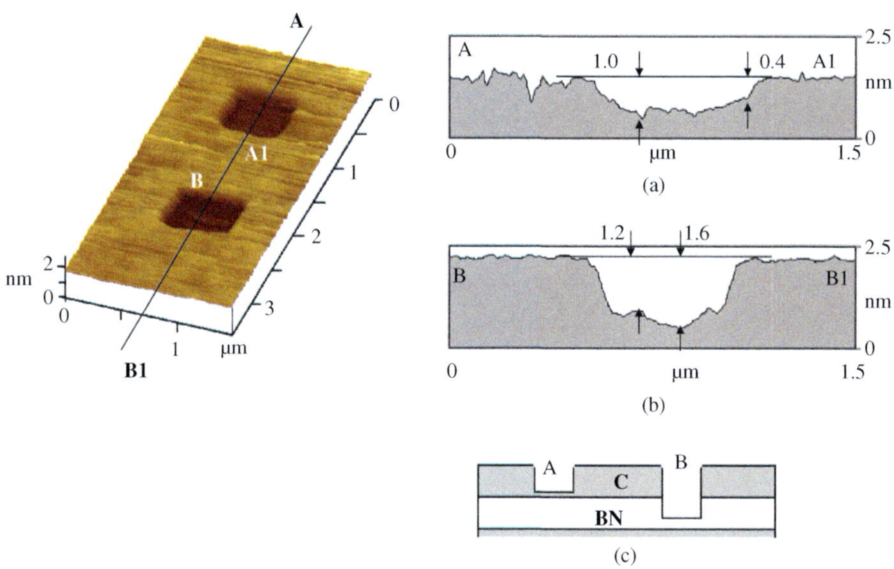

Figure 4.20 AFM image and cross-sectional profiles of the surface topographies obtained at loads of (a) 2100 nN and (b) and 2850 nN. (c) A model of the processed surface of a 2 nm-period film, processed with vibrations.

Figure 4.20 [27]. The processing depths ranged from 0.4 to 1.0 nm along section A–A1 and from 1.2 to 1.0 nm along section B–B1. Figure 4.21 shows the AFM images and cross-sectional profiles of the amplitude for the processed surface [27].

Figure 4.22a shows a model of the boundary in a 2 nm period multilayer film [5, 27]. The multilayer film was prepared by depositing C and BN layers alternately. Compound layers, such as boundaries, were formed between the C and BN layers owing to the

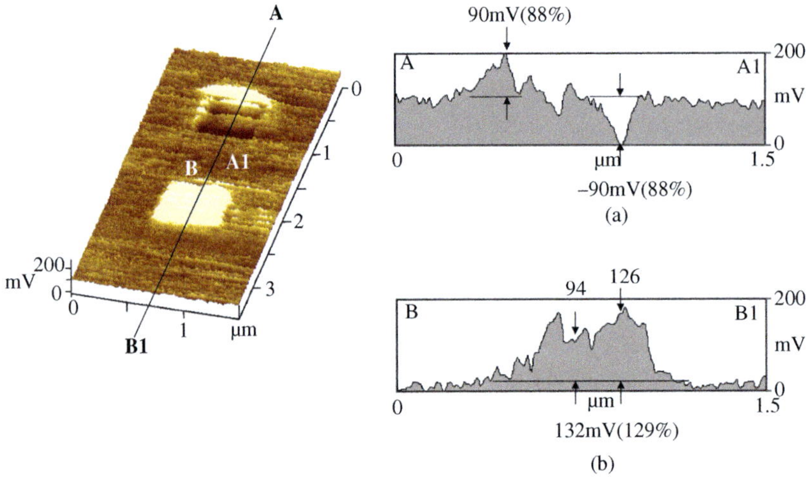

Figure 4.21 AFM image and cross-sectional profiles of the amplitude at loads of (a) 2100 nN and (b) 2850 nN for a 2 nm-period film, processed with vibrations.

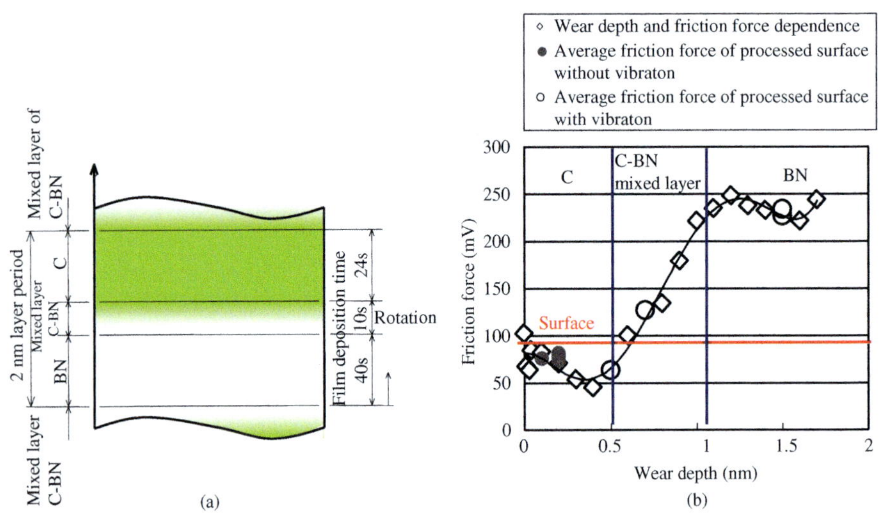

Figure 4.22 Evaluation of the boundary of a nanoperiod multilayer film by force modulation (a) Model of (C/BN)n multilayer film and (b) Relationship between wear depth and friction force.

deposition method used [5, 27]. It was difficult to evaluate the boundaries of multilayer films using conventional surface analysis methods since the boundary thicknesses were only several nanometers long. To improve the smoothness resulting from the surface processing and the processing efficiency, when a sample was processed by sliding a tip subjected to a load along it, vibrations perpendicular to the scanning direction were applied to the sample. The processed part was observed by scanning the sample with the same tip, with a light load and an expanding scanning range. The shape of the processed part and the friction force could be investigated by the lateral force modulation method. As a result, it became possible to control the processing depth using vibration processing. Processing to the depths of the boundaries in C-BN films was realized. Figure 4.22b shows the relationship between the processing depth and friction force. When the processing depth increased from 0.5 to 1.1 nm, the vibration amplitude image (i.e., the friction force) significantly increased. A processing depth of 1.1 nm corresponds to the thickness of the C-BN mixed layer in the C-BN film. At shallow depths, the first layer mostly consisted of C and the friction force was small, owing to the C layer. As the processing depth increased and the BN layer was reached, the friction force increased, owing to the proportion of BN increasing. As shown in Figure 4.22b, the range in which the friction force increased was 0.6 nm, corresponding to the thickness of the C-BN mixed layer, which was formed by alternating between using C and B targets during deposition [5, 27].

To ensure the reliability of magnetic disk devices, improving the nanotribological properties is important. Therefore, it was necessary to clarify the performances of the lubricating and protective films on the disk surface when they are subjected to friction. Thus, the effects of a lubricant on a protective film, such as the effect of a 1-nm-thick lubricating layer, on preventing surface fractures and the effect of supplying a lubricant [25, 26] were evaluated using lateral vibration friction tests.

To investigate the behavior of a lubricant on a perpendicular-recording magnetic disk (see Figure 4.23) [28], scanning friction tests over a square region on a disk were

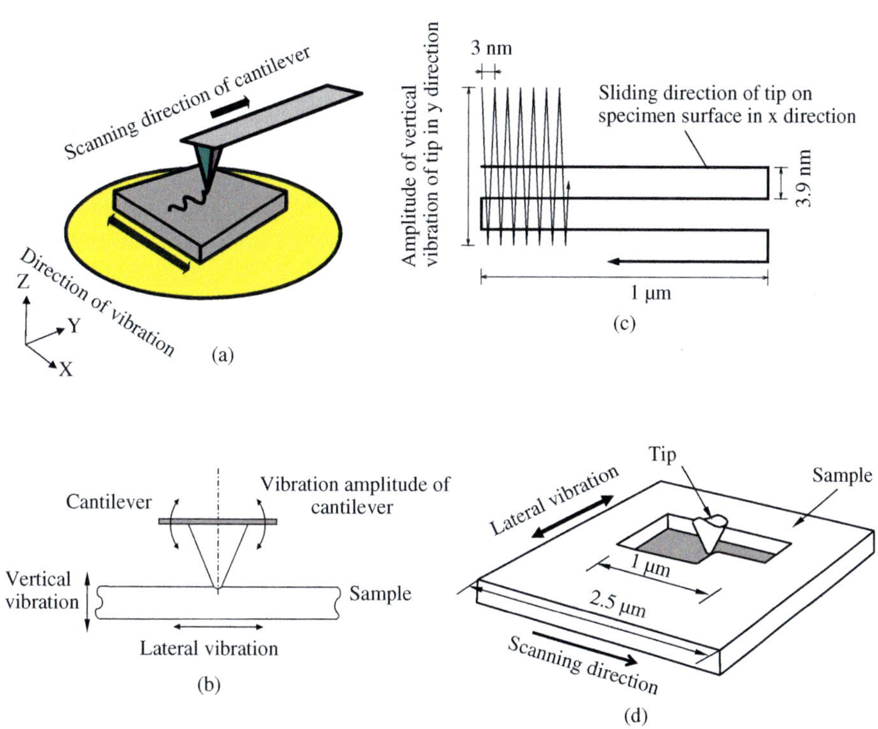

Figure 4.23 Friction wear tests, performed by applying vertical and lateral vibrations during force modulation (a), (b) Force modulation, (c) Schematic of rectangular pattern trajectory of the AFM tip, showing that the tip applies a vertical vibration in the y direction while scanning the sample in the x direction and (d) Friction wear test.

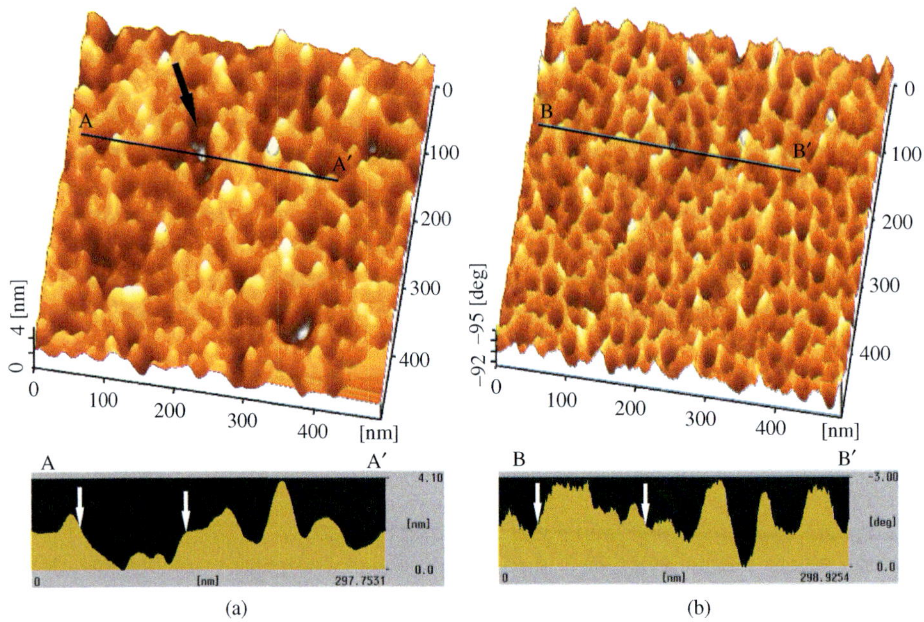

Figure 4.24 Profile AFM images of (a) the shape of the surface and (b) the phase of a perpendicular magnetic disk substrate after dip coating in a PFPE lubricant, obtained with an AFM over an area of 500 nm^2.

performed using a LM-FFM and a viscoelasticity AFM (VE-AFM) in the vertical force mode. The friction forces and viscoelasticities were evaluated from the vibration properties in the lateral and vertical directions [28]. Figure 4.24 shows profile AFM images of the shapes of the surfaces and the phase of a perpendicular-recording magnetic disk, measured with an AFM using force modulation [28]. Crystals were observed as grains on the magnetic layer of the perpendicular-recording magnetic disks, with and without heat curing [28]. The phases in the grooves that formed on the crystal grain layers increased compared with those in the unprocessed areas. This indicated that many lubricant particles remained in the grooves. The wear depth increased owing to the added vibrations. In particular, the wear depth significantly increased upon applying lateral vibrations to the sample where the wear was greatest.

Figures 4.25a and 4.25b show the friction force distributions of the wear area on perpendicular magnetic disks, without and with heat curing, respectively [28]. For the uncured disks (see Figure 4.25c), the friction force in the wear area decreased compared with that in the non-wear area [28]. In the perfluoropolyether (PFPE) lubricant films on the magnetic thin DLC-film-coated disks that were heat-cured, the friction force in the wear areas increased compared with those in the non-wear areas. In addition, the friction force in the wear area markedly increased upon adding lateral vibrations to the sample [28]. It was thus possible to evaluate the quantity of the existing lubricant by measuring the viscoelasticity in the wear area using the vertical force modulation mode.

4.3.4 Evaluation of the Mechanical and Other Physical Properties

When characterizing a surface, it is effective to evaluate the mechanical and other physical properties at the same time. For instance, to clarify the effect of an ultraviolet

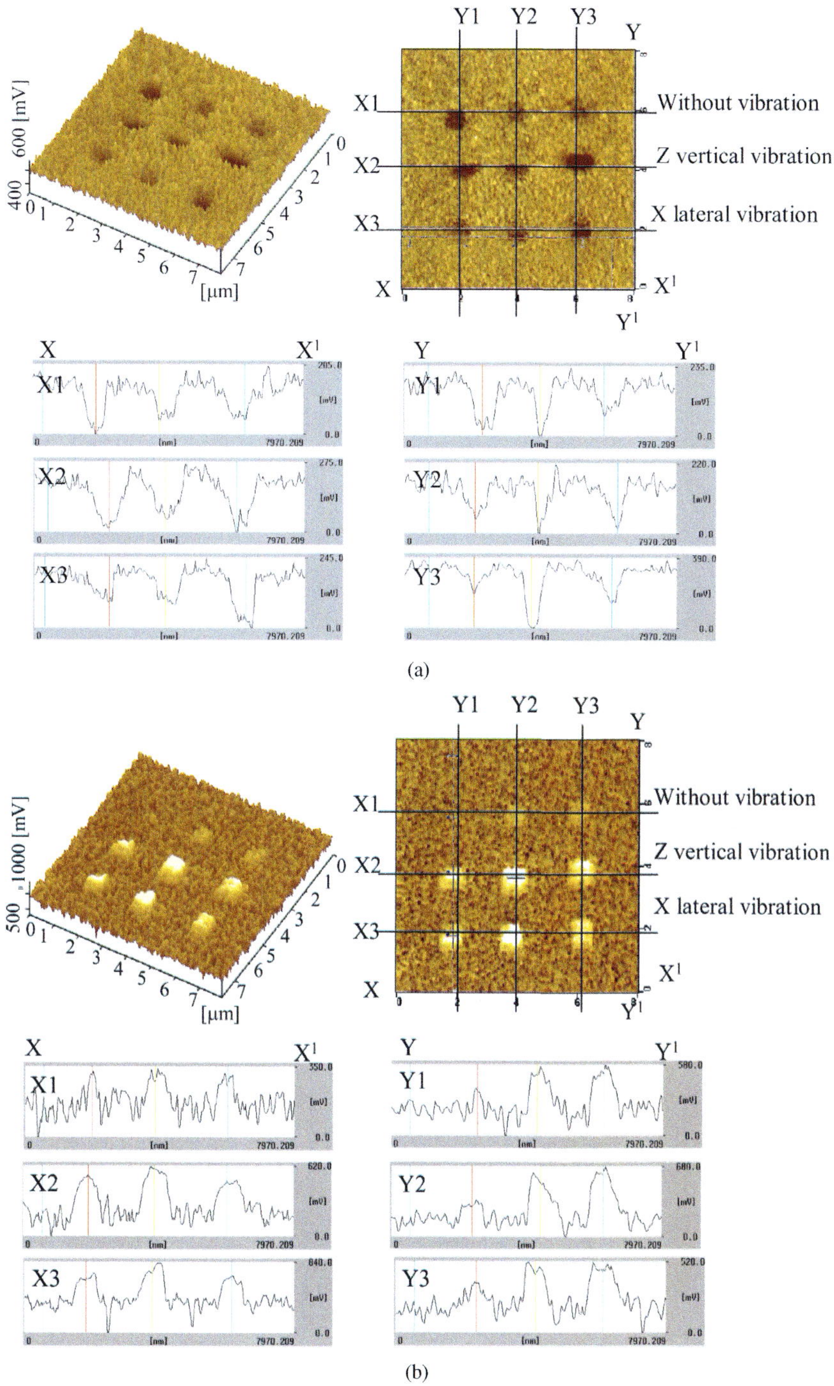

Figure 4.25 AFM images and section profiles of the amplitudes of disks (a) without and (b) with curing, and (c) the amplitude values as a function of the vibrations, obtained using a LM-FFM.

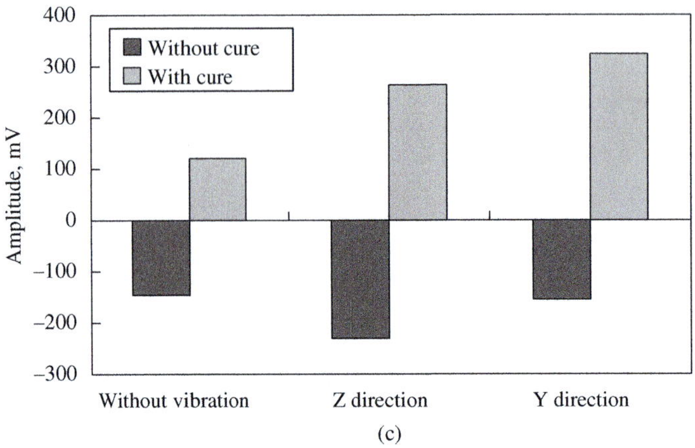

Figure 4.25 (*Continued*)

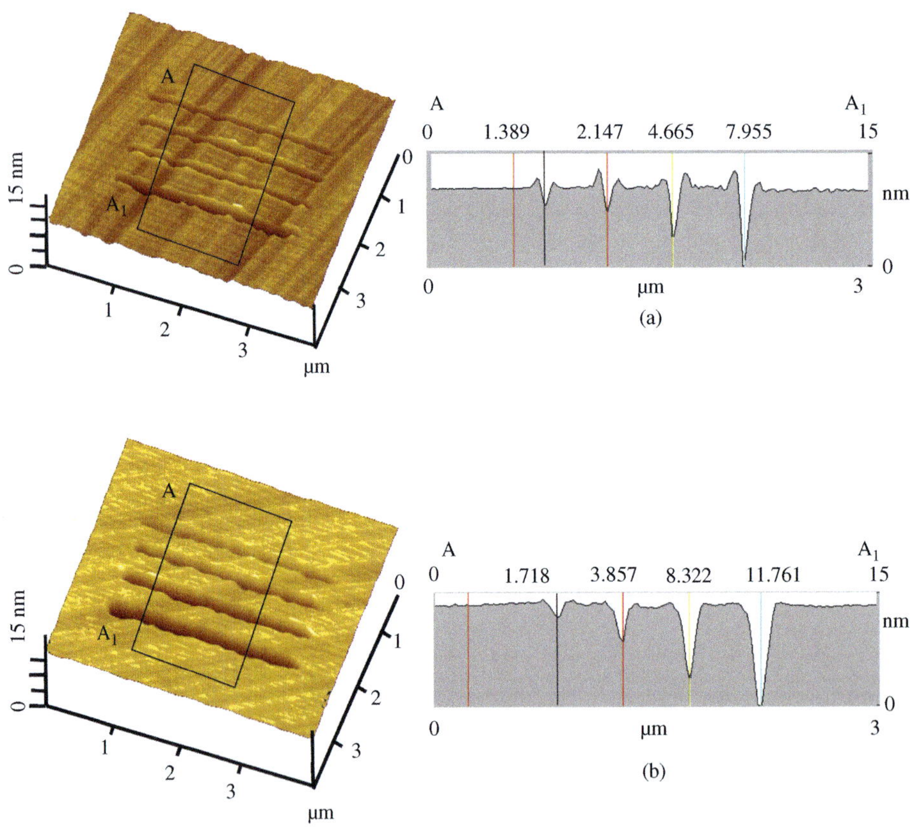

Figure 4.26 Profile and cross-section of the wear marks of (a) UV-irradiated and (b) untreated PFPE films, obtained with a vibration amplitude of 10 nm.

irradiation (UV) curing process on a magnetic disk lubricant, a lateral vibration friction test can be performed with force modulation. The surface damage and lubricant conditions can be evaluated from the friction and current distributions during the lateral vibration friction tests. The wear profile and current AFM images obtained after the wear tests with a vibration amplitude of 10 nm are shown in Figures 4.26 and 4.27 [29]. In UV-irradiated PFPE-coated disks, the current significantly increased at the wear groove that was made during ten scanning cycles, but it decreased at the center of the wear groove. The increase in the current indicates that the wear groove reached the DLC film. However, the PFPE lubricant was removed by scanning with a tip and might have remained at the center of the wear groove, resulting in a decrease in the current. For the untreated PFPE coated disks, the current significantly increased, but slightly decreased at the center of the wear groove after sliding for ten scanning cycles. The increase in the current was thought to occur at the DLC layer and the decrease in the current was caused by the remaining free PFPE lubricant [29].

The microprocessing characteristics of nanoperiod multilayer boron nitride and carbon (C/BN)n films were investigated by employing AFM and a boron-doped (B-doped) CVD (chemical vapor deposition) diamond-coated silicon tip to measure the current. Images of the unprocessed and processed regions were collected [30]. Using a conductive diamond tip, nanoprocessing was carried out by sliding a sample coated with a (C/BN)n film under a load. The topographies and electrical properties of the processed (C/BN)n films were then observed using the same tip. Using a LM-FFM, a 4 nm period (C/BN)n film was processed by sliding tips with loads of 2500, 3000, 3500 and 4000 nN while applying lateral vibrations with an amplitude of 02 nm at a frequency of 1 kHz. Sharp tips with radii of about 30 nm were used in this process.

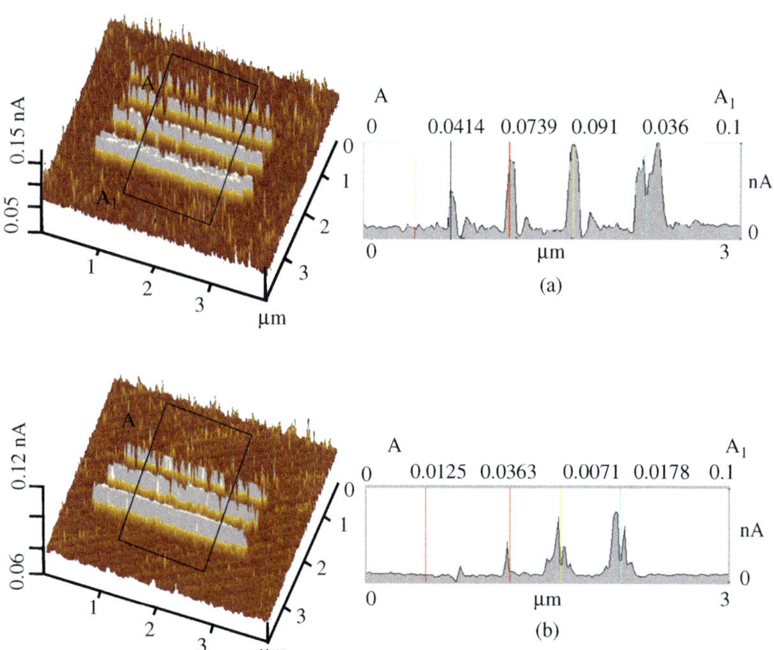

Figure 4.27 Profile and cross-section of the current distribution of (a) UV-irradiated and (b) untreated PFPE films, obtained with a vibration amplitude of 10 nm.

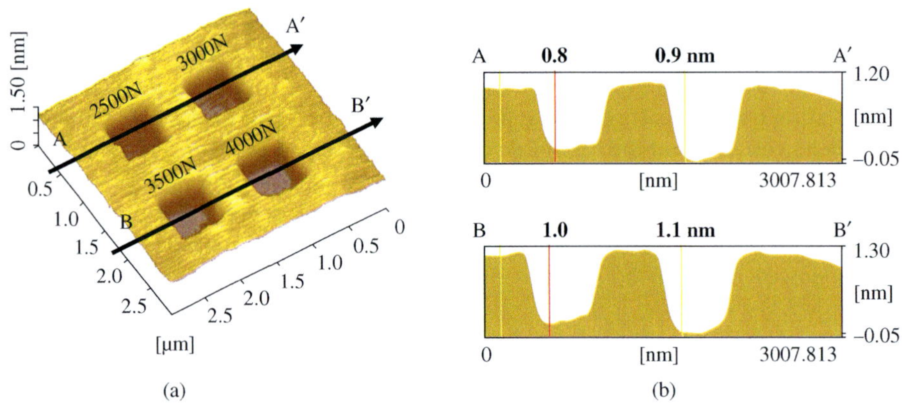

Figure 4.28 (a) Profile and (b) cross-sectional images of a 4-nm-period (C/BN)n film, obtained at loads of 2500, 3000, 3500 and 4000 nN with a vibration amplitude of 20 nm.

Figure 4.28 shows a topographic image of the scanned 4 nm period (C/BN)n film surface [30]. To evaluate the variations in the current and frictional forces over a 4 nm period (C/BN)n film, an attempt to investigate the current and frictional force at the same wear depth as that of a 4 nm period (BN/C)n film was made. Figures 4.29–4.31

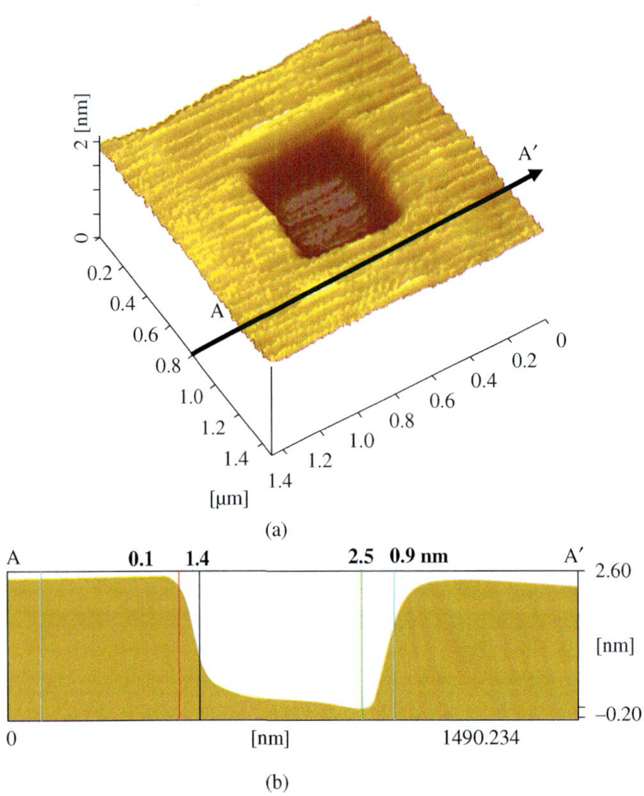

Figure 4.29 (a) Profile and (b) cross-sectional images of the 4-nm-period (C/BN)n topography with a 3500 nN load.

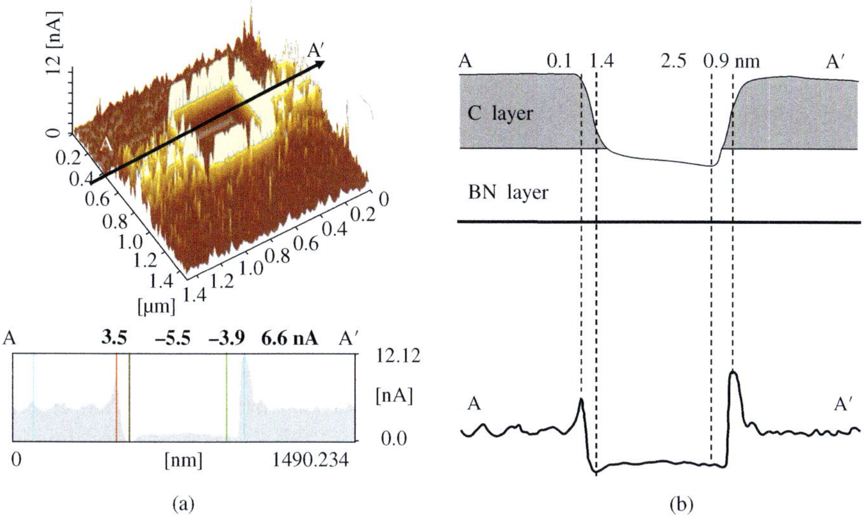

Figure 4.30 (a) Profile and cross-sectional images of the current distribution and (b) a model of the current versus wear track for a 4 nm-period (C/BN)n film with a load of 3500 nN.

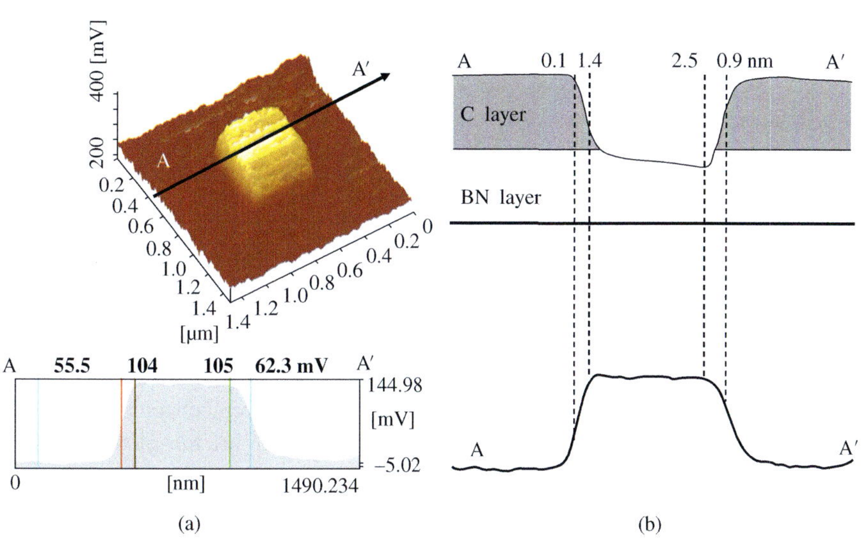

Figure 4.31 (a) Profile and cross-sectional images of the friction distribution and (b) a model of the friction versus wear track for a 4 nm period (C/BN)n film with a load of 3500 nN.

show the topographic, current and friction images of a wear mark made with a load of 3500 nN for a 4 nm period (C/BN)n film. The average mean wear depth was obtained by scanning five different areas on the surface of the same sample [30].

Finally, nanoprocessing is becoming important in nanotechnology [31]. To evaluate the effectiveness of nanoprocessing, surfaces could be effectively observed by evaluating the mechanical and complex properties, including the electrical, chemical, magnetic and other physical properties.

4.4 Conclusions

In summary, using nanoindentation, nanometer wear and force-modulation methods, the hardness, wear and frictional properties of extremely thin films were evaluated and discussed. These results detail the possibilities of what can happen in nano scale evaluation methods, complementing experiments and providing a guide to interpret experimental data. Although nanoindentation cannot confirm which evaluation processes contribute to the experimentally derived hardness determined via nanoindentation, various evaluation methods including nano wear, nanoscratch and nanoprocessing experiments have provided important advances in our understanding of the tribological and mechanical properties of nano scale surface systems.

With the development of nanotechnology, the size of machinery is becoming increasingly small. Owing to the increasing focus on the properties of surfaces, the development of technologies to evaluate the mechanical properties of the surfaces of thin films is becoming indispensable. Future developments are expected as a result of the current research and development.

References

1 Miyake, S, Wang M. Encyclopedia of Nanoscience and Nanotechnology. American Scientific Publishing, vol. 19, 399–451; 2011.
2 Enomoto, Y, Miyake S. Tribology of Thin Films. University of Tokyo Press; 1994.
3 Kaneko, R, Oguchi S, Miyamoto T, Andoh Y, Miyake S. Micro-tribology for magnetic recording. The Society of Tribologists and Lubrication Engineers (STLE), SP-27; 31; 1990.
4 Doerner, MF, Nix WD. A method of interpreting the data from depth-sensing indentation instruments. Journal of Materials Research1(4), 601–609; 1986. DOI: 10.1557/JMR.1986.0601.
5 Miyake, S, Wang M. Evaluation of nanomechanical properties of surfaces. In: Microscopy: Science, Technology, Applications and Education, (eds) A. Méndez-Vilas, J. Díaz, pp. 2013–2021. Formatex Research Center (Pub); 2010.
6 Miyake, S, Saito T, Yu K, Mikami T, Ogata K. Nanometer scale mechanical properties of DLC films deposited by filtered cathodic vacuum arc ion-plating method (FCVA). Surface and Coatings Technology, 55, 669–676; 2004. DOI: 10.4139/sfj.55.669.
7 Farhat, ZN, Ding Y, Northwood DO, Alps AT. Nanoindentation and friction studies on Ti-based nanolaminated films. Surface and Coatings Technology, 89, 24–30; 1997. DOI: 10.1016/S0257-8972(96)02939-8.
8 Miyake, S, Shindo T, Miyake M. Regression analysis of the effect of bias voltage on nano- and macrotribological properties of diamond-like carbon films deposited by a filtered cathodic vacuum arc ion-plating method. Journal of Nanomaterials, vol 2014, article ID 657619; 2014. DOI: 10.1155/2014/657619.
9 Miyake, S, Watanabe S, Miyazawa H *et al.* Improved microscratch hardness of ion-plated carbon film by nitrogen inclusion evaluated by atomic force microscope. Applied Physics Letters 65(25), 3206–3208; 1994. DOI: 10.1063/1.112414.

10 Miyake, S, Kanko R, Kikuya Y, Sugimoto I. Microtribological studies on fluorinated carbon films. Journal of Tribology, 113(2), 384–389; 1991. DOI: 10.1115/1.2920633.
11 Miyake, S. Improvement of mechanical properties of nanometer period multilayer films at interfaces of each layer. J Vac Sci Technol, B21(2), 785–789; 2003. DOI: 10.1116/1.1560217.
12 Miyake, S, Kurosaka W, Oshimoto K. Nanometer scale mechanical properties of extremely thin diamond like carbon films. Tribology: Materials, Surfaces & Interfaces, 3(4) 158–164; 2009. DOI: 10.1179/175158309X12586382418535.
13 Miyake, S, Kawasaki S, Yamazaki S. Nanotribology properties of extremely thin diamond-like carbon films at high temperatures with and without vibration. Wear, 300, 189–199; 2013. DOI: 10.1016/j.wear.2013.01.099.
14 Robertson, J. Ultrathin carbon coatings for magnetic storage technology. Thin Solid Films, 383(1–2), 81–88; 2001. DOI: 10.1016/S0040-6090(00)01786-7.
15 Yamamoto, T, Hyodo H, Tsuchitani S, Kaneko R. Ultrathin amorphous carbon overcoats by filtered cathodic arc deposition. IEEE Transactions on Magnetics, 39(5), 2201–2204; 2003. DOI:10.1109/TMAG.2003.817065.
16 Hyodo, H, Yamamoto T, Toyoguchi T. Properties of tetrahedral amorphous carbon film by filtered cathodic arc deposition for disk overcoat. IEEE Transactions on Magnetics, 37(4), 1789–1791; 2001.
17 Colchero, J, Luna M, Baro AM, Marti O. Lock-In technique for measuring friction on a nanometer scale. Applied Physics Letters 68(20), 2896–2898; 1996. DOI: S0003-6951(96)04220-9.
18 Pfeiffer, O, Bennewitz R, Baratoff A *et al.* Lateral-force measurements in dynamic force microscopy. Physical Review B, 65, 161403(R); 2002. DOI: 10.1103/PhysRevB.65.161403.
19 Kumagai, A, Harada K, Ozawa K, Ishiwata O. Evaluation of tribological properties of a longitudinal magnetic recording medium using lateral modulation friction force microscope. Japanese Journal of Applied Physics 35(12A), 6286–6291; 1996. DOI:10.1143/JJAP.35.6286.
20 Miyake, S, Wang M. Mechanical properties of extremely thin B-C-N protective layer deposited with helium addition. Japanese Journal of Applied Physics 43(6A), 3566–3571; 2004. DOI: 10.1143/JJAP.43.3566.
21 Yamanaka, K, Tomita E. Lateral force modulation atomic force microscope for selective imaging of friction forces. Japanese Journal of Applied Physics 34, 2879–2882; 1995. DOI:10.1143/JJAP.34.2879.
22 Friedenberg, MC, Mate CM. Dynamic visco- elastic properties of liquid polymer films studied by atomic force microscopy. Langmuir , 12(25), 6138–6142; 1996. DOI:S0743-7463(96)00331-9.
23 Miyake, S, Wang M, Saitoh T, Watanabe S. Microtribological properties of B–C–N extremely thin protective films deposited on plasma pretreated magnetic layers. Surface and Coatings Technology, 195(2–3), 214–226; 2005. DOI:10.1016/j.surfcoat.2004.05.037.
24 Radmacher M, Tillmann RW, Gaub HE. 1993. Imaging viscoelasticity by force modulation with the atomic force microscope. Biophysical Journal 1993, 64(3), 735–742. DOI: 10.1016/S0006-3495(93)81433-4.

25 Miyake, S, Takahashi Y, Wang M, Saito T. Durability evaluation of perfluoropolyether-lubricant-coated diamond-like carbon film by the lateral vibration friction test. Journal of Physics D: Applied Physics 38, 2244–2252; 2005. DOI:10.1088/0022-3727/38/13/024.
26 Miyake, S, Wakatsuki Y, Wang M, Matsunuma S. Amplitude dependence of the lateral-vibration wear test for perpendicular recording magnetic disks treated by heat curing. Japanese Journal of Applied Physics 44(5A), 3209–3217; 2005. DOI:10.1143/JJAP.44.3209.
27 Miyake, S, Hashizume T, Wakatsuki Y. Nanoprocessing and evaluation of carbon and boron nitride nanoperiod multilayer films by lateral force modulation method. Surface and Interface Analysis, 38(4), 873–878; 2006. DOI: 10.1002/sia.2192.
28 Miyake, S, Wang M, Ninomiya S. Nano Tribological Properties of Perfluoropolyether-coated Magnetic Disk Evaluated by Vertical and Lateral Vibration Wear Tests. Surface and Coatings Technology 200, 20(21), 6137–6154; 2006. DOI: 10.1016/j.surfcoat.2005.09.038.
29 Miyake, S, Ninomiya S, Wang M. Lubricant state evaluation of ultraviolet-irradiated magnetic disk using lateral friction force modulation microscopy with electric current distribution. Japanese Journal of Applied Physics 44(9), L299–L302; 2005. DOI:10.1143/JJAP.44.L299.
30 Miyake, S, Wang M, Kim J. Investigation of micro-electro-mechanical processing characteristics of layered boron nitride and carbon films, Institute of Physics Publishing, Nanotechnology, 17, 1607 –1614; 2006. DOI:10.1088/0957-4484/17/6/013.
31 Miyake, S. 2005. Nanometer Scale Mechanical Processing. Science of Machines, 57(10), 1011; 2005.

5

Nanoindentation on Tribological Coatings

Francisco J.G. Silva

Ph.D, ISEP School of Engineering, Polytechnic of Porto, Portugal

5.1 Introduction

Extremely hard coatings are widely used in several industrial applications such as machining tools, plastic injection mold inserts, mechanical sliding devices, plastic and aluminum extrusion dies, surgical implants, watches, optics, firearms, among many other parts subject to severe wear.

In the last decades, there is a clear trend to reduce the layers thickness allowing to joint unlike coatings with different characteristics, improving drastically the suitability of the coating set to extremely exigent performances, as is demanding permanently through new solicitations related to machining tools and other advanced applications. Often, customer requirements make necessary to gather in a single coating some different properties, such as hardness, low friction coefficient, stress relief, thermal conductivity, diffusion barrier and corrosion resistance at high temperatures, whose are impossible to congregate on coatings using a single layer. Thus, it is increasingly common to use multi-layer films in order to fulfil all the requirements requested by the market. Figure 5.1 depicts a multilayered system commonly used in tool inserts, allowing obey with all the requirements demanded by this kind of market.

In these cases, the sequence and thickness of each layer play an important role in the global performance of the coating because each layer plays a vital function, as shown in Figure 5.2.

The more demanding applications become more complex, the more intricate become multilayered coatings, assuming configurations as multifaceted as can be observed in Figure 5.3.

Otherwise, it is well known that adhesion between coating and substrate is a critical concern for several applications and, the thicker is the coating, the poorer is the adhesion, limiting by this way the use of thick coatings. Sometimes, depending on the materials metallurgical affinity, it is necessary to apply an intermediary layer in order to improve the adhesion of the remaining coatings to the substrate. Silva [2] used a previous Ni layer with 3 – 4 μm thickness followed by a thick layer of Cu (about 32 μm thickness) and a third layer of Ti (about 2.5 μm thickness) (Figure 5.4) in order to achieve the best conditions to synthetize polycrystalline diamond onto steel substrates, each one with specific functions: Ni layer intends to optimize the adhesion between the steel

Applied Nanoindentation in Advanced Materials, First Edition. Edited by Atul Tiwari and Sridhar Natarajan.

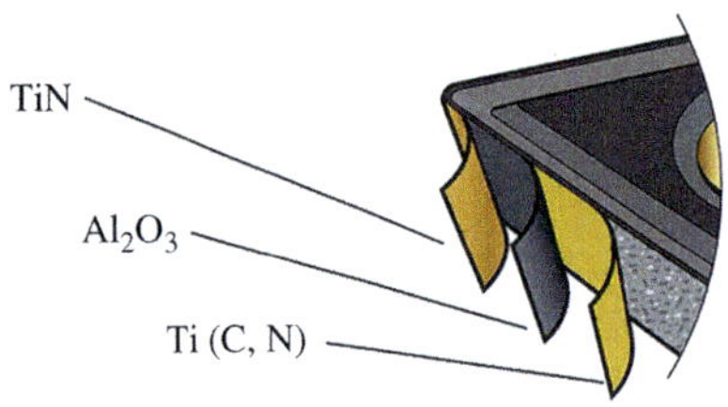

Figure 5.1 Multilayered system commonly used in machining inserts for turning or milling operations. *Source*: SANDVIK 2005, Metalcutting Technical Guide [1].

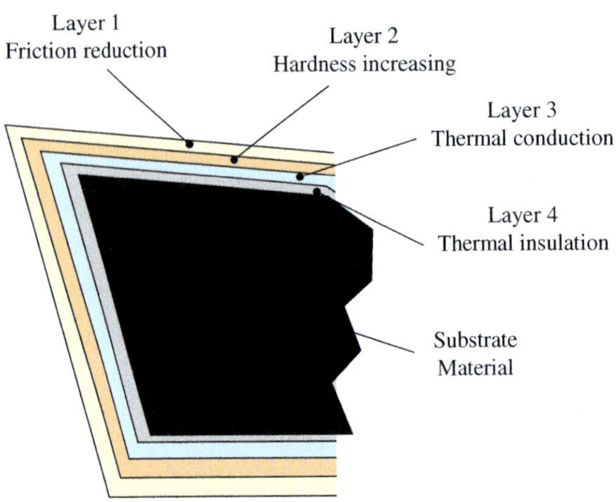

Figure 5.2 Function of each layer in a turning tool insert.

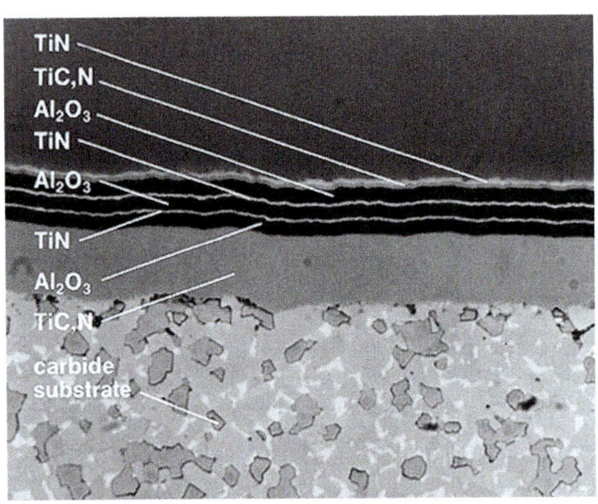

Figure 5.3 Example of a complex multilayered coating system.

substrate and the subsequent layers, Cu layer aims to accommodate stresses generate by the diamond synthesis process (over 900°C) due to different thermal expansion coefficients and the Ti layer means to increase further diamond nucleation.

In the last decades, due to technological advances on the PVD and CVD reactors and corresponding systems of control, as well as several studies carried out about the crack propagation through consecutive layers, it has been created a trend of producing coatings consisting of numerous thin layers, which prevent easy cracking propagation, increasing coating fracture toughness. The production of these coatings is extremely

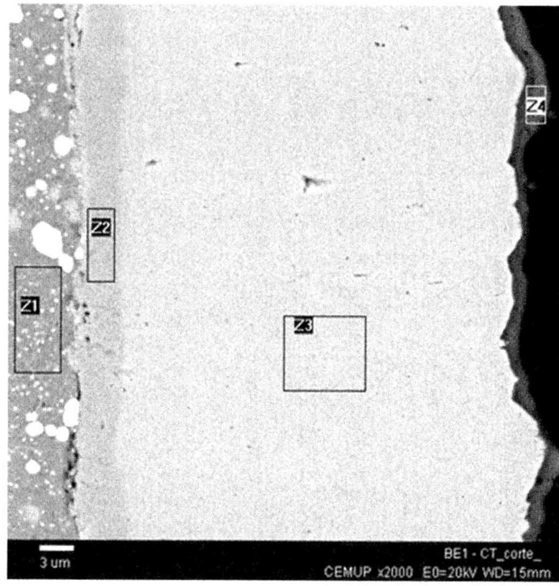

Figure 5.4 Multilayered coating system idealized to accommodate polycrystalline diamond on steel substrates (Z1 – Steel substrate, Z2 – Ni Layer, Z3 – Cu Layer and Z4 – Ti Layer). *Source*: Silva 2004 [2]. Reproduced with permission of Elsevier.

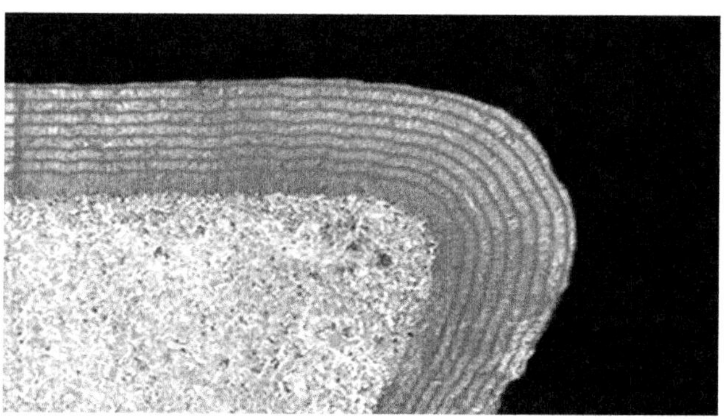

Figure 5.5 Multilayered coating used on tool insert for turning operations. *Source*: Adapted from Benes 2007 [3].

easy by using different targets in the deposition chamber and animating the substrates of rotational motion. The result can be seen in Figure 5.5.

The better is the reactor control systems, the thinner are the produced layers, reaching easily to the nanoscale. The use of nanostructured coatings becomes usual in the last years for several applications, with significant advantages in terms of mechanical properties and tribological behavior. In Figure 5.6 can be seen a nanostructured TiB_2 coating used to increase the abrasion wear resistance of molds used for injection of glass fiber reinforced plastics.

Hence, the use of even more thin layers as part of relative thin coatings introduces new challenges in terms of analysis and characterization. Regarding tribological applications, hardness is probably the most important property to analyze, but, if the layers are very thin, the influence of the previous layers must be experienced in the last layer. The same is true when are being used very thin coatings, where the substrate should

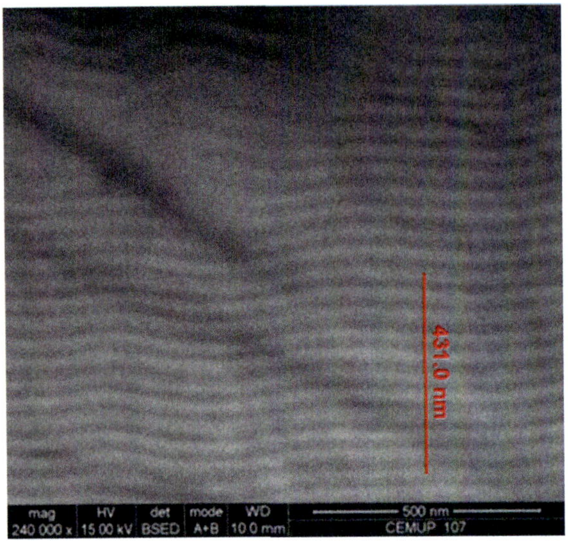

Figure 5.6 TiB_2 nanostructured coating used on moulds for injection of reinforced plastics, consisting of successive layers with different TiB_2 density and period of about 43.1 nm. *Source*: Martinho 2011 [4]. Reproduced with permission of American Scientific Publishers.

influence the hardness values obtained for the coating. It is well known that indentation depth should not exceed one tenth of the coating thickness, in order to avoid influence of the substrate or adjacent previous layers. Thus, the indentation depth penetration must be reduced regarding the thickness decrease promoted on the layers that researchers need to characterize. This requirement conduces to new challenges: the load applied must be lower and the indenter displacement control should be accurate in conformity, allowing measure accurately depth penetrations with resolutions of about 1 nm. Features of recent Nano-Hardness Testers (NHT) include force on the indenter applied by an electromagnetic actuator, whilst the displacement is measured through a capacitive system, giving a force resolution of 10 μN and displacement resolution less than 1 nm, allowing any kind of indenter geometry and the possibility to characterize relative soft coatings such as MoS_2 films. Complimentary studies should carried out by high resolution scanning force microscopy on particular imprint location, obtaining valuable additional information about the indentation effect and subsequent material response.

Depending on coating material and deposition or synthesis parameters, films can present different kinds of morphology. As it is well known, PVD and CVD high deposition rates are usually demanded, in order to become the coatings commercially attractive. Despite several techniques used in order to reduce the coating surface roughness, usually, the greater is the deposition rate, the higher is the coating surface roughness. For tribological applications, the morphology of the coatings surface is very important, however, most of the commercialized coatings present levels of roughness not negligible. Studies carry out allowed to conclude that surface roughness has a significant influence on both the nanohardness and elastic modulus of thin films, when analyzed through nanoindentation tests [5]. In order to minimize the influence of surface morphology, the indentation depth should be much greater than the characteristic size of surface roughness, implying a convenient selection of the indenter size. Distinct indentations can be observed in Figure 5.7 [6], showing the difference between the imprints in smooth and rough coatings. Regarding that surface roughness does not depend on the layer thickness, but the indentation depth cannot exceed 10% of the layer

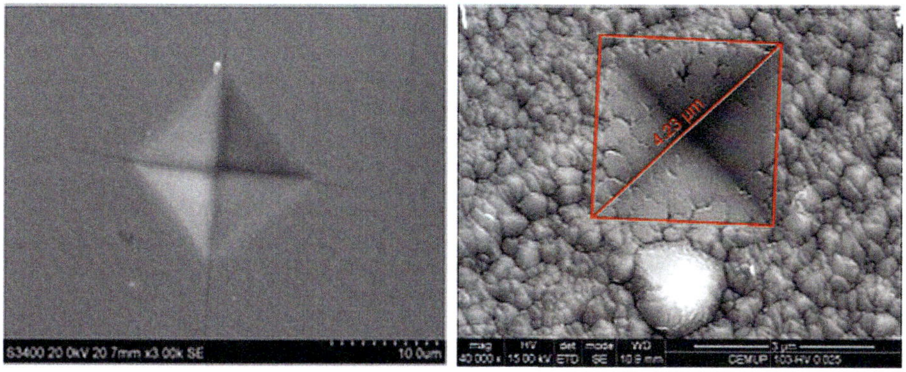

Figure 5.7 Microindentations performed on titanium nitride [6] smooth film and TiAlSiN rough coating.

thickness [7–9], it can easily realize that it is necessary to control the film morphology in order to carry out characterizations by nanoindentation. Others factors also are referred as affecting the coating nanoindentation results, such as sample tilt, sample surface roughness, compliance of the sample, pile-up and residual stress of the films [10].

Despite this limitation, nanoindentation remains a very useful technique which allows a expeditiously access to some of the key properties of coatings usually used for tribological applications, such as hardness, Young's modulus and fracture toughness, among others. Firstly performed for the evaluation of Young's modulus and hardness of thin films and small material volumes, rapidly it starts to be used for a wide-ranging of material mechanical behavior, comprising residual stresses, fracture, elastic and plastic deformation, time-dependent behavior, dislocation behavior and the onset of plastic deformation [11].

Thus, it can be observed that nanoindentation is particularly suitable for extremely thin tribological layers or coatings whose must be characterized and its thickness avoids the use of medium or high loads, promoting high depth imprints. Due to the broad range of coating thickness used for tribological applications, a wide range of loads should be applied, from the nanoscale to the macro scale. ISO 14577-1 standard [12] defines what can be considered as nanoindentation, microindentation and macroindentation, through the depth of the imprint caused and load used, as follows:

- Nanoindentation: $h_{máx} \leq 200$ nm;
- Microindentation: $h_{máx} \geq 200$ nm and $F < 2$ N;
- Macroindentation: $2 \text{ N} \leq F \leq 30$ kN.

ISO 14577-4 [13] was especially addressed for coatings and thin films, as a consequence of the INDICOAT European Project conducted by National Physics Laboratory [14], since ISO 14577-1 was not developed with its focus on coatings and thin films. In 2007, ASTM E 2546-7 [15] was introduced in the U.S.A. covering from the nano to the macro range indentation tests, under the denomination 'Standard Practice for Instrumented Indentation Testing'. This standard also includes some annexes about instruments verification, standard reference blocks and indenter requirements. As an extension of the ISO standards, ISO/TR 29381:2008 'Metallic materials – measurement of mechanical properties by an indentation test – Indentation tensile properties' was lately published [16]. This new document launches methods for the evaluation of tensile properties of metallic materials, such as true stress-strain curves, using instrumented indentation testing. However, the behavior of viscoelastic materials tested by indentation was not covered yet and a new standard is being presented, under proposal of the National Physics Laboratory once again, being denominated as ISO TC 164/SC3 [17]. The work of Versailles Project on Advanced Materials and Standards (VAMAS) also contributes to the development of the ISO 13557 with the European pre-standard ENV WI 152, denominated 'Determination of hardness and Young's modulus by instrumented indentation testing' [18], which is important for some tribological coatings.

5.2 Relevant Properties on Coatings for Tribological Applications

Tribological coatings are widely used for ensure better surface wear resistance relatively to bulk material. Indeed, due to economical purposes or needing different properties, bulk material does not present particularly good properties of wear resistance, requiring a coating in order to improve this property. Sometimes, bulk materials need to present properties those are difficult to match with extremely high hardness required by tribological applications, such as high toughness and/or fatigue resistance. In these cases, the bulk material is selected in order to comply with the main requirements demanded by the application and the surface is treated as another system which requires special attention due to the sliding contact or erosion effects. Otherwise, economical aspects dictate the rules, requiring the application of less noble bulk material and requiring a special material on the surface, where the wear will occurs.

Regarding the PVD and CVD coating techniques, properties can be drastically different by a slight changing in the deposition parameters used. Effectively, just a complete knowledge on the coating properties allows to closely tailoring the films in order to achieve the desirable surface behavior. Main properties required to tribological coatings are the hardness, morphology and friction coefficient. However, related to these properties, there are other ones whose should be analyzed in order to compare and predict their tribological behavior relatively to another coatings which behavior is already known. Tensile properties, residual stress and adhesion between coating and substrate or between layers, are other properties extremely important to predict the wear behavior of the coatings.

5.3 How can Nanoindentation Help Researchers to Characterize Coatings?

Nanoindentation method lets to measure and record continuously the applied load, depth penetration and time during the test, which becomes it very suitable for studying the mechanical properties of the coatings. Furthermore, the effects of substrate and film can be evaluated together when higher displacements are used [19, 20]. Additionally, nanoindentation technique also allows the opportunity of experimentally access to properties which would be extremely difficult to attain using other methods, such as the coating Young's modulus and residual or microstructural stress, the last ones very usual due to thermal gradients during the synthesis or deposition process [19, 21, 22].

Nanoindentation measurement consists of two steps, the so-called loading stage and the relaxation one. Throughout the loading stage, a normal indentation load obliges the diamond indenter to penetrate into the sample. This load is progressively applied and at the same time the indentation depth is evaluated. When the load is removed (relaxing stage), due to the resulting material plastic deformation, there is a residual depth of penetration, depending on the indenter geometry and the applied load. Attending the maximum penetration depth, remaining depth and a number of further parameters related to the nanoindentation procedure, the Young's modulus can be approximately determined by an analytical evaluation method. It is usually assumed that the behavior of an elasticplastic sample can be satisfactorily described through the overlay of a perfectly elastic material behavior with a perfectly plastic one. This consideration simplifies the nanohardness measurement evaluation results, enabling the application of analytical equations to estimate the thin hard coatings Young's modulus, however, without defining their material stress-strain courses.

Introduced about 1990, nanoindentation was initially perceived as a useful technique with the main purpose of determining the micro or ultra-microhardness of bulk materials or thin coatings. Since then, it has been necessary to study the repeatability and accuracy of the measurements, more reliable procedures and finally, the standards above referred. However, in the last two decades, other important properties were capable of being quantified or analyzed by this way, as stated above. The following sections will describe how the nanoindentation can help the researchers to easily characterize tribological coatings. Many authors have been proving that nanoindentation technique is very suitable also for testing ultra-hard coatings such as synthetic diamond and c-BCX films, among other equally ultra-hard materials [22].

Coating and substrate constitute a composite system, being crucial to know the energy repartition during the indentation process. By this way, we can find the intrinsic properties of the film or the properties of the entire system.

Regarding that films usually are very thin, both force and displacement resolution are critical factors in this kind of technique, implying an adequate choice of the equipment used for each case.

The relation between the film thickness and both the coating properties and the substrate properties, as well as the contact size, definitively influence the behavior observed on the test. If a hard coating has been deposited on a softer substrate, plastic deformation of the system can be expected and radial cracks on the film can occur as result of the bent suffered by the coating and plastic deformations imposed to the substrate under high load. Through-thickness cracks will be more pronounced if a pile-up effect is felt in the substrate due to work hardening high level. When a soft coating is deposited on a harder substrate, little or no fracture is expected but the film is crushed under the indenter, being extruded from the indenter tip to the border, as depicted in Figure 5.8 [23].

The use of low loads is very attractive for thin coatings, due to the maximum indentation depth allowed. However, when low loads are being used, the response of hard coatings can be perfectly elastic, giving rise to loading and unloading curves very close to each other. This effect is more pronounced when blunt tip has been used than when we use sharp tip. Moreover, the behavior of the substrate influences the

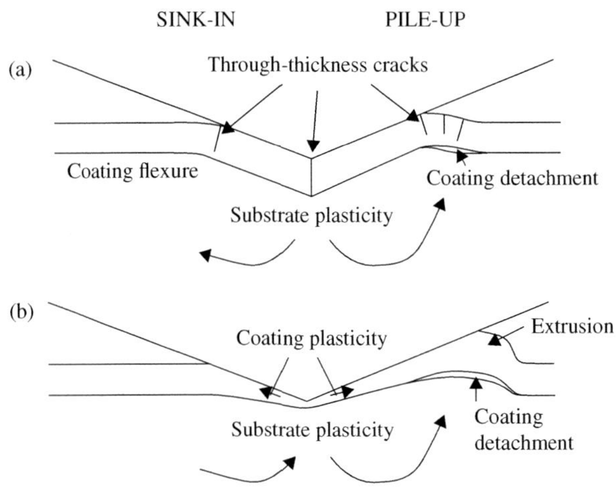

Figure 5.8 Different behaviors of the coatings subjected to nanoindentation: (a) hard coating on a soft substrate and (b) soft coating on a hard substrate.

curves shape, showing higher displacement when a sharp indenter is used, which can be achieved by substrate plastic deformation, coating through-thickness fracture or coating detachment effect. However, these effects promote different curve behavior, because the coating fracture induces a sharp transition whereas a slow but sure change in the loading is usually related to substrate plastic deformation. In spite of a deep analysis of the phenomena related to unusual curves behavior, microscopy analysis will be a useful tool letting to observe what happened during the nanoindentation test.

The early stages of the load-displacement curve largely depend on the load, indenter geometry, as well as indenter and coating roughness. When a Berkovich indenter and extremely low penetration depths are being used, the displacement is nearly proportional to the load. This effect can be attributed to the interaction between the coating and indenter roughness, keep it up until indenter penetration depth surpasses about five times the combined roughness present in the contact. Also a transition between elastic and plastic behavior is usual in these curves, which take place within a wide range of loads and displacements but, for extremely hard coatings, it can impossible to observe the entire elastic-plastic behavior using the loads range usually available on the nanoindenters, prompting the need to use large indentations.

However, there are also some limitations because this technique does not provide direct observation of the operative subsurface deformation mechanisms, such as delaminations and crack propagation.

5.3.1 Thin Coatings Nanoindentation Procedures

Indenter geometries generally used on nanoindentation comprise: Vickers (pyramidal with square base), Berkovich (pyramidal with triangular base), cube corner (pyramidal with triangular base) and spherical. Diamond, tungsten carbide, and sapphire are the materials commonly used in the indenters for tribological coatings. Berkovich indenter is regularly used for tribological films characterization, presenting an angle of 65.03° between the axis of the pyramid and three planes corresponding to indenter faces.

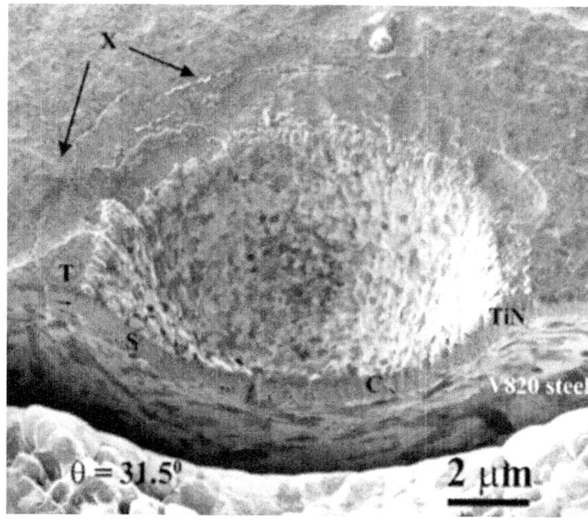

Figure 5.9 Cross-section of a nanoindentation performed with 400 mN using 5 μm radius spherical indenter on a 0.7 μm thick TiN coating deposited on V820 steel substrate. *Source*: MA 2005 [25]. Reproduced with permission of Elsevier.

Berkovich indenters are the most commonly used for hardness and Young's modulus analysis. However, when a more uniform stress field under the indenter is desirable, nanoindentation can be performed using a spherical tipped conical diamond indenter provided with a small tip radius, typically around 5 μm. Cube corner indenter shows a similar geometry but presents a lower angle between the axis and the pyramid faces, 35.26°, which is favorable for thin films due to a larger imprinted area for the same load. When a more uniform stress field under the indenter is desired, sharp indenter should be replaced by spherical indenter, letting easier the interpretation of the microstructural observations [24]. Ma *et al.* [25] carried out some tests on TiN thin coatings using a spherical tipped conical indenter provided with a tip radius of 5 μm and, regardless of stark deformation of the film and substrate, it can be noted that coating preserves its integrity, as can be observed in Figure 5.9, keeping a very good adhesion between coating and substrate, not being visible any evidence of delamination.

The initial calibration of the nanoindentation equipment and indenter is usually carried out using a fused silica as a standard sample due to its elastically isotropic behavior that allows consider its Young's modules constant with depth. Indeed, it is well known that the initial Berkovich indenter tip radius is around 50 nm but there is a degeneration of the tip under work following a ratio of 1 nm per test performed within the first six months of use. Thus, careful indenter tip calibration is crucial in order to achieve accurate nanoindentation results when indentation depths produced are lesser than 100 nm.

Indentation tests on tribological thin films are usually performed at room temperature and there is a period of time in which the peak load is kept constant in order to avoid creep phenomenon, which can undertake values around 5 to 60 seconds, followed by a quick unload (10 seconds) until about 10% of the maximum load, accomplishing a hold period 60 seconds at this load for thermal drift adjustment, usually in the range of 0.001 nm/s to 0.004 nm/s [26]; [27]. Chudoba and Richter [28] have proposed holding times for some usual materials going from 8 seconds to fused silica until 187 seconds to aluminum. The softer is the material tested, the longer is the time of the first holding period. An accurate thermal drift correction is extremely important to make a distinction between depth variations caused by plastic deformation or creep from an apparent

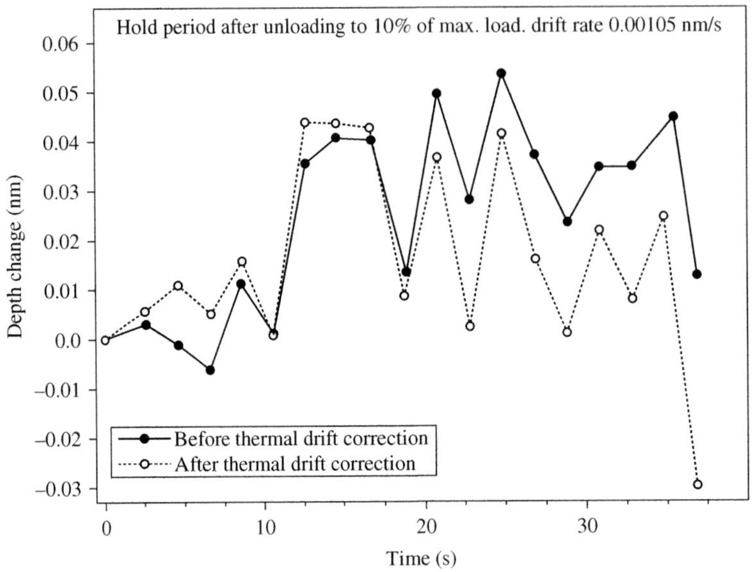

Figure 5.10 Depth variation during the hold period of 60 seconds at 10% of the maximum load of 50 μN, caused by thermal drift when testing DLC coatings with 31.4 nm thickness by nanoindentation. *Source*: Chudoba 2001 [28]. Reproduced with permission of Elsevier.

depth modification as a result of drift effects. This effect cannot be neglected, presenting variations in terms of indentation depth as shown in Figures 5.10 and 5.11.

5.3.2 Hardness Determination

As referred previously, hardness is, probably, the most important property to evaluate in terms of coatings for tribological applications. Also as referred before, film thickness is a variable to take into attention due to substrate or adjacent layers influence. The indentation depth limit of 10% of the layer thickness previously mentioned is a reference but, regarding the INDICOAT report [14] earlier denoted, there is no safe limit to prevent substrate or adjacent layer influence on the hardness measurement by nanoindentation. Some authors tried to avoid that influence using indentation depth between 2–5% [26], both in mono and multilayered films. Otherwise, other authors refer as limit for the nanoindentation depth 20% of the film thickness [29, 30] and the maximum nanoindentation depth can be reach 1000 nm [29], conflicting with the aforementioned recommendation (≤ 200 nm). However, 10% of the coating thickness is usually taken as a rule of thumb, attending that coating is responsible for accommodate elastic and plastic deformations whereas the substrate just cares a small portion of the nanoindentation load.

Regarding the widely used Oliver and Pharr method, hardness (H) calculation is determined by the relationship between the peak load (F_{max}) and the projected area (A), as follows:

$$H = \frac{F_{\max}}{A} \tag{5.1}$$

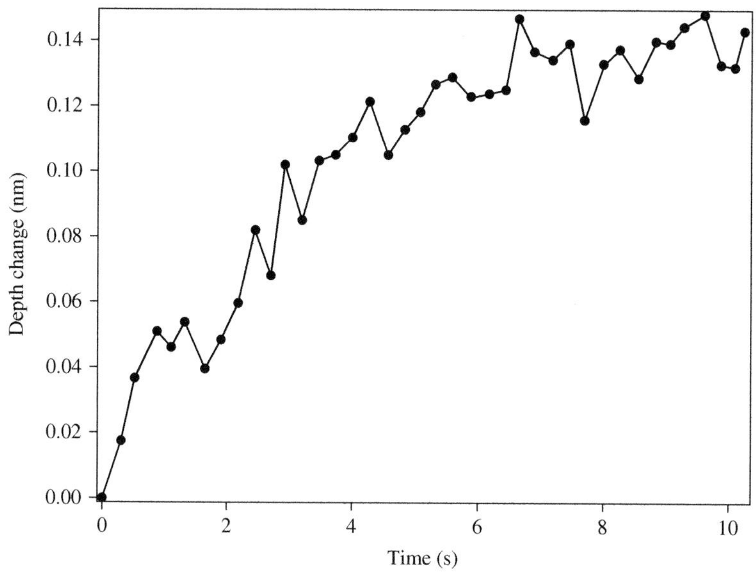

Figure 5.11 Depth variation due to creep effect during the hold time at maximum load of 50 μN after thermal drift adjustment, when testing DLC coatings with 31.4 nm thickness by nanoindentation. *Source*: Chudoba 2001 [28]. Reproduced with permission of Elsevier.

Considering a new Berkovich indenter, we can consider that the contact area is given by the following expression:

$$A_c = 24.5 \cdot d_c \tag{5.2}$$

where d_c is the contact depth, which can be calculated by the following equation:

$$d_c = \varepsilon \frac{F_{máx}}{S_u} \tag{5.3}$$

where ε is a constant which depends on the indenter geometry and S_u corresponds to the unloading curve slope at maximum load. Usually, ε assumes the value of 0.75 for Berkovich indenters.

Nanoindentation tests carried out by Beake *et al.* [26] over mono and multilayered coatings using a Berkovich indenter and 20 mN maximum load, performing 50 indentations and calculating the average values, allows obtain the results presented below in Table 5.1.

As can be observed, the standard deviation is particularly broad regarding the TiCN monolayer coating, which can be attributed to the high roughness of the coating.

Bhushan *et al.* [31] analyzed different coatings by nanoindentation, among them some hard coatings such as SiC, amorphous hydrogenated carbon (a-C:H) and Al_2O_3, and found that film thickness affects severely the hardness and Young's modulus values obtained by nanoindentation. Relatively to the a-C:H films, the hardness value drops from 18 to 14 GPa when the film thickness decreases from 5 μm to 0.4 μm. The same effect was felt in Young's modulus values, beading from 170 to 115 GPa for the same films and thicknesses.

Table 5.1 Hardness and other correlated values regarding some tribological coatings tested by nanoindentation.

Sample	Thickness (μm)	H (GPa)	E_r (GPa)	Load (mN)	Source
TiCN	4.0	29.0 ± 22.8	368 ± 168	20	[26]
Al_2O_3 – TiCN (inner layer)	9.0	14.7 ± 6.6	303 ± 92	20	[26]
TiN – Al_2O_3 – TiCN (inner layers)	12.0	15.3 ± 8.3	375 ± 128	20	[26]
TiN/Ti(C,N)/TiN	1.0/1.5/1.0	> 21	—	16.8	[33]
Synthetic Diamond	0.92 – 4.85	52–95	438–768	3–80	[22]
TiN	0.29	24.9 ± 1.8	265 ± 11	120	[34]
W	0.25	17.9	209	1.7	[34]
TiN + W (alternate)	1.3 + 1.4	24.2 ± 1.4	219 ± 10	120	[34]
TiN + W (alternate)	9.2 + 9.4	21.3 ± 0.7	220 ± 3	120	[34]
TiB_2	2.1	26.82	343.4	50	[35]
TiAlSiN	4.52	19.2	—	50	[36]
TiB_2	2.2	35.7	401.3	50	[37]
DLC	0.5	65	650	200	[29]

Also nanoindentation tests carried out on diamond coatings showed an increasing hardness from 52 to 95 GPa when coatings thickness growths from 0.92 to 4.85 μm [22]. Probably, the usual value assumed for the natural diamond (102 GPa) should be achieved on films over 5 μm thickness. This kind of coatings is not easy to test due to comparable hardness between indenter and coating to test. In this case, some authors [9] are considering the indenter as an elastic body when super-hard coatings are being tested. Effectively, under these circumstances, the indenter cannot be considered as a rigid body and its deformation cannot be ignored. So that check the indenter tip integrity, namely when using a Berkovich indenter, is usual to carry out some nanoindentation tests on copper ($H_{Cu} = 1.1$ GPa and $E_{Cu} = 128.2$ GPa). Due to the high plasticity of the copper, the imprint produced reflects the indenter tip status, acting as a mirror. Further correction should be made as a function of some deformation detected.

Regarding the P - δ curves, some discontinuities on loading can occur, usually denominated by 'pop-ins', whose are commonly related to one or more of the following effects:

- Coating through-thickness fracture;
- Coating delamination;
- Plasticity induced in the substrate close to the coating interface.

The identification of the causes that were at the origin of these phenomena can be explored by microscopy analysis of the deformed area. The size of each pop-in leads to estimate the energy absorbed in each fracture, using nanojoule scale. In certain situations, nanoindentation can promote the coating fracture inducing the formation of separated islands fix to the substrate. Sometimes, when the slopes of both loading and unloading curves begin to get closer to those of the substrate, it means that a transition behavior is occurring and the coating starts failing do not providing the adequate load support, necessary contact stiffness or any important improvement in effective hardness. These phenomena are also observed when Young's modulus are being analyzed by nanoindentation, thus, it will be discussed latter in another section.

Regarding the $P - \delta$ curves, its first stage corresponds to elastic deformation of the substrate/coating system. However, increasing the load, there is a change from elastic to plastic behavior, due to stress exceeding the yield strength of the system, occurring on the substrate or on the coating depending on their properties, indenter geometry and coating thickness, being particularly important when testing films under 1 μm thickness. The stress generated depends on large scale of the indenter tip radius which, as referred above, suffers severe wear during its useful lifetime, starting usually in 100 nm radius but presenting 250 nm radius after some months of use. As expected, the stress field generated by these two different situations leads to very distinct results. Thus, a new and sharp indenter tip is required if the researchers want to induce test thin hard coatings ensuring that contact stress is high enough to induce coating yield before the substrate.

5.3.3 Young's Modulus Determination

In order to obtain the Young's modulus, the unloading portion of the depth–load curve is examined according to a relationship that depends on the contact area:

$$C = \frac{\pi^{0.25}}{2 \cdot E_r \cdot A^{0.5}} \tag{5.4}$$

where C is the contact compliance and E_r is the reduced Young's modulus defined by:

$$\frac{1}{E_r} = \frac{1 - \upsilon_S^2}{E_S} + \frac{1 - \upsilon_i^2}{E_i} \tag{5.5}$$

where υs and υi are the Poisson's ratio for the sample and diamond indenter (0.07), respectively, whereas E_s and E_i are the Young's modulus for the sample and indenter (1141 GPa), respectively. The elastic recovery parameter (ERP) is a useful dimensionless index that is closely related to the ratio between hardness and Young's modulus. It can be defined as follows:

$$ERP = \frac{h_{\max} - h_p}{h_p} \tag{5.6}$$

where h_{max} is the maximum indentation depth and h_p corresponds to plastic indentation depth.

When low loads are applied, the force – displacement behavior is exclusively elastic and usually is in good agreement with the predicted curve for elastic behavior. However, when the load is increased, the loading curve diverges from the predicted curve corresponding to elastic behavior, as shown in Figure 5.12. This comportment can be attributed to the beginning of plastic deformation process. In this regime, some sporadic events occurs (Figure 5.12, P1 and P2 points), so-called 'pop-in' events, showing quick increase in indentation depths which corresponds to small load increments. Usually, these events are correlated with crack nucleation and initial propagation of the coatings [32].

Nanoindentation tests performed on diamond coatings show an increasing on Young's modulus from 438 to 768 GPa when coatings thickness growths from 0.92 to 4.85 μm [21], values that can be compared with 1200 GPa of the natural diamond.

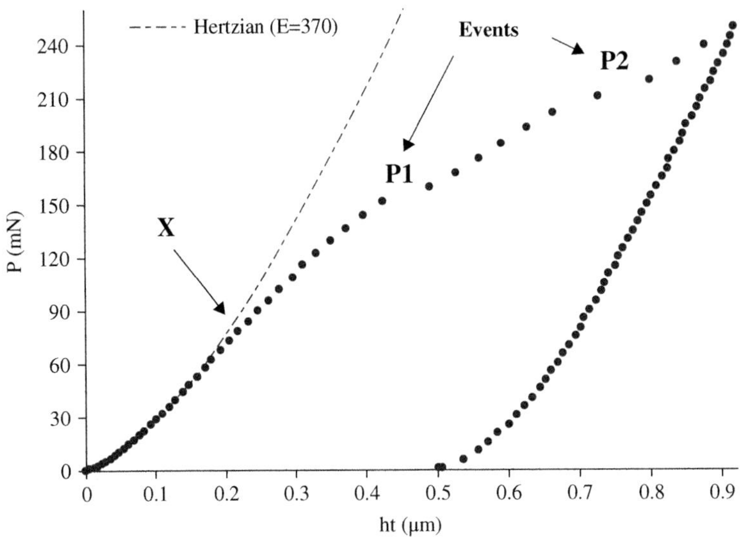

Figure 5.12 Load-displacement curves corresponding to nanoindentations performed on TiN coatings. *Source*: Weppelmann 1994 [32]. Reproduced with permission of Taylor & Francis.

5.3.4 Tensile Properties Determination

Determination of tensile properties with high accuracy by indentation testing presents noteworthy advantages over the traditional tensile test. Nanoindentation uses a small volume of material to perform the tests, thus researchers can get tensile properties of small particles, coatings or surfaces which have been treated (chemically, physically or thermally). Indentation testing machines would be made portable, enabling in situ and on-site testing. Furthermore, nanoindentation testing would be considered as non-destructive and sample preparation requirements are relatively simple. The monitoring process of structures in service would be done easier without the need to extract samples for testing as in the traditional tensile testing methods.

ISO Technical Report 29381 [38] describes as indentation technique can help researchers to get traditional properties usually get through tensile testing. The report refers to three methods and the main requirements, advantages and disadvantages of each one. These methods are as follows: representative stress and strain, Inverse FEA methods, and neural networks. These methods regularly use the instrumented indentation testing cycle, often entitled as "partial unloading" method [39], which applies a progressively increasing force but stops in a series of steps where the force is partially removed to obtain the upper portion of the force removal curve. This portion of the curve is used to obtain the contact stiffness and contact depth at that force. The partial unloading method allows a wide range of indentation sizes to be performed at the same location. In all three methods, the complex stress/strain field beneath the indenter is simplified to a single pair of stress and strain values, often called the "representative" stress and strain.

Determination of the coating tensile properties, such as yield strength and ultimate strength implies, usually, assuming that contact is plastic at full load and von Mises criterion is fulfilled, because if the contact is elastic, von Mises stresses will not reach the material yield stress. Indeed, von Mises stress congregates the whole shearing stresses at the material is subject, displaying its maximum through the effectively shaped indenter at the moment when unloading initiates and allowing to determine the coating yield strength [40, 41].

It is unthinkable to use mechanical conventional methods to measure the yield and ultimate strength of coatings with less than 5 μm thickness. Finite elements methods (FEM) have been used to compute the main mechanical properties such as Young's modulus, yield stress or work hardening exponent, using initially nanoindentation load-displacement curves obtained over bulk materials as experimental data in order to match the results and proceed to model validation [42–44]. Later, these models were extended to coated systems, being nowadays used regularly [45]. Thus, the yield strength of perfectly plastic thin films can easily be assessed by matching the maximum load achieved experimentally in the nanoindentation tests with modeled data regarding the corresponding ratio between coating and substrate elastic properties. The relationship between stress and strain also can be extracted from the nanoindentation tests based on computational analysis performed by FEM, as procedures established by O'Dowd *et al.* [45, 46]. This method uses the information collected from the load-displacement curve and from the measurements made regarding the indentation depression left on the surface after unloading. However, attending possible errors done during the measurement process, this method entails always some uncertainty in the accuracy of the results obtained.

5.3.5 Fracture Toughness in Thin Films

Research about the fracture toughness of thin films has shown that increasing the number of layers, the critical load achieved is larger. This means that multilayer coating system can represent a good way for improving the strength of brittle monolayer coatings. Furthermore, the system of brittle monolayer should be replaced by a multilayered system consisting of alternate brittle and ductile coating layers, improving drastically the global fracture toughness, minimizing or even avoiding by this way the crack nucleation and propagation. Abadias *et al.* have been using Cu as intermediate layer between TiN layers in order to increase the energy dissipation [47, 48]. However, the number of ductile layers should be attained allowing the best compromise between the increasing in fracture toughness and the lost hardness.

Regarding the Li and Bhushan's theory [49], the thin films fracture toughness, K_{IC}, can be calculated using the following expression:

$$K_{IC} = \left[\left(\frac{E}{(1-\upsilon^2 \cdot 2 \cdot \pi \cdot C_g)}\right)\left(\frac{U}{t}\right)\right]^{1/2} \tag{5.7}$$

where E is the elastic modulus, υ represents the Poisson's ratio, $2\pi Cg$ corresponds to the crack length in the film plane, U is the strain energy difference before and after cracking, and t is the layer thickness. A similar approach was also done by the same author in another study [50], conducting to an identical expression, where the facture area (A_{crack}) and the fracture dissipated energy (U_{fr}) are considered, as follows:

$$K_{IC} = \left[\left(\frac{E \cdot U_{fr}}{(1-\upsilon^2 \cdot A_{Crack})}\right)\right]^{1/2} \tag{5.8}$$

Table 5.2 Fracture toughness of different multilayered coatings, performed by nanoindentation.

Sample	Maximum load (mN)	K_{IC} (MPa·m$^{1/2}$)
TiN/TiCN/TiC	16.4	2.18
Al_2O_3 / TiC / TiCN / TiC	18.5	1.74
TiN/TiCN/TiC/TiCN/TiC	35.5	3.40
TiN/TiCN/TiC/TiCN/TiC/TiCN/TiC	56.3	3.90

Source: Kazmanli 1998 [51]. Reproduced with permission of Elsevier.

It is well known that the area under the load/displacement curve corresponds to the work performed by the indenter during the course of elastic-plastic deformation of the film-substrate system. The strain energy released in the ring-like cracking can be calculated from the loading curve. U can be evaluated through the peak loads of the load/displacement curves. Based on the equation above, the first layer fracture toughness of these coating systems can be calculated. The coating layer fracture toughness can be obtained from the load-penetration depth curves, but the variation of the interfacial properties of coating-substrate systems, namely the interfacial properties of the multilayer coatings can scarcely be obtained from the load/penetration depth curves. The changes in slope of plots of load-penetration depth squared can disclose the changes of coating substrate systems [51]. According to *Loubet's* analysis of nanoindentation load-penetration depth curves [52] if the material is being indented, the plastic depth of indentation and the instantaneous elastic contribution can be calculated.

Fracture toughness of different coatings has been analyzed through nanoindentation and the results can be observed in Table 5.2, as follows.

5.3.6 Coatings Adhesion Analysis

Adhesion is a critical issue regarding coating-substrate system. Nanoindentation can be applied to analyze the adhesion behavior, allowing an easy loading configuration and the chance to observe directly the crack formation and film delamination. The nanoindentation takes advantage due to the different reaction of the coating and substrate throughout the relaxation stage. Indeed, during the loading stage, both coating and substrate accomplish the deformation imposed by the indenter. However, when they are unloaded, being the coating usually harder, the elastic recovery is higher, and the adhesion forces between substrate and coating are not strong enough to avoid the delamination.

Depending on the hardness and thickness of the films, maximum load to be used, in order to produce film delamination, must be tested and selected. Elizalde *et al.* [53] carried out nanoindentation tests on Cu - SiO_2 interface, being necessary a normal load of $80 - 120$ mN, using as distance between indentation 3 to 5 μm, depending on the film thickness. After nanoindentation, the cross-section interface must be observed by SEM (Scanning Electron Microscopy), allowing verify its status. Each nanoindentation imprint produces a detachment of the coating, as depicted in Figure 5.13, where x represents the interfacial crack length, $2b$ is the wedge length, u_f expresses the wedge displacement after load removal and d embodies the distance from the indent to the interface of interest which is measured from SEM micrographs.

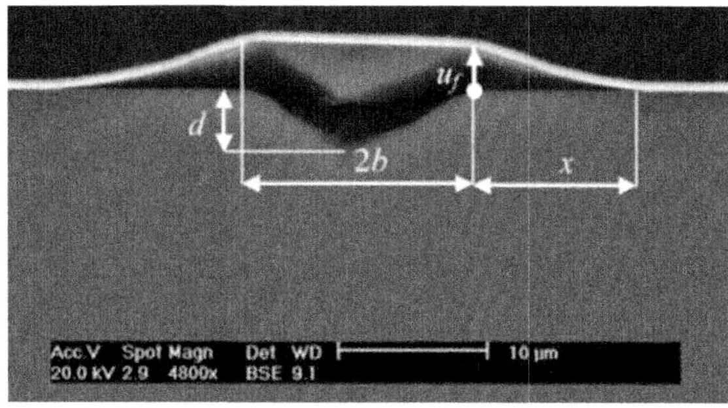

Figure 5.13 Nanoindentation imprint showing film detachment. *Source*: Abadias 2006 [34]. Reproduced with permission of Elsevier.

Based on the variables above mentioned, authors [53] developed a FEM model able to determine the crack propagation, having as input variables the coating thickness, the parameters measured by SEM in the image above (b, x and u_f), yield stress, Young's modulus and Poisson's coefficient. However, the method should not be applied to very thin films, where the roughness avoids a correct reading of the detachment value.

Similar experiences were carried out earlier by other authors [54] on TiN coatings obtained by PVD technique with 1.2 μm thickness. A model was also developed based in FORTRAN to calculate the critical load able to promote the film detachment and the results has been compared with experimental measures also performed, showing a good agreement.

Furthermore, in the last decade some works have been performed in order to understand the mechanics and mechanisms of the crack initiation and propagation on the interface between substrate and coating [10, 55 – 56]. The last one was focused on the effect of the initial position of pre-existing defects on the film – coating interface, whose can lead to easy crack initiation and propagation, masking the adhesion, Young's modulus and hardness results obtained by nanoindentation. The model developed and the results obtained revealed that the interfacial crack growth depends on a large scale of the pre-existing defect position defined in the study. When the defect is considered nearer to the point where shear stress concentration is higher, the crack will initiate quickly. The defect position also affects the load – displacement curve, with repercussions on the results obtained for the hardness and elasticity module.

5.3.7 Stiffness and Other Mechanical Properties

Nanoindentation also has been explored regarding the extraction of other mechanical properties [50]. In 1999, Syed-Asif *et al.* [57] introduced the Dynamic Stiffness Measurement (DSM) method, which has been used to characterize coatings and a wide range of amorphous materials. In 1992, Oliver and Pharr [8] announced the Continuous Stiffness Measurement (CSM) method, which uses a relatively high frequency oscillation of small amplitude superimposed to the signal responsible by the indenter load control. Thus, it becomes possible to get the contact stiffness along the entire displacement curve, improving the study of other properties, such as hardness, elastic modulus, creep or fatigue, as a function of the displacement [50]. Indeed, when the indenter presses the coating and substrate system, the total deformation effect on the surface is not immediately achieved, which can be confirmed keeping the maximum load for a while at the end of the loading cycle and note that the displacement of the indenter increase when the load is kept constant or even reduced. This behavior is directly related to the viscoelastic/

viscoplastic deformation of the system under the stress induced by the indenter. This dynamic behavior can be measured using some variables, such as the storage modulus, E', or the loss modulus, E'', whose are correlated with material elastic properties, energy recovered, material viscosity or energy transformed into heat or friction during indentation.

5.3.8 Simulation and Models Applied to Nanoindentation

After a long work around thin hard coatings, Bouzakis [58] carried out a study in order to develop a model to enable the stress-strain course of the coatings based on Finite Elements Method (FEM). With the aim of increase the calculation accuracy, the traditional Vickers pyramid needed to be replaced by a cone (see Figure 5.14a), allowing the development of the model around a three-dimensional axisymmetric system, considering the edges of the pyramid have a negligible effect on the nanoindentation procedure. The cone was thought keeping in mind that cross-section of the indenter must remain equivalent at the same penetration depth. Similar reasoning was done for the Berkovich indenter, considering a triangular cross-section. The FEM model was established regarding a deformable diamond indenter with a conical shape and based on the above-mentioned assumptions, using contact elements to describe the interface

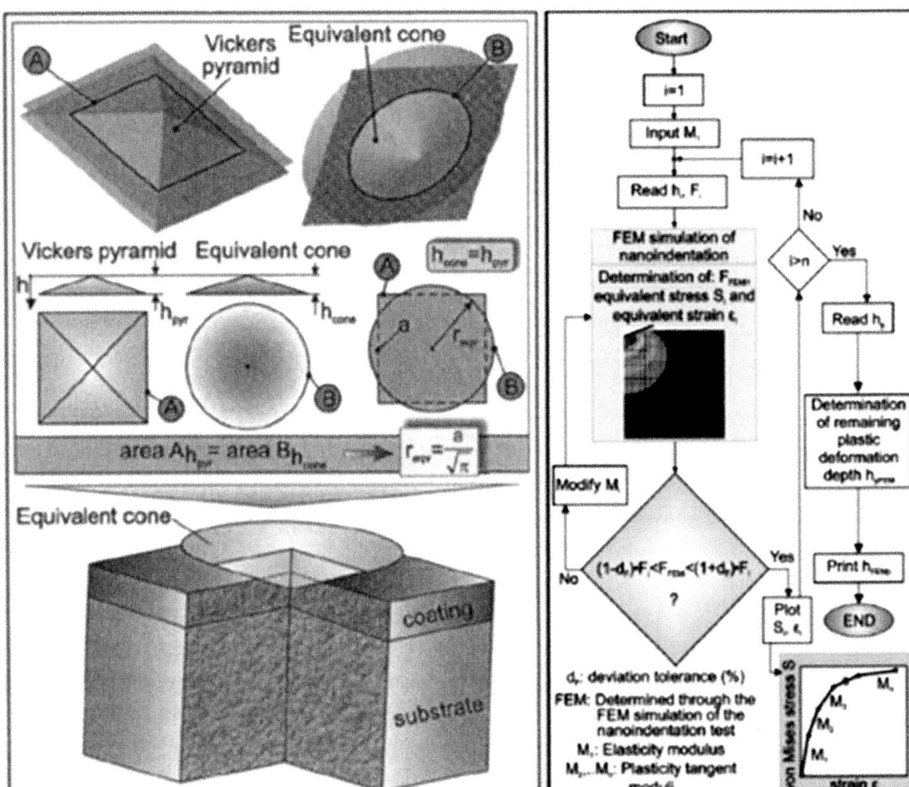

Figure 5.14 Load-displacement curves corresponding to nanoindentations performed on TiN coatings. *Source*: Bouzakis 2001[58]. Reproduced with permission of Elsevier.

between indenter and coating surface and considering that stiffness and coefficient of friction of these elements do not affect the simulation results. Parameters such as substrate and coating properties and penetration depth are kept as variables changeable in order to become the model more flexible and reproducible. Authors [58] took into attention two main hardening rules: isotropic, where the yield remains centered on the tip indenter area and expands in size as a function of the plastic strain, and the kinetic hardening, which assumes that the yield surface rests constant in size and the surface is subjected to progressive yielding. The input data of the model consists of pairs of force – displacement values digitalized during the loading stage, in order to be treated by an algorithm so-called 'SSCUBONI'. Some algorithm conditions were established to filter wrong reading situations, remaining in loop until those conditions are consummate, as shown in the flow chart diagram of the Figure 5.14b. The difference between the calculated h_{FEMp} value due to plastic deformation and the measured hp value, gives us the accuracy of the model calculating the stress – strain curves, being the final presented in the form of a graphical stress-strain curve.

FEM model allows examine the von Mises equivalent stress distribution along the loading stage for each indentation depth. Hence, this model do not avoid the need to perform the indentation test but allows an automatic generation of the load – displacement curves, von Mises stress distribution in the indentation area and hardness and Young's modulus calculations. In this model, it was found that indenter tip radius and coating Poisson's coefficient are the parameters most influents for the model accuracy. Thus, this method should be a very efficient way to illustrate the elastic-plastic deformation behavior of thin tribological coatings.

References

1 SANDVIK. Metalcutting Technical Guide. Sweden: Sandvik Coromant, A73–A74; 2005.
2 Silva FJG, Fernandes AJS, Costa FM, Baptista APM, Pereira E. A new interlayer approach for CVD diamond coating of steel substrates. Diamond and Related Materials, 13, 828–833; 2004. DOI: 10.1016/j.diamond.2003.10.081.
3 Benes, J. Armor for cutting tools – with the right protective coating, tools last longer and cut better (courtesy of diamond tool coating). American Machinist. June 2007, http://customcarbidetools.com/process-solutions/armor-for-tools/ (Accessed 02 April 2014).
4 Martinho RPM, Silva FJG, Alexandre RJD, Baptista APM. TiB_2 nanostructured coating for GFRP injection moulds. Journal of Nanoscience and Nanotechnology, 11, 5374–5382; 2011. DOI: 0.1166/jnn.2011.3772.
5 Jiang, W-G, Su J-J, Feng X-Q. Effect of surface roughness on nanoindentation test of thin films. Engineering Fracture Mechanics, 75 (17), 4965–4972; 2008. DOI: 10.1016/j.engfracmech.2008.06.016.
6 Bailey E, Ray NMT, Hector AL, Crozier P, Petuskey WT, McMillan PF. Mechanical properties of titanium nitride nanocomposites produced by chemical precursor synthesis high-P,T treatment. Materials. 4(10), 1747–1762; 2011. DOI: 10.3390/ma4101747.

7 ASTM. ASTM E-384, Annual Book of Standards 3.01. Philadelphia, PA: ASTM International, p. 469; 1989.
8 Oliver WC, Pharr GM. An improved technique for determining hardness and elastic modulus using load and displacement sensing indentation experiments. Journal of Materials Research. 7, 1564–1583; 1992. DOI: 10.1557/JMR.1992.1564.
9 He JL, Veprek S. Finite element modeling of indentation into super-hard coatings. Surface and Coatings Technology, 163–164, 374–379; 2003. DOI: 10.1016/S0257-8972(02)00628-X.
10 Pejman R, Salehi SH, Salehi M. Numerical study of interfacial crack growth effects on nanoindentation mechanical. Thin Solid Films, 545, 408–413; 2013. DOI: 10.1016/j.tsf.2013.08.033.
11 Lucca DA, Herrmann K, Klopfstein MJ. Nanoindentation: Measuring methods and applications. CIRP Annals – Manufacturing Technology, 59, 803–819; 2010. DOI: 10.1016/j.cirp.2010.05.009.
12 International Organization for Standardization. ISO 14577-1, -2, -3, Metallic Materials – Instrumented Indentation Test for Hardness and Materials Parameters – Part 1: Test Method, Part 2: Verification and Calibration of Testing Machines, Part 3: Calibration of Reference Blocks. Geneve, Switzerland: International Organization for Standardization; 2002.
13 International Organization for Standardization. ISO 14577-4, Metallic Materials – Instrumented Indentation Test for Hardness and Materials Parameters – Part 4: Test Method for Metallic and Non-Metallic Coatings. Geneve, Switzerland: ISO – International Organization for Standardization; 2007.
14 Jennett NM. INDICOAT Final Report – Determination of Hardness and Modulus of Thin Films and Coatings by Nanoindentation. NPL Report MATC (A). Teddington, UK: National Physics Laboratory (NPL); 2001.
15 American Society for Testing and Materials. ASTM E2546-07, Standard Practice for Instrumented Indentation Testing. West Conshohocken, PA, USA: ASTM International; 2007.
16 International Organization for Standardization. ISO TR 29381, Metallic Materials – Measurement of Mechanical Properties by Instrumented Indentation Test – Indentation Tensile Properties. Geneve, Switzerland: International Organization for Standardization; 2008.
17 International Organization for Standardization. ISO/TC 164/SC 3: Hardness Testing – Indentation contact based nano-mechanical measurement of viscoelastic materials. Geneve, Switzerland: International organization for Standardization; 2009.
18 European Committee for Standardization. ENV WI 152: Determination of Hardness and Young's Modulus by Instrumented Indentation Testing. Brussels, Belgium: CEN; 2001.
19 Page TF, Hainsworth SV. Using nanoindentation techniques for the characterization of coated systems: a critique. Surface and Coatings Technology, 61, 201–208; 1993. DOI: 10.1016/0257-8972(93)90226-E.
20 Page TF, Oliver WC, McHargue CJ. The deformation behavior of ceramic crystals subjected to very low load (nano) indentations. Journal of Materials Research, 7, 450–473; 1992. DOI: 10.1557/JMR.1992.0450.

21 Sjostrom H, Hultman L, Sundgren J-E et al. Structural and mechanical properties of carbon nitride CN's (0.2 < x < 0.35) films. Vacuum Science and Technology A, 14, 56–62; 1996. DOI: 10.1116/1.579880.
22 Nitti, MA, Cicala G, Brescia R et al. Mechanical properties of MWPECVD diamond coatings on Si substrate via nanoindentation. Diamond and Related Materials, 20, 221–226; 2011. DOI: 10.1016/j.diamond.2010.12.002.
23 Bull SJ. Nanoindentation of coatings. Journal of Physics D: Applied Physics, 38, 2005; R393–R413, 2005. DOI: 10.1088/0022-3727/38/24/R01.
24 Bradby JE, Williams JS, Leung JW et al. Transmission electron microscopy observation of deformation microstructure under spherical indentation in silicon. Applied Physics Letters, 77, 3749–3751; 2000. DOI: 10.1063/1.1332110.
25 Ma, LW, Cairney JM, Hoffman M, Munroe PR. 2005. Deformation mechanisms operating during nanoindentation of TiN coatings on steel substrates. Surface and Coatings Technology. 192, 2005, 11–18, DOI: 10.1016/j.surfcoat.2004.04.002.
26 Beake BD, Ranganathan N. An investigation of the nanoindentation and nano/micro-tribological behaviour of monolayer, bilayer and trilayer coatings on cemented carbide. Materials Science and Engineering A, 423, 46–51; 2006. DOI: 10.1016/j.msea.2005.11.066.
27 Schwarzer, N, Chudoba T, Richter F. Investigation of ultra-thin coatings using nanoindentation. Surface and Coatings Technology, 200, 5566–5580; 2006. DOI: 10.1016/j.surfcoat.2005.07.075.
28 Chudoba T, Richter F. Investigation of Creep Behaviour under Load During Indentation Experiments and Its Influence on Hardness and Modulus Results. Surface and Coatings Technology, 148, 2–3, 191–198; 2001. DOI: 10.1016/S0257-8972(01)01340-8.
29 Xu Z-H, Rowcliffe D. Nanoindentation on diamond-like carbon and alumina coatings. Surface and Coatings Technology, 161, 44–51; 2002. DOI: 10.1016/S0257-8972(02)00364-X.
30 Zhang TH, Huan Y. Nanoindentation and nanoscratch behaviors of DLC coatings. Composites Science and Technology, 65, 1409–1413; 2005. DOI: 10.1016/j.compscitech.2004.12.011.
31 Bhushan B. Handbook of micro/nanotribology. 2nd edn. Boca Raton: CRC Press; 1999.
32 Weppelmann ER, Hu X-H, Swain MV. Observations and simple fracture mechanics analysis of indentation fracture delamination of TiN films on silicon. Journal of Adhesion Science and Technology, 8, 611–624; 1994. DOI: 10.1163/156856194X00375.
33 Ding J, Meng Y, Wen S. Mechanical properties and fracture toughness of multilayer hard coatings using nanoindentation. Thin Solid Films, 371, 178–182; 2000. DOI: 10.1016/S0040-6090(00)01004-X.
34 Abadias G, Dub S, Shmegera R. Nanoindentation hardness and structure of ion beam sputtered TiN, W and TiN/W multilayer hard coatings. Surface and Coatings Technology, 200, 6538–6543; 2006. DOI: 10.1016/j.surfcoat.2005.11.053.
35 Silva FJG, Casais RCB, Martinho RP, Baptista APM. Role of abrasive material on microabrasion. Wear, 271, 2632– 2639; 2011. DOI: 10.1016/j.wear.2010.11.050.

36 Silva FJG, Martinho RPM, Alexandre RJD, Baptista APM. Wear resistance of TiAlSiN thin coatings. Journal of Nanoscience and Nanotechnology, 12, 9094–9101; 2012. DOI: DOI:10.1166/jnn.2012.6760.
37 Silva FJG, Casais RCB, Alexandre, RJD, Baptista, APM. Mechanical and Tribological Characterization of TiB2 Thin Films. Journal of Nanoscience and Nanotechnology, 12, 9187–9194; 2012. DOI: doi:10.1166/jnn.2012.6759.
38 International Organization for Standardization. ISO TR 29381, Metallic Materials – Measurement of Mechanical Properties by Instrumented Indentation Test – Indentation Tensile Properties. Geneve, Switzerland: ISO, 2008.
39 Field JS, Swain MV. A simple predictive model for spherical indentation. Journal of Materials Research, 8, 2, 297–306; 1993. DOI: 10.1557/JMR.1993.0297.
40 Schwarzer N, Pharr GM. On the evaluation of stresses during nanoindentation with sharp indenters. Thin Solid Films, 469 –470, 194–200; 2004. DOI: 10.1016/j.tsf.2004.08.170.
41 Schwarzer N, Chudoba T, Pharr GM. On the evaluation of stresses for coated materials during nanoindentation with sharp indenters. Surface and Coatings Technology, 200(14–15), 4220–4226; 2006. DOI: 10.1016/j.surfcoat.2005.01.011.
42 Bhattacharya AK, Nix WD. Finite element simulation of indentation experiments. International Journal of Solids and Structures, 24, 881–891; 1988. DOI: 10.1016/0020-7683(88)90039-X.
43 Laursen TA, Simo JC. A Study of the Mechanics of Microindentation Using Finite Elements. Journal of Materials Research, 7, 618–626; 1992. DOI: 10.1557/JMR.1992.0618.
44 Knapp JA, Follstaedt DM et al. Finite-element modeling of nanoindentation. Journal of Applied Physics, 85, 1460–1474; 1999. DOI: 10.1063/1.369178.
45 Tunvisut K, O'Dowd NP, Busso NP. Use of scaling functions to determine mechanical properties of thin coatings from microindentation tests. International Journal of Solids and Structures, 38, 335–351; 2001. DOI: 10.1016/S0020-7683(00)00017-2.
46 Tunvisut K, Busso EP, O'Dowd NP, Brantner HP. Determination of the mechanical properties of metallic thin films and substrates from indentation tests. Philosophical Magazine A, 82 (10), 2013–2029; 2002. DOI: 10.1080/01418610210134332.
47 Abadias G, Tse YY, Michel A et al. Nanoscaled composite TiN/Cu multilayer thin films deposited by dual ion beam sputtering: growth and structural characterization. Thin Solid Films. 443, 166–173; 2003. DOI: 10.1016/S0040-6090(03)00307-9.
48 Abadias G, Tromas C, Tse YY, Michel A. Investigation of the deformation behavior in nanoindented metal/nitride multilayers by coupling FIB-TEM and AFM observations. Materials Research Society Symposium Proceedings, Vol. 880 E, BB 2.8; 2005. DOI: org/10.1557/PROC-880-BB2.8/O3.8.
49 Li X, Diao D, Bhushan B. Fracture mechanisms of thin amorphous carbon films in nanoindentation. Acta Materialia, 45, 4453–4461; 1997. DOI: 10.1016/S1359-6454(97)00143-2.
50 Li X, Bhushan B. A review of nanoindentation continuous stiffness measurement technique and its applications. Materials Characterization, 48(1), 11–36; 2002. DOI: 10.1016/S1044-5803(02)00192-4.
51 Kazmanli MK, Rother B, Urgen M, Mitterer C. Identification of cracks generated by indentation experiments in hard-coating systems. Surface and Coatings Technology, 107(1), 65–75; 1998. DOI: 10.1016/S0257-8972(98)00578-7.

52 Hainsworth SV, Chandler HW, Page TF. Analysis of nanoindentation load-displacement loading curves. Journal of Materials Research, 11, 1987–1995; 1996. DOI: 10.1557/JMR.1996.0250.

53 Elizalde MR, Sánchez JM, Esnaola JMM, et al. Interfacial fracture induced by cross-sectional nanoindentation in metal–ceramic thin film structures. Acta Materialia, 14(51), 4295–4305; 2003. DOI: 10.1016/S1359-6454(03)00256-8

54 Toparli M, Sasaki S. Evaluation of the adhesion of TiN films using nanoindentation and scratch-testing. Philosophical Magazine A, 10(82), 2191–2197; 2002. DOI: 10.1080/01418610208235729.

55 Li W, Siegmund T. An analysis of the indentation test to determine the interface toughness in a weakly bonded thin film coating – substrate system. Acta Materialia, 52, 2989–2999; 2004. DOI: 10.1016/j.actamat.2004.02.046.

56 Xia SM, Gao YF, Bower AF et al. Delamination mechanism maps for a strong elastic coating on an elastic-plastic substrate subjected to contact loading. International Journal of Solids and Structures, 44, 3685–3699; 2007. DOI: 10.1016/j.ijsolstr.2006.10.009.

57 Syed-Asif SA, Wahl KJ, Colton RJ. Nanoindentation and contact stiffness using force modulation with a capacitive load-displacement transducer. Review of Scientific Instruments, 5(70), 2408–2413; 1999. DOI: 10.1063/1.1149769.

58 Bouzakis K-D, Michailidis N, Erkens G. Thin hard coatings stress-strain curve determination through a FEM supported evolution of nanoindentation test results. Surface and Coatings Technology, 142–144, 102–109; 2001. DOI: 10.1016/S0257-8972(01)01275-0.

6

Nanoindentation of Macro-porous Materials for Elastic Modulus and Hardness Determination

Zhangwei Chen

Department of Earth Science and Engineering, Imperial College London, UK

6.1 Introduction

Nanoindentation of homogeneous and elastic-perfectly plastic dense solids is well developed [1]. However, its application in the mechanical characterization of porous materials is not in the same way. Porous materials exhibit very different mechanical behaviours under indentation due to the presence of porous microstructures. For this reason, the true elastic modulus and hardness of a porous material cannot be determined by one-off indentation tests simply at a fixed depth (or load), based on the unloading part of the response curve using the Oliver–Pharr method [2], in which way a fully dense material can be easily characterised. The measurement of mechanical properties of a porous material using nanoindentation requires a series of tests at varied depth and further analytical work.

6.1.1 Nanoindentation Fundamentals for Dense Materials

The nanoindentation technique was first developed in the mid-1970s to measure the hardness of small volumes of material [3]. Significant work has been carried out by Oliver and Pharr [2] with an improved analysis method (i.e. the Oliver–Pharr method) for determining elastic modulus and hardness from the indentation load-displacement data. The technique is also known as depth-sensing indentation and uses very small scale (submicron) tip indenters and control of load and displacement, with high spatial resolution to place the indenters, and provides real-time load-displacement data. Therefore, nanoindentation is an ideal technique to investigate the mechanical properties of small volume materials at under-submicron scales.

Despite the existence of various commercially available apparatuses, the main functional components of a nanoindentation machine are the indenter tip, the sensors and actuators used to apply and measure the load and displacement. The former is usually either a Berkovich tip made of diamond with highly precise three-sided pyramid geometry, or a spherical tip made of diamond or sapphire. Note that compared to sharp indenter tips, such as the Berkovich tip, spherical tips facilitate the distinction from elastic to plastic deformation of materials during indentation due to their less drastic variation of stress (or stress singularity) under loading and hence the unlikelihood of plasticity induced immediately upon contact [4]. The spherical tip is particularly

Applied Nanoindentation in Advanced Materials, First Edition. Edited by Atul Tiwari and Sridhar Natarajan.

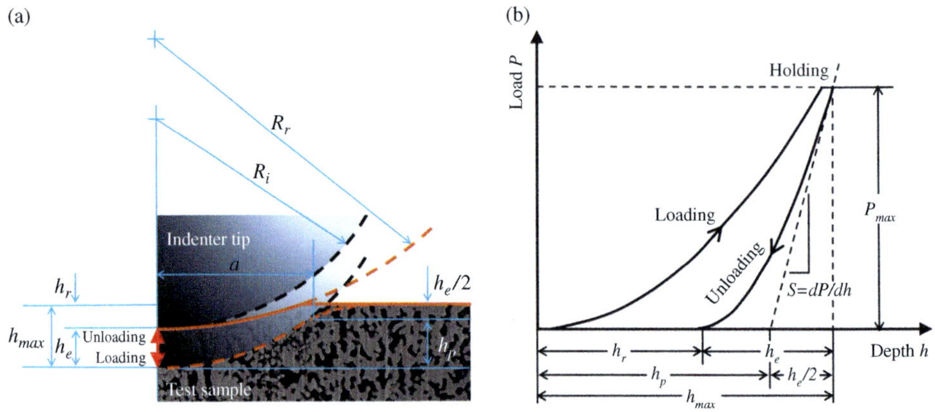

Figure 6.1 (a) Profile and geometries of sample under loading and unloading with a holding period at peak load P_{max} using a rigid spherical indenter tip with radius of R_i; (b) response curve of load versus indent depth and the corresponding geometries. Notice that there is a residual depth h_r with a cross-sectional profile radius R_r.

appropriate for porous materials as the deformation zone can be arranged to be of much greater length scale than the typical length scale of the porous microstructure, e.g. the average pore diameter [5]. They thus can give a result that characterises the long range properties of the porous material. Other benefits of using spherical tips include less sensitivity to surface condition and more accurate resultant hardness [6].

During a typical nanoindentation process, mechanical load and penetration depth (i.e. displacement) are recorded as the indenter tip is pressed into the test material's surface with a predefined loading/unloading rate. A record of these values can be plotted to create a load-displacement curve, which is often referred to as the *P*-*h* curve, such as Figure 6.1(b) depicts. The elastic moduli and hardness of the test material are calculated by applying the Oliver–Pharr method [2] based on the *P*-*h* data obtained. Note that a typical measurement also involves two periods of dwell time at peak load and at the end of unloading, respectively, depending on the testing samples, to accommodate the creep effect and thermal drift. Figure 6.1(a) shows an example of a porous sample's geometry during nanoindentation, assuming that an elastic process takes place with a recovery depth of h_e upon unloading.

In Figure 6.1(b), P_{max} is the peak indentation load, h_{max} is the indenter depth at peak load, h_r is the residual depth of the impression after unloading and S is the initial unloading stiffness, i.e. the slope of the unloading part of the curve ($\mathrm{d}P/\mathrm{d}h$). According to the Oliver–Pharr method, the indentation hardness H (i.e. physically the effective contact pressure) and the reduced elastic modulus E_r of the sample are given by:

$$H = \frac{P_{max}}{A} \tag{6.1}$$

$$E_r = \beta\frac{\sqrt{\pi}}{2}\frac{S}{\sqrt{A}} \tag{6.2}$$

where A is the projected contact area related to the contact depth h_p, and β is a geometry correction factor that depends on the shape of the indenter.

Regarding Equations (6.1) and (6.2), for a Berkovich indenter tip $\beta = 1.034$ and the area function is $A = 24.5h_p^2$, where h_p is the contact depth. Hence Equation (6.2) for the Berkovich tip becomes:

$$E_r = 0.328\,\frac{S}{h_P} = 0.328\frac{1}{h_P}\frac{dP}{dh}\bigg|_{P=P_{max}}^{\text{unloading}} \tag{6.3}$$

For a spherical indenter tip, $\beta = 1$ and $A = \pi a^2$, where a is the radius of the contact circle at maximum depth h_{max} and is given by: $a = \sqrt{2R_i h_P - h_P^2}$, where $h_P = h_{max} - \frac{h_e}{2}$ with reference to Figure 6.1. Hence for the spherical tip the equations become:

$$H = \frac{P_{max}}{\pi a^2} \tag{6.4}$$

$$E_r = \frac{1}{2a}\frac{dP}{dh}\bigg|_{P=P_{max}}^{\text{unloading}} \tag{6.5}$$

The Young's modulus E_s of the sample can then be calculated from the reduced modulus E_r of the indenter-sample system through the following relationship [2]:

$$\frac{1}{E_r} = \frac{1-v_i^2}{E_i} + \frac{1-v_s^2}{E_s} \tag{6.6}$$

where, the subscripts i and s indicate respectively a property of the indenter material and the test sample. A sensitivity study [7] showed that the indentation results depend little on the variation of Poisson's ratio, i.e. no more than 8% of change in sample elastic modulus when varying the Poisson's ratio in the range of 0.2 to 0.4 in the calculation.

6.1.2 Introduction to Porous Materials

Classical porous materials are commonly found in biological systems and natural minerals, including bones and rocks. The development of a great number of industrial porous materials, which are mainly based on ceramics, polymers and other composites, and are in both bulk and film forms, is striving to fulfil demanding requirements of various growing applications, such as for mechanical damping, insulation, catalysis, microelectronics, membranes and bioengineering. As an example, Figure 6.2 shows the porous microstructure features of the top surface and cross-section of an as-sintered cathode material used in SOFCs [5]. It is of great importance to estimate the durability of such materials, which are subject to mechanical stresses arising from environmental instability caused by external mechanical loading and constraints, temperature gradients, flux pressure gradients and thermal expansion coefficients difference.

Pore characteristics (such as pore volume fraction, pore size, pore shape, connectivity and load bearing area) have significant influence on the mechanical properties of the materials. Pore size can be categorised into the nano-scale (<2 nm) and up to the macro-scale (>50 nm). Different fabrication method would result in very different porous features. For example, powder-based processing is the dominant mode of manufacturing on ceramics. It generally involves the consolidation of raw powders by sintering at elevated temperatures which leads to desired porosity for obtaining either key application parameters which increase with increasing porosity such as

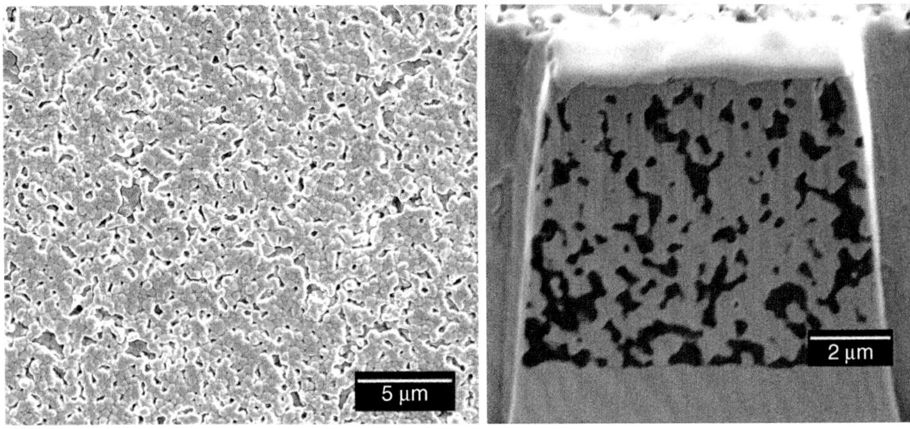

Figure 6.2 Top surface and cross-sectional SEM images of a SOFC cathode film with a highly porous microstructure. *Source*: Chen 2013 [5]. Reproduced with permission of Elsevier.

surface area, or better mechanical performance which can be improved by the opposite (i.e. lower porosity with increased sintering degree) such as strength and elastic modulus. Fabrication methods other than partially sintering of powder, e.g. melt casting, sintering with pore former additives, colloidal processing and lamination etc., would often result in much different microstructure architectures (such as tubular pores resulting from injection moulding and laminar character by rapid prototyping [8]) so that the microstructure dependence of elastic properties would vary significantly. In this case, nanoindentation might not be applicable or the measurement would no longer be reliable, and hence such microstructures with extremely irregular pore shapes are out of the scope of this chapter and are not addressed here.

6.1.3 Studies of Elastic Properties of Porous Materials

In order to characterise the mechanical properties of porous materials, the understanding of their typical elastic properties is a prerequisite. Such properties depend largely on the microstructural parameters of the specimens. Relevant aspects include porosity and geometrical characteristics of the pores, solid phase and the inter-particle necks. Their influences have been widely studied [9–14], despite the fact that the property-microstructure correlation is very complex and there may not have precise definition of some of the features due to the complexity of real microstructures where pores or particles are highly interconnected and vary within a given body so that challenges arise when distinguishing the beginning and end of individual pore/particle [15].

In order to correlate porosity to properties of porous materials, a variety of modelling approaches have been developed, including mainly the following three categories, and most of which treated the porous solids as two-phase microstructures with the second phase being pores in terms of estimating the effective elastic modulus [10, 11].

Empirical and Semi-empirical Methods: A number of equations, either empirical or semi-empirical, have been developed based on fitting of experimentally measured data, generally with the porosity being the main variable accompanied by one or two material-dependent fit constants, such as zero-porosity theoretical elastic modulus. More recently a much thorough overview was given by Pabst *et al.* [16] on most modulus-porosity relations ever proposed, of which the theoretical background, structure and merit were systematically discussed and compared. Examples of these commonly used expressions include the linear relationship first developed by Fryxell and Chandler [17] for polycrystalline BeO modulus-porosity analysis; a simple exponential relationship proposed by Spriggs [18] for fitting experimentally measured modulus of ceramics; and some other non-linear expressions, such as the Coble-Kingery non-linear relation [10] for solids containing isolated pores, the Phani-Niyogi power-law relation [19, 20] to better describe the modulus-porosity data of porous solids over a wide range of porosity and a non-linear relation proposed by Hasselman [21] to overcome the drawback of Spriggs form's resulting modulus not being zero when porosity reaches 100%. However, most of the relations have been rigorously derived only for porous materials with dilute distribution of spherical pores and few of them were related to non-porosity parameters of the microstructures. Moreover, although the models above could be useful in the way of predicting properties within the range of reasonable extrapolation, the understanding of the mechanisms affecting properties was limited.

Mechanistic/micromechanics Based Methods [22]: This category includes the differential [23] and self-consistent methods [24, 25], which focus more on rigorous evaluation of the specific mechanisms, e.g. mechanics, for the porosity effects on the specific property. They generally extend experimentally measured data for microstructures containing a small porosity to higher porosities. Although more insight into mechanisms can be gained by these models, their predictive capabilities are quite limited due to the rigorously idealised the pore characters of spheroidal shapes (e.g. spherical or ellipsoidal) with unspecified stacking. They typically can only treat a very limited number of cases in a narrow way which is quite different to that for real microstructures.

Geometry-based Methods: Particularly those based on the minimum solid area (MSA) [26, 27] by purely geometrical reasoning to approximate the porosity dependence of elastic moduli using load-bearing concepts, of the form $E/E_0 = e^bP$, where E_0 is the dense solid elastic modulus, b is a geometrical constant and P the porosity. However, the expression applies strictly to microstructures with stacking of spherical pores or particles although it has general prediction ability over a range of porosities. In addition, the common drawback of approaches described in the last two categories is that the microstructure corresponding to a particular formula/prediction is not definitely known. Therefore, agreement or disagreement with data can neither confirm nor reject a particular model [28].

Besides these methods mentioned above, the elastic properties could also be precisely estimated, in theory, by computationally solving the equations of elasticity for numerical microstructure models [29]. However, in practice the accuracy of results is greatly limited by large statistical variations and insufficient resolution, which also need further investigation.

Moreover, as mentioned above, there exists a large amount of literature dealing with porosity dependence of elastic modulus of porous materials and results reported show that the ultimate significance of the dependence of modulus on porosity is self-evident. Particularly the elastic modulus of a two-phase (solid/pore) porous microstructure relies primarily on the pore volume fraction of the microstructure. Detailed description on the use of these formulas to estimate/predict elastic modulus of porous solids is out of the scope of this chapter. But rather, the concern is more specifically that how the porous microstructures behave differently under nanoindentation. In particular, the

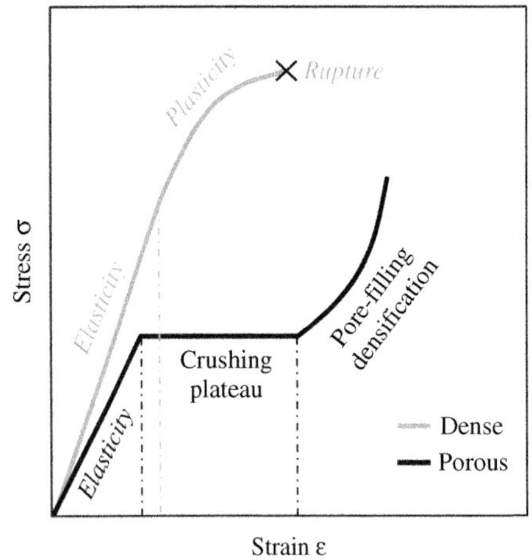

Figure 6.3 The idealised stress-strain curve for indentation of brittle porous solids, compared with that of a dense solid.

effect of porosity on the elastic behaviour of partially sintered ceramic films has been rarely studied. Recent studies of porous films for microelectronic applications using nanoindentation with Berkovich indenters have been reported [30, 31]. However, their methodology is not suitable for typical partially sintered ceramic films with larger pore sizes. Moreover, the effect of porosity variation on elastic modulus and hardness was not considered in those studies. Their films were very smooth, flat and ductile polymers having low porosity (<30%) and extremely small pores in the nm range. Therefore a sharp Berkovich indenter was able to deform a representative volume of the porous films. However, partially sintered ceramic films often feature a relatively rough surface and have pores in the micron range and porosity as high as 50%. Furthermore they are not ductile and have a low elastic limit. The scale of their microstructures requires the use of a blunt indenter in order to deform a representative volume of material in the film. The brittleness also implies a densification mechanism by crushing, which is significantly different from the ductile deformation of polymer films.

Finnie and Patel [32] and Wilsea *et al.* [33] have proposed the idealised stress-strain curve of brittle porous solids under indentation, as shown in Figure 6.3 (with comparison to the schematic curve of fully dense solids) and the idealised behaviour can be described below [33]: 1. Linear elasticity exhibits in a small strain range; 2. Crushing of particle networks (as yielding) occurs beneath the indenter when a critical stress is reached; 3. The stress keeps approximately the same as the crushing of the particle networks proceeds (for a short time); 4. As the crushed particles filled the neighbouring pores, strain hardening takes place with fast increase in compressive stress.

6.2 Nanoindentation of Macro-porous Bulk Ceramics

Porous bulk ceramics are basically ceramics which are partially sintered from powders. The mechanical properties show highly dependency on the porosity obtained by various degree of sintering, as described in the previous section. Conventional mechanical testing methods can be readily applied to determine the mechanical properties of such porous bulk ceramics provided that they can be easily prepared in the desired forms such as bar and rod shapes to meet the testing requirements [13, 34]. In the case when the material is comparatively expensive in cost and only a small amount of material is available or the conventional testing shapes are difficult to fabricate in the lab-scale, nanoindentation can be particularly useful.

However, the effective mechanical properties extraction of porous ceramic materials using nanoindentation has been little studied. As mentioned earlier, porous materials behave mechanically very differently under indentation as compared with fully dense materials due to the existence of porous microstructures. The plastic or permanent deformation mechanisms in porous and dense materials are basically very different as contribution from pore-filling densification would be expected during the nanoindentation into the porous materials. This is due to the networks of bonded ceramic particles being fractured at the necks by the pushing indenter and consolidating eventually, as also observed by many researchers for the indentation of highly porous rocks [35], silica foams [36], and ceramics used in energy technology [5].

Although the Oliver–Pharr method for the calculation of nominal elastic modulus and hardness at each individual depth (or load) stays valid for the indentation of porous bulk ceramics, measurements at a fixed depth (or load) do not suffice for the determination of the effective mechanical properties. Instead, a series of indentation tests with increasing depth (or load) must be performed so that the measured mechanical properties as a function of the varied indentation depth (or load) can be plotted. In this way, the extrapolation of the resulting data points to zero depth can then be executed and the representatively effective mechanical properties are determined with minimal influence from densification.

The recent work performed by Chen *et al.* [5] aimed to understand how the mechanical properties of the ceramic bulk samples ($La_{0.6}Sr_{0.4}Co_{0.2}Fe_{0.8}O_{3-\delta}$, denoted as LSCF) vary with the change of porosity (5%–45%, as a result of different sintering temperatures) as well as other microstructural factors. A spherical indenter was approved more suitable for nanoindentation tests as less sensitivity to surface roughness and pore features was achieved. Densification under the indenter was found to have less influence on the measured elastic modulus than on hardness. The correlation between mechanical properties and porosity was made. Figure 6.4 [5] shows extrapolated lines for the determination of effective mechanical properties of the LSCF bulk samples sintered at different temperatures.

Interestingly, whilst the common elastic sink-in phenomenon [37] (as shown in Figure 6.5) usually found at the edge of the indentation for dense materials appears evident in the porous materials under nanoindentation, the plastic pile-up seems to be rare. Important effect on the calculation of indentation contact area arises when significant pile-up or sink-in takes place, as the former can underestimate the area and the later does the opposite [38, 39]. It is believed that the irreversible nonvolume conserving deformation as a result of pore-filling densification can be a reason responsible for the unlikelihood of pile-up in the indentation of porous materials.

The indentation size effect (ISE), namely the indentation hardness increase with decreasing indentation depth particularly in the sub-micron depth regime, which is well-known for indentation of dense metals [40–42], was not found either in the above study. The greatly smooth nanoindentation response curve shown in Figure 6.6 [5] for the 45% porosity LSCF bulk ceramic sample suggested that discontinuities did not always exhibit in the load-displacement curves for high porosity indentation, and hence

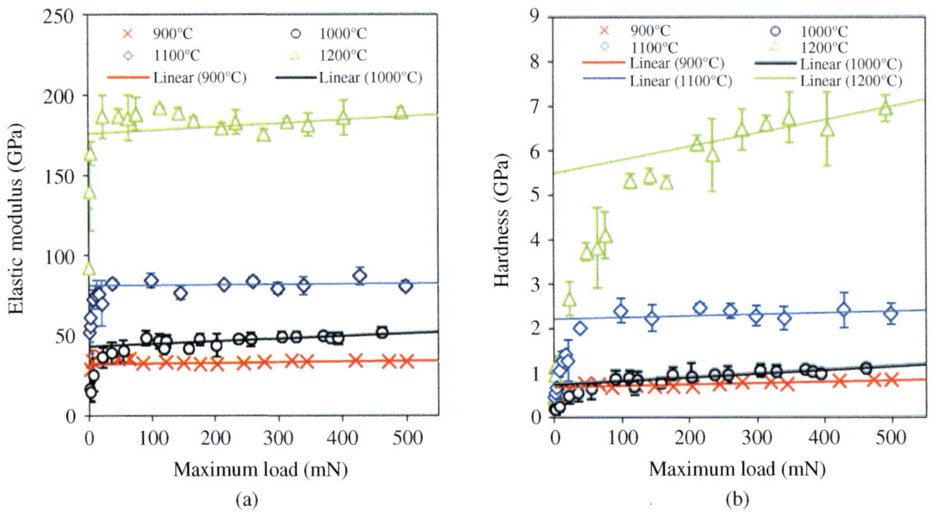

Figure 6.4 Elastic modulus and hardness vs. indentation load for bulk LSCF samples after sintering at 900–1200°C. The solid lines represent extrapolations to zero load. *Source*: Chen 2013 [5]. Reproduced with permission of Elsevier.

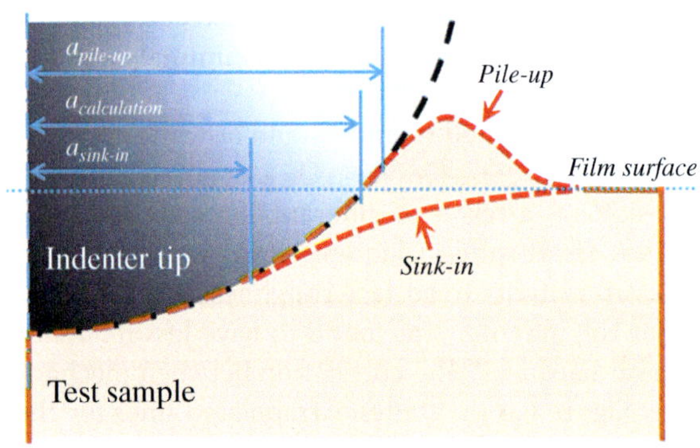

Figure 6.5 Schematic of half-space cross-sectional view of the effect of separated event of pile-up and sink-in on the actual contact area for nanoindentation on a porous bulk material using a spherical indenter.

the de indentation-induced cracking, crushing, densification etc. cannot be readily revealed only by the curve feature.

Using Berkovich nanoindentation, Alao *et al.* [43] has shown the dependency of mechanical properties on the applied loading rate. The pore-filling densification was shown to contribute to the discontinuities on the response curves at high loading rates.

However, most of other studies on porous bulk ceramics used nanoindentation for the characterization of the selective dense part of the samples. For example, Sun *et al.* [44] used nanoindentation to probe solid bridges separating pores in highly porous (i.e. 43%) Ti_3SiC_2 samples. Not surprisingly, the resulting data were qualitatively similar to that of the fully dense bulk. Kaul *et al.* [45] prepared porous SiC with wood precursors and also measured the solid phase elastic modulus and hardness using Berkovich nanoindentation. The porosity, however, was used to assess the elastic modulus by applying conventional elastic modulus and porosity fitting formula with little attention being paid to the evaluation of the nanoindentation behaviour.

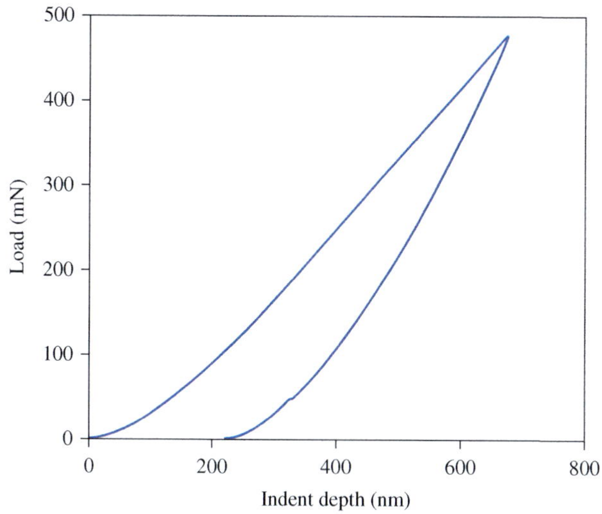

Figure 6.6 A typical nanoindentation curve of a highly porous bulk LSCF sample with a porosity of 45%. *Source*: Chen 2013 [5]. Reproduced with permission of Elsevier.

Some indentation studies have focused on microindentation with larger indentation load and displacement. Latella *et al.* [46] studied the contact damage and fatigue conditions in porous and dense liquid-phase-sintered alumina using spherical indentation. Results showed that increasing porosity induced a transition from an essentially brittle to a quasi-plastic response in the damage mode upon spherical indentation. More recently, Clément *et al.* [4] reported an inverse analysis to validate the spherical microindentation data of highly porous plaster materials, but no further detailed correlation between the microstructures and mechanical data was obtained.

6.3 Nanoindentation of Bone Materials

Unlike the abovementioned porous bulk ceramics fabricated by partial sintering, bone is a naturally growing organ which contains mainly a matrix of mineralised collagen fibrils and pores and has heterogeneous but hierarchical structures. As a load-bearing living component, bone serves to support weight and locomotive muscular activities [47]. Therefore, probing its mechanical properties is particularly essential to clinical purposes.

The elastic properties, hardness and anisotropy of individual bone components at the tissue level with varying microstructures have been studied using nanoindentation [48–52], mostly with sharp Berkovich indenters. These proved the technical feasibility and showed that indentation properties may be quite different from one bone type to another and the values obtained from macroscopic tests, such as bending, are often lower. While some viscoelastic-plastic models [53–55] have been used for Berkovich or spherical indentation, Oliver–Pharr method [2] remains the essential approach for extracting elastic-plastic properties. Bushby *et al.* [56] discussed the concern of possible sensitivity of nanoindentation to the many minute disturbances at the surface and near surface of the bone samples when they are prepared, such as surface roughness, swelling, leaching of bone mineral in solutions and hydration. The nanoindentation data could have significant variation due to the different samples and testing conditions even the starting bone remained the same. Direct comparisons across the different studies seemed to be unlikely. Very recently, Rodriguez-Florez *et al.* [57] conducted a thorough study to evaluate the mechanical properties of the same bones (both wet and dry, and after different embedding processes) using different testing and analysis methods, and thus a direct comparison was made possible. The latest advancement [58, 59] also focused on the incorporation of other experimental techniques to map the nanoindentation data so that the variation of local tissue compositions and the change in mechanical properties can be correlated.

Despite the wide application of nanoindentation in the mechanical characterization of bones, one of the main problems not having been adequately addressed in the above literature is the effect of the porous microstructures on the indentation response. In these studies nanoindentation was basically used as for measuring dense bulk materials, tests only at a fixed depth (or load) were carried out to obtain statistically averaged results. Generally for a highly porous material, a noticeable variation of the measured effective mechanical properties can take place when the indentation depth is changed due to the densification effect of the porous microstructure. Therefore, since the depth-dependent indentation behaviours were neglected, the validity of the results remains suspicious if the study of effective mechanical properties of the porous microstructures was an intention.

6.4 Nanoindentation of Macro-porous Films

The determination of mechanical properties of films using nanoindentation is prevailing due to the advantage that only small volume is required (at under submicron scales) and removal of films from substrates is not needed [60].

Although as mentioned earlier, nanoindentation has been widely used in the mechanical characterisation of films, for porous films deposited on dense substrates the methodology to accurately determine and thus interpret the intrinsic film-only properties has not been well established due to the uncertainties caused by experimental errors [31]. Unlike porous bulk ceramics which are generally free-standing, porous films are typically deposited on dense substrates. As a result, besides the pore-filling densification effect arising from the porous microstructures under indenter (which does not appear in fully dense films), the main difficulty in the nanoindentation of a porous film deposited on a substrate is to minimise the influence from the substrate. Moreover, further surface treatments are sometimes not allowed for the films, thus the surface roughness may also contribute to the uncertainty of the measurements. Figure 6.7 illustrates a plot of measured elastic modulus of a film with 40% porosity deposited on a dense substrate. The trend of the data points with the increase of the indentation depth to thickness ratio (h_{max}/t_f) indicates the essential effects arising from (1) the surface roughness, (2) the pore-filling densification and (3) the substrate. Generally the roughness effect would lead to underestimation of the indentation results and the other two would cause overestimation.

Research is still required to tackle such challenging sources of error, i.e. the influence of the presence of substrates [61, 62], the densification effect upon indentation [63] and

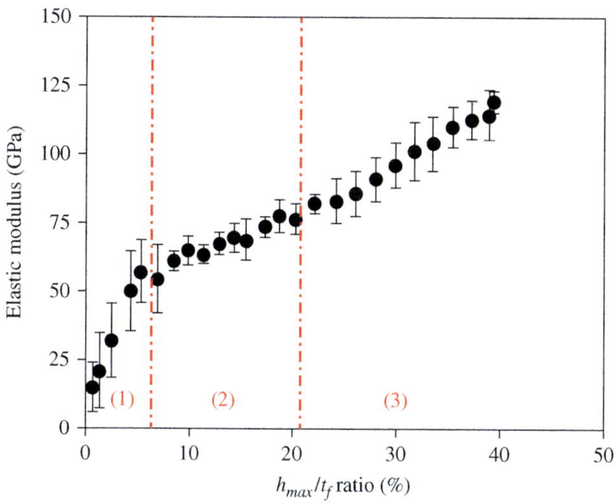

Figure 6.7 Elastic modulus measured using nanoindentation on the LSCF film with 40% porosity deposited on a dense CGO substrate. *Source*: Chen 2013 [5]. Reproduced with permission of Elsevier.

surface roughness of films [64], although extensive effort has been made to understand them [30, 65–67]. In this section, the details are discussed.

6.4.1 Substrate Effect

For a porous film, the elastic modulus measured with an indenter is therefore the long range composite modulus which is affected by its porous microstructure. Indenting to a maximum depth less than 10% of the thickness of a film (namely $h_{max}/t_f < 0.1$) has been empirically considered as a safe condition to avoid effects from substrate and extract intrinsic film properties in routine nanoindentation tests [2], as the substrates could significantly contribute to the contact stiffness of the measurements. Because the elastic deformation field under the indentation normally scales up with the actual indent size and may extend towards the interface with the substrate which then hinders the further extension and results in an overestimation of the modulus if the substrate is stiffer than the film. Xiang *et al.* [31] reported that the effect could be as large as 20%, even if the indentation depth was less than 5% of the film thickness. Although the 'one tenth' rule has no theoretical basis the exact range may be very different depending on the properties of the film and substrate under investigation. It was shown by Chen *et al.* [5] that the effect of the substrate could be neglected for indentation even up to 20% of the film thickness for the porous thin ceramic films under investigation, as Figure 6.8 [5] depicts. However, indenting less than 5% of the film thickness will face the effect arising from film surface roughness.

There are a range of film-substrate systems to consider but typically the discussion in this section focuses particularly on compliant films deposited on stiff substrates (i.e. soft films on hard substrates). Several existing models developed to account for the substrate effect in the nanoindentation measurements often require either the determination of weight factors or the precise prerequisites of the film or substrate properties, such as film thickness, film Poisson's ratio, substrate modulus and Poisson's ratio etc., which are

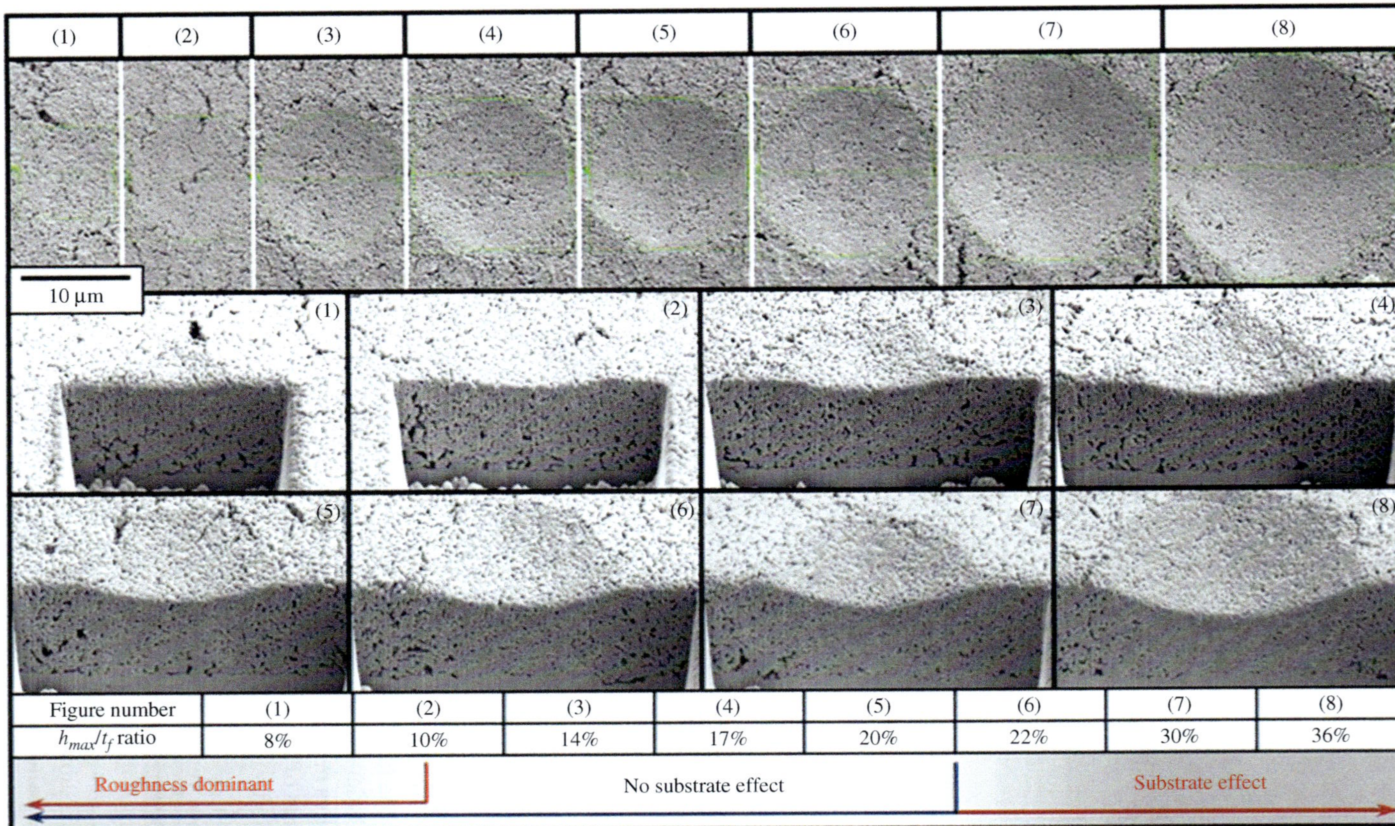

Figure number	(1)	(2)	(3)	(4)	(5)	(6)	(7)	(8)
h_{max}/t_f ratio	8%	10%	14%	17%	20%	22%	30%	36%

Figure 6.8 Top surface and cross-sectional SEM images of indents corresponding to maximum indentation depths from 800 to 3600 nm (i.e. 8% < h_{max}/t_f < 36%), with film being LSCF and substrate being dense CGO. *Source*: Chen 2013 [5]. Reproduced with permission of Elsevier.

Table 6.1 Expressions proposed to account for the substrate effect in the determination of the intrinsic elastic modulus of films deposited on substrates

Oliver and Pharr [2]	$\frac{1}{E_r} = \frac{1-v_i^2}{E_i} + \frac{1-v_f^2}{E_f}$	(6.7)
Doerner and Nix [68]	$\frac{1}{E_r} = \frac{1-v_i^2}{E_i} + \frac{1-v_f^2}{E_f}\left(1-e^{-\frac{\alpha t}{h_{eff}}}\right) + \frac{1-v_s^2}{E_s}\left(e^{-\frac{\alpha t}{h_{eff}}}\right) + b_s$	(6.8)
King [69]	$\frac{1}{E_r} = \frac{1-v_i^2}{E_i} + \frac{1-v_f^2}{E_f}\left(1-e^{-\frac{\alpha t}{a}}\right) + \frac{1-v_s^2}{E_s}\left(e^{-\frac{\alpha t}{a}}\right)$	(6.9)
Saha and Nix [66]	$\frac{1}{E_r} = \frac{1-v_i^2}{E_i} + \frac{1-v_f^2}{E_f}\left(1-e^{-\frac{\alpha(t-h)}{a}}\right) + \frac{1-v_s^2}{E_s}\left(e^{-\frac{\alpha(t-h)}{a}}\right)$	(6.10)
Gao *et al.* [70]	$\frac{1}{G_r} = \frac{1-I_0}{G_s} + \frac{I_0}{G_f};\ E_f = 2G_f(1+v_f)$	(6.11)
Song [71]	$\frac{1}{G_r} = \frac{(1-v_s)(1-v_f)-I_0}{1-[(1-I_1)v_f+I_1v_s]}\left[\frac{(1-I_0)}{G_s}+\frac{I_0}{G_f}\right];\ E_f = 2G_f(1+v_f)$	(6.12)
Hay and Crawford [72]	$\frac{1}{G_r} = \frac{1-I_0}{G_s+FI_0G_f} + \frac{I_0}{G_f};\ E_f = 2G_f(1+v_f)$	(6.13)

not so simple to know. Doerner and Nix [68] proposed an empirical formula based on the indentation depth to film thickness ratio as well as a weight factor, which was only applicable to their own samples described in the paper. King [69] then subsequently modified the formula but still a non-constant weight factor was needed. Later Saha and Nix [66] also improved the King's expression which however resulted in relatively large scatters in the estimation of substrate effect. Some other complex analytical models have been developed based on the work by Gao *et al.* [70] which was originally derived for shear modulus determination with two functions I_0 and I_1 being used to govern the transition in elastic properties from film to substrate and no further adjustable factor was required. The expression was then extended by Song [71] to be more applicable. Very recently, Hay and Crawford [72] have improved the above models by assuming the film and substrate as springs linked in series for assessing the influence of compliant substrates on stiff films. Nevertheless, the proposed expression still requires the knowledge of substrate mechanical properties. The abovementioned expressions are listed in the Table 6.1, compared with the original Oliver–Pharr expression for dense bulk materials.

Note that in this table the symbols in the formulas are typically: E for elastic modulus, v for Poisson's ratio, h_{eff} is effective indentation depth, α is an adjustable parameter, t is the film thickness, G is shear modulus, I_0 and I_1 are weight functions detailed in their respective literature source papers, and the subscripts r, i, f and s represent the symbols for reduced, indenter, film and substrate, respectively.

6.4.2 Densification Effect

Although as mentioned in the introduction section, a number of studies have been carried out to link the elastic modulus with the presence of porosity, the pore-filling densification effect on the nanoindentation measurement of elastic modulus and hardness yet lacks systematic investigation. Studies [5, 30, 73] have revealed the increase of both

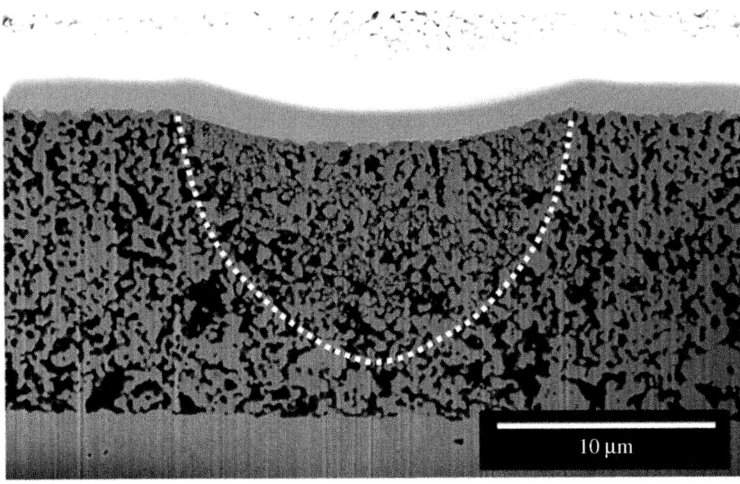

Figure 6.9 Cross-sectional SEM image of the porous LSCF film after indentation, showing crushing densification zone. Note that in the porous film black is pore and grey is the particle networks.

measured elastic modulus and hardness of the test films with the increase of indentation depth, suggesting that the film-only elastic and plastic mechanical properties cannot be directly derived using the classic formulas derived by Oliver and Pharr [2] due to the significant influence of local densification of the porous materials under indentation. As found by Chen *et al.* [5] in the nanoindentation of porous ceramic film (LSCF) deposited on dense substrate (CGO), the film after spherical indentation showed clear permanent densification zone confined in a parabolic shape. As shown in Figure 6.9, it can be readily seen that the densification mechanism was induced by crushing of the particle networks and filling of pores in the porous microstructure, compared to the intact region outside the parabolic curve.

Fleck *et al.* [63] have studied the indentation densification effect of porous solids with known modulus and porosity using FEM and cavity expansion models. Results suggested the indentation pressure increased with decreased initial porosity. By contrast, Chen *et al.* [30] carried out an inverse analysis based on FEM using the Gurson's model [74] to extract the intrinsic mechanical properties from Berkovich nanoindentation of porous materials. They found that densification has a significant impact on the indentation hardness value, while insignificant on indentation modulus. However, in addition to the acquisition of prior knowledge of substrate mechanical properties, the proposed method further required both simulation work and deduction from empirical hardness and modulus mapping, which appeared too much complicated. The applicability of this method in the use of other shapes of indenters might be highly questionable due to their not all being self-similar, compared to the Berkovich indenter used in their study.

It was believed that the morphology (shape, orientation and size) of the interconnected particles and pores in the densified zone can also have significant impact on the determination of the mechanical properties but a quantitative assessment seemed very difficult due to the complexity of the many factors involved [75].

6.4.3 Surface Roughness Effect

Another problem may arise when using the commonly employed 'one tenth' rule for nanoindentation is the effect of surface roughness. The original intention for this rule was to indent films with flat and smooth surfaces [2], or at least the surface roughness effect was not a concern. Ideally, for reliable measurements of the elastic modulus and hardness of the films, uniform thickness with flat and smooth surface should be a pre-requisite. While for a rough film, particularly when the surface roughness is comparable to the indenter size, such an effect must be taken into account.

Here, a case studied by Chen *et al.* [5, 76] of the surface roughness effect on the mechanical property measurement using spherical nanoindentation is described. It can be seen from the measurement result shown previously in Figure 6.7 that, due to the presence of surface roughness the scatter for shallow indent depths less than 10% of the film thickness is particularly large. These shallow indentation depths were comparable to the grain size (270 nm) and the surface roughness (R_a > 402 nm for the film) and resulted in variability in the initial stages on indentation. A cross section through a shallow indent is shown in Figure 6.10 together with the schematic of the spherical indenter tip where it contacts the surface. It can be seen that h_{max} in this case is of similar magnitude to the surface roughness and that the contact is at asperities (outlined in red curves) on the specimen surface. The stresses at these points are much greater than for ideal contact between two smooth surfaces and the displacements are greater.

Due to the facts presented above, surface roughness might have two main effects on the measured results. One was the underestimation of modulus at shallow indentation depth, as shown in Figure 6.7. This underestimation was due to the overestimated contact area, provided that the indentation depth was comparable to the roughness, as also reported by other studies [77, 78]. The measured modulus increased with the indentation depth as the surface roughness effect gradually diminished when an increasing number of the untouched valleys close to the asperities on the surface under compression were touched by the indenter tip. Such effect can be eliminated in data analysis by simply discarding the indentation depths smaller than a critical value, for which the result are less reliable, as suggested by Mencik and Swain [79].

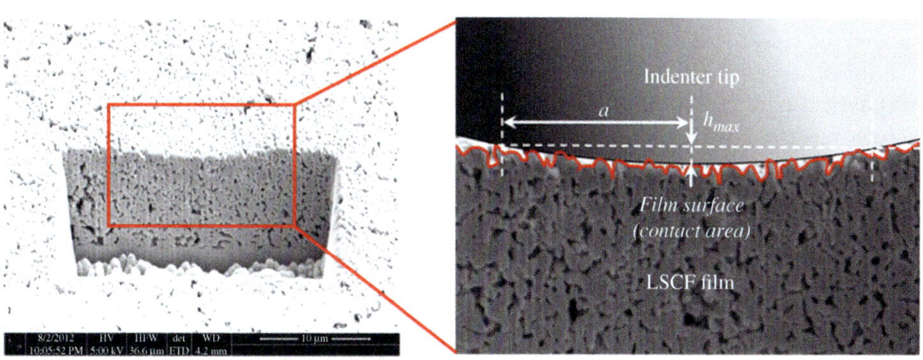

Figure 6.10 Effect of surface roughness on the real contact area with the indenter tip, leading to errors in indentation results. Here the sample shown was sintered at 1000°C.

Therefore, for the very films studied by the authors, indenting deeper than 10% of the film thickness was suggested to be a prerequisite to generate reasonable values close to the film-only properties. Besides, surface roughness could also be responsible for the scatter in the measured data over the whole range of indentation depth as shown in Figure 6.7. Nevertheless, the standard deviations indicate that for the indentation depth in the range of 10–20% of the film thickness, the elastic modulus data were more consistent and reproducible than that for the other depths. As found by the authors in the paper [5], this was particularly true for films sintered at 900 and 1000°C. While for the films sintered at 1100 and 1200°C, the errors were much larger, which were attributed to the much coarser surface features after sintering at higher temperature.

As surface roughness has significant effect on the measured mechanical properties, either a model accounting for the roughness or a larger indenter and deeper indentation should be considered to minimise the surface roughness effect. The influence of the average roughness on the standard deviation of the elastic modulus was studied at different depths for different materials by Quasmi and Delobelle [80] using Berkovich nanoindentation. The relative standard deviation of the elastic modulus was shown to link with the surface roughness to indentation depth ratio via the following expression Equation (6.14), which is similar to the work of Bobji *et al.* [81]:

$$\frac{\sigma_E}{E_{mean}} = \beta\left(\frac{R_{ms}}{h}\right)^n \tag{6.14}$$

where E_{mean} is the average elastic modulus, σ_E is its standard deviation, R_{ms} the average roughness, h indentation depth, β and n are material dependent constants. The study suggested that indenting to a depth ten times larger than the average roughness using a Berkovich indenter could hopefully generate modulus result close to the intrinsic value within an error of 7.5%. Jiang *et al.* [64] investigated the effect of surface roughness on the nanoindentation measured mechanical properties of copper films. They have found that the surface roughness influenced significantly both modulus and hardness of rough thin films and in order to eliminate such an influence, indentation depth much larger than the characteristic size of the surface roughness should be considered. In addition, the authors suggested that indenters with sufficiently large sizes should be used to lessen the sensitivity to the roughness.

On the other hand, the measurement results reported by Chen *et al.* [5] showed that the spherical nanoindentation was less sensitive to the local microstructural features (such as surface roughness and pore size) indicated a clear dependence of measured elastic modulus on the indentation depth, which was largely controlled by the combined effects from surface roughness, densification and substrate. A very simple approach (although empirical) was used by the author to account for the substrate and densification effects for the determination of the film-only nanoindentation modulus and hardness of the various porous films sintered at different temperatures. The approach simply extrapolated the series of resulting data points deduced by the Oliver–Pharr method at different indentation depths to zero indentation depth (or load), by eliminating the data points which are obviously dominated by the effects of both surface roughness and substrate. Very good agreement was reached when comparing the results with the elastic moduli computed for the real 3D microstructures with varied porosities reconstructed using focused ion beam (FIB) tomography. Based on such a comparative examination, the authors suggested that the combination of thorough nanoindentation analyses with FEM simulations of real microstructures could be a much suitable approach for the determination of mechanical properties of the porous thin films.

6.5 Concluding Remarks

Nanoindentation is a powerful tool to characterise the mechanical properties of porous samples in the form as both porous bulk and thin films deposited on substrates. However, when high porosity is included in the films, measures must be taken to account for some challenging issues, such as pore-filling densification effect on the measured properties. For porous thin films deposited on substrates, extra care and efforts should be made to minimise the effects arising from surface roughness and the substrates. In order to extract the detailed film-only mechanical properties a series of nanoindentation tests at different depths (or load) must be conducted. Reasonable values should fall in the comparatively constant range of the measured results as a function of indentation depth (or load). The results affected by surface roughness and substrate should be avoided before the extrapolation to zero indentation depth is performed to eventually account for the pore-filling densification effect. In this way, the reliable determination of the film-only properties can thus be achieved.

Acknowledgements

The author is grateful to Prof. Alan Atkinson and other colleagues in Imperial College London for their assistance.

References

1 Fischer-Cripps AC. Nanoindentation, 3rd edn. New York: Springer; 2011.
2 Oliver WC, Pharr GM. An improved technique for determining hardness and elastic modulus using load and displacement sensing indentation experiments. Journal of Materials Research, 7(6),1564–1583; 1992.
3 Bulychev S, Alekhin V, Shorshorov M, *et al.* Determining Young's modulus from the indentor penetration diagram. Industrial Laboratory, 41(9),1409–1412; 1975
4 Clément P, Meille S, Chevalier J, Olagnon C. Mechanical characterization of highly porous inorganic solids materials by instrumented micro-indentation. Acta Materialia, 61(18):6649–6660; 2013.
5 Chen Z, Wang X, Bhakhri V, Giuliani F, Atkinson A. Nanoindentation of porous bulk and thin films of $La_0.6Sr_0.4Co_0.2Fe_0.8O_{3-\delta}$. Acta Materialia, 61(15), 5720–5734; 2013.
6 Swain MV, Menčík J. Mechanical property characterization of thin films using spherical tipped indenters. Thin Solid Films, 253(1–2), 204–211; 1994.
7 Mesarovic SD, Fleck NA. Spherical indentation of elastic–plastic solids. Proceedings of the Royal Society of London Series A: Mathematical, Physical and Engineering Sciences, 455(1987), 2707–2728; 1999.

8 Reed JS. Introduction to the Principles of Ceramic Processing. Chichester: Wiley; 1988.
9 Rice RW. Porosity of Ceramics: Properties and Applications. CRC Press; 1998.
10 Coble R, Kingery W. Effect of porosity on physical properties of sintered alumina. Journal of the American Ceramic Society, 39(11), 377–385; 1956.
11 Dean EA, Lopez JA. Empirical dependence of elastic moduli on porosity for ceramic materials. Journal of the American Ceramic Society 66(5), 366–370; 1983.
12 Herakovich C, Baxter S. Influence of pore geometry on the effective response of porous media. Journal of Materials Science 34(7), 1595–1609; 1999.
13 Hardy D, Green DJ. Mechanical properties of a partially sintered alumina. Journal of the European Ceramic Society 15(8), 769–775; 1995.
14 Green DJ, Nader C, Brezny R. Elastic behaviour of partially-sintered alumina. In: Sintering of Advanced Ceramics Proc Symposium on Sintering of Advanced Ceramics Cincinnati, Ohio, 2–5 May 1988, pp. 345–356; 1988.
15 Rice R. Porosity of ceramics. In: (ed.) Marcel Dekker, Porosity of ceramics. New York; 1998.
16 Pabst W, Gregorová E, Tichá G. Elasticity of porous ceramics—A critical study of modulus-porosity relations. Journal of the European Ceramic Society 26(7), 1085–1097; 2006.
17 Fryxell R, Chandler B. Creep, strength, expansion, and elastic moduli of sintered BeO as a function of grain size, porosity, and grain orientation. Journal of the American Ceramic Society 47(6), 283–291; 1964.
18 Spriggs R. Expression for effect of porosity on elastic modulus of polycrystalline refractory materials, particularly aluminum oxide. Journal of the American Ceramic Society 44(12), 628–629; 1961.
19 Phani KK, Niyogi SK. Elastic modulus-porosity relation in polycrystalline rare-earth oxides. Journal of the American Ceramic Society 70(12), C-362-C-366; 1987.
20 Phani K, Niyogi S: Young's modulus of porous brittle solids. Journal of Materials Science 1987, 22(1):257–263.
21 Hasselman D. On the porosity dependence of the elastic moduli of polycrystalline refractory materials. Journal of the American Ceramic Society 45(9), 452–453; 1962.
22 Hashin Z. Analysis of composite materials: a survey. Journal of Applied Mechanics 50(3), 481–505; 1983.
23 McLaughlin R. A study of the differential scheme for composite materials. International Journal of Engineering Science 15(4), 237–244; 1977.
24 Budiansky B. On the elastic moduli of some heterogeneous materials. Journal of the Mechanics and Physics of Solids 13(4), 223–227; 1965.
25 Hill R. A self-consistent mechanics of composite materials. Journal of the Mechanics and Physics of Solids 13(4), 213–222; 1965.
26 Rice RW. Evaluation and extension of physical property-porosity models based on minimum solid area. Journal of Materials Science 31(1):102–118.1996.
27 Rice R. Comparison of physical property-porosity behaviour with minimum solid area models. Journal of Materials Science 31(6), 1509–1528; 1996.
28 Roberts AP, Garboczi EJ. Elastic properties of model porous ceramics. Journal of the American Ceramic Society 83(12), 3041–3048; 2000.

29 Poutet J, Manzoni D, Hage-Chehade F *et al.* The effective mechanical properties of random porous media. Journal of the Mechanics and Physics of Solids 44(10), 1587–1620; 1996.
30 Chen X, Xiang Y, Vlassak JJ. Novel technique for measuring the mechanical properties of porous materials by nanoindentation. Journal of Materials Research 21(03), 715–724; 2006.
31 Xiang Y, Chen X, Tsui TY, Jang J-I, Vlassak JJ. Mechanical properties of porous and fully dense low-κ dielectric thin films measured by means of nanoindentation and the plane-strain bulge test technique. Journal of Materials Research 21(02), 386–395; 2006.
32 Finnie I, Patel M. Structural features and mechanical properties of rigid cellular plastics (Rigid cellular plastics mechanical properties based on model assuming pentagonal dodecahedron cell form). Journal of Materials 5, 909–932; 1970.
33 Wilsea M, Johnson KL, Ashby MF. Indentation of foamed plastics. International Journal of Mechanical Sciences 17(7), 457–460; 1975.
34 Nanjangud SC, Brezny R, Green DJ. Strength and Young's modulus behavior of a partially sintered porous alumina. Journal of the American Ceramic Society 78(1), 266–268; 1995.
35 Suarez-Rivera FR, Cook NGW, Cooper GA, Zheng Z. Indentation by pore collapse in porous rocks. In: Proceedings of the 31st US Symposium on Rock Mechanics (USRMS), pp. 671–678, American Rock Mechanics Association; January 1990.
36 Toivola Y, Stein A, Cook RF. Depth-sensing indentation response of ordered silica foam. Journal of Materials Research 19(01), 260–271; 2004.
37 Bolshakov A, Pharr G. Influences of pileup on the measurement of mechanical properties by load and depth sensing indentation techniques. Journal of Materials Research 13(4), 1049–1058; 1998.
38 Tsui T, Oliver W, Pharr G. Influences of stress on the measurement of mechanical properties using nanoindentation: Part I. Experimental studies in an aluminum alloy. Journal of Materials Research 11(03), 752–759; 1996.
39 Bolshakov A, Oliver W, Pharr G. Influences of stress on the measurement of mechanical properties using nanoindentation: Part II. Finite element simulations. Journal of Materials Research 11(03), 760–768; 1996.
40 Stelmashenko N, Walls M, Brown L, Milman YV. Microindentations on W and Mo oriented single crystals: an STM study. Acta Metallurgica et Materialia 41(10), 2855–2865; 1993.
41 Ma Q, Clarke DR. Size dependent hardness of silver single crystals. Journal of Materials Research 10(04), 853–863; 1995.
42 Nix WD, Gao H. Indentation size effects in crystalline materials: A law for strain gradient plasticity. Journal of the Mechanics and Physics of Solids 46(3), 411–425; 1998.
43 Alao A-R, Yin L. Nano-scale mechanical properties and behavior of pre-sintered zirconia. Journal of the Mechanical Behavior of Biomedical Materials 36, 21–31; 2014.
44 Sun Z, Murugaiah A, Zhen T *et al.* Microstructure and mechanical properties of porous Ti3SiC2. Acta Materialia 53(16), 4359–4366; 2005.
45 Kaul VS, Faber KT. Nanoindentation analysis of the elastic properties of porous SiC derived from wood. Scripta Materialia 58(10), 886–889; 2008.

46 Latella BA, OConnor BH, Padture NP, Lawn BR. Hertzian contact damage in porous alumina ceramics. Journal of the American Ceramic Society 80(4), 1027–1031; 1997.
47 Kim D-G, Elias KL. Elastic, viscoelastic, and fracture properties of bone tissue measured by nanoindentation. In: Handbook of Nanomaterials Properties. Springer, pp.1321–1341; 2014.
48 Rho J-Y, Tsui TY, Pharr GM. Elastic properties of human cortical and trabecular lamellar bone measured by nanoindentation. Biomaterials 18(20), 1325–1330; 1997.
49 Turner CH. The elastic properties of trabecular and cortical bone tissues are similar result from two microscopic measurement techniques.Journal of Biomechanics 32, 437–441; 1999.
50 Zysset PK, Edward Guo X, Edward Hoffler C, *et al.* Elastic modulus and hardness of cortical and trabecular bone lamellae measured by nanoindentation in the human femur. Journal of Biomechanics 32(10), 1005–1012; 1999.
51 Chowdhury S, Thomas V, Dean D *et al.* Nanoindentation on porous bioceramic scaffolds for bone tissue engineering. Journal of Nanoscience and Nanotechnology 5(11), 1816–1820; 2005.
52 Tricoteaux A, Rguiti E, Chicot D *et al.* Influence of porosity on the mechanical properties of microporous β-TCP bioceramics by usual and instrumented Vickers microindentation. Journal of the European Ceramic Society 31(8),1361–1369; 2011.
53 Olesiak SE, Oyen ML, Ferguson VL. Viscous-elastic-plastic behavior of bone using Berkovich nanoindentation. Mechanics of Time-Dependent Materials 14(2), 111–124; 2010.
54 Isaksson H, Nagao S, MaŁkiewicz M *et al.* Precision of nanoindentation protocols for measurement of viscoelasticity in cortical and trabecular bone. Journal of Biomechanics 43(12), 2410–2417; 2010.
55 Menčík J, He LH, Swain MV. Determination of viscoelastic–plastic material parameters of biomaterials by instrumented indentation. Journal of the Mechanical Behavior of Biomedical Materials 2(4), 318–325; 2009.
56 Bushby A, Ferguson V, Boyde A. Nanoindentation of bone: Comparison of specimens tested in liquid and embedded in polymethylmethacrylate. Journal of Materials Research 19(01), 249–259; 2004.
57 Rodriguez-Florez N, Oyen ML, Shefelbine SJ. Insight into differences in nanoindentation properties of bone. Journal of the Mechanical Behavior of Biomedical Materials 18, 90–99; 2013.
58 Oyen ML, Ferguson VL, Bembey AK *et al.* Composite bounds on the elastic modulus of bone. Journal of Biomechanics 41(11), 2585–2588; 2008.
59 Kavukcuoglu N, Patterson-Buckendahl P, Mann A. Effect of osteocalcin deficiency on the nanomechanics and chemistry of mouse bones. Journal of the Mechanical Behavior of Biomedical Materials 2(4), 348–354; 2009.
60 Oliver WC. Measurement of hardness and elastic modulus by instrumented indentation: Advances in understanding and refinements to methodology. Journal of Materials Research 19(1), 3; 2004.
61 Chen X, Vlassak JJ. Numerical study on the measurement of thin film mechanical properties by means of nanoindentation. Journal of Materials Research 16(10), 2974–2982; 2001.

62 Bhattacharya AK, Nix WD. Analysis of elastic and plastic deformation associated with indentation testing of thin films on substrates. International Journal of Solids and Structures 24(12), 1287–1298; 1988.

63 Fleck NA, Otoyo H, Needleman A. Indentation of porous solids. International Journal of Solids and Structures 29(13), 1613–1636; 1992.

64 Jiang W-G, Su J-J, Feng X-Q. Effect of surface roughness on nanoindentation test of thin films. Engineering Fracture Mechanics 75(17), 4965–4972; 2008.

65 Tsui TY, Vlassak J, Nix WD. Indentation plastic displacement field: Part I. The case of soft films on hard substrates. Journal of Materials Research 14(06), 2196–2203; 1999.

66 Saha R, Nix WD. Effects of the substrate on the determination of thin film mechanical properties by nanoindentation. Acta Materialia 50, 23–38; 2002.

67 Bouzakis K-D, Michailidis N, Hadjiyiannis S *et al.* The effect of specimen roughness and indenter tip geometry on the determination accuracy of thin hard coatings stress–strain laws by nanoindentation. Materials Characterization 49(2), 149–156; 2002.

68 Doerner MF, Nix WD. A method for interpreting the data from depth-sensing indentation instruments. Journal of Materials Research 1(04), 601–609; 1986.

69 King R. Elastic analysis of some punch problems for a layered medium. International Journal of Solids and Structures 23(12), 1657–1664; 1987.

70 Gao H, Chiu C-H, Lee J. Elastic contact versus indentation modeling of multilayered materials. International Journal of Solids and Structures 29(20), 2471–2492; 1992.

71 Song H. Selected mechanical problems in load and depth sensing indentation testing. PhD thesis, Rice University, Houston, USA; 1999.

72 Hay J, Crawford B. Measuring substrate-independent modulus of thin films. Journal of Materials Research 26(06), 727–738; 2011.

73 Volinsky AA, Vella JB, Gerberich WW. Fracture toughness, adhesion and mechanical properties of low-k dielectric thin films measured by nanoindentation. Thin Solid Films 429(1), 201–210; 2003.

74 Gurson AL. Continuum theory of ductile rupture by void nucleation and growth: Part I—Yield criteria and flow rules for porous ductile media. Journal of Engineering Materials and Technology 99(1), 2–15; 1977.

75 Vanstreels K, Wu C, Gonzalez M *et al.* Effect of pore structure of nanometer scale porous films on the measured elastic modulus. Langmuir 29(38), 12025–12035; 2013.

76 Chen Z, Wang X, Giuliani F, Atkinson A. Surface quality improvement of porous thin films suitable for nanoindentation. Ceramics International 40(3), 3913–3923; 2014.

77 Rico A, Garrido MA, Otero E, Rodriguez J. Roughness effect on the mechanical properties of ceramic materials measured from nanoindentation tests. Key Engineering Materials 333, 247–250; 2007.

78 Walter C, Mitterer C. 3D versus 2D finite element simulation of the effect of surface roughness on nanoindentation of hard coatings. Surface and Coatings Technology 203(20), 3286–3290; 2009.

79 Mencik J, Swain M. Errors associated with depth-sensing microindentation tests. Journal of Materials Research 10(6), 1491–1501; 1995.
80 Qasmi M, Delobelle P. Influence of the average roughness Rms on the precision of the Young's modulus and hardness determination using nanoindentation technique with a Berkovich indenter. Surface and Coatings Technology 201(3), 1191–1199; 2006.
81 Bobji M, Biswas S. Estimation of hardness by nanoindentation of rough surfaces. Journal of Materials Research 13(11), 3227–3233; 1998.

7

Nanoindentation Applied to DC Plasma Nitrided Parts

Silvio Francisco Brunatto[1] *and Carlos Maurício Lepienski*[2]

[1] *Plasma Assisted Manufacturing Technology and Powder Metallurgy Group, Department of Mechanical Engineering, Universidade Federal do Paraná, Curitiba, Paraná, Brazil*
[2] *Department of Physics, Universidade Federal do Paraná, Curitiba, Paraná, Brazil*

7.1 Introduction

Cold plasmas [1–4] and their applications were strongly studied, researched and developed throughout the last century, and it seems to be also a basic tendency for the 21st century. This is confirmed nowadays by the intense use of direct current (DC) plasmas for materials processing, comprising different techniques, as for example, sputtering as a deposition process [5, 6], etching [7–9], PVD [10], CVD [11], sintering [12–14], surface texturing [15], cleaning [16], carburizing [17–19], besides nitriding [20–24]. The importance of the DC plasma nitriding technique in the metallurgical-mechanical field is shown by its very well-established use in industry, and different modes in promoting nitrogen surface alloying of metallic materials, as for example, in plasma nitriding itself [21, 23, 25], in post-discharge [3], in cathodic or active screen cage and/or in floating potential [26–29], in hollow cathode discharge [30, 31], or even in anode [32–34]. So, distinct nitriding mechanisms have been studied, which lead the part surfaces to present different characteristics and properties. In this entry special attention is given to the DC power supply nitriding, which is an important plasma excitation technique. The term plasma nitriding is generally used here referring to this case, i.e. for parts acting as cathode, anode or under floating potential, which are nitrided by using DC power supply. Other plasma excitation techniques using radio frequency, plasma immersion ion implantation (PIII), and/or microwave power supply are not considered here, as a whole.

Plasma nitriding, like other plasma-assisted thermochemical treatments, is typically a diffusion-controlled process [24]. In this case, the type of the (metallic) material to be treated, the mode by which nitriding is carried out, and the adequate choice of processing parameters are key points for the plasma-surface interaction, and thus for the final surface characteristics of DC plasma nitrided parts.

Different results for the nitrided layer growth kinetics have been obtained depending on the above mentioned variables. As verified for distinct thermochemical process and treated materials [17, 19, 24, 35, 36], incubation times (defined as the time necessary to lead the nitrided layer to be formed) can eventually be present. Regarding the original state of the part surface, the presence of stable native oxide can also infer

Applied Nanoindentation in Advanced Materials, First Edition. Edited by Atul Tiwari and Sridhar Natarajan.

important changes in nitriding process. In low-temperature plasma nitriding kinetics of martensitic stainless steel, as for example, incubation time has been assumed as the time necessary to remove the passive oxide layer from the surface of samples, which enables the nitrogen to diffuse into the substrate [24].

The original state of the part surface can be strongly modified during plasma nitriding treatment, especially for parts acting as cathode, as a result of the plasma species bombardment. So, the last aspect to be introduced here refers to two important phenomena in plasma nitriding, which are directly related to the plasma (fast) species bombardment, and thus the plasma-surface interaction, namely the part heating and sputtering in surfaces. As is known, heating is crucial to perform thermochemical treatments, since they are diffusion-controlled, as previously mentioned. In this case, depending on the type of material to be nitrided, composition gradients for nitrogen can be present all over the nitrided layer [37]. This is the typical case of plasma nitrided layers achieved in stainless steels, as will be discussed below. In addition, the crystallographic orientation of planes and directions present in the material surface can also infer important changes for anisotropic hardness and indentation modulus results, and thus on the monotonic mechanical behavior of the treated material, as verified in [38–41]. In addition, both the nitrogen solubility and diffusivity in iron alloys are strongly dependent on the crystal structure (lattice), as for example the b.c.c., and the f.c.c. iron structures [42]. On the other side, sputtering during plasma nitriding can lead finishing aspects of the parts to be modified, resulting in important changes in the roughness of the original surface. Therefore, all these above-mentioned aspects are to be discussed here, especially considering that the nanoindentation technique is strongly sensitive to surface aspects. In this entry, examples (case studies) comprising plasma nitriding of stainless steels, niobium, and titanium are considered, aiming to bring to the reader aspects and challenges of the nanoindentation technique applied to plasma nitrided parts characterization.

On the other hand, hardness tests, such as the Vickers and Knoop, have been the most used to determine the mechanical properties of nitride layers. They are very powerful techniques to investigate deep nitrided layers obtained by long-term nitriding, performed at high nitriding temperatures (on the range of 500–580°C, for carbon and low-alloy steels; and higher than 580°C for different metals, like niobium and titanium for example). It is to be noted that such conditions have shown to be very harmful, especially for the case of stainless steels nitriding. This is due to the characteristics of the obtained nitride layer, which is composed by different nitrides (iron nitrides and chromium nitrides), resulting in a very hard layer, of brittle behavior, and low resistance to corrosion. In order to overcome these problems, which is specific for stainless steels, long-term and high-temperature nitriding has been substituted by new process conditions, resulting in the termed 'low-temperature plasma nitriding' processes. Differently, in this case, shallow layers constituted of nitrogen-supersaturated solid solution phases are formed. Such relatively new processes on nitriding are made at low temperatures (on the range of 300–400°C), implying on small deep layers, for which traditional microhardness tests cannot correctly determine their hardness values. Similar result occurs for plasma nitrided parts of high N-affinity metals, like niobium, and titanium. Both these metals form very thin nitrogen-modified layers, as it is shown ahead, in the case studies, thus the mechanical behavior characterization of such materials, from conventional hardness tests, is not the most indicate testing procedure.

So, nanoindentation [43, 44] shows up to be the most suited technique to characterize surface hardness for all above-mentioned material cases, since it is a powerful technique to investigate mechanical properties at shallow depths on materials surface, like those attained in plasma nitriding of advanced materials.

Another point to be referred, before presenting the nanoindentation power for plasma nitrided parts, is related to the eventual presence of nitrogen concentration gradients along the nitrided layer. This result mainly occurs in the formation of nitrogen-expanded phases, like those observed in low-temperature plasma nitriding of stainless steels. It is to be remembered that, as the nitrogen diffuses into the original substrate, its concentration tends to be decreased from the surface in contact with the gas phase (plasma) into the bulk, resulting in nitrogen gradients distributed all over the nitrided layer, in the treated sample. In addition, nitrogen can also be present in the nitride precipitates form, for different phases and stoichiometries. All these aspects supposedly imply gradients for mechanical properties along the nitrided layer, depending on the measurement depth from the surface. In short, nanoindentation analysis (hardness and elastic modulus) only from the top surface is good enough to evaluate the behavior of nitrided complex layers subjected to external mechanical agents.

Regarding the nitrided surfaces, the most important aspects related to nanoindentation can be listed as follows:

- It is a very powerful technique to investigate mechanical properties of thin layers;
- It is an easy method to determine differences in mechanical properties induced by different plasma nitriding conditions, by comparing untreated to treated part surfaces;
- Gradients on mechanical properties can be easily investigated by evaluation of hardness and the elastic modulus via nanoindentation technique on cross-sections of the parts even for small depth modified layers; and
- The possibility to apply a large matrix of indentations to investigate the top surface at small penetration depths, by using low loads to evaluate anisotropic aspects modified by nitriding processes. This is a very important characteristic that should be explored for a good mechanical evaluation of the nitriding process, in function of the existent phases and different crystalline orientations at the surface grains (crystals). In this case, conjugation with other characterization techniques, like electron backscatter diffraction (EBSD), provides powerful synergy aiming to obtain the closest correspondence from diffusion and hardening (strengthening) processes promoted by the nitrogen-alloying in solid solution at the grains present in treated surface.

Additionally, surface roughness is perhaps the major problem against the universal application of nanoindentation to investigate the mechanical behavior of plasma nitrided layers, mainly for parts processed in cathode. As previously indicated, plasma nitrided surfaces tend to present increased roughness, which are caused by the plasma species bombardment of the part surface, leading the sputtering phenomenon to occur [21, 23, 45–47]. Surface roughness degradation has also been observed in studies using the surface etching effect caused by sputtering in glow discharge cathode surfaces [48]. Even in the active screen plasma nitriding of parts working at floating potential, slight surface morphology changes by sputtering and re-condensation mechanism can be present. In, this case iron nitride polygonal (nano) particles with normal distribution have been observed in the entirety of the nitrided sample surfaces [26, 49].

Other important surface finishing aspect that can result in difficulties to evaluate the relative hardening, attained by surface treatment, derives from the mechanical polishing of cross-sectioned surfaces, usually prepared by conventional metallographic techniques. Surface deformation caused by standard mechanical polishing is also common reason for the poor quality of the EBSD patterns [50], despite that for some metallic materials like tantalum, mechanical polishing using very fine Al_2O_3 can provide a surface smooth enough for high quality EBSD patterns [51, 52]. The standard mechanical polishing procedure generally induces the increase of the dislocation density in the surface of metals, considering that hardness is a measure of plastic deformation. All these above mentioned problems are treated in details ahead, at the case studies section.

In the next sections, the reader is introduced to topics comprising basic aspects of DC plasma nitrided parts. Theoretical aspects and the main characteristics of parts processed in linear glow discharges, and bases of the DC abnormal regime glow discharge, which is the regime used in plasma nitriding are also shown. Case studies are presented, for which the main surface characteristics related to nanoindentation characterization and its challenges, are treated and discussed, aiming to bring to the reader an overview on nanoidnentation technique applied to DC plasma nitrided parts.

7.2 Basic Aspects of DC Plasma Nitrided Parts

DC abnormal glow discharges for plasma nitriding purposes can be easily obtained in a discharge chamber, by using gas mixtures of $N_2 + H_2$ (eventually +Ar). The glow discharge (plasma) is achieved by applying a potential difference between electrodes placed in a low pressure gas environment. Charged particles (ions and free electrons present in gas) are accelerated by the obtained electric field. Successive collisions with neutral atoms and molecules lead the gas to be ionized. Electron impact ionization/excitation inelastic collisions originate a bright-aspect (glow) discharge, characterizing the formation of plasma [1–4]. The abnormal glow discharge, established according to the current-voltage characteristic, is important in plasma nitriding, since it makes possible to obtain homogeneous treatments in plasma-assisted materials processing, by covering completely all cathode surfaces. For additional details about abnormal glow discharge regime, the reader should see references [1, 15].

7.2.1 The Potential Distribution for an Abnormal Glow Discharge

The representation of a typical abnormal (linear) glow discharge is shown in Figure 7.1. In this case, the plasma (characterized by the glow region) is obtained between the cathode and the anode from an electric field imposed by an external plasma excitation source (usually a DC square waveform pulsed power supply). The cathode is negatively biased at a specified (supplied) potential V_{sup}, according to the power to be transferred to plasma, necessary to heat the part to nitriding temperature. The anode is grounded and presents a potential of 0 V. This electrode configuration is the same used for plasma nitriding of parts acting as cathode (see Figure 7.3a).

The plasma is approximately equipotential, and it is the most positive region of the discharge, presenting plasma potential (V_{pl}) [1]. Both the potential falls of the cathode and anode sheaths, thus the electrical fields in the considered system, tend to repel electrons and accelerate positive ions to the respective electrodes. Ions and fast neutrals

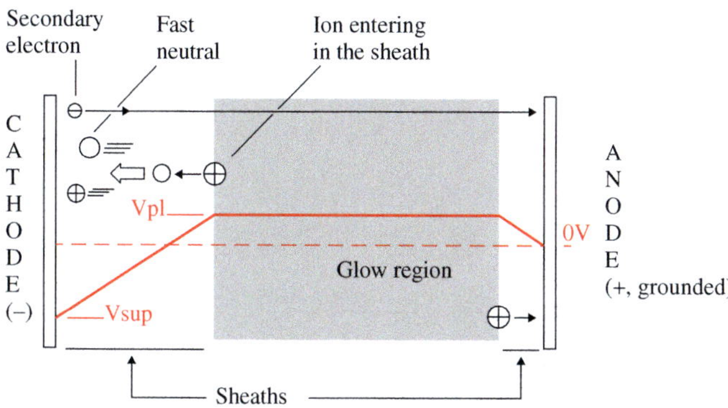

Figure 7.1 Schematic representation of the linear glow discharge in abnormal regime, and the potential distribution along the discharge (in red). Also shown is the charge change collision in the cathode sheath. *Source*: Chapman 1980 [1]. Reproduced with permission of John Wiley & Sons.

(produced by charge change collision in the cathode sheath) of nitrogen bombarding the part surface are mainly responsible for the heating and sputtering effects in the plasma-surface interface (see Figure 7.1). In charge change collision, ion entering (accelerated by the electrical field) in the sheath collides with slow neutrals, changes occur resulting in fast neutrals and slow ions (which are just accelerated by the remaining field) [1]. The higher the cathode fall (potential), keeping the other parameters unaltered, the higher is the bombarding energy of ions and fast neutrals colliding with the cathode surface, being the applied pulse voltage and the switched-on time (duty cycle) user-controlled parameters aiming to provide adequate plasma-surface interaction of the parts during pulsed DC plasma nitriding (in practice, referred to the part heating and sputtering phenomena).

7.2.2 Plasma-surface Interaction in Cathode Surface

Figure 7.2 shows the main events of the plasma-surface interaction, occurring in cathode surface, as result of the plasma species bombardment (ions and fast neutrals). These species bombarding the surface promote sputtering and heating, can be reflected, and can lead electrons and surface atoms to be torn.

The electrons torn from the original surface are strongly accelerated into the glow discharge by the electrical field present in the cathode sheath. Such electrons, termed 'secondary electrons' travelling across the glow region in direction to the anode are mainly responsible for the plasma (glow discharge) maintenance, by increasing the electron impact ionization/excitation collision rates with slow neutrals of the glow region. On the other hand surface atoms are torn, when an energy threshold is overcome, by the momentum transfer from plasma species bombarding the surface. This phenomenon, generically termed 'sputtering' [1], is present in different DC abnormal glow discharge processes, such as those aiming at the deposition of thin films in substrates (parts), as previously indicated [1, 5, 6]. Sputtering is also present in plasma nitriding and usually leads the roughness of part surface to be increased.

In plasma nitriding process, sputtering is directly related to the power transferred to plasma (thus leading the part to be heated and held at the nitriding temperature), and discharge parameters like the pulse voltage, gas mixture, and pressure. In generic terms, the sputtering phenomenon in plasma nitriding is increased by using higher pulse

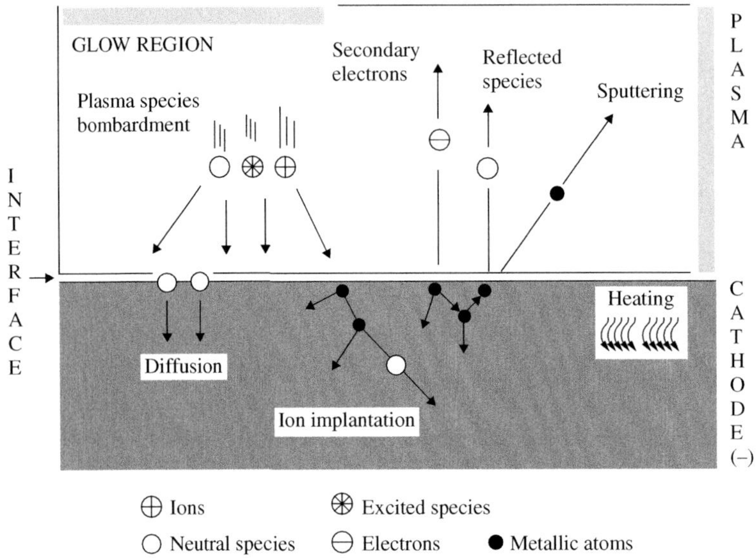

Figure 7.2 Plasma-surface interaction in cathode surface.

voltages, higher N_2 (and also Ar, if used) vol% than H_2 contents in gas mixture, and lower pressures. On the other hand, the part heating in plasma nitriding is intensified by using higher pulse voltages, higher N_2 (and also Ar, if used) vol% contents in gas mixtures, and higher pressures. The nitrided layer growth by nitrogen atoms diffusion into the part surface is dependent on the temperature attained in the part. Finally, for DC power supply, which is the subject of this entry, the power transferred to plasma is controlled by adjusting the pulse period, the pulse voltage, and the switched-on time (the duty cycle). For additional details of plasma-surface interaction, the reader should see references [1, 12, 13, 15, 24].

7.2.3 Electrical Configuration Modes in DC Plasma Nitriding

The great flexibility of the DC plasma nitriding technique is confirmed by the many modes by which the nitriding is performed, for different electrical configurations in the discharge. As previously explained, DC plasma nitriding can be carried out for parts acting as cathode, anode or under floating potential. The main differences are in the way by which the part is heated, the nitrogen-surface alloying occurs, and the energy is transferred to the part [12, 15].

The first nitriding mode considered in this entry is the plasma nitriding for parts acting as cathode, here termed *cathode configuration* (see Figure 7.3a).

In this case, heating to the nitriding temperature is only due to the part surface bombardment by the ions and fast neutrals (plasma species). This is the most used mode as for research purposes as for industrial applications. It is to be noted that in industry, the plasma nitriding furnace walls ('hot wall' discharge chamber) usually present heating components, such as electrical resistances. This design is used aiming to provide temperature homogenization all over the furnace loading, which is important for large scale production, and it can lead the sputtering effects to be reduced at the treated surface.

The second (termed *floating configuration*), and the third (*anode configuration*) nitriding modes considered in this entry occur for parts under floating potential (see

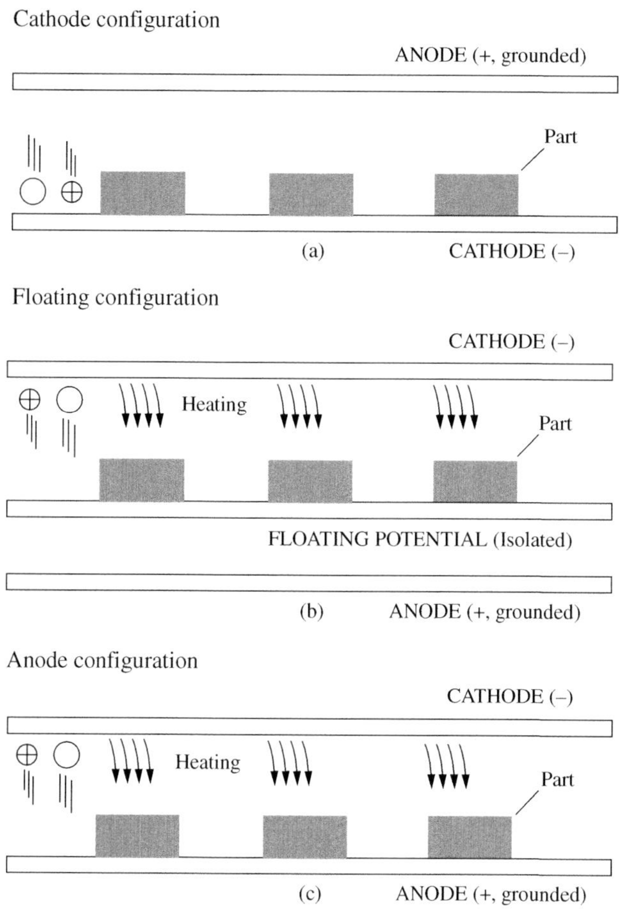

Figure 7.3 Schematic representation of plasma nitriding for part: (a) acting as cathode (cathode configuration); (b) under floating potential (floating configuration); and (c) acting as anode (anode configuration).

Figure 7.3b), and acting as anode (see Figure 7.3c), respectively. In both these cases, the part is not subjected to bombardment of ions and fast neutrals, thus sputtering tends to be suppressed. On the other hand, the part heating is due to the heat transferred by radiation from the cathode under plasma species bombardment [26–29, 32–34], or by means of the use of hot wall discharge chambers.

7.3 Basic Aspects of Nanoindentation in Nitrided Surfaces

Nanoindentation bases are now very well established and known [43, 44, 53, 54]. Recent achievements and new trends for the use of this nanomechanical analysis technique has comprised the characterization of nanostructured surfaces, and even nanoparticulate materials [55, 56]. Therefore, the presentation of full details about obtaining hardness (H) and elastic modulus (E) from force (load) versus displacement curves is not considered here. Notwithstanding, some important aspects to a better understanding by the reader are presented. At first, it is worth mentioning that the authors call nanoindentation testing, in this entry, not only the instrumented indentation performed up to 200 nm, but all instrumented indentation tests carried out even if penetration depths on the order of few micrometers are attained.

Mechanical characterization of nitrided layers from nanoindentation depends on a good understanding about the actual role of different variables, such as roughness, gradient properties, grain anisotropy, mechanical polishing, and existence of compressive stresses at the surface, besides others. All these aspects are important and need to be taken into account to select adequate testing parameters and limits, and fundamental, to perform a correct evaluation of nanoindentation data, aiming at a good mechanical properties characterization for nitrided parts.

Nanoindentation is based on the correct determination of the applied load and the displacement of the indenter into the sample surface as a function of time. In this case, the data of the force versus indenter displacement curve are processed in conjunction with mechanical–mathematical models, aiming to determine hardness and elastic modulus properties. Small loads, which can range from some μN up to hundreds of mN, are usually used. In this way, if the first (ideal) contact of the indenter with the sample is not correctly evaluated, all the other contact depth values will be incorrect. Different types of equipment use different methods to define the first contact from the penetration with the sample. For very flat surfaces presenting very low roughness, automated testing processes are good enough to determine adequate and reproducible reference zero depth. This condition is illustrated in Figure 7.4(a). An adequate zero depth reference is crucial for good properties evaluation in nanoindentaion techniques, considering the force versus indenter displacement curve, which are the data used in the Oliver and Pharr model [43], aiming to determine both properties. However, careful analysis and adequate corrections are necessary if rough surfaces are subjected to nanoindentation testing (see Figure 7.4b, c, d).

In the worst cases, useful values for hardness and/or elastic modulus can be impossible to obtain from automated processes, even by using machine algorithms, due to the large scattering on obtained data values.

As shown in Figure 7.4, different tip-surface approaches can be evidenced for different surfaces. Therefore, the correct determination of the first contact is something difficult for rough surfaces. The adequate first contact (relative to a real zero depth) does not always correspond to the first point at which the indenter touches the surface. Indenter sliding on surface asperities, as shown in Figure 7.4(c), can lead to a not valid first contact at the indenter penetration, resulting in an unreal increase of the penetration depth. Thus an inaccurate hardness/elastic modulus evaluation usually occurs due to the undesirable/unexpected indenter movement just after the indenter touches the rough surface, leading to an untrue indenter penetration.

On the other hand, indentations performed on the top region of asperities, as shown in Figure 7.4(b, and d), will correspond to higher penetrations than those obtained for a flat surface. For these cases, lower hardness/elastic modulus values will be obtained, since the mechanical behavior of the substrate surface is influenced by the uncommon deformation of the asperity compared to a flat surface. This effect also causes a decrease of both the properties measured by nanoindentation at shallow penetration depths. Contrarily, if the indenter touches two or more asperities at the same time, the effected

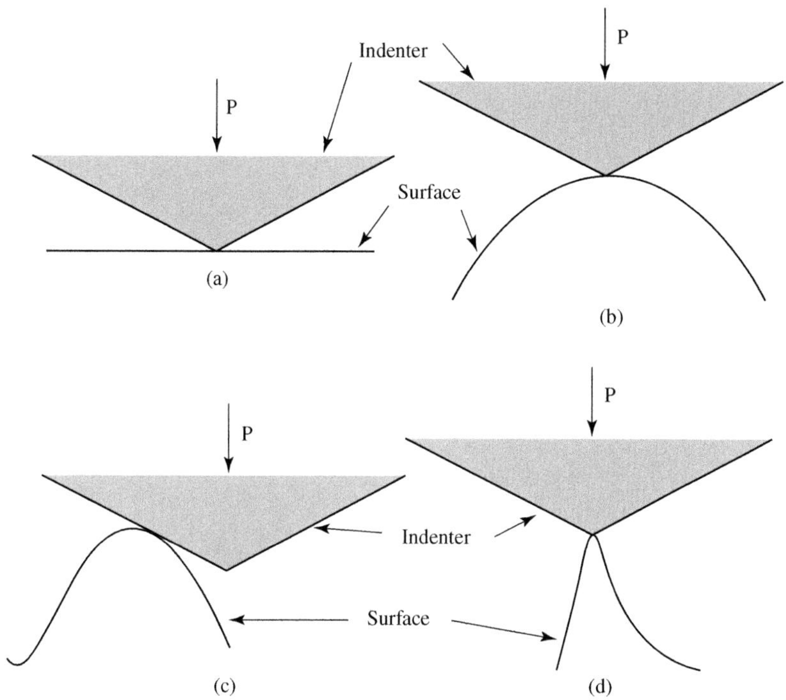

Figure 7.4 Different tip-surface approach: a) no asperities present (flat surface); b) asperities larger than tip diameter; c) tip sliding at the asperity; and d) asperity curvature diameter in the same order than tip diameter. *Source*: de Souza 2006. Reproduced with permission of Materials Reseach - Brazil..

penetration can be lower than that verified for a flat surface, and thus the obtained hardness value shall be higher than the expected.

It is to be noted that distinct techniques can be employed to surpass such difficulties. In some equipments, the first contact of the indenter with the sample surface is specified by analyzing the ratio $\Delta P/\Delta h$ (ΔP: load increment; and Δh: displacement increment), which is termed 'contact stiffness'. In this case, the first contact is defined when the $\Delta P/\Delta h$ ratio value surpasses a predefined value, during the indenter approximation, by applying small and few loadings in order to guide the indenter into the surface, as the contact stiffness is measured.

Alternative method has been proposed by de Souza *et al.* [57], by analyzing the load (P) versus displacement (h) data aiming to redefine an initial penetration depth (termed *zero depth*), considering the analysis of the contact stiffness. Since nitrided surfaces are very hard, the value of contact stiffness is high, as the true surface of the nitrided part is achieved. Then, due to the surface roughness, the proposed method consists in defining a new contact stiffness ($\Delta P/\Delta h$) limit, which is much higher than that defined by the equipment algorithm, thus making it possible to determine the new zero depth surface contact reference. In addition, considering that the load versus displacement curve should respect the relation $P = Ah^m$ (A, m: constants), for which the m value must be close to 2, according to Oliver and Pharr [43]. This procedure is illustrated in Figure 7.5.

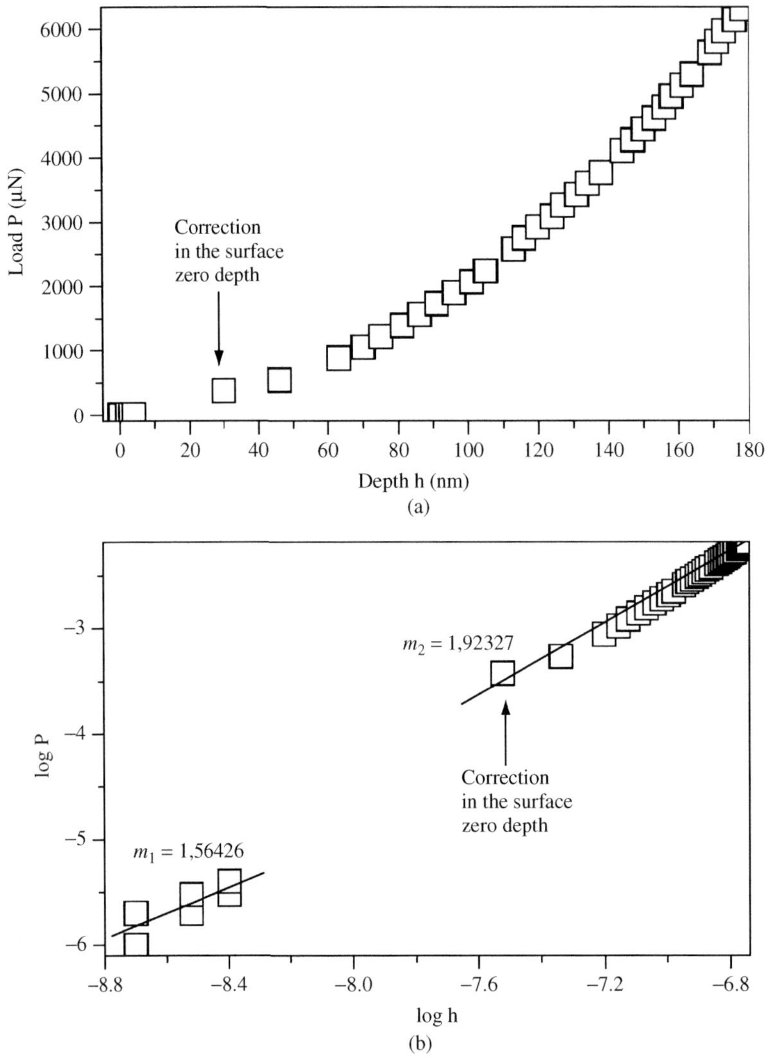

Figure 7.5 (a) Initial portion of the loading curve; and (b) the same curve linearized, for a nanoindentation test of the sample treated at 60% N_2/40% H_2 atmosphere and 600°C. The arrows indicate the same experimental point in both plots. *Source*: de Souza 2006. Reproduced with permission of Materials Reseach - Brazil.

As an example, it is indicated in Figure 7.5(a) the surface zero depth obtained by using the machine algorithm (relative to 0 nm depth in the x-axis), and the new surface zero depth obtained by using this alternative method, which is shifted to 30 nm depth (being the new zero depth reference). On the other hand, in Figure 7.5(b), one can see that the fitted curve by using the data obtained from this new zero reference presents a constant $m = 1.92$. So, the application of this procedure can transform useless data into comprehensible values for hardness and elastic modulus. In that work, comparative values of hardness obtained by standard application of the Oliver and Pharr method [43] are presented, considering small contact stiffness to define first contact from a new one based on these alternative approaches. The new values were compared with data obtained from Vickers hardness, which is based on image analysis, by using higher loads. The comparison has shown that the alternative method leads to consistent hardness values, being an important alternative procedure for characterization of rough nitrided surfaces. By applying this methodology, Souza and co-authors [57] were able to overcome the very high roughness attained in titanium surfaces nitrided in the 700–900°C temperature range, aiming to obtain most reasonable hardness values of Ti nitride, as presented later.

On the other hand, as previously commented, it is well known that mechanical polishing of metals can induce mechanical deformation, causing the dislocation density of the metal to be increased, or even martensitic transformation induced by deformation to occur in some steels, as a consequence of the work hardening at the first atomic layers or even at the first microns below the surface. Results obtained from nanoindentation tests on such surfaces show increase in hardness for shallow penetrations and hardness decrease for deep tip penetrations, as it is shown ahead in the case study for niobium parts. Evaluations for relative hardness increments achieved by comparing the mechanical properties obtained from treated and untreated materials are frequent in this case, if it is not carefully performed, leading to untrue and incorrect results. In untreated austenitic stainless steels, for example, hardness of mechanically polished surfaces at depths of 200 nm can be about 50% higher than that verified for higher penetrations, on the order of 1500–2000 nm. Adequate electrochemical or chemo-mechanical polishing, after mechanical polishing, should be used to remove the work hardened layers, thus leading the characterization of new surface to show hardness values much more close to the true substrate bulk hardness.

7.4 Examples of Nanoindentation Applied to DC Plasma Nitrided Parts

The glow discharge (plasma) parameters such as the gas mixture, gas flow rate, pressure, the pulse voltage, duty cycle, current, nitriding time and temperature can be easily modified and adjusted in accordance with the desired characteristics and properties of the treated surface. On the mechanical point of view, nitriding is usually employed aiming to improve the wear resistance (for different mechanisms like cavitation-erosion [20, 21], besides abrasion, adhesion, and others) of the treated material. Fatigue resistance is also generally increased, as a result of the compression stress state generated in the nitrided surface. In [58], for example, it is stated that the hard nitriding layer has effectively protected the surface from crack initiation, thus causing the cyclic properties to be improved. In addition, improvements in corrosion resistance can also be eventually achieved. But, in this case, besides the discharge parameters, the result is strongly dependent on the type of the part (substrate) material to be nitrided. For stainless steels [59–61], comprising an important class of high-alloy steels, the use of nitriding temperatures (T) (usually, for $T > 400$°C, associated with relatively long times) high enough to promote chromium carbide precipitation in nitrided layer tends to lead to strong detriment of the steel corrosion resistance, as a whole [21]. Differently, for parts manufactured from iron, carbon steels, and low-alloy steels [62], the nitriding application generally results in an increase in corrosion resistance. The other class of high-alloy steels, the tool ones [62], is also considered in this entry, in this case for plasma nitrided H13 tool steel parts [37].

Aspects related to the metal-nitrogen solubility, affinity, and diffusivity are also worth stressing here. In short, the behavior of the nitrided layer growth is strongly influenced by the metal type or alloying elements present in the substrate material. For the purpose of a better understanding, in generic terms, materials to be nitrided can be separated in three distinct classes, as follows:

- *Iron, carbon steels, and low-alloy steels*
  The nitriding process is typically based on the Fe-N equilibrium system. As both the nitrogen solubility and affinity in Fe matrix are relatively low, the nitrided case is usually constituted of a 'compound (or white) layer' + 'diffusion layer'. Depending on the discharge parameters, mainly the gas mixture ($N_2 + H_2$) composition used in plasma, the compound layer can present one or two iron nitride phases, in the case Fe_4N and/or $Fe_{2\text{-}3}N$. As an example, low N_2 vol% content gas mixtures (<5% N_2) can even cause the white layer formation to be avoided. Naturally, the diffusion layer depth is a function of the N diffusivity inside the iron matrix;
- *High-alloy steels*
  Nitriding aspects for this material class are separately treated, considering the two types of steels, in the case the stainless and tool steels. For the case of stainless steels (which present Cr contents higher than 12 wt% in solid solution) nitriding is mainly based on Cr-Fe-N ternary equilibrium system. Depending on the discharge parameters, the compound layer can be preferentially formed by N-expanded phases (i.e. α'_N = N-expanded martensite, and/or γ_N = N-expanded austenite, depending on the steel matrix [22, 24]) or even presenting chromium nitride phase (i.e. CrN [21, 24]). Due to the high Cr-N affinity and the presence of high Cr contents in such alloys, the obtained nitrided layers are comparatively thinner than those obtained for iron, carbon steels, and low-alloy steels, for similar treatment conditions. Questions to be answered aiming to reveal the morphology, phase structure, atomic ordering and chemical state of N-expanded phases have been the subject of different works [22, 63]. In a recent study of 304L austenitic stainless steel low-temperature plasma nitriding [63], the use of global and local probes advanced characterization techniques has made it possible for various authors to interpret that though the structure of the γ_N phase appears to be XRD homogeneous, it can be microscopically heterogeneous, acting as a heterogeneous highly strained material with nanometric CrN precipitates embedded in Fe_4N-like matrix.

  On the other hand, for the case of tool steels, decarburizing can occur in plasma nitrided surfaces, since the hydrogen species present in nitriding atmospheres, associated with the sputtering phenomenon, are prompt to react with the carbon present in steel. In low-temperature plasma nitriding of tool steels as shown in [37] for example, too thin ε-$Fe_{2\text{-}3}N$ layer has been formed at the outmost sample surface, showing increased diffusion layers for longer nitriding times, as expected for diffusion-controlled process; and finally,
- *High N-affinity metals*
  This is the case of niobium (Nb) [23, 64], and titanium (Ti) [57], as for example, which are also considered in present entry. The strongest N-affinity for both these metals,

related to the previous materials, result in nitrided layers (constituted of NbN and/or Nb_2N [23], and TiN [57] phases, respectively), which are comparatively much thinner than those obtained for Fe-based, and Fe-Cr-based alloys, for similar treatment conditions.

In sequence, examples of the use of nanoindentation technique are presented for different materials presenting different crystalline lattices (i.e., f.c.c., t.c.c., b.c.c., and h.c., such as austenitic stainless steel, martensitic stainless steel, niobium, and titanium, respectively), aiming to stress the main aspects and challenges of the application of nanoindentation to DC plasma nitrided parts.

7.4.1 Mechanical Polishing: Nanoindentation in Niobium

Nanoindentation of cross-sectioned surfaces of nitrided materials is a common procedure to investigate the nitrogen effect in depth. The surface finishing aspect for the nanoindentation test is performed normally by mechanical polishing of cross-sectioned surfaces, usually prepared by conventional metallography techniques. Such procedures always induce the increase of the dislocation density in the surface of the metals. Considering that hardness is a measure of plastic deformation it will be affected by mechanical polishing. So, two distinct situations can be stressed:

i) *For nanoindentation applied on a polished cross-section of untreated (non-nitrided) metal*: the increase of the dislocation density caused by mechanical polishing promotes an increase in hardness due to the dislocation density increment, when compared to the unaffected metal bulk region, which presents a lower dislocation density. Consequently, nanoindentation at shallow depths, in this case, always presents higher hardness than that observed for deeper indenter penetrations. In order to avoid the work hardening effect induced by mechanical polishing, it is possible, in some conditions, to perform chemical or electrochemical polishing, normally after the mechanical polishing, aiming to eliminate the surface region for which the dislocation density was increased; and
ii) *For nanoindentation applied on a polished cross-section of treated (nitrided) metal*: in this case, the hardness increment induced by mechanical polishing in the nitrided layer, at the cross-section of the part, is much less pronounced (for the case of nitrided layers composed of nitrogen-expanded phases) or even practically non-existent (for the case of nitrided layers constituted of high-N-affinity-alloying-elements nitrides), by comparing to the original metal (the nitrogen-unalloyed substrate bulk). This is due to fact that the nitrided surface presents a higher strengthening level, since the presence of nitrides and/or N-expanded phases makes it difficult to increase the dislocation density in the same proportion as that observed for an unalloyed metal.

In Figure 7.6 a typical result is shown for instrumented indentation hardness in a mechanically polished surface of non-nitrided niobium (corresponding to the first case, for which nanoindentation is applied on a polished cross-section surface of a non-nitrided metal). It can be observed for shallow depths (about 100 nm) an increase on the order of three times in hardness regarding the substrate bulk hardness, supposedly attained for higher depths (about 4000 nm).

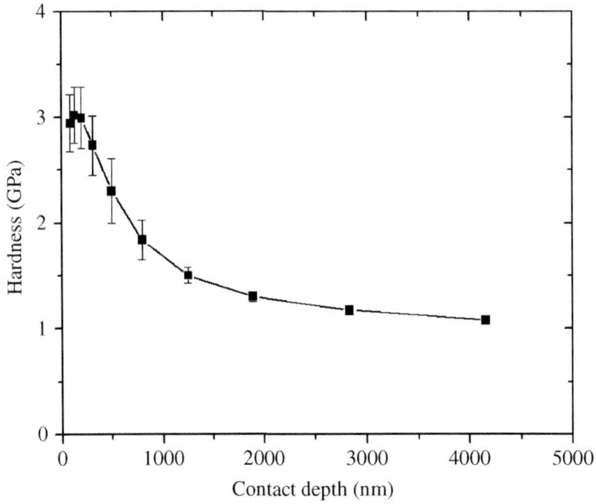

Figure 7.6 Instrumented indentation hardness in mechanically polished surface of niobium. *Source*: Courtesy of Ricardo Kertscher.

This result confirms the great sensitivity of the nanoindentation technique to characterize work hardening process attained by mechanical polishing in ductile metals. It also illustrates the possibility in making the wrong comparative analysis, as for example, aiming to determine the hardness increment achieved by the surface treatment, by comparing the hardness of a nitrided and non-nitrided part surface. In this case, one could suppose that the hardness of the non-nitrided surface would be about 3 GPa, when in truth it was about 1 GPa only, corresponding to the material bulk hardness. In order to avoid this misunderstanding, it is suggested that the highest loads available in the nanoindentation test equipment be used, aiming to determine the actual hardness value. As previously indicated, polishing procedure improvement, using the electrochemical (polishing) techniques just after mechanical polishing, can lead to good results. On the other hand, for the second case (nanoindentation applied on a polished cross-section surface of a nitrided metal), the above-mentioned suggestions are also valid. But in this case, adequate choice of a non-nitrided region is to be performed, as distant as possible from the nitrided layer, aiming to avoid the nitrogen-alloyed region, in the diffusion layer. In addition, by using the electrochemical polishing technique, it is important to keep in mind that different etching effects can occur for both the nitrogen-alloyed and non-alloyed regions, since a secondary effect of roughness could appear in the nitrided layer region. In most cases, it is difficult to obtain adequate results for electropolished surfaces presenting gradients as in composition as in properties.

7.4.2 Surface Roughness: Nanoindentation in DC Plasma Nitrided Parts

7.4.2.1 Nanoindentation in DC Plasma Nitrided Niobium

The effect of roughness on the hardness measured by the nanoindentation technique for DC plasma nitrided niobium parts was observed by Borcz and co-authors [23]. In this work, plasma nitriding of pure niobium was performed in a pulsed DC square

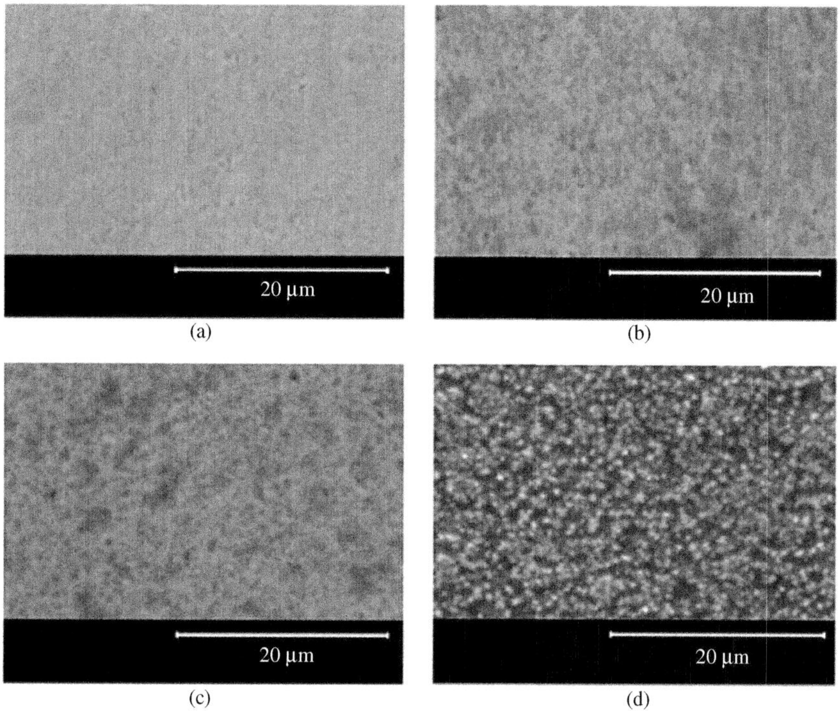

Figure 7.7 Surface aspect of niobium samples nitrided at: a) 500; b) 750; c) 915; and d) 1080°C [23]. *Source*: C. Borcz 2013 [23]. Reproduced with permission of Elsevier.

waveform power supply, for parts acting as cathode (see Figure 7.3a), at a voltage of 675 V. So, changes on the material surface characteristics due to both the heating and physical–chemical reactions caused by the ions and fast neutrals bombardment during the treatment were investigated. Nitriding was successfully performed at the temperatures of 500, 750, 915 and 1080°C, for 2 h nitriding time. As diffusion-controlled process, low nitriding temperatures (<750°C) tend to result in very thin nitrided layers, but presenting lower surface roughness. Otherwise, the use of high nitriding temperatures (>750°C) has lead to the obtaining of thicker nitrided layers, showing incremented surface roughness. In this work the same alternative method described by Souza and co-authors [57], which is presented ahead, was used to investigate the hardness of the treated surfaces as a function of the indenter penetration depth for the different studied conditions.

The surface aspect of niobium parts nitrided at different temperatures are shown in Figure 7.7. Low roughness aspect can be observed for parts treated at 500°C (Figure 7.7a). On the other hand, rougher surface is noted for parts nitrided at 1080°C (Figure 7.7d). In this case, the surface morphology remembers the presence of atom islands growing at the surface, due to sputtering and re-condensation phenomena, resulting in the niobium nitride formation at the treated surface.

The roughness profiles of the parts nitrided at 500 and 1080°C are shown in Figure 7.8(a, b), respectively. The great number of roughness peaks presenting positive values ranging from 200 to 600 nm (Figure 7.7b) agrees with the presence of

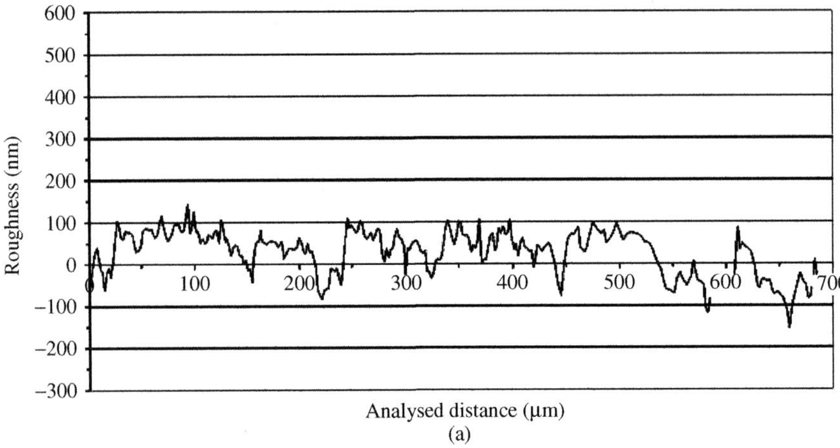

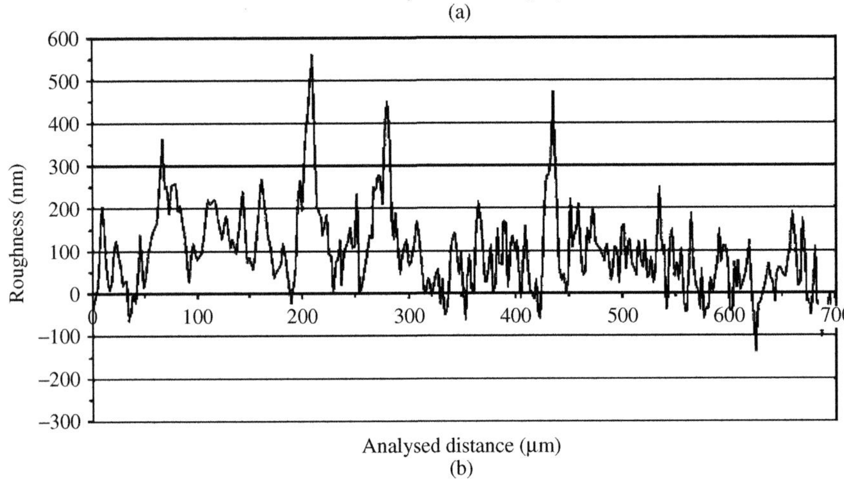

Figure 7.8 Roughness profile for surfaces plasma nitrided at: a) 500; and b) 1080°C [23]. *Source*: C. Borcz 2013 [23]. Reproduced with permission of Elsevier.

sub-micrometric atom islands at the surface (Figure 7.7d), as discussed in [23]. These results clearly show the effects of the ions and fast neutrals bombardment on the treated surfaces. Comparatively, the strong surface roughness increase evidenced in the niobium part nitrided at 1080°C can be attributed to the higher pulse switched-on time (duty cycle) necessary to perform the part heating up to the nitriding temperature, since both the used pressure and pulse voltage were the same, thus leading the bombarding energy of the plasma species to be maintained practically unaltered.

Figure 7.9 presents the curves of nitrided niobium surface hardness as a function of the indenter displacement depth for the different studied conditions shown in Figure 7.7.

Results were determined using a MTS Systems Nanoindenter XP device with Berkovich diamond tip indenter. It can be observed that values in the order of 21 GPa for penetration depths about 50, and 125 nm were attained at 750 and 915°C nitriding temperature, respectively. It is to be noted that the maximum hardness expected for niobium nitride phases can range from 30 to 45 GPa, as presented in [23]. For shallow indenter displacements, in these cases, smaller measured hardness values

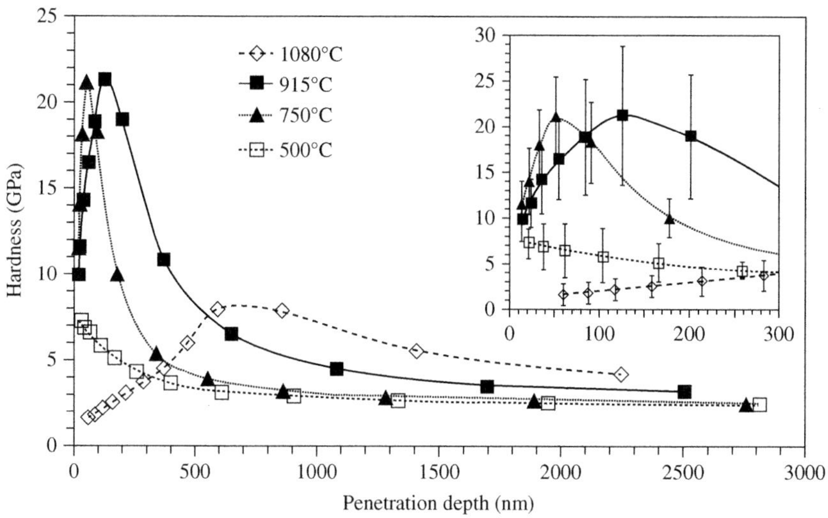

Figure 7.9 Hardness of the treated surfaces as a function of the indenter penetration depth for the different studied conditions. *Source*: C. Borcz 2013 [23]. Reproduced with permission of Elsevier.

are due to the surface roughness, despite the use of the alternative method [57]. In such conditions, the first contact of the indenter supposedly did not occur on a flat surface (see Figure 7.4), but was performed on the asperities of the nitrided surfaces. Thus, the load versus displacement curve is then shifted to higher penetrations depths and the estimated contact depth is wrong. It is to be noted that, even by applying the alternative method proposed by Souza and co-authors [57], the adopted procedure was not sufficient to determine true surface hardness values for the sample treated at 1080°C, as a consequence of the strong increment of the surface roughness after the nitriding treatment. Contrarily, for the part nitrided at 500°C, for which low roughness was attained, this problem effectively did not occur, and hardness measurements close to the true value were supposedly obtained. Nevertheless despite the correct values of hardness not being obtained, the measured values still can be useful to comparatively analyze the mechanical behavior of the DC plasma nitrided surfaces.

Finally, from the Figure 7.9 results, despite that true hardness values cannot be obtained, it can be inferred that deeper nitrided layers were obtained by increasing the temperature from 500 to 1080°C, by analyzing only the final part of the hardness profiles. This assertion is based on the fact that nitrided layer thickness, not higher than 2000 nm, can be indirectly estimated from nanoindentation results, considering that hard layers built into the soft substrate surface tends to present lower hardness values, after the maximum one, when the penetration depth is higher than 10% of the hard layer thickness [23, 55, 65]. This estimation, used to show the effect of ductile metal bulks on the hard layer thickness, made possible to predict nitrided layer thickness on the order of 250 and 600 nm for the parts nitrided at 750 and 915°C, respectively. In addition, results for the sample nitrided at 1080°C would only indicate that the hard layer has greater thickness. But it is worth mentioning that, in this case, it is not possible to infer if the obtained nitrided layer has higher hardness than those verified for the lower nitriding temperatures.

7.4.2.2 Nanoindentation in DC Plasma Nitrided Titanium

Souza and co-authors [57] have reported a work describing the application of nanoindentation to determine mechanical properties of nitrided titanium. Nitriding for parts acting as cathode (see Figure 7.3a) was performed in a conventional DC plasma device. Depending on the applied voltage, current and gas composition, different nitrided surfaces were obtained. Samples were nitrided at 600°C, 700°C, 800°C, and 900°C, for 3 h treatment time. Hardness and elastic modulus profiles were determined by a MTS Systems Nanoindenter XP device, for loads ranging from 12.5 to 400 mN. In order to have representative values, 40 indentations, per sample, separated 50 μm to each other were performed, using a Berkovich indenter. In order to reach deeper regions (substrate), Vickers microhardness measurements were also performed in a Carl Zeiss MHP-160 device, with loads ranging from 500 to 1600 mN. Surface roughness profiles were obtained by a VEECO/SLOAN 'Dektak3' profile meter. Table 7.1 shows some data of the obtained roughness, as presented in [57], indicating the strong increment of the surface roughness *Ra* as the nitriding temperature is increased.

Application of the standard nanoindentation procedure [43, 53], commonly used for low roughness flat surfaces, in the present case, for which high roughness levels were attained, did not give useful results for hardness and elastic modulus. On the other hand, application of the alternative method has assured the obtainment of more reliable values. Figure 7.10 shows results for hardness and elastic modulus attained by using the alternative method for surface definition on rough nitrided surfaces, as described in [57] and previously presented in section 7.3.

The hardness data obtained by applying the alternative method for surface definition in nanoindentation tests were compared with the Vickers hardness values attained at higher loads. The nanoindentation and Vickers hardness values agree very well for higher loads (see Figure 7.11), and the same tendency is shown with the hardness decrease, as penetration depth is increased.

By applying the contact stiffness analysis to correct the most probable zero surface contact point, surface hardness on the order of 23 GPa maximum, depending on the nitriding condition, was measured. In the same way, elastic modulus changing from 136 to 300 GPa was determined. These corrected values are in agreement with the literature data [57]. The results presented in both Figures 7.10 and 7.11, for DC plasma nitrided titanium parts, confirm the power of the alternative method illustrated here. For additional details of the alternative methodology, the reader can refer to [57].

Table 7.1 Some nitriding parameters and surface average roughness (Ra). All titanium samples were plasma nitrided for 3 hours.

Nitriding Temperature (°C)	Atmosphere	Roughness R_a (nm)
Untreated	—	19 ± 9
600	60% N_2/ 40% H_2	23 ± 6
700	60% N_2/ 40% H_2	69 ± 9
900	60% N_2/ 40% H_2	235 ± 62

Source: de Souza 2006 [57]. Reproduced with permission of Materials Reseach - Brazil.

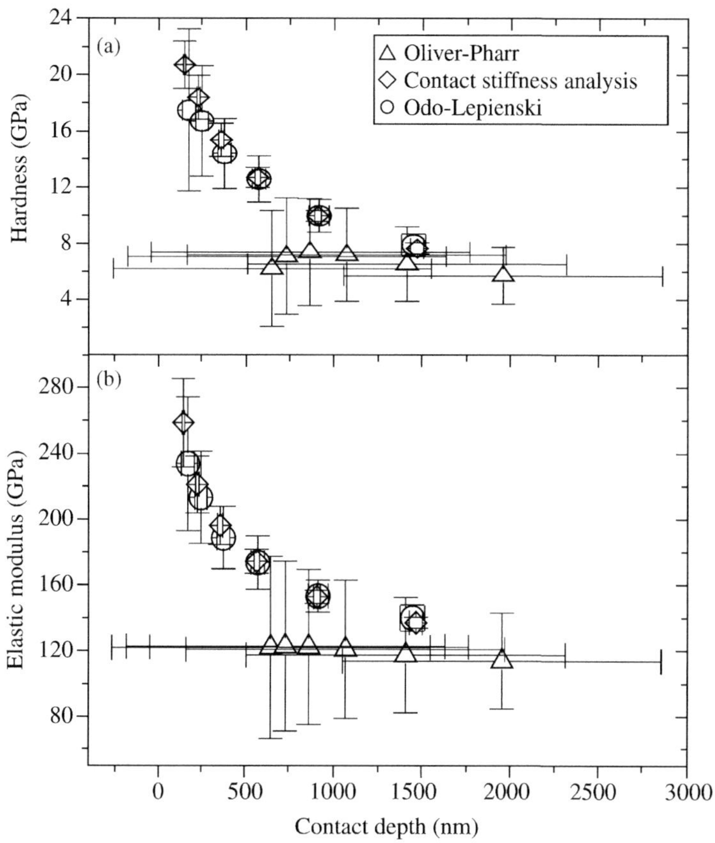

Figure 7.10 (a) Hardness; and b) elastic modulus vs. contact depth for titanium nitrided at 60% N_2/40% H_2 atmosphere and 800°C: (b) not corrected (as obtained for the nanoindenter algorithm), (c) according to the contact stiffness analysis correction and (d) the Odo-Lepienski method. *Source*: de Souza 2006 [57]. Reproduced with permission of Materials Reseach - Brazil.

7.4.2.3 Nanoindentation in DC Plasma Nitrided Martensitic Stainless Steel

In [21], plasma nitriding of CA-6NM martensitic stainless steel, for parts acting as cathode (see Figure 7.3a), was performed for high H_2 content gas mixtures comprising 5, 10, and 20 vol% N_2, at 500°C, for a time of 2 h. Roughness change relative to the untreated surface initial condition promoted by the sputtering effect, due to ions and fast neutrals bombardment, was characterized. Hardness and elastic modulus profiles were determined by a MTS Systems Nanoindenter XP device, for loads up to 400 mN, using a Berkovich tip indenter. Results indicate that the higher the N_2 content in the plasma gas mixture the higher is the attained surface roughness. Roughness Ra of 0.02, 0.13, 0.21, and 0.24 μm, and roughness Rz of 0.19, 1.03, 1.34 and 1.67 μm for an untreated surface in its initial condition, and plasma nitrided at 5, 10, and 20 vol% N_2 was obtained, respectively. This effect is explained by the higher amount of heavy species (in the case when nitrogen is compared to hydrogen) as the N_2 content is increased in the plasma gas mixture, leading to a more effective sputtering at the nitrided surface [21]. On the other

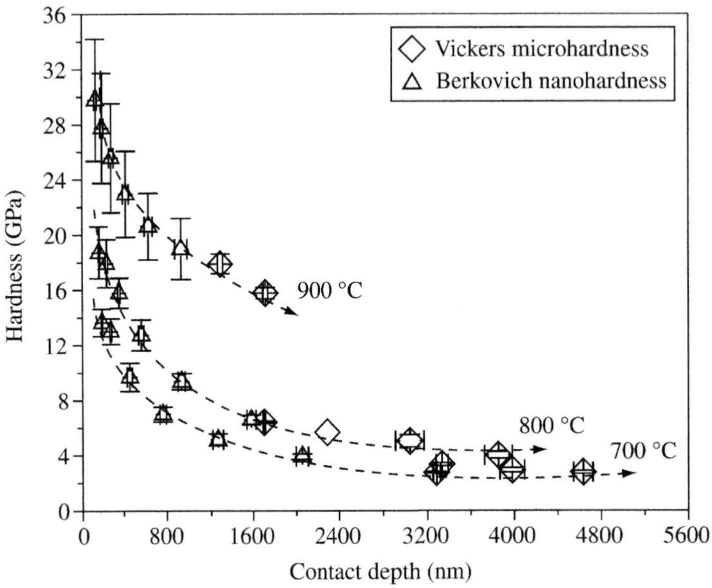

Figure 7.11 Hardness vs. contact depth for titanium samples nitrided at 700°C, 800°C and 900°C and 20% N_2 /80% H_2 atmosphere, corrected by the contact stiffness analysis (Δ). Vickers microhardness values are also shown (◇). The lines are a guide for the eyes. *Source*: de Souza 2006 [57]. Reproduced with permission of Materials Reseach - Brazil.

hand, the large dispersion observed for the measurements in the hardness versus displacement curve is an additional example of the surface roughness effect on the nanoindentation characterization. Hardness results ranging from 10 to 16 GPa confirm the strong strengthening effect of the nitrogen in high-chromium steels, which is the case of the martensitic stainless steel studied in [21]. Regarding the elastic modulus results, average value on the order of 250 GPa was observed for all the studied surfaces. In the referred work, better cavitation erosion resistance was verified for samples treated at 5% N_2, which was attributed to the finer and more homogeneous distribution of the phases present in nitrided layer. For additional details, the reader can consult reference [21].

7.4.3 Nitrogen-concentration Gradients: Nanoindentation in DC Plasma Nitrided Tool Steel

Mechanical properties can vary along the nitrided layer depending on the presence of nitrogen concentration gradients within the layer. The difference in hardness caused by the presence of nitrogen-concentration gradients can successfully be investigated, by analyzing the properties along the nitrided layer cross-section, from nanoindentation characterization. In the Zagonel and co-authors work [37], plasma nitriding of AISI H13 hot work tool steel parts was carried out in a DC plasma device, for parts acting as cathode (see Fiure 7.3a). Nitrided layers presenting thickness on the order of 50 μm were obtained at 400°C temperature, and times up to 36 h, aiming to determine how the nitrogen diffusion, precipitation, phase transformation and hardening take place in the treated steel. Nanoindentation tests were performed using a Berkovich indenter, and hardness values were estimated from loading/unloading curves. Measurements were performed on polished cross-sections and on the surface of the samples. The position of each indentation within the nitrided layer from surface was determined a posteriori by scanning electron microscopy (SEM). The composition depth profiles were analyzed on mirror polished samples cross-sections by electron probe microanalysis (EPMA), using the wavelength dispersive spectrometer technique (WDS). For additional procedure details the reader can consult [37]. Maximum contents on the order of 16 wt% N was obtained near the surface, independently of the nitriding time. Nitrogen presence in the nitrided layer was observed for depths of 43, 50, and 73 μm, using 9, 16, and 36 h nitriding times, respectively [37]. Hardness tests show that the material hardness (considering the bulk hardness of 7.5 GPa) increases linearly with the nitrogen concentration, reaching up to 14.5 GPa in the surface, independently of the treatment time, while the Young modulus remains essentially unaffected. It was found a linear relation between hardness and N concentration along the nitrided layer, in the case: $H - H_0 = 0.73$ [N] (where H: hardness expressed in GPa; H_0: the bulk hardness; and [N]: nitrogen content expressed in atomic %) [37]. Despite that this result is valid for AISI H13 steel nitrided at 400°C, the obtained linear relation can give good insight into the present subject, which could be extended to other situations, in a first approach, by comparing maximum and minimum hardness values on the sample nitrided layer cross-section.

7.4.4 Crystallographic Orientation: Nanoindentation in DC Plasma Nitrided Austenitic Stainless Steels

Stinville and co-authors [38] have applied nanoindentation aiming to investigate the anisotropy modification caused by the nitriding process on different grains of an AISI 316L austenitic stainless steel. Plasma nitriding was performed at 400°C, for 8 = h, under floating potential (see Figure 7.3b), conditions for which nitrided layer thickness on the order of 10 μm can be expected. In this case, the indentation curves were analyzed using the method proposed by Oliver and Pharr [43]. The sample surface was analyzed by EBSD technique, and the chosen region was then mapped by a large nanoindentation tests matrix aiming to determine the hardness and elastic modulus as a function of crystal orientation of the analyzed grains. In order to avoid roughness and work hardening of the studied surface, an electrochemical polishing process was performed previously to EBSD analysis and nanoindentation tests. They found hardness varying according to the anisotropy (or orientation) factor A_{hkl}, which for cubic crystal it is equal to $(h^2k^2+k^2l^2+l^2h^2)/(h^2+k^2+l^2)^2$, with limiting values of 1/3 for <h h h> orientation type, and 0 for <h 0 0> orientation type [38]. So, for the <1 0 0> direction, $A_{100} = 0$, while for the <1 1 1> direction, $A_{111} = 1/3$. From the obtained results, untreated steel has shown for $A \cong 0$, hardness H close to 1.8 GPa, while for $A \cong 1/3$, hardness H is close to 2.1 GPa. The elastic modulus E has also shown similar behavior, varying from ~190 GPa, for $A \cong 0$, to ~215 GPa, for $A \cong 1/3$ [38]. The most important result reported by Stinville and co-authors [38], by analyzing the nitrided 316L steel surface, is that both the hardness and indentation modulus exhibit an inverted anisotropy compared to the untreated surface. This result is likely to be associated with the presence of the N atoms in interstitial sites, supposedly inducing lattice expansion, thus developing high compressive stresses and activating plasticity mechanisms in the nitrided layer [38]. The higher hardness and elastic modulus values were obtained for A ≅ 0, while the lower ones were observed for grains oriented with A ≅ 1/3, at the sample surface. According to the results of Stinville and co-authors [38], the surface hardness enhancement after nitriding depends on the grain orientation (the $H_{<111>}$ is multiplied by ~8.5, and the $H_{<001>}$ is multiplied by ~6.5). Finally, it is concluded that plasma nitriding changes the anisotropy of hardness and the indentation modulus in the surface of 316L stainless steel. In this case, the <0 0 1> direction, which is the softest direction of 316L, becomes the hardest one after nitriding [38]. As a practical point of view, changes in anisotropy should be considered in studies of nitrided layers presenting nitrogen concentration gradients, since gradients can infer changes on the hardness versus displacement curve, as previously presented.

In another work of Tromas and co-authors [39], the same research group has expanded the analysis of anisotropy change by investigating the influence of nitrogen concentration on the local mechanical properties, by repeating the indentation matrices in the same area after successive partial removals of the nitrided layer, thus as a function of depth from the nitrided surface. Selective chemo-mechanical polishing procedure using neutral colloidal polishing slurry was employed to successively expose several regions at 5, 7, 8.5, 10.5, 13.2, and 19.4 μm depths from the surface, along the nitrided layer [39]. This method made possible to avoid work hardening effect, leading the obtained final RMS roughness within a grain to be smaller than 2 nm. They concluded that despite the fact that both hardness and elastic modulus qualitatively depict the nitrogen profile within the nitrided layer, their precise dependence on the nitrogen concentration remains unclear [39]. According to the authors [39], nitrogen profiles as well as the changes induced by nitriding (residual stresses, lattice rotations) are grain-orientation-dependent. Thus, experimental observations in individual grains cannot be described taking into account merely orientation effects. Finally, the authors suggest that the primary effect of the experimentally observed orientation dependence is a nitrogen concentration effect [39].

7.5 Conclusion

In this entry, the reader was introduced to topics comprising basic aspects of DC plasma nitrided parts. Theoretical aspects and the main characteristics of parts processed in linear glow discharges, and bases of the DC abnormal regime glow discharge, were also shown. Case studies were presented, for which the main surface characteristics related to nanoindentation characterization and its challenges were treated and discussed. Different examples of DC plasma nitriding technique application in niobium, titanium, and stainless steel parts were presented, comprising important aspects related to the mechanical properties characterization by nanoindentation. The main challenge was to obtain reliable hardness and elastic modulus data. The surface roughness due to ions and fast neutrals bombardment was treated and discussed. Aspects comprising the occurrence of anisotropy, and nitrogen concentration gradients in DC plasma nitrided layers were also illustrated. Finally, the power of the nanoindentation technique was proved aiming to determine mechanical properties of nitrided layers achieved in the DC plasma nitriding process.

Acknowledgements

The authors wish to express their thanks to CNPq and CAPES Brazilian Agencies by supporting this work.

References

1 B. Chapman. Glow Discharge Processes; John Wiley & Sons, New York, USA; 1980.
2 A. v. Engel. Ionized Gases, 2nd edn. American Institute of Physics, New York, USA; 1994.
3 A. Ricard. Reactive Plasmas, 1st edn; Société Française du Vide, Paris, France; 1996.
4 J.R. Roth. Industrial Plasma Engineering, vol. 1: Principles. Institute of Physics, London, UK; 1995.
5 A. Eltayeb, R.K. Vijayaraghavan, A. McCoy, *et al.* Control and enhancement of the oxygen storage capacity of ceria films by variation of the deposition gas atmosphere during pulsed DC magnetron sputtering. Journal of Power Sources, Volume 279, 94–99; 2015.
6 B. Laïk, S. Bourg, J.-P. Pereira-Ramos, *et al.* Electrochemical reaction of lithium with ruthenium nitride thin films prepared by pulsed-DC magnetron sputtering. Electrochimica Acta, 164, 12–20; 2015.
7 L. Liang, Y. He, H. Song, X. Yang, X. Cai. Effect of placement of aluminium foil on growth of etch tunnels during DC etching. Corrosion Science, 79, 21–28; 2014.
8 J.Y. Shin, K.H. Choi, K.H. Noh, D.K. *et al.* Selective etching of GaAs over $Al_0.2Ga_0.8As$ semiconductor in pulsed DC BCl_3/SF_6 plasmas. Thin Solid Films, 521, 245–248; 2012.
9 H.C. Barshilia, A. Ananth, J. Khan, G. Srinivas. Ar + H_2 plasma etching for improved adhesion of PVD coatings on steel substrates. Vacuum, 86(8), 1165–1173; 2012.
10 Hanizam, H. Soufhwee A.R, K.Anuar A.R *et al.* The effect of pulse DC and DC substrate bias during in situ cleaning PVD process on surface roughness. Procedia Engineering, 53, 562–568; 2013.
11 A. Shanaghi, A.R. Rouhaghdam, S. Ahangarani, *et al.* Effects of duty cycle on microstructure and corrosion behavior of TiC coatings prepared by DC pulsed plasma CVD. Applied Surface Science, 258(7), 3051–3057; 2012.
12 A.N. Klein, R.P. Cardoso, H.C. Pavanati *et al.* DC plasma technology applied to powder metallurgy: an overview. Plasma Science and Technology, 15(1), 70–81; 2013.
13 S.F. Brunatto, I. Kühn, J.L. Muzart. Surface modification of iron sintered in hollow cathode discharge using an external stainless steel cathode. Journal of Physics D: Applied Physics, 38, 2198–2203; 2005.
14 S.F. Brunatto, J.L. Muzart. Influence of the gas mixture flow on the processing parameters of hollow cathode discharge iron sintering. Journal of Physics D: Applied Physics 40, 3937–3944; 2007.
15 S.F. Brunatto, R.P. Cardoso, A.N. Klein, J.L. Muzart. Sintering and surface texturing: direct current-coupled plasma-assisted parts manufacturing. In: Encyclopedia of Iron, Steel, and their Alloys, pp. 3194–3207. Taylor & Francis Group, New York, published online 30 Mar 2016.

16 M.A. Santos, H.R. Silva, A.M. Maliska, A.N. Klein, J.L. Muzart. Oil removal and nitriding of sintered parts using a DC discharge. Surface and Coatings Technology, 195(2–3), 198–203; 2005.
17 C.J. Scheuer, R.P. Cardoso, M. Mafra, S.F. Brunatto. AISI 420 martensitic stainless steel low temperature plasma assisted carburizing kinetics. Surface and Coatings Technology, 214, 30–37; 2013.
18 C.J. Scheuer, R.P. Cardoso, F.I. Zanetti *et al.* Low-temperature plasma carburizing of AISI 420 martensitic stainless steel: Influence of gas mixture and gas flow rate. Surface and Coatings Technology, 206, 5085–5090; 2012.
19 S.F. Brunatto, R.P. Cardoso, C.J. Scheuer. DC low-temperature plasma carburizing of martensitic stainless steel. In: Encyclopedia of Iron, Steel, and their Alloys. Taylor and Francis Group, New York, 2015.
20 A.N. Allenstein, C.M. Lepienski, A.J. Buschinelli, S.F. Brunatto. Improvement of the cavitation erosion resistance for low-temperature plasma nitrided CA- 6NM martensitic stainless steel. Wear, 309, 159–165; 2014.
21 A.N. Allenstein, C.M. Lepienski, A.J. Buschinelli, S.F. Brunatto. Plasma nitriding using high H_2 content gas mixtures for a cavitation erosion resistant steel. Applied Surface Science, 277, 15–24; 2013.
22 A.N. Allenstein, R.P. Cardoso, K.D. Machado *et al.* Strong evidences of tempered martensite-to-nitrogen-expanded austenite transformation in CA-6NM steel. Materials Science and Engineering A, 552, 569–572; 2012.
23 C. Borcz, C.M. Lepienski, S.F. Brunatto. Surface modification of pure niobium by plasma nitriding. Surface and Coatings Technology, 224, 114–119; 2013.
24 R.P. Cardoso, S.F. Brunatto, C.J. Scheuer. Stainless steel: Low-temperature nitriding kinetics. In: Encyclopedia of Iron, Steel, and their Alloys, pp. 283–3293. Taylor and Francis Group, New York, published online 30 Mar 2016.
25 E. Roliński. Plasma-assisted nitriding and nitrocarburizing of steel and other ferrous alloys. Chapter 11 in: Thermochemical Surface Engineering of Steels: Improving Materials Performance (eds E.J. Mittemeijer, M.A. Somers), pp. 413–457. Elsevier, 2015.
26 A. Saeed, A.W. Khan, F. Jan, M. *et al.* Validity of 'sputtering and re-condensation' model in active screen cage plasma nitriding process. Applied Surface Science, 273, 173–178; 2013.
27 R.R. de Sousa, F.O. de Araújo, L.C. Gontijo *et al.* Cathodic cage plasma nitriding (CCPN) of austenitic stainless steel (AISI 316): Influence of the different ratios of the (N_2/H_2) on the nitrided layers properties. Vacuum, 86(12), 2048–2053; 2012.
28 K.J.B. Ribeiro, R.R.M. de Sousa, F.O. de Araújo *et al.* Industrial application of AISI 4340 steels treated in cathodic cage plasma nitriding technique. Materials Science and Engineering: A, 479(1–2), 142–147; 2008.
29 L.Wang, Y. Li, X. Wu. Plasma nitriding of low alloy steels at floating and cathodic potentials. Applied Surface Science, 254(20), 6595–6600; 2008.
30 K. Nikolov, K. Köster, P. Kaestner *et al.* Strip hollow cathode method for plasma thermochemical treatment for surface modification of thin metal strips: Plasma nitriding of austenitic stainless steel sheets for bipolar plates. Vacuum, 102, 31–37; 2014.

31 M. Benda, J. Musil. Plasma nitriding enhanced by hollow cathode discharge: a new method for formation of superhard nanocomposite coatings on steel surfaces. Vacuum, 55(2), 171–175; 1999.
32 Y. Li, L. Wang, J. Xu, D. Zhang. Plasma nitriding of AISI 316L austenitic stainless steels at anodic potential. Surface and Coatings Technology, 206(8–9), 2430–2437; 2012.
33 M. Benda, J. Vlcek, J. Musil. Anodic plasma nitriding with a molybdenum cathode. Vacuum, 46(1), 43–47; 1995.
34 Y. Li, L. Wang, L. Shen *et al.* Plasma nitriding of 42CrMo low alloy steels at anodic or cathodic potentials. Surface and Coatings Technology, 204(15), 2337–2342; 2010.
35 M. Keddam, M. Kulka, N. Makuch *et al.* A kinetic model for estimating the boron activation energies in the FeB and Fe_2B layers during the gas-boriding of Armco iron: Effect of boride incubation times. Applied Surface Science, 298, 155–163; 2014.
36 C. Sudha, R. Anand, V.T. Paul *et al.* Nitriding kinetics of inconel 600. Surface and Coatings Technology, 226, 92–99; 2013.
37 L.F. Zagonel, J. Bettini, R.L. Basso *et al.* Nanosized precipitates in H13 tool steel low temperature plasma nitriding. Surface and Coatings Technology 207, 72–78; 2012.
38 J.C. Stinville, C. Tromas, P. Villechaise, C. Templier. Anisotropy changes in hardness and indentation modulus induced by plasma nitriding of 316L polycrystalline stainless steel. Scripta Materialia, 64(1), 37–40; 2011.
39 C. Tromas, J.C. Stinville, C. Templier, P. Villechaise. Hardness and elastic modulus gradients in plasma-nitrided 316L polycrystalline stainless steel investigated by nanoindentation tomography. Acta Materialia, 60(5), 1965–1973; 2012.
40 J.C. Stinville, J. Cormier, C. Templier, P. Villechaise. Monotonic mechanical properties of plasma nitrided 316L polycrystalline austenitic stainless steel: Mechanical behavior of the nitrided layer and impact of nitriding residual stresses. Materials Science and Engineering A, 605, 51–58; 2014.
41 J.C. Stinville, J. Cormier, C. Templier, P. Villechaise. Modeling of the lattice rotations induced by plasma nitriding of 316L polycrystalline stainless steel. Acta Materialia, 83, 10–16; 2015.
42 W. Hume-Rothery. The structures of alloys of iron. Pergamon Press, Oxford; 1966.
43 W.C. Oliver, G.M. Pharr. An improved technique for determining hardness and elastic modulus using load and displacement sensing indentation experiments. Journal of Materials Research, 7(6), 1564–1583; 1992.
44 A.C. Fischer-Cripps. Nanoindentation, 3rd edn. Springer, New York, 2011.
45 S. Cucatti, E.A. Ochoa, M. Morales *et al.* Effect of bombarding steel with Xe^+ ions on the surface nanostructure and on pulsed plasma nitriding process. Materials Chemistry and Physics, 149–150, 261–269; 2015.
46 D.R. Irala, L.C. Fontana, J.C. Sagás, H.S. Maciel. The effects of plasma nitriding pretreatment in steel substrates on the photocatalytic activity of TiO_2 films. Surface and Coatings Technology, 240, 154–159; 2014.
47 F. Borgioli, E. Galvanetto, T. Bacci. Influence of surface morphology and roughness on water wetting properties of low temperature nitrided austenitic stainless steels. Materials Characterization, 95, 278–284; 2014.

48 V. Brackmann, V. Hoffmann, A. Kauffmann *et al.* Glow discharge plasma as a surface preparation tool for microstructure investigations. Materials Characterization, 91, 76–88; 2014.

49 A. Nishimoto, K. Nagatsuka, R. Narita, H. *et al.* Effect of the distance between screen and sample on active screen plasma nitriding properties. Surface and Coatings Technology, 205, Supplement 1, S365–S368; 2010.

50 J. Guo, S. Amira, P. Gougeon, X.G. Chen. Effect of the surface preparation techniques on the EBSD analysis of a friction stir welded AA1100-B4C metal matrix composite. Materials Characterization, 62(9), 865–877; 2011.

51 F.X. Lin, A. Godfrey, D.J. Jensen, G. Winther. 3D EBSD characterization of deformation structures in commercial purity aluminum. Materials Characterization, 61(11), 1203–1210; 2010.

52 M.A. Wall, A.J. Schwartz, L. Nguyen. A high-resolution serial sectioning specimen preparation technique for application to electron backscatter diffraction. Ultramicroscopy, 88(2), 73–83; 2001.

53 W. Oliver, G. Pharr. Measurement of hardness and elastic modulus by instrumented indentation: advances in understanding and refinements to methodology. Journal of Materials Research, 19, 3–20; 2004.

54 A. Tiwari. Nanomechanical Analysis of High Performance Materials, Springer, New York; 2014.

55 A. Tiwari, L.H. Hihara. Nanoindentation and morphological analysis of novel green quasi-ceramic nanocoating materials. Progress in Organic Coatings, 77(7), 1200–1207; 2014.

56 M. L. Palacio, B. Bhushan. Depth-sensing indentation of nanomaterials and nanostructures. Materials Characterization, 78, 1–20; 2013.

57 G.B. de Souza, C.E. Foerster, S.L. da Silva, C.M. Lepienski. Nanomechanical properties of rough surfaces. Materials Research, 9(2), 159–163; 2006.

58 H. Mayer, R. Schuller, M. Fitzka *et al.* Very high cycle fatigue of nitrided 18Ni maraging steel sheet. International Journal of Fatigue, 64, 140–146; 2014.

59 Folkhard, E. Welding metallurgy of stainless steels. Wien Springer Verlag: New York; 2012.

60 J.C. Lippold, D.J. Kotecki. Welding metallurgy and weldability of stainless steel. John Wiley and Sons: New Jersey; 2005.

61 K.H. Lo, C.H. Shek, J.K. Lai. Recent developments in stainless steels. Materials Science and Engineering Review, 65, 39–104; 2009.

62 K.-E. Thelning. Steel and its heat treatments, 2nd edn. Butterworth, London; 1984.

63 A. Martinavicius, G. Abrasonis, A.C. Scheinost *et al.* Nitrogen interstitial diffusion induced decomposition in AISI 304L austenitic stainless steel. Acta Materialia, 60, 4065–4076; 2012.

64 W.D. Wilkinson. Fabrication of Refractory Metals. Gordon and Breach Science Publishers New York; 1970.

65 C.E. Foerster, A. Assmann, S.L. da Silva *et al.* AISI 304 nitrocarburized at low temperature: Mechanical and tribological properties. Surface and Coatings Technology, 204, 3004–3008; 2010.

8

Nanomechanical Properties of Defective Surfaces

Oscar Rodríguez de la Fuente

Departamento de Física de Materiales, Universidad Complutense, Madrid, Spain

8.1 Introduction

Real surfaces, unlike those commonly used in surface science approaches, tend to contain a large number of imperfections: vacancies, adsorbed hydrocarbons, steps and facets, water adlayers, emerging dislocations etc. All of them are elements of what we usually name *real surfaces*, and can be intrinsic or introduced by external manipulation. As Bowden and Tabor stated [1], 'Putting two solids together is rather like turning Switzerland upside down and standing it on Austria.' This statement is a powerful visualization of the real situation in terms of surface topography. Real surfaces, far from the atomically flat terraces separated by single steps used with surface science techniques, contain mounds and hills, grain boundaries and protrusions, and so the mechanical contact of two solids usually takes place between the asperities at both sides. This dissimilarity between the conditions at the *laboratory* and those for real surfaces is usually called *materials gap* in surface science, and it is a concept relevant to many areas (heterogeneous catalysis is a recurrent example [2]). The consequences of the existence of a materials gap wouldn't be important except for the proven fact that surface defects not only may influence physico-chemical properties of solids, but actually control some of them.

8.1.1 The Role of Surface Defects in Plasticity

For the case of the bulk, the role of defects in crystals has been long studied, and textbooks on solid state physics usually address one or more chapters to the subject [3]. The situation is rather different for the case of surface defects. While they are located in a very specific part of the solid, several physico-chemical processes take place totally or partially at surfaces, and thus the state of the surface and the density and type of defects contained in it play a role. Heterogeneous catalysis, as already mentioned, is the paradigmatic example because of the so-called *active sites*, where reactions are often considered to take place, and which have been usually identified with surface defects [4]. But other properties, such as wettability, also strongly depend on the presence of surface defects.

Mechanical properties should not be an exception. Gilman already reported in 1961 [5] that stress-strain curves in some crystals are sensitive to the conditions at the crystal surfaces. Plasticity is, in essence, the creation and movement of defects (most of them

Applied Nanoindentation in Advanced Materials, First Edition. Edited by Atul Tiwari and Sridhar Natarajan.

already present in the material) and the expression of their mutual interactions. The state of the surface should determine its plastic response, mostly in the region close to the surface and also in nanosized objects, where a relevant part of the atoms are located *at* or *close to* the surface, and where the volume is so small that the probability of finding pre-existing dislocations is reduced. In this case, incipient plasticity is ruled by the nucleation of new defects, some of which may take place at the surface. Indeed, nanosized materials show plastic properties qualitatively different from those of bulk materials [6, 7], and the state of the surface also has to do with it. All the previous statements hold also true for the case of very localized applied stresses, which is the case of nanoindentation processes. So, either in the case of nanosized objects or for very localized deformations in macroscopic materials, the state of the surface morphology can be conclusive in determining the plastic response. This assertion is most easily visualized for the case of friction, either in the wear or non-wear regime, where the friction from the protruding surface defects adds to the atomic friction.

8.1.2 Experimental Techniques for Visualization and Generation of Surface Defects

The lesser knowledge on surface defects, when compared to the bulk, is probably linked to the absence, for decades, of suitable experimental techniques capable of locating and characterising surface defects at the atomic level. The level of the knowledge in the case of the bulk of solids is higher, probably, because of the existence of experimental techniques suitable for exploring the bulk in real space (mainly TEM). The development of scanning probe microscopies (mainly scanning tunnel microscopy (STM) and atomic force microscopy (AFM) paved the way to the characterization of surface defects and contributed to understand their role in physico-chemical properties (Figure 8.1).

Together with those experimental techniques suitable for exploring surfaces in real space at the atomic level, adequate techniques aimed to introduce surface defects in a controlled way is mandatory as well. Real surfaces are defective *per se*, but the heterogeneity of their defects is too high to understand their role profoundly. To separately understand the contribution of each type of defect to a given process, ways to fabricate surface defects are needed. As we will see, a surface step is the most typical and best-studied surface defect. The role of an isolated single step can be studied in surfaces

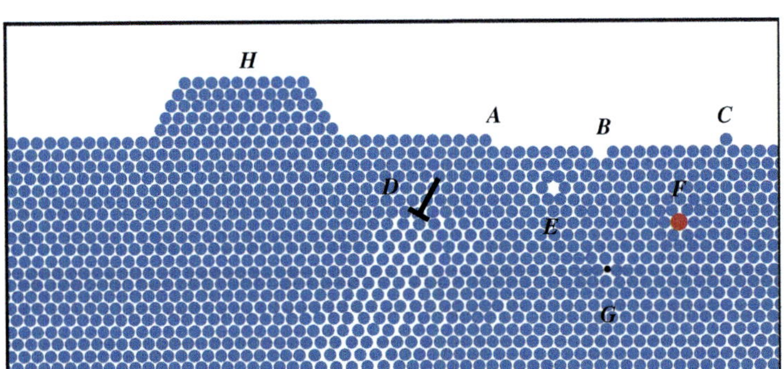

Figure 8.1 Scheme of a defective surface with some of the possible types of surface defects most frequently encountered. A: surface step, B: surface vacancy, C: adatom, D: sub-surface dislocation, E: vacancy, F: substitutional atom, G: interstitial, H: asperity.

of single crystals if they are prepared carefully. Usually, it is not a challenge to obtain atomically flat terraces separated by single atomic steps. If one is more concerned about the role of multiple steps, an approximation can be that of using vicinal surfaces. These are surfaces of single crystals miscut a few degrees off a high symmetry direction. This misalignment can be adjusted, at the surface, by a succession of atomic steps, regularly separated. The distance between the steps can be adjusted through the miscut angle. But for this purpose, the bunching or aggregation of the steps must be avoided.

Another way of introducing surface defects is through low energy ion bombardment. The overall process can be rather complex, as well as the morphology of the modified surface and the nature of the generated defects. But, by adjusting the experimental parameters (temperature, dose, ion energy, etc.) one can explore a large set of configurations: from a surface with a low concentration of dispersed surface vacancies to a structured surface, formed by nanometric dots or mounds evenly distributed [8]. If the energy of the incident ions is of the order of a few keV, most of the energy is deposited in the first few nanometers below the surface, generating atomic displacements in a sequence of events. Atomic sputtering or ejection of atoms out of the solid also occurs. After the recombination of the generated vacancies and interstitials, the process may end up with a few surface vacancies, which is the case of most metallic surfaces when irradiation is carried out at room temperature. The diffusion and clustering of these vacancies form vacancy islands, and when these are large enough, new vacancy islands can nucleate and grow inside the formers. In this way, starting from a flat surface, a surface with multiple exposed levels (with mounds and holes) can be created if the ion dose is large enough.

Irradiation with other particles (photons, electrons etc.) can be also utilized to generate surface defects, but their use is much more restricted. Photons have a long penetration depth and their damage is not, in principle, so localized in the vicinity of the surface. Electrons with energies of the order of a few hundreds of electronvolts have, in contrast, a penetration depth of a few nanometers, but their use for defect generation is limited, in principle, to insulators. Another way to introduce defects in the vicinity of a surface is by performing an indentation in it. A set of dislocations is created in the vicinity. Controlling the temperature, on the other hand, is a way to control the density of vacancies.

8.1.3 Approaches to Study and Probe Nanomechanical Properties

Thanks to the work carried out in the last 20 years in the field of nanomechanical properties, much has been learnt about incipient plasticity and nanomechanical properties. Most of the works performed in this field have used nanoindentation as the experimental tool to perform the mechanical tests. The rise in the use of instrumented nanoindenters, with very high both spatial and load resolutions, together with the use of microscopic techniques with unprecedented resolution, has pushed the limits. A special mention deserves the use of the AFM both as a nanoindenter and microscope. It is indeed designed to apply specific mechanical loads on a surface. Its main advantage is the chance to visualize the surface before and after the application of the deformation.

Regarding simulations or theoretical procedures, approaches based in continuum elasticity have been used to describe the role of surface defects on nanomechanical properties. Despite the atomic nature of solids, continuum mechanics works reasonably well even at a scale of a few atomic distances and in the vicinity of defects (dislocations, for instance). Continuum approaches are not much affected by the discreetness of solids in the bulk. Following this approach, contact processes between defective or rough surfaces have been described, also, using continuum contact mechanics. However, atomic scale roughness of small asperities at the surfaces leads to dramatic deviations from continuum theories. Luan and Robbins nicely disclosed this idea in [9]. They showed, comparing continuum and atomistic approaches, that the former underestimate contact areas and yield stress σ_y. As for nanotribology, they also showed that continuum approaches overestimate friction and contact stiffness.

Discrete dislocation dynamics (DDD) approaches have also dealt with incipient plasticity at rough or defective surfaces [10]. However, these models need constitutive relations (with activation energies, critical shear stresses etc.), and at present no surface nucleation criterion has been fully developed. Atomistic simulations are, in this respect, more convenient. As they consider the atomic nature of solids, they describe in a more natural way any plastic process involving limited volumes or plasticity being initiated or affected by sites with atomic dimensions. The main disadvantage of atomistic simulations is, still, the restricted sizes and simulation times they are limited to.

8.2 Homogeneous and Heterogeneous Dislocation Nucleation

Three-dimensional dislocations (as opposed to vacancies or even two-dimensional dislocations) are not defects in thermal equilibrium at a given temperature. The application of a mechanical stress is needed for a dislocation to nucleate, either homogeneously or heterogeneously, shearing the crystal lattice. Dislocation nucleation is a thermally activated and stress-assisted process. The probability of nucleation of a dislocation can be described as:

$$\dot{n} = \dot{n}_0 \exp\left(-\frac{\varepsilon - \sigma V}{k_B T}\right) \tag{8.1}$$

where n_0 is the attempt frequency, $\varepsilon - \sigma V$ is the effective activation energy (an activated energy ε assisted by a stress σ applied to a volume V), k_B is the Boltzmann constant and T is the temperature.

A complete knowledge of surface plasticity must necessarily differentiate homogeneous and heterogeneous dislocation nucleation. Homogeneous nucleation refers to the nucleation of the defect in the perfect crystal. In heterogeneous nucleation, the dislocation will nucleate at a previously existing defect. The mechanisms and energetics involved in the two types of processes are very different, and so the plastic response of the material is profoundly influenced by the existence or dominance of either type of defect nucleation.

8.2.1 Homogeneous Dislocation Nucleation

It is well known that the theoretical shear strength of a crystal, estimated by Frenkel [11] as $\tau_c = G/2\pi$ (where G is the shear modulus of the crystal), is much lower in real macroscopic solids. The maximum stress attained in a solid before deforming plastically is typically around 100-1000 times lower than the estimated theoretical value τ_c. It is also well known that the reason for this behaviour is the pre-existence of dislocations in the bulk of the crystal. Indeed, this was one of the reasons for seriously considering their existence before they were visualized. The theoretical estimation τ_c considers that crystals irreversibly deform through shear forces sliding adjacent crystallographic planes. But it is well known that even well annealed crystals contain a large amount of dislocations (10^6–10^8 dislocations/cm^2), and that putting them into motion imposes plastic flow (permanent deformation). And it is also well established that this process is energetically much more convenient than sliding atomic planes in perfect crystals. Thus, the investigation of the homogeneous nucleation of a dislocation (nucleation of a dislocation somewhere inside the perfect crystal) is not an attainable goal for a macroscopic crystal subject to an extensively distributed external stress.

In contrast, nanoindentation experiments in well-prepared surfaces are a very convenient tool to prove incipient plasticity related to the nucleation of *new* dislocations in the solid. The volume affected by the nanoindentation is sufficiently small and the applied stress is so local that the affected solid can be regarded, most of the times, as a perfect crystal without pre-existing dislocations. Force versus penetration curves during nanoindentation, if performed with sufficient resolution, show an initial elastic response (usually fit to a hertzian response), followed by a series of discrete events, called bursts or *pop-ins*. These discontinuities in the load-penetration curve have been associated to the nucleation of dislocations below the indenter. The fact that the first pop-in in the *F(z)* curves occurs when (as estimated from elasticity theory) the applied shear stress is close to the theoretical shear strength τ_c is a clear suggestion of the nucleation of a new dislocation by shear displacement of atomic planes.

In this way, values of the applied stress close to the theoretical strength τ_c have been attained, and it has soon become clear that plasticity associated to nanoindentation experiments can be related to the nucleation of new dislocations rather than to the flow of pre-existing ones. In nanoindentations at flat surfaces, the probability of finding heterogeneous sources is low. And for a long time it has been considered that the nucleation of dislocations under these conditions is of the homogeneous type.

Calculations based on continuum elasticity theory can show that the stress applied on a flat surface by a spherical indenter is far from being uniform. The shear stress, as a function of the depth z is given by [12]:

$$\sigma = -\frac{p_0}{a}\left(z - \frac{z^2}{\sqrt{a^2+z^2}}\right) \tag{8.2}$$

where p_0 is the maximum pressure attained below the indenter and a is the half width of the area of mechanical contact. From Equation (8.2), the shear stress (ultimately responsible for dislocation nucleation) reaches a maximum at $z = 0.78\,a$. Thus, the place where a dislocation is expected to nucleate lies below the surface, and not at the contact point between the indenter and the surface. However, and despite the long range of elastic fields, Equation (8.2) shows that the maximum stress is attained relatively close to the contact area. Due to the nature of the process (a hidden structural process at the atomic level below a surface subject to high mechanical stresses) obtaining experimental evidences of these predictions has been hard.

In this sense, simulations have helped a lot in discerning the mechanisms of homogeneous dislocation nucleation. Two-dimensional bubble rafts [13, 14] have shown this fact in a rather precise way. Also atomistic simulations (mainly molecular dynamics simulations) have shown that the nucleation of the seed dislocation takes place, approximately, at the expected depth and shear stress [15]. However, atomistic simulations also present interesting peculiarities related to the crystal anisotropy and not predicted by continuum elasticity theory. A molecular dynamics simulation in an f.c.c. crystal during a purely homogeneous process shows the nucleation of two or more dislocation loops below the surface, which soon emerge to the surface and glide into the bulk, evolving into complex structures [15]. But the initial seed loops are located off the indenter axis, as expected, on the other hand, if one takes into account crystalline anisotropy.

The identification of a reliable criterion that describes those conditions leading to homogeneous dislocation would be of prime importance. This criterion for nucleation would be incorporated to the constitutive laws of those continuum models working at length scales approaching the nanoscale (where, for instance, Frank-read sources are not available and homogeneous nucleation must be considered). Homogeneous dislocation nucleation has been usually described, in a first approximation, via the critical resolved shear stress through the Schmid law. The Schmid factor defines, essentially, the stress applied or resolved on the slip plane and slip direction of the crystal. However, progresses beyond the consideration of the Schmid factors would improve the theory. In this respect, the identification of non-Schmid factors has been acknowledged in simple f.c.c. systems [16]. Non-Schmid factors may refer, for instance, to the stress applied perpendicular to the slip plane. It can be easily understood that, for the case of an f.c.c. metal, if the {111} interplanar distance changes, this fact may affect the slip of these {111} atomic planes (through atomic friction). It is now well accepted that homogeneous dislocation nucleation depends on both the Schmid (in-plane) factor and the non-Schmid (normal) factor. Other more complex criteria have been also developed, such as the so-called Λ-criterion [17], which extends the notion of theoretical strength to a non-uniform loading. It is based on taking second-order derivatives of the free energy, and elastic instabilities leading to dislocation nucleation can be predicted.

8.2.2 Heterogeneous Dislocation Nucleation

There are not many doubts that, when probing small volumes with nanoindentations, *new* dislocations nucleate under the mechanical action of the indenter. However, the exact nature and location of the nucleation are not so well known and it is indeed a subject of debate nowadays. We will see in the next sections that surface defects (for instance, surface steps) are preferential places for dislocation nucleation, and thus for incipient dislocation activity. This has been clearly shown in a few cases. Important doubts remain, though. An open question is how often dislocation nucleation processes at real surfaces are indeed heterogeneous. In other words, one could wonder whether nucleation processes inside the *perfect* crystal, below the surface, are really homogeneous.

In this respect, a very interesting contribution has been carried out by Schuh *et al.* [18–20]. They have performed nanoindentations in Pt single crystals at different temperatures, obtaining decreasing values of the yield point for increasing temperatures (see Figure 8.2). This result points to a significant contribution of thermal energy to the onset of plasticity. Based in Equation (8.1), accomplishing many indentations at different temperatures and introducing statistical methods to quantitatively evaluate the data,

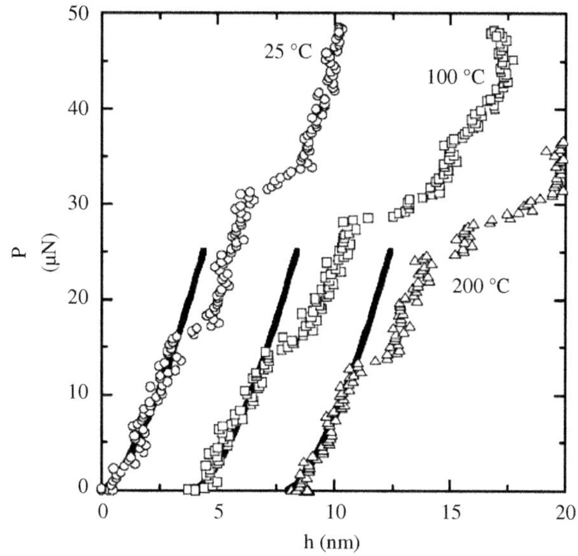

Figure 8.2 Representative force-penetration curves performed at various temperatures on Pt(110). The yield point (where the experimental data deviate from the expected hertzian elastic behavior, black line) decreases with higher temperatures. *Source*: Mason 2006 [18]. Reproduced with permission of the American Physical Society.

the authors have suggested that incipient plasticity involves heterogeneous nucleation sites. This conclusion has been reached after having obtaining an activation volume $V \approx 10$ Å^3, which corresponds to the approximate volume of a vacancy in the bulk. This and other data obtained from the analysis are not, according to the authors, compatible with the homogeneous nucleation of a dislocation loop as anticipated by atomistic simulations. The authors suggest that heterogeneous dislocation nucleation at subsurface vacancies or at surface asperities could be responsible for incipient plasticity. Paul *et al.* have arrived at similar conclusions [21]. By performing nanoindentations in Au(111), combining scanning tunnel microscopy and atomic force microscopy, they have claimed that the forces measured at the first pop-in correspond to stresses much lower than the theoretical shear strength. These results again suggest that dislocations are heterogeneously nucleated.

In the last years, the fast development of focussed ion beams (FIB) has allowed the scientific community to perform mechanical tests in the micro- and nanostructures fabricated with them. In the field of mechanical properties, the investigation of deformation tests in pillars with small dimensions is an important assessment of the nucleation type. Micro- and nanopillars are, in principle, free of defects in the bulk (unless those introduced during the sculpting process), but have a large surface-to-volume fraction. Also, the stress field during compression tests in pillars is rather uniform, and dislocation nucleation can occur anywhere in the pillar, including free surfaces and surface defects. This is a difference with respect to nanoindentation experiments, where the stress state in nonuniform, and the stress is, in principle, focused underneath the surface (although the existence of defects with a reduced activation energy or acting as stress concentrators may change this situation). Bei *et al.* [22] have been able to distinguish between homogeneous and heterogeneous dislocation nucleation by experimentally measuring different strengths (elastic limits) in the same material (a Mo alloy), either in the form of singe crystalline surface or micropillar. While the authors have found that the critical shear stress is around $\sim G/8$ in nanoindentation tests at the surface, they have measured a value around ~$G/26$ when compressing micropillars. The authors attribute this lower shear stress to the fact that in micropillars the dislocations nucleate heterogeneously as a half or quarter dislocation loop at the surface or at the corner of a surface, respectively. It is also interesting to note that the nucleation of a dislocation at the surface can be considered as a kind of heterogeneous nucleation event.

8.3 Surface Steps

Surface steps are, probably, the paradigmatic surface defects. While obvious, it is important to remind that surface roughness, either at the micron or nanometre scales, is often accommodated with surface steps: an upward deviation from perfect flatness is, at the atomic scale, a succession of ascending steps. Due to their ubiquity in real surfaces, their easy visualization in single-crystals with near-field microscopies, and the relatively good knowledge attained about their physical properties, they are relatively well understood. Surface steps have quite often been a benchmark for the study of surface physicochemical properties. As for the case of catalysis, where steps have been often designated as the *active sites* for certain surface chemical reactions [23], they might be also considered, in certain cases, as active sites for the initiation of crystal plasticity.

The role of surface steps on the nanomechanical properties of surfaces has been widely studied experimentally and theoretically with both continuum and atomistic approaches. A most simplistic approach considers that when a dislocation half loop is nucleated at a surface step, the collapse of a portion of the step reduces its total energy (see Figure 8.3). If γ_L is the line tension of the step (energy per unit length) and l is the length of the part of the step collapsed by the applied force, the energy of the surface will be reduced by $\Delta U_{STEP} = -\gamma_L\, l$. While the energy term associated with the creation of the half loop is present both for the flat and stepped surfaces (and considered to be

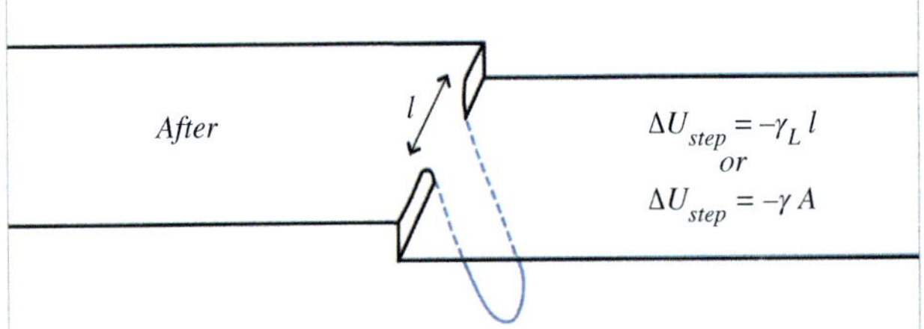

Figure 8.3 Scheme of a surface step before (top) and after (bottom) a partial collapse resulting from an applied force. A dislocation half loop is nucleated at the step, and the energy of the step is reduced by ΔU_{STEP}. The former can be described in terms of step line energy γ_L or step surface energy γ . In any case, ΔU_{STEP} is an energetically favourable term.

equal, in a first approximation) this energetically favourable term ΔU_{STEP} will be only present for the case of the stepped surface. If the step is indeed a multiple step formed by consecutive atomic steps (ledge) or by a facet, a second approach, considering not the line energy but the surface energy γ of the step, could be regarded. In this case $\Delta U = -\gamma A$, where A is the area of the collapsed step.

Surface steps may influence nanomechanical properties in several ways. They can affect the plastic response i) through their elastic field, acting as stress concentrators, ii) because of their geometry, reducing the contact area and iii) because of their atomistic nature, since atoms at the steps are less coordinated and thus easier to deform.

Since the work of Marchenko and Parshin (MP) [24], it is well known that surface steps interact elastically. In the same way as atoms at free surfaces tend to relax downwards, compensating the lack of neighbours, atoms at the steps also relax, but in a more complex way [25]. Unlike for dislocations, where the Burgers vector determines the displacement field through the elastic constants, there is no equivalent relationship specifying the strain field of a step in terms of its height and orientation, and quite often atomistic calculations are needed [26]. It is often considered that atoms at the upper part of the step tend to relax towards the bulk, while those at the lower part (corner) tend to relax upwards (out of the surface). This can be viewed as a pair of opposite forces (dipole), producing a relaxation field, which penetrates well into the bulk. The MP prediction of the interaction is repulsive and depends on the square of the inverse of the interstep distance ($E \sim d^{-2}$). While many experimental studies confirm the MP prediction, others have shown the possibility of attractive interactions. These interactions stabilize the step distribution at surfaces, and thus may play a decisive role in surface morphology (especially in vicinal surfaces [27, 28]). Moreover, electrostatic interactions originated from the incomplete screening of step atoms must be also considered, and have been in fact locally measured with AFM [29]. The conclusion is that the elastic fields created by the steps are complex, depend on the specific system, and can play a role on nanomechanical properties through the addition of stress to the externally applied stress.

Surface steps also tend to reduce the real contact area. If the centre of a spherical indenter contacts a single step, just the upper terrace will start severely deforming, while the lower terrace will keep untouched in the very first stages of mechanical contact. The applied force will be then distributed just in the contact areas, and the consequence is a stress enhancement in these areas. Finally, because step atoms have a larger number of missing neighbours than the rest of the surface atoms, a lower stiffness and easier deformation (also plastic) can, in principle, be expected.

8.3.1 Studies on Surface Steps

Surface steps and their role in surface plasticity have been broadly studied under every type of approximation, both theoretically (with continuum elasticity models, discrete dislocation dynamics, atomistic simulations etc.) and experimentally. There are hints in the past suggesting, indirectly, an important role of surface steps as dislocation nucleation centres. One of the first experimental mechanical tests, performed locally on single steps, came from Kiely *et al.* [30]. Short before this work, other authors [31, 32] had performed nanoindentations in well-prepared single crystal surfaces, showing that the yield strengths of flat, defect-free surfaces, were comparable to the ideal strength of a crystal. In their work, Kiely *et al.* [30] gave one step further and carried out nanoindentations on passivated Au surfaces at different distances from atomic steps. The study showed that when the indenter is positioned directly on the step, the yield stress σ_y is 30%–45% lower than the value of σ_y in flat terraces. The authors also reported that the effect of the step is non-local, i.e. no direct contact of the indenter with the step is necessary to significantly reduce the yield stress. An interaction of the stress field of the indenter with the stress field of the step was then suggested. This investigation was one of the first experimental studies of a differentiated plastic response of a surface in the presence of defects. In a previous investigation [32], also in Au, a difference in the *F(z)* curves when they were performed on step bunches was also reported. The authors suggested the possibility that dislocations would preferentially nucleate at the steps.

Atomistic simulations of metallic surfaces with single steps [33] indicate that the presence of a step indeed exerts an influence on the plastic response, lowering the yield strength. Dislocation half loops are observed to nucleate at the steps. But the study also suggests that the direct mechanical contact of the indenter with the step is needed. An interesting conclusion of this work, which has to do with the concept of the real contact area, is that the value of the mean pressure beneath the indenter estimated from the hertzian approximation, $\sigma_y = F_y/(\pi a_y^2)$, is no longer valid, either for very sharp tips or at the very initial stage of the contact. a_y, the contact radius, is not well defined in defective surfaces due to the discontinuity in the position of surface atoms at both sides of the step.

Similar conclusions have been reached with the quasicontinuum (QC) method [34]. The QC method consists in performing atomistic simulations in regions with high deformation gradient, close to the indentation point, while a continuum treatment is considered in the regions with a slow varying deformation. This combined method allows, in principle, for an increase in the simulation size. QC simulations also show [34] the need for a real contact of the indenter with the step in order to perceive a reduction in the yield strength, which can be as large as 70% of the yield strength of the perfect surface. Cyclic loading indentation experiments using a Bragg-Nye bubble raft model [14] also show that when the surface contains steps, these act as dislocation nucleation centres which are emitted into the bulk. Starting from a flat surface, dislocation nucleation is homogeneous and takes place in the crystal below the indenter. But the surface steps formed during the first indentations act as heterogeneous nucleation centres for further indentations, confining all dislocation nucleation activity to the surface in a few cycles.

So, both experimental and simulation works show that, under an indenter load, steps at metal surfaces play an active role by decreasing the elastic threshold and assisting the nucleation of dislocations. However, what happens when the surface has a high density of steps and multiple dislocations nucleate and grow simultaneously? Collective effects based on interactions involving pre-existing defects can lead to hardening phenomena [35, 36], and this issue can have profound implications in plasticity models. Following these ideas, we carried out a study [37] of the incipient plasticity in a Au(788) vicinal surface, miscut 3.5° off the [111] direction, where surface steps are evenly separated by flat (111) terraces just 3.9 nm wide. Steps run parallel to the [0-11] direction in this vicinal surface. When comparing the *F(z)* curves on this surface with those of the flat (111) surface, qualitatively different behaviours are observed. While the penetration curves in the flat Au(111) surface show a good hertzian fit up to the elastic limit, (where a clear pop-in is observed), for the case of the vicinal surface there is a continuous deviation

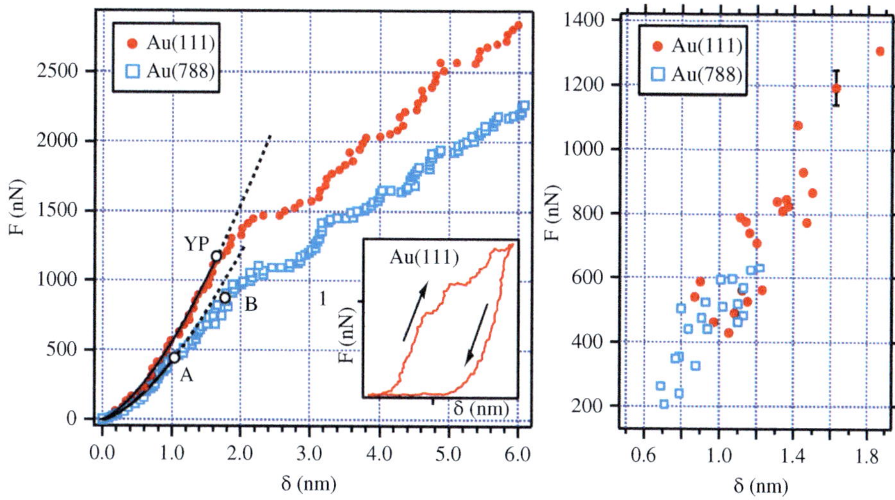

Figure 8.4 (Left) Representative force-penetration experimental curves for a flat Au(111) surface (red) and a stepped Au(788) surface (blue). The hertzian fits are shown for both curves, which terminate at point YP at the flat surface and at point A at the stepped surface. (Right) Distribution of termination points YP and A for the flat and stepped surfaces, respectively. *Source*: Navarro 2008 [37]. Reproduced with permission of American Physical Society.

(without pop-in) from the elastic fit at lower loads (see Figure 8.4). At higher loads, a clear excursion is also observed in the stepped surface. Taking the respective yield points F_Y for each surface, and in terms of the Hertz model, we obtained the maximum resolved shear stress $\tau = 0.31\ p_0$, where $p_0 = [(6F_y\ E^{*2})/(\pi^3 R^2)]^{1/3}$ (E^* is the reduced Young modulus [37]). Using the corresponding values, we obtain $\tau_{111} = 2.1$ GPa and $\tau_{788} = 1.6$ GPa. The value for the flat surface agrees with the theoretical shear strength of a perfect Au crystal. Thus, regularly stepped surfaces yield at smaller loads and without abrupt excursions in the load. Steps seem to lower and *soften* the elastic-plastic trasition.

We carried out, in parallel, atomistic simulations of the process to gain information of the plastic events occurring below the surface, especially during the intermediate stage in the stepped surfaces (between the soft deviation from the elastic response and the first abrupt pop-in). The simulations qualitatively reproduce the behaviour observed experimentally. By comparing experiments and simulations, we conclude that during the continuous and soft deviation of the $F(z)$ indentation curve, dislocation half loops nucleate at the steps under a lower resolved shear stress than the required critical stress at the flat surface. Collective dislocation behaviour leading to hardening processes do not seem to be present, at least under the parallel step configuration studied. The dislocations nucleated at the steps are reversible and disappear reabsorbed by the steps if the load is removed. This uncommon intermediate stage could be termed as *quasiplastic*, since defects (dislocations) are nucleated, but they are not irreversible, no permanent deformation is achieved and no pop-in is observed. Other authors [38] have also reported this intermediate regime of reversible plasticity, where the defects heal upon withdrawal of the indenter. In this respect, some studies argue that substantial dislocation activity can be detected before the first pop-in appears in the load curves [39, 40]. These results suggest that in defective surfaces a certain plastic activity, through heterogeneous dislocation nucleation, exists before any abrupt discontinuity in the load-displacement curves appears.

Of course, other stress states different to that produced by an indenter can be present at surfaces. That is the case of uniaxial stress perpendicular to the step, which has been extensively studied by Beauchamp and collaborators [41–46]. In-plane stresses are present, for instance, in heteroepitaxial thin films, and their study is technologically relevant. The electronic or optoelectronic response, as well as the degradation of the film depend on the existence or control of thin film misfit dislocations. Using atomistic simulations for Al, it has been observed [41] that a uniaxial stress favours dislocation nucleation at the steps well below the theoretical strength. A combination of a continuum study with atomistic simulations show [42] that the stress concentration along a slip plane below a surface step (the *precursor* of the dislocation nucleation) is well explained by the conjugate effect of (i) the elastic stress induced by the step and (ii) the shear component of the external applied stress. Similar conclusions have been reached for the case of semiconductor Si surfaces, with atomistic simulations either using semiempirical potentials [43] or with *ab initio* calculations [44]. Temperature is shown to reduce the strength limit in strained Al [45], and it is interesting to note that at high temperatures, dislocations in Si are replaced by melting/amorphization zones in the slip zones below the surface [43]. The influence of the step height has been also studied in Al and Cu [46]: the nucleation strain decreases for steps being higher than double or triple steps.

Incipient plasticity from surface steps has also been studied with continuum elasticity models. Using a formulation of the Peierls-Navarro dislocation model, it has been claimed [47] that the atomic scale surface step reduces the critical stress required for dislocation nucleation at the defect-free surface by nearly an order of magnitude as the height of the step increases. Other works [35, 36], combining micro-mechanical dislocation models with atomistic simulations, have studied a possible contact hardening behaviour, whose origin is the existence of various slip planes under the steps. Depending on the slip planes activated, dislocations can slip away into the bulk or interact with each other giving rise to a hardened layer. Rough surface contacts, at a small scale, can lead to dislocation segregation and to the formation of a tensile sub-layer. Other studies [48] have addressed the loss of contact in the vicinity of a step and the singular, non-homogeneous contact pressure distribution generated beneath. The singularity produced by the step induces a shear stress concentration, which eases incipient plasticity in stepped surfaces. Some of these models are continuum-based but developed with a low commitment to atomistic information. However, the limitations of these models for the correct analysis of an atomic-scale process, such as dislocation nucleation, are unclear.

8.4 Subsurface Defects

Besides surface steps, whose role in the initiation of plastic activity has been presented in the previous section, other types of imperfections can also influence surface plasticity. To mention a few, vacancies, interstitials, impurities and clusters of any of the former point defects or pre-existing dislocations could be as well sites for heterogeneous dislocation nucleation. In particular, the influence of equilibrium or non-equilibrium point defects on the mechanical response of materials is an old problem. Strengthening effects due to Cottrell atmospheres and dislocation climb are clear examples. The question here is whether subsurface point defects (as well as extended defects such as clusters or dislocations) can influence the plastic response of the surface.

8.4.1 Sub-surface Vacancies

Vacancies are defects in thermal equilibrium in crystals: at any temperature, a given density of vacancies is energetically favourable because of the entropic contribution reducing the free energy $F = U\text{-}TS$. Their ubiquity justifies the study of their possible influence on mechanical properties. The works by Schuh and collaborators [18–20] have introduced the idea that the first pop-in in $F(z)$ load curves, commonly associated to homogeneous dislocation nucleation below the indenter, had indeed an heterogeneous origin, perhaps boosted by subsurface vacancies. These should feel the influence of the elastic field produced by the indenter, and thus be in the proximity of the contact area. While Equation (8.1) gives the nucleation rate for a process both thermally and mechanically activated in an element of volume under a constant stress, the global rate over a given volume Ω is given by:

$$\dot{N} = \dot{n}_0 \exp\left(-\frac{\varepsilon}{k_B T}\right) \cdot \iiint_{\Omega} \exp\left(\frac{\sigma V}{k_B T}\right) d\Omega \tag{8.3}$$

But vacancy concentration at Pt at the temperatures considered (between 20°C and 200°C) is too low to deem pre-existing vacancies beneath the indenter. However, an out-of-equilibrium concentration in the single crystal and also vacancy migration could be considered. Vacancies tend migrate to regions with high compressive stress and the dislocation nucleation process would be then diffusion-limited. However, the introduction of vacancy migration does not fit well into the model developed [18], and can be discarded. So, according to the authors, two plausible mechanisms remain: (i) heterogeneous nucleation at subsurface vacancies, present at a high non-equilibrium density or (ii) activation of dislocation sources such as surface steps or asperities.

Atomistic simulations, combined with experiments, have been performed in Ni and Fe to study the effect of the presence of vacancies during nanoindentation [49, 50]. The conclusion is that the presence of vacancies seems to lower the yield point, acting as direct or indirect heterogeneous dislocation sources. The authors propose that dislocations do not necessarily need to nucleate at the vacancies, and that they might be influencing the nucleation process through their strain field. The introduction of other types of defects (di-vacancies and stacking fault tetrahedra) below Cu surfaces also shows a reduction of the maximum shear stress [51]. Interestingly, the position of the maximum shear stress seems to be less temperature-dependent in the presence of defects.

8.4.2 Sub-surface Impurities and Dislocations

Atomistic simulations under uniaxial strain in Cu crystals with Ag impurities have shown [52] a decrease in the yield point with the presence of the substitutional atoms and with the increase of temperature. Ag atoms lower the stacking fault energy, due to the local strain they produce in the lattice, and act as sources for partial dislocation nucleation. Other elements similar to Ag, such as Au and Ni, should also reduce the yield strain. Also, a similar behaviour might be expected under the stress state produced by a nanoindentation. Very similar results have been obtained for the case of Sb-substituted Cu crystals [53]. These examples are interesting for they somehow oppose to the well known strengthening effect of impurities, which is well known in metallurgy and quite often encountered in macroscopic samples.

A most interesting type of subsurface defect is a pre-existing dislocation. Plastic activity, from the most incipient stages to plastic flow or work hardening, is so intimately related to dislocations, that studying their role during nanoindentations seems mandatory. In microindentation tests, size effects are well known, and depend on the shape and size of the indenter and also on the indentation depth. In general, indentation size effects (ISE) indicate that the smaller the local contact, the larger the locally observed hardness, and this is a consequence of Taylor hardening below the microindenter. The hardness H for a given penetration depth h has been shown to depend as $H = H_0 + C/h$ where C is a constant and H_0 is the macroscopic hardness measured with high probing loads. Both pre-existing dislocations (statistically stored dislocations, SSD) and dislocations created to accommodate plastic deformation (geometrically necessary dislocations, GND) contribute to the Taylor hardening, via dislocation-dislocation interactions, augmenting the measured hardness:

$$H = H_{\text{Peierls}} + \alpha G b \sqrt{\rho_{\text{SSD}} + \rho_{\text{GND}}(h)} \tag{8.4}$$

H_{Peierls} is the hardness of the material without dislocations (mainly derived from the Peierls stress), α is a constant aggregating other constants, G is the shear modulus and b is the representative Burgers vector of the material. ρ_{SSD} is the density of the pre-existing dislocations (before deforming) and ρ_{GND} is the density of those dislocations generated during deformation, which in turn depends on the depth of the indentation h. When the range of nanoindentation is approached (the main topic of this book) the situation qualitatively changes and Equation (8.4) needs to be reconsidered. As already stated, ρ_{SSD} is low or close to cero, and ρ_{GND} is also small if the plastic deformation is close to its incipient stages. Between the microscopic indentation range (relatively well described by models considering ISE) and the nanometric range (where dislocations are presupposed to be non-existing) there may be intermediate cases where just a few dislocations can be present in the neighbourhood of a nanoindentation.

In this respect, Lodes *et al.* [54] have performed nanoindentations in CaF_2 single crystals in regions where larger indentations were previously done. In this way, the nanoindentations were carried out in the presence of a controlled number of pre-existing dislocations. The authors, by combining experimental measurements with molecular dynamics simulations, have observed that plasticity initiates by the movement of previous dislocations in the vicinity. Also, the first pop-in takes place at lower load values and the transition between the elastic regime and the plastic stage is smoother when nanoindenting in the plastic zone. In this sense, the hardness of the surface is reduced with respect to the undeformed surface, and Taylor hardening is only observed in areas with high dislocation density. Exploring the same ideas but using a different protocol, Montagne *et al.* [55] have controllably created dislocations in MgO with crystal cleavage to later perform nanoindentations (Figure 8.5). They have found that pre-existing dislocations lower the pop-in load and promote dislocation nucleation. The authors reveal and distinguish pre-existing and newly-formed dislocations by a double etching method. They have found that, surprisingly, the dislocations

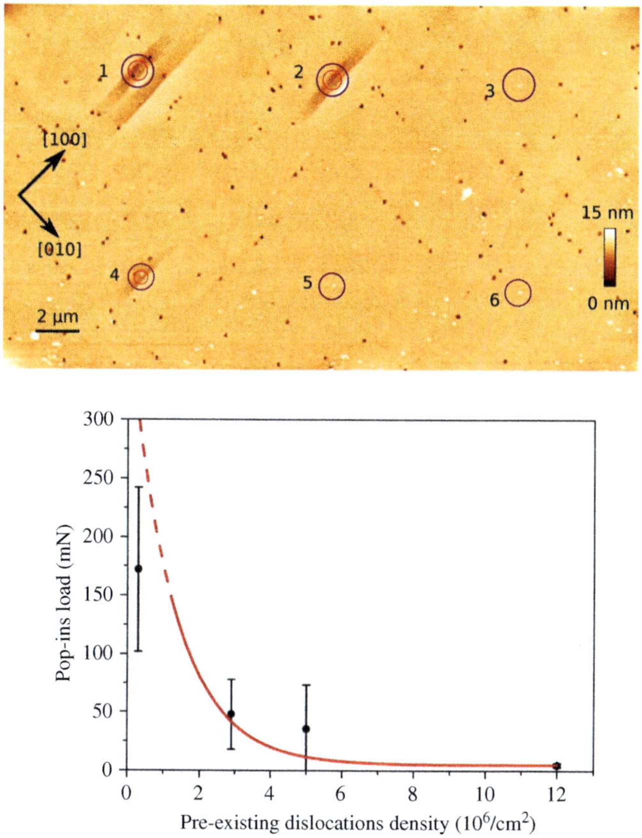

Figure 8.5 (Top) AFM image showing six nanoindentations in a MgO single crystal. The contact points are represented by a purple circle, and those indentations where the plastic yield has been reached show an additional red circle. All the pits correspond to pre-existing dislocations. (Bottom) Load at the yield point versus density of pre-existing dislocations. *Source*: Montagne, 2013 [55]. Reproduced with permission of Elsevier.

nucleated during nanoindentation are not originated from a dislocation source or from pre-existing dislocations. They propose instead that they are homogeneously nucleated in the perfect crystal but assisted by the stress of the pre-existing dislocations.

8.5 Rough Surfaces

Searching for an increasing degree of complexity in structural surface defects, a rough surface obviously represents a more complicated and heterogeneous state than, for instance, a stepped surface. By rough surface we refer here to a surface formed by nanometric mounds and holes distributed along all directions. It is clear in this case that the surface contains steps, accommodating the roughness, distributed along all possible directions. Depending on the history of the rough surface (i.e. the growth or modification method), describing its state and quantifying its response can be rather complex. First, if the roughness has been induced by low energy ion bombardment, the existence of subsurface defects resulting from the incoming ions (interstitials, vacancies or dislocation loops) may complicate the interpretation of the mechanical response. Second, in the case of two very rough surfaces contacting with a low real contact surface (the Austria–Switzerland analogy [1] mentioned in the introduction), it seems rather obvious that highly localized stresses appear at the contact points. The quantification of this stress is a difficult task since the real contact area is not well known. Despite, some efforts have been done to study nanoplasticity in rough surfaces.

Surface roughness is, at the atomic level, constituted by many surface steps aligned along the two or three equivalent directions (depending on the surface symmetry). In the case of vicinal surfaces, all incipient dislocation half loops are nucleated in the same slip plane (determined by the unique step direction), reducing the chance for dislocations to interact. In contrast, in rough surfaces, no specific slip plane is preferred and all can be active in a spatial range close to the surface. This would allow, in principle, that dislocation loops nucleated along different slip planes might intersect and form locks, possibly hardening the surface [56]. Thus, a mechanical softening behaviour of rough surfaces, as measured in stepped surfaces, cannot be ascertained *a priori*.

Shibutani *et al.* [57] have used electropolished rough Al single crystals to probe their nanomechanical properties with indentation. The duration of electropolishing determines the surface roughness, which is smaller for longer times. The authors have found that the higher the roughness, the lower is the applied force to produce the first burst in the load-displacement curve. Also, they have found a linear correlation between the applied load and the width of the first burst or pop-in: the higher the amount of load needed to produce the first pop-in, the higher the width of the burst. It seems that, in this case, surface roughness tends to reduce the needed applied force to reach plasticity.

As already mentioned, one of the most extended ways to modify surfaces is low energy ion bombardment. In this regard, we have performed nanoindentation experiments on ion-eroded Au(001) surfaces [58] at different ion fluxes. Bombarded surfaces are rich in surface and sub-surface defects. At the surface, steps are aligned along different crystallographic directions, as opposed to the case of stepped vicinal surfaces. Also, sub-surface defects, such as displaced atoms (interstitials), vacancies or implanted ions (Ar^+ in our case), may play a role in the mechanical response of the surface, as they play in the case of bulk mechanical properties. Indeed, the controlled introduction of point defects in solids is a technical process aimed at hardening materials. The interaction of dislocations with point defects is one of the most studied subjects in the field of mechanical properties of solids [59], usually limiting dislocation flow and, consequently, augmenting the hardness of the solid. Hence, the expected role of each type of defect (either at the surface or below it) might have opposite contributions to the hardness in ion-bombarded surfaces. We have shown that ion-eroded Au surfaces, bombarded under high ion fluxes, show a reduced onset of surface plasticity (see Figure 8.6). These experimental observations are supported by molecular dynamics (MD) simulations [60], which show that nanoindentations at surface protrusions do not exhibit any sign of lock formation from the interaction of dislocations from different slip systems. In fact, MD simulations show a mechanical softening due to surface roughness. Also, MD simulations show a slight increase of the hardness with the presence of interstitials. But the experiments show that the mechanical softening contribution of the roughened surface dominates over the possible hardening trend of point defects in ion-bombarded Au surfaces.

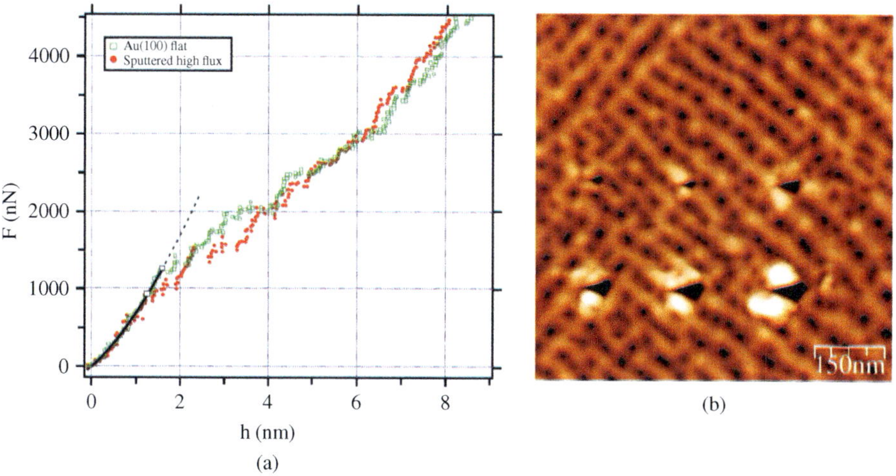

Figure 8.6 (Left) Representative force versus penetration curves performed in flat (green) and ion-bombarded (red) Au(100) surfaces. The hertzian fit is also represented, which terminates at the yield point of each surface. (Right) AFM image of a matrix of several nanoindentations (performed with increasing maximum applied load from top to bottom and from left to right) on a Au(100) surface. This surface has been previously nanostructured with low energy ion bombardment at high ion doses and fluxes. *Source*: Rodríguez 2013 [63]. With permission of IOP.

Ziegenhain *et al.* [60] have performed nanoindentation simulations on adatom islands in Cu surfaces. They have corroborated the general result (i.e. surface defects weaken the surface reducing the yield point), and have also studied the dependency of the plastic response on the size of the island (relative to the size of the indenter). Large islands behave like flat surfaces, for there is no direct contact of the indenter with the island steps. Smaller islands immediately disappear in the form of prismatic dislocation loops emitted into the bulk. Indentation proceeds then as for a flat surface. And in between, in intermediate adatom islands, dislocations are nucleated in some of the steps, and plastic activity remains localized under the island.

We can end up referring to the plastic response of a rough surface of an oxide, TiO_2. By far, most of the experiments and simulations on nanomechanical properties of defective surfaces have been accomplished in metals. These are most often encountered as structural materials and, from the point of view of their understanding, they are simpler than other types of solids. Oxides, for instance, are a very important class of materials, but the complexity of their interactions and the nature of their defects makes it difficult to interpret the experiments. We have tested in the past nanomechanical properties of TiO_2 [61]. Under the action of an ion beam, we have observed that a TiO_2(110) surface exhibits a complex and interesting behaviour, which can include an structural and chemical transformation under very high ion doses [62]. We have also studied the nanomechanical response of TiO_2(110) surfaces modified with Ar^+ beams [63]. After bombardment, this surface is topographically rough and chemically reduced, because O atoms have a higher sputtering yield than Ti atoms. In contrast with the situation found in metallic surfaces, such as in Au [58], we have not found any noticeable changes in the nanomechanical properties of ion-bombarded TiO_2 compared to the pristine surface. Probably, opposite effects cancel out. This is an indication of how the mechanical response of oxides can be quite different to that of metals also at the nanoscale.

8.6 Conclusions

Surface defects definitely influence, to a greater or lesser extent, the incipient mechanical response of a surface. The physics of the process mainly relies on the heterogeneous character of dislocation nucleation. Defects, in general, tend to lower the load threshold needed to enter the plastic range and to soften the transition. Also, reversible plastic events have been identified at surface defects, i.e. nucleation of dislocations which disappear at the surface once the load is removed. Most interestingly, the role of defects is probably more relevant than it was ordinarily believed in the past. Some works have remarked that the first plastic events at flat surfaces, usually reported as homogeneous (for the surfaces were considered to be defect-free), have been in the last years identified as of the heterogeneous type.

Most of the knowledge and works reviewed here are a representative set of the efforts aimed to understand how surface plasticity can be tuned through surface defects. With the size of many devices and physical systems being continuously reduced, the controlled and purposed generation of surface defects, planned to tailor the mechanical response, will probably become increasingly important. Together with the control of the microstructure and the size and shape of the material, the control of surface defects will probably be another strategy to enhance the mechanical properties of the material. In this sense, much further work is needed.

Acknowledgements

I would like to thank all my former and present collaborators, who have helped me over the last years, to learn what I know about surface nanomechanics. And definitely, I would like to dedicate this work to Professor Juan Rojo, my advisor and friend, from whom I have learnt so much and such good things.

References

1 F.P. Bowden, D. Tabor. The Friction and Lubrication of Solids. Oxford, Oxford University Press; 1950.
2 G.A. Somorjai, Y. Li. Introduction to Surface Chemistry and Catalysis, John Wiley and Sons, 2010.
3 C. Kittel. Introduction to Solid State Physics. Wiley; 2004.
4 H.S. Taylor. A theory of the catalytic surface. Proceedings of the Royal Society of London A, 108, 105; 1925.
5 J.J. Gilman. The mechanism of surface effects in crystal plasticity. Philosophical Magazine, 6, 159; 1961.
6 T. Zhu, J. Li. Ultra-strength materials. Progress in Materials Science, 55, 710; 2010.

7 J.R. Greer, and J.T.M. De Hosson. Plasticity in small-sized metallic systems: Intrinsic versus extrinsic size effect. Progress in Materials Science, 56, 654; 2011.
8 E. Carrasco, O. Rodríguez de la Fuente, M.A. González, J.M. Rojo. Characterising and controlling surface defects. European Journal of Physics, 40, 421; 2004.
9 B. Luan, M.O. Robbins. The breakdown of continuum models for mechanical contacts. Nature, 435, 929; 2005.
10 L. Nicola, A.F. Bower, K.S. Kimet *et al.* Surface versus bulk nucleation of dislocations during contact. Journal of the Mechanics and Physics of Solids, 55, 1120; 2007.
11 J. Frenkel. Zur Theorie der Elastizitätsgrenze und der Festigkeit kristallinischer Körper. Zeitschrift für Physik, 37, 572; 1926.
12 K.L. Johnson. Contact Mechanics. Cambridge, Cambridge University Press; 1985.
13 A. Gouldstone, K.J. van Vliet, S. Suresh. Nanoindentation: Simulation of defect nucleation in a crystal. Nature, 411, 656; 2001.
14 K.J. van Vliet, S. Suresh. Simulations of cyclic normal indentation of crystal surfaces using the bubble-raft model. Philosophical Magazine A, 82, 1993; 2002.
15 C.L. Kelchner, S.J. Plimpton, J.C. Hamilton. Dislocation nucleation and defect structure during surface indentation. Physical Review B, 58, 11085; 1998.
16 M.A. Tschopp, D.E. Spearot, D.L. McDowell. Atomistic simulations of homogeneous dislocation nucleation in single crystal copper. Modelling and Simulation in Materials Science and Engineering, 15, 693; 2007.
17 J. Li, T. Zhu, S. Yip, K.J. van Vliet, S. Suresh. Elastic criterion for dislocation nucleation. Materials Science and Engineering A, 365, 25; 2004.
18 J.K. Mason, A.C. Lund, C.A. Schuh. Determining the activation energy and volume for the onset of plasticity during nanoindentation. Physical Review B, 73, 054102; 2006.
19 C.A. Schuh, J.K. Mason, A.C. Lund. Quantitative insight into dislocation nucleation from high-temperature nanoindentation experiments. Nature Materials, 4, 617; 2005.
20 A.C. Lund, A.M. Hodge, C.A. Schuh. Incipient plasticity during nanoindentation at elevated temperatures. Applied Physics Letters, 85, 1362; 2004.
21 W. Paul, D. Oliver, Y. Miyahara, P.H. Grütter. Minimum threshold for incipient plasticity in the atomic-scale nanoindentation of Au(111). Physical Review Letters, 110, 135506; 2013.
22 H. Bei, Y.F. Gao, S. Shim *et al.* Strength differences arising from homogeneous versus heterogeneous dislocation nucleation. Physical Review B, 77, 060103; 2008.
23 I. Palacio, J.M. Rojo O. Rodríguez de la Fuente. Surface defects activating new reaction paths: formation of formate during methanol oxidation on Ru(0001). ChemPhysChem, 13, 2354; 2012.
24 V.I. Marchenko, A.Y. Parshin. Elastic properties of the crystal surface. Soviet Physics: Journal of Experimental and Theoretical Physics, 52, 129; 1980.
25 R.V. Kukta, A. Peralta, D. Kouris. Elastic interaction of surface steps: effect of atomic-scale roughness. Physical Review Letters, 88, 186102; 2002.
26 L.E. Shilkrot, D.J. Srolovitz. Elastic field of a surface step: Atomistic simulations and anisotropic elastic theory. Physical Review B, 53, 11120; 1996.
27 G. Prevot, P. Steadman, S. Ferrer. Determination of the elastic dipole at the atomic steps of Pt(977) from surface X-ray diffraction. Physical Review B, 67, 245409; 2003.

28 G. Prevot, B. Croset. Revisiting elastic interactions between steps on vicinal surfaces: The buried dipole model. Physical Review Letters, 92, 256104; 2004.
29 J.Y. Park, G.M. Sacha, M. Enachescu *et al.* Sensing dipole fields at atomic steps with combined scanning tunneling and force microscopy. Physical Review Letters, 95, 136802; 2005.
30 J.D. Kiely, R.Q. Hwang, J.E. Houston. Effect of surface steps on the plastic threshold in nanoindentation. Physical Review Letters, 81, 4424; 1998.
31 T.A. Michalske, J.E. Houston. Dislocation nucleation at nano-scale mechanical contacts. Acta Materialia, 46, 391; 1998.
32 S.G. Corcoran, R.J. Colton, E.T. Lilleodden, W.W. Gerberich. Anomalous plastic deformation at surfaces: Nanoindentation of gold single crystals. Physical Review B, 55, 16057; 1997.
33 J.A. Zimmerman, C.L. Kelchner, P.A. Klein *et al.* Surface step effects on nanoindentation. Physical Review Letters, 87, 165507; 2001.
34 D. Shan, L. Yuan, B. Guo. Multiscale simulation of surface step effects on nanoindentation. Materials Science and Engineering A, 412, 264; 2005.
35 H.H. Yu, P. Shrotriya, Y.F. Gao, K.S. Kim. Micro-plasticity of surface steps under adhesive contact: Part I – Surface yielding controlled by single-dislocation nucleation. Journal of the Mechanics and Physics of Solids, 55, 489; 2007.
36 Y.F. Gao, H.H. Yu, K.S. Kim. Micro-plasticity of surface steps under adhesive contact: Part II – Multiple dislocation mediated contact hardening. Journal of the Mechanics and Physics of Solids, 56, 2759; 2008.
37 V. Navarro, O. Rodríguez de la Fuente, A. Mascaraque, J.M. Rojo. Uncommon dislocation processes at the incipient plasticity of stepped gold surfaces. Physical Review Letters, 100, 105504; 2008.
38 G. Ziegenhain, H.M. Urbassek. Reversible plasticity under nanoindentation of atomically flat and stepped surfaces of f.c.c. metals. Philosophical Magazine Letters, 89, 717; 2009.
39 J. Knap, M. Ortiz. Effect of indenter-radius size on Au(001) nanoindentation. Physical Review Letters, 90, 226102; 2003.
40 A. Minor, S.A. Syed Asif, Z. Shan *et al.* A new view of the onset of plasticity during the nanoindentation of aluminium. Nature Materials, 5, 697; 2006.
41 S. Brochard, P. Beauchamp, J. Grilhé. Dislocation nucleation from surface steps: Atomistic simulation in aluminium. Philosophical Magazine, 80, 503; 2000.
42 S. Brochard, P. Beauchamp, and J. Grilhé. Simulations of dislocation nucleation from atomic size surface steps and grooves. Materials Science and Engineering A, 309, 456; 2001.
43 J. Godet, L. Pizzagalli, S. Brochard, P. Beauchamp. Theoretical study of dislocation nucleation from simple surface defects in semiconductors. Physical Review B, 70, 054109; 2004.
44 J. Godet, S. Brochard, L. Pizzagalli *et al.* Dislocation formation from a surface step in semiconductors: An ab initio study. Physical Review B, 73, 092105; 2006.
45 P. Hirel, S. Brochard, L. Pizzagalli, P. Beauchamp. Effects of temperature and surface step on the incipient plasticity in strained aluminium studied by atomistic simulations. Scripta Materialia, 57, 1141; 2007.
46 S. Brochard, P. Hirel, L. Pizzagalli, J. Godet. Elastic limit for surface step dislocation nucleation in face-centered cubic metals: Temperature and step height dependence. Acta Materialia, 58, 4182; 2010.

47 C. Li, G. Xu. Critical conditions for dislocation nucleation at surface steps. Philosophical Magazine, 86, 2957; 2006.
48 M.R. Shankar. Surface steps lead to heterogeneous contact mechanics and facilitate dislocation nucleation in nanoindentation. Applied Physics Letters, 90, 171924; 2007.
49 I. Salehinia, V. Perez, D.F. Bahr. Effect of vacancies on incipient plasticity during contact loading. Philosophical Magazine, 92, 550; 2012.
50 E.K. Njeim, D.F. Bahr. Atomistic simulations of nanoindentation in the presence of vacancies. Scripta Materialia, 62, 598; 2010.
51 I. Salehinia, D.F. Bahr. Mechanical behavior of FCC single crystals at finite temperatures in the presence of point defects. Materials Science and Engineering A, 588, 340; 2013.
52 N. Amigo, G. Gutiérrez, M. Ignat. Atomistic simulation of single crystal copper nanowires under tensile stress: Influence of silver impurities in the emission of dislocations. Computational Materials Science, 87, 76; 2014.
53 R.K. Rajgarhia, D.E. Spearot, A. Saxena. Heterogeneous dislocation nucleation in single crystal copper-antimony solid-solution alloys. Modelling and Simulations in Materials Science and Engineering, 17, 055001; 2009.
54 M.A. Lodes, A. Hartmaier, M. Göken, K. Durst. Influence of dislocation density on the pop-in behavior and indentation size effect in CaF_2 single crystals: Experiments and molecular dynamics simulations. Acta Materialia, 59, 4264; 2011.
55 A. Montagne, V. Audurier, C. Tromas. Influence of pre-existing dislocations on the pop-in phenomenon during nanoindentation in MgO. Acta Materialia, 61, 4778; 2013.
56 A. Gannepalli, S.K. Mallapragada. Atomistic studies of defect nucleation during nanoindentation of Au(001). Physical Review B, 66, 104103; 2002.
57 Y. Shibutani, A. Koyama. Surface roughness effects on the displacement bursts observed in nanoindentation. Journal of Materials Research, 19, 183; 2004.
58 V. Navarro, O. Rodríguez de la Fuente, A. Mascaraque, J.M. Rojo. Plastic properties of gold surfaces nanopatterned by ion beam sputtering. Journal of Physics: Condensed Matter, 21, 224023; 2009.
59 J.P. Hirth and J. Lothe, Theory of Dislocations, 2nd edition. New York, John Wiley and Sons; 1982.
60 G. Ziegenhain, H.M. Urbassek. Nanostructured surfaces yield earlier: Molecular dynamics study of nanoindentation into adatom islands. Physical Review B, 81, 155456; 2010.
61 V. Navarro, O. Rodríguez de la Fuente, A. Mascaraque, J.M. Rojo. Reduced hardness at the onset of plasticity in nanoindented titanium dioxide. Physical Review B, 78, 224107; 2008.
62 B.M. Pabón, J.I. Beltrán, G. Sánchez-Santolino *et al.* Formation of titanium monoxide (001) single-crystalline thin film induced by ion bombardment of titanium dioxide (110). Nature Communications, 6, 6147; 2015.
63 O. Rodríguez de la Fuente, M.A. González-Barrio, V. Navarro *et al.* Surface defects and their influence on surface properties. Journal of Physics: Condensed Matter, 25, 484008; 2013.

9

Viscoelastic and Tribological Behavior of Al_2O_3 Reinforced Toughened Epoxy Hybrid Nanocomposites

Mandhakini Mohandas[1] *and Alagar Muthukaruppan*[2]

[1] *Centre for Nanoscience and Technology, Anna University, Chennai, India*
[2] *Polymer Composite Laboratory, Departnment of Chemical Engineering, Anna University, Chennai, India*

9.1 Introduction

The polymeric materials usually exhibit mechanical strength, light weight, ease of processing, versatility and low cost, together with acceptable thermal and environmental resistances, in order to be suitable for tribological applications they should also exhibit good abrasion resistance and wear resistance. This target is not easy to achieve, since the viscoelasticity of polymeric materials makes the analysis of the tribological features and the processes involved in such phenomena are quite complicated. Indeed, it is well-known that an improvement of the mechanical properties can be effectively achieved by including 'small' inorganic particles in the polymer matrices [1]. For applications taking place in hard-working conditions, such as slide bearings, the development of composite materials, which possess a high stiffness, toughness and wear resistance, becomes crucial. On the one hand, the extent of the reinforcing effect depends on the properties of composite components, and on the other hand it is strongly affected by the microstructure represented by the filler size, shape, homogeneity of distribution/dispersion of the particles within the polymer, and filler/matrix interface extension. This latter plays a critical role, since the composite material derives from a combination of properties, which cannot be achieved by either of the components alone. Thus it is generally expected that the characteristics of a polymer, reinforced with a certain volume fraction of particles having a certain specific surface area, are more strongly influenced when very small particles (nanofillers), promoting an increased interface within the bulk polymer, are used [2–4]. However, this happens only when a high dispersion efficiency of the nanoparticles within the polymer matrix is assessed: indeed, nanoparticles usually tend to agglomerate because of their high specific surface area, due the adhesive interactions derived from the surface energy of the material. In particular, the smaller the size of the nanoparticles, the more difficult is the breaking down of such agglomerates, so that their homogeneous distribution within the polymer matrix is compromised. As a consequence, the development of nanocomposites showing high tribological features requires a deep investigation into their micro-to-nanostructure, aiming to find synergistic mechanisms and reinforcement effects exerted by the nanofillers. In particular, applications such as brake pads or clutches usually require a high friction coefficient

Applied Nanoindentation in Advanced Materials, First Edition. Edited by Atul Tiwari and Sridhar Natarajan.

and, at the same time, a low wear resistance. However, in the case of gears or bearings, acting as smooth metallic counterparts under dry sliding conditions the development of polymer composites having low friction and wear properties are strongly needed.

Hence in the present study, the two component of C4 e-BMI toughened epoxy matrix have been developed to obtain excellent toughness in parallel with mechanical properties. Hence an attempt has been aimed at clarifying the effects of nanoalumina and to investigate the effect of crosslinking density on the wear of C4 e-BMI toughened epoxy alumina nanocomposites and also to investigate the effect of functionalized alumina concentration on wear behavior of e-BMI toughened epoxy nanocomposites.

9.2 Experimental

9.2.1 Materials

Epoxy resin (Diglycidyl ether of bisphenol A, DGEBA) LY556 having an epoxy equivalent of 180–190, with the viscosity of about 10,000 cP and 4.4-diaminodiphenylmethane (DDM), was obtained from Ciba-Geigy Ltd., India. The alumina nanoparticles of size 17 nm were purchased from Sigma-Aldrich. C4-ether-linked bismaleimide (C4 e-BMI) was synthesized from maleic anhydride as reported [5].

The FTIR measurements were taken by a Nicolet Avatar 320 Fourier transform infrared spectrometer at room temperature (25°C) in the range of 4,000–400 cm^{-1}. The dynamic mechanical analysis (DMA) was carried out on SII Nanotechnology (Model DMS-6000 Japan) in a three point bending configuration, at a temperature range of 30–250°C using a specimen with the dimensions of 55 mm x 100 mm x 3 mm. The frequency used is 1.0 Hz, and the rate is 2°C min^{-1} for cured epoxy matrices. Nanoindentation experiments were performed to measure the hardness and Young's moduli of the samples. In these experiments, we have performed the nanoindentation testing using Agilent Nano Indenter G200. A Berkovich (three-side pyramidal) diamond indenter was used throughout the experiments in order to determine the nanohardness of specimens. To determine sliding wear, scratches along the samples were performed on the wear path length of 200 μm using a wear load of 50 mN and wear test velocity of 50 μm s^{-1}. Samples with grooves obtained from sliding wear determination were coated with gold. The mode of wear was investigated in detail by analyzing the sample surfaces using FESEM apparatus.

3-Glycidoxypropylmethoxysilane (GPS) was used as the coupling agent to functionalize the particles as per the procedure reported [6, 7] and the schematic representation is depicted in Figure 9.1.

Transmission electron microscopy is very useful for the characterization of filler structures. The TEM images obtained for nano- Al_2O_3 and silane-treated nano- Al_2O_3 powders are compared in Figure 9.2. Comparison of TEM images (a) and (b) show that the surface modification of the silane-treated nano- Al_2O_3 particles dramatically affects the aggregate particle size.

The 15:85 C4 e-BMI/DGEBA-reinforced Al_2O_3 samples were prepared by blending Al_2O_3 and C4 e-BMI in DGEBA at about 60°C through mechanical stirring. The modified resin was mixed with a stoichiometric amount of DDM and stirred continuously until a homogeneous solution was obtained. The mixtures were cured in an aluminium

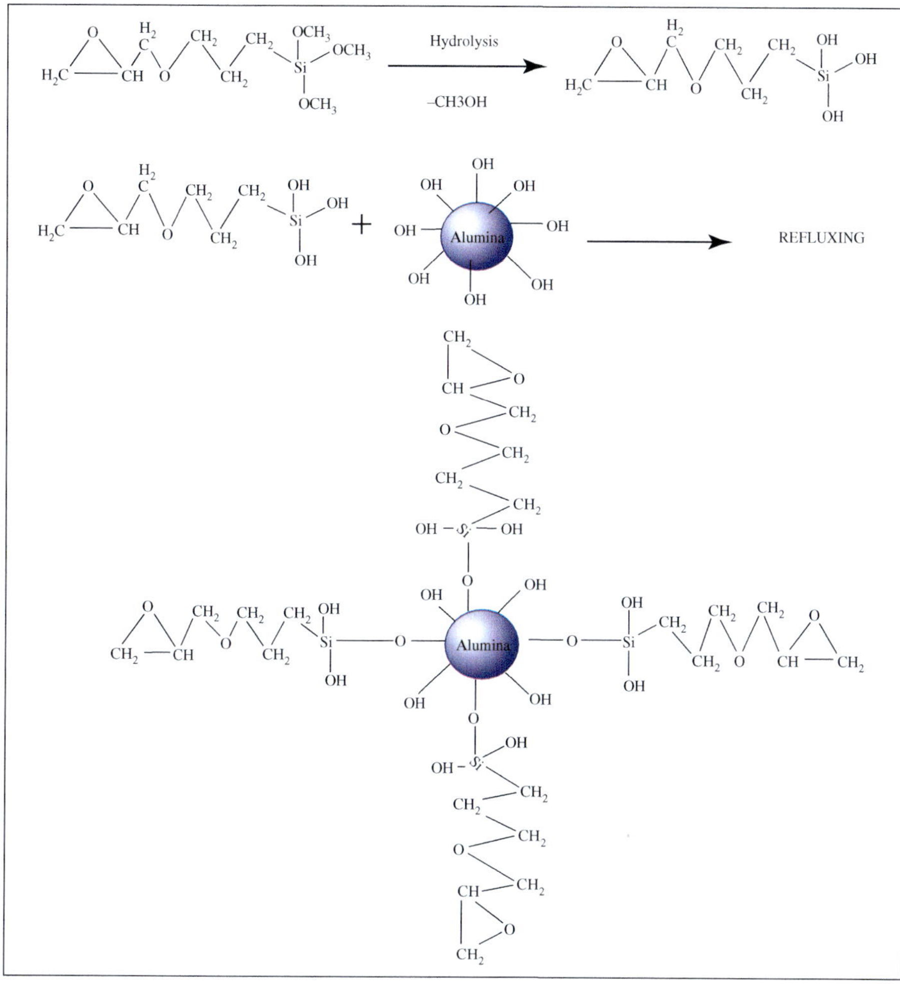

Figure 9.1 Functionalization of alumina particles with GPS.

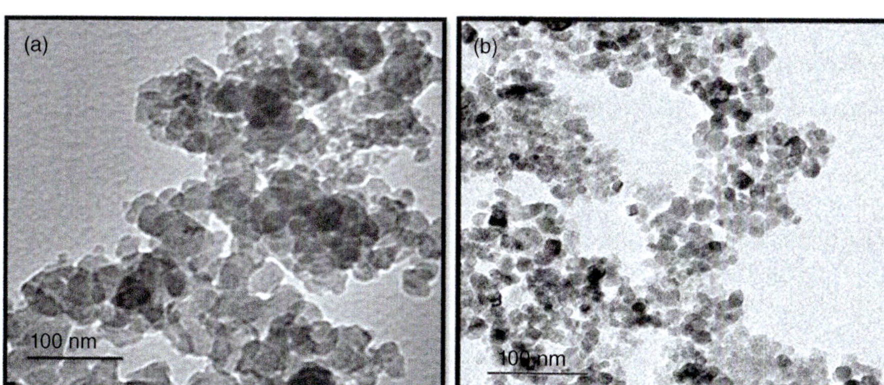

Figure 9.2 TEM micrograph of (a) nano- Al_2O_3 and (b) silane-treated nano-Al_2O_3.

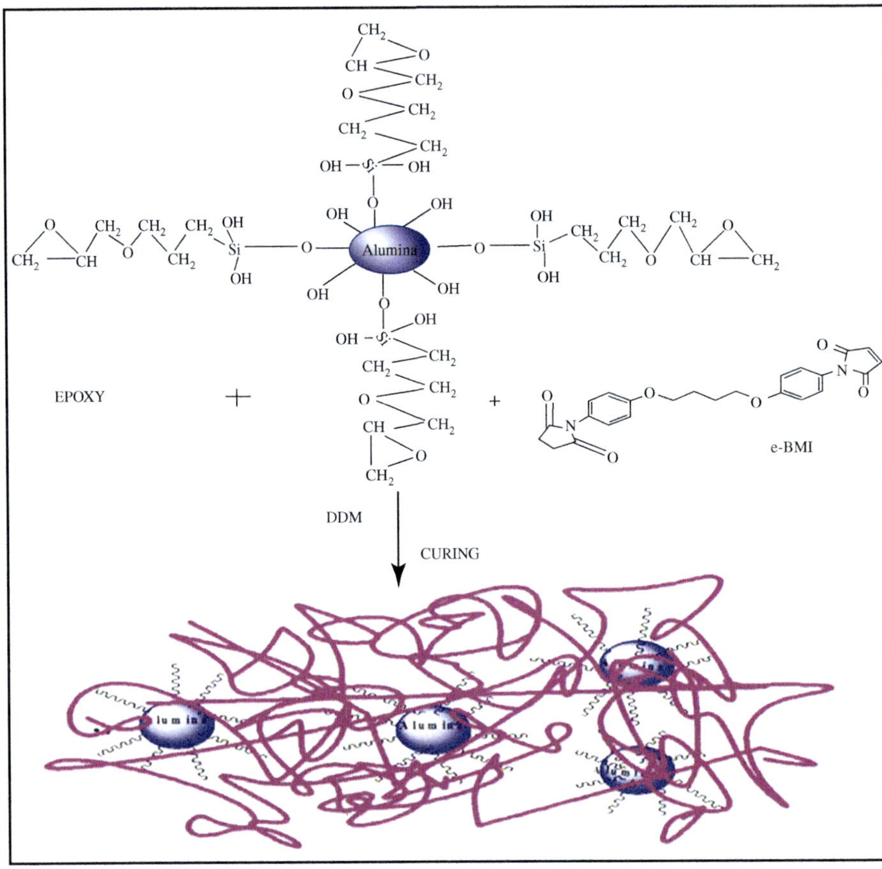

Figure 9.3 Preparation of C4 e-BMI/DGEBA/Al_2O_3 nanocomposites.

mould after being degassed under vacuum for 10 min at room temperature. The liquid blends were cured at 120°C for 2 h and post cured at 180°C for 3 h to attain the complete curing reaction and to allow the samples to cool gradually to room temperature, and the schematic representation is shown in Figure 9.3.

9.2.2 FTIR Analysis

The FTIR spectrum of uncoated particles in Figure 9.4a shows a broad peak at around 3293 cm^{-1} which is due to the hydroxyl groups on the surface of the alumina particles [8, 9]. The FTIR spectrum of GPS-coated alumina particles (Figure 9.4b) indicates an additional peak at 1110.5 cm^{-1}, which is absent in the FTIR spectrum of uncoated alumina particles. This peak is attributed to Si–O–Al bond formation, the confirmation of which is explained as follows.

The peak at 1046 cm^{-1} represents the presence of Si–O–Si bonds, whereas a peak at 1110.5 cm^{-1} corresponds to asymmetrical stretching of Si–O in Si–O–H. The peaks at 2928 and 2880 cm^{-1} correspond to the asymmetrical and symmetrical stretching of CH_3 and CH_2 groups. The peaks at 1200.5 cm^{-1} and 918 cm^{-1} correspond to the asymmetrical aliphatic C–O stretching and epoxide ring vibrations respectively. When the

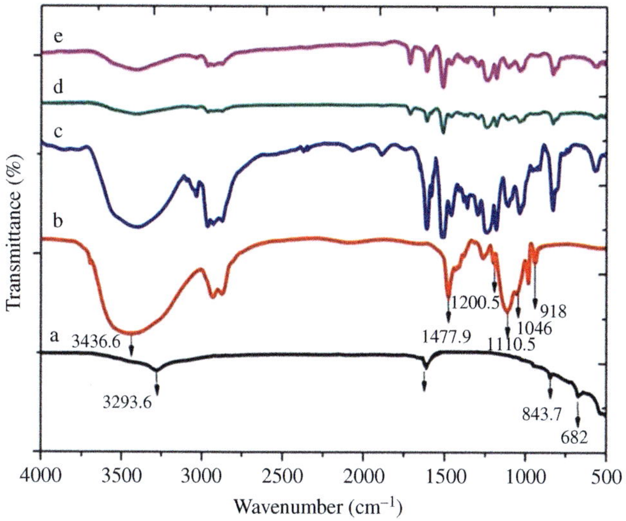

Figure 9.4 FTIR spectra of C4 e-BMI/Al_2O_3-toughened epoxy nanocomposites.

GPS-coated particles are dispersed in the epoxy resin and cured, the epoxide end of the GPS is involved in the crosslinking with the epoxy. Thus, a strong covalent bond was formed between the epoxy and alumina nanoparticles through GPS functionalization, and it is clearly shown in Figure 9.4c, d and e for 1 and 5 wt% GPS-treated alumina epoxy nanocomposites. The schematic diagram of the strong covalent bond formed between the epoxy and alumina nanoparticles through GPS functionalization is shown in Figure 9.3.

9.2.3 Results and Discussion

The morphology and roughness of the samples were characterized by SEM and optical profiler. Optical surface roughness profiles of the neat epoxy, C4 e-BMI-toughened epoxy and alumina-reinforced C4 e-BMI-toughened epoxy nanocomposite samples are shown in Figure 9.5. The average roughness values of all the samples were below 400 nm, indicating that the surfaces of all samples were relatively smooth. To verify whether the particles were properly dispersed in the epoxy matrix, SEM images were taken for each weight percentage. Figure 9.6a–d shows the SEM images of nanocomposite having 1, 3, 5 and 10 wt% GPS-coated alumina nanoparticles, respectively. The nanocomposites with 1, 3 and 5 wt% GPS-alumina showed enhanced tribological properties due to uniform distribution of the alumina particles. However, negative results were observed with 10 wt% alumina reinforcement due to insufficient dispersion and large agglomerates of alumina nanoparticles [10].

In order to get a more clear view of the actual size of the particles, TEM images were taken. Figure 9.7a–c shows the dispersion state of nano- Al_2O_3 content of 3, 5 and 10 wt% in C4 e-BMI-toughened epoxy matrix. The GPS-coated Al_2O_3 particles were homogenously mixed within epoxy resin, and individually isolated particles without any agglomeration were observed up to 5 wt% beyond which the particles started to agglomerate. The GPS had an epoxide functional group, which was compatible with

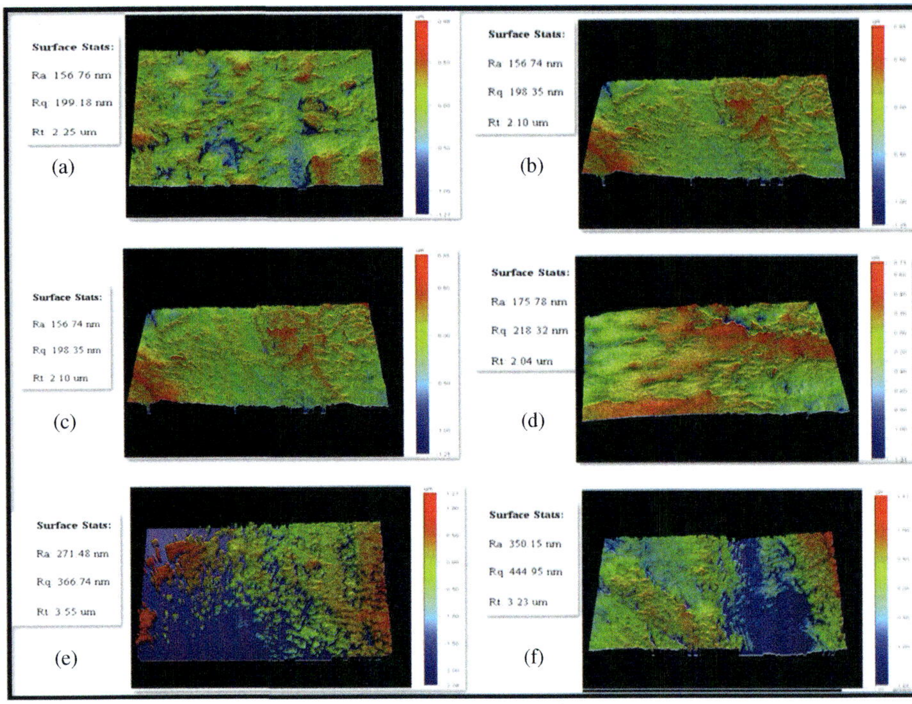

Figure 9.5 The optical surface profile images of (a) neat epoxy, (b) 15 wt% alumina (c) 15 wt% C4 e-BMI/epoxy system/1 wt% GPS-alumina (d) 15 wt% C4 e-BMI/epoxy system/3 wt% GPS (e) 15 wt% C4 e-BMI/epoxy system/5 wt% GPS-alumina C4 e-BMI/epoxy system and (f) 15 wt% C4 e-BMI/epoxy system/10 wt% GPS-alumina.

the epoxy resin. The epoxide groups at the other end of silane could react with the epoxy during the curing process. The dispersion of Al_2O_3 nanoparticles is thus enhanced by the silane coating. In effect, the interacting force between the Al_2O_3 nanoparticles and epoxy matrix is improved because the formation of chemical bonding between the Al_2O_3 nanofillers and the epoxy matrix is enhanced.

9.2.3.1 Viscoeleastic Properties

9.2.3.1.1 Storage Modulus

The results of DMA, which was performed to evaluate viscoelastic properties such as storage modulus and glass transition temperature (T_g) of the composites, are presented in Table 9.1. In the case of the 10 wt% C4 e-BMI-toughened epoxy matrices, low storage modulus is exhibited as shown in Figure 9.8, which may be due to the flexibility brought about by the ether linkage.

However, the storage modulus is seen to increase with an incremental concentration of the 15 wt% C4 e-BMI, and this is ascribed to the presence of entangled networks and a high level of compatibility between interpenetrated networks of epoxy and C4 e-BMI, which also fills the defects existing in the single polymer. The interfacial thickness is expected to be enhanced with an increase in interpenetration which consequently

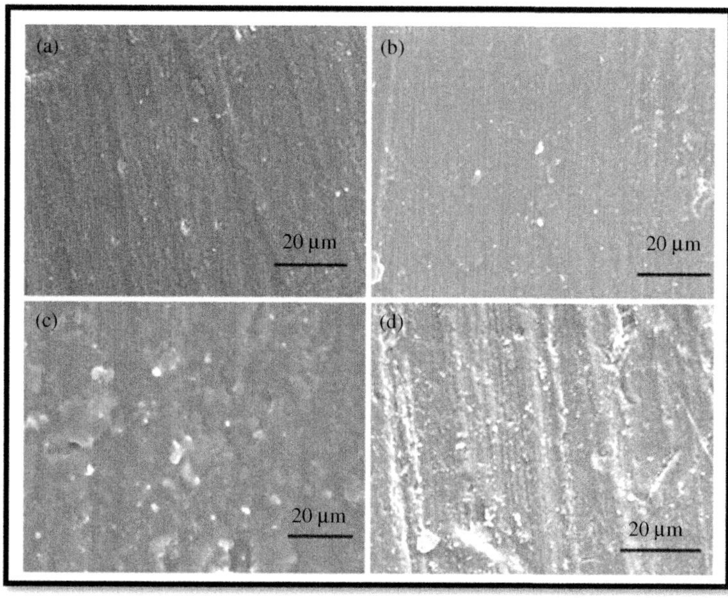

Figure 9.6 SEM images of C4 e-BMI-toughened epoxy matrices containing (a) 1 wt% GPS, (b) 3 wt% GPS Al_2O_3, (c) 5 wt% GPS Al_2O_3 and (d) 10 wt% GPS Al_2O_3.

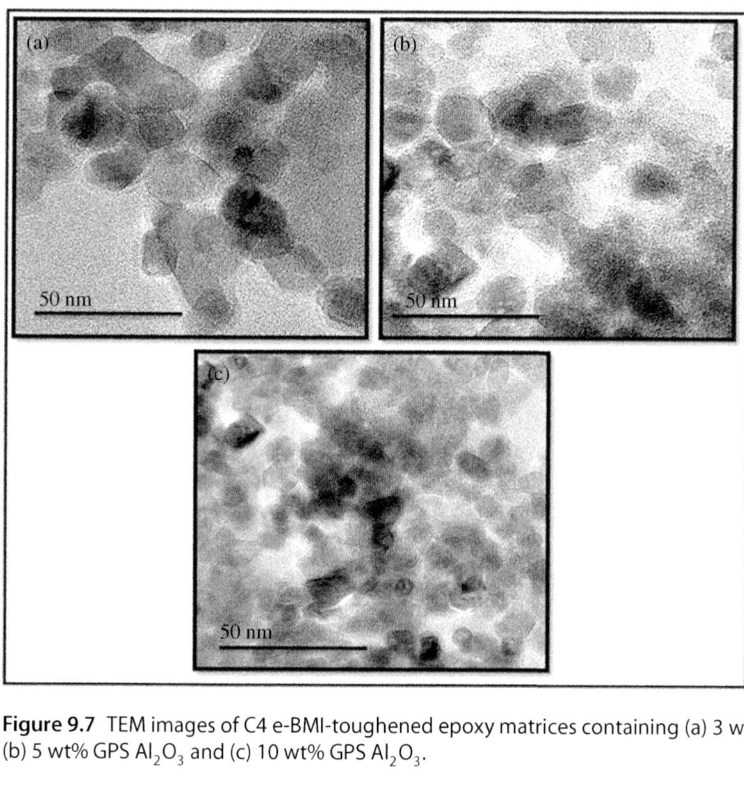

Figure 9.7 TEM images of C4 e-BMI-toughened epoxy matrices containing (a) 3 wt% GPS Al_2O_3, (b) 5 wt% GPS Al_2O_3 and (c) 10 wt% GPS Al_2O_3.

Table 9.1 Effect of crosslinking on tribological behaviour.

Epoxy + C4 e-BMI + Al_2O_3	Elastic modulus (GPa)	Hardness (GPa)	Wear track deformation (μ2)	M_C ($g\ mol^{-1}$)	$\nu_e \times 10^{21}$ Chains m^{-3}
100/00/00	3.0	0.13	115.4	609.4	1.29
100/15/00	3.9	0.18	113.22	829.0	0.82
100/15/01	3.9	0.19	97.50	847.8	0.91
100/15/05	4.2	0.22	78.00	600.0	1.30
100/15/10	3.7	0.19	109.80	696.0	1.10

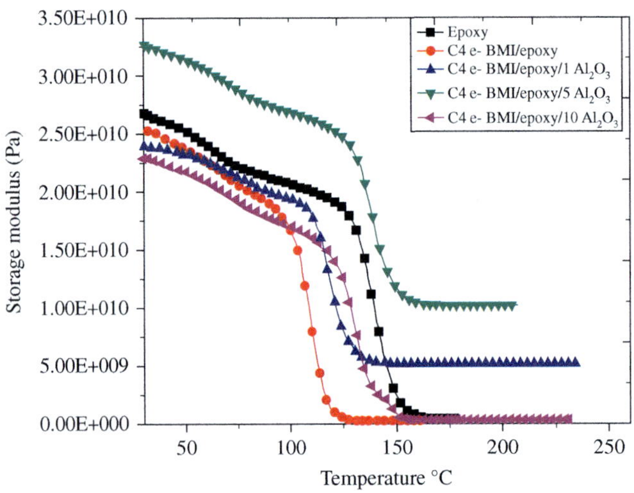

Figure 9.8 Storage modulus of the C4 e-BMI/epoxy/Al_2O_3 composite systems.

facilitated the transfer of stress between the components. Thus, the 15 wt% C4 e-BMI brings about an improvement in interfacial adhesion and consequently an increase in the storage modulus. Since the 15 wt% C4 e-BMI epoxy composite proves to be a synergistic system with enhanced toughness, the study was further extended to analyse the nanohardness and wear studies of the composite by reinforcing nanoalumina. At low weight percentage (1 wt%) of alumina particles, the nanocomposite exhibits a 10% increase in storage modulus over the pure epoxy matrix; the strengthening being mainly due to the random dispersion of harder particles called dispersion hardening (a term often used for precipitate hardening metals and alloys). When the particle content is increased to 5 wt%, the interparticle distance between the particles is so small that the particle– particle interaction becomes important and hence the reinforcement is greatly increased by 70%. The enhancement of storage modulus may arise from a reaction of the surface epoxy ring with the amine group in the hardener. During the curing process, the hardener DDM reacts not only with the epoxy resin, but also with the epoxy groups on Al_2O_3, forming covalent bonds between fillers and matrices which ultimately resulted in effective mechanical load transfer from the epoxy polymer matrix to the fillers, such that a higher storage modulus is achieved with silanized Al_2O_3/C4 e-BMI-toughened epoxy nanocomposites. At higher concentration of 10 wt% nanoalumina, the filler particles are dispersed irregularly into the epoxy matrix and the void of the crosslinking is increased. Hence, the segmental mobility rises, which leads to a decrease in the storage modulus at the same temperature range.

9.2.3.1.2 Glass Transition Temperature

The loss tangent values of the C4 e-BMI-toughened epoxy composites are shown in Figure 9.9. The 10 wt% C4 e-BMI-toughened epoxy exhibited two relaxation peaks at 128° and 138°C, while the representative IPN 15 wt% C4 e-BMI-toughened epoxy showed a single T_g at 119.1°C. It is presumed that, before the curing process, the C4 e-BMI/epoxy is homogeneous at the molecular level. However, as the curing proceeds, the molecules grow in size and the entropy contribution to the free energy of mixing decreases as a result of which the phase separation tends to occur in 10 wt% C4 e-BMI/epoxy. The formation of interpenetration and entanglement between two discrete networks of C4 e-BMI/epoxy, particularly in a densely cross-linked network, may counter this tendency and essentially enhanced the phase homogeneity in the case of the 15 wt% C4 e-BMI/epoxy. The 15 wt% of C4 e-BMI-toughened matrices have low T_g, in spite of bearing a storage modulus which is comparable with that of the neat epoxy matrix. This may be attributed to the well-known steric hindrance of the well-known network interlock that makes the curing reactions more intricate, and hence, the uncured portion of the epoxy and C4 e-BMI would probably act as a plasticizer and lowers the T_g of each IPN [11].

On increasing the alumina content from 1 to 5 wt%, the T_g showed an increasing trend increase and this is attributed to the small surface-to-surface distances ('matrix bridges') which restricted the chain mobility of epoxy segments resulting from good adhesion between the nanoparticles and the surrounding polymer matrix where the hard nanoparticles incorporated into the polymer acted as additional virtual network

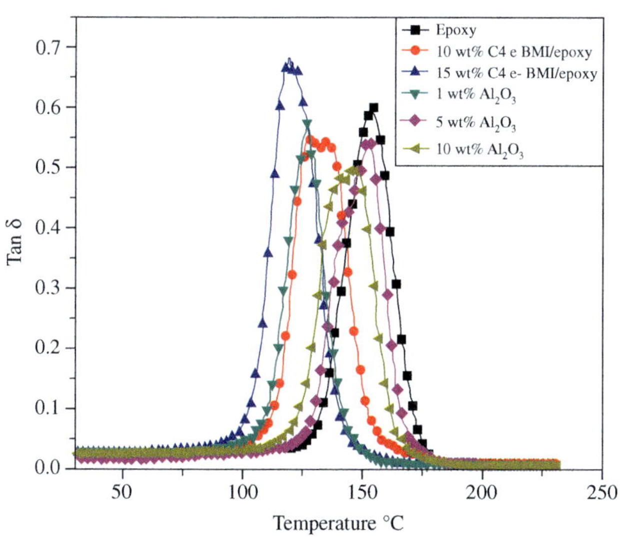

Figure 9.9 Glass transition temperatures of the C4 e-BMI/Al_2O_3 –toughened epoxy nanocomposite systems.

nodes [12]. However, a dual- T_g response is indicated at the highest particle loading of 10 wt%, and the failure to observe dual-T_g character at lower loadings may be a reflection of the high degree of compatibility inherent to the combination of epoxy/C4 e-BMI matrices with the GPS modified nano-Al_2O_3 [13]. The reduction in tan δ peak is consistent with the increase in T_g as it is seen in the DMA measurements may be related to the interface layer effect where the polymer chains are being tied down [14] on to the surface of nanoalumina at low level of loading namely 1 and 5 wt%. On the other hand, when the alumina nanoparticles loading is enhanced to 10 wt%, the T_g decreases. However, it can be inferred that the main important factors that affect T_g are degree of particle dispersion and curing conditions.

9.2.3.2 Hardness and Modulus by Nanoindentation

9.2.3.2.1 Elastic Modulus

The nanoindentation technique was employed to evaluate the material's mechanical properties, including elastic modulus and hardness. The nanoindentation tests started with a surface approaching segment and, once the surface of the material was detected, the indenter was brought into contact with the material at a constant strain rate of 0.05 s^{-1} until maximum load 50 mN was achieved. In the Nano Indenter G200, the mechanism for 'pressing' is electromagnetic. Thus, the instrument is fundamentally force-controlled. The force on the indenter is held constant for a dwell time at the peak force. The indenter was then withdrawn from the surface at the same rate until 10% of the maximum load was reached, and a constant strain rate was chosen to load on the samples in order to avoid any strain hardening effect on the measurements. At least 10 indents were performed on each sample, and the distance between the indentations was 100 μm to avoid an interaction and the average denotes the nanohardness and elastic modulus.

The values of hardness and Young's modulus of all the samples are given in Table 9.1. The elastic modulus of the C4 e-BMI-toughened epoxy composite increased significantly in the presence of C4 e-BMI. For instance, the elastic modulus of neat mechanical interlocking within the matrix and allow the transfer of the effective load from the matrix to the Al_2O_3 nanoparticles [15]. However, at 10 wt% of Al_2O_3, there was a competition between reinforcing effect of nano- Al_2O_3 and mobility of the polymer molecules as a result, the antiplasticization effect which comes into play cannot suppress the liquid-like behaviour due to a very low crosslink density of the network and hence the elastic modulus goes down. This evidences that increasing the reinforcement rate up to 5 wt% has a hardening effect beyond which it starts to decrease in the nanocomposites studied epoxy is 3 GPa and the corresponding C4 e-BMI-toughened epoxy matrix showed an elastic modulus of 3.9 GPa which is about 23% greater than the initial value. This is due to the rigidity contributed by hard bismaleimide skeleton, which increased the modulus [16]. The Young's modulus of the GPS Al_2O_3 /C4 e-BMI-toughened epoxy nanocomposites increased significantly in the presence of Al_2O_3. For example, the addition of 1 and 3 wt% of Al_2O_3 displayed no massive change in elastic modulus, nanohardness and showed properties similar to that of C4 e-BMI-toughened epoxy composites. On the other hand, at 5 wt% Al_2O_3 content, the effect of chemical interaction compatibility between the epoxy organic phases and the silane-functionalized Al_2O_3 inorganic phase is enhanced and hence the stiffness of the loaded Al_2O_3 together with the interfacial interaction between the organic and inorganic phases is believed to be the cause of improvement in elastic modulus. Secondly, the Al_2O_3 may form mechanical interlocking within the matrix and allow the transfer of the effective load from the matrix to the Al_2O_3 nanoparticles. However, at 10 wt% of Al_2O_3, there was a competition between reinforcing effect of nano-Al_2O_3 and mobility of the polymer molecules as a result, the antiplasticization effect which comes into play cannot suppress the liquid like behaviour due to a very low crosslink density of the network and hence the elastic modulus goes down. This evidences that increasing the reinforcement rate up to 5 wt% has a hardening effect beyond which it starts to decrease in the nanocomposites studied.

9.2.3.2.2 *Hardness by Nanoindentation*

The values of hardness of neat epoxy show a value of 0.13 GPa; however, the hardness of the C4 e-BMI-toughened epoxy composite increased significantly to 0.18 GPa. This is because of the presence of high performance hard filler like bismaleimide, which prevented the penetration of the indenter into the matrix. The higher hardness of Al_2O_3 filled C4 e-BMI/epoxy nanocomposite is due to the intrinsic hardness behaviour of Al_2O_3 nanoparticles which imposed better resistance to penetration under indentation [17]. It is evident that up to 5 wt% reinforcement of Al_2O_3 nanoparticles, there was more uniform distribution of alumina nanoparticles which contributed to the improvement in hardness and the decreasing trend in hardness indicates the inability of the nanocomposite to withstand the applied flexural load exerted by the indenter at 10 wt% of Al_2O_3. Thus, significant deformation occurs relatively easily, especially in the case of the nanocomposite reinforced with more than 10 wt% Al_2O_3. Based on the above observation, the 10 wt% Al_2O_3 containing matrix system did not effectively improve hardness of the nanocomposite due to the presence of agglomeration of Al_2O_3 nanoparticles.

9.2.3.2.3 *The Hardness Profile*

The hardness behaviour was studied on the neat epoxy, 15 wt% C4 e-BMI-toughened epoxy and 15 wt% C4 e-BMI/ Al_2O_3 -toughened epoxy nanocomposites systems, and the hardness profile could be generally divided into three regions, as shown in Figure 9.10.

In region I, i.e. from the surface to approximately below 1000 nm, the hardness dramatically decreased followed by a slight increase with increasing indentation depth. The initial decrease in the hardness may be due to the indentation size effect (ISE) or some uncertainties existing at the very surface layer of the material [18] . Then, the hardness profile increased with the indentation depth or displacement, and it may be due to the presence of dominant amorphous population and inhomogeneous distribution of crystalline morphology which slightly increases along the indentation direction (into the bulk), paving way for slight increment of the hardness. The region II, which lies between 1250–2000 nm, showed a stabilized hardness profile of the composite system. This region is considered as the 'transition zone', which indicates the microstructural changes from the amorphous 'skin' to crystalline 'core' in the composite samples. In region III, i.e. from 2000 nm onwards, the hardness again shows a steady increase with indentation displacement. It is believed that the inner (or 'core') part undergoes a slow/long cooling process (i.e. having much time to crystallize), thus having higher degree of crystallinity than the outer portion, and consequently leading to increment of the hardness with indentation depth [18].

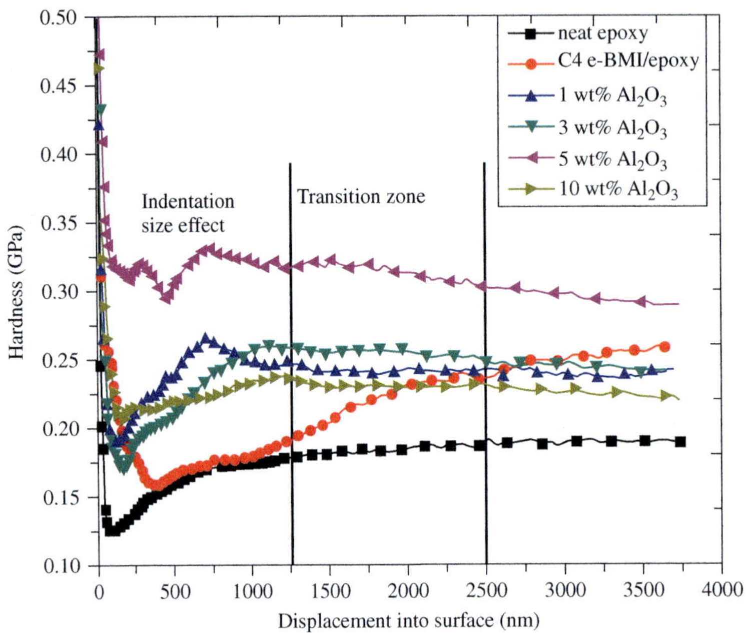

Figure 9.10 Hardness profile of 15 wt% C4 e-BMI/Al_2O_3 -toughened epoxy nanocomposite systems.

9.2.3.2.4 *Effect of Crosslinking on Tribological Behaviour*

It can be seen from the experimental data that a correlation exists between the observed elastic modulus and hardness with crosslinking density. Both elastic modulus and hardness increased with crosslinking density. This is because during crosslinking, the weak van der Waal's force between the polymeric chains is replaced by the strong covalent bonds and this ultimately reduces the molecular mobility of chains between the junctions and also the number of conformations that a polymer can adopt thereby enlightening higher elastic modulus and hardness with respect to crosslink density [19].

9.2.3.2.5 *Profiles of the Wear Track and Wear Performance*

The surface profile across the groove formed after applying a load of 5 mN along the same groove for 100 cycles is shown in Figure 9.11a–f. A closer observation of the wear profile of neat epoxy (Figure 9.11a) indicates that the wear cycle follows the original surface profile, and from the wear cycle tests, it is inferred that the wear track deformation is 115.4 μ^2 . The wear profile of C4 e-BMI- toughened epoxy matrix (Figure 9.11b) indicates that the grooves on the initial surfaces are filled with transferred matrix as a result of which the groove is decreased and wear track deformation also decreased to 113.22 μ^2. This is further evidenced by SEM in Sect. 9.2.3.2.6. The groove surfaces of all the Al_2O_3 filled C4 e-BMI-toughened epoxy nanocomposite are shown in Figure 9.11c–f, and they exhibited wave like pattern along the sliding direction and showed cracks, as a consequence of stiffness imparted by the Al_2O_3 filler incorporation. A closer look at the wear tracks revealed wedge formation at the end of the groove. Thus, the wear track profile results provided us with the clear mechanism of wear in epoxy composites when the indenter slides through the surface; it resulted in the formation of top ridges along the

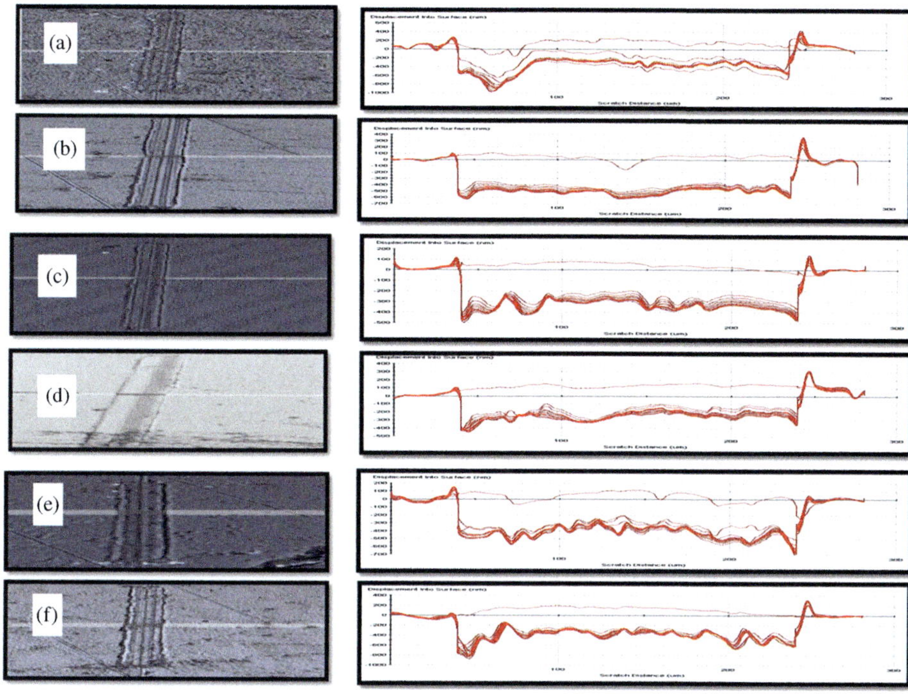

Figure 9.11 The surface profile across the groove of (a) neat epoxy. (b) 15 wt% C4 e-BMI/epoxy system. (c) 15 wt% C4 e-BMI/epoxy system/1 wt% GPS-functionalized alumina. (d) 15 wt% C4 e-BMI/epoxy system/3 wt% GPS-functionalized alumina. (e) 15 wt% C4 e-BMI/epoxy system/5 wt% GPS-functionalized alumina and (f) 15 wt% C4 e-BMI/epoxy system/10 wt% GPS-functionalized alumina.

sides of the ploughed grooves, and they are also removed as chips or developed a wedge on the front as in the case of 1, 3 and 5 wt% Al_2O_3 filled C4-e-BMI-toughened epoxy nanocomposite and hence the wear track deformation decreased to 97.5, 78.6, 78.11 μ^2, respectively. Unexpectedly at 10 wt% Al_2O_3 loading, the wear profile changed to aggressive abrasive behavior and hence the wear track deformation increased to 109.8 μ^2 showing poor wear performance of the nanocomposite.

9.2.3.2.6 FESEM of the Wear Surfaces

The role of nanoparticles in the reduction of wear rate and the wear mechanism taken place is studied in detail by analyzing the wear surface by field emission scanning electron microscope (FESEM) and the wear pattern of each sample is presented in Figure 9.12. The wear surface of neat epoxy matrix shows that fatigue is the main wear mechanism involved, on the otherhand combined fatigue and delamination ccurred together in some cases and it is reported elsewhere [20–23]. The magnified microstructures and indepth analysis of the neat epoxy wear surface (Figure 9.12a) showed wear platelets together with chunks of debris loosely smeared on the wear surface. These platelets are a result of the generation and propagation of surface and subsurface cracks. The small-sized platelets and debris found on the wear surface refer to the brittle nature of the epoxy resin. The wear surface of C4 e-BMI-toughened neat

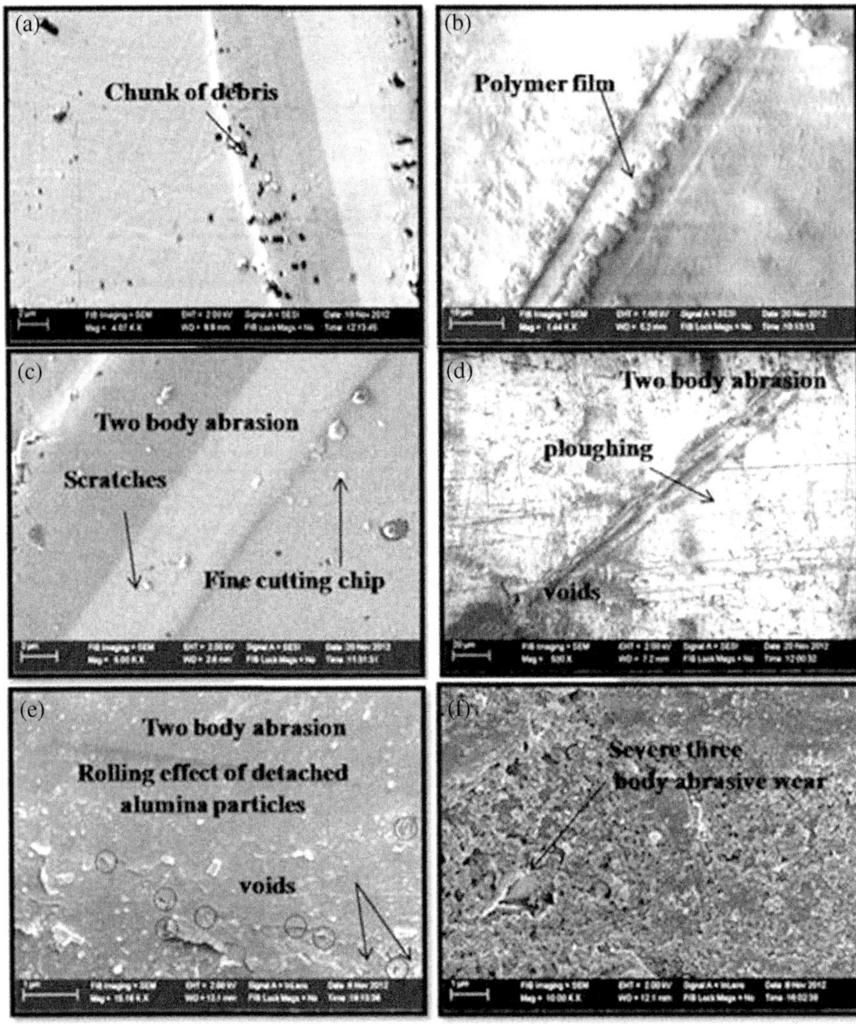

Figure 9.12 SEM images of worn surfaces of C4 e-BMI/Al_2O_3 toughened epoxy nanocomposite systems.

epoxy matrix presented in Figure 9.12b represents an adhesive wear, and the reason for the wear behaviour is as follows. When an indenter slides against the surface of the composites, the actual contact is between asperities of the two surfaces. The physical interactions at these contact regions possibly create a junction due to the heating at the interface, and this heat is high enough to cause melting or softening of the sliding polymer surface [24]. As the wear sliding proceeded, wear fragments were detached from the surface resulting in adhesive wear if the bonding at the interface of the adhering asperities is greater than the strength of the asperities. Ultimately, this led to the formation of film or lumps of polymer on the slider contact surface, and this polymer film is clearly visible in the FESEM and strongly provides a proof for adhesive wear [25]. The key aspect of abrasive wear relates to cutting or ploughing of the surface by harder particles or asperities. These cutting points may either be embedded in the counterface or lie loose within the contact zone. The former case is commonly called two-body abrasion and is seen in the wear of 1, 3 and 5 wt% Al_2O_3 reinforced nanocomposite evidenced by the FESEM images (Figure 9.12c, d) with the presence of scratches, ploughing caused by the asperities, scoring marks on the wear surface and the debris produced by abrasion frequently take on the appearance of fine cutting chips similar to those produced during machining, although at a much finer scale, and the latter, three-body abrasion took place in 10 wt% Al_2O_3-reinforced nanocomposite. In the case of three-body abrasion, the free abrasive readily penetrated the polymeric surface, which acted as an emery cloth resulting in increasing the wear of counter surface [26]. The FESEM image of wear surface of 5 wt% Al_2O_3 reinforced C4 e-BMI-toughened epoxy nanocomposite (Figure 9.12e) is rough with few voids. This is due to the detachment of spherical Al_2O_3 particles from the matrix during the wear process. Though the wear surface is rough, the wear resistance of the nanocomposite is good based on the wear track deformation. This might be due to the positive rolling effect [27] where the separated particles acted as rolling balls, during sliding; a rolling effect of nanoparticles has reduced the shear stress [17]. The matrix damages in the interfacial region were reduced by this rolling effect, and this effect can change sliding friction into rolling friction, between the indenter and the sample. Thus, it can be deduced that Al_2O_3 not only acted as a load carrier in the nanocomposite but also improved the bonding between the transfer film and the steel counterface. This promising activity of rolling particles is justified from the decreasing trend in the wear track deformation. When particle content is high (10 wt%), the particles started to agglomerate severely leading to a number of loosened clusters of particles in the tribo-system, which could be easily removed during sliding wear conditions. Ultimately, the composites would be easily damaged under repeated frictional stress (Figure 9.12 f). The wear surface became more rugged as more particles were detached from the matrix under wear load, and thus, it is easier for the nanoparticles to aggregate together. Consequently, the aggregations with irregular shapes caused relative severe three-body abrasive wear between the indenter and the sample leading to poor wear resistance. The abrasive wear caused by the hard wear debris enhanced the wear of the nanocomposite block.

9.3 Conclusion

When the nano-Al_2O_3 particle content was increased to 5 wt%, the particle interaction became efficient enough to contribute to 70% increment in the storage modulus. The addition of 5 wt% nano-Al_2O_3 particle increased the hardness of C4 e-BMI-toughened epoxy nanocomposites to 0.22 GPa from 0.13 GPa for neat epoxy matrix. The wear track deformation decreased to 78 μ^2 from 115.4 μ^2 for neat epoxy. Based on microscopic observations of the worn out surfaces, a positive rolling effect of the debris (matrix and nanoparticles) between the sample and the counterface led to remarkable reduction in wear track. This effect became more remarkable when the nano-Al_2O_3content was 5 wt%.

The wear resistance of C4 e-BMI/epoxy nanocomposites considerably increased with increasing content of nano- Al_2O_3. Thus, the improvement in the tribological behaviour of nano- Al_2O_3-filled C4 e-BMI nanocomposite is related to the improved characteristics of the transfer film. An optimum wear resistant composition was found to be with 5 wt% nano-Al_2O_3 particles. The main wear mechanism of the nanocomposite changed from the severe fatigue (for neat epoxy matrix) to two-body abrasion, and finally, three-body abrasion wear as the nano-Al_2O_3 concentration was increased. It can be briefly stated that 5 wt% nanoalumina incorporated C4 e-BMI-toughened epoxy nanocomposite showed the best balance of viscoelastic and thermal properties due to surface functionalization and homogeneous dispersion of alumina nanoparticles in the polymer matrix.

References

1 Dasari A, Yu ZZ, Mai YK. Fundamental aspects and recent progress on wear/scratch damage in polymer nanocomposites. Materials Science and Engineering Review 63, 31–80; 2009.
2 Christian J, Schwartz S. Studies on the tribological behavior and transfer film–counterface bond strength for polyphenylene sulfide filled with nanoscale alumina particles. Wear 237, 261; 2000.
3 Karger Kocsis J, Zhang Z. Structure-property relationships in nanoparticle/semicrystalline thermoplastic composites, In: Balta Calleja, J.F., Michler, G., (eds) Mechanical Properties of Polymers Based on Nanostructure and Morphology. CRC Press, 547–596; 2005.
4 Sawyer WG, Freudenberg K, Bhimaraj P, Schadler L. A study on the friction and wear behavior of PTFE filled with alumina nanoparticles. Wear 254, 573–580; 2003.
5 Ashok Kumar A, Alagar M, Rao R. Synthesis and characterization of siliconized epoxy-1, 3-bis (maleimido) benzene intercross linked matrix materials. Polymer 43, 693–702; 2002.
6 Guo Z, Pereira T, Choi O *et al.* Surface functionalized alumina nanoparticle filled polymeric nanocomposites with enhanced mechanical properties. Journal of Material Chemistry 16, 2800–2208; 2006.
7 Zunjarrao SC, Singh RP. Characterization of the fracture behavior of epoxy reinforced with nanometer and micrometer sized aluminum particles. Composite Science and Technology 66, 296–305; 2006.
8 Lu S, Hongyu J, Zhang H, Wang X. Wear and mechanical properties of epoxy/SiO_2-TiO_2 composites. Journal of Material Science 40, 2815–2818; 2005.
9 Parfitt GD. Surface chemistry of oxides. Pure and Applied Chemistry 48, 415–418; 1976.
10 Rong MZ, Zhang MQ, Zheng YX *et al.* Structure property relationships of irradiation grafted nano-inorgnic particle filled polypropylene composites. Polymer 42, 167–183; 2001).
11 Lin MS, Liu CC, Lee CT Toughened interpenetrating polymer network materials based on unsaturated polyester and epoxy. Journal of Applied Polymer Science 72 585–592; 1999.
12 Wang QH, Xu JF, Shen, WC, Xue QJ. The effect of nanometer SiC filler on the tribological behavior of PEEK. Wear 209, 316–321; 1997.

13 Wetzel B, Haupert F, Qiu Zhang M. Epoxy nanocomposites with high mechanical and tribological performance. Composite Science and Technology 63, 2055–2067; 2003.
14 Omrani A, Simon LC, Rostami AA. The effects of alumina nanoparticle on the properties of an epoxy resin system. Materials Chemistry and Physics 114, 145–150; 2009.
15 Lu M, Lau KT, Tam W Y, Liao K Enhancement of Vickers hardness of nano-clay supported nanotube reinforced novel polymer composite. Carbon 44, 383; 2006.
16 Qin H, Mather PT, Baek JB, Tan LS. Modification of bisphenol-A based bismaleimide resin (BPA-BMI) with an allyl-terminated hyper branched polyimide (AT-PAEKI). Polymer 47, 2813–2821; 2006.
17 Kurahatti RV, Surendranathan AO, Srivastava S *et al.* Role of zirconia filler on friction and dry sliding wear behavior of bismaleimide nanocomposites. Materials Design 32, 2644–2649; 2011.
18 Shen L, Phang IY, Chen L, Liu T, Zeng K. Nanoindentation and morphological studies on nylon 66 nanocomposites. I. Effect of clay loading. Polymer 45, 3341–3349; 2004.
19 Li F, Larock R. Synthesis, structure and properties of new tung oil-styrene-divinly benzene copolymers prepared by thermal polymerization. Biomacromolecules 4, 1018–1025; 2003.
20 Bonfield W, Edwards BC, Markham AJ, White JR. Wear transfer films formed by carbon fibre reinforced epoxy resin sliding on stainless steel. Wear 37, 113–121; 1976.
21 Zhang MQ, Rong MZ, Yu SL *et al.* Effect of particle surface treatment on the tribological performance of epoxy based nanocomposites. Wear 253, 1086–1093; 2002.
22 Bassani R, Levita G, Meozzi M, Palla G. Friction and wear of epoxy resin on inox steel: Remarks on the influence of velocity, load and induced thermal state. Wear 247, 125–132; 2001.
23 Durand JM, Vardavoulias M, Jeandin M. Role of reinforcing ceramic particles in the wear behaviour of polymer-based model composites. Wear 181, 833–839; 1995.
24 Briscoe BJ, Stuart BH, Sebastian S, Tweedale PJ The failure of poly (ether-ether-ketone) in high speed contacts. Wear 162, 407–417; 1993.
25 Sviridenok AI, Bely VA, Smurugov VA, Savkin VG. A study of transfer in frictional interaction of polymers. Wear 25, 301–308; 1973.
26 Myshkin NK, Petrokovets MI, Kovalev AV. Tribology of polymers: adhesion, friction, wear, and mass transfer. Tribology International 38, 910–921; 2005.
27 Chang L, Zhang Z. Tribological properties of epoxy nanocomposites. I. Enhancement of the wear resistance by nano-TiO_2 particles. Wear 258, 141–148; 2005.

10

Nanoindentation of Hybrid Foams

Anne Jung, Zhaoyu Chen and Stefan Diebels

Applied Mechanics, Saarland University, Saarbrücken, Germany

10.1 Introduction

10.1.1 Motivation

The increasing demand for lightweight and energy absorbing materials leads not only to searching new materials but also to developing new design concepts, in which already well-known materials will be used as special structures. In this context, foams and other cellular structures are known to combine many interesting physical and mechanical properties [1, 2]. Hence, metal foams have become of interest for researchers in both scientific and industrial applications. They consist of a three-dimensional network of stochastically distributed pores and mimic the construction elements of natural load-bearing materials like bones, wood or coral [1, 2]. In analogy to these materials, they offer a high stiffness to weight ratio, which is why they are used as lightweight construction materials [2–5]. Due to their porous structure, by the successive buckling of pores, metal foams undergo large deformations at a nearly constant stress level, which is known as plateau stress [2–5]. This is the reason why metal foams can be used as crash and energy absorber in the automotive and aerospace industry [2–5]. As well as for blast alleviation or ballistic impacts [6–8].

Foams are divided into open- and closed-cell foams. The following work only deals with open-cell foams. Open-cell metal foams are commonly synthesized with polymer foams by an investment casting process [2, 4]. It is a manufacturing method without the direct foaming of the metal. Polymer foams, e. g. polyurethane foams, are used as a template. The pores are infiltrated with a slurry of a heat resistant material like mullite and dried subsequently. After the curing of the slurry, the polymeric template foam is removed by a thermal treatment. The remaining voids, which replicate the structure of the template foam, represent the casting mold for the foams. Molten metal is then cast into the mold. After the curing, the mold is removed by pressurized water, leaving behind nothing but the metal foam, being an exact replicate of the polymer foam. Despite being present in some niche markets, metal foams have not fully succeeded commercially yet. The reason for this is that, due to the low density ($\rho_{Al} = 2.70$ g cm^{-3}), low melting point (about 660 °C), and low costs, most of the open-cell foams are made of aluminum. A big disadvantage in comparison to other metals is the low tensile

Applied Nanoindentation in Advanced Materials, First Edition. Edited by Atul Tiwari and Sridhar Natarajan.

strength of about 300 MPa for $AlSi_7Mg_{0.3}$, which is the most common alloy for open-cell metal foams. Nickel, for example, offers much better mechanical properties. The tensile strength for polycrystalline nickel lies in the order of 400 MPa. For nanocrystalline materials, the yield strength is much higher than for polycrystalline materials. Based on the Hall-Petch relationship [9, 10], the yield strength is inversely proportional to the grain size. Hence, in the case of nanocrystalline nickel, the tensile strength rises up to 800 MPa. A big disadvantage of most of the materials is their high density due to having a higher tensile strength than aluminum. Foams made of steel ($\rho_{Fe} = 7.86$ g cm^{-3}) or nickel ($\rho_{Ni} = 8.91$ g cm^{-3}) are three to four times heavier than aluminum foams without their mechanical performance being improved by the same factor. A further disadvantage are the high melting points with 1536 °C for steel and 1455 °C for nickel, making the production process much more expensive not only due to the costs of nickel or steel but also due to the high energy costs and the handling. Due to the production using the casting process, it is not possible to produce foams made of nanocrystalline metals. The reason, why metal foams are not extremely often used is not only the low mechanical strength, but also the bad reproducibility of their mechanical properties due to defects in the microstructure like curvatures in the struts, the build-up of thin membranes rather than open porosity or the formation of blow holes during the casting process.

A common design strategy in materials science to achieve mechanical properties that cannot be offered by a bulk material is to make a composite. The coating of materials to change the surface properties like hardness, wear resistance, etc., of a material is also a kind of composite formation. In order to overcome the disadvantages of aluminum foams and to bring them into use, a combination of the three abovementioned design strategies in materials sciences (cellular materials, nanotechnology and composites) was used resulting in the so-called hybrid metal foams. Hybrid metal foams in this case are normal metal foams that have been finished with a coating in order to improve the mechanical properties. Electrodeposition is an important non-line-of-sight deposition technique for the coating of materials, which is not only applicable to planar parts but also to complexly shaped three-dimensional components. Due to the large surface inner surface of foams and hence, their large surface to volume ratio, coating of foams does not only change the surface properties, but can also be used to tailor the mechanical properties of the composite foam, producing hybrid foams with superior properties.

The first attempt to produce nanocrystalline coated metal foams was done in 2008 by Boonyongmaneerat *et al.* [11]. They wanted to strengthen open-cell aluminum foams by means of electrodeposition of a nanocrystalline NiW alloy. Boonyongmaneerat *et al.* obtained positive effects of the hard coating on the stiffness, strength, and energy absorption capacity of the foams, but, due to the mass increase by the coating, these enhancements were not reflected by the specific properties. In 2009, Bouwhuis *et al.* [12] worked on nickel coated aluminum foams. They were able to confirm the finding of Boonyongmaneerat *et al.* and performed a detailed study on the deformations and damage mechanisms as well as on the coating thickness distribution of coated struts in foams. Bouwhuis *et al.* came to the conclusion that, due to the electromagnetic shielding effect of the foam microstructure, it is not possible to produce homogeneously coated foams via electrodeposition. Based on these shielding effects, the foams show up to ten times higher coating thicknesses for struts near the geometrical surface of the foam compared to the struts in the foam center. Jung *et al.* discerned in 2010 that the reason for the non-homogeneous coating is not only the electromagnetic shielding effect, but also, by a large amount, the mass transport limitation during the electrodeposition [13, 14]. They developed a semi-non-destructive technique for the determination of the coating thickness homogeneity based on measurements of the remanent magnetic flux density distribution of nano-nickel coated foams [15, 16]. Hence, they were able to improve the coating homogeneity up to 80% by developing a new deposition setup. Due to this much more homogeneous coating, for the nano-Ni/Al hybrid foams of Jung *et al.* there was not only an increase in the absolute mechanical properties but also a significant enhancement of the specific properties, despite of the increasing mass. Some recent work by Sun *et al.* (17, 18] confirmed these findings for copper/aluminum hybrid foams using the same deposition setup as described by Jung *et al.* [13, 14]. As demonstrated above, the field of nanocrystalline coated metal foams and nano hybrid foams respectively is a very new research area, in which various studies still have to be performed to earn a better understanding of these highly innovative materials.

Metal foams in general are microheterogeneous materials. The structural constitution of such materials can be divided into three hierarchical levels. The macroscale comprises the complete sample, the mesoscale single pores or a small number of pores, and the microscale only single struts. On each scale, the scale below is regarded as a continuum. In macroscopic tests, after a first linear elastic region, metal foams show a distinct flow plateau with nearly constant stress. This plateau in the stress-strain-curve arises due to the local bending and buckling of single struts. Hence, the global, macroscopic properties of metal foams mainly depend on two factors, on the one hand on the mechanical properties of the strut material, and one the other hand on the microstructure (shape, size and distribution of pores) itself [19–21]. A detailed knowledge of these two factors is crucial for the reliable material simulation of such materials [22–24]. Based on the manufacturing process, the material properties of single struts of metal foams significantly differ from the mechanical properties of the bulk materials [19, 20]. For pure foams, due to the casting process, the cooling behavior of a foam is much faster than of a cast massy part. This leads to different grain structures and grain sizes. A significant effect of the ratio of strut thickness to grain size can be observed. Whereas in a bulk material, several hundred grains can be found in a sample dimension, there a only several grains over the cross section of a strut of the foam. Hence, micromechanical tests on single struts are crucial for the development of reliable material models for cellular materials like metal foams. The determination of micromechanical parameters like Young's modulus and yield strength can be made by means of nanoindentation [20, 25–28] as well as using microtensile tests [29–35]. Up to now, only a few works have been done in this field of the micromechanical testing of cellular materials. Within this context, most of the literature deals with trabecular bones. For hybrid metal foams, it is much more difficult to determine the mechanical parameters. The mechanical properties of the electrodeposited coating do not only depend on the deposition process, but also on the foam structure. As mentioned above, the electrodeposition of metal foams is strongly limited by mass diffusion problems [12, 13]. This mass transport limitation has a significant effect on the grain size and grain structure of the deposited metal and promotes the formation of nanocrystallinity by inhibiting the surface diffusion of as-deposited metal ions and hence inhibiting the grain growth for the benefit of nucleation.

In the case of nano-coated hybrid metal foams, nanoindentation is the only method to determine the exact micromechanical properties of the single constituents like the foam and coating of the composite. The following section will give an overview on the nanoindentation of metals and especially of metal foams.

10.1.2 State of the art of Nanoindentation of Metal and Metal Foam

Nanoindentation, also known as depth-sensing indentation (DSI), uses the continuously sensed indentation depth combined with the measured applied force to determine the elastic modulus and hardness according to the Oliver & Pharr method [36, 37]. Since Nanoindentation can accurately measure the mechanical properties of extremely small volume materials at the micro- and nanoscale without removing the substrate, it has become a fairly popular testing technique for thin films and coatings. During the long-term evolution, the indentation technique, which has its origins in Mohs' hardness scale of 1822 [38], has been transformed into a fine multifunctional research tool that is suitable for revealing various physical structures and measuring the mechanical, thermal, and electrical behaviors of quite different materials varying from soft biological to superhard diamond-like materials. The evolution has been caused by the increase of high-resolution, digital controlling techniques and the development of the corresponding analysis methods of the original data.

First of all, a quick review will be made of the existing methods and approaches to determine different properties varying from elastic, viscoelastic up to elastic-plastic or elastic-viscoplastic behavior in the literature. It is well known that the elastic contact theory was created by Hertz in solving the problem of the elastic contact of two paraboloids [39]. This contact solution gives a simple relation between load, displacement, radius of a sphere, and the elastic moduli of the bodies in contact with the assumption of a small strain. Based on the Hertz contact theory, the analytical solution for a rigid axisymmetric body of arbitrary shape indenting into a half-space was derived by Sneddon [40, 41]. The basic theory of linear viscoelastic contact problems was developed by Lee and Radok [42] in 1960. The method is called the correspondence principle and extends the elastic contact theory to linear viscoelastic models by using some mathematical transformation, e. g. the Laplace transformation. This method is restricted to a monotonically increasing contact area. Later Ting [43, 44] and Graham [45, 46] proposed a more general solution. In 1985, Johnson [47] summarized the correspondence analysis of spherical indentation replacing the elastic constants with the Boltzmann viscoelastic hereditary integral operators. It is known that a nonlinear behavior of materials, e.g. elastic-plastic or elastic-viscoplastic, is very difficult to analytically describe. Therefore, it is not surprising that the biggest part of analytical studies involves the linear elastic behavior [48]. The most often used Oliver & Pharr procedure was developed in 1992 [36] and refined in 2004 [37] to determine the hardness and Young's modulus of elastic-plastic material by fitting the unloading load-displacement curve with the Galin solution [41]. Although the Oliver and Pharr procedure has been used in most available indentation instruments, problems of inaccurate determination due to the inaccuracy of defining the initial slope of the unloading curve and indentation depth as well as due to the neglect of the real indenter shape [48–54] still remain unsolved. The method of determining yield stress consists of the Tabor-Johnson approach [47, 53], in accordance with the relation between the hardness H, the yield stress σ_Y, and the representative strain ε_R: $H = C\,\sigma_Y\,(\varepsilon_R)$ [55–58]. In the case of elastic-viscoplastic behavior, the information of a single experiment is insufficient for a simultaneous determination of the yield stress and the strengthening parameter. Therefore, various methods of determining plastic properties using independently gained information with different indenter are considered in [58–61]. The use of the finite element method (FEM) to compute the boundary value problem of the contact mechanics allows scientists to take into account the indentation process non-linearity, real indenter geometry, and behaviors (constitutive laws) of the material [48]. The comparison of the experimental data and the computational results forms the basis of the inverse problem, in which the parameters of the materials' constitutive laws can be identified based on the indentation curves. The inverse method is increasingly applied in the indentation not only of metals [62–66], but also of polymers and biomaterials [67–71] using various numerical optimization strategies. The analysis approaches to determine the elastic, viscoelastic, elastic-plastic, and elastic-viscoplastic properties using nanoindentation, as given in various literatures, are presented in Table 10.1.

Table 10.1 Analysis approaches to determine material properties using nanoindentation in various literatures.

Properties	Methods	Determination	Assumption	References
Elastic	Hertz contact solution	E, H[a)]	Contact of two paraboloids, small strain	[39]
	Sneddon's solution	E, H	Rigid axisymmetric indenter, half-space	[40, 45]
	Inverse method	E, hyperelasticity	—	[68, 70, 72]
Viscoelastic	Correspondence principle	Linear viscoelastic operators	Incompressible half-space, nondecreasing contact radius, small deformation	[42]
	General methods	Linear viscoelastic operators	Small deformation	[43–45]
	Inverse method	Various linear or non-linear viscoelastic models	—	[67–70, 73, 74]
Elastic-plastic	Oliver &Pharr method	E, H	No rate-dependent deformation, only elastic recovery during unloading	[36, 37]
	Tabor-Johson approach	σ_Y	Relation: $H = C\sigma_Y(\varepsilon_R)$	[52–54]
	Inverse method	Various constitutive laws	—	[60, 75]
Elastic-viscoplastic	Analytical methods	Yield stress, strengthening parameters	Using independent indentation data with different indenters	[55–58]
	Inverse method	Various consitutive models	—	[61–63, 68, 76]

a) E and H refer to the elastic modulus and hardness.

Secondly, the developed analysis methods used to evaluate the mechanical properties that have been discussed above stimulate wide-range applications in terms of very small volume materials with nanoscale structures. The micromechanical properties as well as volume fractions of metal foams, which have microstructural material phases and very small dimension cell walls, have been obtained by means of nanoindentation [19, 21, 77–79]. The measurements cannot be carried out by traditional macroscale methods due to the small dimensions, low local bearing capacity, and local yielding and bending of the cell walls. However, the load-displacement curves obtained by nanoindentation in the nanometer range can overcome these problems [77, 78] and the used methodology can be effectively used to characterize the mechanical properties of the cell walls of any cellular material [51]. Structural materials such as cellular metal or polymer foams, trabecular bones, and porous ceramics, exhibit several types of heterogeneity at the micro- and nanoscales. It is hard to rigorously define their volumes and microstructural distribution during nanoindentation. Hence, nanoindentation is often used combined with complementary techniques, e. g. scanning electron microscopy (SEM), atomic force microscopy (AFM), and electron backscatter diffraction (EBSD) to separate and to characterize the material phases and the microstructural distribution [8]. The complicated microstructures of trabecular bones and porous ceramics have been treated by a combination of nanoindentation and a small-scale tensile test in [30, 79]. The microstructure of these materials could then be reconstructed using X-ray microtomography in combination with the specimen loading [30]. The texture and grain growth of Ni/Al-hybrid metal foams were investigated with the EBSD technique working within an SEM environment, while the local hardness was determined using nanoindentation [8]. Furthermore, the size-dependent mechanical properties of nanoporous gold were investigated using a combination of nanoindentation, column microcompression, and molecular dynamics simulations [80]. As a result of the high precision at a small-scale, nanoindentation has also been employed by chemists to evaluate the mechanical properties and intermolecular interactions of the crystals of organic, inorganic, and metal-organic framework (MOF) [80, 82–84]. For example, the stiffness of different types of bonds in MOFs and the elastic modulus can be explored with the aid of the nanoindentation technique [84]. Nanoindentation could be used to probe the properties of microstructural constitutes, and in turn, for the assessment of structural inhomogeneities such as intergrowth of secondary phases or polymorphs [84]. Systematic experiments have been performed to investigate the rate sensitivity of the deformation in fully dense nanocrystalline Ni using nanoindentation and tensile testing. Results from both types of tests reveal that the strain-rate sensitivity is a strong function of the grain size [85]. The role of oxide phases in association with AlN particles is investigated to understand the mechanical behavior of AlN/Al metal matrix nanocomposites using a nanoindentation tester. load-displacement profiles reveal the distribution of oxide phases along with the AlN particles and their effect on the indent penetration [86]. For polycrystalline materials, the influence of the grain boundary on the determined hardness using nanoindentation has been discussed in [87–89]. The majority of the authors have observed a hardening when the indentation approached the boundary. This effect was explained by a hindering of the slip transfer to the neighboring grain [90]. A number of studies on the nanoindentation of different coated surfaces or thin-film systems show that the hardness and Young's modulus of films depend on various factors such as yield stress and the elastic modulus of the

Table 10.2 Classification of the applications of nanoindentation.

Classification	Material	Measurements	References
Bulk	Metals: Fe, Zn, Cu, Ni	Equivalent stress-strain curve, E, yield stress, tangent modulus	[52, 53, 55, 56, 101]
	Steel S690QL, austenitic steels	Viscoplasticity model of Perzyna-type, stress-strain	[59, 79]
	Polymers: silicon rubber, PU, natural rubber	Hyperelasticity, viscoelasticity, adhesion between indenter and polymers	[64, 67, 68, 74]
Coatings/thin films	TiAIN/TiN gradient coating, AlTiN films	Influence of gradient layer on strain distribution, influence of nitrogen partial	[96, 100]
	Graded hybrid PTFE/steel thin film	Relation between E, H, and chemical structure	[99]
	Quasicrystalline AlCuFe coatings	Stress-strain behavior, plastic flow	[57]
	Ti-C thin film on oxidized silicon and fused quartz substrates	E, H, plastic deformation	[98]
	Al, gold, and platinum thin film/sapphire	Film thickness-dependent plastic deformation	[95]
Nanocomposites, nanowire, nanocrystalline	Nano-layered Al-Sic composite	Microstructure, E, H, influence of layer thickness and substrate on E and H	[102]
	AlN/Al metal matrix nanocomposites	Influence of AlN and oxide phase on E and H	[86]
	Metallic glass nanowires	Viscosity at high temperatures	[97]
	Organic, inorganic, and metal-organic framework (MOF)	Evaluating the mechanical properties and intermolecular interactions of crystals	[81–84]
	Fully dense nanocrystalline Ni	Grain size-dependent strain-rate sensitivity	[85]
Foam	Ni/Al-hybrid metal foam	Local hardness	[8]
	Trabecular bone, porous Au	Elastic-plastic material model and comparison with tensile test, strain field in microstructure, size effect	[30, 31, 80]
	Al-alloy foam, AlSiCuMg alloy foams	Submicron E and H, effective elastic model, volume fraction of microstructure	[19, 21, 77, 78]

substrate, film/substrate adhesion, film thickness, and indenter geometry [91–95]. The evaluation of the mechanical properties of nanocomposites sputtered thin films was usually carried out using nanoindentation [96–98]. The nanomechanical characterization of functionally graded hybrid polymer/metal thin films was performed using nanoindentation to connect the local hardness and elastic modulus with the chemical composition, chemical bonds and structures [99]. Also, the influence of the gradient coating on strain distributions in the area of the indenter/coating was investigated using nanoindentation [100]. A classification and conclusion of the applications of nanoindentation are presented in Table 10.2 according to the review of the literature.

Finally, an important factor that should be mentioned is the pile-up phenomenon, which is observed in many indentations of elastic-plastic materials and which leads to a larger contact area than estimated by this method. As a consequence, both the determined hardness and elastic modulus are overestimated, sometimes as much as by 50% (Bolshakov and Pharr 1998). The fundamental material parameters affecting the pile-up are the ratio of the effective modulus to the yield stress, E^*/σ_y, and the work hardening behavior (Oliver and Pharr 2004). In general, the pile-up is the greatest in materials with a large E^*/σ_y and with little or no capacity for work hardening. The capacity for work-hardening inhibits the pile-up because, if the material at the surface adjacent to the indenter hardens during the deformation, it constrains the upward flow of material to the surface. Some methods have been developed to identify the pile-up behavior and to correct its influence using experimentally measurable parameters, e. g. penetration depth [102] and indentation [103].

10.2 Sample Material and Preparation

10.2.1 Al Material and Coating Process

In this study, nano-Ni/Al hybrid metal foams produced by direct current electrodeposition of nanocrystalline nickel will be investigated using nanoindentation. For the synthesis of this composite material, open-cell aluminum foams $AlSi_7Mg_{0.3}$, produced by CellTec Materials, Dresden, Germany) with a pore size of 10 ppi were used as substrate and cathode respectively, in the electrodeposition process. In accordance with Jung *et al.*, a special cathode–anode arangement [13, 14] was used for the plating process. Nickel with an approximated coating thickness of 150 μm and an average grain size, determined by means of the X-ray diffraction (XRD) of 50 μm nickel, was deposited from a commercial nickel sulfamate electrolyte with a nickel content of 110 g/l at a temperature of 50 °C and a pH of 3.8. The exact plating procedure with the necessary pretreatment to prevent the dissolution of the aluminum foam in the acid electrolyte has been distributed by Jung *et al.* [13, 104, 105]. In the literature, it has been demonstrated that the nanocrystalline coating leads to superior mechanical properties in comparison to pure aluminum foams [11, 12, 14]. But the texture and micromechanical testing of the coating has been analyzed only once [8]. Resulting from a growth-determined deposition mechanism, electrochemically produced bulk nickel typically shows a columnar grain growth with a growth direction perpendicular to the substrate surface [106–109]. This leads to an almost perfect texturing of the coating in the direction of the (001)-axis, which is parallel to the growth direction [107, 110]. Normally, such a strong anisotropic grain structure has a weakening effect on the material. Using electron backscatter diffraction (EBSD) measurements [8], it has been discovered that a perpendicular grain growth for coating of metal foams appears, but based on the complex-shaped three-dimensional surface of the foams with countless curvatures, a structurally induced isotropic material behavior of the hybrid foams can be observed. Directly at the Ni/Al interface, a layer with very fine grains of less than 1 μm has been detected. Due to the columnar grain growth of the electrodeposited nickel, the directly following region consists of more or less elongated grains. In the following, nanoindentation will be performed on both the aluminum struts and nickel coating in order to determine the local hardness and yield strengths.

10.2.2 Sample Preparation for Nanoindentation

The objective of the nanoindentation on the hybrid Ni/Al nano-coated foams is to measure the local mechanical behaviors of both the Ni coatings and Al substrates taking into account the influence of the hybrid interface. Hence, the sample preparation has to be carefully performed in order to get a smooth working surface of both the Ni coatings and Al substrates at the nanoscale. Firstly, a mechanical polishing using emery paper in grit sizes ranging from 320 grit to 1200 grit (for P-Grade ranging from P360 to P4000) has been carried out to remove the coarse grain. Afterwards, the surfaces with a very low roughness have been further polished down to 250 nm with diamond slurry and then electro-polished with a 1 M sulfuric acis-methoanol solution [111] to remove the mechanical damage on the surfaces. A very smooth surface of the Ni coatings with a maximum topographic asperity about 3 nm was obtained (Figure 10.1a, b).

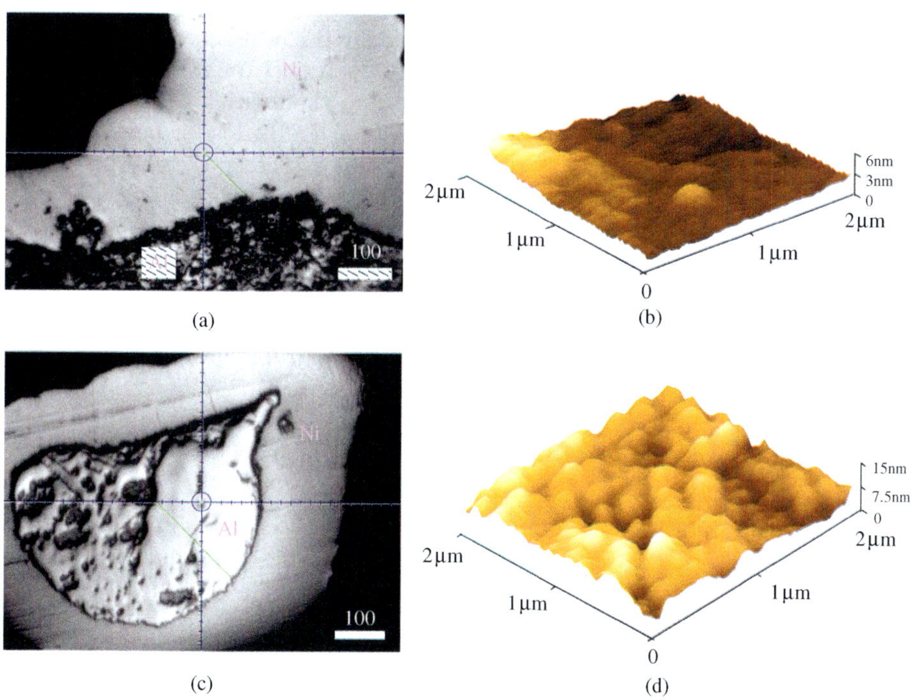

Figure 10.1 The images after polishing the specimen: 2D image using electro microscopy of Ni(a) and Al(c); 3D topography from in situ scanning probe microscopy (SPM) image mode using a Berkovich indenter tip of Ni(b) and Al(d).

However, the quality of the Al surface decreased during the electro-polishing. Therefore, the second sample, which has been firstly mechanically polished using emery paper and diamond slurry, was further polished with Oxide Polishing Suspension (OPS). The OPS is a colloidal silica suspension with a pH of 9.8 and a grain size of about 40 nm. The final obtained Al sample surface is of good quality with a maximum topographic asperity of about 8 nm (Figure 10.1c, d). All the samples were cleaned by being submerged for ten minutes in an ethanol ultrasonic bath after each polishing.

10.3 Nanoindentation Experiments

10.3.1 Experimental Setup

The nanoindentation of Ni/Al hybrid foams in this work was performed with a TI 900 TriboIndenter® of Hysitron Inc., MN, USA[1]. This TriboIndenter® is a fully automated and powerful indentation instrument, because it is a quantitative system with an internal analysis procedure and it can provide in-situ topographical images of the sample surface before and after a test. In the quasi-static mode, the TriboIndenter R has a force noise floor better than 100 nN and a displacement noise floor less than 0.2 nm. The indenter tip is a very important part of the nanoindentation test. The geometry and the material it is made from can dramatically influence the raw data. In this study, a diamond Berkovich indenter with a semi-angle of 65.27 ° is chosen considering the high stiffness and the nano- or micro-scale grain size of the Ni nanocoatings. The thermal drift refers to a change in the dimensions of the instrument due to the heat generation from the electronic devices and the plastic deformation [38]. To correct the thermal drift, it is common to measure the displacement increment during a hold series of the load. However, even for some metals, creep may be prominent, when the load is held constant and the depth readings increase due to the superposition of creep and thermal drift. Hence, the experiments were performed with a loading history as shown in Figure 10.2 a) by means of force control. A maximum force of 2000 μN was superpositioned on the indenter with a loading rate of 500 μN/s and the force was then completely removed with the same rate during the unloading part. During this procedure, the force was held once at the maximum value for 2s and once at a small value of 250 μN for 1s in order to check the creep behavior and the thermal drift, respectively. As shown in Figure 10.3, an indentation pattern used in this study is a 3 × 3 matrix with an interval of 5 μm. The directions of the arrows display the indentation path. The intervals should be large enough in order to avoid any influence from the neighboring indentations.

10.3.2 Results and Discussion

First of all, the contributions of the creep behavior and the thermal drift on the increment of the displacement will be discussed. As shown in Figure 10.2a, the creep behavior is more likely to be aroused by the holding at the maximum load, whereas the thermal drift can be captured from the hold period data at a low value of load, where creep

---

1 The nanoindentation experiments were performed using the device possessed by the Chair of Experimental Science of the Materials Science at Saarland University under the direction of Prof. C. Motz.

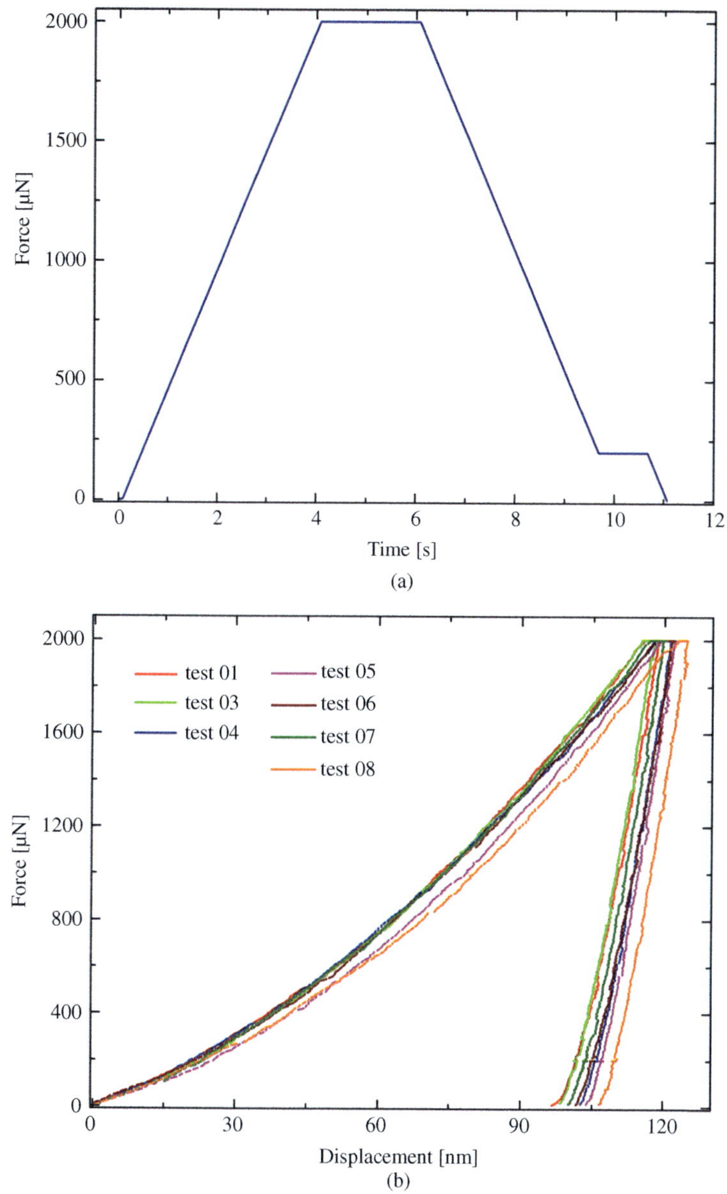

Figure 10.2 The loading history using a force controlled mode (a), and the corresponding force-displacement curves of a set of indentation points (b).

within the material is less likely to occur [38]. The force-displacement data plotted in Figure 10.2b show a negligible and an undiscoverable increment due to the creep and thermal drift, respectively. Hence, in this study, the Ni coating of the hybrid foam mainly behaves in an elastic-plastic way. As we have discussed in the first section, metal foams in general are micro-heterogeneous materials because of the inhomogeneous microstructure originating mainly in the manufacturing process. Hence, nanoindentation is applied

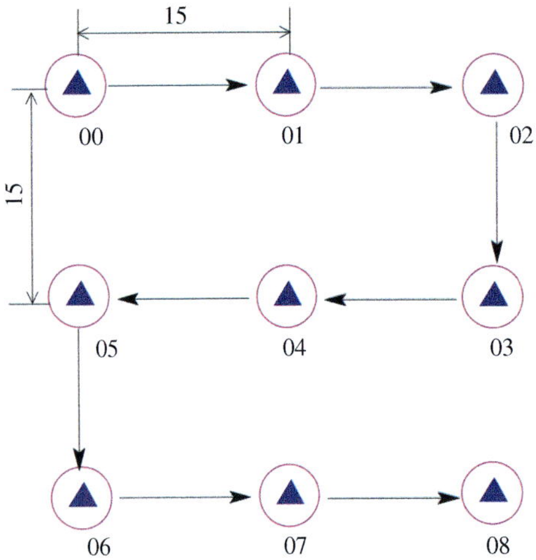

Figure 10.3 Indentation pattern (automatic way).

Table 10.3 The local Young's modulus and hardness of Ni coating determined by nanoindentation.

		line1	line2	line3	line4
Pos1	E (GPa)	150.6±3.6	207.7±8.9	209.7±3.8	209.9±6.4
	H(GPa)	2.9±0.2	4.8±0.3	4.2±0.2	4.0±0.2
Pos2	E (GPa)	176.3±6.4	213.1±3.7	205.9±8.4	217.4±5.3
	H(GPa)	3.6±0.4	4.1±0.4	4.1±0.3	3.7±0.4
Pos3	E (GPa)	195.4±6.6	229.5±3.8	224.0±3.5	216.2±6.8
	H(GPa)	3.0±0.3	4.2±0.1	3.9±0.4	4.1±0.2
Pos4	E (GPa)	204.2±6.2	227.9±8.9	239.3±9.0	243.0±4.0
	H(GPa)	3.4±0.6	4.2±0.7	4.6±0.4	4.5±0.6
Mean	E (GPa)	181.6±23.8	219.6±10.8	219.7±15.2	221.6±14.6
	H(GPa)	3.2±0.3	4.3±0.3	4.2±0.3	4.1±0.3

here to capture the microscale local mechanical properties of the strut material and the nanocrystalline nickel coating (Table 10.3). According to the microstructural analysis using the electron backscatter diffraction (EBSD) technique in our previous work, the grain growth direction in the coating part is always perpendicular to the Ni/Al hybrid interface [8]. In the nanoindentation tests, as shown in Figure 10.4, four lines are chosen as testing zones including two lines perpendicular to the Ni/Al interface (line1 and line3) and the other two lines inclined to the interface at an angle of 45 ° and 60 °, respectively. The main objective is to check the correlation between the mechanical behaviors and the grain growth direction of the nanocrystalline coating. Hence, four testing positions with a gradient distance (listed in the table in Figure 10.4) to the Ni/Al interface are considered in each test zone to enrich the obtained local mechanical properties, in which the influence of the hybrid interface is involved. The indentation is repeated nine times on each testing position according to the indentation pattern as was shown in Figure 10.3 for the sake of reproducibility.

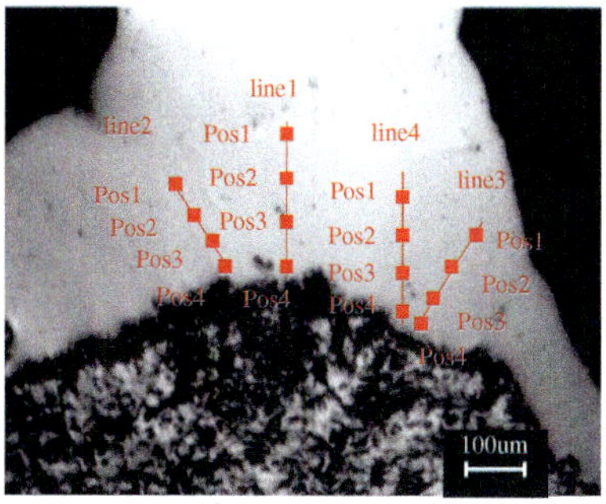

line	Pos1	Pos2	Pos3	Pos4
line1	257.0um	169.4um	66.8um	17.9um
line2	192.9um	155.5um	110.2um	43.2um
line3	207.4um	168.5um	97.2um	30.2um
line4	203.7um	151.2um	104.3um	45.6um

Figure 10.4 Four test lines are chosen on the pre-treated Ni part in nanoindentation with four positions in each line. The distances between the chosen position and the interface boundary are listed on the right-hand side.

The force-displacement data are plotted in Figure 10.5a, c, e and g, which are obtained by means of nanoindentation on all test zones of the Ni coating part. The corresponding determined elastic moduli and the hardness according to the Oliver & Pharr method are plotted in Figure 10.5b, d, f, h, and the values are listed in Table 10.4. It is clearly shown that the scattering of the force-displacement curves of nine repeated indentations within

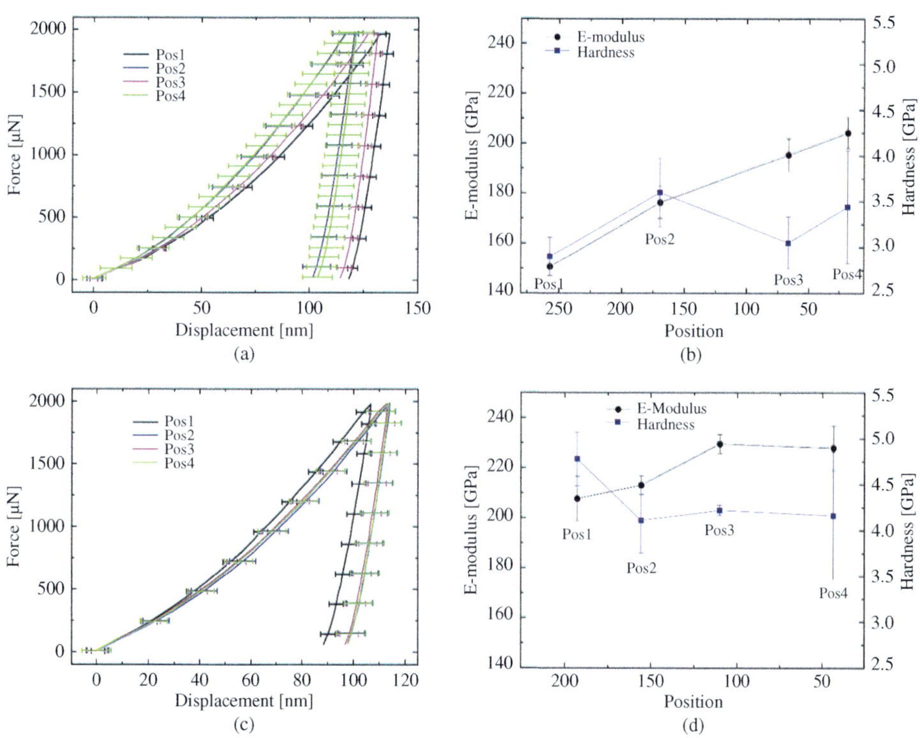

Figure 10.5 The measurements of Ni at the test line1 (a, b), line2 (c, d), line3 (e, f), line4 (g, h): load-displacement data (a, c, e, g) and the determined Young's modulus (E-modulus) and hardness (b, d, f, h) on different positions using nanoindentation.

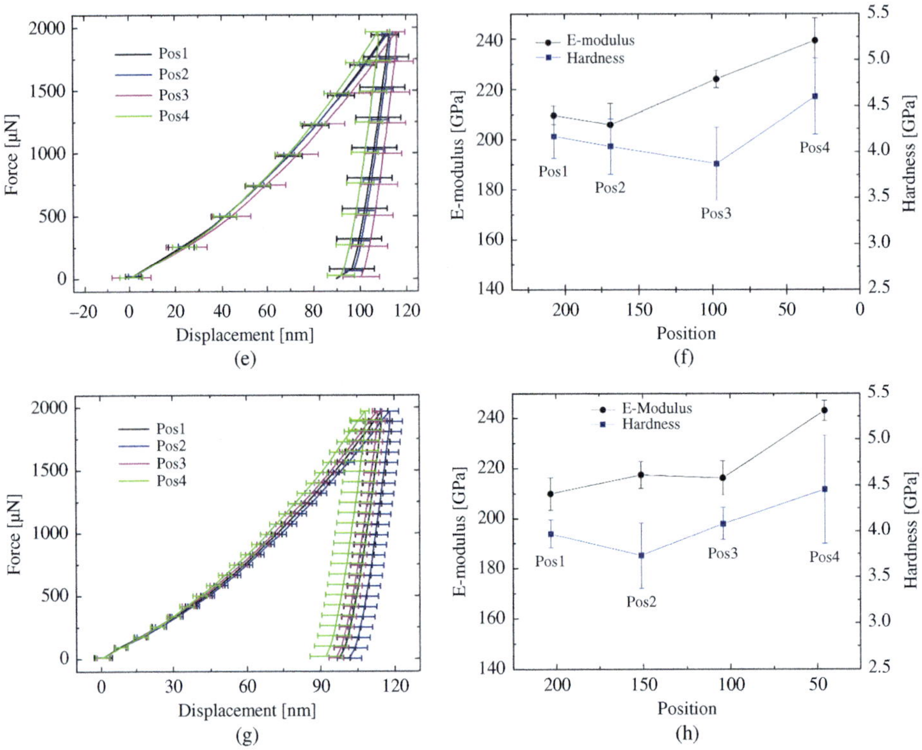

Figure 10.5 *(Continued)*

Table 10.4 The local Young's modulus and hardness of Al substrate determined by nanoindentation.

		Pos1	Pos2	Pos3	Pos4	Mean
line1	E (GPa)	102.6±9.4	103.6±3.5	96.5±5.9	97.1±9.2	100.0±3.7
	H(GPa)	1.5±0.1	1.4±0.04	1.4±0.1	1.4±0.2	1.4±0.1
line2	E (GPa)	95.8±2.6	93.6±4.1	88.1±3.9	—	92.5±4.0
	H(GPa)	1.3±0.2	1.4±0.2	1.3±0.2	—	1.3±0.1

each test position depends on the distance of the test position to the hybrid interface. A greater scattering is displayed in all the four test zones, if the indentation position is closer to the interface boundary. Therefore, it is proven by the indentation tests that the local inhomogeneity of the nanocrystalline Ni coating at the microscale becomes stronger due to the influence of the hybrid interface.

This expanded inhomogeneity can be connected to the different grain sizes generated during the manufacturing process of the Ni/Al hybrid foam, which have been observed in our previous EBSD measurement [8]. Directly at the Ni/Al interface boundary, a layer of very fine grains can be observed and the grain size increases along the direction outward the interface boundary. Hence, if an indentation pattern of 3 × 3 with an interval of

5 μm is applied, a higher probability of variation in the indentation on the grains and on grain boundary could be met in the area close to the interface boundary than in the range further away from the boundary. It is well known that the indentation response of the crystalline metal strongly depends on the grain boundary [87–89]. Both the measured Young's modulus and hardness (Table 10.3) are plotted as a function of the distance of the indentation position to the hybrid interface in Figure 10.5b, d, f, h for the four test lines, respectively. As shown in Figure 10.5b), the elastic modulus measured in test line1 (b)) increases almost linearly from 150.6 GPa in the bulk of Ni coating to 204.2 GPa near the interface. The determined moduli in test line2 (d)), line3 (f)) and line4 (h)) almost lie in the same region, which is approximately 210 GPa, in the Ni bulk and rise up to approximately 240 GPa near the interface. Similar to the trend shown in Figure 10.5 b), linear relationships between the modulus and the distance are also displayed excluding the Pos3 in line2, Pos2 in line3 and Pos3 in line4. It is proved by these results that the hybrid Al/Ni interface has a certain influence on the locally measured elastic modulus determined by nanoindentation. The modulus has the maximum value near the interface and the minimum value at the test position further away from the interface. Furthermore, an approximately linear function can be used to describe the relationship between the measured modulus and the distance of the indentation position to the interface boundary. The cause of the internal discrepancy that the measured moduli in test line1 are much lower than the values of the corresponding test positions in the other three test lines should be further studied with the EBSD technique in a future work. In contrast to the determined modulus, there is no distinct relationship between the measured hardness and the distance of the indentation position from the interface. The internal scattering of the hardness measured at Pos4 near the interface is greater than the global scattering of test line1, line2 and line4. From this point of view, the determination of the hardness depends more strongly on the local inhomogeneity than the determination of the modulus. The local inhomogeneity, including the local surface topography and microstructure, can have a strong influence on the determination of the contact area. In the Oliver & Pharr method [36, 37], the hardness is a function of the contact area whereas the modulus is a function of the area's square root. However, if the hardness tested on the Ni bulk and the hardness near the interface are compared, the hardness of the latter is larger. The result goes well with the findings in the previous EBSD measurements [8].Very small globular grains are found at and near the interface and large elongated grains in the bulk due to the columnar growth. The hardness is known to depend strongly on the grain size, according to the Hall-Petch formula [9, 10]. This formula reveals that finer grains are associated with more grain boundaries and result in a greater mechanical strength.

The local elastic modulus and hardness of the Al substrate are of particular interest because of the observed strong influence of the hybrid interface on the measured mechanical properties of the Ni coating. Two test lines on the pre-treated Al part are taken with four indentation positions in each line as shown in Figure 10.6.

The indentation positions are chosen on the smooth part with a gradient distance away from the interface boundary in order to investigate the effect of the hybrid boundary on the measured modulus and hardness. The scattering of the force-displacement data in Figure 10.7 is much smaller compared to the case of Ni coating. The experimental data on Pos4 in test line2 is excluded from Figure 10.7d because of its extremely large difference from other data. The data show that a maximum displacement about

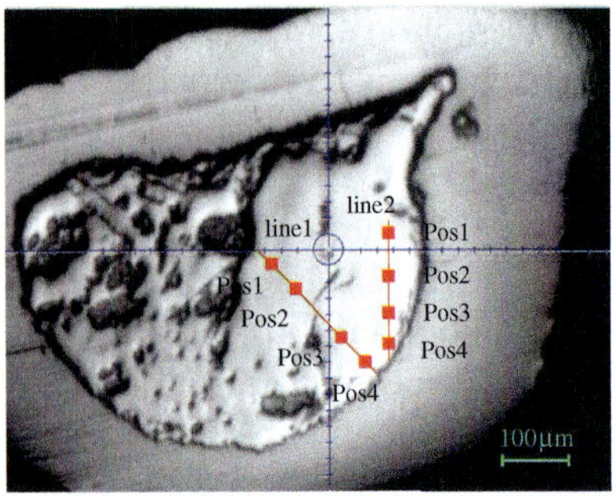

Line	Pos1	Pos2	Pos3	Pos4
line1	237.6 μm	181.5 μm	115.7 μm	49.7 μm
line2	218.3 μm	179.2 μm	114.1 μm	21.3 μm

Figure 10.6 Two test lines are chosen on the pre-treated Al part in nanoindentation with four positions in each line. The distances between the chosen position and the interface boundary are listed on the right-hand side.

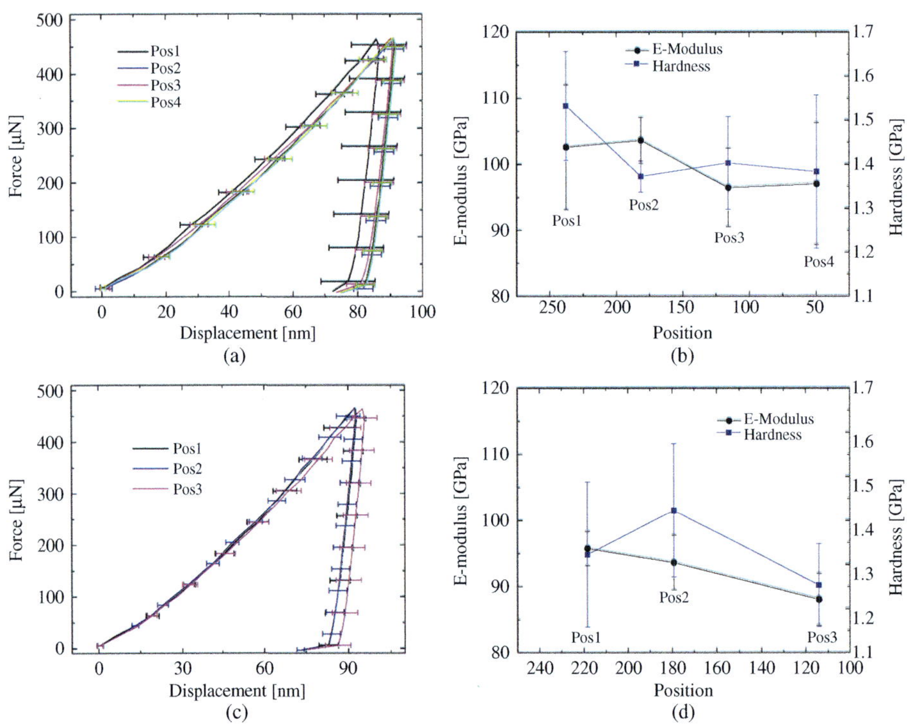

Figure 10.7 The measurements of Al at the test line1 (a, b) and line2 (c, d): load-displacement data (a, c) and the Young's modulus (E-modulus) and hardness (b, d) on different positions determined by nanoindentation.

30 nm is obtained under the maximum load of 500 μN and a very large scattering is involved. The determined modulus and hardness fall in the range of the corresponding values of the Ni coating. A reason could be that Pos4 is distanced only 21.3 μm from the interface boundary and a pattern of 3 × 3 with interval of 5 μm is used. There is a great possibility that the indenter may penetrate the interface or its boundary and hence the influences of the hybrid interface and the Ni coating can be involved. This

reason is of particular interest and should be further proven in the following work by the microstructure around the boundary using the EBSD technique.

In Figure 10.7b and d, the divergence between the average values of the four test positions and the scattering within the individual test positions are considered a global and local discrepancy, respectively. It is clearly shown that both the measured modulus and the hardness of the Al alloy strut have an extremely large local discrepancy compared to the global one. The very slight global discrepancy suggests that the Al alloy strut still possesses a homogeneous mechanical property at the microscale. The relatively large local discrepancy could be caused by the silicone precipitations and the grain boundaries. As can be seen from the element-mapping image in [8], silicone precipitations, whose modulus and hardness are both expected to be much larger than the Al alloy, is randomly distributed in the Al strut. The measured modulus and hardness of the indentation points on or near the silicone precipitation should be larger than those obtained from the indentation points on the Al strut, which are further away from the precipitation. Therefore, from a statistical point of view, the local divergence due to the strong influence of the silicone precipitation can disappear in the average values of the different test positions. This interpretation is also suitable for the influence of the grain boundary on the measured local and global modulus and hardness.

10.4 Conclusions and Outlook

The strengthening of metal foams by means of electrodeposition of coatings such as nanocrystalline nickel is an important method to improve the mechanical properties of metal foams as it leads to tremendous increases in stiffness and in the energy absorption capacity [8, 12, 14]. In order to develop new cellular hybrid materials for energy absorbers and lightweight components, both experimental and simulation approaches to characterize the behaviors of the hybrid foam structures are necessary. For an efficient modeling approach, multiscale numerical models are needed. Such models require the determination not only of the global material behavior of the hybrid foams but also of the local behavior of single hybrid struts. Based on the thin coating and the small strut dimensions, nanoindentation is the only method for the direct experimental determination of the local properties of the coating and the aluminum alloy strut itself. The nanoindentation experiments in this work show very reproducible Young's moduli and hardnesses for the aluminum struts, but there are strong position dependent changes in the properties of the nickel coating. These changes are associated with position-dependent variations in the grain structure of the nickel coating, which have been determined in a previous work [8]. In general, the nickel coating has twice the modulus and three times the hardness of the aluminum struts. The results show furthermore a large scattering in the local nickel modulus and hardness directly at the Ni/Al interface. This is based, on the one hand, on the nine points measuring pattern. As a result of this broad pattern, some of the indents were performed in Al and some in Ni if the location of the pattern was located at the Ni/Al interface. On the other hand, the large scattering at the interface is caused by the interface itself. Due to the different hardnesses of nickel and aluminum, it is not possible to polish both materials with the same intensity. Aluminum is softer than nickel. This difference in hardness leads to a small step at the contact zone of nickel and aluminum. In conclusion to the fact that the global mechanical behavior of foams is based on the buckling and bending of the struts of the foams, with the help of the very local nanoindentation experiments, the increased global strength and energy absorption capacity of hybrid metal foams can be explained. The three times harder nickel coating improves the bending stiffness of the struts and hence the global mechanical properties of the Ni/Al hybrid foams. As the highest stresses in compression and tension arise during the bending at the outer fibre of a strut, the hardness increase of the nanocrystalline nickel coating leads to a weight-optimized composite strut, whereas the heavier nickel is only needed in the outer regions of the strut cross section. Due to the local changes in the properties of the nickel coating and the strong scattering at the Ni/Al interface, for a clear identification and a reliable explanation of the measured local properties, it is necessary for future work to couple EBSD measurements with nanoindentation experiments to determine the local microstructure and local properties. As mentioned before, multiscale simulation is need to simultaneously study the nano- or microcellular structure and macroscale properties of the hybrid foam materials. However, the locally measured Young's modulus and hardness are not enough to predict the nano- and micro-behaviors of the cellular Ni/Al hybrid material in the multiscale numerical model. Constitutive models with identified parameters are required to describe the mechanical responses. Hence, in the future, characterizing the local behaviors of the Ni/Al hybrid foam using constitutive models will be treated with the FEM based inverse method combined with the nanoindentation experimental data, which has been developed in our previous work [64, 68].

Acknowledgements

The authors are grateful to the DFG (German Science Foundation-Deutsche Forschungsgemeinschaft) for the financial support through the grant number Di 430/20-1. The very helpful discussion with M.Sc M. Zamanzade at the Chair of Materials Science and Methods at Saarland University during nanoindentation tests is greatly appreciated.

References

1 Ashby M. The mechanical properties of cellular solids. Metall Trans A, (1983) 14(9): 1755–1769.
2 Ashby M, Evans A, Fleck NA, *et al.* Metal Foams: A Design. Butterworth-Heinemann, Oxford; 2000.
3 Banhart J. Manufacturing Routes for Metallic Foams. JOM; 2000.
4 Banhart J. Manufacture, characterisation and application of cellular metals and metal foams. Progress in Materials Science (2001) 46(6): 559–U3.
5 Banhart J, Baumeister J, Weber M. Aluminium foams for transport industry. Materials Design (1997) 18: 221
6 Abrate S. Localized impact on sandwich structures with laminated facings. Applied Mechanics Review (1997) 50(2): 69–82
7 Destefanis R, Schfer F, Lambert M, Faraud M. Selecting enhanced space debris shields for manned spacecraft. International Journal of Impact Engineering (2006)

33(1–12): 219–230, DOI: 10.1016/j.ijimpeng.2006.09.065, Impact Proceedings of the 2005 Symposium.

8 Jung A, Diebels S, Koblischka-Veneva A, *et al.* Microstructural analysis of electrochemical coated open-cell metal foams by EBSD and nanoindentation. Advanced Engineering Materials (2014a) 16: 15–20.

9 Hall E. The deformation and ageing of mild steel: Discussion of results. Proceedings of the Physical Society of London B, (1951) 64: 747–753.

10 Petch NJ. The cleavage strength of polycrystals. Journal of Iron and Steel Research International (1953) 174: 509–517.

11 Boonyongmaneerat Y, Schuh C, Dunand D. Mechanical properties of reticulated aluminum foams with electrodeposited Ni-W coatings. Scripta Materialia (2008) 59(3): 336–339, DOI:10.1016/j.scriptamat.2008.03.035

12 Bouwhuis B, McCrea J, Palumbo G, Hibbard G. Mechanical properties of hybrid nanocrystalline metal foams. Acta Materialia (2009) 57(14): 4046–4053, DOI: 10.1016/j.actamat.2009.04.053

13 Jung A, Koblischka MR, Lach E, Diebels S, Natter H. Hybrid metal foams: mechanical testing and determination of mass flow limitations during electroplating. International Journal of Material Science (2012) 2(4): 97–107.

14 Jung A, Natter H, Diebels S, *et al.* Nanonickel coated aluminum foam for enhanced impact energy absorption. Advanced Engineering Materials (2011) 13(1–2): 23–28, DOI: 10.1002/adem.201000190.

15 Jung A, Klis D, Goldschmidt F. Experiments, modell and simulation of the magnetic behavior of inhomogeneously coated nickel/aluminium hybrid foams. Journal of Magnetism and Magnetic Materials (accepted).

16 Jung A, Natter H, Hempelmann R, *et al.* Study of the magnetic flux density distribution of nickel coated aluminum foams. Journal of Physics: Conference Series, IOP Publishing, (2010a) 200: 082011.

17 Sun Y, Burgueo R, Vanderklok A, *et al.* Compressive behavior of aluminum/copper hybrid foams under high strain rate loading. Materials Science and Engineering A, (2013) 592: 111–120.

18 Sun Y, Burgueo R, Wang W, Lee I. Effect of annealing on the mechanical properties of nano-copper reinforced open-cell aluminum foams. Materials Science and Engineering A, (2014) 613: 340–351.

19 Hasan M, Kim A, Lee HJ. Measuring the cell wall mechanical properties of al-alloy foams using the nanoindentation method. Composite Structures (2008) 83: 180–188.

20 Lee HJ, Kim A, Hasan M, Choen S, Lee. The constitutive behavior of metallic foams using nanoindentation technique and fe modelling. Key Engineering Materials (2005) 297–300: 1050–1055.

21 Kim A, Hasan M, Lee HJ, Cho S (2005) Characterization of submicron mechanical properties of Al-alloy foam using nanoindentation technique. Materials Science Forum 475–479: 4199–4202.

22 Christensen R. Mechanics of low density materials. Journal of the Mechanics and Physics of Solids (1986) 34: 563–578.

23 Gibson L, Ashby M, Schajer G, Robertson C. The mechanics of two-dimensional cellular materials. Proceedings of the Royal Society of London, Ser A, (1982) 382: 25–42.

24 Meguid S, Cheon S, El-Abbasi N. Fe modelling of deformation localization in metallic foams. Finite Elem Anal Des (2002) 38: 631643.
25 Kralik V, Nemecek J. Micromechanical properties of porous material based on metal foam. Chemicke Listy (2011) 105: 672–675.
26 Markaki AE, Clyne TW. The effect of cell wall microstructure on the deformation and fracture of aluminium-based foams. Acta Materialia (2001) 49: 1677–1686.
27 Rho J, Kuhn-Spearing L, Zioupos P. Mechanical properties and the hierarchical structure of bone. Med Eng Phys (1998) 20: 92–102.
28 Rho JY, Tsui T, Pharr G. Elastic properties of human cortical and trabecular lamellar bone measured by nanoindentation. Biomaterials (1997) 18: 1325–1330.
29 Branca FP, Marinozzi F. Microtensile device for stress: Elongation tests on nonstandard specimens. Review of Scientific Instruments (2000) 71(6): 2526–2531.
30 Jiroušek O, Jandejsek I, Vavřík D. Evaluation of strain field in microstructures using micro-ct and digital volume correlation. Journal of Instrumentation (2011) 6(1): 1–7.
31 Jiroušek O, Němeek J, Kytýř D, *et al.* Nanoindentation of trabecular bone-comparison with uniaxial testing of single trabecula. Chemicke Listy (2011) 105: 668–671.
32 Lucchinetti E, Thomann D, Danuser G. Micromechanical testing of bone trabeculae potentials and limitations. Journal of Materials Science (2000) 35(24): 6057–6064.
33 Nazarian A, Stauber M, Müller R. Design and implementation of a novel mechanical testing system for cellular solids. Wiley InterScience (2005) 73(2): 400–411.
34 Ryan SD, Williams JL. Tensile testing of rodlike trabeculae excised from bovine femoral bone. J Biomechanics (1989) 22(4): 351–355.
35 Zhou J, Allameh S, Soboyejo WO. Microscale testing of the strut in open cell aluminum foams. Journal of Materials Science (2005) 40(2): 429–439.
36 Oliver WC, Pharr GM. An improved technique for determining hardness and elastic modulus using load and displacement sensing indentation experiments. Journal of Materials Research (1992) 7: 1564–1583.
37 Oliver WC, Pharr GM. Measurement of hardness and elastic modulus by instrumented indentation: Advance in understanding and refinements to methodology. Journal of Materials Research (2004) 19: 3–20.
38 Fischer-Cripps AC. Nanoindentation. Springer-Verlag, New York; 2004.
39 Hertz H. Über die Berührung fester elstischer körper. Journal für die Reine und Angewand Mathematik (1881) 92: 156–171
40 Sneddon IN. Boussinesq's problem for a rigid cone. Mathematical Proceedings of the Cambridge Philosophical Society (1948) 44: 492–507.
41 Sneddon IN. The relation between load and penetration in the axisymmetric Boussinesq problem for a punch of arbitrary profile. International Journal of Engineering Science (1965) 3: 47–57.
42 Lee EH, Radok JRM. The contact problem for viscoelastic bodies. Journal of Applied Mechanics (1960) 30: 438–444.
43 Ting TCT. The contact stresses between a rigid indenter and a viscoelastic half-space. Journal of Applied Mechanics (1966) 33: 845–854.
44 Ting TCT. Contact problems in the linear theory of viscoelasticity. Journal of Applied Mechanics (1968) 35: 248–254.
45 Graham GAC. The contact problem in the linear theory of viscoelasticity. International Journal of Engineering Science (1965) 3: 27–46

46 Graham GAC. The contact problem in the linear theory of viscoelasticity when the time dependent contact area has any number of maxima and minima. International Journal of Engineering Science (1967) 5: 495–514.
47 Johnson KL. Contact Mechanics. Cambridge, Cambridge University Press; 1985.
48 Kushch VI, Dub SN. The assessment of elasto-plastic properties of materials from nanoindentation and computer modeling. 1. state-of-the-art of the problem (review of the literature). Journal of Superhard Materials (2012) 34: 149–155.
49 Balasundaram K, Cao YP, Raabe D. Identifying the limitation of Oliver and Pharr method in characterizing the viscoelastic-plastic materials with respect to indenter geometry. In: Nano, Properties MMM, under Extreme Environments B, (eds) A. Misra and T. J. Balk and H. Huang *et al.* Materials Research Society Symposium Procedings vol. 1137. Warrendale, PA, 1137-EE10-23; 2009.
50 Ma D, Ong C, Wong S. New relationship between Young's modulus and nonideally sharp indentation parameters. Journal of Materials Research (2004) 19: 2144–2151.
51 Cheng YT, Cheng CM. Relationship between contact stiffness, contact depth, and mechanical properties for indentation in linear viscoelastic solids using axisymmetric indenters. Structural Control and Health Monitoring (2006) 13: 561–569.
52 Chudoba T, Richter F. Investigation of creep behaviour under load during indentation experiments and its influence on hardness and modulus results. Surface and Coatings Technology (2001) 148: 191–198.
53 Tabor D. Hardness of Metals. Oxford: Oxford University Press; 1951.
54 Kim JY, Lee BW, Read D, Kwon D. Influence of tip bluntness on the size-dependent nanoindentation hardness. Scripta Materialia (2005) 52: 353–358.
55 Backes B, Durst K, Gken M. Determination of plastic properties of polycrystalline metallic materials by nanoindentation: Experiments and finite element simulations. Philosophical Magazine (2006) 86: 5541–5551.
56 Dao M, Chollacoop N, Van Vliet K, *et al.* Computational modeling of the forward and reverse problems in instrumented sharp indentation. Acta Materialia (2001) 49: 3899–3918.
57 Ogasawara N, Chiba N, Chen X. Representative strain of indentation analysis. Journal of Materials Research (2005) 20: 2225–2234.
58 Bucaille JL, Stauss S, Felder E, Michler J. Determination of plastic properties of metals by instrumented indentation using different sharp indenters. Acta Materialia (2003) 51: 1663–1678.
59 Cao Y, Qian X, Lu J, Yao Z. An energy-based method to extract plastic properties of metal materials from conical indentation tests. Journal of Materials Research (2005) 20: 1194–1206.
60 Milman Y, Lotsko D, Dub S, *et al.* Mechanical properties of quasicrys-talline Al-Cu-Fe coatings with submicron-sized grains. Surface and Coatings Technology (2007) 201: 5937–5943.
61 Swaddiwudhipong S, Tho K, Liu Z, Zeng K. Material characterization based on dual indenters. International Journal of Solids and Structures (2005) 42: 69–83.
62 Hartmann S, Gibmeier J, Scholtes B. Experiments and material parameter identification using finite elements, uniaxial tests and validation using instrumented indentation tests. Experimental Mechanics (2006) 46: 5–18.
63 Huber N, Nix WD, Gao H. Identification of elastic-plastic material parameters from pyramidal indentation of thin films. Proceedings of the Royal Society of London Series A: Mathematical, Physical and Engineering Sciences (2002) 458: 1593–1620.

64 Huber N, Tyulyukovskiy E. A new loading history for identification of viscoplastic properties by spherical indentation. Journal of Materials Research (2004) 19: 101–113.
65 Klötzer D, Ullner C, Tyulyukovskiy E, Huber N. Identification of viscoplastic material parameters from spherical indentation data. part II: Experimental validation of the method. Journal of Materials Research (2006) 21: 677–684.
66 Tyulyukovskiy E, Huber N. Identification of viscoplastic material parameters from spherical indentation data. Part I: Neural networks. Journal of Materials Research (2006) 21: 664–676.
67 Chen Z, Diebels S, Peter NJ, Schneider AS. Identification of finite viscoelasticity and adhesion effects in nanoindentation of a soft polymer by inverse method. Computational Materials Science (2013) 72: 127–139.
68 Chen Z, Scheffer T, Seibert H, Diebels S. Macroindentation of a soft polymer: Identification of hyperelasticity and validation by uni/biaxial tensile tests. Mechanics of Materials (2013) 64: 111–127.
69 Chen Z, Diebels S. Parameter re-identification in nanoindentation problems of viscoelastic polymer layers: small deformation. Journal of Applied Mathematics and Mechanics/Zeitschrift für Angewandte Mathematik und Mechanik (ZAMM) (2013) 93: 88–101.
70 Le Saux V, Macro Y, Bles G, *et al.* Identification of constitutive model for rubber elasticity from micro-indentation tests on natural rubber and validation by macroscopic tests. Mechanics of Materials (2011) 43: 775–786.
71 Rauchs G, Bardon J. Identification of elasto-viscoplastic material parameters by indentation testing and combined finite element modelling and numerical optimization. Finite Elements in Analysis and Design (2011) 47: 653–667.
72 Chen Z, Diebels S. Nanoindentation of hyperelastic polymer layers at finite deformation and parameter re-identification. Archive of Applied Mechanics (2012) 82: 1041–1056.
73 Chen Z, Diebels S. Numerical investigation of nanoindentation of viscoelastic polymer layers and parameters re-identification. Proceedings in Applied Mathematics and Mechanics (2011) 11: 765–766.
74 Rauchs G, Bardon J, Georges D. Identification of the material parameters of a viscous hyperelastic constitutive law from spherical indentation tests of rubber and validation by tensile tests. Mechanics of Materials (2010) 42: 961–973.
75 Rauchs G. Optimization-based material parameter identification in indentation testing for finite strain elasto-plasticity. ZAMM Zeitschrift für Angewandte Mathematik und Mechanik (2006) 86: 539–562.
76 Constantinescu A, Tardieu N. On the identification of elastoviscoplastic constitutive laws from indentation tests. Inverse Problems in Engineering (2001) 9: 19–44.
77 Gibson LJ, Ashby MF. Cellular Solids, 2nd edn. Cambridge University Press; 1997.
78 Němeek J, Králík V, Vondřejc J. A two-scale micromechanical model for aluminium foam based on results from nanoindentation. Computers and Structures (2013) 128: 136–145.
79 Haušild P, Nohava J, Materna A. Identification of stress-strain relation of austenitic steels by instrumented indentation. Chemicke Listy (2011) 105: 676–679.
80 Biener J, Hodge A, Hayes J, *et al.* Size effects on the mechanical behavior of nanoporous Au. Nano Letters (2006) 6: 2379–2382.

81 Karunatilaka C, Buar DK, Ditzler L, *et al.* Softening and hardening of macro- and nano-sized organic cocrystals in a single-crystal transformation. Angewandte Chemie: International Edition (2011) 50: 8642–8646.
82 Ramamurty U; Jang JI. Nanoindentation for probing the mechanical behavior of molecular crystals-a review of the technique and how to use it. Crys-tEngComm (2013) 16: 12–23.
83 Tan J, Cheetham A. Mechanical properties of hybrid inorganic-organic framework materials: Establishing fundamental structure-property relationships. Chemical Society Reviews (2011) 40: 1059–1080.
84 Varughese S, Kiran M, Ramamurty U, Desiraju G. Nanoindentation in crystal engineering: Quantifying mechanical properties of molecular crystals. Angewandte Chemie: International Edition (2013) 52: 2701–2712.
85 Schwaiger R, Moser B, Dao M, Chollacoop N, Suresh S. Some critical experiments on the strain-rate sensitivity of nanocrystalline nickel. Acta Materialia (2003) 51: 5159–5172.
86 Fale S, Likhite A, Bhatt J. Nanoindentation, compressive and tensile deformation study of in-situ AlAlN metal matrix composites. Transactions of the Indian Institute of Metallurgy (2014) DOI: 10.1007/s12,666–014–0445–0.
87 Iglesias R, Leiva E. Two-grain nanoindentation using the quasi-continuum method: Two-dimensional model approach. Acta Materialia (2006) 54: 2655–2664.
88 Nix W, Greer J, Feng G, Lilleodden E. Deformation at the nanometer and micrometer length scales: Effects of strain gradients and dislocation starvation. Thin Solid Films (2007) 515: 3152–3157.
89 Soer W, Aifantis K, De Hosson J. Incipient plasticity during nanoindentation at grain boundaries in body-centered cubic metals. Acta Materialia (2005) 53: 4665–4676.
90 Golovin Y. Nanoindentation and mechanical properties of solids in submi-crovolumes, thin near-surface layers, and films: A review. Physics of the Solid State (2008) 50: 2205–2236.
91 Bhattacharya A, Nix W. Analysis of elastic and plastic deformation associated with indentation testing of thin films on substrates. International Journal of Solids and Structures (1988) 24: 1287–1298.
92 Bull S. Nanoindentation of coatings. Journal of Physics D: Applied Physics (2005) 38: R393–R413.
93 Burnett P, Rickerby D. The mechanical properties of wear-resistant coatings. i: Modelling of hardness behaviour. Thin Solid Films (1987) 148: 41–50.
94 Nix W. Mechanical properties of thin films. Metallurgical Transactions A, (1989) 20: 2217–2245.
95 Ohmura T, Matsuoka S, Tanaka K, Yoshida T. Nanoindentation load-displacement behavior of pure face centered cubic metal thin films on a hard substrate. Thin Solid Films (2001) 385: 198–204.
96 Cai F, Zhang S, Li J, Chen Z, Li M, Wang L. Effect of nitrogen partial pressure on Al-Ti-N films deposited by arc ion plating. Applied Surface Science (2011) 258: 1819–1825.
97 Nakayama K, Yokoyama Y, Xie G, *et al.* Metallic glass nanowire. Nano Letters (2008) 8: 516–519.

98 Raman K, Kiran M, Ramamurty U, Rao G. Structure and mechanical properties of tic films deposited using combination of pulsed DC and normal DC magnetron co-sputtering. Applied Surface Science (2012) 258: 8629–8635.

99 Nunes J, Piedade A. Nanoindentation of functionally graded hybrid polymer/metal thin films. Applied Surface Science (2013) 284: 792–797.

100 Szparaga, Ratajski J, Bartosik P (2013) Strain field analysis in nanoindentation test of gradient coatings. Archives of Materials Science and Engineering 64: 219–227.

101 Pelletier H, Gauthier C, Schirrer R. Friction effect on contact pressure during indentation and scratch into amorphous polymers. Materials Letters (2009) 63: 1671–1673.

102 Deng X, Cleveland C, Karcher T, *et al.* Nanoindentation behavior of nanolayered metal-ceramic composites. Journal of Materials Engineering and Performance (2005) 14: 417–423.

103 Bolshakov A, Pharr GM. Influences of pileup on the measurement of mechanical properties by load and depth sensing indentation techniques. Journal of Materials Research (1998) 13: 1049–1058.

104 Cheng YT, Cheng CM. Relationships between hardness, elastic modulus, and the work of indentation. Applied Physics Letters (1998) 73: 614–616.

105 Jung A, Natter H, Hempelmann R, Lach E. Metal foams, (2009) ep 2261398 a1.

106 Jung A, Natter H, Hempelmann R, Lach E. Metal foams, (2010) wo 2010/142436 a1.

107 Amblard J, Froment M, Maurin G, *et al.* Nickel electrocrystallization-from nucleation to textures. Electrochimica Acta (1983) 28: 909–915.

108 Amblard J, Epelboin I, Froment M, Maurin G. Inhibition and nickel electrocrystallization. Journal of Applied Electrochemistry (1979) 9: 233–242.

109 Bicelli L, Kozlov V. Texture formation of electrodeposited fcc metals. Materials Chemistry and Physics (2003) 77(1): 289–293, DOI 10.1016/S0254-0584(02)00004-4.

110 Pangarov N. On the crystal orientation of electrodeposited metals. Electrochimica Acta (1964) 9(6): 721–726.

111 Vicenzo A, Cavallotti P. Structure and electrokinetic study of nickel electrodeposition. Russian Journal of Electrochemistry (2008) 44(6): 716–727, DOI 10.1134/S1023193508060128.

112 Barnoush A, Zamanzade M. Effect of substitutional solid solution on dislocation nucleation in Fe3Al intermetallic alloys. Philosophical Magazine (2012) 92: 3257–3268.

11

AFM-based Nanoindentation of Cellulosic Fibers

Christian Ganser[1,2] *and Christian Teichert*[1,2]

[1] *Institute of Physics, Montanuniversitaet Leoben, Austria*
[2] *Christian Doppler Laboratory for Surface Chemical and Physical Fundamentals of Paper Strength, Graz University of Technology, Austria*

11.1 Introduction

In the last two decades, nanomechanical characterization of various kinds of cellulose fibers became possible due to the advent of modern nanoindentation (NI) [1] and atomic force microscopy (AFM) [2] related techniques. Investigated materials range from raw wood fibers [3], pulp fibers for paper production [4–6], and man-made cellulose fibers such as viscose and lyocell [7, 8]. The reasons for determining the materials' mechanical properties are also diverse and could stem from the wish to deepen the understanding of wood [3, 9] or paper mechanics [5, 10], or to quantify differences of several treatments in paper production [6] or man-made fiber production [7, 11].

The mechanical characterization by AFM based nanoindentation (AFM-NI) presented in this chapter was carried out on pulp and viscose fibers. Here, the goal was to understand the bond between two paper fibers in more detail. Although paper is known and produced for thousands of years, the mechanisms holding the material together are still elusive. Nevertheless, five mechanisms that could contribute to the strength of a bond between two paper fibers have been identified [12]:

- hydrogen bonds
- van der Waals interactions
- Coulomb interactions
- interdiffusion of cellulose molecules
- mechanical interlocking.

It is, however, unknown to which extent these mechanisms contribute and if there are additional ones present. Also, capillary bridges [13, 14] and microcompressions induced by stresses during drying of the fibers [15, 16] are likely to have an effect on the bond strength.

From the five bonding mechanisms listed above, numbers one to four would lead to a stronger bond when the area in molecular contact is maximized. The fifth mechanism, on the other hand, depends on the surface structure and its features [17]. To gain understanding of these mechanisms, the mechanical properties of the paper fibers' surfaces and their morphology needs to be known on the nanometer scale – the scale on

Applied Nanoindentation in Advanced Materials, First Edition. Edited by Atul Tiwari and Sridhar Natarajan.

which bond formation takes place. Moreover, bonds are initially formed between highly swollen fibers in an aqueous environment and it is usually insufficient to consider the properties of dried fibers. Both, the mechanical parameters and the surface morphology will change drastically, and to some degree irreversibly [18], after water is removed from the fiber.

Since the majority of the fiber-fiber bonding mechanisms listed above relies on a high area in molecular contact, a measure for the fibers' ability to conform to each other is needed. Indentation hardness and Young's modulus are such measures. On the nanometer scale, usually classical NI is employed to determine these properties. Due to the rough nature of paper fibers, the application of this method is limited and likely to result in a large scattering of the data. Therefore, AFM-NI is used in this work to approximate these material parameters. By employing a setup to control the environment's water content, the effect of water uptake is studied beginning with the fully dried state until highly swollen materials. While the goal is ultimately to understand the importance of fiber swelling during bond formation in paper, a comprehensive mechanical characterization of cellulose fibers is achieved as well. The results obtained in this chapter allow not only a comparison of the dry and wet states of cellulose fibers, but indicate the transition from one to the other. Further, the effectiveness of industrially available ways to influence the swelling behavior is demonstrated with viscose fibers.

To successfully apply AFM-NI for mechanical characterization of cellulosic fibers, the method has been expanded by taking the viscoelasticity upon unloading into account [19]. Further, an improved method to directly detect the indenter's geometry and thus its area function by AFM imaging of a calibration grid is introduced. This easy imaging of the indenter allows also for checking its degree of contamination and employing a cleaning procedure, which is described in this chapter. Additionally, a setup to control relative humidity was developed and is also detailed in this chapter.

The presented AFM-NI studies on paper fibers lead to the result that in the presence of water, their hardness H is decreased by a factor of about 100 compared to the dried state. Simultaneously, the reduced modulus E_r of paper fibers decreases by a factor of approximately 100. The characterization of H and E_r of modified swollen viscose fibers proved that the mechanical properties of these fibers can be tuned from soft and compliant with a high swelling potential to hard and stiff with low swelling.

11.2 Experimental

11.2.1 AFM Instrumentation

The atomic force microscope used for all experiments in this chapter was an Asylum Research MFP-3D (Santa Barbara, CA), equipped with a planar closed-loop sample scanner. Its maximum x–y scan size is 90 μm × 90 μm, and the z-range is 15 μm. In this AFM system, the cantilever is tilted by 11°, which is the case for most commercial AFMs.

Probes For AFM-NI, ND-DYIRS full diamond probes (Advanced Diamond Technologies, Romeoville, IL) were used. These probes have an aluminum reflective coating at the backside of the cantilever and feature a regular, four-sided pyramid as a tip. The angles between the pyramidal faces are about 75°, and the apex radius is 15 nm in average and guaranteed to be less than 50 nm. Spring constants were determined to be between 45 N m^{-1} and 60 N m^{-1}, as described below. The cantilever is about 135 μm long, 33 μm wide, and 3 μm thick, according to the probes' datasheet.

In addition to ND-DYIRS probes, also ND-DTIRS full diamond probes were utilized for AFM-NI. These probes have a 3-sided pyramidal tip with an angle of the side planes of about 60° and an angle between the front plane and the back edge of 120°. The apex radius is the same as the one of ND-DYIRS probes. Also most parameters are in accordance with ND-DYIRS probes, only the cantilever length is 120 μm. Since the manufacturer stopped to produce ND-DTIRS probes after first results were obtained, all more recent investigations were carried out using ND-DYIRS probes.

The large opening angles of both tip types are an asset when the area function needs to be determined, as is described later on. A high opening angle is also beneficial when comparing obtained results with those of classical NI.

Calibration of Cantilever Spring Constants In order to measure forces with an AFM, the cantilever's spring constant k as well as the optical lever sensitivity S_{OLS} have to be determined precisely. The sensitivity is the ratio between the signal at the diode V_{PD} and the cantilever's actual deflection z_d

$$V_{PD} = S_{OLS} \times z_d. \tag{11.1}$$

It is calibrated by pressing the cantilever on a hard substrate and recording V_{PD} as a function of the piezo movement z. If the substrate does not deform, obviously, $z = z_d$ and V_{PD} plotted versus z_d is a straight line with slope S_{OLS}. Note that this method implies a calibrated z-piezo or an active z-sensor, otherwise the deflection z_d cannot be determined. A silicon wafer with native oxide can be used as the hard substrate. To verify if a silicon wafer is appropriate, the calibration of S_{OLS} has been cross-checked on a diamond coated fused silica substrate where no difference was detectable. The spring constant is determined after S_{OLS} is known using the thermal sweep method [20]. There, the spring constant is extracted from the thermal spectrum of a free cantilever.

Controlled Relative Humidity To control relative humidity φ_r inside the fluid cell, the setup presented schematically in Figure 11.1 is employed. This setup is based on one that was developed to measure pull-off forces in humid environments with AFM [21]. A flow

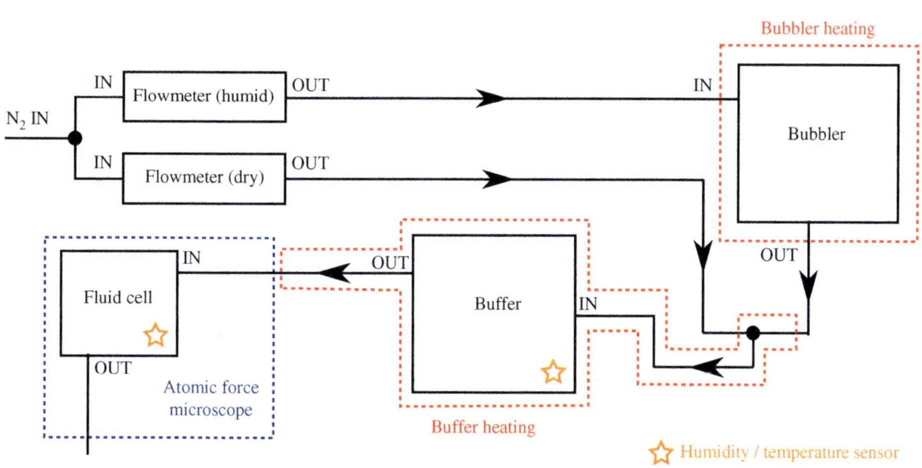

Figure 11.1 Setup to control relative humidity within the AFM's fluid cell as a schematical representation.

of dry nitrogen is divided into two lines with adjustable flow rates. One line is kept dry, whereas the other one is lead through a bubbler where the nitrogen is saturated with water ($\varphi_r = 1$). Since the bubbler is – due to the fact that water is evaporated – most likely the coldest spot in the system, φ_r of the humid line will decrease below 1 outside the bubbler. Heating the bubbler above room temperature ensures that φ_r cannot drop below its initial value. Beyond the bubbler, the dry and saturated nitrogen lines are merged together, leading to a mixture with a water content according to the flow rates of the two lines. In front of the fluid cell, a buffer volume is located which absorbs fluctuations in relative humidity. This is necessary, because the fluid cell has a volume of only 10 ml. Since temperatures of up to 35°C are observed in the fluid cell, the buffer as well as the tubes between bubbler and fluid cell are also heated.

Separate heating circuits for buffer and bubbler result in a high flexibility in the temperature profile and allow reaching very high values in relative humidity inside the fluid cell. For example, without heating the buffer, the maximum φ_r was ≈ 0.8, whereas by heating it, a relative humidity of up to $\varphi_r = 0.95$ can be generated inside the fluid cell. Note that at $\varphi_r = 0.95$ a temperature decrease from 30° to 29.11° would result already in condensation. Condensation inside the fluid cell can be a problem, since the sample will then be subject to bulk water and not only humid air. Also, the humidity sensor might be covered with water droplets, causing short-circuits, false readings, and – in the worst case – permanent damage to the sensor.

Sensirion SHT21 humidity sensors (Staefa, Switzerland) are positioned inside the fluid cell as well as in the buffer. The sensors record the relative humidity with a tolerance of $\Delta\varphi_r = \pm 0.03$ in the φ_r interval [0.2; 0.8] which increases linearly to $\Delta\varphi_r = \pm 0.05$ until $\varphi_r = 0.0$ and $\varphi_r = 1.0$, respectively. Additionally, the temperature is recorded with a maximum tolerance of 0.4 °C between 5 °C and 60 °C.

Fluid Cell For AFM measurements in water as well as experiments under controlled relative humidity, an Asylum Research fluid cell was employed. The setup to control relative humidity requires in- and outlets for nitrogen and connections to the humidity sensor. Therefore, the fluid cell body drawn in Figure 11.2a was used in a closed configuration, in order to establish a stable environment. The cell's interior was sealed completely (Figure 11.2c), allowing only the passing of the controlled nitrogen flow through the open ports.

Samples completely submerged in water were investigated using the portless fluid cell body, presented in Figure 11.2b. The reason for this is that the lack of ports and any additional openings reduces the risk of water leaking from the cell. To permit a large range of movement and to allow easy access to the sample, the open configuration of the fluid cell was chosen for experiments in water. Only a drop of water with a diameter of about 10 mm was added to the sample. After approaching with the cantilever, a water meniscus between the sample and the cantilever holder will form and produce a stable environment (see Figure 11.2d).

11.2.2 AFM-based Nanoindentation

AFM based nanoindentation is carried out using the Asylum Research MFP-3D AFM. First, the surface is scanned in tapping mode to locate sufficiently flat regions. Then, force vs. distance (P – z) plots were recorded in contact mode at a desired position. For all measurements, the load schedule featured a hold time $t_{h,c}$ at maximum load and a second hold $t_{h,t}$ towards the end of the unloading segment at $0.05P_{max}$. The parameters were

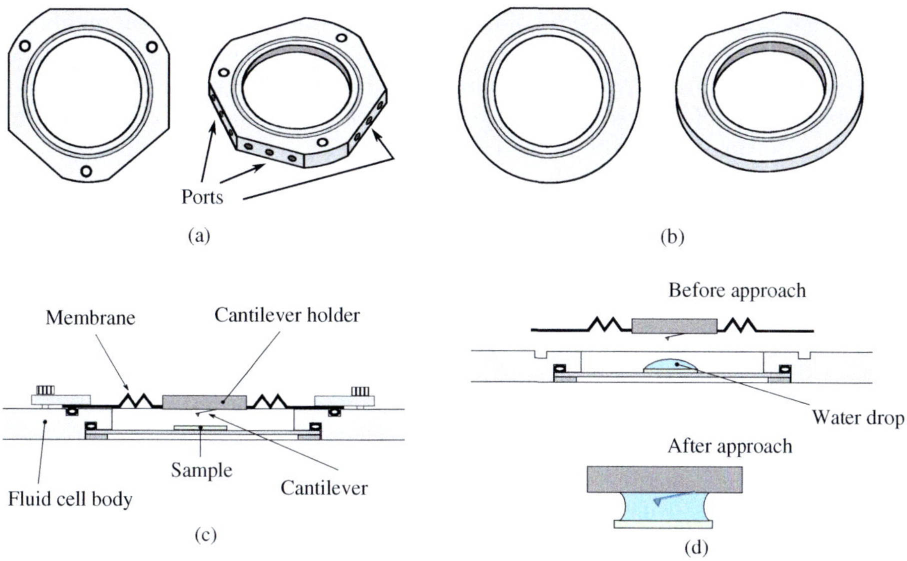

Figure 11.2 Schematic of (a) the fluid cell body with connection ports and (b) the body without ports. Both bodies can be operated in (c) the closed and (d) the open configuration.

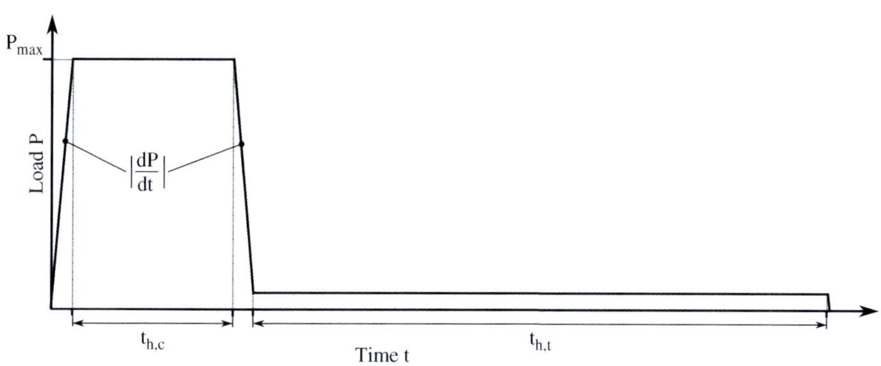

Figure 11.3 Sketch of the load schedule used for AFM-NI. See text for details.

maximum load $P_{max} = 10$ μN, loading and unloading rate $|dP/dt| = 10$ μN s^{-1}, and hold times $t_{h,c} = 10$ s, and $t_{h,t} = 30$ s. A schematic representation of the used load schedule is shown in Figure 11.3.

The recorded $P - z$ plots are then converted to force vs. indentation depth ($P - z_i$) plots by subtracting the cantilever's deflection z_d from the distance z. Further evaluation of the $P - z_i$ plots follows the method of Oliver and Pharr [1] with modifications to yield improved results on viscoelastic materials [19]. However, the basic equations do not change and allow to calculate hardness H and reduced modulus E_r according to:

$$H = \frac{P_{max}}{A(z_c)} \tag{11.2}$$

and

$$E_r = \frac{\sqrt{\pi}}{2} \cdot \frac{S}{A(z_c)}. \tag{11.3}$$

The difference to the original method of Oliver and Pharr lies in the calculation of the unloading stiffness S and thereby also the contact depth z_c, which is needed to calculate the area function at contact depth $A(z_c)$. Due to creep effects during unloading, the measured stiffness S_m is different from the true one. Improved values for H and E_r are gained by correcting the unloading stiffness with the creep rate $\dot{z}_{\text{creep}}$ at the end of the holding time $t_{h,c}$ and the unloading rate $|\dot{P}|$ according to [19]:

$$S = \left[\frac{1}{S_m} + \frac{\dot{z}_{\text{creep}}}{|\dot{P}|}\right]^{-1}. \tag{11.4}$$

According to Equations (11.2) and (11.3), it is of fundamental importance to correctly determine the area function A(z). This means, an exact knowledge of the indenter's geometry has to be obtained. A method using AFM to image the three-dimensional shape of the indenter is described in the next section.

Note that due to the bending of the cantilever in AFM-NI, a lateral force component will be introduced during indentation. This effect leads to results deviating from those of true NI. However, in the past it was verified that AFM-NI is feasible on soft materials such as polymers [22–24]. On rough, soft samples such as paper fibers, the availability of recording topography images in tapping mode [25] without damaging the surface is a further advantage. By acquiring the samples' morphology, it is possible to select flat areas for indentation, which should decrease the scattering of the data due to surface roughness [10].

Determination of the Area Function The area function A(z) of an indenter describes the projected area at a certain distance z from the tip apex. In classical NI, this function is usually calibrated by indenting into a material with varying indentation depths [1]. In AFM-NI, however, a direct imaging of the indenter's geometry is easily achieved by scanning the tip across a grid with sharp spikes, a so-called tip characterizer. If the opening angle of the spikes is less than the indenter's opening angle, the indenter is imaged, as is drawn schematically in Figure 11.4. This is commonly known as tip-sample dilation [26] and is sometimes misleadingly called "convolution".

To image the tip, the commercially available TGT01 calibration grid (NT-MDT, Moscow, Russia) was employed. On this grid, spikes with an opening angle of 30° and a tip radius of 10 nm are arranged periodically. Scanning this grid yields already satisfying information on the indenter's three-dimensional shape. To improve the result even more, the obtained image is eroded with the modeled spike geometry (24-sided pyramid with an opening angle of 30° and an apex radius of 10 nm) using built-in functions of the AFM image analysis software Gwyddion [27]. Erosion, in this case, denotes the inverse operation of dilation. The area function is calculated by intersecting a horizontal plane

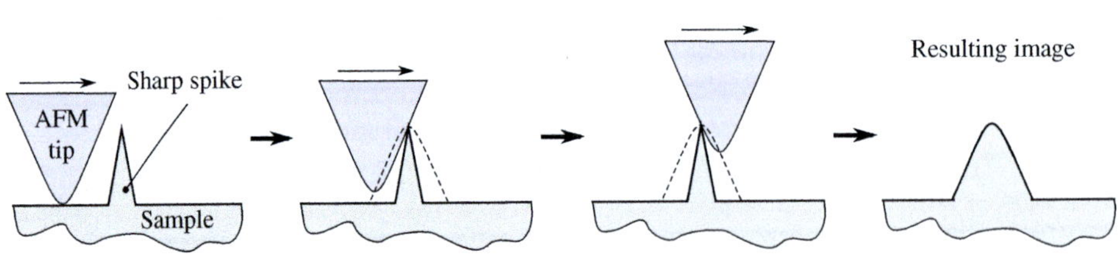

Figure 11.4 The principle of tip-sample dilation. If the opening angle of the sharp spike is lower than the one of the tip, the resulting image will actually represent the tip.

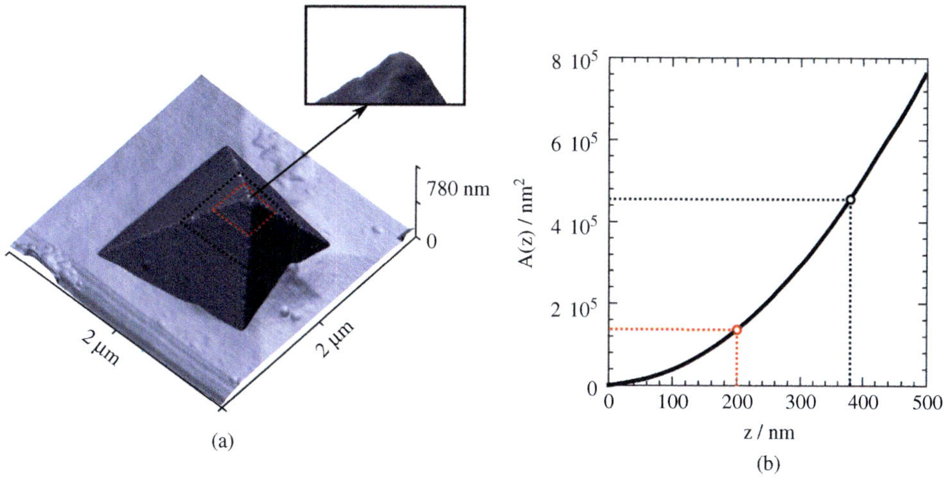

Figure 11.5 (a) A 3D representation of the 4-sided pyramid tip (ND-DYIRS) and (b) its corresponding area function. The areas represented by quadrilaterals in (a) correspond to the circles with the respective color in (b). A finite tip apex is emphasized by the inset in (a).

with the tip geometry at any distance z from the apex (see Figure 11.5a) and counting the pixels within the intersection. Since the physical dimensions of one pixel are known, A(z) is the direct result and displayed in Figure 11.5b. This procedure will henceforth be called the dilation method, according to the principle of which the tip's geometry is recorded.

Area Function of a Tilted Ideal Four-sided Pyramid By assuming that the tip is an ideal, regular four-sided pyramid with a square base, a half opening angle β, and tilted by the angle γ, its area function can be calculated analytically. Due to the cantilever's tilt, the tip becomes asymmetric (with respect to the surface) with a front opening angle of $\beta + \gamma$ and a back opening angle of $\beta - \gamma$. Geometrical considerations will then yield that the area function A(z) is given by:

$$A(z) = C_2^{(t)} z^2 . \tag{11.5}$$

The coefficient $C_2^{(t)}$, where the superscript (t) marks a theoretically obtained value, depends on β and γ in the following form:

$$C_2^{(t)} = [\tan(\beta + \gamma) + \tan(\beta - \gamma)] \sin(\beta) \left[\frac{1}{\cos(\beta + \gamma)} + \frac{1}{\cos(\beta - \gamma)} \right] . \tag{11.6}$$

Imaging the four-sided ND-DYIRS pyramidal AFM tips as described above, yielded that the opening angle $\beta = 37.2° \pm 0.6°$. From the asymmetry, the cantilever's tilt angle can also be measured and amounted to $\gamma = 10.8° \pm 0.5°$. Both values are averages obtained from 14 different tips and are presented with their respective standard deviations. Evaluating Equation (11.6) using the measured β and γ values yields:

$$A(z) = 2.54 z^2 . \tag{11.7}$$

Here, it was assumed that γ does not depend on the cantilever's deflection z_d. This is justified when the cantilever length $L << z_d$. In this work, the maximum load $P_{max} = 10\ \mu N$ resulted in a deflection of $z_d \approx 200$ nm, by assuming a spring constant of

$k = 50\ N\ m^{-1}$. For a cantilever length L of 135 μm, it is true that L << z_d and the deflection would change γ by approximately 0.1°. Thus, it is safe to assume that γ is constant.

Area Function of a Tilted Pyramid Obtained by the Dilation Method According to Equation (11.5), the area function is a parabola. However, determining A(z) of ND-DYIRS tips experimentally, using the dilation method, a function that is best fitted by the model:

$$A(z) = C_1^{(e)} z + C_2^{(e)} z^2 + C_3^{(e)} z^3 \tag{11.8}$$

is obtained, with (e) marking experimentally obtained parameters. By evaluating 53 area functions of all together 14 tips of the same type, the coefficients amount to $C_1^{(e)} = (248 \pm 98)\,\text{nm}$, $C_2^{(e)} = 2.49 \pm 0.36$, and $C_3^{(e)} = (2.64 \times 10^{-4} \pm 1.02 \times 10^{-3})\,\text{nm}^{-1}$. The linear term in Equation (11.8) can be explained most likely by a non-ideal geometry with characteristics of a paraboloid of rotation. An ideal paraboloid of rotation has an area function that depends linearly on z. In a similar way, the cubic term does also account for imperfections in tip geometry and improves the fit. It is reassuring to observe that $C_2^{(e)}$ and $C_2^{(t)}$ are very similar and the tip geometry should thus be well described by a tilted four-sided pyramid.

Area Function from SEM Images While it is in principle possible to image a tip with scanning electron microscopy (SEM), it is not a preferred method. One reason is that using SEM is time consuming. The tip has to be introduced into the SEM system, imaged at different angles at a sufficiently high resolution, and then transferred back to the AFM system. Any additional step, where the AFM probe is handled manually increases the risk of breaking it. When using the dilation method, only the sample has to be replaced by a characterization grid and scanned subsequently. However, the most important reason, why the use of SEM to collect an indenter's area function is discouraged is the linear term in Equation (11.8). While it is in principle possible, it would be practically unfeasible to generate a 3D image out of a series of SEM scans. The effort of measurement and evaluation would simply be too high for a characterization which should be carried out before the start of every experiment.

Additionally, there are the resolution limits of SEM which come into account when the first 100 nm of the tip are of interest. Therefore, instead of reconstructing the entire tip, one would measure the angles of the tip's side planes. From this, it is possible to derive an analytical expression for the area function by considering the tip geometry, similar to, e.g., Equation (11.5). The problem of this approach was already discussed: While the theoretical area function is described by a purely parabolic relation, a linear term is introduced in the experimental area functions by imperfections such as curve side faces or a tip apex with a finite radius. Obviously, the linear term of A(z) will become negligible at high indentation depths. For the four-sided pyramidal tips (ND-DYIRS) considered above, this would mean an indentation depth of approximately 1.6 μm is necessary for the linear term to be less than 5% of the total area. Since most of the indentations presented in this chapter were between 40 nm (dry cellulose) and 500 nm (highly swollen cellulose), the linear term of Equation (11.8) could never be neglected.

Instead of simply measuring the angles of the tip's side planes from an SEM image, a method was proposed where the widths of the side planes are measured within the SEM image and used to calculate the area at different heights [6]. Then, the data are fitted

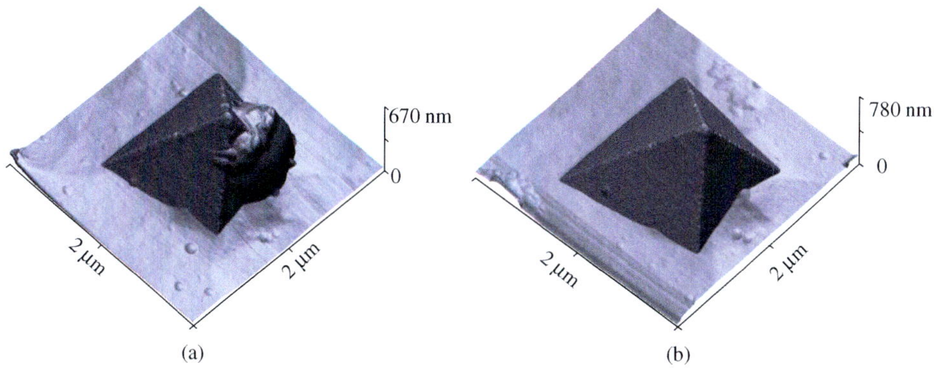

Figure 11.6 Resulting 3D shape of a contaminated (a) and cleaned (b) full diamond tip, obtained by scanning across a spike of the TGT01 calibration grid.

with a second order polynomial. This method will most likely yield better results than calculating the area function from the measured angles, however, it suffers the same shortcomings for low indentation depths. Therefore, the dilation method was chosen here to characterize the tip shape for AFM-NI.

Tip Cleaning While performing indentations, the tip might pick up loose nanoscale debris from the surface under investigation. In fact, this should be a dynamic process of picking up and losing contaminations during the progress of the experiment. It can hardly be avoided but seems of minor concern during indentation, since no degenerative change in hardness or reduced modulus was observed. However, if the tip is used again, the determination of its area function A(z) will be disturbed by the debris attached to the tip. In Figure 11.6a, a contaminated tip is displayed after using it for a series of indents. The easiest way to clean such a contaminated tip is to indent it into an epoxy layer with a thickness of at least several hundred micrometer. This layer was prepared by placing a drop of two component epoxy glue (UHU Endfest 300) on a steel plate with a diameter of 18 mm, covering most of the plate's surface. The material was then cured at 100°C for approximately 10 min in an oven. Performing indents on the epoxy surface while the cantilever is excited at its resonance frequency and repeating this at three to five different spots leads to a perfectly clean tip, as is presented in Figure 11.6b.

11.2.3 Comparison with Results of Classical NI

To check if AFM-NI with the procedure outlined previously does yield meaningful results, a comparison to literature values [7] was performed. For this purpose, viscose fibers (Kelheim Fibres GmbH, Kelheim, Germany) were embedded in epoxy resin and microtome cut to access their cross-sections. Then, H and E_r were determined by using a three-sided pyramid as well as a four-sided pyramid as indenters. The results are presented in Table 11.1. Apparently the tip shape has a significant influence on H and E_r, which is true for the AFM-NI values as well as the ones obtained by NI [7]. While the results from AFM-NI do not completely coincide with the results from NI, they are undoubtedly in a close range. Considering the difference in tip geometries and the fact that fibers investigated by NI are from a different producer, this is a remarkable finding and proves the usefulness of AFM-NI for soft materials.

Table 11.1 Comparison of AFM-NI results obtained from viscose fiber cross-sections with literature values.

Source	Tip	Fiber producer	H (MPa)	E_r (GPa)	Number of indents
This work[a)]	3-sided pyramid	Kelheim Fibres	160 ± 16	7.8 ± 1.3	24
This work	4-sided pyramid	Kelheim Fibres	239 ± 8	12.6 ± 2.1	15
[7]	Berkovich	Lenzing	≈230	≈10	3
[7]	cube-corner	Lenzing	≈350	≈6.5	5

a) Published in [10].

11.2.4 Sample Preparation

Pulp Fibers The investigated pulp fibers were unbleached, unrefined, industrial kraft pulp fibers. The pulp is a mixture of spruce and pine fibers with a diameter between 20 μm and 30 μm and a length of approximately 3 mm to 5 mm. All samples arrived as never-dried pulp and were stored in a refrigerator to keep them in their initial state and to inhibit the growth of mold on the samples.

The pulp fibers were placed on top of a drop of uncured nail polish [28, 29] to ensure that the fibers cannot move vertically during indentation, as is presented schematically in Figure 11.7a. Then the nail polish was left to cure at room temperature for at least 12 h. If a fiber had to be investigated in its never-dried state, a sequence of preparation steps as indicated in Figure 11.7b was employed. First, small water droplets were placed outside the nail polish, without touching it. The pulp fiber was then placed on the nail polish in such a way that its ends were sticking into the water droplets, as drawn in Figure 11.7b. It was found that the best fixation of the fibers is achieved when the nail polish could cure for several minutes in air before placing the whole sample in distilled water. Nail polish will then also cure in water within 12 h, which was the time the fibers were stored in water to ensure complete swelling.

Viscose Fibers Viscose fibers were classical (or regular) viscose with a typical diameter of 20 μm to 30 μm, cut to a length of 4 mm as well as viscose fibers with a rectangular cross-section which are 200 μm wide and about 5 μm thick. The viscose fiber samples were prepared in the same way as dried pulp fibers. All samples characterized in their fully swollen state were stored in distilled water for at least 12 h before AFM-NI characterization.

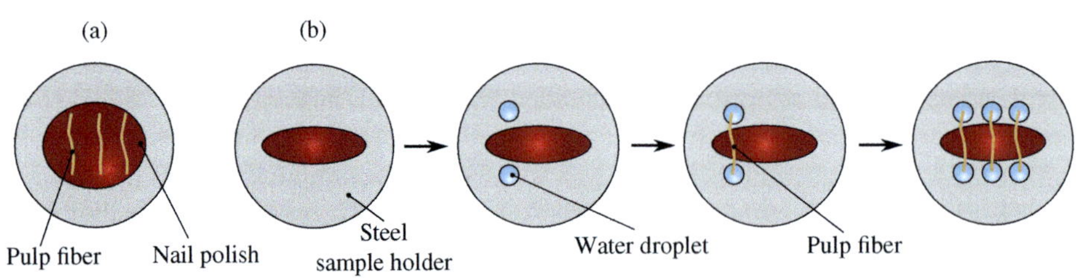

Figure 11.7 Schematical illustration of the preparation of (a) dried and (b) wet fibers for AFM-NI.

11.3 Mechanical Properties of Cellulose Fibers

11.3.1 Pulp Fibers

The mechanical properties of pulp fibers were characterized as humidity dependent H and E_r, using AFM-NI and the setup to control relative humidity inside the AFM's fluid cell. Hardness and reduced modulus are presented as a function of relative humidity φ_r, including the fully wet state. In order to exclude an influence of the substrate on the measured properties, a model developed for classical NI on coated substrates was applied [30]. The hardness and reduced modulus of the nail polish in the dried state were measured with a three-sided pyramid and found to be 100 MPa and 1.5 GPa, respectively. By assuming a fiber thickness of 4 μm, no significant influence on H and E_r was detectable [10]. Since no swelling of nail polish in water was observed, this fact is also true for investigations of fully swollen materials in water.

For these AFM-NI investigations, softwood kraft pulp fibers were glued on a drop of nail polish after they were dried in air, as is described in section 11.2.4. It is known that pulp fiber surfaces exhibit a rough and wrinkled surface topography [31]. Tip-sample dilation will, therefore, obscure parts of the surface which are then also not accessible for AFM-NI. Figure 11.8 depicts this effect. In an area such as region I in Figure 11.8a, no AFM-NI can be performed. Only areas such as region II will be flat enough to place indents and extract useful nanomechanical information. An actual AFM topography image with severe dilation effects at the left border is presented in Figure 11.8b. The central region of the image, however, was sufficiently flat to place indents.

AFM-NI investigations on kraft pulp fibers were performed by using three-sided pyramidal tips (ND-DTIRS) as indenters. This revealed that H and E_r of pulp fibers are decreasing with increasing relative humidity. The results are visualized in Figure 11.9. Both, hardness and the reduced modulus seem to follow an almost linear relationship with relative humidity. A linear least squares fit for H and E_r with a linear model was performed for the results. The hardness is then:

$$H(\varphi_r) = (270 \pm 1)\,\text{MPa} - (243 \pm 1)\,\text{MPa} \times \varphi_r. \tag{11.9}$$

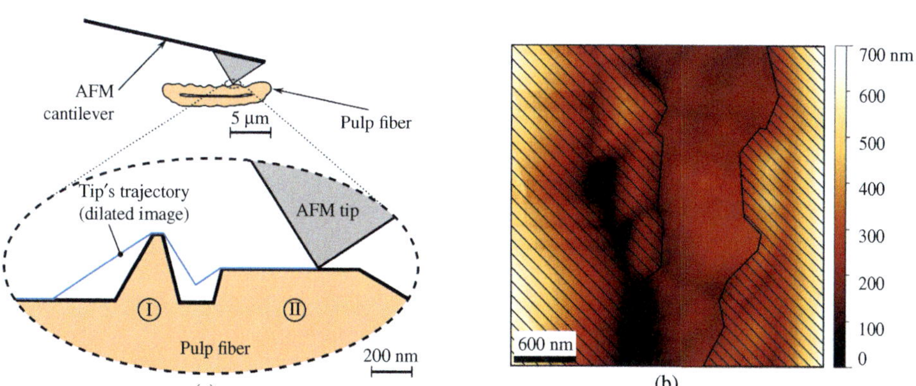

Figure 11.8 The effect of tip-sample dilation on topography images recorded on pulp fiber surfaces with AFM-NI probes. (a) Schematic illustration and (b) dilation (marked by the shaded areas) on an actual AFM topography image. *Source*: Ganser 2014 [10]. Reproduced with permission of De Gruyter.

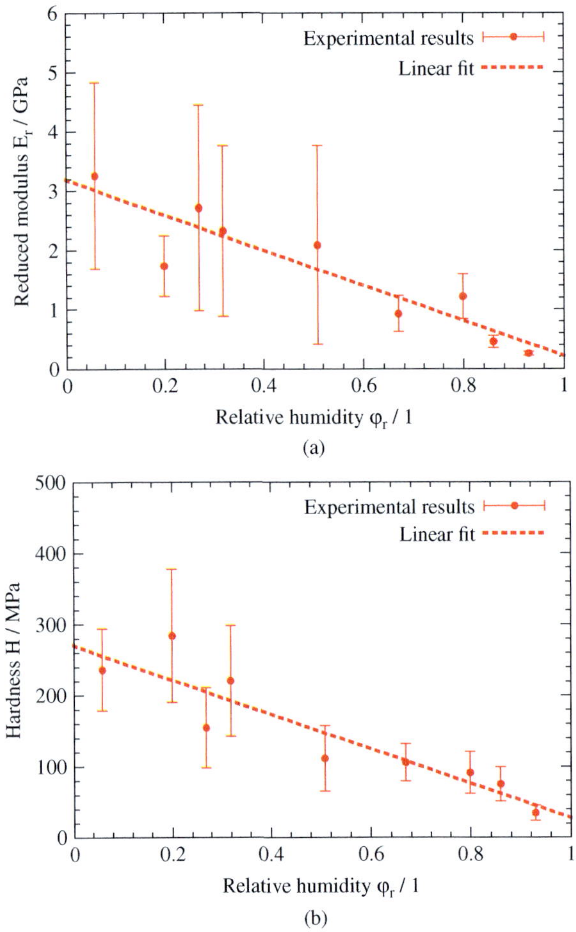

Figure 11.9 Hardness (a) and reduced modulus (b) of pulp fibers as a function of relative humidity, determined by using a three-sided pyramidal tip. The data points are averages of 10 to 30 individual indents from one to two fibers. *Source*: Ganser 2014 [10]. Reproduced with permission of De Gruyter.

Similarly, the fit for the reduced moduli resulted in:

$$E_r(\varphi_r) = (3.2 \pm 1.4)\,\text{GPa} - (3.0 \pm 2.4)\,\text{GPa} \times \varphi_r. \tag{11.10}$$

According to the errors of the parameters in Equation (11.9) and Equation (11.10) it is confirmed that the hardness follows the linear model to a higher extent than the reduced modulus. The mechanical properties of fully swollen fibers were also determined, and resulted in $H_{wet} = (2.7 \pm 0.7)$ MPa, and $E_{r,wet} = (32 \pm 6) \times 10^{-3}$ GPa. Both quantities are decreased by a factor of about 100 compared to the fully dried state.

The elastic modulus of softwood pulp fibers in the fully wet state using AFM based methods, has been reported several times in the literature. By applying loads in the order of approximately 50 nN, values for E_r of $\approx 10 \times 10^{-3}$ GPa [4] and $\approx 5 \times 10^{-3}$ GPa, with a maximum value of 50×10^{-3} GPa [5] were determined. In both cases, fully elastic behavior was assumed. More recently, an AFM-NI approach similar to the one presented here was also applied to investigate softwood pulp in the wet state, where a reduced modulus of $E_r = (182 \pm 50) \times 10^{-3}$ GPa and a hardness of $H = (140 \pm 38)$ MPa was found [6]. In all three studies, standard AFM tips – designed for topography measurements – were used. The first of the listed works relied on an estimation of the tip radius [4], the second assumed a perfectly conical indenter [5] and the third employed a tip characterization using SEM [6]. In summary, literature values for E_r of wet kraft pulp fibers are reported to be several times lower or higher as the values presented here. These discrepancies could be caused by the different ways of taking the tip geometry into account and also by the applied evaluation procedures, which assumed either elastic or elastic-plastic behavior. Additionally, there is an effect of the tip shape itself, which is clearly visible in Table 11.1.

Using four-sided tips, a significant pile-up effect on pulp fibers as well as on viscose fibers was detected at low values of relative humidity. To counter such effects, a manual evaluation of the single indents can be employed, e.g. after [32], and will be discussed elsewhere [33].

11.3.2 Swollen Viscose Fibers

Viscose fibers can act as a model system for paper fibers and have the advantage that they are initially chemically pure. This fact allows to selectively enable certain mechanisms, such as Coulomb interactions, by adding modifiers to the fibers' raw material [34].

The viscose fibers investigated here were modified with carboxymethyl cellulose (CMC) and cationic starch (CS). Both additives were introduced into the fibers during the production process. Modification with CMC leads to anionic fibers whereas CS produces cationic fibers. The details of the modifications as well as the sample designations are given in Table 11.2.

All AFM-NI experiments here were performed using the four-sided pyramidal tips ND-DYIRS. It is, therefore, not easy to compare with results from the swollen pulp fibers discussed earlier. The benchmark experiments carried out on viscose fiber cross-sections (Table 11.1), however, suggest that H and E_r obtained by four-sided tips

Table 11.2 Sample designations and amount of modifications present in viscose fibers. The amount of COOH- and N-groups correlate to the amount of CMC and CS, respectively.

Sample designation	Fiber type	Modification type	COOH % w/w	N % w/w
cRF	classical	none	0.77	-
cAF1		anionic	2.60	-
cAF2			3.78	-
cCF1		cationic	-	0.26
cCF2			-	0.13
rRF	rectangular	none	0.52	-
rAF		anionic	2.54	-
rCF		cationic	-	0.25
rACF		an./cat.	1.42	0.31

Source: Weber 2013 [34]. Reproduced with permission of Springer.

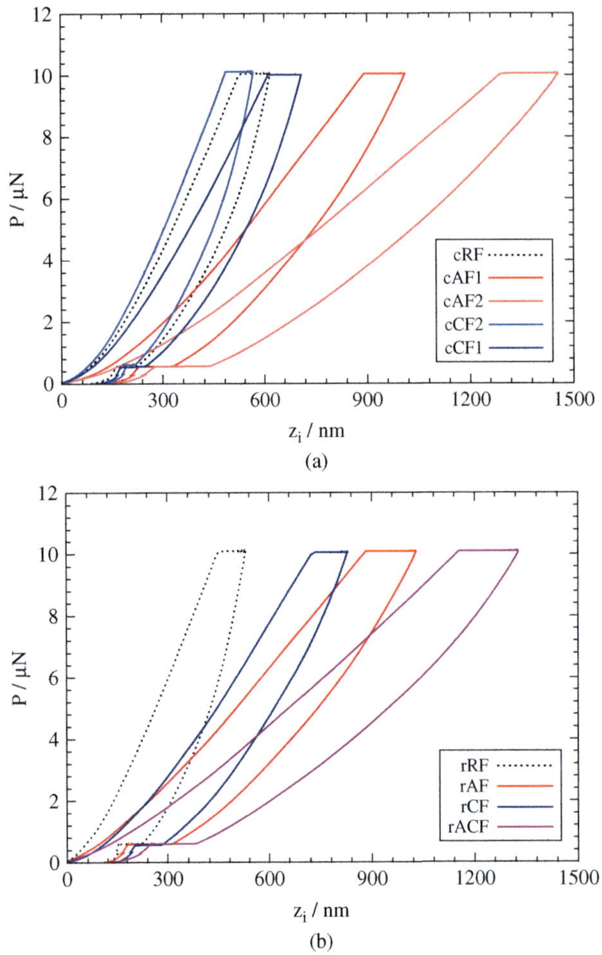

Figure 11.10 Exemplary P − z_i plots recorded on (a) classical and (b) flat, rectangular viscose fibers with different modifications incorporated.

are 50% to 60% higher than H and E_r measured by three-sided tips. The viscose fibers were embedded in nail polish in the fully swollen state.

In Figure 11.10, P − z_i plots of the modified fibers are compared. Classical viscose fibers (Figure 11.10a) and flat viscose fibers (Figure 11.10b) respond in a similar manner to the additives. It is clearly visible that all modifications have a significant effect on the fibers' mechanical properties.

Hardness and reduced modulus obtained from all samples are visualized for classical and rectangular fibers in Figure 11.11 and Figure 11.12, respectively. Unmodified classical fibers exhibit a hardness of about 14 MPa with a reduced modulus of 51×10^{-3} GPa. The addition of CMC to the fiber (cAF1) leads to a decrease of H by a factor of two and a reduction of E_r by a factor of almost three. If more CMC is added (cAF2), the hardness is decreased by a factor of three and the reduced modulus by a factor of six compared to cRF. cAF2 exhibits the lowest values for hardness and reduced modulus of

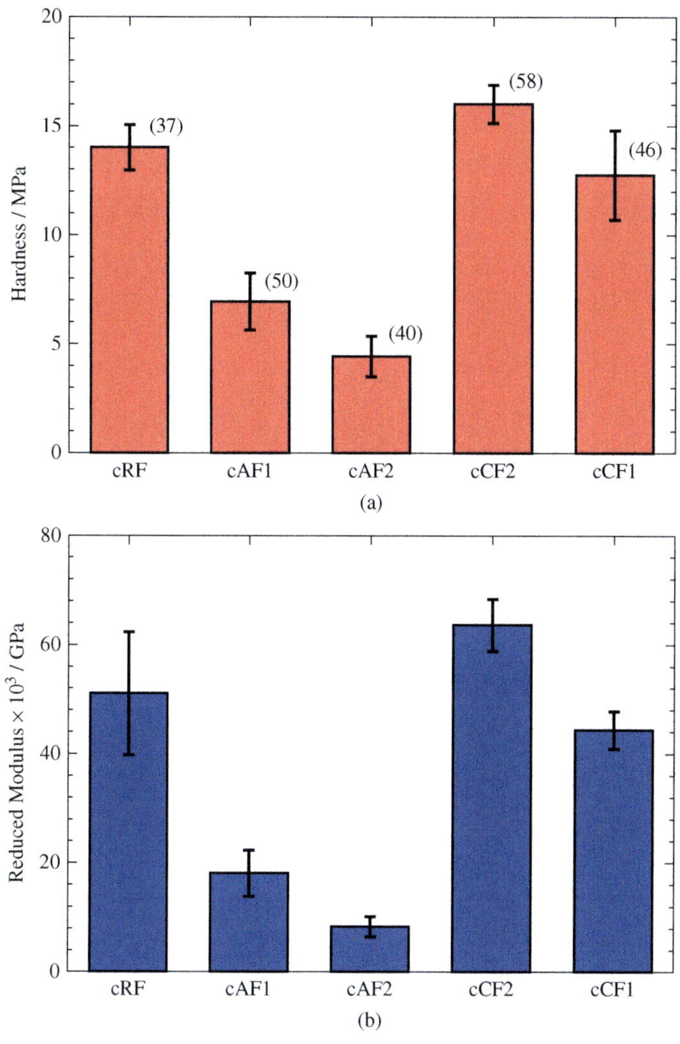

Figure 11.11 AFM-NI results of swollen classical viscose fibers as (a) hardness and (b) reduced modulus. The number of averaged indents are given in brackets. *Source*: Ganser 2014 [11]. Reproduced with permission of ICE Publishing.

all the viscose fibers investigated. The incorporation of CMC into the fiber is, therefore, decreasing both parameters drastically. CS, on the other hand leads to a slight increase in H and E_r at low concentrations (cCF2) and to a decrease for higher concentrations (cCF1). All differences are significant under a Welch test at a significance level of 0.05 [11].

The values of H and E_r of the flat reference fiber (rRF) are similar to the ones observed for the classical fiber without modifications. rAF, which contains an amount of CMC comparable to that of cAF1, exhibits a hardness and reduced modulus which are almost identical to the values found for cAF1. Only the cationic fiber rCF differs significantly from its counterpart cCF1. However, H and E_r of rCF are ranged between these values

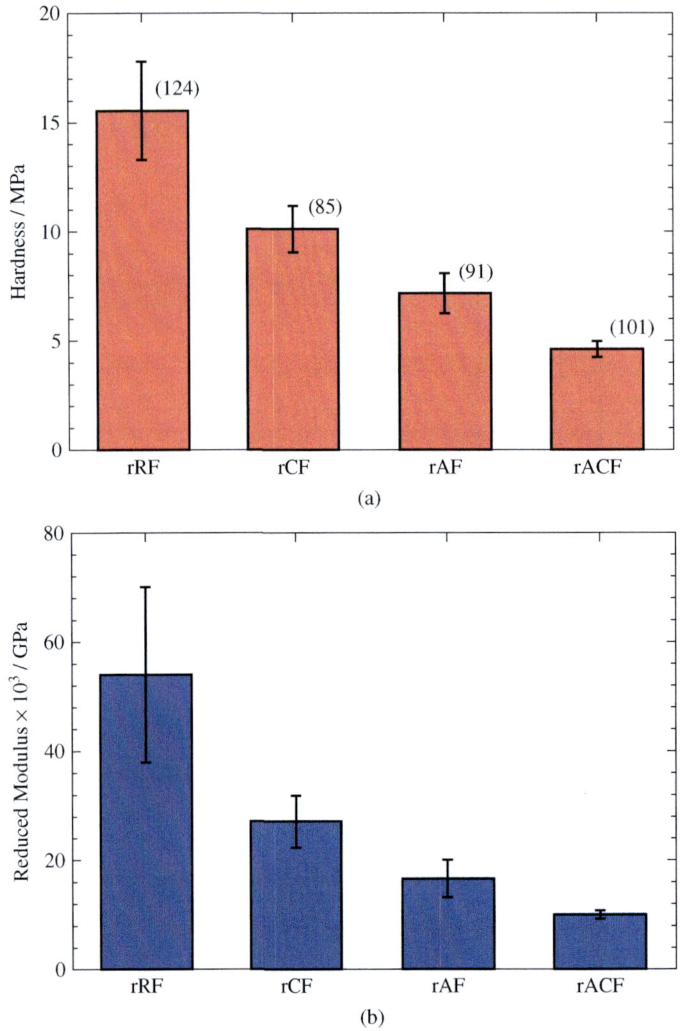

Figure 11.12 AFM-NI results of swollen rectangular viscose fibers as (a) hardness and (b) reduced modulus. The number of averaged indents are given in brackets. *Source*: Ganser 2014 [11]. Reproduced with permission of ICE Publishing.

from rRF and rAF, which is also the case for cCF1 and, hence, leads to the conclusion that rCF fits in principle into the series. Slight variations during the production process could explain the discrepancy between cCF1 and rCF, however, the amount of CS in both fibers is approximately the same.

An influence of incorrect sample preparation – i.e., gluing the fibers on the nail polish – is unlikely. H and E_r of both samples are averages of at least three different fibers and feature only a low standard deviation and are, therefore, well reproducible. The mechanical properties of rACF are similar to those of cAF2, but there is no direct comparison for rACF possible, since no classical viscose fibers with both anionic and cationic additives were prepared.

A different study by Weber *et al.* dealt with the swelling behavior of the classical viscose fibers listed in [34]. There, the geometric swelling was measured by employing an optical microscope and quantified as the swelling ratio $s_r = d_s/d_0$, with d_s and d_0 denoting the diameter – or width for rectangular fibers – of the swollen and dried fiber, respectively. In Figure 11.13, H and E_r of classical viscose fibers are plotted against the swelling ratio. Apparently, cRF, cCF1, and cCF2 swell by approximately the same amount, but differ significantly in their mechanical properties. Both anionic fibers cAF1 and cAF2 clearly swell more than the reference fibers, which is in good agreement with their decreased H and E_r values. It seems that there is a definite relation between the geometrical swelling and the mechanical properties, when inspecting Figure 11.13. Such a behavior is to be expected, since geometrical swelling is nothing more than an increase

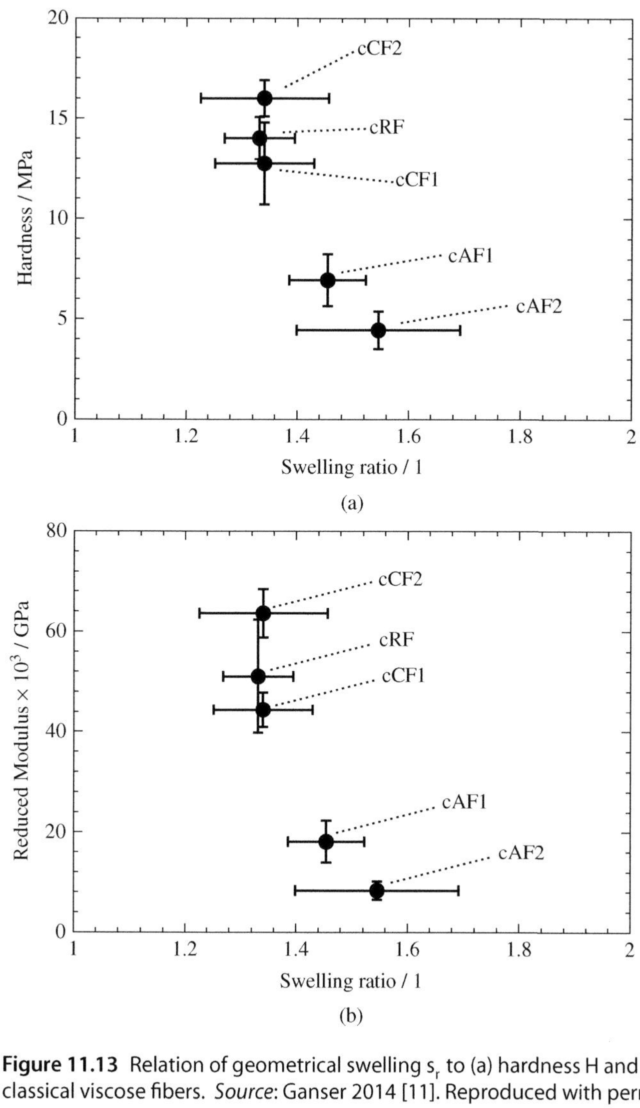

Figure 11.13 Relation of geometrical swelling s_r to (a) hardness H and (b) reduced modulus E_r of classical viscose fibers. *Source*: Ganser 2014 [11]. Reproduced with permission of ICE Publishing.

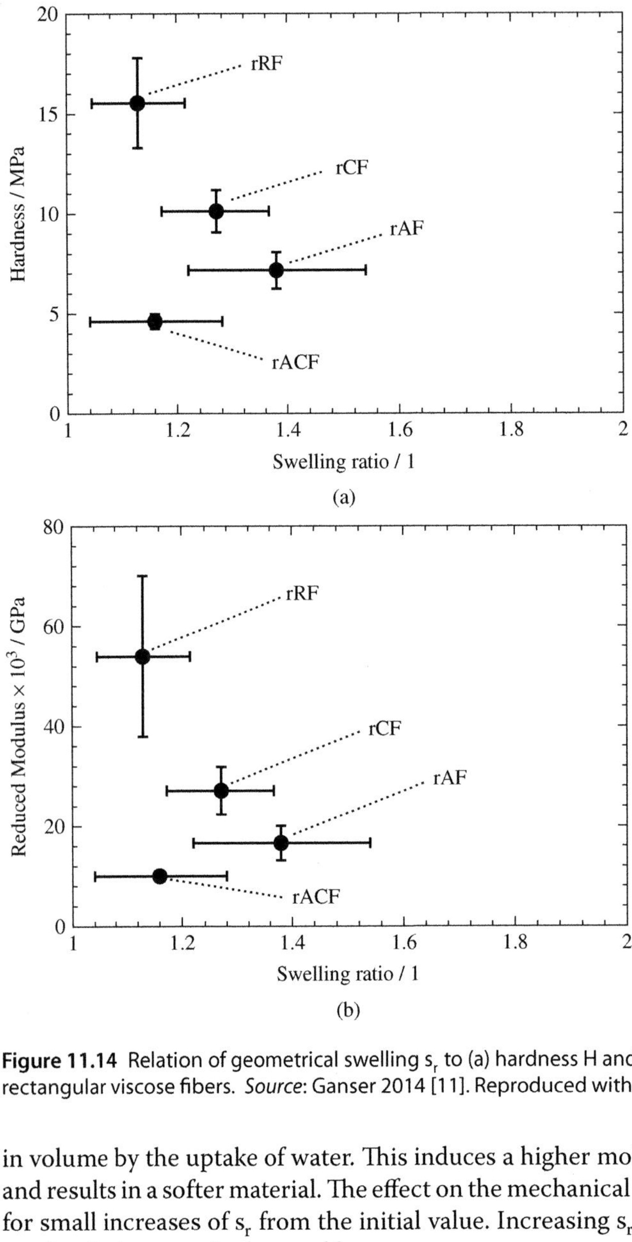

Figure 11.14 Relation of geometrical swelling s_r to (a) hardness H and (b) reduced modulus E_r of rectangular viscose fibers. *Source*: Ganser 2014 [11]. Reproduced with permission of ICE Publishing.

in volume by the uptake of water. This induces a higher mobility of the cellulose chains and results in a softer material. The effect on the mechanical properties seems to be large for small increases of s_r from the initial value. Increasing s_r further, seems to affect the mechanical properties less and less.

To prove the effectiveness of the additives further, also the swelling of flat, rectangular viscose fibers was characterized [11]. The results are presented in Figure 11.14 and follow in principle the trend observed for classical viscose fibers.

A first finding is that the unmodified reference rRF as well as the cationic fiber rCF swell less than their classical viscose counterparts. Only rAF is comparable to its classical equivalent. This discrepancy might be caused by the method of detecting the fibers' widths. Flat fibers have a width to thickness ratio of about 40, enabling them to twist easily. Such twists cannot always be detected during the swelling experiment via the optical microscope and will lead to a corruption of the measured widths. Additionally, an anisotropy in swelling might be present, causing the fibers to swell differently in thickness and width. This, however, cannot be detected by the method used here.

Another interesting observation is that rACF swells approximately by the same amount as the unmodified fibers, while being the softest and most compliant rectangular fiber. To compare rACF with the other fibers, the amounts of incorporated CMC and CS have to be considered. In rACF, less CMC than in rAF is present, whereas the amount of CS is larger than in rCF. If the additives would act synergetically, one would expect a slight increase in swelling accompanied by a moderate reduction in H and E_r. Obviously, this is not the case. There is no information available on how exactly CMC and CS will interact in the cellulose matrix, but it can be expected that both will attract each other and agglomerate due to their opposite charges. This would lead to an effective charge compensation leaving only a small net charge – positive or negative, depending on the absolute amounts – which will naturally have a low effect on the geometrical swelling. The large decrease in H and E_r could be caused by the agglomerated additives which act as impurities and disrupt the cellulose-cellulose interactions locally. To some extent, this effect is expected to be present in all modified viscose fibers investigated here [11].

11.4 Conclusions and Outlook

AFM based nanoindentation was employed here as an alternative to classical nanoindentation for soft samples with a rough surface morphology. On viscose fiber cross-sections, AFM-NI yields results comparable to classical NI, proving an overall usefulness of the method. The great advantage of AFM-NI over NI on soft and rough samples is the possibility of collecting topography images to select surface regions suitable for indentation and thus decrease the scattering of the results.

Results obtained on paper fibers and viscose fibers proved the applicability of AFM-NI for these materials. The evolution of H and E_r with increasing relative humidity was recorded for paper fibers and followed an almost linear trend until $\varphi_r \approx 0.95$. Swollen viscose fibers measured in water revealed their tunability in terms of swelling and mechanical properties by certain additives, which can be – to some extent – also adjusted independently.

Future extensions of AFM-NI could also provide a means to measure viscoelastic properties of cellulose fibers. This can be achieved by either AFM based dynamic mechanical analysis (DMA) [35, 36] or detecting the creep rate as a function of the applied load, as has been done with classical nanoindentation [8].

Acknowledgments

The financial support of Mondi, Frantschach, and the Federal Ministry of Economy, Family and Youth and the National Foundation for Research, Technology and Development, Austria, is gratefully acknowledged. We also acknowledge the fruitful collaboration and discussions with R. Schennach, U. Hirn, F. J. Schmied, W. J. Fischer, F. Weber, C. Czibula, I. Bernt, and S. Rohm.

References

1 Oliver WC, Pharr GM. Improved technique for determining hardness and elastic modulus using load and displacement sensing indentation experiments. Journal of Materials Research 7(6), 1564–1583; 1992.
2 Binnig G, Quate CF, Gerber C. Atomic force microscope. Physical Review Letters 56(9), 930–933; 1986.
3 Gindl W, Gupta HS, Schöberl T et al. Mechanical properties of spruce wood cell walls by nanoindentation. Applied Physics A: Materials Science and Processing 79(8), 2069–2073; 2004.
4 Nilsson B, Wågberg L, Gray D. Conformability of wet pulp fibres at small length scales. Sundsvall: Mitthögskolan, Fibre Science Communication Network; 2000.
5 Chhabra N, Spelt JK, Yip CM, Kortschot MT. An investigation of pulp fibre surfaces by atomic force microscopy. Journal of Pulp and Paper Science 31(1), 52–56; 2005.
6 Yan D, Li K. Conformability of wood fiber surface determined by AFM indentation. Journal of Materials Science 48(1), 322–331; 2013.
7 Gindl W, Konnerth J, Schöberl T. Nanoindentation of regenerated cellulose fibres. Cellulose 13(1), 1–7; 2006.
8 Lee S-H, Wang S, Pharr GM et al. Mechanical properties and creep behavior of lyocell fibers by nanoindentation and nano-tensile testing. Holzforschung 61(3), 254–260; 2007.
9 Wimmer R, Lucas BN, Tsui TY, Oliver WC. Longitudinal hardness and Young's modulus of spruce tracheid secondary walls using nanoindentation technique. Wood Science and Technology 31(2), 131–141; 1997.
10 Ganser C, Hirn U, Rohm S et al. AFM nanoindentation of pulp fibers and thin cellulose films at varying relative humidity. Holzforschung 68(1), 53–60; 2014.
11 Ganser C, Weber F, Czibula C et al. Tuning hardness of swollen viscose fibers. Bioinspired, Biomimetic and Nanobiomaterials 3(3), 131–138; 2014.
12 Lindström T, Wågberg L, Larsson T. On the nature of joint strength in paper: a review of dry and wet strength resins used in paper manufacturing. 13th Fundamental Research Symposium, pp. 457–562; 2005.
13 Van de Ven TGM. Capillary forces in wet paper. Industrial Engineering and Chemical Research 47(19), 7250–7256; 2008.
14 Tejado A, van de Ven TGM. Why does paper get stronger as it dries? Materials Today 13(9), 42–49; 2010.
15 Haslach HW. A model for drying-induced microcompressions in paper: buckling in the interfiber bonds. Composites Part B 27(1), 25–33; 1996.
16 Vainio A, Paulapuro H. Interfiber bonding and fiber segment activation in paper. BioResources 2(3), 442–458; 2007.
17 Schmied FJ, Teichert C, Kappel L, Hirn U, Bauer W, Schennach R. What holds paper together: Nanometre scale exploration of bonding between paper fibres. Scientific Reports 3, 2432; 2013.
18 Fernandes Diniz JMB, Gil MH, Castro J. Hornification: its origin and interpretation in wood pulps. Wood Science and Technology 37(6); 489–494; 2004.
19 Tang B, Ngan AHW. Accurate measurement of tip–sample contact size during nanoindentation of viscoelastic materials. Journal of Materials Research 18(05), 1141–1148; 2003.

20 Hutter JL, Bechhoefer J. Calibration of atomic-force microscope tips. Review of Scientific Instruments 64(7), 1868–1873; 1993.
21 Çolak A, Wormeester H, Zandvliet HJW, Poelsema B. Surface adhesion and its dependence on surface roughness and humidity measured with a flat tip. Applied Surface Science 258(18), 6938–6942; 2012.
22 Clifford CA, Seah MP. Quantification issues in the identification of nanoscale regions of homopolymers using modulus measurement via AFM nanoindentation. Applied Surface Science 252(5), 1915–1933; 2005.
23 Tranchida D, Piccarolo S, Soliman M. Nanoscale Mechanical Characterization of Polymers by AFM Nanoindentations: Critical Approach to the Elastic Characterization. Macromolecules 39(13), 4547–4556; 2006.
24 Jee A-Y, Lee M. Comparative analysis on the nanoindentation of polymers using atomic force microscopy. Polymer Testing 29(1), 95–99; 2010.
25 Zhong Q, Inniss D, Kjoller K, Elings VB. Fractured polymer/silica fiber surface studied by tapping mode atomic force microscopy. Surface Science Letters 290(1–2): L688–L692; 1993.
26 Villarrubia JS. Morphological estimation of tip geometry for scanned probe microscopy. Surface Science 321(3), 287–300; 1994.
27 Nečas D, Klapetek P. Gwyddion: an open-source software for SPM data analysis. Central European Journal of Physics 10(1), 181–188; 2012.
28 Schmied FJ, Teichert C, Kappel L, Hirn U, Schennach R. Joint strength measurements of individual fiber-fiber bonds: An atomic force microscopy based method. Review of Scientific Instruments 83(7), 73902–739028; 2012.
29 Fischer WJ, Zankel A, Ganser C et al. Imaging of the formerly bonded area of individual fibre to fibre joints with SEM and AFM. Cellulose 21(1), 251–260; 2013.
30 Korsunsky AM, McGurk MR, Bull SJ, Page TF. On the hardness of coated systems. Surface and Coatings Technology 99(1–2), 171–183; 1998.
31 Persson BNJ, Ganser C, Schmied F et al. Adhesion of cellulose fibers in paper. Journal of Physics Condensed Matter 25(4), 045002; 2013.
32 Kese KO, Li Z-C, Bergman B. Method to account for true contact area in soda-lime glass during nanoindentation with the Berkovich tip. Materials Science and Engineering A: Structure 404(1), 1–8; 2005.
33 Ganser C, Kreiml P, Morak R, Weber F, Paris O, Schennach R, Teichert C. The effects of water uptake on mechanical properties of viscose fibers. Cellulose 22(4), 2777–2786; 2015.
34 Weber F, Koller G, Schennach R, Bernt I, Eckhart R. The surface charge of regenerated cellulose fibres. Cellulose 20(6), 2719–2729; 2013.
35 Mahaffy RE, Park S, Gerde E, Käs J, Shih CK. Quantitative analysis of the viscoelastic properties of thin regions of fibroblasts using atomic force microscopy. Biophysical Journal 86(3), 1777–1793; 2004.
36 Foschia R, Jobin M, Hengsberger S. Local dynamic mechanical analysis. Micron 40(1), 51–55; 2009.

12

Evaluation of Mechanical and Tribological Properties of Coatings for Stainless Steel

A. Mina[1], *J.C. Caicedo*[1], *W. Aperador*[2], *M. Mozafari*[3] *and H.H. Caicedo*[4,5]

[1] *Tribology, Powder Metallurgy and Processing of Solid Recycled Research Group, Universidad del Valle, Cali, Colombia*
[2] *Department of Engineering, Universidad Militar Nueva Granada, Bogotá-Colombia*
[3] *Bioengineering Research Group, Nanotechnology and Advanced Materials Department, Materials and Energy Research Center (MERC), Tehran, Iran*
[4] *Department of Anatomy and Cell Biology, University of Illinois at Chicago, USA*
[5] *National Biotechnology and Pharmaceutical Association, Chicago, USA*

12.1 Introduction

Stainless steel (316L SS) is commonly used as standard orthopedic implants due to its proper mechanical, biocompatibility and corrosion resistant characteristics in the body [1]. However, it has been reported that after long-term implantations, 316L SS may potentially cause allergic reactions due to the release of nickel (Ni), molybdenum (Mo) and chromium (Cr) ions [2, 3]. As an innovative strategy, surface modification of this class of metallic implants with ceramic coatings represents the most common solution to change the bulk properties of the metal substrate, and also improve the mechanical and tribological behavior of the surface into interface bone/implant [4–7]. In recent studies, hydroxyapatite (HA), compounds of HA and β–tricalcium phosphate (β–TCP), HA and chitosan and β–TCP and chitosan have been frequently examined as coating materials on metallic substrates for biomedical applications due to their favorable biocompatibility, mechanical, and tribological characteristics [8–10].

It is known that the chemical composition of β–TCP is similar to the mineral phase of natural bone, for that reason it is widely applied in biomedical industry in different applications. However, β–TCP coatings are the newest application trying to improve the biocompatibility of metallic implants [11–13]. On the other hand, chitosan is a natural cationic polysaccharide produced by alkaline N-deacetylation of chitin, a constituent of the exoskeleton from crabs and squid. It is widely used in biomedical applications due to its bioactive characteristics that promote cell proliferation [14]. The combination of β–TCP and chitosan as a bioactive compound can potentially improve the bone fixation and accelerate bone growth. So the electrochemical deposition of biomaterials on metallic substrates has recently provided greater advantages in biomedical applications [2]. Many bioactive compounds that accelerate bone growth and improve bone fixation have been used as coating materials. The incorporation of these bioactive coatings to metallic implants represents great advances in the development of orthopedic devices. The coatings can be applied into the substrates by a variety of techniques, such as

Applied Nanoindentation in Advanced Materials, First Edition. Edited by Atul Tiwari and Sridhar Natarajan.

sputtering [15], thermal evaporation [16], chemical vapor deposition [17], spin coating [18], spray pyrolysis [19], sol–gel deposition [20] and electrochemical deposition [21]. The electrochemical deposition has superior advantages compare to other techniques due to its simplicity, low cost and room temperature operation [22]. Furthermore, electrochemical deposition of biomaterials can provide functionalized coating implants by modifying experimental solutions and conditions [23].

In many applications, the time-dependent properties of β-TCP make it an ideal coating for implants, which supplies a temporary support for bone ingrowth, and eventually is replaced by natural tissue [24]. Among several clinical applications involving β-TCP, posterior spinal fusion and cranio-maxillofacial applications have attracted considerable attentions [25]. Calcium phosphate-based bioceramics with improved mechanical properties and controlled resorbability can assist in designing optimal biodegradable bone substitutes for spinal fusion and craniomaxillofacial applications. The use of such bone implants also avoids the second surgery required for autograft harvesting [26].

The initial response of the body to the biomaterials depends on the surface characteristics that directly influence the biocompatibility by affecting the surface charge, surface topography, and etc. [27]. In addition, some mechanical and tribologial properties are also of importance (hardness and friction coefficient). For example, materials used as catheters need excellent tensile and mechanical characteristics. Therefore, both the biocompatibility and the mechanical properties of biomaterials need to be improved [28]. In this work, β–TCP/Ch coatings were deposited on the 316L SS substrates by using cathodic electro-deposition technique in order to obtain a possible functionalized coating implant. The influence of the chitosan percentage in the coating structures, which improved the mechanical and tribological response was characterized and discussed in current research. In this sense, the β–TCP/Ch coatings could be considered as promising materials for biomedical industry in temporal invasive prosthesis applications as they can effectively improve the bone/prosthesis interfaces.

12.2 Experimental Details

The β–TCP/Ch coatings were deposited via cathodic electro-deposition on AISI 316L SS substrates (12 mm of diameter and 5 mm of thickness). The electro-deposition technique was performed using the following conditions: current density of 260 mA, agitation velocity of 250 rpm, pH electrolyte of 10.4, and a temperature deposition of 60°C. The cylindrical 316L SS substrates were prepared by grinding to 600 grit water proof silicon carbide papers, then they were activated by superficial electrochemical etching using an acid solution of 1:1:1 $HCl:H_2SO_4:H_2O$, to improve the adhesion characteristics in the substrate-coating interfaces. Finally, the substrates were cleaned in an ultrasonic bath with acetone and rinsed in de-ionized water in order to remove the organic contaminants.

The electro-deposition process was performed using an electrolyte compound of two solutions. The first one was β-TCP dissolved in an ethanol-water 1:3 solution; and the second one was carried out with chitosan solution dissolved in acetic acid 2%. The structural characterization was performed by X-ray diffraction (XRD) Bruker D8 Advance with Cu cathode and scintillation detector using the setup 0–20 and making a sweep of 20° to 80° with a step of 0.01 degrees and a time per step of 2 seconds with Cu Ka radiation ($\lambda = 1.5405$ Å). The diffraction patterns were obtained as a function of increase in the chitosan concentration from β-TCP/Ch coatings, these patterns were further analyzed to determinate the changes in the crystal lattice arrangements. In addition, the different lattice arrangements of coating were compared and the changes in preferential peaks were determinate due to the inclusion of chitosan. The increase of intensity as a function of chitosan weight percentage was determined after to be applied the full width at half maximum (FWHM) criterion for the XRD patterns [29]. The surface chemical characterization of the β-TCP-Ch coatings were performed using a scanning electron microscope (SEM) JEOLJSM-6490LV to determinate the different phases on the surface coating. Energy Dispersive X-ray (EDX) analysis was carried out to investigate the distribution of the different elements in the composites and to evaluate the changes of chemical composition as a function of chitosan weight percentage.

Atomic Force Microscopy (AFM) was used for a quantitative study of the surface morphology in β-TCP/Ch coatings using an AFM Asylum Research MFP-3D® and calculated by a Scanning Probe Image Processor (SPIP®), which is the standard program for processing and presenting AFM data. Therefore, this software has become the de-facto standard for image processing at the nanoscale. The mechanical properties were defined using a Nanoindentation test, by using a Nanoindenter Ubi1-Hysitron to obtain elastic-plastic properties of the coatings. A three-side Berkovich diamond indenter with a tip radious of about 100 nm was used. After nanoindentation measurements, the load-penetration depth curves were performed to calculate the hardness and elastic modulus through the Oliver–Pharr method [30]. Tribological characterization was developed with a Microtest, MT 400 98 tribometer, using a 6 mm diameter 100Cr6 steel-ball-like counter body slide. The load applied was 5 N with a total running length of 1000 m (around 29 000 cycles) and 0.10 m/s. A scratch test was used to evaluate the adhesion strength between the β-TCP/Ch coatings and the 316L SS substrates. This technique was performed with a scratching speed of 4 mm/min and a continuous load rate of 1 N/s. The scratch test was also used to calculate the critical load (L_c) as a function of chitosan percentage, and determine the cohesive failure of the coatings.

12.3 Results and Discussion

12.3.1 Crystal Lattice Arrangement of β-TCP/Ch Coatings

To study of the crystalline structure of the prepared β-TCP/Ch coatings, XRD technique was used to obtain the diffraction patterns of each coating. Figure 12.1 shows the diffraction patterns of β-TCP peaks at low angles, preferential orientations (0018), (1118) and (0502) for 2θ = (43.55°, 47.06° and 50.68°), respectively. The results indicate that the preferential orientations, corresponding to the ceramic phase of the coating layers, are in agreement with the Joint Committee on Powder Diffraction Standards (JCPDS) file number 01-070-2065 from the International Centre for Diffraction Data (ICDD). Figure 12.1 also exhibits the changes in rhombohedral configurations of β-TCP as a function of chitosan percentage. From this figure, the compression stress was also identified due to the changes in the peak morphology with chitosan percentage increasing [31]. Moreover, the diffraction patterns show the influence between the peak intensity and the chemical composition for β-TCP/Ch coatings. In this study, a displacement and widening of the preferential peaks was observed for all the samples.

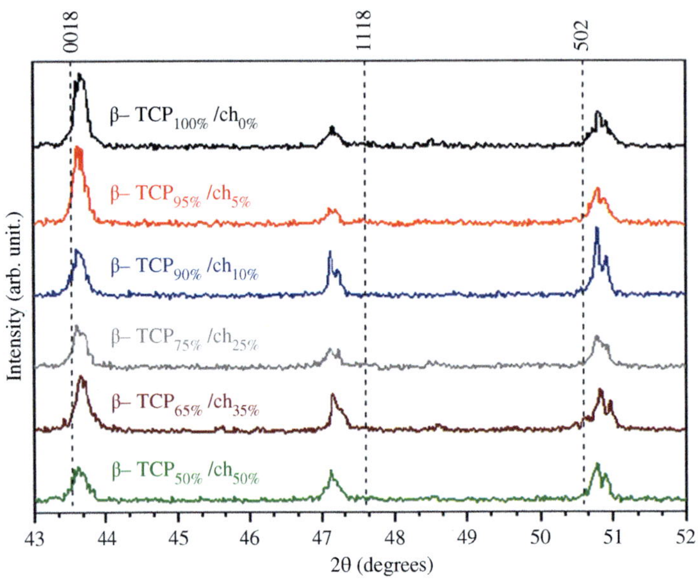

Figure 12.1 XRD patterns of β-TCP/Ch coatings deposited with different chemical compositions. The diffraction patterns show rhombohedral configurations corresponding to β-TCP and also a displacement and widening of the preferential peaks that indicate the compression stress due to the increase in the chitosan percentage.

The diffraction patterns exposed in Figure 12.1 show a variation in the intensity of the preferential peaks oriented (0018) at $2\theta = 43.6°$ as a function of chitosan percentage, which is in an agreement with the results reported by other authors [10, 31]. The variations in XRD patterns were appreciated using the full width at half maximum (FWHM) criterion for the patterns. Therefore, the crystal structure deformation data suggests an inversely proportional relationship between the chitosan percentage and the peak intensity. Analyzing the diffraction patterns in Figure 12. 2 shows a variation in the intensity of the preferential peaks oriented (0018) at $2\theta = 43.6°$ as a function of chitosan percentage in the coatings.

The crystallographic behavior described above is shown in Figure 12.2a where it is analyzed the shape and size variation associated to preferential peaks. Furthermore, the same shape and size variation as a function of chitosan percentage was observed in Figure 12.2b. This figure also exhibits a decreasing in peak area and peak height for the preferential peak (0018) which can be contributed to the changes in the internal residual compression stress [10, 23, 31]. Consequently, it was possible to observe that the peak variation suggests an inverse relationship between the chitosan percentage and the peak intensity. The observation of this peak variation demonstrates a crystal structure deformation due to the compression stress applied by chitosan molecules in the β-TCP structure.

12.3.2 Surface Coating Analysis

To analyse the surface characteristics of different β-TCP/Ch coatings SEM was used. Figure 12.3 exhibits SEM micrographs of different coatings as a function of chemical

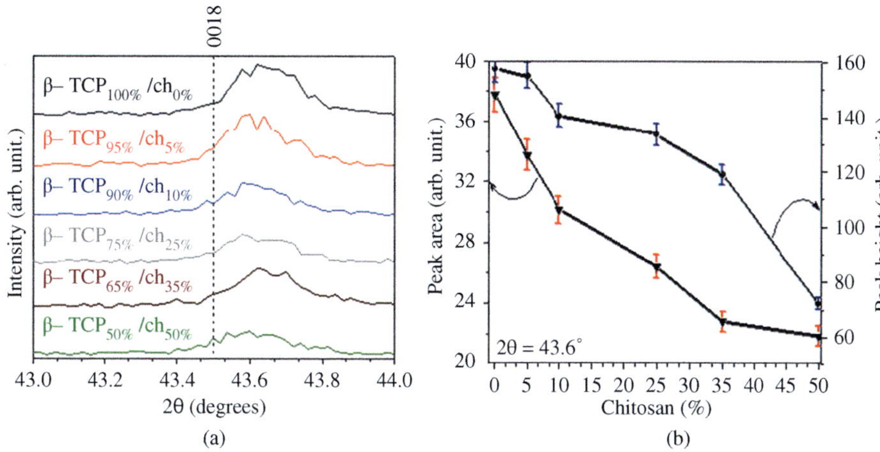

Figure 12.2 XRD patterns of different β-TCP/Ch coatings: (a). full width at half maximum (FWHM) variations as a function of chitosan percentage change and (b) peak area and peak height changes of preferential peak (0018) as a function of chitosan percentage. The curves show the compression stress effect in crystal structure due to the changes in chitosan percentage; it was evidenced by the decrease of peak area and peak height while the chitosan percentage increased.

compositions. The micrographs show the changes in the surface morphology as a function of chitosan percentage, which is in accordance with previous publications [10, 30, 32]. Therefore, Figure 12.3a presents a fibrillar structure matrix associated to the β-TCP particles, this figure also demonstrated irregular dispersed particles associated to the particles which were not dissolved in the electrolyte solution deposited on the surface coating by ionic exchange from (Ca_2+ and $PO_{4\text{-}3}$).

Figure 12.3b exhibits the surface coating when it has 5% of chitosan, it showed that the number of particles increased associated with the polymerization of chitosan molecules which are not dissolved in the saturated electrolyte. Also, it was possible to observe that fibrillar structure matrix has disappeared, and it was replaced by a matrix with angular particles dispersed on the surface coating without defined orientation. This effect visually demonstrates the functionalization effects between the β-TCP particles and Ch molecules [24, 26].

On the other hand, Figure 12.3 shows that the thickness of the deposited particles decrease as a function of the increasing of chitosan percentage (Figure 12.3b–e). Moreover, increasing of the chitosan percentage influences the deposited particles on the surface to be thinner than the other coatings with lower chitosan percentage and presents an angular shape similar to needle shape.

Furthermore, these results exhibit progressive increase in the percentage and the size of particles deposited on the surface coating. The chitosan particles have a tendency to homogenize eliminating the surface area of the needle-like structure. In this way, Figure 12.6f (a typical micrograph of surface coating with 50% of chitosan) shows a surface without needle-like structures, probably due to increasing of irregular particles.

Figure 12.3e shows a typical micrograph of the surface coating with 50% of chitosan, this micrograph presents a surface without needles structures probably due to the increasing of irregular particles observed in the last figures (Figure 12.3b–e). Allotropic changes in the surface morphology of β-TCP/Ch coatings were described previously,

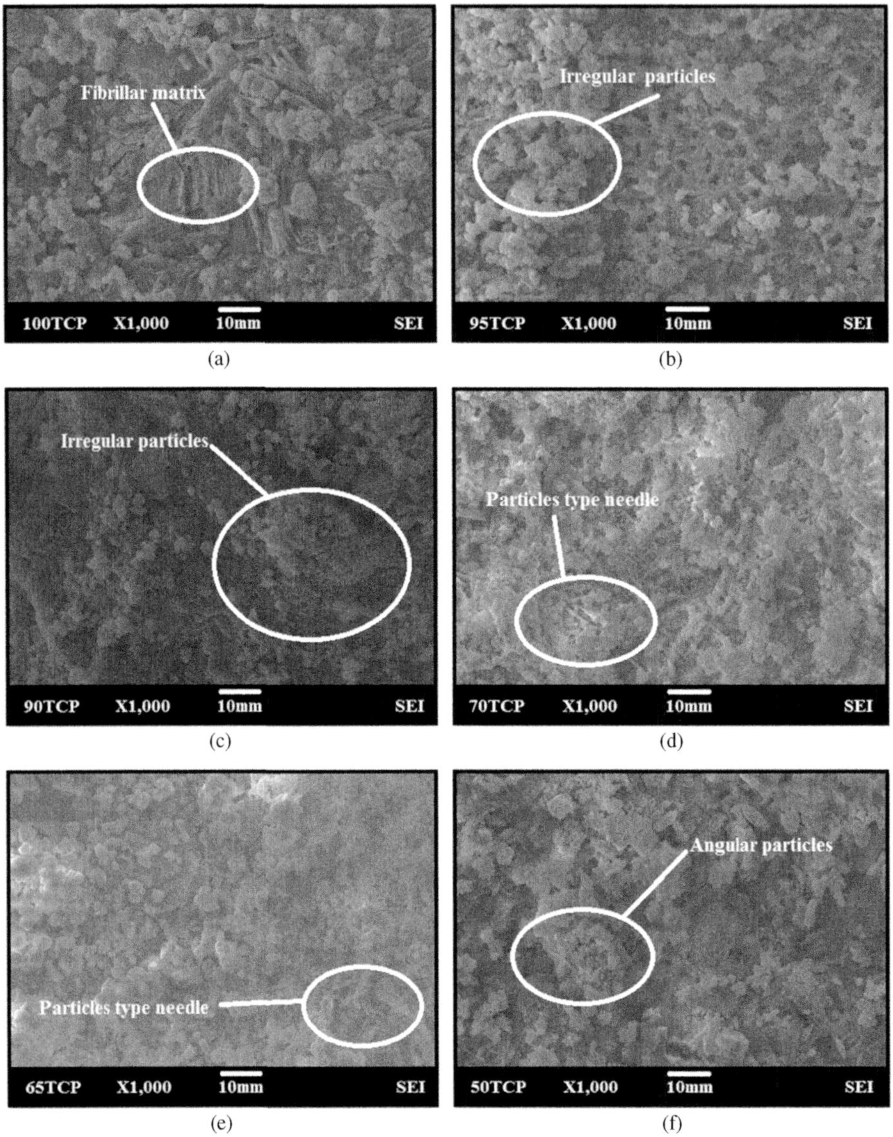

Figure 12.3 SEM micrographs of different β-TCP/Ch coatings: (a) β-$TCP_{100\%}/Ch_{0\%}$, (b) β-$TCP_{95}\%/Ch_{5}\%$, (c) β-$TCP_{90}\%/Ch_{10}\%$, (d) β-$TCP_{90}\%/Ch_{25}\%$, (e) β-$TCP_{65}\%/Ch_{35}\%$, and (f) β-$TCP_{50}\%/Ch_{50}\%$. The irregular particles are related with the particles without a defined shape; Particles type needle are elongated particles and angular particles are agglomerated particles with angular shapes.

which can be related to the changes in the crystallographic arrangement presented in the Figure 12.1 and Figure 12.2.

12.3.3 Morphological Analysis of the β-TCP-Ch Coatings

AFM analysis was used to study the surface morphology of different β-TCP/Ch coatings. It is known that the surface morphology and roughness of coatings have a very important role on the final characteristics of the implants, since the reactions at the tissue/coating interface is directly related to the surface morphology. Taking this note into account, different β-TCP/Ch coatings as function of chitosan percentage were analysed in an area of 20 μm × 20 μm for each sample with a z-scale around 86.30 ± 2.0 μm.

The morphology and the root-mean-square roughness of the β-TCP/Ch coatings are shown in Figure 12.4, it suggests that increasing the chitosan percentage can cause major changes in the surface topography and roughness. In this way, by increasing the chitosan percentage, the surface morphology changed and the surface roughness decreased.

The AFM micrograph shown in Figure 12.4a was obtained from the β-TCP coating surface, which shows rounded particles overlapped between them, taking into account that the scale of measurements is clear evidence that this coating is obtained from nanopowder. In addition, a qualitative analysis indicates that the sample contains predominantly

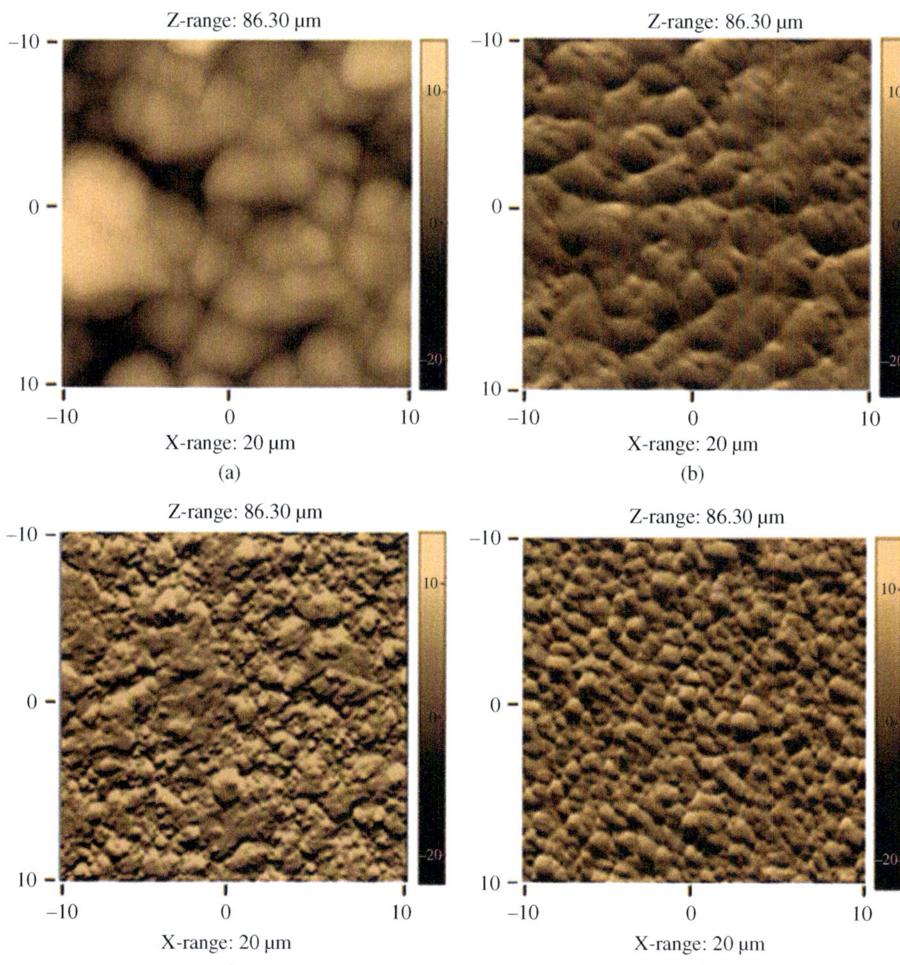

Figure 12.4 Representative AFM images of β-TCP/Ch coatings on the 316L SS substrates: (a) β-TCP100%/Ch0%; (b) β-TCP95%/Ch5%; (c) β-TCP65%/Ch35%; and (d) β-TCP50%/Ch50% coatings.

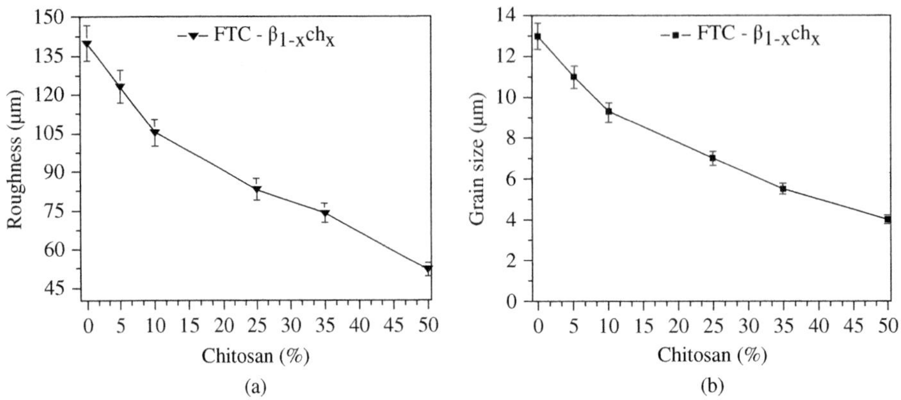

Figure 12.5 Morphological analysis of β-TCP/Ch coatings on the 316L SS substrates: (a) Roughness as a function of chitosan percentage curve, and (b) Grain size as a function chitosan percentage.

particles of almost the same size and shape similar to the results reported by previous publications [33].

The allotropic changes in the surface morphology of the β-TCP/Ch coatings were described previously, indicating that these changes are related to the changing in the size and distribution of the particles. It is possible to infer that the changes in the crystallographic arrangement obtained by XRD analysis are related to the growth, size and distribution of the particles, at the same time the allotropic changes of the surface morphology is related to the changes in the surface topography observed by the SEM micrographs.

Applying statically analysis on AFM images using a Scanning Probe Image Processor (SPIP®), it was possible to measure the morphological changes of the β-TCP/Ch coatings. Figure 12.5 shows the roughness and grain size trend as a function of chitosan percentage. Similar to the results from the SEM analysis it was possible to observe a decrease of coating roughness by further increasing the chitosan content. The previous published studies indicated that the values of grain size are closely related to the values of roughness, therefore, a decrease in the surface roughness implies a decrease in the grain size [34]. This effect is evidenced in Figure 12.5 which shows the decrease of roughness and grain size as a function of increasing of the chitosan amount. Finally, it is important to remark that the surface morphology results are related to the mechanical and tribological behaviour. Therefore, the grain size reduction increases the toughness and hardness values. Also a smooth surface with a superficial homogeneity and low friction coefficient can help to increase the wear resistance [34, 35].

12.3.4 Mechanical Properties

Hardness and elastic modulus

The load-displacement indentation curves of the β-TCP/Ch coatings were obtained using the standard Berkovich indenter in a nanoindentation test. The load-displacement indentation as a function of chitosan percentage is shown in Figure 12.6, which exhibits an elastic-plastic behavior of the coatings. The displacement of the indenter at maximum load was analyzed for each coating. Figure 12.6 shows that the values of

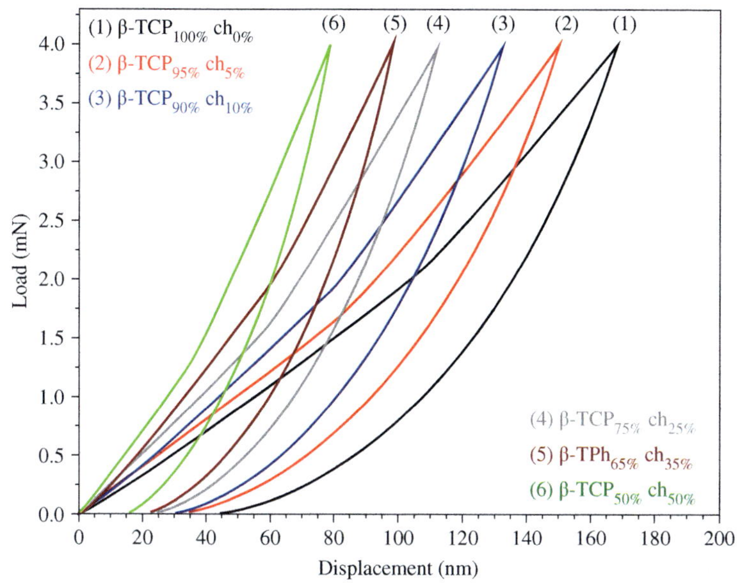

Figure 12.6 Nanoindentation test results for the β-TCP/Ch coatings: load-displacement indentation curves as a function of increasing the chitosan percentage.

displacement decrease as a function of increase in the percentage of chitosan content. The lower values of displacement indicate an increase of penetration resistance, which implies a possible mechanical properties improvement.

The XRD patterns indicated that the compression stresses applied by chitosan inclusion in crystallographic structure of β-TCP associated with the deformation in the preferential peaks. This deformation in β-TCP structure suggest a kind of hardening by deformation due to a possible inclusion of chitosan in interstitial zones of β-TCP that can increase the value of the lattice parameter Burgers vector (this can produce a hardening). In this way, the improvement in mechanical properties could be related to the structural changes occurring by inclusion and increase of chitosan in the crystallographic arrangement of β-TCP.

It was possible to determinate the values of hardness (H) and elasticity modulus (E) using the indentation method developed by Oliver and Pharr [36]. Figure 12.7a shows an increase of hardness while increase the chitosan percentage by using Equation (12.1) (which relate hardness with the ratio between maximum applied load and contact area):

$$H = \frac{P_{\max}}{A(h_C)} \tag{12.1}$$

where P_{max} is the maximum applied load and A(hc) is the area associated to the surface where the load is applied. This equation is according with the Oliver–Pharr method [30, 36]. Figure 12.7 shows the increasing of mechanical properties as a function of chitosan percentage. As mentioned before, this increase of E and H could be related with little changes in the crystallography orientation observed in the XRD patterns (Figure 12.1) associated with the changes in the area and height of the central peak.

Hajek and co-workers [37] proposed an expression to calculate the elastic recovery for the coatings. In this research, Figure 12.8 shows the plastic deformation resistance

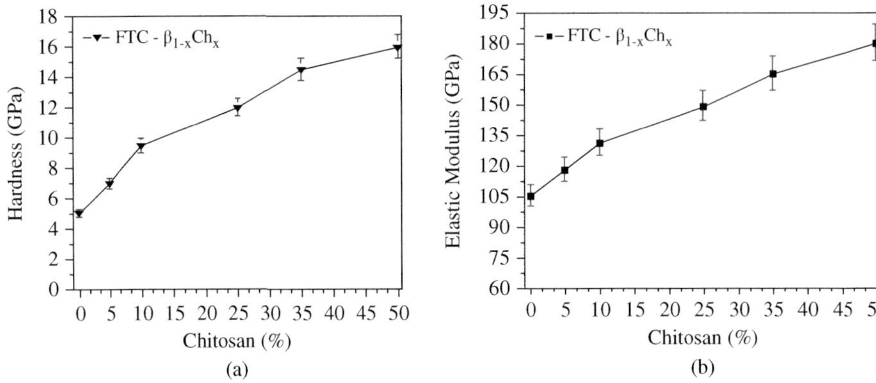

Figure 12.7 Mechanical properties of the β-TCP/Ch coatings: (a) Hardness values of the β-TCP/Ch coatings as a function of chitosan percentage and (b) Elastic modulus of the β-TCP/Ch coatings as a function of chitosan percentage.

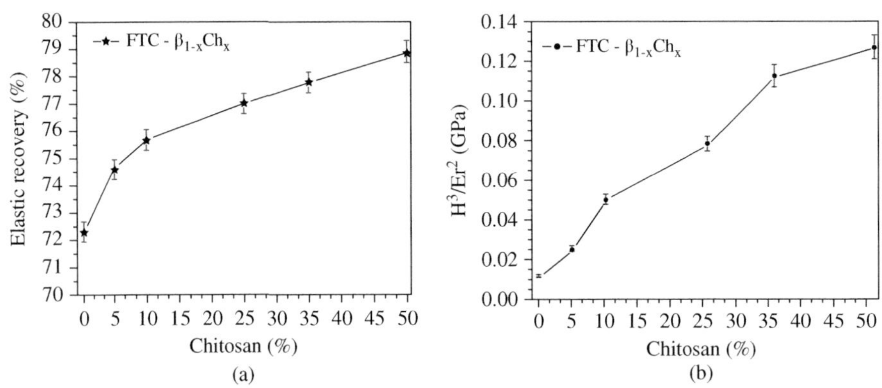

Figure 12.8 Mechanical properties: (a) Elastic recovery as a function of chitosan percentage, and (b) Plastic deformation resistance as a function of increasing in the chitosan percentage.

(H3/E2) and elastic recovery (R). The obtained results here are in agreement with other authors [38–42]. The elastic recovery for the β-TCP/Ch coatings was calculated by using the following Equation [36]:

$$R = \frac{\delta_{\max} - \delta_p}{\delta_{\max}} \tag{12.2}$$

where $\delta_{\max}$ is the maximum displacement and δ_p is the residual or plastic displacement. The equation data was taken from the load-penetration depth curves of indentations for each coating according to the results obtained from nanoindentation test and exhibited in Figure 12.6. The best mechanical properties were reached by the coating with the highest amount of chitosan content. Figure 12.8 shows an inverse proportional relation between the mechanical properties and chitosan percentage. In this sense it was possible to observe that the elastic recovery and plastic deformation resistance values increase as a function of chitosan amount increase, similar to the results analyzed in elastic modulus and hardness.

12.3.5 Tribological Properties

Pin-On-Disk Analysis

The friction coefficient values were obtained from pin-on-disk test, in this way, the tribological properties were analyzed into tribological pair between the 316L SS substrate and β-TCP/Ch coatings system and a 100Cr6 steel balls. Figure 12.9a shows the friction coefficient as a function of sliding distance for the substrate and coatings. The curves displayed in Figure 12.9a showed two different stages; the first stage is called the running-in period, where the friction coefficient began at low levels; due to an interferential friction mechanism it is caused by few contacts between the steel ball and the coating surface through the roughness tips in both counterparts. As slide distance increase, the friction coefficients increase too due to the formation of wear debris by cracking of roughness tips on both counterparts according to the tribological model proposed by Archard [43].

The second stage, also called steady state is a zone of the curve where the roughness of tribological pair is smoothed and the friction coefficient value is stabilized. In this stage, the surface coatings are molded by the surface of a 100Cr6 steel ball, since the ball surface is harder than the coating surface. Finally, it is possible to suppose that the friction coefficients are closely related to the slide distance, and by increasing the slide distance, the friction coefficient decreases [44, 45].

A good combination between elastic-plastic properties (H, Er) and surface properties (Ra) could possibly guarantee a good tribological response. In fact, the tribological properties depend mainly on the elastic-plastic properties, thus, while the chitosan percentage increases H and Er, the roughness decreases. Therefore, the values of friction coefficient decrease as is shown in Figure 12.9b. As it was explained above the chitosan amount and friction are closely related, so Figure 12.9b shows the friction coefficient as a function of chitosan percentage increase curve, this curve shows that μ decreases when the chitosan percentage increases.

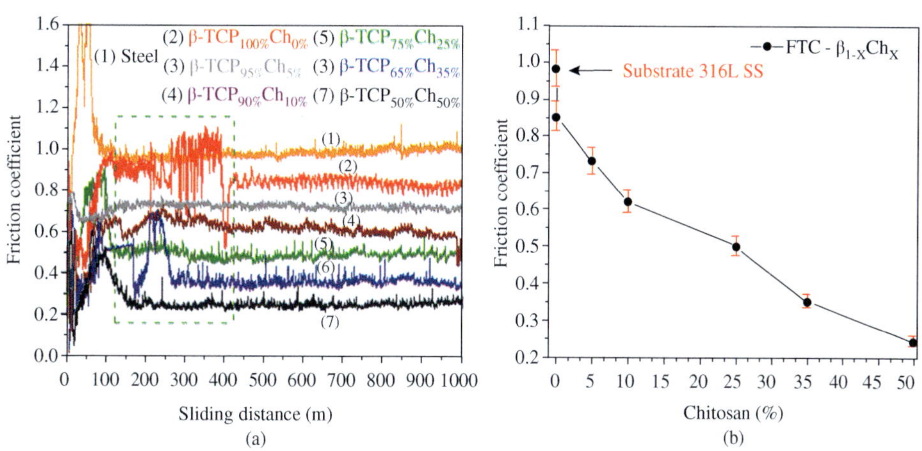

Figure 12.9 Tribological results of the 316L SS substrates coated with β-TCP/Ch coatings: (a) friction coefficient as a function of sliding distance, and (b) friction coefficient as a function of chitosan percentage. The curves explain an inversely proportional relationship between the chitosan amount and friction coefficient.

12.3.6 Surface Wear Analysis

After a sliding distance of 1000 m in the pin-on-disk tests, SEM analysis was used to evaluate the wear properties of the surfaces. Figure 12.10 shows the micrographs of the surface wear for all the coating samples at 1000× and their wear mechanism, demonstrating a continuous and smooth wear track, with a depth below the film thickness.

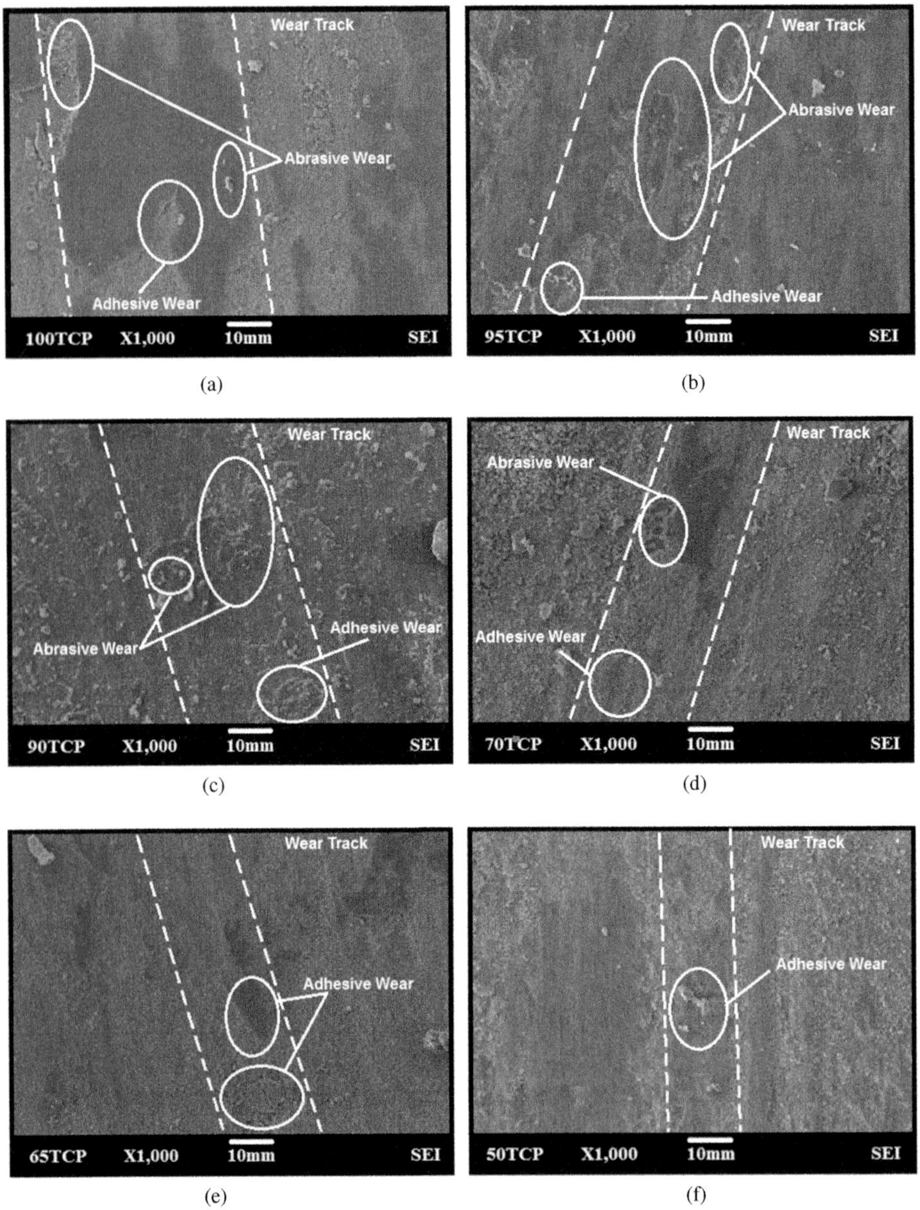

Figure 12.10 (a-f) SEM micrographs for the wear tracks, evidencing the changes in the wear mechanism (abrasive y adhesive wear) as a function of chitosan percentage for all the β-TCP/Ch coatings.

Therefore, the abrasive wear is presented in each micrograph, which is related to the generation of loose wear debris corresponding to rupture, cohesive failure and delamination of the coatings due to the interaction between the slider pair (steel ball) and the β-TCP/Ch coating surfaces.

Thus, the wear debris interact again with both surfaces, generating grooves and scratches on the coating surfaces. In this sense, high friction coefficient, and other wear mechanisms associated with the mechanical fatigue due to cyclical loads of pin on the surface is evidenced by cracks. Figure 12.10a–f clearly show the effects of chitosan percentages on the wear mechanism of the coatings. Therefore, it can be concluded that by increasing the chitosan percentage, a decrease in the trace of abrasive particles and cracks could be observed. The decreasing of the wear mechanism (adhesive and abrasive failure) for the β-TCP/Ch coatings is associated with the morphological surface evolution exhibited in the AFM results (Figure 12.5) together with the evolution of the mechanical properties evidenced in Figure 12.7 and Figure 12.8.

12.3.7 Adhesion Behaviour

Scratch-test technique was carried out to measure the adherence strength of the β-TCP/Ch coatings. To identify the adhesion properties of the coatings, it is important to consider a lower critical load (Lc1). It is defined as the load where the first cracks occurred (cohesive failure) and the highest critical load (Lc2) where occurs the first delamination at the edge of scratch (adhesive failure) [46]. Lc1 and Lc2 values correspond to the tribological zones where the friction is independent on the applied load [47]. The critical load (Lc1 and Lc2) values for all the β-TCP/Ch coatings are shown in Figure 12.11, which shows the critical load values experimentally determined from the friction coefficient as a function of applied load.

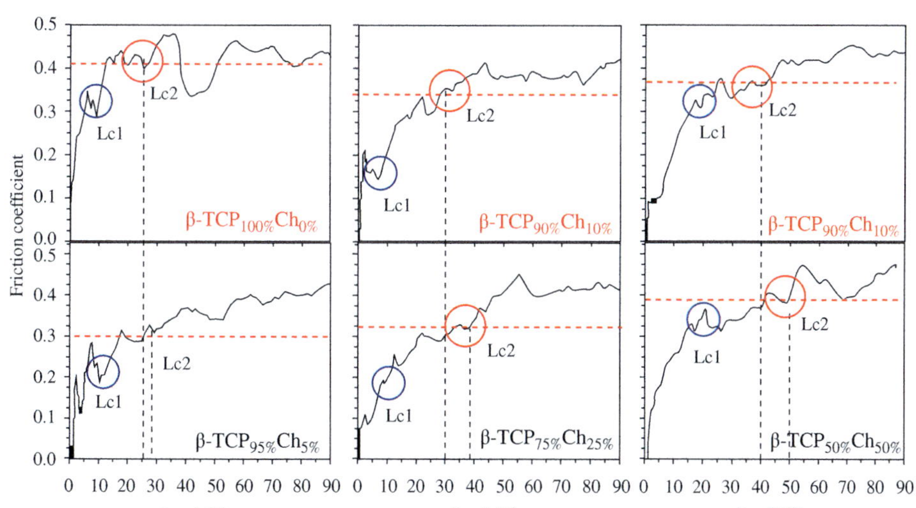

Figure 12.11 Tribological results for the friction coefficient curves versus the applied load; showing the cohesive (Lc1) and adhesive (Lc2) failure mode for the β-TCP/Ch coatings as a function of increasing in the chitosan percentage.

From Figure 12.11 it was possible to analyze the real adhesive response associated with the β-TCP/Ch coatings, hence, the curves of friction coefficient as a function of load showed that the values of Lc1 are similar in Figure 12.11a–d. However, for the coatings with higher percentages of chitosan (35% and 50%), the critical load associated with the cohesive failure is higher, as shown in Figure 12.11e and 12.11f. On the other hand, the critical load is associated with the adhesive failure, which is closely related to the chitosan percentage. In this way, the friction coefficient curves as a function of load showed that Lc2 increased when the chitosan percentage increased. Moreover, the improvement in the adhesion behavior of the β-TCP/Ch coatings could be related to the tribological analysis showed in Figure 12.9 from pin-on-disk results. Similarly, the improvement in the mechanical behavior exhibited in Figure 12.7 and Figure 12.8, could be also associated to the changes in the crystal lattice arrangement studied in Figure 12.1 and Figure 12.2 and the evolution in the topography and morphological surface observed by SEM and AFM techniques, respectively.

As mentioned before, the chitosan percentage is influenced on the values of Lc2, Figure 12.12 shows the Lc2 curve as a function of chitosan percentage. This curve indicates that the adhesive critical load of the coatings increase by further increasing the chitosan percentage, in another word, a higher stress should be applied to reach the delamination of the coatings. From the data reported by other authors, it is possible to conclude that the tribological results were governed by the elastic-plastic properties of the β-TCP/Ch coatings and adhesion corresponding to the interface (β-TCP/Ch coatings–substrate) because the β-TCP/Ch systems have a relatively high elastic recovery (R%) values [47]. As can be seen in Figure 12.8a, the critical load changes occur due to the increasing of plastic deformation resistance (H3/E2) [47] (showed

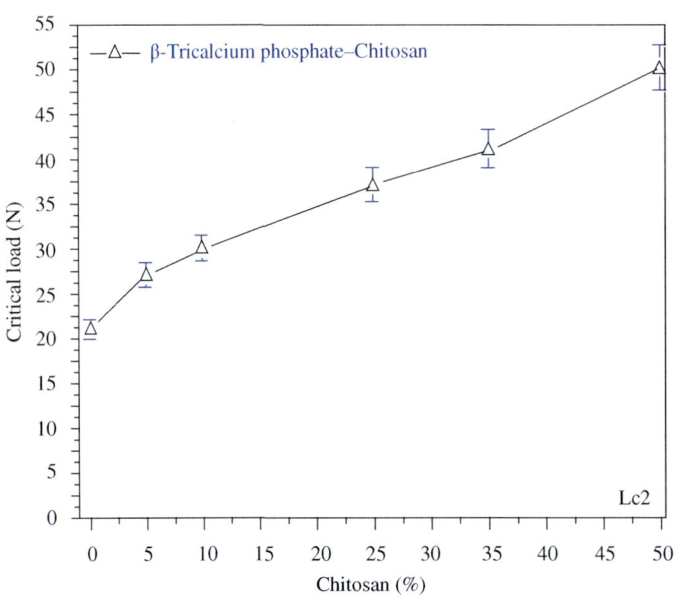

Figure 12.12 Critical load associated to the adhesion failure (Lc2) as a function of chitosan percentage for all the coatings: (β -TCP100%/Ch0%, β TCP95%/Ch5%, β -TCP90%/Ch10%, β -TCP90%/Ch25%, β -TCP65%/Ch35%, and β -TCP50%/Ch50%).

in Figure 12.8b), which generates opposition to indenter penetration producing an opposition to coating deformation. Thus, the last tribological and mechanical properties are related to a rhombohedral arrangement with the obtained results for all the coatings.

12.4 Conclusions

In this study, different types of β-TCP/Ch coatings were deposited via an electrodeposition method. It is shown that the preferential orientation for rhombohedral arrangement were (0018), (1118) and (0502) orientations for 2θ = (43.55°, 47.06° and 50.68°), respectively. From the SEM micrographs, it was possible to observe the topography changes on the surface coatings in relation with different chitosan percentages. These changes were associated with the results of the morphological analysis of β-TCP/Ch coatings performed by using AFM analysis, which showed the decrease of grain size and roughness as a function of chitosan percentage.

Different mechanical properties such as hardness were obtained for the β-TCP/Ch coatings. It was also determined that the elastic modulus (E_r) and elastic recovery increased as function of increasing the chitosan percentage. These results were expected taking into account the changes in preferential peaks of XRD that suggested a deformation hardening.

The tribological behavior was studied, in which a low friction coefficient was obtained in the tribological system (steel pin and β-TCP/Ch coatings) for all the coatings. This study suggested an improvement in tribological properties of the coatings evidencing an increase of adhesion resistance as function of chitosan increase. Therefore, the improvement in tribological behavior could be expected taking into account the improvement of the elastic-plastic behavior exposed by the β-TCP/Ch coatings by increasing the chitosan percentage.

From the SEM results it was possible to identify the wear mechanisms in each coating (abrasive and adhesive wear) with failure by fatigue evidenced by micrographs which are reduced when the chitosan percentage is increased. In this sense it is possible to observe a decrease in the evidence of wear mechanisms due to the improvement in elastic-plastic behavior of the coatings, which open up new features for the future of surgical implants.

Acknowledgements

This research was supported by El patrimonio Autónomo Fondo Nacional de Financiamiento para la Ciencia, and financial support from the Universidad Militar Nueva Granada, contract number ING-1775-2015.

References

1 Ratner BD, Hoffman AS, Schoen FJ, Lemons JE (eds). Biomaterials Science an Introduction to Materials in Medicine, 2nd edn. Elsevier Academic Press, San Diego, p. 137; 2004.

2 Karim, B, Jean J, Mainard D, *et al.* Patrick. Evaluation of the effect of three surface treatments on the biocompatibility of 316L stainless steel using human differentiated cells. Biomaterials 17, 491; 1996.
3 Salahinejad E, Hadianfard MJ, Macdonald DD *et al.* In vitro electrochemical corrosion and cell viability studies on nickel-free stainless steel orthopedic implants. PLoS ONE 8, 1–8; 2013.
4 Mozafari M, Salahinejad E , Macdonald DD *et al.* Multilayer bioactive glass/zirconium titanate thin films for bone tissue engineering and regenerative dentistry. International Journal of Nanomedicine 8, 1665–1672; 2013.
5 Naghib SM, Ansari M, Pedram A *et al.* Bioactivation of 304 Stainless Steel Surface through 45S5 Bioglass Coating for Biomedical Applications. International Journal of Electrochemical Science 7, 2890–2903; 2012.
6 Salahinejad E, Hadianfard MJ, Macdonald DD *et al.* Surface modification of stainless steel orthopedic implants by means of sol-gel derived $ZrTiO_4$ and $ZrTiO_4$–PMMA thin films. Journal of Biomedical Nanotechnology 9, 1327–1335; 2013.
7 Cross MJ, Parish EN. A hydroxyapatite-coated total knee replacement: prospective analysis of 1000 patients. Journal of Bone Joint Surgery, 87(8), 1073–6; 2005.
8 Jinno T, Davy DT, Goldberg VM. Comparison of hydroxyapatite and hydroxyapatite tricalcium-phosphate coatings. Journal of Arthroplasty 17 (7), 902–909; 2002.
9 Pang X, Zhitomirsky I. Electrophoretic deposition of composite hydroxyapatite $CaSiO_3$- Chitosan composite coatings. Colloid and Interface Science 330, 323–329; 2009.
10 Peña D. Estupiñan H. Cordoba H. Vasquez Comportamiento electroquímico en ringer's de recubrimientos de HAP obtenidos por electrodeposición sobre acero inoxidable 316LVM anodizado. Revista Facultad de Ingeneria, University of Antioquia, Medellín, Colombia 54, 15–23; 2010.
11 Abe Y, Kokubo T, Yamamuro T. Apatite coating on ceramics, metals and polymers utilizing a biological process. Materials Science: Materials in Medicine 1, 233; 1990.
12 Pighinelli L, Kucharska M. Properties of microcrystalline chitosan-calcium phosphate complex composite. Journal of Biomaterials and Nanobiotechnology 4, 20–29; 2013.
13 Ding S-J. Preparation and properties of chitosan/calcium phosphate composites for bone repair. Dental Materials Journal 25, 706–712; 2006.
14 Kucharska M, Niekraszewicz A, Wisniewska-Wrona M *et al.* Preparation and estimation of chitosan usable dressing forms. Progress in Chemistry and Application of Chitin and Its Derivatives 3, 69–72; 2003.
15 Bamwenda G.R, Sayama K, Arakawa H. The effect of selected reaction parameters on the photoproduction of oxygen and hydrogen from a WO_3–Fe $_2$+–Fe_3+ aqueous suspension. Journal of Photochemstry and Photobiology A: Chemistry 122, 175; 1999.
16 Bohnke O, Bohnke C, Robert G. Electrochromism in WO_3 thin films. I. $LiClO_4$-propylene carbonate-water electrolytes. Solid State Ionisation 6, 121; 1982.
17 Davazoglou D, Donnadieu A, Donnadicu A. Study on the optical and electrochromic properties of polycrystalline WO_3 thin films prepared by CVD. Solar Energy Materials 71, 379; 1988.
18 Salahinejad E., M.J. Hadianfard, D.D. Macdonald *et al.* Zirconium titanate thin film prepared by an aqueous particulate sol-gel spin coating process using carboxymethyl cellulose as dispersant. Materials Letters 88 5–8; 2012.

19 Hurdich R. Electrochromism in hydrated tungsten-oxide films. Electron Letters 11, 142; 1975.
20 Lee K.D. Deposition of WO thin films by the sol–gel method. Thin Solid Films 302, 84; 1997.
21 Monk PMS, LS Chester. Electro-deposition of films of electrochromic tungsten oxide containing additional metal oxides. Electrochimica Acta 38, 1521; 1993.
22 Inamdar AI, Mujawar SH, Ganesan V, Patil PS. Surfactant-mediated growth of nanostructured zinc oxide thin films via electrodeposition and their photoelectrochemical performance. Nanotechnology 19, 325706; 2008.
23 Castro EM, Estupiñan HA, Peña DY. Obtención de recubrimientos de fosfatos de calcio sobre Ti6Al4V ELI mediante electrodeposición catódica. Mechanical properties of hydroxyapatite and tricalcium phosphate. Scientia et Technica 13, 36; 2007.
24 Metsger DS, Rieger MR, Foreman DW. Mechanical properties of hydroxyapatite and tricalcium phosphate. Journal of Materials Science: Materials Medicine 10, 9–17; 1999.
25 Guo X, Lee L, Law LP *et al.* Recombinant human bone morphogenetic protein-4 (rhBMP-4) enhanced posterior spinal fusion without decortication. Journal of Orthopaedic Research 20, 740–746; 2002.
26 Banerjee SS, Tarafder S, Davies NM *et al.* Understanding the influence of MgO and SrO binary doping on the mechanical and biological properties of β-TCP ceramics. Acta Biomaterialia 6, 4167–4174; 2010.
27 Den Braber ET, de Ruijter JE, Ginsel LA. Quantitative analysis of fibroblast morphology on microgrooved surfaces with various groove and ridge dimensions. Biomaterials 17, 2037–2044; 1996.
28 Zhang M, Li XH, Gong YD *et al.* Properties and biocapability of biodegradable composites based on PHB-PHV copolymer. Biomaterials 23, 2641–2648; 2002.
29 Martínez R, Estupiñan H, Peña D, Mohan P. Efecto del chitosán en el comportamiento electroquímico de recubrimientos de fosfatos de calcio. Scientia et Technica 36, 36; 2007.
30 Oliver WC, Pharr GM. An improved technique for determining hardness and elastic modulus using load and displacement sensing indentation experiments. Journal of Materials Research 7–6, 1564–1583; 1992.
31 Jiawei W. Electrolytic deposition of calcium phosphate/chitosan coating on titanium alloy: growth kinetics and influence of current density, acetic acid, and chitosan. Journal of Biomedical materials Research Part A. 76, 503–511; 2006.
32 Tejada A, Piña C, Martínez S, Ávila G. Cerámicas Bioeutecticasw-tcp. Revista Mexicana de Física 5(2), 187–192; 2004.
33 Tomoaia G, Mocanu A, Vida-Simiti I *et al.* Silicon effect on the composition and structure of nanocalcium phosphates: in vitro biocompatibility to human osteoblasts. Materials Science and Engineering: C 37, 37–47; 2014.
34 Kim GS, Lee SY, Hahn JH. Synthesis of CrN/AlN superlattice coatings using closed-field unbalanced magnetron sputtering process. Surface and Coatings Technology, 171, 91–95; 2003.
35 Zhang J, Liu W, Schnitzler V *et al.* Calcium phosphate cements for bone substitution: chemistry, handling and mechanical properties. Acta Biomaterialia 10, 1035–1049; 2014.

36 Cabrera G, Caicedo JC, Amaya C *et al.* Enhancement of mechanical and tribological properties in AISI D3 steel substrates by using a non-isostructural CrN/AlN multilayer coating. Materials Chemistry and Physics 125, 576–586; 2011.

37 Hajek VV, Rusnak K, Vlcek J *et al.* Tribological study of CN x films prepared by reactive dc magnetron sputtering. Wear 213, 80–86; 1997.

38 Caicedo JC, Amaya C, Yate L *et al.* TiCN/TiNbCN multilayer coatings with enhanced mechanical properties. Applied Surface Science 256, 5898–5904; 2010.

39 Kim GS, Lee SY, Hahn JH. Synthesis of CrN/AlN superlattice coatings using closed-field unbalanced magnetron sputtering process. Surface and Coatings Technology 171, 91–95; 2002.

40 Burnett PJ, Rickerby DS. The relationship between hardness and scratch adhesion. Thin Solid Films 154, 403–416; 1987.

41 Caicedo JC, Bejarano G, Gómez ME *et al.* Nanostructured multilayers of TiN/ZrN obtained by magnetron sputtering. Physica Status Solidi 4(11), 4127–4133; 2007.

42 Romero J, Lousa A, Esteve J, Martínez E. Nanoindentation stress-strain curves as a method for thin-film complete mechanical characterization: application to nanometric CrN/Cr multilayer coatings. Applied Physics A: Materials Science and Processing 77, 419–427; 2003.

43 Archard JF. Contact and rubbing of flat surfaces. Experiences from scratch testing of tribological PVD coatings. Journal of Applied Physics 24, 981;1953.

44 Nledengvist P, Hogmark S. Experiences from scratch testing of tribological PVD coatings. Tribology International 30, 507–516; 1997.

45 Ipaz L, Caicedo JC, Esteve J *et al.* Improvement of mechanical and tribological properties in steel surfaces by using titanium–aluminum/titanium–aluminum nitride multilayered system. Applied Surface Science 258, 3805–3814; 2012.

46 Bull SJ, Rickerby DS, Matthews A *et al.* The abrasive wear resistance of sputter ion plated titanium nitride coatings. Surface and Coatings Technology 36, 503–517; 1988.

47 Lackner JM, Major L, Kot M. Modelling of nanoindentation to simulate thin layer behaviour. Bulletin of the Polish Academy of Sciences: Technical Sciences 59(3), 343–355; 2011.

13

Nanoindentation in Metallic Glasses

Vahid Nekouie, Anish Roy and Vadim V. Silberschmidt

Wolfson School of Mechanical, Electrical and Manufacturing Engineering, Loughborough University, Leicestershire, UK

13.1 Introduction

13.1.1 Motivation

Amorphous alloys, or metallic glasses (MGs), are engineering materials with unique physical, chemical and mechanical properties such as a high ratio of elastic limit to the Young's modulus, high fracture toughness (when compared to their crystalline counterparts), corrosion resistance and biocompatibility thanks to absence of a long-range order in their atomic structure and a lack of defects such as dislocations. These unique properties make them an ideal candidate for modern high-tech applications such as MEMS (micro-electromechanical systems), miniaturised biomedical devices, implants, etc. Metallic glasses were first reported in 1960s; they were processed using solid-state amorphisation [1] via hydrogen absorption [2, 3], mechanical alloying [4], or heat treatment enabling anomalous diffusion in crystalline bi-layers [5]. From the late 1980s, metallic glasses were manufactured in a variety of multi-component alloys with cooling rates less than $100\,K\,s^{-1}$ and thickness of several centimetres. Over the last 40 years, critical casting thickness has increased by more than three orders of magnitude, and high quantity of metallic glasses was produced.

Typically, inorganic glasses are brittle at room temperature, exhibiting a smooth fracture surface as a result of mode-I brittle fracture. In metallic glasses, formation and evolution of localised shear bands is a primary deformation mechanism resulting in a brittle response in the macroscale. However, there is convincing experimental evidence indicating that metallic glasses demonstrate ductility in micro- and nano-scale. Interestingly, at the small length scale shear bands act as carriers of ductility. Thus, understanding formation, evolution and propagation of shear bands is crucial for understanding mechanics of metallic glasses.

A number of mechanistic theories have been proposed to describe plastic flow and deformation behaviour of metallic glasses. Plastic deformation in MGs was described using concepts of a free volume and shear transformation zones (STZs) proposed in the Argon and Spaepen's model, based on motion of atoms [6, 7]. With respect to plastic deformation in metallic glasses, STZs occur at high free volume sites and then percolate to form shear bands along the direction of maximum shear stress. At the macro-scale, the deformation mechanism of metallic glasses based on the free-volume and STZs

Applied Nanoindentation in Advanced Materials, First Edition. Edited by Atul Tiwari and Sridhar Natarajan.

theory can occur homogeneously or inhomogenously depending on the levels of strain rate, temperature and applied stress. Schuh and Nieh [8] developed a deformation map with inclusion of strain-rate contours, where the homogeneous regime is divided into 'elastic', 'Newtonian', and 'non-Newtonian' sub-regions. A transition from the Newtonian sub-region to non-Newtonian one occurs at $10^{-5}\,s^{-1}$; the non-Newtonian flow is observed at strain rates below this value. However, it is important to mention that at high enough shear rates, the non-Newtonian flow as well as shear localisation – can happen at high temperatures even in the supercooled region. The deformation map developed by Schuh *et al.* [9] does not consider the evolution of glass structure during deformation and cannot be used to explain the fracture behaviour. However, this map can be employed to compare mechanical responses of different metallic glasses at the same absolute temperature (e.g. room temperature) but at different homologous temperatures.

13.1.2 Nanoindentation Studies of Metallic Glasses

During the 1970s and 1980s, it was difficult to measure the mechanical properties of metallic glasses due to limitation of small specimen sizes produced. Thus, standard microhardness measurement was a common technique to assess mechanical strength of thin ribbons of metallic glass. In recent years, depth-sensing indentation experiments have been used to characterise mechanical properties across various length scales. Nanoindentation experiments play a major role in analysis and elucidation of the mechanisms of plastic flow in metallic glasses as this technique can be used to gain insight into the nuances of deformation their behaviour. The underlying reason for ubiquitous use of the indentation technique is its relative experimental simplicity. Experiments can be performed several times on a single specimen including the possibility to probe different volumes of materials via an appropriate choice of load and tip geometry [10].

In indentation experiments, an applied load and a penetration depth into the studied material are simultaneously recorded, and a load-penetration depth curve is produced (see Figure 13.1). This curve contains a wealth of information related to the deformation

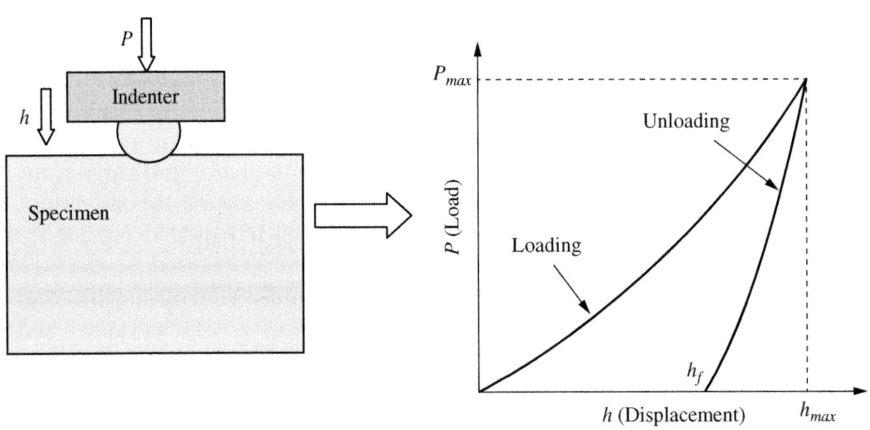

Figure 13.1 Schematic of indentation of bulk materials. *Source*: Guo 2010 [11]. Reproduced with permission of ORBi.

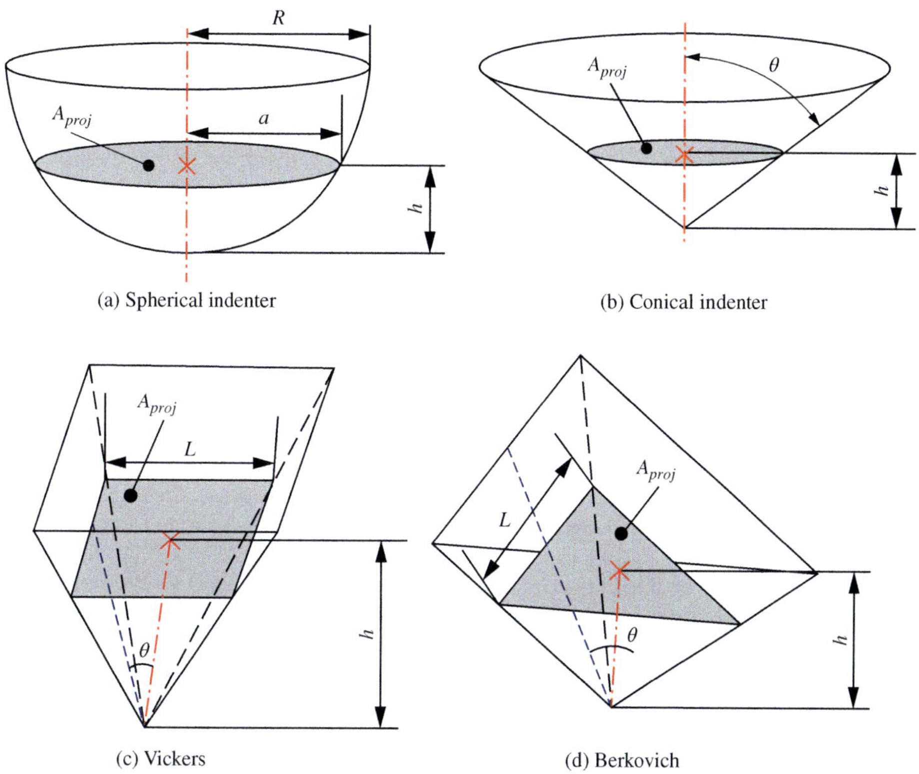

Figure 13.2 Geometries of indenters used in instrumented indentation. *Source*: Fischer-Cripps 2001 [13]. Reproduced with permission of Cambridge University Press.

behaviour of materials and can be used to determine mechanical properties such as hardness and elastic modulus [12]. Indenters are classified based on their shape and geometry. Among the most popular, as shown in Figure 13.2, used in experiments and analysis are conical, spherical, Berkovich, cube-corner indenter and Vickers. Spherical indenters (Figure 13.2a), have been widely used for brittle materials where a small deformation is preferable. In spherical indentation, the contact stresses are initially small and produce only elastic deformation; later, a transition from elastic to plastic deformation occurs, allowing assessment of yielding and work-hardening of the indented material. Sharp indenters such as Vickers and Berkovich are preferred for indentation at nano-scale, as minimising the effect of friction. In contrast to them, cube-corner indenters have relatively small face angles, useful in producing small, well-defined cracks around imprints in brittle materials. These cracks can be used to estimate fracture toughness at a very small scale. For the cube-corner indenters, the projected area of contact is the same as that for the Berkovich indenter.

Due to their specific mechanical behaviour, the initiation of localised plastic deformation in MGs is captured in nanoindentation be analysis of pop-in events. The pop-in behaviour depends on the type of metallic glass and its free-volume content [14] as well as the specifics of experimental procedures such as the indenter's tip shape and applied loading rates. Generally, the size of pop-ins increases with the depth of

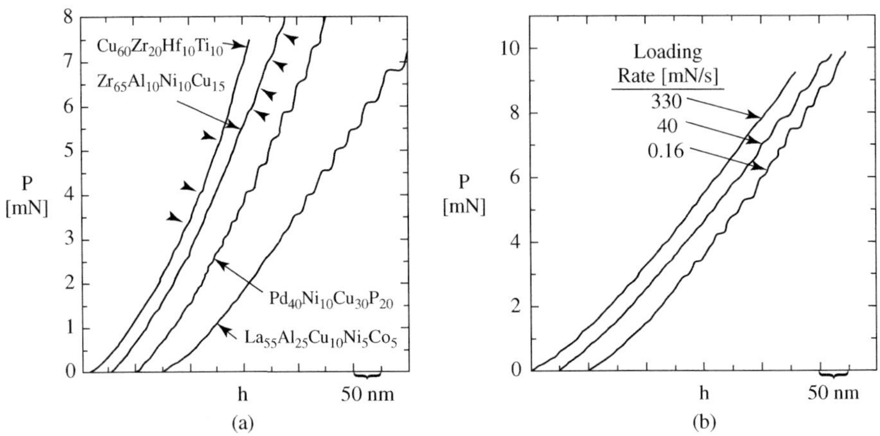

Figure 13.3 (a) Example of indentation P-h curves for different MGs, showing discrete pop-ins, or flow serration. (b) Example of rate-dependence of serrated flow beneath nanoindenter in Pd-based metallic glass. *Source*: Schuh 2004 [8]. Reproduced with permission of Cambridge University Press.

indentation based on the strain accommodated by the indenter [14, 15]. For instance, the cube-corner indenter produces larger pop-in events than the Berkovich tip thanks to its lower included angle and higher generated stress [16].

As shown in Figure 13.3, the pop-in behaviour differs with composition of MGs. For instance, Pd-based glasses show sharp and well defined pop-ins, whereas Zr-based metallic glasses demonstrate ripples rather than displacement bursts. This variation from glass to glass is expected because of their varied chemical nature and different states of short- or medium-range ordering. A large number of studies showed that the pop-in behaviour was changed in MG with the same composition when varying the deformation rate [9, 14]. As illustrated in Figure 13.3, low indentation rates show more displacement burst compared to high indentation rates, due to the activation of single shear as a result of plastic deformation [8].

In other studies, an intrinsic change in plasticity of MGs was reported at high homologous temperatures (T/T_g), as the degree of flow serration was significantly reduced. Furthermore, there was an evidence of creep during unloading or constant-load hold with an increasing penetration depth when metallic glasses were indented at high homologous temperatures [17, 18].

More recently, a flattened indenter tip was used to perform micro-compression tests on micro-pillars patterned with Focused Ion Beam (FIB). Micro-compression of metallic glasses helps to run uniaxial testing with high resolution without imposing strain gradients to study the effect of individual shear banding [19–21]. Nevertheless, a size-dependent behaviour was reported by many researchers in micro-compression studies in MG micropillars with a reported correlation between a size and mechanical properties such as maximum plastic strain before failure and yield. However, other studies concluded differently (Table 13.1). This experimental technique suffers from several challenges, for instance, imperfect pillar geometry, such as a tapering effect, and sensitivity to pillar alignment with regard to the indenter face [22].

Table 13.1 Literature on size effect of strength in metallic glasses [23].

Reference	BMGs	Strength size-dependence
[50]	Mg-based	Dramatic increase
[21]	Zr-based	Dramatic increase
[51]	Pd-based	Slight reduction
[52]	Pd-based	Slight reduction
[53]	Zr-based	No change
[54]	Zr-based	Small increase
[55]	Zr-based	Yes
[56]	Zr-Cu-based	No change

Source: Greer 2011 [23]. Reproduced with permission of Elsevier.

13.1.2.1 Pile-up and Sink-in

Two main modes of deformations are observed in indentation. The loaded material is displaced with compression along the radial direction, imparting further displacement perpendicular to the indent surface. These combined movements force the material around the indenter either to pile up (upwards) or sink-in (downwards) [24]. Both phenomena are schematically illustrated in Figure 13.4. The extent of these changes depends on the E/σ_Y and strain–hardening properties of the materials [25]. Alcala *et al.* [26] investigated the piling-up and sinking-in using Vickers and spherical indentation tests. The study demonstrated that sinking-in predominated in materials

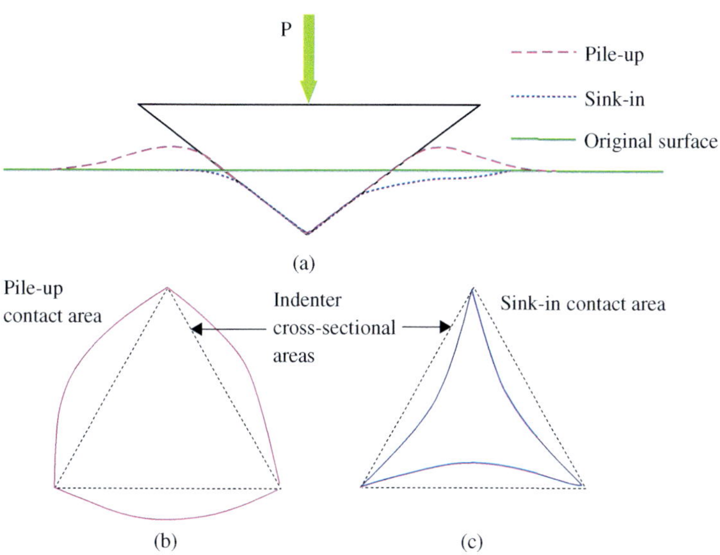

Figure 13.4 Schematic of pile-up and sink-in effects in indentation. *Source*: Alcala 2000 [26]. Reproduced with permission of Elsevier.

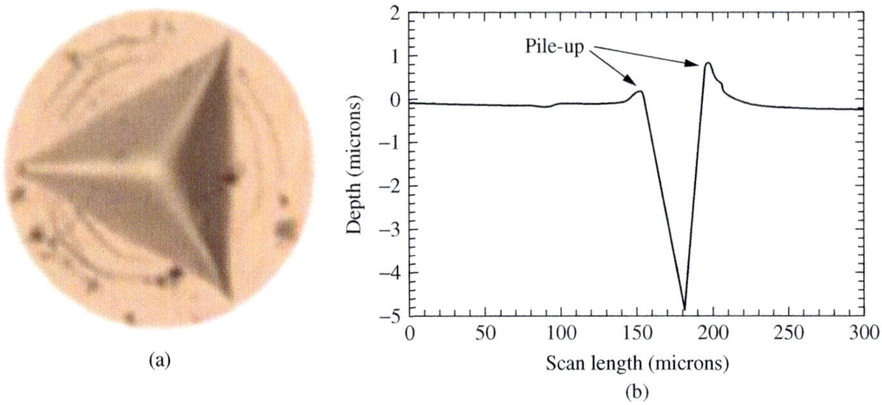

Figure 13.5 (a) Pile-up around indenter in MG. (b) Pile-up profile around indenter. *Source*: Yang 2007 [15]. Reproduced with permisison of AIP Publishing LLC.

with a strain-hardening modulus larger than 0.2. Circular patterns around the indent representing layers of overlapped displaced materials are not considered as cracks. The non-strain hardening features of metallic glasses were further proved by an upward flow of the material (i.e. pile up) along the indenter's face. Metallic glasses seem to deform in an elastic-perfectly plastic manner (Figure 13.5).

The piling-up and sinking-in phenomena are of interest, since they affect the accurate determination of contact areas and, thus, the assessment of hardness and the elastic modulus. For instance, sink-in patterns reduce the contact area, while pile-up patterns increase it, and neglecting them can result in significant errors, when extracting properties from the experimental data. Fischer-Cripps [13] argued that the errors could be as high as 60%. An Oliver and Pharr approach takes into account the sink-in effect but not the pile-up effect; thus, Martin and Troyon [27] pointed out that it can grossly underestimate the true contact area. To determine the actual contact area, different techniques have been suggested. Atomic force microscopy (AFM) was recently employed to determine the actual contact area for Berkovich indenter tips, assuming simplified pile-up geometry. However, this procedure considered a residual imprint, i.e. an accurate continuous measurement of the elastic recovery was not possible. To overcome this limitation, numerical techniques such as a finite-element (FE) method can provide a helpful alternative in development of accurate models for indented materials.

Therefore, in indentation tests on metallic glasses, the pile-up height should be measured, and added to the measured indentation depth to calculate a representative contact area using the standard Oliver–Pharr scheme. This method is usually employed to measure hardness and to characterise the indentation size effect (ISE). Another important factor is the effect of residual stress on morphology of pile-up in indentation as reported in the literature for different types of materials, using both experiments and finite element modelling [28, 29]. The results showed that the amount of free volume in the materials and the stress state beneath the indenter change the pile-up morphology and formation: a large amount of free volume can induce stable and homogeneous flow in nanoindentation at low temperature [30]. As shown in Figure 13.6, there is a confined plastic zone due to compressive stress, leading to extensive pile-up formation. In a tensile state, there was no indication of pile-up formation. It was observed in tests that

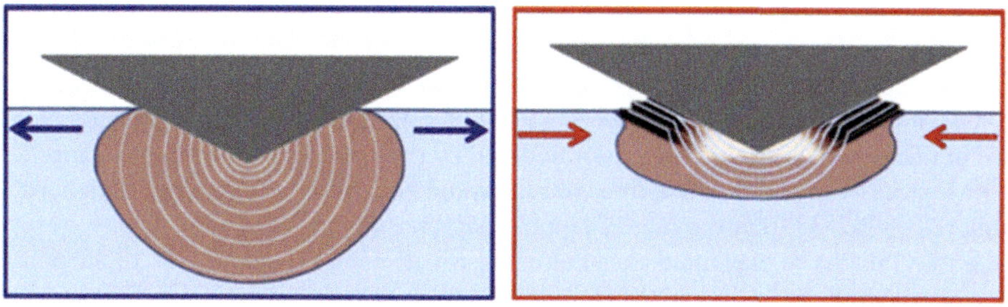

Figure 13.6 Effect of tensile (a) and compressive (b) residual stress on plastic-zone size, pile-up behaviour and shear-band activity. *Source*: Haag 2014 [30]. Reproduced with permission of Elsevier.

activation of shear bands on the shear plane underneath the indenter was increased under tension and decreased under compression.

13.1.2.2 Indentation Size Effect

Homogeneous and isotropic materials are expected to possess unique values of hardness and modulus. However, experimental results show a variation in hardness and modulus with indentation depth. It is now well known that size plays a significant role in the overall mechanical response of material. The apparent variation in the perceived material properties such as hardness and elastic modulus with the depth of indentation h and indenter radius R is known as *indentation size effect* (ISE) [31].

A number of studies argued that ISE depended on the level of strain and the magnitude of strain gradient in the loaded material. The characteristic features of plastic deformation in conventional metals are formation, motion and storage of dislocations. The level of indentation hardness of such materials is observed to increase with a decrease in the indentation depth owing to the nucleation of dislocations in the plastic zone. In conventional metals, the most popular strain-gradient-based ISE mechanism was proposed by Nix and Gao [32], who assumed a density of dislocation together with a Taylor's dislocation strengthening model employing a concept of geometrically necessary dislocations (GNDs). The depth dependence on hardness can be obtained from this model in the following form:

$$\frac{H}{H_0} = \sqrt{1 + \frac{h^*}{h}}, \qquad (13.1)$$

where H is the hardness for the given indentation depth h, H_0 is the hardness in the limit of the infinite depth and h^* is the characteristic length depending on the indenter shape, shear modulus and H_0. Experimental results for cold-worked polycrystalline copper and single-crystal silver showed that ISE is more pronounced in metals with a low value of intrinsic hardness. Hence, little and significant ISEs are expected for soft and hard crystalline materials, respectively. Since the Nix–Gao model is based on dislocation strengthening, it is expected that ISE does not occur in materials with a non-crystalline (amorphous) structure because of the absence of dislocations and strain hardening. However, some researchers reported ISE-like behaviours in Zr-, Pd- and Fe-based metallic glasses [33, 34]. There are different hypothesises described ISE mechanisms in MGs. Firstly, Lam and Chong [35] and, more recently, Yang [36] used the Nix–Gao model to describe ISE in MGs based on the strain-gradient plasticity. The concept of non-crystalline flow defects including shear clusters or excess free volume were used instead of dislocations. An increase in these defects at low indentation depths is mainly responsible for ISE in MGs, as the increase in dislocations is in metals in the Nix–Gao model. Hence, it is claimed that plasticity in metallic glasses is induced by a strain gradient, which should be accommodated by these defects. However, this model has drawbacks as an increase in a free volume would lead to softening rather than hardening in MGs [37, 38] that the model cannot characterise.

Secondly, ISE can be explained based on a strain-softening behaviour [39] and dependency of hardness on in indentation-imposed strain as proposed by Van Steenberge *et al.* [34]. Vincent *et al.* [40] conducted micro-hardness tests on Zr-based metallic glass at four different loads of 1, 2, 3 and 5 N and nano-indentation experiments with a maximum load 80 mN. The hardness values from nanoindentation were higher than those reported from micro-indentation. In addition, a decrease in hardness with increase in the load level was observed. This phenomenon can be described based on [31], where metallic glasses showed strain softening in plastic deformation rather than strain hardening due to shear localisation. In shallow indentations, a plastic zone is quite small to have sufficient STZs and, hence, shear bands operate at limited locations resulting in an increase in hardness. For large penetration depths, a continuous process of accumulation of excess free volume during deformation can induce strain softening, and this softening is the source of ISE. Hence, a relationship between hardness and a strain rate based on the classical metallic flow equations developed by Spaepon [7] and Argon [6] was introduced:

$$H\left(\approx 3\sqrt{3}\,\tau\right) \propto \sinh^{-1}\left(\frac{\alpha\dot{\gamma}}{C_f}\right), \tag{13.2}$$

where τ is the shear stress, α is a constant, $\dot{\gamma}$ is the shear strain rate and C_f is the concentration of flow defects. According to this equation, the strain rate reduces in indentation tests under a constant loading rate, and the high indentation strain rate at low indentation depth can induce an increase in hardness.

However, both stress and strain should not vary significantly in penetration of a sharp indenter according to the classical contact mechanics. Jang and co-workers [41] systematically studied ISE in a Zr-based BMG through nanoindentation with Berkovich indenters with different angles from 35° to 70°. As shown in Figure 13.7, the variation of hardness changed with indentation depth. The sharper indenters produced smaller contact radius and, thus a localised highly-stressed or plastic zone, which led to lower activities of STZs. Lower activation of STZs resulted in higher hardness values, as confirmed in Figure 13.7. The authors showed that the mentioned hypothesis cannot be used to explain the angle effect on hardness as the use of sharp indenters resulted in higher hardness value.

The phenomenon of ISE can also be explained by occurrence of STZs [9]. In this hypothesis, the STZ population controls the indentation size or the volume of indentation-induced elastic/plastic deformation. In low-depth indentation, a highly stressed volume beneath the indenter is too small to generate a sufficient number of STZs; thus, the shear bands are forced to operate in a confined zone, leading to higher hardness values for lower loads. In contrast, in high-depth indentation, a large volume of plastic zone is formed resulting in activating a large population of STZs and shear bands. The STZ hypothesis can be used to describe the angle effect on hardness, as higher hardness is expected for sharp indenters.

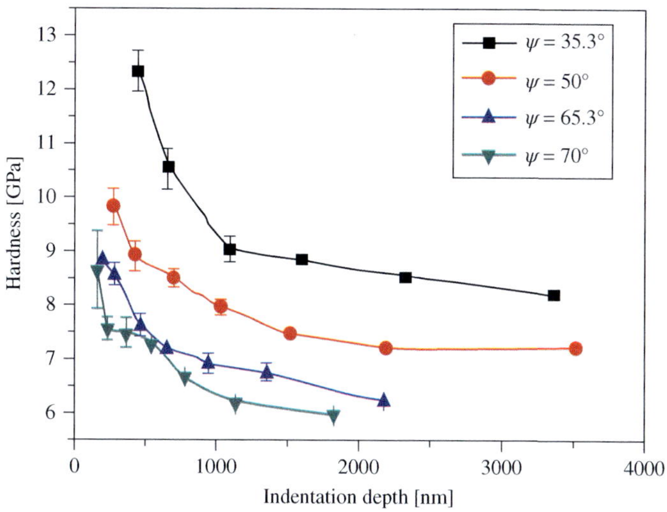

Figure 13.7 Analysis of hardness data according to the Nix–Gao model: plots of hardness *vs.* penetration depth. *Source*: Jang 2011 [41]. Reproduced with permission of Elsevier.

Choi et al. [31] studied the existence of ISE in Zr-based BMGs by employing spherical nanoindentation experiments with varying tip radii (R) of 2.91, 5.75 and 31.5 μm. The load, at which a first pop-in event occurred in Zr-based MGs, was measured for three tip radii and in two different structural states: as-cast and structurally relaxed. The experimental results showed the existence of ISE, with shear yield stress increasing with a decrease in the tip radius. Figure 13.8 shows variations in the average values of τ_{max} with$R$ for both as-cast and annealed specimens, where τ_{max} increases with reducing R for both states. Larger volumes of metallic glass were affected as R grew, increasing the

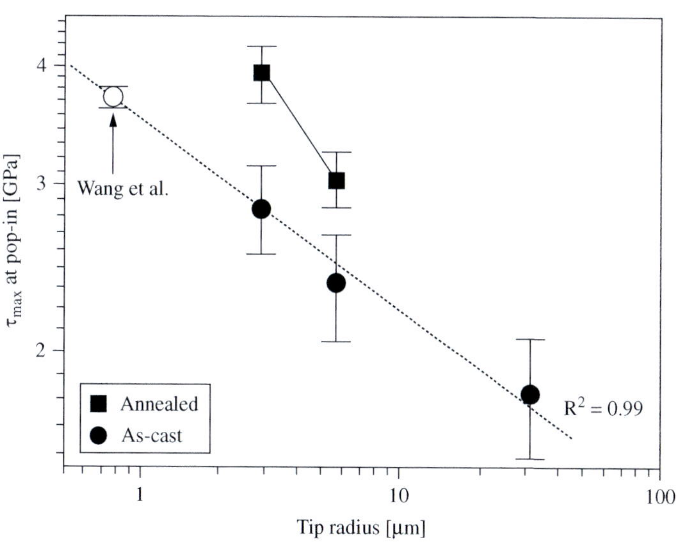

Figure 13.8 Variation in maximum shear stress at first pop-in with tip radius for as-cast and annealed Zr-based metallic glass.

possibility of finding weaker regions in the bulk materials; as a result, lower stress was needed to initiate plasticity. Additionally, structural relaxation decreased availability of potential sites for STZ operation with annihilation of free volume in the metallic glass; thus, the annealed specimen exhibited higher τ_{max} at the first pop-in.

Nachum and Greer [42] studied four types of metallic glasses including two-Pd based and two Zr-based MGs using nanoindentation with a spherical indenter tip radius of 10 μm and collected experimental data from the literature detailing indentation with such tips (180 nm $< R <$ 31.5 μm) to investigate the relation between the indenter's tip radius and indentation pressure at yield. Clear signs of ISE were found as normalised yield pressure increased at the elastic limit when smaller tip radii were used: it was around three times higher than that in the macroscopic regime using the Hertzain equation to calculate yield pressure and stress. The increases in yield pressure and yield strain were independent of glass composition. The experiments demonstrated that yield pressure reached maximum when the tip radius decreased to 10 nm and below due to fewer structural defects. Greer and co-workers [42] showed that the magnitude of the indentation size effect was inversely correlated with yield strain.

Still, existence of ISE is debated. Haung *et al.* [43] studied two types of Ti-based metallic glasses (ribbon-shaped and in a bulk form) to investigate the ISE effect. Ti-based ribbon specimens demonstrated that of lower hardness compared to bulk ones because of a higher amount of free volume. Hardness was observed to decrease, in both ribbon and bulk MGs, with increasing in the indentation depth. However, the perceived ISE disappears when the pile-up effect is accounted for with the Oliver–Pharr analysis.

13.2 Experimental Studies

In this section, Nanotest Platform III used for our nanoindentation experiments is explained in details. The calibration measurements required to ensure correct functioning of the instrument including frame compliance, cross-hair calibration and indenter area function are also discussed. A basic analysis for extraction of mechanical properties, such as hardness and an elastic modulus for a spherical indenter, are also described. A Zr-Cu-based metallic glass with a nominal composition $Zr_{48}Cu_{36}Al_8Ag_8$ (at%) was characterised using the indentation technique. Single-loading and incremental loading-unloading experiments were performed to investigate initial elastic deformation and hardening/softening phenomena, respectively. Additionally, a first pop-in stress was examined by means of the latter method; serration flow behaviour for a varying loading rate and a localised heating mode were also analysed.

13.2.1 Nano Test Platform III Indentation System

The NanoTest Platform III system, manufactured by Micro Materials Wrexham UK, was used for the indentation experiments. Its sample stage can be manipulated using precise DC motors having a displacement resolution of 17.3 nm in the *x*, *y* and *z* directions. The NanoTest system was supplied with a high-resolution microscope and a microscope monitor, assisting inexact positioning of indents in a specimen. The high-resolution microscope had magnifications of 4×, 10×, 20× and 40×. The 'focal plane' is a plane

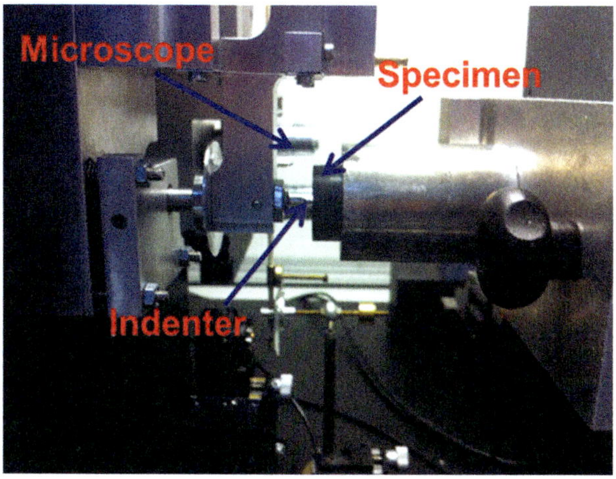

Figure 13.9 Experimental arrangement for indentation test of MG.

parallel to the y-z plane of the sample holder, in which the specimen's surface was in focus at the highest magnification. Whenever the specimen was changed, it was brought to the focal plane before starting an indentation test. The 'measurement plane' was a plane parallel to the y-z plane of the sample holder, in which the surface of the sample was approximately 50 μm away from the tip of the indenter. The experimental equipment used in our study of metallic glass sample is shown in Figure 13.9. The nano head consists of a pendulum with an indenter and provides a load range of 0.1–500 mN for nanoindentation and 1–20 N for micro-indentation with a resolution of 0.1 μN. As can be seen in Figure 13.9, the specimen was mounted on the sample stage, having its surface vertical and parallel to the microscope's viewing field and the indenter tip.

13.2.2 Calibration

A nanoindentation system requires calibration (based on calibration standards) to ensure that the instrument is working correctly. Three calibration measurements are required for most nanoindentation systems including frame compliance, indenter area function and cross-hair, or targeting alignment. It is expected that calibration standards show no pile-up or sink-in behaviour and very little creep. Standards should be economical and easily obtainable in an optically flat or highly polished condition. It is desirable that the standard materials show little or no ISE. Table 13.2 illustrates material properties of used calibration standard materials for a diamond indenter.

Table 13.2 Material properties of commonly used calibration standard material for diamond indenter.

Material	Elastic modulus, E	Poisson's ratio, ν	Hardness	Reduced modulus, E_r
Fused Silica	72 GPa	0.17	8.8 GPa	0.75
Tungsten	412 GPa	0.28	~6.6 GPa	0.75

13.2.2.1 Frame Compliance

A magnitude of depth measured in the indentation process includes the depth of penetration of the indenter into the specimens along with any displacement of the instrument due to deflection [44]. The system's compliance C_f is defined as deflection of the instrument divided by the load. This compliance includes compliance of the loading frame, the indenter shaft and the specimen's mount. The unloading stiffness dp/dh measured in indentation comes from both the responses of the specimens and the instrument. The compliance of the indenter's material 1/S is included in the reduced modulus E_r, where the contact stiffness S is given a rearrangement of:

$$\frac{dh}{dp} = \frac{1}{S} + C_f, \tag{13.3}$$

Calibration of the frame compliance is required to determine an area function, and the standard material should have a low ratio of hardness to elastic modulus. Hence, a large and stiff contact can be easily obtained with a minimum error of the indenter area function. Tungsten is ideal for this as it demonstrates isotropic elasticity under indentation.

13.2.2.2 Cross-hair Calibration

Cross-hair calibration is performed any time when an indenter tip is exchanged in the system. This calibration process involves characterising precisely a movement of the sample stage in a way that the region of the sample at the microscope's cross-hair is re-positioned exactly in front of the indenter. When the cross-hair calibration is completed correctly, it is possible to position imprints at a chosen location with an accuracy of about 1 μm. Hence, the only requirement of the calibration standard for this test is that it readily holds a clear indentation impression. It is necessary that the calibration standard is very hard and stiff for the indenter area function calibration as loads at very low indentation depth are still large enough to be well within the force resolution of the indentation system. Thus, fused silica with superior elastic isotropy is appropriate for this calibration.

13.2.2.3 Indenter Area Function

In indentation testing, a contact area at plastic depth $h_p$is found from geometry. The area for each type of indenter can be measured following Table 13.2. The contact area is considered for the ideal and undamaged indenter. Crystal anisotropy of diamonds indenters can also affect the expected shape of indents. Thus, it is necessary to apply a correction factor to the equations shown in Table 13.3 in order to measure the actual contact area. The correction factor is a ratio of the actual contact area to the real contact area at a depth of h_p.

To calculate the area function, a series of indentation is performed at varying maximum loads on a standards test specimen, which elastic modulus E and Poisson's ratio v are known. Then, the actual contact area A is used to determine elastic properties of the specimen. A typical force-displacement curve, which was used to calculate the elastic moduli and hardness of metallic glasses, is shown in Figure 13.10a.

When a flat specimen is loaded with a spherical indenter, there may be an initial elastic response at low loads followed by elastic–plastic deformation of the specimen's material at higher loads. As shown in Figure 13.10, the maxium penetration depth of a spherical indenter beneath the original specimen's free surface is h_t at full load P_{max}. When the laod is removed, (and assuming no reverse plasticity), the unloading is elastic, and at

Table 13.3 Projected areas, intercept corrections and geometry correction factors for various types of indenters.

Indenter type	Projected area	Semi-angle θ	Effective cone angle α	Intercept factor	Geometry correction factor β
Sphere	$A \approx 2\pi R h_p$	N/A	N/A	0.75	1
Berkovich	$A = 3\sqrt{3}h_p^2\tan^2\theta$	65.3°	70.2996°	0.75	1.034
Vickers	$A = 4h_p^2\tan^2\theta$	68°	70.32°	0.75	1.012
Knoop	$A = 2h_p^2 \tan\theta_1 \tan\theta_2$	$\theta_1 = 86.25°$ $\theta_2 = 65°$	77.64°	0.75	1.012

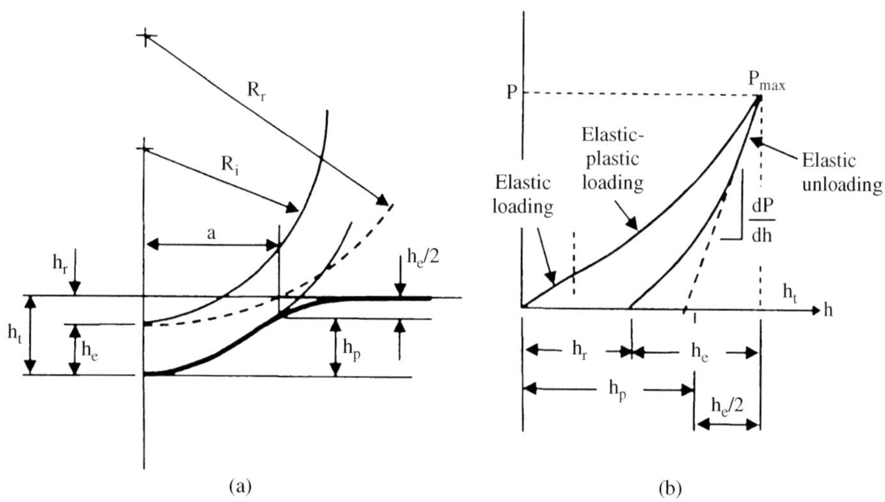

Figure 13.10 (a) Geometry of loading-induce impression of radius R_r with a rigid indenter radius R_i. (b) Load- displacement curve for specimen loaded with spherical indenter showing both loading and unloading responses. *Source*: Fischer-Cripps 2001 [13]. Reproduced with permission of Cambridge University Press.

the complete unloading, there is a residual impression of depth h_r .As can be seen in Figure 13.10, the elastic depth h_e is the unloading distance from h_t to h_r, $h_e = h_t - h_r$; according to the Hertz equation:

$$P = \frac{4}{3}E_r R^{1/2} h_e^{\frac{3}{2}} \tag{13.4}$$

where R is the tip radius and E_r is the reduced modulus depending on the deformation of the material and the diamond tip, determined from:

$$\frac{1}{E_r} = \frac{1-\nu^2}{E} + \frac{1-\nu_i^2}{E_i} \tag{13.5}$$

Here E and E_i are the modulus of the specimen and indenter, respectively, ν and ν_i are the Poisson's ratios of the specimens and indenter, respectively. Calculation of the elastic moduli with Equation 13.5 assumed a spherical diamond tip with the elastic modulus of 1.141 TPa and the Poisson's ratio of 0.07. The well-established method proposed by Oliver and Pharr [16] is used to calculate elastic properties of materials in nanoindentation tests. In this method, the initial portion of the unloading curve is employed to determine the plastic depth h_p. The derivative of the unloading curve $\frac{dP}{dh}$ at the point of initial unloading h_t can be calculated as:

$$S(h_t) = \frac{dP}{dh} = 2E_r R^{\frac{1}{2}} h_e^{\frac{1}{2}} \tag{13.6}$$

dP/dh can be used to determine the contact stiffness S indicating the material's resistance to elastic deformation. By substituting Equation (13.6) into Equation (13.4):

$$P = \frac{2}{3}\frac{dp}{dh}h_e \tag{13.7}$$

Then,

$$h_e = \frac{2}{3}P\frac{dh}{dP} \tag{13.8}$$

Based on the Hertz equations for spherical indenter, the depth of the contact area h_a beneath the specimen free surface is half of the elastic displacement h_e:

$$h_a = \frac{h_e}{2} \tag{13.9}$$

Hence

$$h_a = \left[\frac{3}{4}\right]\frac{P_t}{\frac{dP}{dh}}. \tag{13.10}$$

From 13.10, the radius of the circle of contact can be extracted from geometrical considerations:

$$a \approx \sqrt{2R_i h_p} \tag{13.11}$$

E_r can also be calculated from the slope of the initial unloading:

$$E_r = \frac{dP}{dh}\frac{1}{2a} = \frac{dP}{dh}\frac{\sqrt{\pi}}{\sqrt{A}}\frac{1}{2}. \tag{13.12}$$

where $A = \pi a^2$ is the area of contact. These parameters were computed from the unloading part of the force-displacement curves with Nanoindenter II software, using a well-established method [16].

According to definition, hardness is the mean pressure under the indenter at maximum depth. In the indentation process with the spherical indenter tip, hardness can be calculated from the corresponding P-h curves [13]:

$$H = \frac{P}{2\pi R h_p}. \tag{13.13}$$

Plastic displacement: h_p can be found from:

$$h_p = h_i - \varepsilon \times \frac{P}{S}. \tag{13.14}$$

where h_i is the recorded indenter displacement, ε is the shape function, which is 0.75 for a spherical indenter, and S is the stiffness obtained from the unloading curve of the indentation test.

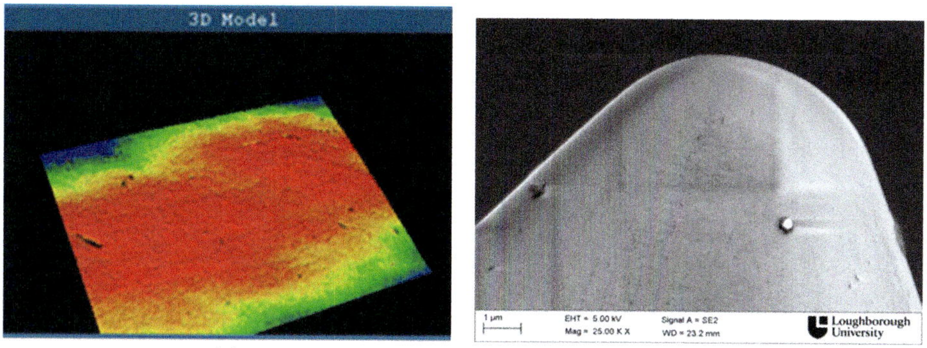

(a) (b)

Figure 13.11 (a) Roughness micrograph of surface of MG specimen showing roughness less than 5 nm. (b) SEM image of indenter tip with nominal tip radius of 5 μm.

13.2.3 Experimental Procedure

A Zr-Cu-based MG alloy $Zr_{48}Cu_{36}Al_8Ag_8$ (at %) was critically evaluated. The alloy was initially produced by arc-melting on a water-cooled copper mould under a high-purity argon atmosphere at IFW Dresden, Institute for Complex Materials, Germany. The ingot was re-melted at least three times to obtain homogeneity. Furthermore, using the alloy ingot rectangular beam-shaped samples with a cross-section of 10 mm × 2 mm and a nominal length of 80 mm were prepared by suction casting into a copper mould. The amorphous nature of the supplied beam samples was initially investigated using X-ray diffraction (XRD), by Bruker D-8 Diffractometer, Bruker AXS GmbH, Karlsruhe, Germany using Cu-kα radiation with wave length of 1.5406 Å. The entire sample were analysed at a step size of 0.02° in 2θ and a step time of 1 s; XRD patterns were recorded over for 2θ in the range of 20–90°. No traces of crystalline phase were found in the XRD spectrum, with only broad amorphous peaks present.

The specimens for indentation were cut from the supplied MG beams with a low-speed diamond disc (1000 rpm) under constant water irrigation to minimise the undesired mineral formation on the specimen's surface. Prior to indentation, the specimens were polished to a mirror finish with average roughness of around 5 nm (Figure 13.11a). All the tests on MG specimens to measure the elastic modulus and hardness were performed at room temperature (23°C). Both load- and displacement-controlled nanoindentation tests were performed. The holding (dwell) time was settled as 60 s to allow negligible creep displacement. A thermal shift correction was used so that the absolute values of a thermal drift rate were usually well below 0.05 nm/s. In the present study to capture the first pop-in, a spherical indenter with a radius of 5 μm represented a conical shape with a spherical tip (Figure 13.11b). Obviously, only the indenter tip was used to penetrate the specimen surface in indentation tests.

13.2.4 Results and Discussion

To investigate initial elastic deformation, a maximum load of 10 mN was applied with a loading rate of 0.1 mN/s to the specimens in the nanoindentation test. The inset image of Figure 13.12 shows that there was no disparity between loading and unloading at a maximum load of 3 mN, indicating an overall elastic response of the material under

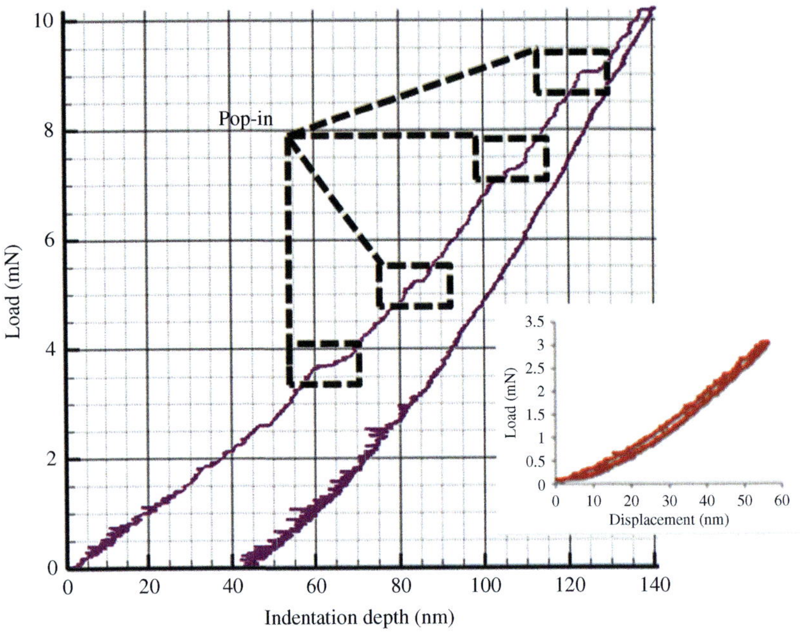

Figure 13.12 Typical load-displacement response of $Zr_{48}Cu_{36}Al_8Ag_8$ at loading rate 0.1 mN/s showing initial deformation.

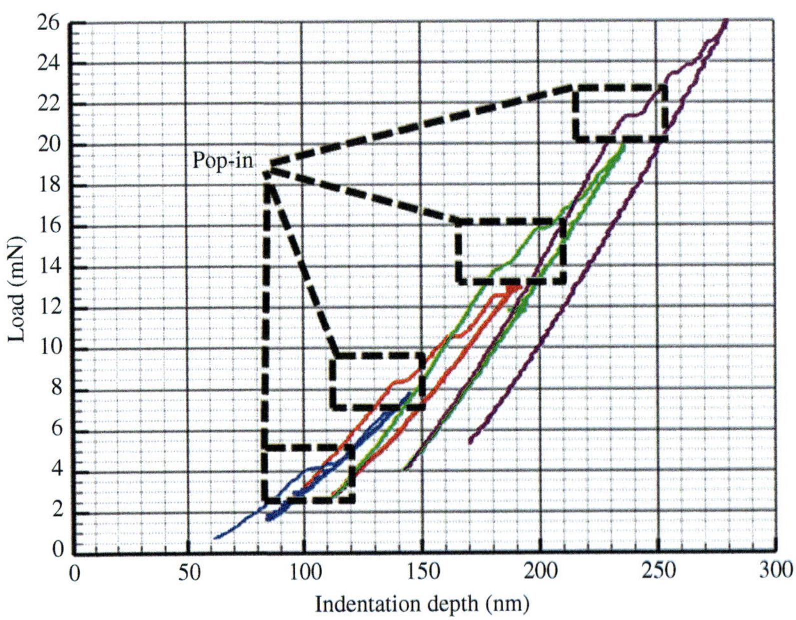

Figure 13.13 Indentation load-depth plots for $Zr_{48}Cu_{36}Al_8Ag_8$ under incremental loading-unloading nanoindentation at load rate of 0.1 mN/s.

nanoindentation. Presented load-displacement plots (Figures 13.12–13.14) are a typical example for several tests, showing closely correlated pop-in events. A significant plastic event was observed at around 4 mN (see Figure 13.12), which was considered to be the first pop-in event. It is associated with initiation and propagation of individual shear bands underneath the indenter tip. Below 4 mN, the loading and unloading responses are similar, with no obvious disparity between the two, implying pure elastic deformation. As can be seen in Figure 13.12, other pop-in events were found to occur at ~6 mN, ~8 mN and ~9 mN.

The incremental loading-unloading experiments were performed to investigate hardening or softening effect on yield stress, hardness or elastic modulus as this technique allows for measurements of hardness and modulus at different indentation depths for the same location in the specimen. First, such experiments were performed using a spherical tip with 5 μm diameter in the load range from 7 mN to 27 mN with the same loading rate –0.1 mN/s. Unloading was performed with the rate of 1 mN/s down to a prescribed limit of 20% of the load peak in the previous cycle. As shown in Figure 13.13, pop-in events in each reloading stage took place at higher levels than those in the previous cycle, demonstrating an increase in the first pop-in from ~4 mN to ~21 mN. As mentioned earlier, the pop-in events depend on composition and free-volume content of MGs as well as on parameters of the experimental procedure such as a tip shape and applied loading rates. The maximum shear stress (τ_{max}) at the first-pop-in corresponds to the critical shear strength of the onset of plasticity in the indented material.

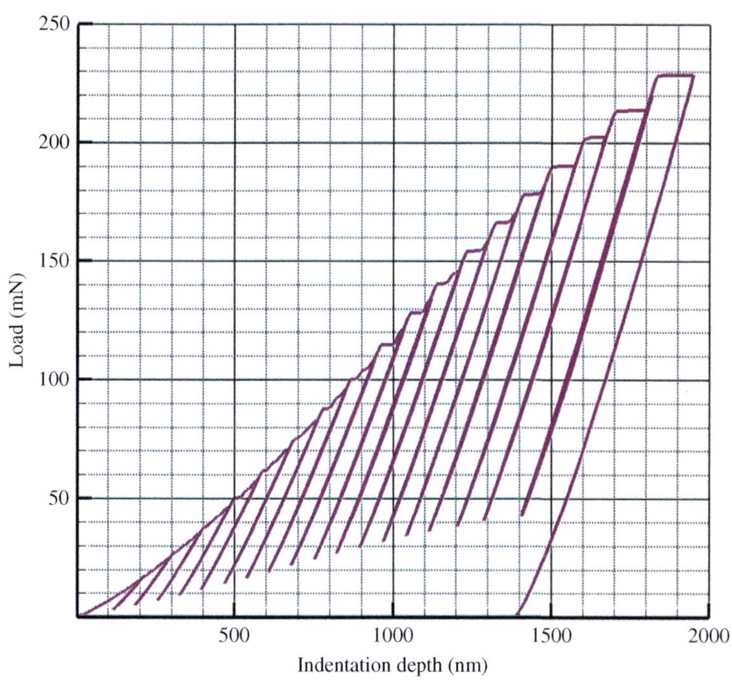

Figure 13.14 Indentation load-displacement curve of Zr-based MG for incremental loading-unloading at loading rate of 2 mN/s.

In spherical indentation, τ_{max} occurs at a distance of approximately half the contact radius, directly below the rotational axis of the contact and can be calculated as:

$$\tau_{max} = 0.31\left(\frac{6E_r^{\ 2}}{\pi^3 R^2}P_I\right)^{1/3} \tag{13.15}$$

where P_I is the load at the first pop-in, R is the indenter radius and E_r is the reduced modulus, which was around 94 GPa. There was a significant rise (some 50%) in the average value of τ_{max} as first pop-in load increased from ~4 mN to ~15 mN. Initially, the maximum shear stress was approximately 2 GPa for in the former load, increasing to around 3.45 GPa at the latter one. This is the evidence of the hardening effect caused by multiple loading-unloading experiments. The reason for this phenomenon is explained below.

Subsequently, multiple unloading-reloading experiments were performed using a loading rate of 2 mN/s to investigate the variation of hardness and elastic modulus with depth in MG specimens. The maximum applied indentation depth ranged between 200 nm and 2000 nm and 20 partial unloads down to 20% of the peak load at each step were applied in these tests. As discussed previously, a large plastic zone is formed under the indenter tip in indentation [35, 36]. The plastic zone contains a high density of shear bands, which is ideal for investigation of deformation-induced hardening and softening effects [36]. Our results show the dependence of hardness on the penetration depth indicating a work-hardening phenomenon in metallic glass, especially at nano-scale (see Figure 13.14). It is clear from Figure 13.14 that the onset of yielding upon each reloading always took place at a higher load than in the previous cycle, apparently suggesting the hardening effect. In addition, the hardness increased from 3.7 GPa to 5.7 GPa with a sharp enhancement in the first 10 cycles and a slight rise in the subsequent 10 cycles. This can be described employing ideas of a free-volume evolution proposed originally by Spaepen [7]. According to this model, the change of free volume (which is a thermally activated process) is a competition between the free-volume accumulation in a stress-biased system and the free-volume annihilation by local atom rearrangements. During the unloading process, the latter is the dominant process thanks to a sudden reduction of the applied stress. This results in relaxation of the free volume, leading to an immediate arrest of propagating shear bands and a significant temperature decrease inside shear bands because of their arrest. As a result, hardness increases due to decreasing temperature and free volume [45]. At the next reloading step, previously generated shear bands remains arrested resulting in an increase in hardness. However, with the onset of initiation and propagation of multiple shear bands, the increased free volume in the propagating shear bands causes a reduction of hardness, known as the *recovery phenomenon*. The measured hardness is a combined response of propagating and arrested shear bands as well as the undeformed region in the material [14].

To study rate sensitivity of $Zr_{48}Cu_{36}Al_8Ag_8$ metallic glass, nanoindentation tests were carried out at various loading rates (0.5 mN/s and 2 mN/s) with the maximum load of 100 mN (Figure 13.15). A serrated flow was observed at the lower loading rate that weakened and eventually disappeared with an increasing strain rate. This phenomenon suggests that deformation of the Zr-Cu-based metallic glass depends on the strain rate. In addition, as can be seen in Figure 13.15, there is no evidence of creep. Schuh *et al.* [8, 9] also confirmed the absence of pop-in events at high loading rates. The elastic modulus E and hardness H were obtained from the corresponding P-h curves based on the Oliver

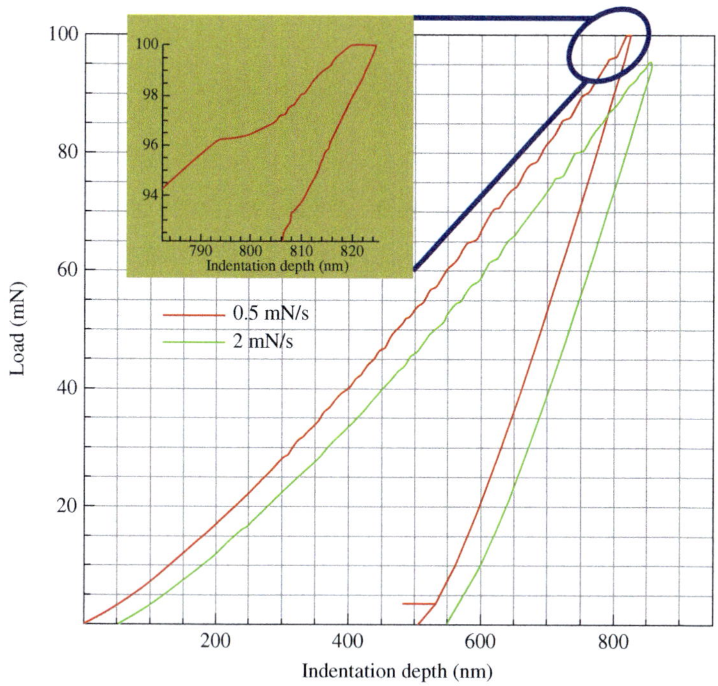

Figure 13.15 Load-displacement curves for two strain rates. The inset is an enlarged portion showing no creep behaviour.

Table 13.4 Characteristics value at different strain rates with maximum depth control of 900 nm under nanoindentation experiments.

Load rate / mN/s	Elastic modulus (*E*)/GPa	Hardness (*H*)/GPa	Maximum load/mN	Maximum depth/nm
0.5	95.84 ± 1.25	5.58 ± 0.22	100	824
2	92.00 ± 2.75	5.29 ± 0.35	96	808

and Pharr method [46] and are summarized in Table 13.4. Apparently, both the hardness and modulus values in the metallic glass are higher at the lower loading rate. The magnitude of the loading rate and diffuse relaxation of the atomic structure after STZ initiation influence the severity of plastic flow localisation and the shear bands nucleation rate [47].

The diffuse relaxation time is somewhat higher at low loading rates, reducing the free volume content thus making MG denser and, hence, requiring higher stresses for STZ activation.

A constant loading rate can be converted into a representative strain rate using the following equation:

$$\dot{\varepsilon} = \frac{\dot{P}}{2P}. \tag{13.16}$$

This is a more suitable parameter to describe inhomogeneous deformation according to the classic free volume theory [46].

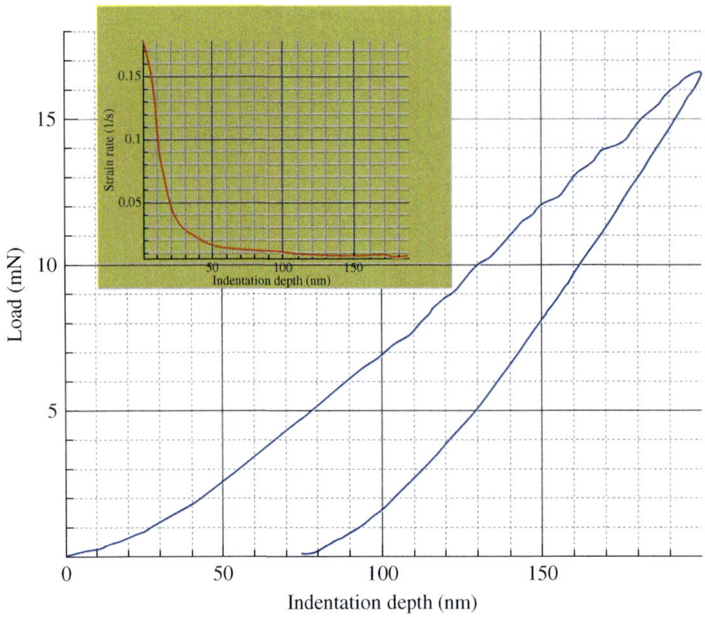

Figure 13.16 Representative P-h curve for loading rate of 2 mN/s demonstrating discrete serrated flow (The inset shows a conversion from the loading rate to the strain rate.)

Figure 13.16 presents a P-h curve obtained with the loading rate of 2 mN/s for a termination load of 16 mN, with the inset presenting the corresponding conversion from the loading rate to the strain rate. The discrete serrated flow can be observed for indentation depths in excess of 50 nm, where the strain rate is below 0.02 s^{-1}. In other words, lower strain rates promote more pronounced serrations, and high rates supress the serrated flow. Hence, the pop-in events for this metallic glass, demonstrated as the stepped P-h curve show significantly strong rate sensitivity. The serration flow behaviour can be analysed using the notions of free volume and localised heating mode.

Several researchers showed that temperature rises (from 0.1 K to ~ 1000 K) occur inside shear bands. Such temperature rises resulted in appearance of droplets on fracture surfaces. Therefore, it is important to study the mechanism of shear-band softening to understand whether the temperature rise or/and free volume increase leads was the underpinning mechanism. Both mechanisms cause a lower viscosity of metallic glasses resulting in shear-band propagation and formation. Eshelby and Pratt [48] proposed the following equation to predict a temperature rise inside a shear band:

$$\Delta T = \frac{\tau \dot{\gamma}_b h}{2\pi k}\sqrt{\frac{\pi k \Delta \gamma}{C_p \dot{\gamma}_b}}. \tag{13.17}$$

Here, shear yield stress τ is 925 MPa for the studied MG, which is half of compression yield strength (1850 MPa), assuming an angle of 45° to the loading axis. The shear strain rate $\dot{\gamma}_b$ inside the shear band is estimated to be 10^3 s^{-1}. The shear band thickness$h$, thermal conductivity k and specific heat capacityC_p are ~ 10 nm, 20 W(mK) and 0.33 J/(gK), respectively. The average displacement burst Δh for each serration is 2 nm. Plastic deformation was considered to proceed in the plane of the shear band; the shear

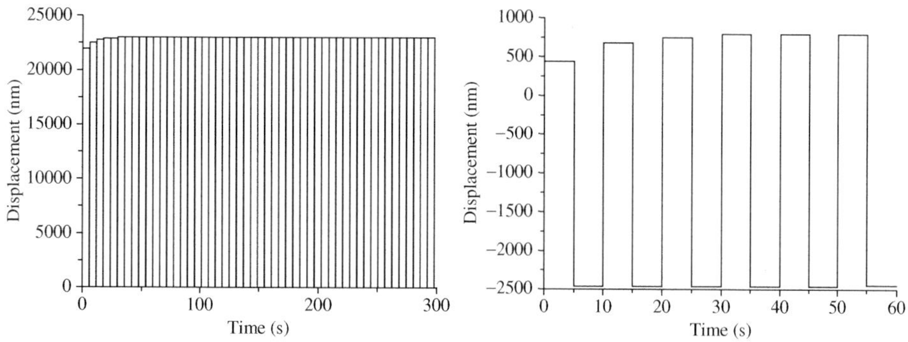

Figure 13.17 Typical load-displacement plots for Zr-based MG at loading rate of 0.1 mN/s (a) pure elastic deformation; (b) initial plastic deformation.

strain $\Delta\gamma = \Delta h/h$ can be calculated as 0.20. Hence, the temperature rise is around 0.04 K after substituting the above data into Equation 13.17, which is similar to experimental measurements of the local temperature rise under quasi-static loading [49]. It was shown that the temperature rise is negligible at low strain rates; therefore we may conclude that the onset of serration is controlled by the free-volume creation and annihilation leading to formation of shear bands. Under high-strain-rate deformation the temperature rise may occur and affect shear bands propagation.

To study high strain rates, an impact regime of nanoindentation was implemented employing a pendulum-based impact test with the nano-test system using a diamond spherical probe of radius $R = 5$ μm. The operating principles of the experimental set are outlined in [10]. This regime can be used for analysis of low cycling fatigue, work hardening and dynamic hardness. In this experiment, the pendulum is moved away from the specimen by a known distance and then released to produce a single impact. Successive impacts can be performed at a single point until failure occurs. In most cases, an initial period is characterised by damage generation, in which cracks develop and expand, but in which no appreciable increase in penetration depth in observed. Figures 13.17a and b illustrate that there is an evidence of strain hardening as there is a significant increase in displacement (penetration depth) in the first 30 seconds, where the cracks coalesce, and a slight enhancement afterwards. Additionally, these results indicate that the Zr-Cu-based MG was an elastic-perfectly plastic material since the material is linear-elastic and, as the load increases to the yield point, its stress remained constant and strain increased.

13.3 Conclusions

This chapter has reviewed the theory behind the analysis techniques used for extracting magnitudes of the elastic modulus and hardness from nanoindentation test data for metallic glasses using a spherical tip. The calibration measurements required to ensure correct performance of the instrument including frame compliance, cross-hair calibration and indenter area function were also discussed.

An instrumented nanoindentation setup was utilised to reveal a mechanical response of $Zr_{48}Cu_{36}Al_8Ag_8$ metallic glass. An increasing yield strength of the Zr-Cu-based metallic glass was investigated by recourse to spherical indentation experiments employing the incremental loading technique. The load at the first pop-in was converted into the maximum shear stress underneath the indenter with the help of Hertz's spherical contact mechanics. The experimental results showed the hardening behaviour, with the first pop-in load increasing with each load increment. The incremental loading technique was also used to investigate the dependence of hardness on penetration depth, and this similarly indicated that the work hardening phenomena in the metallic glass, especially at nanoscale.

A discrete serrated flow was observed at low strain rates ($0.02\ s^{-1}$); however, high rates suppressed this type of flow. As a result, the pop-in events for this metallic glass manifested as a stepped P-h curve showed significantly strong rate sensitivity. The serration flow behaviour was analysed using the concept of free volume and localised heating. It was found that localised shear deformation was dominated by an increase in the free volume. The temperature rise, calculated as 0.04 K for the low strain rate, had a negligible effect on the deformation mechanism.

References

1 Johnson WL. Thermodynamic and kinetic aspects of the crystal to glass transformation in metallic materials. Progress in Materials Science 30(2), 81–134; 1986.
2 Yeh X, Samwer K, Johnson W. Formation of an amorphous metallic hydride by reaction of hydrogen with crystalline intermetallic compounds-A new method of synthesizing metallic glasses. Applied Physics Letters 42(3), 242–243; 1983.
3 van Diepen A, Buschow K. Hydrogen absorption in $CeFe_2$ and $ThFe_3$. Solid State Communications 22(2), 113–115; 1977.
4 Aning A, Wang Z, Courtney T. Tungsten solution kinetics and amorphization of nickel in mechanically alloyed Ni-W alloys. Acta Materialia 41(1), 165–174; 1993.
5 Schwarz R, Johnson W. Formation of an amorphous alloy by solid-state reaction of the pure polycrystalline metals. Physical Review Letters 51(5), 415; 1983.
6 Argon A. Plastic deformation in metallic glasses. Acta Materialia 27(1), 47–58; 1979.
7 Spaepen F. A microscopic mechanism for steady state inhomogeneous flow in metallic glasses. Acta Materialia 25(4), 407–415; 1977.
8 Schuh C, Nieh T. A survey of instrumented indentation studies on metallic glasses. Journal of Materials Research 19(1), 46–57; 2004.
9 Schuh CA, Hufnagel TC, Ramamurty U. Mechanical behavior of amorphous alloys. Acta Materialia 55(12), 4067–4109; 2007.
10 Gouldstone A, Chollacoop N, Dao M. *et al.* Indentation across size scales and disciplines: Recent developments in experimentation and modeling. Acta Materialia 55(12), 4015–4039; 2007.
11 Guo W. On the influence of indenter tip geometry on the identification of material parameters in indentation testing. Acta Materialia 34(12), 3015–3039; 2010.
12 Cheng Y, Cheng C. Scaling, dimensional analysis, and indentation measurements. Materials Science and Engineering 44(4), 91–149; 2004.
13 Fischer-Cripps A. Use of combined elastic modulus in the analysis of depth-sensing indentation data. Journal of Materials Research 16(11), 3050–3052; 2001.

14 Schuh C, Nieh T. A nanoindentation study of serrated flow in bulk metallic glasses. Acta Materialia 51(1), 87–99; 2003.
15 Yang B, Wadsworth J, Nieh T. Thermal activation in Au-based bulk metallic glass characterized by high-temperature nanoindentation. Applied Physics Letters 90(6), 061911; 2007.
16 Eswar Prasad K, Ramamurty U. Effect of temperature on the plastic zone size and the shear band density in a bulk metallic glass. Materials Science and Engineering: A, 535, 48–52; 2012.
17 Wei B, Zhang T, Li W *et al.* Indentation creep behavior in Ce-based bulk metallic glasses at room temperature. Materials Transactions 46(12), 2959–2962; 2005.
18 Zhang L, Wei B, Xing D *et al.* The characterization of plastic deformation in Ce-based bulk metallic glasses. Intermetallics 15(5), 791–795; 2007.
19 Schuster BE, Wei Q, Ervin MH *et al.* Bulk and microscale compressive properties of a Pd-based metallic glass. Scripta Materialia 57(6), 517–520; 2007.
20 Shan Z, Li J, Cheng Y *et al.* Plastic flow and failure resistance of metallic glass: Insight from in situ compression of nanopillars. Physical Review:B, 77(15), 155419; 2008.
21 Lai Y, Lee C, Cheng Y *et al.* Bulk and microscale compressive behaviour of a Zr-based metallic glass. Scripta Materialia 58(10), 890–893; 2008.
22 Greer AL, Cheng YQ, Ma E. Shear bands in metallic glasses. Materials Science and Engineering R: Reports 74(4), 71–132; 2013.
23 Greer JR, de Hosson JTM. Plasticity in small-sized metallic systems: Intrinsic versus extrinsic size effect. Progress in Materials Science 56(6), 654–724; 2011.
24 Abu Al-Rub RK. Prediction of micro and nanoindentation size effect from conical or pyramidal indentation. Mechanics of Materials 39(8), 787–802; 2007.
25 Taljat B, Pharr G. Development of pile-up during spherical indentation of elastic–plastic solids. International Journal of Solids and Structures 41(14), 3891–3904; 2004.
26 Alcala J, Barone A, Anglada M. The influence of plastic hardening on surface deformation modes around Vickers and spherical indents. Acta Materialia 48(13), 3451–3464; 2000.
27 Martin M, Troyon M. Fundamental relations used in nanoindentation: Critical examination based on experimental measurements. Journal of Materials Research 17(09), 2227–2234; 2002.
28 Keryvin V. Indentation of bulk metallic glasses: Relationships between shear bands observed around the prints and hardness. Acta Materialia 55(8), 2565–2578; 2007.
29 Durst K. Göken M, Pharr GM. Finite element simulation of spherical indentation in the elastic-plastic transition. Zeitschrift für Metallkunde 93(9), 857–861; 2002.
30 Haag F, Beitelschmidt D, Eckert J, Durst K. Influences of residual stresses on the serrated flow in bulk metallic glass under elastostatic four-point bending : A nanoindentation and atomic force microscopy study. Acta Materialia 70, 188–197; 2014.
31 Choi I, Zhao Y, Kim Y *et al.* Indentation size effect and shear transformation zone size in a bulk metallic glass in two different structural states. Acta Materialia 60(19), 6862–6868; 2012.
32 Nix WD, Gao H. Indentation size effects in crystalline materials: A law for strain gradient plasticity. Journal of the Mechanics and Physics of Solids 46(3), 411–425; 1998.

33 Concustell A, Sort J, Alcalá G *et al.* Plastic deformation and mechanical softening of Pd40Cu30Ni10P20 bulk metallic glass during nanoindentation. Journal of Materials Research 20(10), 2719–2725; 2005.
34 van Steenberge N, Sort J, Concustell A *et al.* Dynamic softening and indentation size effect in a Zr-based bulk glass-forming alloy. Scripta Materialia 56(7), 605–608; 2007.
35 Lam DC, Chong AC. Model and experiments on strain gradient hardening in metallic glass. Materials Science and Engineering: A, 318(1), 313–319; 2001.
36 Yang B, Riester L, Nieh T. Strain hardening and recovery in a bulk metallic glass under nanoindentation. Scripta Materialia 54(7), 1277–1280; 2006.
37 Bei H, Xie S, George EP. Softening caused by profuse shear banding in a bulk metallic glass. Physical Review Letters 96(10), 105503; (2006).
38 Yoo B, Park K, Lee J *et al.* Role of free volume in strain softening of as-cast and annealed bulk metallic glass. Journal of Materials Science 24(04), 1405–1416; 2009.
39 Jang J, Yoo B, Kim Y *et al.* Indentation size effect in bulk metallic glass. Scripta Materialia 64(8), 753–756; 2011.
40 Vincent S, Murty B, Kramer M, J. Bhatt. Micro and nano indentation studies on $Zr_{60}Cu_{10}Al_{15}Ni_{15}$ bulk metallic glass. Materials Design 65, 98–103; 2015.
41 Jang J, Yoo B, Kim Y *et al.* Indentation size effect in bulk metallic glass. Scripta Materialia 64(8), 753–756; 2011.
42 Nachum S, Greer A. Indentation size effect in metallic glasses: Mean pressure at the initiation of plastic flow. Journal of Alloys and Compounds. 615, S98–S101; 2014..
43 Huang Y, Sun Y, Shen J. Tuning the mechanical performance of a Ti-based bulk metallic glass by pre-deformation. Intermetallics 18(11), 2044–2050; 2010.
44 Fischer-Cripps A. A review of analysis methods for sub-micron indentation testing. Vacuum 58(4), 569–585; 2000.
45 Yang B, Morrison ML, Liaw PK *et al.* Dynamic evolution of nanoscale shear bands in a bulk-metallic glass. Applied Physics Letters 86(14), 141904; 2005.
46 Cheng L, Jiao Z, Ma S *et al.* Serrated flow behaviors of a Zr-based bulk metallic glass by nanoindentation. Journal of Applied Physics 115(8), 084907; 2014.
47 Bhattacharyya A, Singh G, Prasad KE *et al.* On the strain rate sensitivity of plastic flow in metallic glasses. Materials Science and Engineering: A, 625, 245–251; 2015.
48 Eshelby J, Pratt P. Note on the heating effect of moving dislocations. Acta Metallurgica 4(5), 560–562; 1956.
49 Flores KM, Dauskardt RH. Enhanced toughness due to stable crack tip damage zones in bulk metallic glass. Scripta Materialia 41(9), 937–943; 1999.
50 C.J. Lee, J.C. Huang, T.G. Nieh. Sample size effect and microcompression of Mg65Cu25Gd10 metallic glass. Applied Physics Letters, 91(16): 161913; 2007.
51 C.A. Volkert, A. Donohue, F. Spaepen. Effect of sample size on deformation in amorphous metals. Journal of Applied Physics 103(8) 083539; 2008.
52 B.E. Schuster, Q. Wei, T.C. Hufnagel, K.T. Ramesh. Size-independent strength and deformation mode in compression of a Pd-based metallic glass. Acta Materialia, 56(18): 5091–100; 2008.
53 A. Dubach, R. Raghavan, J.F. Loffler, *et al.* Micropillar compression studies on a bulk metallic glass in different structural states. Scripta Materialia, 60(7): 567–570; 2009.
54 D. Jang, J. R. Greer. Transition from a strong-yet-brittle to a stronger-and-ductile state by size reduction of metallic glasses. Nature Materials, 9: 215–219; 2010.

55 A. Bharathula, S.-W. Lee, W.J. Wright, K.M. Flores. Compression testing of metallic glass at small length scales: effects on deformation mode and stability. Acta Materialia, 58(17): 5789–5796; 2010.
56 J. T. De Hosson. Advances in transmission electron microscopy: In situ straining and in situ compression experiments on metallic glasses. Microscopy Research and Technique, 72(3): 250–260; 2009.

Part II

14

Molecular Dynamics Modeling of Nanoindentation

C.J. Ruestes[1,2], *E.M. Bringa*[1,2], *Y. Gao*[3] *and H.M. Urbassek*[3]

[1] *Facultad de Ciencias Exactas y Naturales, Universidad Nacional de Cuyo, Mendoza, Argentina*
[2] *CONICET, Mendoza, Argentina*
[3] *Physics Department and Research Center OPTIMAS, University Kaiserslautern, Germany*

14.1 Introduction

Indentation is an experimental technique commonly used to gather information about the elastic modulus and hardness of materials [1, 2]. At the submicron scale, nanoindentation has provided insights into a broad range of material properties: Weppelmann and co-workers [3] used the technique to study indentation cracking of brittle thin films on brittle substrates; Volinsky, Vella and Gerberich [4] studied fracture toughness, adhesion and mechanical properties of dielectric thin films by nanoindentation; Yang, Riesler and Nieh [5] focused on strain hardening and recovery in a bulk metallic glass under nanoindentation, Bendavid, Martin and Takikawa [6] used the technique to evaluate phase transformations on the deposition and modification of titanium dioxide thin films produced by filtered arc deposition; Sargent and Ashby [7] studied creep while Veprek [8] focused on superhard materials; Lu, Mai and Shen [9] offered a review on the origin of superhardness in nanocomposite coatings; Kumar, Van Swygenhoven and Suresh [10] included the technique in their thorough review of the mechanical behavior of nanocrystalline metals and alloys; the technique is even applicable to biomaterials, where Cuy and co-workers [11] used nanoindentation to map the mechanical properties of human enamel. The combination of the modern experimental testing method with the Oliver-Pharr [12] analysis has led to a widespread utilization of nanoindentation testing.

In the quest to understand deformation mechanisms at the atomic scale during nanoindentation, atomistic simulations have emerged as a fantastic tool to 'probe' materials [13]. Since the landmark contribution by Kelchner *et al.* [14], molecular dynamics (MD) simulations have been extensively applied to study plasticity mechanisms during indentation processes. Molecular dynamic simulations allow to track the indentation process and offer the advantage of obtaining an atomistic picture of the generation, reaction and annihilation of dislocations, twinning, extension of the plastic zone generated, etc. Such simulations have nowadays been performed for many materials, in particular f.c.c. and b.c.c. metals [15–38]. Contributions to the study of h.c.p. metals are, however, still rare [39]. Also simulation studies of non-metals are not

Applied Nanoindentation in Advanced Materials, First Edition. Edited by Atul Tiwari and Sridhar Natarajan.

frequent; we mention the indentation studies of covalently bound materials such as Si [40–42], of the ionic compound CaF_2 [43], of ceramics like SiC [44–48]. A relevant exception is 2D graphene, for which several studies are available [49–51].

MD has played an important role in identifying the processes of dislocation nucleation and slip allowing to correlate experimentally measured features in the load-displacement curves with their atomistic origin [52]. On the one hand, MD simulations offer the possibility to model surfaces free of ambient water layers, oxides, hydrocarbons, etc. On the other hand, the mismatch in length and time scales is a serious barrier to overcome in the task of relating experimental results and simulations; MD can simulate samples of the order of tens of nm both in size and depth while traditional nanoindentation is used to probe samples with lateral sizes in the order of micrometers and hundreds of nm in depth. A similar mismatch can be found for timescales where an MD simulation would take a few nanoseconds for a test that a commercial nanoindenter would perform in a much longer time of seconds.

As Paul, Oliver and Grütter pointed out recently [53], matching experimental length scales to MD simulations is a challenge since it requires proper characterization of nanometer-size indenters, the ability to measure nanonewton forces, and the capability to image residual impressions with nanometer resolution. Their article focuses on recent experimental progress in closing this length scale gap, particularly in the application of scanning probe microscopy and field ion microscopy. In situ electron microscopes equipped with picoindenters in their sample holders have the potential to bridge this length scale gap as well [54–60]. In addition, advances in electron backscatter diffraction (EBSD) allow measurements of geometrically necessary dislocation densities which can also be compared to atomistic results [61].

We present in section 14.2 the basics of the method of MD simulation and explain the issues to consider when applying it to nanoindentation. A reliable simulation requires the use of realistic and validated interatomic interaction potentials (section 14.3). A short section on the elastic regime (section 14.4) is followed by the central sections 14.5 and 14.6 which describe the onset of plasticity and its evolution, respectively. An outlook summarizing the role of MD simulation in the context of other modeling schemes for nanoindentation concludes this chapter.

14.2 Methods

Atomistic simulations can be performed using a standard MD package such as the open-source LAMMPS code [14, 62] or DL Poly [63, 64]. The chosen integration timestep for the simulations should be of the order of 0.1–3 fs. Figure 14.1 shows a schematic representation of the MD simulation system and a loading profile. It illustrates the configuration of the nanoindenter and the substrate. The suggested simulation consists of three parts: (i) nanoindentation, during which the indenter is pushed in the $-z$ direction into the substrate; (ii) a holding stage in which the indenter position remains fixed; and (iii) retraction of the indenter: the indenter is moved out of the substrate until no force is exerted on the substrate.

The nanoindenter can have one of several shapes: (i) spherical; (ii) conical; (iii) cylindrical; (iv) pyramidal. The preferred shape for MD studies is a spherical tip, the reason being that almost all experimental indenters end up with a smoothing radius at the very tip of them. Even though the pyramidal Berkovich indenter is the most common

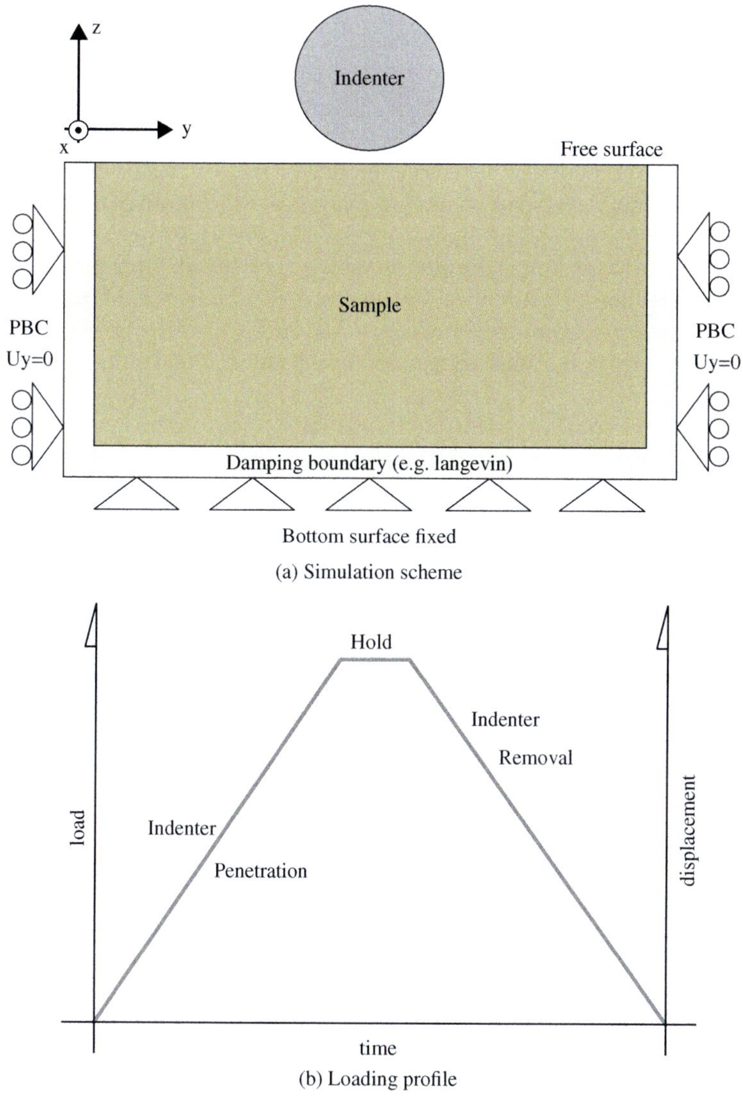

Figure 14.1 Setup of the simulation system (a) and loading profile (b). The substrate has thermostatting and rigid zones at its boundaries.

indenter used in nanoindentation experiments because it maintains self similarity, such an indenter has a rounded tip with a radius of up to 150 nm. Cube corner nanoindentation tips can have a radius as small as 40 nm. Therefore, it can be expected that the spherical tip used in MD studies renders the same plasticity mechanisms produced by a Berkovich tip or a cube corner tip prior to significant effects produced by the pyramidal planes. A thorough discussion on indentation tips used in MD simulations is presented in section 14.2.1.

In principle, the substrate can be represented by any material provided the interaction between atoms can be accurately modeled by an interatomic potential. As examples of b.c.c. metals, Fe can be described by the Mendelev potential [65] and Ta can be described by Dai *et al.* potential [66]. As examples of f.c.c. metals, Al can be described using the potential developed by Mendelev *et al.* [67], while Cu can be modeled using the Mishin *et al.* potential [68]. The sample to be indented can be single crystalline or polycrystalline. Typical state-of-the-art sample dimensions are 20–50 nm in each direction. The simulation box must be large enough to avoid the generated defects (e.g. dislocations) to reach the boundaries of the simulation box. A few atom layers at the bottom of the sample are fixed in order to suppress any rigid- body movement of the substrate. Typically, another few layers at the bottom, as well as the outermost layers of the substrate in lateral directions are kept at a fixed temperature by a thermostat (e.g. a Langevin thermostat with a viscosity set to maximize damping of any elastic waves reaching the boundary). In addition, periodic boundary conditions are applied in the lateral x and y directions. The top surface is free.

Before performing nanoindentation simulations, substrates have to be relaxed until all stress components have reached values $< 10^{-5}$ GPa [27]. The center of the indenter is placed at a certain height above the substrate surface such that all indenter atoms are outside the cut-off range of the interaction potential between the tip and the sample. After that nanoindentation proceeds with the indenter moving with a constant velocity v. In the holding stage the indenter is kept in position for a certain time, typically in the order of a tenth of the indentation time, in order to allow dislocation motion to decrease significantly. After the holding stage, the indenter is retracted from the substrate at a similar velocity. The simulation is performed in the so-called displacement-controlled way; that is, at every MD time step Δt, the indenter proceeds rigidly by a path length $v \cdot \Delta t$. The indenter penetrates the substrate until a desired depth is reached. A thorough discussion on indentation penetration rate used in MD simulations is presented in section 14.2.2.

Often, indentation simulations are carried out at zero temperature; the low temperature eases the detection of defects. At room temperature, many materials will show – even quantitatively – a similar behavior as at 0 K. For example Gao *et al.* [69] reported on the differences found in Fe indentation between 0 K and 300 K; at the higher temperature the hardness decreases by 12% while the dislocation network increases by 5.5%. Note that at the high penetrations speeds that are adopted in MD, thermal processes are very unlikely to occur, even at room temperature.

Simulations at elevated temperature are rare. Alcalá *et al.* [34] performed MD simulations of Ta nanoindentation for temperatures in the range of 77–900 K. They report systematic changes in the defect nucleation stage; the load drops associated with defect nucleation occur at smaller pressures and decrease in height with increasing temperature.

14.2.1 The Indentation Tip

In nanoindentation MD simulations, the election of the method to simulate the indenter tip can play a major role on the plasticity mechanisms that take place in the sample. For instance, one typical way to model the indenter tip is to use a 'smooth spherical tip' represented by a repulsive potential such that the spherical indenter exerts a force of magnitude:

$$F(r) = K(r - R)^2 \quad for \quad r \leq R,\ F(r) = 0 \text{ otherwise.} \tag{14.1}$$

This method was introduce by Kelcher, Plimpton and Hamilton [14] in their study of dislocation nucleation and defect structure during surface indentation of Au(111) back in 1998 and since then, it has been extensively used in MD studies [15, 16, 27, 36, 37]. 'Smooth tips' tend to nucleate dislocations homogeneously, that is within a region of defect-free crystalline material, once a critical resolved shear stress is reached or in a region of maximum shear gradient [70]. Should dislocations be produced by heterogeneous nucleation by this type of indenter, it would result from point defects [71], surface steps [15, 72] or thermally activated transient effects [73].

An alternative method is the use of atomistically rough indenters [74]. Considering atomic length scales, it is reasonable to expect that ledges provided by an atomistically rough indenter will produce heterogeneous nucleation of dislocations. The crystalline nature of the indenter leads to roughness in its outer surface and the discrete changes in the outer surface have shown to locally increase the contact pressure [74] and a continuum mechanics description of the contact no longer applies. To reveal the consequences of the use of an atomically rough indenter in favour of a spherical 'smooth tip' the same simulation must be carried out with both stepped and smooth indenters. Such an example can be found in Wagner *et al.* [73] where the authors compare the effect of temperature on the nucleation of dislocations both under the use of stepped and smooth indenters. Examples on the use of atomistically rough indenters in MD simulations can be found in [35, 75–77] to cite a few. The indenter is often modeled as a rigid atomistic sphere or an atomistic cone, made of carbon atoms arranged in a diamond lattice to mimic the structure of diamond tips. This in turn leads to the need of an interatomic potential not only capable of modelling the interactions between atoms in the sample, but also between these atoms and those of the C tip. The usage of an atomistic tip with mobile atoms adds additional complexity to the problem and is generally avoided.

14.2.2 Control Methods Used in Experiment and in MD Simulations

In nanoindentation experiments, the onset of plasticity and other plastic processes are often detected as 'pop-ins'; these are excursions in depth at constant load in the load-displacement curve. For defect-free crystals, such events are often linked to dislocation nucleation at or just below the surface [70, 78, 79]. Pop-ins can be the result of the nucleation of dislocations (both homogeneous and heterogeneous), the propagation of existing dislocations, or even of a phase transformation of the substrate material, depending on the applied stress.

Depending on the control method employed for the nanoindentation experiment, the pop-ins show up differently in a load-displacement curve. Typically, nanoindentation tests are carried out in the so-called load-controlled mode, with the pop-in appearing as a plateau in the load-displacement curve, the reason being that under this setup the load is increased at a constant rate and when a pop-in occurs as a result of a plastic process taking place, the same load is maintained as the indenter penetrates suddenly into the material. In a condition diametrically opposed to the latter, nanoindentation tests carried out in the displacement-controlled mode would show a pop-in as a sudden load drop in the load-displacement curve, the reason being that under this setup the displacement is increased at a constant rate and the load is just a measure of the material reaction; hence as a noticeable plastic process takes place the pop-in unloads the contact without further penetration of the indenter.

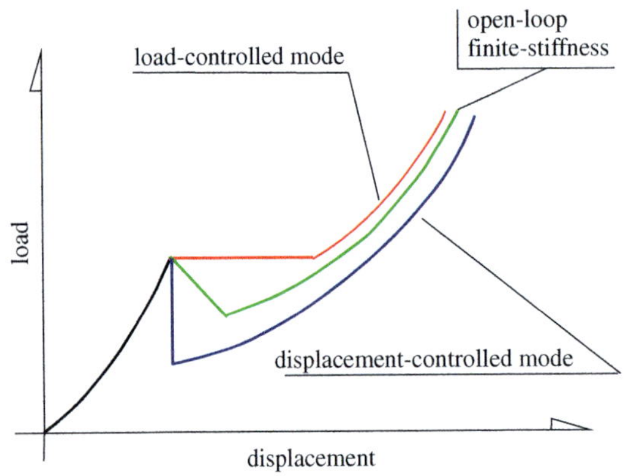

Figure 14.2 Control methods for experimental nanoindentation. MD simulations can produce each of the shown pop-ins.

While displacement-controlled indentation is less common, it can indeed be achieved in an interfacial force microscope as well as in some commercial nanoindenters. In between these two modes, atomic-force-microscope (AFM) based indentation offers the possibility to impart a finite stiffness between the indenter and the substrate; hence any noticeable plastic process would produce a pop-in as a simultaneous drop in force with an increase in penetration depth. This indentation mode has been referred as 'open-loop finite-stiffness' [53]. The three modes are depicted in Figure 14.2.

MD simulations of nanoindentation are usually performed by simulating the displacement-controlled penetration of the indenter; here the displacement is controlled by the penetration rate. If the penetration rate is low enough, once a pop-in event takes place it will show up as a sudden load drop since it will unload the contact much faster than the next indenter penetration step. If the penetration rate is high, the indenter penetration step can match the deformation of the material due to the pop-in and the event will be seen as a plateau in the load-displacement curve. Penetration rates in between those cases would produce pop-ins similar to those described above as the 'open-loop finite-stiffness' indentation mode. MD simulations of nanoindentation would offer, in principle, the possibility to mimic these three modes by making the adjustment of a single parameter, the penetration rate. However, in order to avoid large strain-rate effects, MD simulations should be performed at the lowest penetration rate allowed by the computational resources available. Indentation rate effects are discussed in detail in section 14.5.3.

14.2.3 Penetration Rate

A serious problem of the application of MD simulation to real experiments consists in the disparate indentation velocities and corresponding time scales of MD (nanoseconds) and experiment (seconds). The slowest simulation that has been performed up to now was by Alcalá *et al.* [34] who indented with a speed of 0.004 m/s, spanning times of hundreds of ns. Still this is three orders of magnitude faster than in typical experiments.

For smaller indentation velocities, thermal processes may play an increasingly important role. A number of techniques have been developed which incorporate (slow) thermally activated processes – these techniques are summarized as accelerated molecular dynamics (aMD) techniques. This development started with three hallmark papers by Voter *et al.* [80–82]; in those days important applications lay in the field of surface science such as surface diffusion. Recently these techniques were also applied to tribological studies where the atomic stick-slip behavior of a tip sliding on a single-crystalline surface was studied. Velocities of the order of mm/s and below were reached [83, 84]. These studies find that conventional MD applies to an 'athermal' regime where thermal processes play no significant role, while aMD allows to introduce thermal effects. To our knowledge, no aMD study of nanoindentation has been published up to now.

14.3 Interatomic Potentials

A thorough literature review reveals that there is no source of interatomic potentials specifically fitted for nanoindentation studies. When studying f.c.c. and b.c.c. metals, potentials of the embedded-atom-method (EAM) type [85] are used extensively; this is caused, among other reasons, by the good balance between accuracy and computational cost. Even when choosing a widely used potential, the MD practitioner should confirm that the interatomic potential chosen provides in the pressure and temperature range to be applied (i) a stable crystal structure and (ii) that it correctly reproduces not only zero-temperature properties like the elastic moduli but also also predicts a reasonable evolution of such properties with pressure and temperature. Regarding plasticity, two further important issues to be verified are the structure of dislocation cores [86] which will affect dislocation formation and mobility, and the generalized planar fault energy since it is considered as the primary intrinsic material property affecting the twinning propensity [87], as seen in section 14.3.2 below. Last but not least, the MD practitioner must verify the absence of unphysical behavior such as the existence of non-valid slip systems or solid-solid phase transitions not predicted by the phase diagram [88].

14.3.1 Elastic Constants

Taking as an example metals with cubic symmetry, the elastic behavior is described completely by their elastic constants, C_{11}, C_{12}, and C_{44}. As for polycrystalline samples, the following isotropic properties can be defined based on the elastic constants. The bulk modulus, B, is given by:

$$B = \frac{C_{11} + 2C_{12}}{3}. \tag{14.2}$$

The average shear modulus G is defined as the arithmetic mean over the Voigt- and Reuss-averaged shear moduli:

$$G = \frac{1}{2}(G_{\text{Voigt}} + G_{\text{Reuss}}), \tag{14.3}$$

where G_{Reuss} [89] is:

$$G_{\text{Reuss}} = \frac{5(C_{11} - C_{12})C_{44}}{4C_{44} + 3(C_{11} - C_{12})}, \tag{14.4}$$

and G_{Voigt} [90] is:

$$G_{\text{Voigt}} = \frac{C_{11} - C_{12} + 3C_{44}}{5}. \tag{14.5}$$

Isotropic materials are completely determined by these two elastic constants, G and B. From these, further quantities can be derived. The Poisson ratio, ν, is defined as:

$$\nu = \frac{3B - 2G}{2(G + 3B)}, \tag{14.6}$$

the Young modulus, E, is:

$$E = 2G(\nu + 1) \tag{14.7}$$

and the elastic anisotropy is described by the Zener anisotropy factor, X, as:

$$X = \frac{2C_{44}}{C_{11} - C_{12}}. \tag{14.8}$$

However, when dealing with single crystals, the properties become dependent on the orientation of the single crystal. Using the elasticity equations, the orientation-dependent Young moduli for (100), (110), and (111) single crystals can be calculated by means of the following equations [91]:

$$\frac{1}{E_{100}} = \frac{C_{11} + C_{12}}{(C_{11} + 2C_{12})(C_{11} - C_{12})}, \tag{14.9}$$

$$\frac{1}{E_{110}} = \frac{4C_{44}(C_{11} + C_{12}) - 2C_{44}(C_{11} + 2C_{12}) + (C_{11} + 2C_{12})(C_{11} - C_{12})}{4C_{44}(C_{11} + 2C_{12})(C_{11} - C_{12})}, \tag{14.10}$$

$$\frac{1}{E_{111}} = \frac{6C_{44}(C_{11} + C_{12}) - 4C_{44}(C_{11} + 2C_{12}) + (C_{11} + 2C_{12})(C_{11} - C_{12})}{6C_{44}(C_{11} + 2C_{12})(C_{11} - C_{12})}. \tag{14.11}$$

Ziegenhain *et al.* [27] produced an orientation dependent evaluation of a potential for Cu and another for Al. Ruestes *et al.* [36] presented a thorough evaluation of three Ta potentials, including predictions from Equations (14.2) through (14.11).

Given the large stresses produced by nanoindentation, it is important to establish whether the potential offers an adequate pressure dependence of the elastic constants; this is shown in Figure 14.3 for two typical potentials used in Ta simulations. The extended Finnis-Sinclair potential by Dai *et al.* [66] behaves smoothly under pressure and in the range considered here, while it can be seen that the EAM potential by Li *et al.* [92] performs poorly at pressures that can be reached during nanoindentation. Discontinuities or steep variations in the derivative of the elastic constants can lead to artificial effects during loading, such as artificial micro-structures [36] and phase transitions [93] as well as soft phonon modes [94]. The removal of the indenter might lead to tensile stresses, and this should also be adequately described by the interatomic potential. Even if the pressure in a certain region stays low, it is often the case that atomic pressures can reach both compressive and tensile values of up to tens of GPa, and the potential should be able to handle those pressure ranges properly.

14.3.2 Generalized Stacking Fault Energies

In f.c.c. metals plasticity often occurs by the emission of partial loops, i.e. partial dislocations bounding a stacking fault, which are absent for b.c.c. metals. This behavior may be rationalized by considering the generalized stacking fault energy (GSFE) [95, 96]. Figure 14.4 shows the basic difference between the GSFE curves for a typical f.c.c. metal, Cu, and a typical b.c.c. metal, Ta. The curves show the energy necessary to produce an (unstable) stacking fault by shifting half the crystal along a lattice plane by a

Figure 14.3 Pressure dependence of the elastic constants C_{ij} of Ta for (a) the extended Finnis-Sinclair potential by Dai *et al.* [66], and (b) the EAM potential by Li *et al.* [92]. *Source*: Li 2003. Reproduced with permission of American Physical Society.

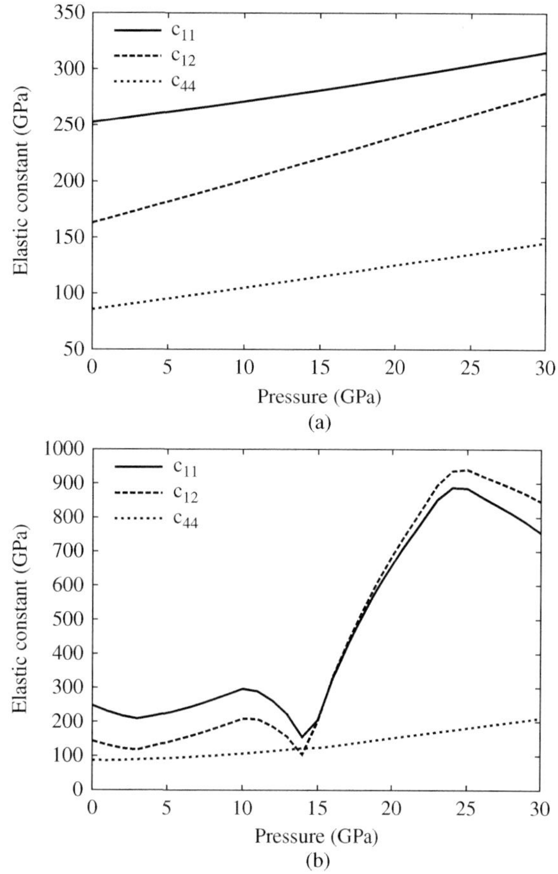

vector $\boldsymbol{f}$ in the direction of the Burgers vector $\boldsymbol{b}$ of a dislocation. For Cu, Figure 14.4(a) considers a shift of the (111) planes along the $[11\bar{2}]$ direction; the GSFE curve shows a minimum at the Burgers vector of a partial dislocation, $b=\frac{1}{6}[11\bar{2}](111)$, with an energy of only around 40 mJ/cm^2 above the ideal lattice. This characterizes the formation of stacking faults such as they are produced abundantly in nanoindentation of f.c.c. crystals. B.c.c. metals, on the other hand, do not exhibit local minima for the relevant $\langle 111 \rangle$ shear directions [97], and hence stacking fault formation is not expected. Still GSFE surface calculations for b.c.c. metals along the $\langle 111 \rangle$ slip directions should be performed for the potential used since the calculations can unveil unphysical behavior, as demonstrated by Ruestes *et al.* [36].

The importance of the generalized twinning fault energy (GTFE) surface in plasticity of b.c.c. nanowires was recently discussed by Cao [98] and Ruestes *et al.* [36]. In these contributions, the twinning fault (TF) energy is defined as the energy measured at a displacement of b/3 for a full dislocation operating in the $\{112\}\langle 111 \rangle$ system. Unstable and stable twinning fault energies ((γ_{UTF} and γ_{TF}), respectively) are the energy maximum and minimum along the GTFE curve, the twin migration (TM) energy is the difference between UTF and TF energies, $(\gamma_{UTF} - \gamma_{TF})$, and the twin propensity is given by the ratio

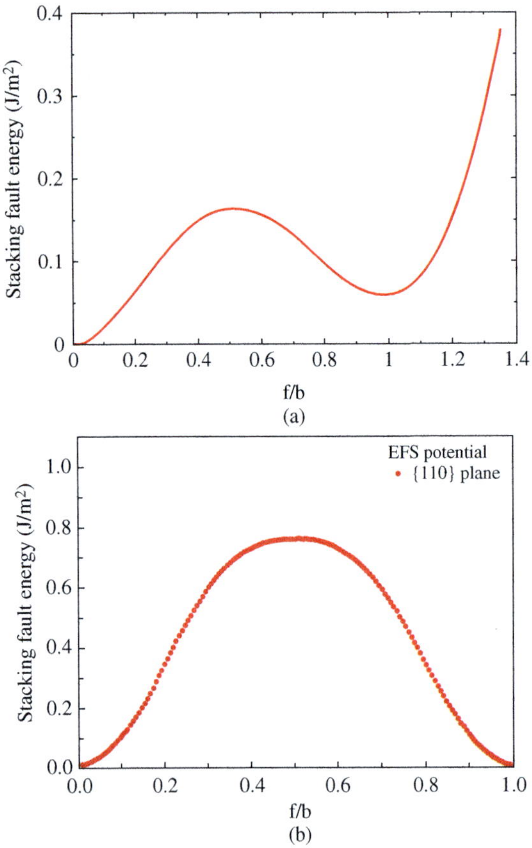

Figure 14.4 Generalized stacking fault energy. (a) An f.c.c. metal, Cu, shows a stable minimum for $\boldsymbol{b} = \frac{1}{6}\{11\bar{2}]\}$. Data obtained for the (111) plane and an EAM potential [68], after [95]. (b) A b.c.c. metal, Ta, does not develop stable stacking faults along the [111] direction. A stacking fault with $\boldsymbol{b} = \frac{1}{2}\{111]\}$ reproduces the lattice. Data obtained for the (110) plane, after [36]; the extended Finnis-Sinclair potential by Dai *et al.* [66] has been used.

$\gamma_{UTF}/\gamma_{USF}$. A twin propensity ratio below one usually indicates that twinning is dominant over subsequent slip.

Pressure might modify the GSFE surface [99] as well as the GTFE, and therefore thorough studies are needed to identify possible pressure-induced changes.

14.4 Elastic Regime

For typical MD simulations, the relevant geometrical parameters can be schematically seen in Figure 14.5. Besides the radius R of the indenter, the contact radius a_c and the displacement (or indentation depth) d_c characterize the indent.

When studying the elastic interaction between a sphere of radius R and an elastic isotropic solid, Hertz [100] found that the force F imparted by the sphere perpendicular to the surface is related to the displacement d_c into the surface:

$$F = \frac{4}{3} E^* R^{1/2} d_c^{3/2}, \tag{14.12}$$

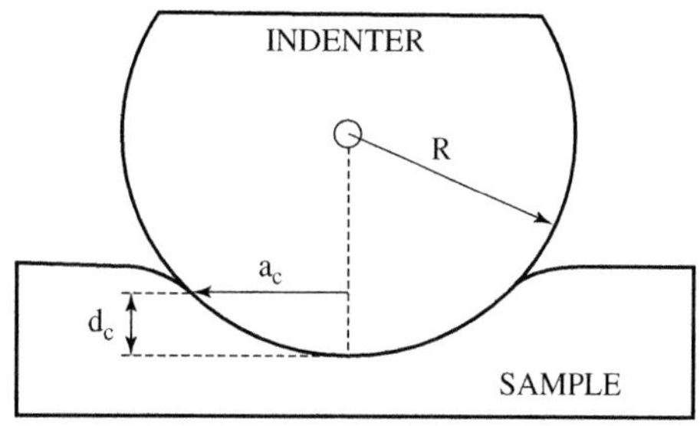

Figure 14.5 Relevant geometrical parameters.

where E^* is often called the 'combined indentation modulus' or the 'reduced indentation modulus' of the system, and is given by:

$$\frac{1}{E^*} = \frac{1-\nu^2}{E} + \frac{1-\nu'^2}{E'}. \tag{14.13}$$

Here E, ν, E' and ν' are the Young modulus and Poisson ratio of the surface and indenter, respectively; see section 14.3.1. E^* is the one and only parameter describing the materials elastic response. In the Hertzian treatment, the indenter is frictionless, that is there are no forces in the direction tangential to the tip. Within the Hertzian model, the mean (averaged) contact pressure amounts to:

$$p_{mean} = \frac{3}{4\pi}\frac{1}{E^*}\sqrt{\frac{d_c}{R}}. \tag{14.14}$$

The Hertz treatment is considered exact for $d_c \ll R$ and an elastically isotropic solid. Several authors [27, 101, 102] have shown that when probing single crystals, the Hertzian laws hold with an indentation modulus that depends on the surface orientation:

$$\frac{1}{E^*} = \frac{1-(\nu_{ijk})^2}{E_{ijk}}. \tag{14.15}$$

The above equations are often used to calculate Hertzian loading curves for comparison purposes to loading curves obtained by MD. We note, however, that elasticity might break down at the nanoscale [74].

14.5 The Onset of Plasticity

14.5.1 Evolution of the Dislocation Network

A particular benefit of MD simulation is that it allows to identify in detail the stages during which plasticity forms and develops in the substrate under the penetrating indenter. For the case of f.c.c. metals, this has been documented by van Vliet *et al.* [17] and later by Lee *et al.* [105]. We show an example in Figure 14.6, where we correlate the excursions in the load-depth curve with the development of plasticity inside the material. At the first load drop (from point A to B) suddenly a strongly defective region develops under the indenter; close to the indenter the defect structure cannot be analyzed, but two small

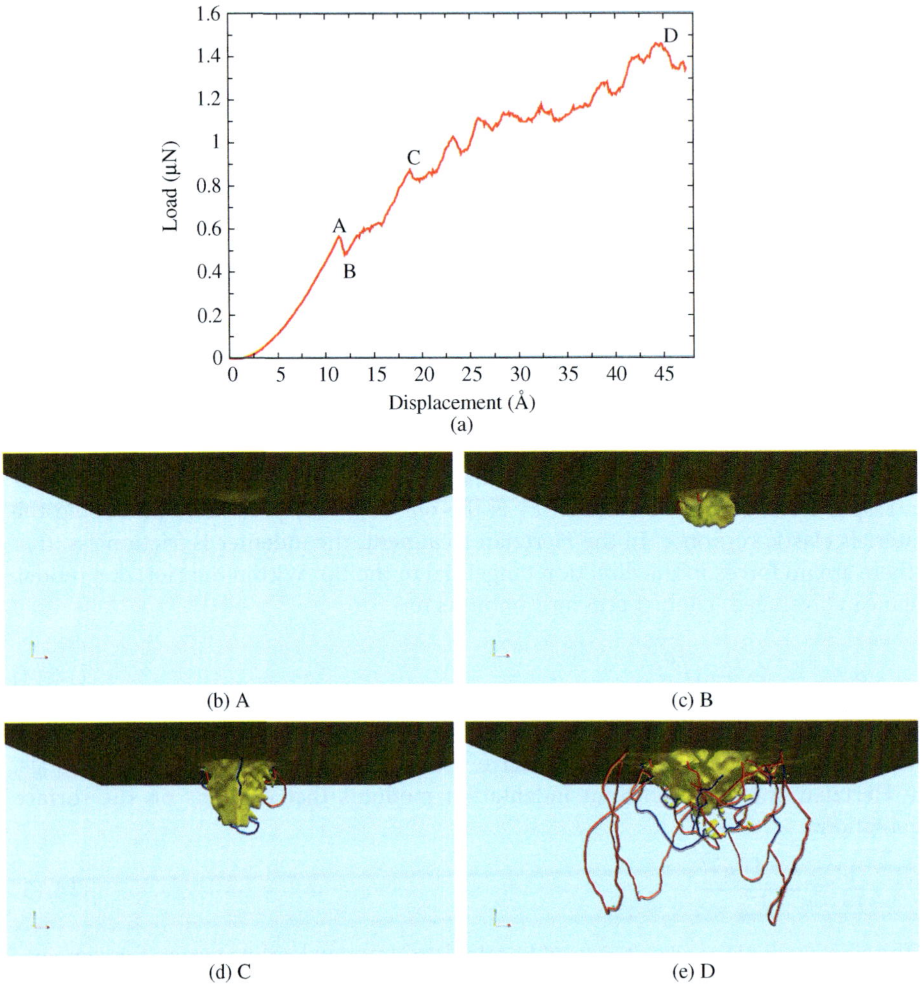

Figure 14.6 Indentation with a rigid atomistic indenter (radius $\boldsymbol{R} = \mathbf{4.28}$ nm, consisting of 57 903 C atoms arranged in a diamond lattice) into a (100) Fe surface with a velocity of 10 m/s. The substrate contains $\mathbf{4.3 \times 10^6}$ atoms to provide enough space for the development of dislocations. Substrate temperature was set to 0 K to ease detection of lattice defects. In the load-displacement curve (a) several points are marked. These are shown successively in (b)–(e). Yellow: deformed surface including unidentified defects. Dislocation lines with Burgers vector $\boldsymbol{b} = \frac{1}{2} < 111 >$ are shown in red, those with $\boldsymbol{b} =< 100 >$ in blue. Green arrows indicate direction of $\boldsymbol{b}$. Dislocations are detected using the DXA algorithm [103]. Visualization has been prepared using Paraview [104].

dislocations are seen with Burgers vector $\boldsymbol{b} = \frac{1}{2} < 111 >$. This demonstrates that load drops are directly connected to the generation of dislocations. Upon further indentation, at point C, these dislocations have grown and also new dislocations have formed. In b.c.c. Fe, besides the energetically most favourable $\boldsymbol{b} = \frac{1}{2} < 111 >$ dislocations, also dislocations with Burgers vector $\boldsymbol{b} =< 100 >$ form. Finally, at point D we observe the fully developed plastic zone. Note the extended shear loops reaching far out of from the indenter into the substrate. Triple dislocation junctions have developed, typically by the reaction of two $\boldsymbol{b} = \frac{1}{2} < 111 >$ dislocations with one $\boldsymbol{b} =< 100 >$. The quantitative characteristics of the plastic zone developing under the indenter will be discussed in section 14.6.

14.5.2 Contact Area and Hardness

The quantification of the parameters necessary to describe the nanoindentation process using MD typically proceeds as follows. When the indenter is pressed into the material, the resulting reaction force F is calculated based on the interactions tracked by the MD code. The determination of the contact radius a_c, see Figure 14.5, needs some care [27].

In spite of its importance, there is not a unique definition of the contact area. For macroscopic length scales, there are at least two concepts usually taken into account: the Brinell contact area which represents the convex contact surface, and the Meyer contact area which represents the projection of the convex surface onto the initial surface plane [91]. Considering the Hertzian treatment, the contact pressure is the substrate response in the normal direction and hence, the Meyer definition should be followed. However, in contrast to continuum approaches, atomistic simulations are based on the discreteness of matter and hence, 'holes' may appear between the atoms. This introduces systematic errors, particularly in the elastic regime, where as elastic deformation proceeds, the atoms separate more and more, leading to an increase in the size of the holes with increasing deformation. As a result the area may be underestimated and, as a consequence, the contact pressure may be overestimated.

Furthermore we note that on the atomistic scale, different textures provide different areal densities of surface planes. When compared to dense surfaces, such as the (111), typically rough surfaces, such as the (110), end up with an underestimation of the area, leading to a distortion in the determination of the contact area.

The contact pressure or hardness, H, is given by:

$$H = \frac{F}{A} = \frac{F}{\pi a_c^2}, \tag{14.16}$$

where F is the reaction force and A is the contact area. Of particular interest is the contact pressure at the onset of plasticity. We denote it as the pop-in pressure, σ_p, and is detected in the simulation as the contact pressure, at which plastic deformation is first noticeable. The maximum resolved shear stress is related to the mean contact pressure and reaches a maximum of $\tau_{max} \sim 0.465 p_{mean}$ slightly below the contact surface [78]. Schuh *et al.* [71] demonstrated that the first pop-in event requires a large critical stress and the maximum shear stress, τ_{max}, must be compared with the theoretical shear stress threshold for homogeneous nucleation of dislocations ($\tau_{th} = G/2\pi$). While we would expect τ_{max} to be lower than τ_{th}, a value above the theoretical threshold for homogeneous nucleation is possible, because of pressure hardening effects [27]. These are attributed to the pressure dependence of the shear modulus, which increases with pressure, hence increasing the threshold. Also, Li *et al.* [70] pointed out that dislocation nucleation might be driven by stress gradients, not stress maxima. Recently, Kang *et al.* [106] reported a complex behavior of the shear stress needed to move a dislocation, depending on the exact angle of the applied shear, with mobility showing a faceted behavior. It should be noticed that the Orowan value ($G/2\pi$) is only a first-order estimate of the theoretical strength, and that more exact calculations predict values that are significantly different [91].

Table 14.1 Hardness measurements (GPa) on Ta (100), (110) and (111), for four different indenter diameters D (nm) using the extended Finnis-Sinclair potential and the same ***d/D*** ratio.

D nm	(100) GPa	(110) GPa	(111) GPa
8	19.9	19.5	14.8
12	17.7	19.3	14.2
16	16.3	18.2	13.9
20	16.6	—	—

It is common practice in experiments to indicate pop-in events in terms of the dimensionless ratio a_c/D, where D is the tip diameter. To cite an example, for the $\{001\}$ Ta surface, experiments and previous simulations indicate that $a_c/D \sim 0.19 \pm 0.01$ at 293 K [34]. In MD simulations, slight variations are expected in such values based on the interatomic potential used.

Hardness estimations can be derived from MD simulations using Equation (16) and are presented in Table 14.1 for (100), (110) and (111) Ta single crystals. There is a decrease in hardness with the diameter of the indenter D. This size dependent behavior is similar to the indentation size effect reported by Swadener *et al.* [107]. Hardness decreases as the indenter diameter increases, without influence on the penetration depth or contact radius.

14.5.3 Indentation Rate Effect

It is common practice to perform MD simulations at different penetration velocities to assess the effect of the indentation rate on the onset of plasticity. Figure 14.7 presents an example for comparison purposes. In this case a Ta (100) single crystal modeled by the extended Finnis-Sinclair potential was indented using a 10 nm radius tip and two penetration velocities were used separately to establish the effect of strain rate on the indentation. The chosen velocities were 34 m/s (~1/100 of the bulk sound velocity) and 3.4 m/s (~1/1000 of the bulk sound velocity). As one might expect, the critical load to

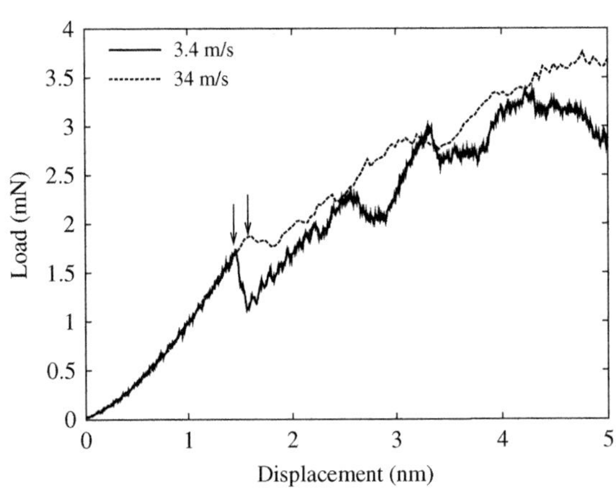

Figure 14.7 Indentation velocity dependence of the load-displacement curves for a typical MD simulation. As the penetration rate decreases, the first pop-in event becomes more noticeable, with a marked reduction in load at the onset of plasticity. Arrows indicate the first pop-in events for the two indentation velocities explored.

Figure 14.8 Indenter-size dependence of the loading curves for a Ta (100) defect-free single crystal. The legend indicates the radius of the spherical tip used for the simulations, using an indenter velocity of 34 m/s.

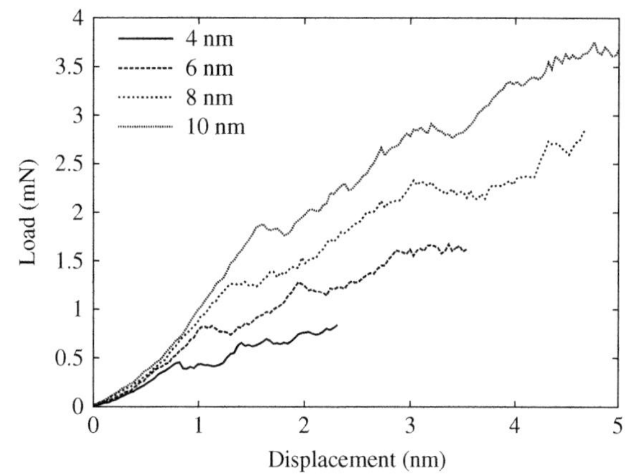

induce plasticity depends on the indentation velocity. The critical load is lowered by ≈ 10 % for a tenfold decrease in penetration velocity. As shown by Ruestes and co-workers [36], the analyst can extract an exponent m characterizing the indentation-rate sensitivity of the contact pressure, σ_p, at which plastic deformation is first noticeable (pop-in pressure) via the relation:

$$\sigma_p = C v^m, \tag{14.17}$$

For this example one obtains $m = 0.04$.

We note that the range of strain rates sampled by MD simulations of nanoindentation can lead to complex effects on strength [108], and plasticity [109], including modification of the dominant plasticity mechanisms [110].

14.5.4 Tip Diameter Effect

To probe the response of the material to variations in the tip dimensions, MD practitioners should perform a systematic set of nanoindentation simulations varying the indenter tip diameter. Such an example can be found in the contribution by Ruestes *et al.* [36] where the authors study the response of Ta single crystals when probed using spherical indenters with diameters ranging from 8 to 20 nm. As shown by the authors, under certain circumstances, the combination of indenter diameter, penetration rate and interatomic potential used can lead to significant differences in the plastic response of the substrate. Systematic studies are mandatory to identify and exclude simulation artefacts. The load-displacement curves presented in Figure 14.8 are an excerpt of the data presented by Ruestes and co-workers [36] for the extended Finnis-Sinclair potential showing that as the mean pressure for pop-in is a constant, the point at which the onset of plasticity occurs increases with increasing indenter diameter.

14.6 The Plastic Zone: Dislocation Activity

14.6.1 Face-centered Cubic Metals

The nucleation and structure of dislocations in f.c.c. metals under indentation was first studied by Kelchner *et al.* [14] and later by Van Vliet *et al.* [17] and Lee *et al.* [105].

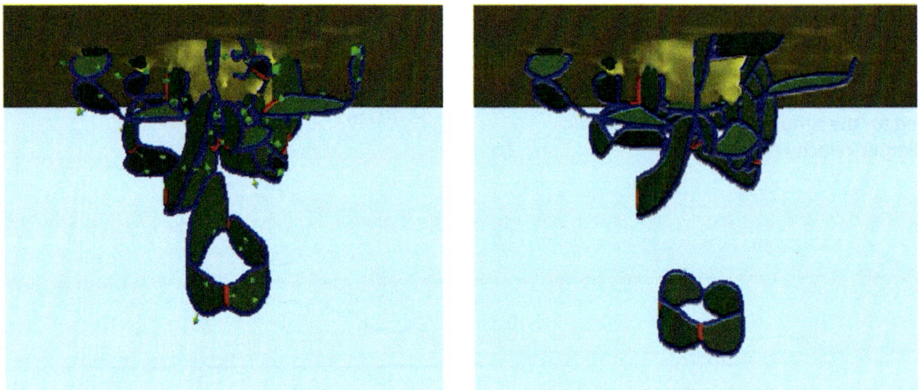

(a) (b)

Figure 14.9 Indentation of a rigid atomistic indenter (radius R=2.14 nm, consisting of 7248 C atoms arranged in a diamond lattice) into a (110) Cu surface with a velocity of 20 m/s at 0 K. The substrate contains 4.1 x 10^{16} atoms; only the zone close to the indenter is shown. Between the indentation depth of (a) 21.0 Å and (b) 21.4 Å, a prismatic loop detaches from the complex dislocation network developing under the indenter. Yellow: deformed surface including unidentified defects. Dislocation lines with Burgers vector $\boldsymbol{b} = \frac{1}{6} < 112 >$ are shown in red, those with $\boldsymbol{b} = \frac{1}{2} < 110 >$ in blue. Green arrows indicate direction of $\boldsymbol{b}$. Dark green areas denote stacking faults. Dislocations are detected using the DXA algorithm [103]. Visualization has been prepared using Paraview [104].

The primary glide system in this material class is the $< 110 > \{111\}$ system. Hence immediately after nucleation, shear loops are formed which attach to the indenter surface. Since dislocations in f.c.c. metals dissociate and form partials these shear loops are accompanied by stacking fault planes. Figure 14.9 shows a typical plastic zone developing in f.c.c. metals after indentation, including shear loops which react to form prismatic loops.

Prismatic loops transport material away from the surface into the substrate. If glide vectors $< 110 >$ are available that lie parallel to the surface (such as for a (111) surface) V-shaped loops are formed at the surface that are free to glide out of the high-stress indentation zone.

14.6.2 Body-centered Cubic Metals

Alcalá and co-workers [34] have shown that the nucleation and propagation of twin and stacking fault bands driven by a combination of shear stresses and pressure are responsible for nanocontact plasticity in b.c.c. Ta. They also propose that dislocations appear after a thermally assisted twin annihilation and they report that this mechanism is common to other b.c.c. metals like niobium, iron and tungsten.

Recently, Remington *et al.* [37] presented a comprehensive nanoindentation study and inspired by the american cowboys proposed a 'lasso' mechanism for the formation of prismatic loops based on nanoindentation MD simulations. Using the EAM potential by Ravelo *et al.* [111], the authors performed MD simulations of nanoindentation of Ta single crystals along the three principal crystallographic orientations. They reported that after the formation and annihilation of planar defects similar to the ones reported by Alcalá *et al.* [34], shear loops form and propagate along $< 111 >$ directions,

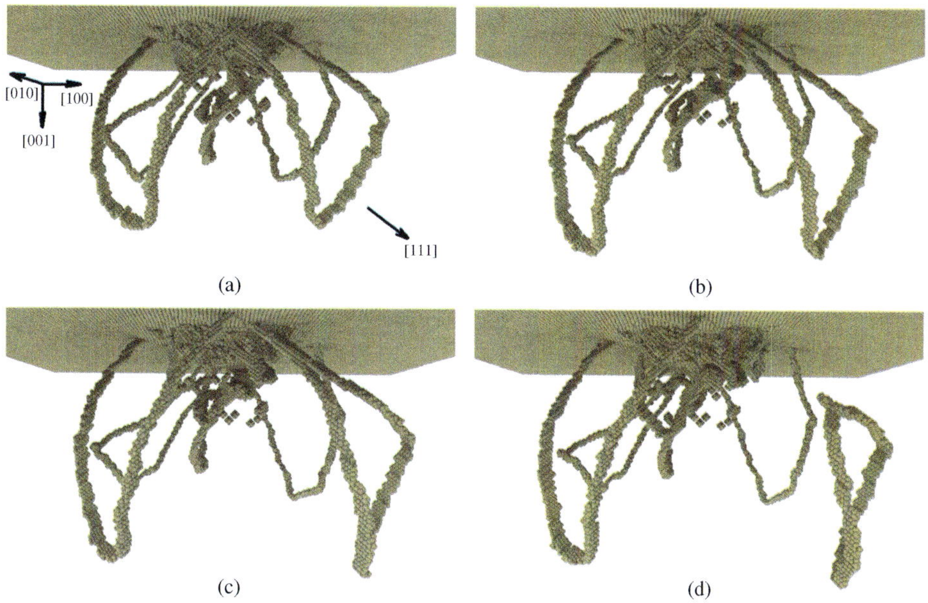

Figure 14.10 Example of the 'lasso' mechanism as proposed by Remington and co-workers [37]. The process of prismatic loop formation along the {111} direction can be seen in this sequence (a) - (d): the shear loops nucleated during the indentation process expand into the material by the advancement of their edge components, while screw components of the loop cross-slip. As they approach they annihilate to produce prismatic loops. MD nanoindentation simulation of (100) Ta at 300 K using an indenter size of 8 nm radius and a penetration velocity of 34 m/s.

consistent with b.c.c. slip systems (Figure 14.10a). These shear loops increase in extension by the advance of their edge components (Figure 14.10b), while screw components resulting from that advance undergo limited cross-slip leading to a curve in the edge front as the screw segments, being of parallel Burgers vector but dislocation lines with opposite signs, attract each other (Figure 14.10c) and, after a certain time, cancel and lead to a pinching off event producing a prismatic loop;(Figure 14.10d). This mechanism was verified for each of the three crystallographic orientations, namely {100}, {110} and {111}. The reader is referred to reference [37] for a thorough explanation with schematic drawings. Lodes *et al.* [43] presented results indicative of the same mechanism detailed by Remington and co-workers [37] when studying the influence of the dislocation density on the pop-in behavior and the indentation-size effect in cubic ionic CaF_2 single crystals, where an incomplete 'lasso' and a prismatic loop can be seen.

Hagelaar *et al.* [35] observed prismatic loop formation in their atomistic simulations of the formation and destruction of nanoindentation contacts in tungsten. In that study they verified the presence of such loops when studying the indentation along the {111} direction and associated its generation with shear stresses in their atomistic indenter.

14.6.3 Quantification of Dislocation Length and Density

In nanoindentation modelling, dislocations are typically divided into two groups: Geometrically-necessary dislocations (GNDs) which are those related to strain gradients, such as the ones commonly produced by nanoindentation, and statistically stored dislocations (SSDs) which account for dislocations that do not produce strain gradients, such as prismatic loops and dislocation dipoles [112, 113].

Dislocation generation and multiplication can be tracked and quantified by computing the total dislocation length. The values obtained can then be compared to the GND and SSD length estimates.

In 2010, Hua and Hartmaier [114] presented an analysis method to quantify the Burgers vector of dislocations in atomistic ensembles and for the calculation of GNDs and SSDs. The authors applied their proposed method to the analysis of an atomistic nanoindentation simulation. A slight variation of this method was recently used by Gao *et al.* [38] and Ruestes *et al.* [36] for systematic studies of nanoindentation simulations. First of all, the method makes use of the evaluation of the local plastic slip from atomistic data. This can be accomplished by using the slip vector analysis [15] as it was done by Hua and Hartmaier [114], or by using the Burgers circuit method [103] as it was done by Urbassek and co-workers [36, 38].

The Burgers circuit method is the preferred method by the authors as it is fully implemented in the dislocation extraction algorithm (DXA) [103], which produces a discretized line representation of the dislocations present in the sample. Using the DXA output, Gao and co-workers prepared a code programmed ad-hoc to implement the idea of Hua and Hartmaier [114], that is defining hemispherical shells and by using the DXA output, classifying the dislocation contribution to each shell and finally computing the dislocation density associated to each shell. The local dislocation density, $\rho_d(r)$, is computed as the total local dislocation length divided by the volume of the shell. Typical results are presented in Figure 14.11 showing the computed local dislocation density as a function of the distance to the center of the spherical indenter. The dislocation density decays from a maximum of around 10^{17} m^{-2} to less than 10^{16} m^{-2} over a distance of the order of the indenter radius. This agrees with earlier models by Durst *et al.* [115, 116] showing that the plastic deformation occurs in a volume whose radius is around 2 – 3 times the radius a_c. As known from previous experimental results by Chiu and Ngan [117], the dislocation density, ρ_d, varies with depth. A dislocation density of $\sim 10^{16}$ m^{-2} corresponds to a dislocation separation of $\sim$ 10 nm; for a $\sim$ 10 nm indenter diameter, the mean spacing between dislocations would be of the same order as the indenter itself.

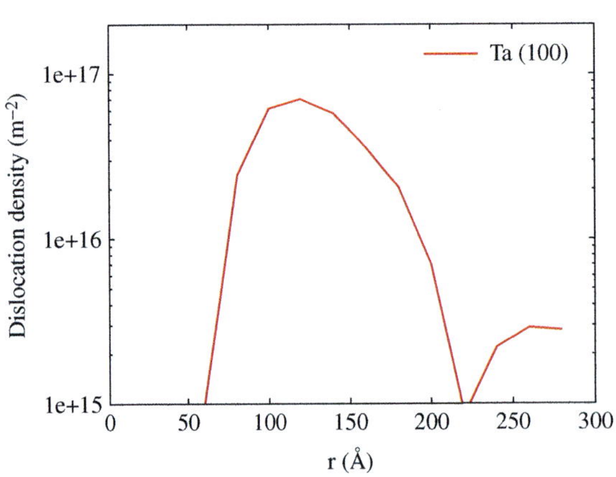

Figure 14.11 Radial dislocation density profile for Ta (100) using an extended Finnis-Sinclair potential and a 20 nm diameter indenter. The increase near 25 nm is due to the presence of detached prismatic loops.

By incorporating a GND density definition (see [114]) and taking the statistically stored dislocation density as the difference between the total dislocation density and the GND density, both GNDs and SSDs can be estimated and their evolution determined. This is very useful for the modeling community as local information about dislocation densities provides useful information to fill the gap across scales between atomistic models and continuum descriptions of plasticity in nanoindentation.

14.6.4 Pile-up

During the analysis of nanoindentation experiments, the residual pile-ups at the indenter site are often explored by AFM or similar techniques [79, 118–120]. The pile-up patterns can also be extracted from MD simulations as shown in Figure 14.12 for the simulation of a b.c.c. metal indented with a 16 nm diameter tip.

Metals with b.c.c. structure with dislocation-mediated plasticity accommodate plastic strain by nucleation and propagation of dislocations in < 111 > slip systems. As a direct consequence, {100} oriented single crystals are expected to produce four pile-ups in a 4-fold symmetry and with a certain degree of anisotropy consistent

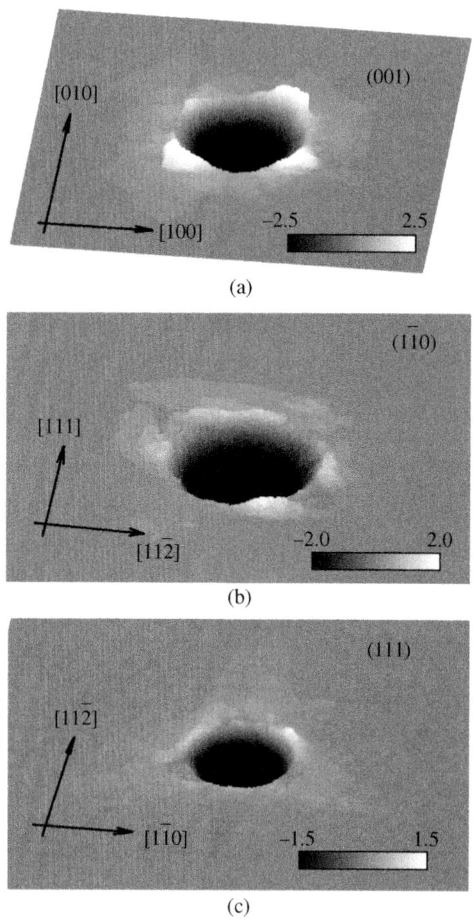

Figure 14.12 Topography of imprints produced by a 16 nm diameter indenter on the three simply indexed Ta surfaces. Gray-scale bar shows pile-up height in nm. (a) Ta (100), (b) Ta (110), (c) Ta (111).

with random fluctuations in the lattice. This can be seen in our simulation of a {100} Ta single crystal (Figure 14.12). Biener *et al.* [79] experimentally observed a 4-fold symmetry and anisotropy of the pile-up pattern when studying {100} Ta under a spherical indenter. In a recent study of the plastic anisotropy of tungsten single crystals under spherical micro-indentation, Yao and co-workers [120] experimentally found the same pile-up anisotropy as shown in Figure 14.12. For a {110} oriented surface, the slip in < 111 >direction produces a 2-fold symmetry of the pile-up. Finally, for a {111} surface, the < 111 > slip produces a 3-fold symmetry of the pile-up due to dislocation activity in $(11\bar{2})$, $(1\bar{2}1)$ and $(\bar{2}11)$ planes. The height of the pile-ups is determined by net material transport towards the surface on the < 111 > directions being activated.

While crystal plasticity models implemented in finite element method (FEM) calculations are also able to predict pile-ups with the same geometry, they lack the typical anisotropy found in experiments, leading to perfectly symmetric pile-ups, i.e. equivalent in all planes [118, 119]. Instead, MD simulations include thermal and stress fluctuations that lead to slightly different hillocks on the surface.

In a short conclusion, the analysis of the residual imprint in MD simulations can be used much like AFM images of experimental nanoindentation are used, as it has been presented in this section.

14.6.5 Geometrically-necessary Dislocations and the Identification of Intrinsic Length-scales from Hardness Simulations

The indentation size effect, which has often been observed in micro- and nanoindentation studies, shows up as an increase in hardness with decreasing depth of penetration [107]. It has been often treated using the concept of GNDs associated with strain gradients. In order to identify the intrinsic material length parameter associated to GNDs, we will revisit a simple procedure presented by Nix and Gao [121], Swadener *et al.* [107] and Al-Rub and Voyiadjis [122].

The fundamental parameters for indentation testing are the applied load F, the residual contact radius of the indentation imprint a_c, the residual indentation depth d_c, and the profile diameter of the residual indentation, here taken equal to the indenter diameter D, for simplicity.

The plastic zone is assumed to extend in a hemispherical region around the indent; its radius, R_{pl}, is included in Figure 14.13. The ratio of $f = R_{pl}/a_c$ is called the plastic-zone size factor. Experiments [115] and MD simulations [38] find that f is in the region of 2–3, for many f.c.c. and b.c.c. metals.

Considering the scheme presented in Figure 14.13, it is common practice to assume that the plastic deformation is accommodated by circular loops of GNDs with Burgers vector normal to the plane of the sample surface. The determination of GNDs under conical and spherical indenters has been extensively treated [107, 121, 123, 124]. As the indenter is forced into the substrate of a single crystal, the permanent shape change at the surface must be accommodated by GNDs. For a conical indenter with an angle θ with respect to the sample surface, Nix and Gao [121] derived a model where the profile of the residual pit was described by a cone with a well-defined slope:

$$\tan(\theta) = \frac{d_c}{a_c} = \frac{b_G}{L_G}. \tag{14.18}$$

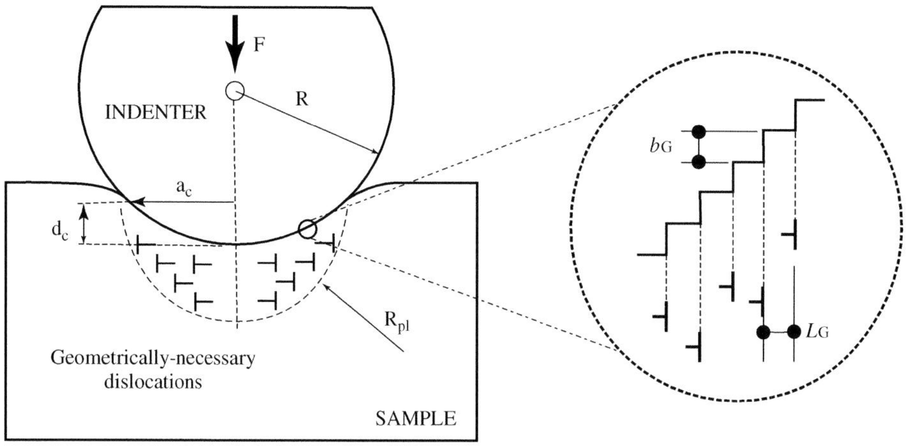

Figure 14.13 Specimen is indented by a rigid spherical indenter, producing an arrangement of circular loops of GNDs during the process. The slope of the pit profile can be related to the separation between slip events and the Burgers vector of the dislocations.

The slope of the residual pit is related to the spacing between individual slip steps corresponding to GNDs, hence resulting in:

$$L_G = \frac{b_G a_c}{d_c}, \tag{14.19}$$

where L_G *is* the spacing between individual slip steps on the indentation surface.

As for spherical indenters, it is common practice to assume that the profile of the residual pit in the unloaded condition can be described by a paraboloid defined as [125]:

$$w_r = -d_c + r^2/D \quad for \quad 0 \leq r \leq a_c, \tag{14.20}$$

By taking the derivative of Equation (20), the slope of the pit profile can be represented by:

$$\left|\frac{dw}{dr}\right| = \frac{2r}{D} = \frac{b_G}{L_G} \quad or \quad L_G = \frac{b_G D}{2r}, \tag{14.21}$$

This expression also shows that the GND loops are unequally spaced along the surface of the material, as opposed to the result for a conical indenter, Equation (14.19). If the pit profile can be represented by a paraboloid, then the dislocation loops are closely spaced near the indenter tip and the spacing increases as one moves away from it. This is supported by TEM experimental observations by Chiu and Ngan who reported a maximum of GNDs at the tip of the indenter when investigating indentation plastic zones in Ni_3Al (Cr, B) single crystals.

The total length λ of the injected dislocation loops can be expressed in terms of L_G in such a way that its differential between r and $r + dr$ is:

$$d\lambda = 2\pi r \frac{dr}{L_G}, \tag{14.22}$$

which after integration gives the total dislocation length depending on the form of L_G.

Since it is assumed that the injected loops remain within a hemispherical volume V defined by the contact radius a_c around the indentation profile and the plastic-zone factor f:

$$V = \frac{2}{3}\pi(f a_c)^3, \tag{14.23}$$

Typically, the plastic-zone factor can be taken as equal to 1.9 after Durst *et al.* [115] and Gao *et al.* [38].

The GND density (ρ_G) during indentation is determined by:

$$\rho_G = \frac{\lambda}{V}. \tag{14.24}$$

Hence for a conical indenter, Nix and Gao [121] determined $\rho_G = (3/2b d_c)\tan^2\theta$; later Al-Rub and Voyiadjis determined $\rho_G =(2/b_G D)$ for a spherical indenter producing a paraboloid pit profile [122].

As pointed out by Ruestes *et al.* [36] the choice of f at this nano-scale is not beyond questioning. Recently, a detailed study of the determination of the plastic zone size factor f at this scale was presented, finding that, after the indenter retraction, f assumes values between 2 and 3, not depending systematically on crystal structure, surface orientation, indenter size and penetration depth [38]. Densities observed in the MD simulations suggest that there is a significant number of SSDs. The relative proportions of GNDs and SSDs seem to vary with the radial distance.

The cross-section of the pit is also of interest to both experimentalists and analysts in the nanoindentation field since the intrinsic length scales of strain gradient plasticity theory can be derived from it [122].

Recently Faghihi and Voyiadjis [126] proposed two physically-based models to capture the influence of temperature and rate on the indentation size effect for b.c.c. metals by considering different expressions of the GND density. In that work, the authors present an equation for the material length scale of single-crystalline materials (see Equation 13 in [126]) which, for niobium, gives a value of 2.94 nm, that is around 10 times the magnitude of the Burgers vector.

14.7 Outlook

Indentation can be studied by a broad band of simulation techniques, ranging from macroscopic finite-element methods over mesoscopic dislocation-density based models to atomistic MD simulations [127–130]. New developments in software and hardware should allow atomic-scale simulations of more complex materials, including nanocrystals with a distribution of grain boundaries and surface roughness, samples with pre-existing defects like dislocations [131], impurities, and precipitates, and samples where indentation stresses might lead to changes in chemical bonding. Connection to experiments is improving but continues to be challenging, even for simple materials like Au [132].

Classical continuum models assume that plastic deformation occurs by the motion of pre-existing dislocations at a stress level corresponding to the yield strength of the material. The fulfillment of this assumption is quite easy when dealing with metals at the macroscale, but nanoscale experiments and simulations have shown that due to the high stress levels attained during nanoindentation, homogeneous nucleation of dislocations can be expected. As a general rule, any atomistic process with a length scale shorter than the length scale of the smallest heterogeneity considered in the macroscale model is, by definition, neglected. Therefore, MD simulation emerges as a valuable tool to provide insights to such plasticity mechanisms.

As for every nanomechanical test, the main goal pursued when performing nanoindentation testing and simulations is to gain information on local material properties together with fundamental deformation mechanisms. The multiscale nature of nanoindentation becomes evident as the typical macroscale response of the load-displacement curve can be correlated to experimental characterization carried out on transmission electron microscopes (TEM) and scanning electron microscopes. This multiscale nature also applies when treating nanoindentation [130], especially in simulation techniques where atomistic models can shed light on the details of defect nucleation and their propagation, informing mesoscale models such as dislocation density models which in turn can be used to improve continuum mechanical approaches, revealing the interplay between mechanical loading and atomistic scale processes across length scales [128].

While each of these methods has its own virtues, the MD technique is certainly the appropriate method to describe processes occurring at the atomic scale. We showed in this review how detailed information can be obtained about the onset of plasticity, the evolution of the dislocation network and the role of twins in the emerging plasticity. Modern analysis algorithms allow us to determine the structure and lengths of dislocations and thus to analyze the size and dislocation density of the plastic zone. It seems safe to foresee that such structural information will provide useful input both in the analysis of experiments and as the atomistic basis for multiscale models that describe nanoindentation at larger spatial scales and longer time scales.

Acknowledgements

CJR and EMB acknowledge support by CONICET, PICT-2015-0342, PICT-PRH-0092 and SeCTyP UNCUYO. YG and HMU acknowledge support by the Deutsche Forschungsgemein-schaft via the Sonderforschungsbereich 926.

References

1 A. C Fischer-Cripps. Nanoindentation. 2nd edn, Mechanical engineering series. Springer, New York; 2004.

2 R. W. Armstrong, W. L. Elban, S. M. Walley. Elastic, plastic, cracking aspects of the hardness of materials. International Journal of Modern Physics B 27, 1330004; 2013.

3 E. Weppelmann, M. Wittling, M. V. Swain, D. Munz. Indentation cracking of brittle thin films on brittle substrates, in: R. C. Bradt, D. P. H. Hasselman, D. Munz *et al.* (eds), Fracture Mechanics of Ceramics. Springer US, Boston, pp. 475–486; 1996.

4 A. A. Volinsky, J. B. Vella, W. W. Gerberich. Fracture toughness, adhesion and mechanical properties of low-k dielectric thin films measured by nanoindentation. Thin Solid Films 429(1–2), 201–210; 2003.

5 B. Yang, L. Riester, T. Nieh, Strain hardening and recovery in a bulk metallic glass under nanoindentation. Scripta Materialia 54(7), 1277–1280; 2006.
6 A. Bendavid, P. Martin, H. Takikawa, Deposition and modification of titanium dioxide thin films by filtered arc deposition. Thin Solid Films 360(1–2), 241–249; 2000.
7 P.M. Sargent, M.F. Ashby. Indentation creep. Materials Science and Technology 8(7), 594–601; 1992.
8 S. Veprìcek, The search for novel, superhard materials. Journal of Vacuum Science and Technology A: Vacuum, Surfaces, and Films 17(5), 2401; 1999.
9 C. Lu, Y.-W. Mai, Y.-G. Shen. Recent advances in understanding the origin of superhardness in nanocomposite coatings: A critical review. Journal of Materials Science 41(3), 937–950; 2006.
10 K. Kumar, H. Van Swygenhoven, S. Suresh. Mechanical behavior of nanocrystalline metals and alloys. Acta Materialia 51(19), 5743–5774; 2003.
11 J.L. Cuy, A.B. Mann, K.J. Livi *et al.* Nanoindentation mapping of the mechanical properties of human molar tooth enamel. Archives of Oral Biology 47(4), 281–291; 2002.
12 W.C. Oliver, G.M. Pharr. An improved technique for determining hardness and elastic modulus using load and displacement sensing indentation experiments. Journal of Materials Research 7(6), 1564–1583; 1992.
13 S.B. Sinnott, S.-J. Heo, D.W. Brenner *et al.* Computer simulations of nanometer-scale indentation and friction, in: Springer Handbook of Nanotechnology. Springer, pp. 955–1011; 2010.
14 C. Kelchner, S. Plimpton, J. Hamilton. Dislocation nucleation and defect structure during surface indentation. Physical Review B, 58(17), 11085–11088; 1998.
15 J.A. Zimmerman, C.L. Kelchner, P.A. Klein *et al.* Surface step effects on nanoindentation. Physical Review Letters 87(16), 165507; 2001.
16 E.T. Lilleodden, J.A. Zimmermann, S.M. Foiles, W.D. Nix. Atomistic simulations of elastic deformation and dislocation nucleation during nanoindentation. Journal of the Mechanics and Physics of Solids 51(5), 901–920; 2003.
17 K. Van Vliet, J. Li, T. Zhu *et al.* Quantifying the early stages of plasticity through nanoscale experiments and simulations. Physical Review B, 67(10): 104–105.
18 X.-L. Ma, W. Yang. Molecular dynamics simulation on burst and arrest of stacking faults in nanocrystalline Cu under nanoindentation. Nanotechnology 14(11), 1208; 2003.
19 T. Zhu. Predictive modeling of nanoindentation-induced homogeneous dislocation nucleation in copper. Journal of the Mechanics and Physics of Solids 52(3), 691–724; 2004.
20 H. Liang, C. Woo, H. Huang *et al.* Crystalline plasticity on copper (001),(110), and (111) surfaces during nanoindentation. CMES-Computer Modeling in Engineering and Sciences 6(1), 105–114; 2004.
21 J. Biener, A. M. Hodge, J. R. Hayes *et al.* Size effects on the mechanical behavior of nanoporous au. Nano Letters 6(10), 2379–2382; 2006.
22 A. Asenjo, M. Jaafar, E. Carrasco, J. Rojo. Dislocation mechanisms in the first stage of plasticity of nanoindented Au (111) surfaces. Physical Review B, 73(7), 075431; 2006.

23 T. Tsuru, Y. Shibutani. Anisotropic effects in elastic and incipient plastic deformation under (001),(110), and (111) nanoindentation of Al and Cu. Physical Review B, 75(3), 035415; 2007.
24 S.-P. Ju, C.-T. Wang, C.-H. Chien *et al.* The nanoindentation responses of nickel surfaces with different crystal orientations. Molecular Simulation 33(11), 905–917; 2007.
25 E. Carrasco, O. Rodríguez De La Fuente, J. M. Rojo. Dislocation emission at the onset of plasticity during nanoindentation in gold. Philosophical Magazine 88(3), 281–296; 2008.
26 G. Ziegenhain, H. M. Urbassek. Effect of material stiffness on hardness: A computational study based on model potentials. Philosophical Magazine 89(26), 2225–2238; 2009.
27 G. Ziegenhain, H. M. Urbassek, A. Hartmaier. Influence of crystal anisotropy on elastic deformation and onset of plasticity in nanoindentation: A simulational study. Journal of Applied Physics 107(6), 061807; 2010.
28 C. Begau, A. Hartmaier, E. George, G. Pharr. Atomistic processes of dislocation generation and plastic deformation during nanoindentation. Acta Materialia 59(3), 934–942; 2011.
29 Y. Liang, Q. Wang, N. Yu *et al.* Study of dislocation nucleation mechanism in nanoindentation process. Nanoscience and Nanotechnology Letters 5(5), 536–541; 2013.
30 I. Salehinia, D. F. Bahr. The impact of a variety of point defects on the inception of plastic deformation in dislocation-free metals. Scripta Materialia 66(6), 339–342; 2012.
31 I. Salehinia, S. Lawrence, D. Bahr. The effect of crystal orientation on the stochastic behavior of dislocation nucleation and multiplication during nanoindentation. Acta Materialia 61(5), 1421–1431; 2013.
32 I.-C. Cheng, A. M. Hodge. Strength scale behavior of nanoporous Ag, Pd and Cu foams. Scripta Materialia 69(4), 295–298; 2013.
33 N. Naveen Kumar, R. Tewari, P. Durgaprasad *et al.* Active slip systems in b.c.c. iron during nanoindentation: A molecular dynamics study. Computational Materials Science 77, 260–263; 2013.
34 J. Alcalá, R. Dalmau, O. Franke *et al.* Planar defect nucleation and annihilation mechanisms in nanocontact plasticity of metal surfaces. Physical Review Letters 109(7), 075502; 2012.
35 J. Hagelaar, E. Bitzek, C. Flipse, P. Gumbsch. Atomistic simulations of the formation and destruction of nanoindentation contacts in tungsten. Physical Review B 73(4),045425; 2006.
36 C. Ruestes, A. Stukowski, Y. Tang *et al.* Atomistic simulation of tantalum nanoindentation: Effects of indenter diameter, penetration velocity, and interatomic potentials on defect mechanisms and evolution. Materials Science and Engineering: A 613, 390–403; 2014.
37 T. Remington, C. Ruestes, E. Bringa *et al.* Plastic deformation in nanoindentation of tantalum: A new mechanism for prismatic loop formation. Acta Materialia 78, 378–393; 2014.

38 Y. Gao, C. J. Ruestes, D. R. Tramontina, H. M. Urbassek. Comparative simulation study of the structure of the plastic zone produced by nanoindentation. Journal of the Mechanics and Physics of Solids 75, 58–75; 2015.
39 A. V. Verkhovtsev, A. V. Yakubovich, G. B. Sushko *et al.* Molecular dynamics simulations of the nanoindentation process of titanium crystal. Computational Materials Science 76, 20–26; 2013.
40 D. E. Kim, S. I. Oh. Atomistic simulation of structural phase transformations in monocrystalline silicon induced by nanoindentation. Nanotechnology 17, 2259; 2006.
41 K. Mylvaganam, L. C. Zhang, P. Eyben *et al.* Evolution of metastable phases in silicon during nanoindentation: mechanism analysis and experimental verification. Nanotechnology 20, 305705; 2009.
42 S. Goel, N. H. Faisal, X. Luo, J. Yan, A. Agrawal. Nanoindentation of polysilicon and single crystal silicon: Molecular dynamics simulation and experimental validation. Journal of Physics D 47, 275304; 2014.
43 M. Lodes, A. Hartmaier, M. Göken, K. Durst Influence of dislocation density on the pop-in behavior and indentation size effect in CaF_2 single crystals: Experiments and molecular dynamics simulations. Acta Materialia 59(11) 4264–4273; 2011.
44 I. Szlufarska, A. Nakano, P. Vashishta. A crossover in the mechanical response of nanocrystalline ceramics. Science 309(5736), 911–914; 2005.
45 I. Szlufarska, R. Kalia, A. Nakano, P. Vashishta. Atomistic mechanisms of amorphization during nanoindentation of SiC: A molecular dynamics study. Physical Review B 71, 174113; 2005, http://doi.org/10.1103/PhysRevB.71.174113 doi:10.1103/PhysRevB.71.174113.
46 H-P. Chen, R. K. Kalia, A. Nakano *et al.* Multimillion-atom nanoindentation simulation of crystalline silicon carbide: Orientation dependence and anisotropic pileup. Journal of Applied Physics 102, 063514; 2007.
47 I. Szlufarska, R. K. Kalia, A. Nakano, P. Vashishta. A molecular dynamics study of nanoindentation of amorphous silicon carbide, Journal of Applied Physics 102 (2007) 023509.
48 M. Mishra, I. Szlufarska. Possibility of high-pressure transformation during nanoindentation of SiC. Acta Materialia 57(20), 6156–6165; 2009; dx.doi.org/10.1016/j.actamat.2009.08.041 .
49 C. Lee, X. Wei, J. W. Kysar, J. Hone. Measurement of the elastic properties and intrinsic strength of monolayer grapheme. Science 321(5887), 385–388; 2008.
50 X. Tan, J. Wu, K. Zhang *et al.* Nanoindentation models and young's modulus of monolayer graphene: A molecular dynamics study. Applied Physics Letters 102(7), 071908; 2013.
51 Z. Sha, Q. Wan, Q. Pei *et al.* On the failure load and mechanism of polycrystalline graphene by nanoindentation. Scientific Reports 4: 7437.
52 A. Gouldstone, N. Chollacoop, M. Dao *et al.* Indentation across size scales and disciplines: Recent developments in experimentation and modeling. Acta Materialia 55(12), 4015–4039; 2007.
53 W. Paul, D. Oliver, P. Grütter. Indentation-formed nanocontacts: an atomic-scale perspective. Physical Chemistry Chemical Physics 16(18), 8201–8222; 2014.

54 B. Bhushan, A. V. Kulkarni, W. Bonin, J. T. Wyrobek. Nanoindentation and picoindentation measurements using a capacitive transducer system in atomic force microscopy. Philosophical Magazine A 74(5), 1117–1128; 1996.
55 B. Bhushan. Nano-to microscale wear and mechanical characterization using scanning probe microscopy. Wear 251(1), 1105–1123; 2001.
56 M. Jin, A. Minor, D. Ge, J. Morris. Study of deformation behavior of ultrafine-grained materials through in situ nanoindentation in a transmission electron microscope. Journal of Materials Research 20(07), 1735–1740; 2005.
57 A.M. Minor, S.S. Asif, Z. Shan *et al.* A new view of the onset of plasticity during the nanoindentation of aluminium. Nature Materials 5(9), 697–702; 2006.
58 R. Ghisleni, K. Rzepiejewska-Malyska, L. Philippe *et al.* In situ SEM indentation experiments: Instruments, methodology, and applications. Microscopy Research and Technique 72(3), 242–249; 2009.
59 J.T.M. de Hosson. Advances in transmission electron microscopy: in situ straining and in situ compression experiments on metallic glasses. Microscopy Research and Technique 72(3), 250–260; 2009.
60 H.D. Espinosa, R. Bernal, T. Filleter. In situ TEM electromechanical testing of nanowires and nanotubes. Small 8(21), 3233–3252; 2012.
61 J. Jiang, T. Britton, A. Wilkinson. Measurement of geometrically necessary dislocation density with high resolution electron backscatter diffraction: Effects of detector binning and step size. Ultramicroscopy 125, 1–9; 2013.
62 S. Plimpton. Fast parallel algorithms for short-range molecular dynamics. Journal of Computational Physics 117(1), 1–19; 1995.
63 I.T. Todorov, W. Smith, K. Trachenko, M.T. Dove. Dl_poly_3: new dimensions in molecular dynamics simulations via massive parallelism. Journal of Materials Chemistry 16(20), 1911–1918; 2006.
64 D. Kilymis, J.-M. Delaye. Nanoindentation studies of simplified nuclear glasses using molecular dynamics. Journal of Non-Crystalline Solids 401, 147–153; 2014.
65 M.I. Mendelev, S. Han, D.J. Srolovitz *et al.* Development of new interatomic potentials appropriate for crystalline and liquid iron. Philosophical Magazine 83, 3977; 2003.
66 X.D. Dai, Y. Kong, J.H. Li, B. X. Liu. Extended Finnis-Sinclair potential for b.c.c. and f.c.c. metals and alloys. Journal of Physics: Condensed Matter 18(19), 4527–4542; 2006.
67 M. Mendelev, M. Kramer, C. Becker, M. Asta. Analysis of semi-empirical interatomic potentials appropriate for simulation of crystalline and liquid Al and Cu. Philosophical Magazine 88, 1723; 2008.
68 Y. Mishin, M.J. Mehl, D.A. Papaconstantopoulos *et al.* Structural stability and lattice defects in copper: Ab initio, tight-binding, and embedded-atom calculations. Physical Review B, 63, 224106; 2001.
69 Y. Gao, C.J. Ruestes, H.M. Urbassek. Nanoindentation and nanoscratching of iron: Atomistic simulation of dislocation generation and reactions. Composites and Materials Science 90, 232–240; 2014.
70 J. Li, K.J. van Vliet, T. Zhu *et al.* Atomistic mechanisms governing elastic limit and incipient plasticity in crystals. Nature 418(6895), 307–310; 2002.

71 C.A. Schuh, J.K. Mason, A.C. Lund. Quantitative insight into dislocation nucleation from high-temperature nanoindentation experiments. Nature Materials 4(8), 617–621; 2005.
72 G. Ziegenhain, H.M. Urbassek. Nanostructured surfaces yield earlier: Molecular dynamics study of nanoindentation into adatom islands. Physical Review B, 81, 155456; 2010.
73 R. Wagner, L. Ma, F. Tavazza, L. Levine. Dislocation nucleation during nanoindentation of aluminum. Journal of Applied Physics 104(11), 114311; 2008.
74 B. Luan, M.O. Robbins. The breakdown of continuum models for mechanical contacts. Nature 435(7044), 929–932; 2005.
75 T-H. Fang, C-I. Weng, J-G. Chang. Molecular dynamics analysis of temperature effects on nanoindentation measurement. Materials Science and Engineering: A, 357(1), 7–12; 2003.
76 C-L. Liu, T-H. Fang, J-F. Lin. Atomistic simulations of hard and soft films under nanoindentation. Materials Science and Engineering: A, 452, 135–141; 2007.
77 S. Goel, S.S. Joshi, G. Abdelal, A. Agrawal. Molecular dynamics simulation of nanoindentation of Fe_3C and Fe_4C. Materials Science and Engineering: A, 597, 331–341; 2014.
78 D. Lorenz, A. Zeckzer, U. Hilpert *et al.* Pop-in effect as homogeneous nucleation of dislocations during nanoindentation. Physical Review B, 67(17), 172101; 2003.
79 M. Biener, J. Biener, A. Hodge, A. Hamza. Dislocation nucleation in b.c.c. Ta single crystals studied by nanoindentation. Physical Review B, 76(16), 165422; 2007.
80 A.F. Voter. Hyperdynamics: Accelerated molecular dynamics of infrequent events. Physical Review Letters 78, 3908–3911; 1997.
81 A.F. Voter. Parallel replica method for dynamics of infrequent events. Physical Review B, 57, R13985–R13988; 1998, http://doi.org/10.1103/PhysRevB.57.R13985 doi:10.1103/PhysRevB.57.R13985.
82 M.R. Sorensen, A.F. Voter. Temperature-accelerated dynamics for simulation of infrequent events. Journal of Chemical Physics 112, 9599; 2000.
83 W.K. Kim, M.L. Falk. Accelerated molecular dynamics simulation of low-velocity frictional sliding. Modeling and Simulation in Materials Science and Engineering. 18, 034003; 2010.
84 Q. Li, Y. Dong, D. Perez *et al.* Speed dependence of atomic stick-slip friction in optimally matched experiments and molecular dynamics simulations. Physical Review Letters 106, 126101; 2011, https://doi.org/10.1103/PhysRevLett.106.126101 doi:10.1103/PhysRevLett.106.126101.
85 M.S. Daw, M.I. Baskes. Embedded-atom method: Derivation and application to impurities, surfaces, and other defects in metals. Physical Review B, 29(12), 6443; 1984.
86 N. Gunkelmann, EM. Bringa, K. Kang *et al.* Polycrystalline iron under compression: Plasticity and phase transitions. Physical Review B, 86(14), 144111; 2012.
87 Y. Zhu, X. Liao, X. Wu. Deformation twinning in nanocrystalline materials. Progress in Materials Science 57(1), 1–62; 2012.
88 J.A. Martinez, D.E. Yilmaz, T. Liang *et al.* Fitting empirical potentials: Challenges and methodologies. Current Opinion in Solid State and Materials Science 17(6), 263–270; 2013.

89 A. Reuss, Berechnung der FließŸgrenze von Mischkristallen auf Grund der Plastizitätsbedingung für Einkristalle. Zeitschrift für Angewandte Mathematik und Mechanik (ZAMM) 9(1), 49–58; 1929.
90 W. Voigt. Lehrbuch der Kristallphysik (mit Ausschluß der Kristalloptik). BG Teubner; 1928.
91 M.A. Meyers, K.K. Chawla. Mechanical Behavior of Materials. Cambridge University Press, Cambridge; 2009.
92 Y. Li, D. Siegel, J. Adams, X.-Y. Liu. Embedded-atom-method tantalum potential developed by the force-matching method. Physical Review B, 67(12), 125101; 2003.
93 D. Tramontina, P. Erhart, T. Germann *et al.* Molecular dynamics simulations of shock-induced plasticity in tantalum. High Energy Density Physics 10, 9–15; 2014.
94 G. Kimminau, P. Erhart, E. M. Bringa *et al.* Phonon instabilities in uniaxially compressed f.c.c. metals as seen in molecular dynamics simulations. Physical Review B, 81(9), 092102; 2010.
95 G. Ziegenhain, A. Hartmaier, H. M. Urbassek. Pair vs many-body potentials: Influence on elastic and plastic behavior in nanoindentation of f.c.c. metals. Journal of the Mechanics and Physics of Solids 57(9), 1514–1526; 2009.
96 V. Vitek. Intrinsic stacking faults in body-centred cubic crystals. Philosophical Magazine 73, 773; 1968.
97 M. Duesbery, V. Vitek. Plastic anisotropy in b.c.c. transition metals. Acta Materialia 46(5), 1481–1492; 1998.
98 A. Cao. Shape memory effects and pseudoelasticity in b.c.c. metallic nanowires. Journal of Applied Physics 108(11), 113531; 2010.
99 G.S. Fanourgakis, V. Pontikis, G. Zérah. Phase stability and intrinsic stacking faults in aluminum under pressure. Physical Review B, 67(9), 094102; 2003.
100 H.R. Hertz, Ueber die Berührung fester elastischer Körper. Journal für die Reine und Angewandte Mathematik. 92, 156–171; 1882.
101 J.J. Vlassak, W.D. Nix. Indentation modulus of elastically anisotropic half spaces. Philosophical Magazine A, 67(5), 1045–1056; 1993.
102 J.J. Vlassak, W. Nix. Measuring the elastic properties of anisotropic materials by means of indentation experiments. Journal of the Mechanics and Physics of Solids 42(8), 1223–1245; 1994.
103 A. Stukowski, K. Albe, Extracting dislocations and non-dislocation crystal defects from atomistic simulation data. Modelling and Simulation in Materials Science and Engineering 18(8), 085001; 2010.
104 A. Henderson. Paraview guide, a parallel visualization application. Kitware Inc.; 2007, http://www.paraview.org
105 Y. Lee, J.Y. Park, S.Y. Kim *et al.* Atomistic simulations of incipient plasticity under al(111) nanoindentation. Mechanics of Materials 37, 1035; 2005.
106 K. Kang, V.V. Bulatov, W. Cai. Singular orientations and faceted motion of dislocations in body-centered cubic crystals. Proceedings of the National Academy of Sciences 109(38), 15174–15178; 2012.
107 J.G. Swadener, E.P. George, G.M. Pharr. The correlation of the indentation size effect measured with indenters of various shapes. Journal of the Mechanics and Physics of Solids 50(4), 681–694; 2002.

108 B.A. Remington, P. Allen, E.M. Bringa *et al.* Material dynamics under extreme conditions of pressure and strain rate. Materials Science and Technology 22(4), 474–488; 2006.

109 P. Cao, X. Lin, H. S. Park. Strain-rate and temperature dependence of yield stress of amorphous solids via a self-learning metabasin escape algorithm. Journal of the Mechanics and Physics of Solids 68, 239–250; 2014.

110 D. Warner, W. Curtin, S. Qu. Rate dependence of crack-tip processes predicts twinning trends in f.c.c. metals. Nature Materials 6(11), 876–881; 2007.

111 R. Ravelo, T. Germann, O. Guerrero *et al.* Shock-induced plasticity in tantalum single crystals: Interatomic potentials and large-scale molecular-dynamics simulations. Physical Review B, 88(13), 134101; 2013.

112 A. Arsenlis, D. Parks. Crystallographic aspects of geometrically-necessary and statistically-stored dislocation density. Acta Materialia 47(5), 1597–1611; 1999.

113 H. Gao, Y. Huang. Geometrically necessary dislocation and size-dependent plasticity. Scripta Materialia 48(2), 113–118; 2003.

114 J. Hua, A. Hartmaier. Determining burgers vectors and geometrically necessary dislocation densities from atomistic data. Modelling and Simulation in Materials Science and Engineering 18(4), 045007; 2010.

115 K. Durst, B. Backes, M. Göken. Indentation size effect in metallic materials: correcting for the size of the plastic zone. Scripta Materialia 52(11), 1093–1097; 2005.

116 K. Durst, B. Backes, O. Franke, M. Göken. Indentation size effect in metallic materials: Modeling strength from pop-in to macroscopic hardness using geometrically necessary dislocations. Acta Materialia 54(9), 2547–2555; 2006.

117 Y.L. Chiu, A.H.W. Ngan. A TEM investigation on indentation plastic zones in Ni_3Al(Cr, B) single crystals. Acta Materialia 50(10), 2677–2691; 2002.

118 O. Casals, J. Očenášek, J. Alcala. Crystal plasticity finite element simulations of pyramidal indentation in copper single crystals. Acta Materialia 55(1), 55–68; 2007.

119 O. Casals, S. Forest. Finite element crystal plasticity analysis of spherical indentation in bulk single crystals and coatings. Computational Materials Science 45(3), 774–782; 2009.

120 W. Yao, C. Krill, B. Albinski *et al.* Plastic material parameters and plastic anisotropy of tungsten single crystal: a spherical micro-indentation study. Journal of Materials Science 49(10), 3705–3715; 2014.

121 W. D. Nix, H. Gao. Indentation size effects in crystalline materials: A law for strain gradient plasticity. Journal of the Mechanics and Physics of Solids 46(3), 411–425; 1998.

122 R.K. Abu Al-Rub, G.Z. Voyiadjis. Analytical and experimental determination of the material intrinsic length scale of strain gradient plasticity theory from micro-and nano-indentation experiments. International Journal of Plasticity 20(6), 1139–1182; 2004.

123 M.S. De Guzman, G. Neubauer, P. Flinn, W. D. Nix. The role of indentation depth on the measured hardness of materials. Materials Research Society Symposium Proceedings 308, 613; 1993.

124 N. A. Stelmashenko, M. G. Walls, L. M. Brown, Y. V. Milman. Microindentations on W and Mo oriented single crystals: an STM study. Acta Metallurgica et Materialia 41(10), 2855–2865; 1993.

125 S. Kucharski, Z. Mroz. Identification of plastic hardening parameters of metals from spherical indentation tests. Materials Science and Engineering: A, 318(1), 65–76; 2001.

126 D. Faghihi, G. Z. Voyiadjis. Determination of nanoindentation size effects and variable material intrinsic length scale for body-centered cubic metals. Mechanics of Materials 44, 189–211; 2012.

127 H-J. Chang, M. Fivel, D. Rodney, M. Verdier. Multiscale modelling of indentation in f.c.c. metals: From atomic to continuum. Comptes Rendus Physique 11, 285; 2010.

128 P.S. Engels, C. Begau, S. Gupta *et al.* Multiscale modeling of nanoindentation: From atomistic to continuum models, in: A. Tiwari (ed.), Nanomechanical Analysis of High Performance Materials, no. 203 in Solid Mechanics and Its Applications, p. 285. Springer; 2014.

129 J. Knap, M. Ortiz. An analysis of the quasicontinuum method. Journal of the Mechanics and Physics of Solids 49(9), 1899–1923; 2001.

130 J. Knap, M. Ortiz. Effect of indenter-radius size on au (001) nanoindentation. Physical Review Letters 90(22), 226102; 2003.

131 A. Hasnaoui, P. Derlet, H. Van Swygenhoven. Interaction between dislocations and grain boundaries under an indenter–a molecular dynamics simulation. Acta Materialia 52(8), 2251–2258; 2004.

132 W. Paul, D. Oliver, Y. Miyahara, P. H. Grütter. Minimum threshold for incipient plasticity in the atomic-scale nanoindentation of au (111). Physical Review Letters 110(13), 135506; 2013.

15

Continuum Modelling and Simulation of Indentation in Transparent Single Crystalline Minerals and Energetic Solids

J.D. Clayton[1], *B.B. Aydelotte*[2], *R. Becker*[1], *C.D. Hilton*[3] *and J. Knap*[4]

[1] *Impact Physics, RDRL-WMP-C, US ARL, APG, MD, USA*
[2] *Lethal Mechanisms, RDRL-WML-H, US ARL, APG, MD, USA*
[3] *Oak Ridge Institute for Science and Education, US ARL, APG, MD, USA*
[4] *Computational Sciences, RDRL-CIH-C, US ARL, APG, MD, USA*

15.1 Introduction

Models of indentation provide insight into material behaviours not available from standard indentation experiments alone. Traditional indentation experiments—whether conducted at nano-, micron-, or milli-meter scales—provide indenter force versus displacement data and enable the possibility of post-mortem surface characterization. Details regarding the material response beneath the indented surface are usually not recorded or even accessible. In contrast, models of the indentation process may provide descriptions of the response of finite volumes of the indented material, including the local sub-surface response. Model results can supplement experimental data, providing insight into the most likely operative deformation mechanisms and their driving forces such as local stress components associated with the onset of defect generation or motion.

In the context of crystalline solids of interest in the present chapter, deformations consist of elasticity (i.e., mechanically reversible motions of atoms) [1–3], plasticity resulting from the glide of mobile dislocations [3, 4], and deformation twinning resulting primarily from the collective motions of partial/twinning dislocations [3, 5, 6]. Classical analytical solutions to indentation problems are restricted to relatively simple geometries (e.g., spherical, wedge, or conical indenters) and to relatively simple material constitutive relations (e.g., isotropic linear elasticity or rigid-perfect plasticity) [7–11]. Such solutions are not usually physically relevant for describing the response of single crystalline materials of present interest, which tend to display highly anisotropic and highly nonlinear elastic-inelastic deformation mechanisms, as will become evident in the context of results shown later. Purely atomistic simulations of indentation processes [12, 13] are restricted regarding maximum system sizes (e.g., nano-meter regimes) and time scales [e.g., very high loading rates for molecular dynamics (MD) simulations]. In contrast, modelling techniques invoked in the present work (consisting of continuum nonlinear elasticity, continuum crystal plasticity, and phase field theory) enable numerical simulations of processes spanning length scales from nano-meters to milli-meters and time scales commensurate with real quasi-static and dynamic indentation experiments.

Applied Nanoindentation in Advanced Materials, First Edition. Edited by Atul Tiwari and Sridhar Natarajan.

Atomic-continuum multi-scale simulation methods [14, 15], though not discussed further in the present work and often restricted to constant-temperature (e.g., 0 K) simulations, may also alleviate length scale restrictions and can be highly suitable for modelling indentation processes.

The remainder of this chapter consists of three major sections. The first of these sections develops a general continuum theory of material behaviour applicable to the anisotropic mechanical response of single crystals that may undergo large elastic, plastic, and/or twinning deformations. This theory, which essentially merges previous nonlinear elastic-plastic [16, 17] and nonlinear elastic-twinning [18, 19] models, invokes a continuum slip system-level representation of crystal plasticity and a phase field representation of deformation twinning, with state variables and order parameters associated with cumulative slip and twinning transformations, respectively. Spatial gradients of these field variables, physically representative of dislocation densities and twin boundaries, provide a realistic context for regularization of the constitutive model via material length scales, facilitating mesh-size independent numerical solutions of boundary value problems.

Particular assumptions reduce the general theory to forms used previously for modelling indentation in transparent single crystals [16–19]. The general framework of section 15.2 is similar to a phase field-crystal plasticity model reported recently elsewhere for magnesium single crystals [20, 21]. The present contribution, more applicable to stiffer non-metals with a relatively large ratio of shear to bulk modulus [22], explicitly considers nonlinear elastic response, while theory in [20] applies linear elasticity which tends to be appropriate for ductile metals only when pressures are relatively low. Phase field theory for twinning enables prediction of twin morphologies without the need for user-prescribed shearing resistances to twinning deformation typically invoked in pseudo-slip continuum representations [3, 23]. A phase field description of dislocation slip is not invoked herein because resolution of individual dislocations in phase field theory [24–26] restricts such simulations to regions too small or refined in length scale for the present applications.

The second major section of this chapter describes application of the constitutive theory towards modelling and simulation of indentation of oriented single crystals of the energetic (i.e., explosive) material cyclotrimethylene trinitramine, abbreviated as RDX (Research Development Explosive). Single crystals of adequate purity are transparent, and the crystal structure of the ambient (α) phase is orthorhombic. Twinning is not observed in this particular energetic material, unlike others [27], so the phase field component of the general theory mentioned above is not invoked for this application. However, a number of possible slip systems have been proposed in the literature [28–32]. The present results and analysis, which extend those documented previously [16, 17] and include new comparisons with experimental and atomic simulation data, provide further insight into magnitudes of driving forces needed to initiate slip, as well as suitability of various sets of anisotropic elastic constants reported in the atomic modelling community [33, 34] and experimental literature [35–39].

The third major section of this chapter describes application of the constitutive theory towards modelling and simulation of indentation of oriented single crystals of the common mineral calcite, which are transparent and of rhombohedral (i.e., trigonal) symmetry under ambient conditions. Calcite twins readily when subjected to concentrated loading including indentation, and twinning may be reversible for loads not exceeding a threshold magnitude [5, 40]. At room temperature, dislocation slip is very limited if existent [40, 41]. Therefore, the crystal plasticity component of the general theory is not invoked, and the constitutive model reduction for nonlinear elasticity and twinning is used [18, 19, 42], with a variational approach suitable for quasi-static indentation at constant ambient temperature. Previously published results focused on two-dimensional (2D) simulations of wedge indentation [18]; the present paper newly considers fully three-dimensional (3D) simulations of spherical indentation and includes comparison with recent experimental data and observations [43].

Notations of vector calculus and continuum mechanics are used. Vectors and higher-order tensors are written in bold-italic font, and scalars and scalar components are written in italic font. Components are referred to a fixed Cartesian frame of reference.

15.2 Theory: Material Modelling

A general continuum theory is set forth in section 15.2.1, which encompasses both slip and twinning behaviours in a way not modelled before (crystal plasticity + phase field). Specialization to slip only and to twinning on a single system follow in respective sections 15.2.2 and 15.2.3. Familiarity of the reader with fundamentals of nonlinear continuum mechanics, nonlinear elasticity, and finite plasticity is assumed [1–4].

15.2.1 General Multi-field Continuum Theory

Let $\boldsymbol{x} = \boldsymbol{x}(\boldsymbol{X}, t)$ denote motion of a material particle. The deformation gradient is decomposed multiplicatively as:

$$\boldsymbol{F} = \nabla \boldsymbol{x} = \boldsymbol{F}^E \boldsymbol{F}^\eta \boldsymbol{F}^\gamma \tag{15.1}$$

where $\nabla = \partial/\partial \boldsymbol{X}$ is the material gradient, $\boldsymbol{F}^E$ is thermoelastic deformation, $\boldsymbol{F}^\eta$ is twinning deformation, and $\boldsymbol{F}^\gamma$ is deformation due to plastic slip (i.e., due to dislocation glide). The spatial velocity gradient is:

$$\boldsymbol{L} = \partial \boldsymbol{v}/\partial \boldsymbol{x} = \dot{\boldsymbol{F}}^E \boldsymbol{F}^{E-1} + \boldsymbol{F}^E \boldsymbol{L}^\eta \boldsymbol{F}^{E-1} + \boldsymbol{F}^E \boldsymbol{L}^\gamma \boldsymbol{F}^{E-1} \tag{15.2}$$

where superposed dots denote material time derivatives, $\boldsymbol{v} = \dot{\boldsymbol{x}}$, and contributions from twinning and slip are, respectively:

$$\boldsymbol{L}^\eta = \dot{\boldsymbol{F}}^\eta \boldsymbol{F}^{\eta-1} = \sum_i \dot{\varphi}^i \gamma^T \boldsymbol{s}^i \otimes \boldsymbol{m}^i, \quad \boldsymbol{L}^\gamma = \boldsymbol{F}^\eta \dot{\boldsymbol{F}}^\gamma \boldsymbol{F}^{\gamma-1} \boldsymbol{F}^{\eta-1} = \sum_j \dot{\gamma}^j \boldsymbol{s}^j \otimes \boldsymbol{m}^j \tag{15.3}$$

Regarding twinning kinematics, index i runs over the number of twin systems, $\varphi^i = \varphi(\eta^i) : [0,1] \to [0,1]$ is an interpolation function, γ^T is the twinning eigenshear for a complete structural transformation that depends on crystallography of the material under consideration, and $\eta^i \in [0,1]$ is the order parameter for twin system i. At material point $\boldsymbol{X}$, the parent crystal is characterized by $\eta^i(\boldsymbol{X}) = 0 \forall i$; a single fully formed twin is characterized by $\eta^i(\boldsymbol{X}) = 1$ for a single system i; and twin boundaries obey $\eta^i(\boldsymbol{X}) \in (0,1)$ for at least one twin system i. Regarding plastic slip kinematics, index j runs over the number of slip systems, and γ^j is the cumulative slip on system j. In each case, $\boldsymbol{s}$ and $\boldsymbol{m}$ are the orthonormal direction of shear and the normal to the habit plane or slip plane, respectively. The equation for the plastic velocity gradient can be expanded to distinguish among contributions from slip in parent and twinned regions, with $\boldsymbol{s}$ and $\boldsymbol{m}$ in the latter updated via the rotation/reflection appropriate to a given twinning mode [3, 6, 23]. Thermoelastic deformation and volume change are quantified via:

$$\boldsymbol{C}^E = (\boldsymbol{F}^E)^{\mathrm{T}}\boldsymbol{F}^E, \quad J = \det \boldsymbol{F} = \sqrt{\det \boldsymbol{C}^E} \tag{15.4}$$

with the T superscript denoting transposition and noting that twinning and slip are isochoric.

Let θ denote temperature. The Helmholtz free energy density per unit reference volume is of the general functional form:

$$\psi = \psi(\boldsymbol{C}^E, \eta^i, \nabla\eta^i, \gamma^j, \nabla\gamma^j, \theta) \tag{15.5}$$

Dependence of energy on the order parameters and their spatial gradients (taken here with respect to material coordinates) enables quantification of twin boundary surface energy [18, 19, 42]. Dependence of free energy on slip and slip gradients enables quantification of residual elastic energies associated with statistically stored and geometrically necessary dislocation densities in an approximate sense (more precisely, geometrically necessary dislocations can be associated with the curl of plastic deformation [3, 44]). Cauchy stress and entropy, respectively, obey the usual finite strain thermoelastic relations:

$$\boldsymbol{\sigma} = 2J^{-1}\boldsymbol{F}^E(\partial\psi/\partial\boldsymbol{C}^E)\boldsymbol{F}^{E\,\mathrm{T}}, \quad S = -\partial\psi/\partial\theta \tag{15.6}$$

Kinetic equations are of the same general form as (2.5). For example, letting M^{ik} denote mobility coefficients, twinning can be prescribed to obey the Ginzburg-Landau or Allen-Cahn type kinetics [42]:

$$\dot{\eta}^i = -\sum_k M^{ik}(\partial\psi/\partial\eta^k - \nabla\cdot\partial\psi/\partial\nabla\eta^k) \tag{15.7}$$

Letting τ^j denote the resolved shear stress that is work conjugate to the slip rate on system j, plasticity kinetics can be expressed in the general form:

$$\dot{\gamma}^j = \dot{\gamma}^j(\tau^j, \eta^i, \nabla\eta^i, \gamma^j, \nabla\gamma^j, \theta) \tag{15.8}$$

with more detailed expressions to follow later.

15.2.2 Crystal Plasticity Theory

Particular assumptions can be used to reduce the general theory of section 15.2.1 to nonlinear crystal plasticity theories similar to those described in [4, 45, 46] and implemented in [16, 17]. These assumptions are that twinning is absent and that gradient contributions to slip resistance are not included in the thermodynamic potentials or kinetic equations. Equations (15.1)–(15.3), (15.5), and (15.8) then reduce to:

$$\boldsymbol{F} = \nabla\boldsymbol{x} = \boldsymbol{F}^E\boldsymbol{F}^\gamma \tag{15.9}$$

$$\boldsymbol{L} = \partial\boldsymbol{v}/\partial\boldsymbol{x} = \dot{\boldsymbol{F}}^E\boldsymbol{F}^{E-1} + \boldsymbol{F}^E\boldsymbol{L}^\gamma\boldsymbol{F}^{E-1} \tag{15.10}$$

$$\boldsymbol{L}^\gamma = \dot{\boldsymbol{F}}^\gamma\boldsymbol{F}^{\gamma-1} = \sum_j \dot{\gamma}^j\boldsymbol{s}^j \otimes \boldsymbol{m}^j \tag{15.11}$$

$$\psi = \psi(\boldsymbol{C}^E, \gamma^j, \theta) \tag{15.12}$$

$$\dot{\gamma}^j = \dot{\gamma}^j(\tau^j, \gamma^k, \theta) = \dot{\gamma}_0 \left| \frac{\tau^j}{g^j} \right|^m \mathrm{sgn}(\tau^j) \tag{15.13}$$

with $\dot{\gamma}_0$ a reference slip rate (material constant) and m describing strain rate sensitivity. Slip system shearing resistance $g^j(\gamma^k, \theta)$ potentially depends on temperature and cumulative slip on up to all available slip systems, allowing for thermal softening and strain hardening, for example. In [16, 17, 45], a particular form of stress-strain law for nonlinear elasticity invoking a logarithm of the right thermoelastic stretch tensor [i.e., $(\ln \boldsymbol{C}^E)/2$]) is used that approximately satisfies the first of (15.6) when the material is anisotropic.

15.2.3 Phase Field Theory for Twinning

Particular constitutive assumptions can be used to reduce the general theory of section 15.2.1 to nonlinear phase field models for twinning implemented in [18, 19, 42, 47]. These assumptions are that plastic slip due to dislocation glide is absent and that only one twin system is active. Further assume isothermal and quasi-static conditions, such that a variational approach can be used, with kinetic equations replaced by Euler-Lagrange equations obtained from free energy minimization [42]. Equations (15.1), (15.3) and (15.5) are then replaced with:

$$\boldsymbol{F} = \nabla \boldsymbol{x} = \boldsymbol{F}^E \boldsymbol{F}^\eta \tag{15.14}$$

$$\boldsymbol{F}^\eta = 1 + \varphi(\eta)\gamma^T \boldsymbol{s} \otimes \boldsymbol{m} \tag{15.15}$$

$$\psi = \psi(\boldsymbol{C}^E, \eta, \nabla\eta) \tag{15.16}$$

Free energy per unit reference volume is prescribed of the following functional form, with W the elastic strain energy density, a double-well potential with respect to the order parameter for twinning η, and isotropic twin boundary surface energy:

$$\psi = W[\boldsymbol{C}^E(\boldsymbol{F}, \eta)] + A\eta^2(1-\eta)^2 + \kappa|\nabla\eta|^2 \tag{15.17}$$

The kinetic equation for twinning transformation in (2.7) is replaced with the equilibrium equation [18, 42]:

$$(\partial W/\partial\eta)|_F = -A\eta(2\eta^2 - 3\eta + 1) + 2\kappa\nabla^2\eta \tag{15.18}$$

Material constants A and κ are related to surface energy Γ and unstressed twin boundary thickness (regularization width) l via [18, 42]:

$$A = 12\Gamma/l, \quad \kappa = 3\Gamma l/4 \tag{15.19}$$

Previous work has invoked several different nonlinear elastic potentials W incorporating Green-Lagrange strain for anisotropic elasticity [42, 47] and neo-Hookean theory for compressible isotropic elasticity [18, 19]. Reduction of the nonlinear theory to geometrically and materially linear elasticity is also documented elsewhere [18, 42]. Interpolation functions used previously for twinning kinematics include a cubic polynomial [18, 42] and a Fermi–Dirac exponential function with constant $k > 0$ [19]:

$$\varphi(\eta) = 3\eta^2 - 2\eta^3, \quad \varphi(\eta) = 1/[1 + e^{-2k(\eta - \frac{1}{2})}] \tag{15.20}$$

15.3 Application: Indentation of RDX Single Crystals

Energetic molecular crystals are found in various military and industrial systems involving explosives and propellants. For example, certain kinds of plastically bonded explosive (PBX) consist of energetic single crystals embedded in a waxy binder phase [48] (see Figure 15.1, left). Defects in energetic materials are thought to affect their initiation sensitivity. Local stresses may concentrate in the vicinity of cracks, pores, or lattice defects, including dislocations and point defects, which may, in turn, affect initiation of reactions associated with burning or detonation [49]. In plastically anisotropic single crystals, sufficient availability of slip systems associated with glissile dislocations may lower peak stresses and decrease sensitivity to shock initiation [50].

The focus of this section is the mechanical behavior of the energetic material cyclotrimethylene trinitramine ($C_3H_6N_6O_6$), typically referred to as RDX (Research Development Explosive). Single crystals of RDX belong to an orthorhombic space group with eight molecules per unit cell. Dislocations in RDX have been characterized using etch pit [51] and X-ray topographic [52, 53] techniques. Likely slip systems for dislocation glide in RDX have been suggested upon analysis of anisotropic hardness profiles [28] and nano-indentation experiments [29, 30]. The latter experiments also provide an estimate of the critical resolved shear stress associated with slip initiation, thought to be on the order of the theoretical strength of the crystal (i.e., $G_0/10 - G_0/20$, with G_0 a representative ambient elastic shear modulus), which corresponds physically to homogeneous dislocation nucleation in an otherwise perfect crystal. Inelastic behaviors of RDX single crystals have also been probed using shock physics experiments [50, 54] and molecular dynamics simulations [31, 32, 55].

Continuum crystal plasticity theory permits mesoscale modeling of materials behavior at length scales larger than feasible using atomic or molecular models, but with greater physical fidelity than that afforded by macroscopic elastic-plastic models that omit description of anisotropy and slip system activity. Grain interactions can be studied in direct numerical simulations via finite element models, wherein each crystal of a polycrystal is resolved geometrically [45, 46]. Crystal plasticity models have been implemented elsewhere to study shock loading of the energetic materials cyclotetramethylene tetranitramine (HMX) [27] and pentaerythritol tetranitrate (PETN) [56, 57]. A crystal plasticity model for RDX [16, 17] is utilized in the present work. The right side of Figure 15.1 shows single crystals of low defect content and high purity that can be grown to several cm in size.

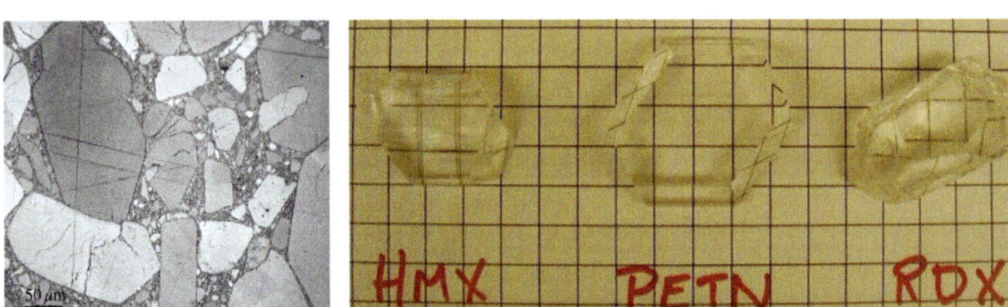

Figure 15.1 Polycrystal with binder (left, PBX-9404) and single crystals (right, 1 cm grid) [courtesy D. Hooks, LANL].

15.3.1 Review of Prior Work

The single crystal elastic-plastic model developed in [16, 17] extends the general framework of Becker [45] first formulated for cubic crystals loaded to possibly very high pressures. This model, which follows the constitutive equations outlined here in section 15.2.2, was applied to crystals with orthorhombic symmetry characteristic of RDX. Anisotropic elastic constants and pressure-dependent compressibility were obtained from the experimental literature [35, 36, 58]. Six slip systems (from four different families of systems) were implemented following analysis of indentation loading profiles and surface impressions [28–30]. The model was applied to study indentation (with a spherical indenter) of (001), (021), and (210) faces of single crystals of RDX. The diamond indenter had an initial (undeformed) radius of 1482 nm; properties for the nonlinear elastic model used for diamond are tabulated in [17]. Frictionless contact was assumed, and quasi-static implicit finite element calculations were performed using the ALE3D multi-physics code [45].

Results from simulations were compared with numerical predictions from nonlinear isotropic elasticity and analytical isotropic elastic (i.e., Hertzian) solutions [11]. Critical shear strengths associated with slip initiation were assigned upon comparison of results with load excursion data from experiments [29, 30]. Constant (i.e., non-hardening or perfectly plastic) shear strengths g^j were prescribed for each family of slip systems under isothermal conditions, though values were permitted to vary among families in some exploratory simulations [17]. Bulk slip system activity was quantified, and residual surface slip activity was compared with experimental observations.

Results in [16, 17] showed significant influences of elastic anisotropy and elastic nonlinearity on force-displacement data. Model predictions for initial elastic response using constants measured with resonant ultrasound spectroscopy [35] agreed with experimental force-displacement data for indentation on (001), (021), and (210) planes. Model predictions for initial elastic response using constants measured with Brillouin scattering [36] were found to be in reasonable agreement with experiments for indentation on (210), but were stiffer than experiments for indentation on (001) and (021). Orientation (001) was predicted to be elastically most compliant with respect to indentation force, in agreement with experiments.

Model predictions of force for larger indentation depths, wherein local plastic slip was substantial, tended to exceed experimental values regardless of which set of elastic constants [35, 36] was used. The constant strength (i.e., perfectly plastic) slip model implemented in [16, 17] was unable to replicate nearly horizontal steps in indentation force observed in experiments. Such steps were thought most likely to correspond to cleavage fractures observed in experiments [29] (at the surface) or material failure not directly observed (sub-surface), either of which could contribute to discrepancies between simulations and experiments, as discussed later. Other potential reasons for differences include uncertainties in true indenter tip geometry, and to discrete slip events of width too fine to be captured by a conventional continuum slip model. Dependencies of shear strength and shear stiffness on slip history and pressure were omitted; incorporation of such physics, for example, as suggested by atomic simulations [33], might provide improved agreement.

Critical shear strengths associated with plastic yielding were estimated to be on the order of $G_0/20$. Simulations suggested that slip planes (010) and {011} contain active systems for indentation on (001), with slip on system (010)[001] dominating the inelastic response; experimental surface observations [28–30] confirm that these and {021} slip planes may also be active. Simulations suggested that slip planes (010) and {021}, and to a lesser extent {011}, are active at the specimen surface for indentation on (210); these same planes have also been confirmed as active in experiments [28–30]. Simulations suggest that planes (010), (011), and (021) contain active systems for indentation on (021); particular slip planes active for this orientation have not been reported from experimental observations. The present simulations suggest that the single slip system (010)[001] provided the largest contribution to the inelastic material response (i.e., the post-yield force versus displacement curve) for indentation on (001) and (021) planes, while five slip systems {021}[100], {011}[100], and (010)[100] all contributed to inelastic response for indentation on (210) planes. Plastic deformation and hysteresis were more extensive for indentation on (021) and (210) than for indentation on (001). Since much plastic deformation occurs within the bulk of the material, and since different slip mechanisms may be prominent at the surface and in the bulk, these results offer new insight into inelastic mechanical behaviour of RDX single crystals not available from experimental observations of residual surface topography alone.

As shown on the left side of Figure 15.2, superior agreement with experimental load excursion data was obtained using the softer elastic constants of [35] rather than the stiffer constants of [36]. Shown on the right side of each part of Figure 15.2 is total cumulative slip on all systems, $\sum \gamma^k$. The lack of symmetry in plastic deformation contours is a result of elastic and plastic anisotropy and the limited number of geometrically independent available systems in RDX.

15.3.2 New Results and Analysis

Additional simulations not reported in [16, 17] have been completed on RDX faces of crystallographic orientations (001), (021), and (210). These new simulations were conducted with elastic constants calculated from recent atomic simulations [34]. Elastic constants are compared in Table 15.1 (B'_0 is the pressure derivative of the ambient bulk modulus). Experimental values have been converted from isentropic (measured) to isothermal constants using standard thermodynamic formulae [3]. The same indentation geometry discussed in section 15.3.1 and in [16, 17] was simulated using the ALE3D code, using the aforementioned RDX crystal plasticity model which contained six slip systems among four families for α-RDX.

Simulation results are comparable to the pre-existing results from [16, 17], as shown in Figure 15.3. For indentation onto the (001) plane, results obtained using atomic constants are similar to those obtained using constants from [36] and are too stiff. For indentation onto (021) and (210) planes, similarly to those obtained using constants from [35], atomic elastic constants provide suitable accuracy up to excursion from elastic behaviour. As with the previous results, agreement with experimental data [29] diverges with increasing load, after excursion into the inelastic regime. This suggests that the model may lack some requisite physics for fracture or perhaps other slip systems [54] not included in the implementation. Cauchy pressures are shown on the right side of Figure 15.3, here corresponding to elastic constants of [36]. At an indentation depth of 200 nm, maximum pressures are substantial, on the order of ¼ the ambient bulk modulus B_0. Results obtained using the constants of [35] are anticipated to be very similar to

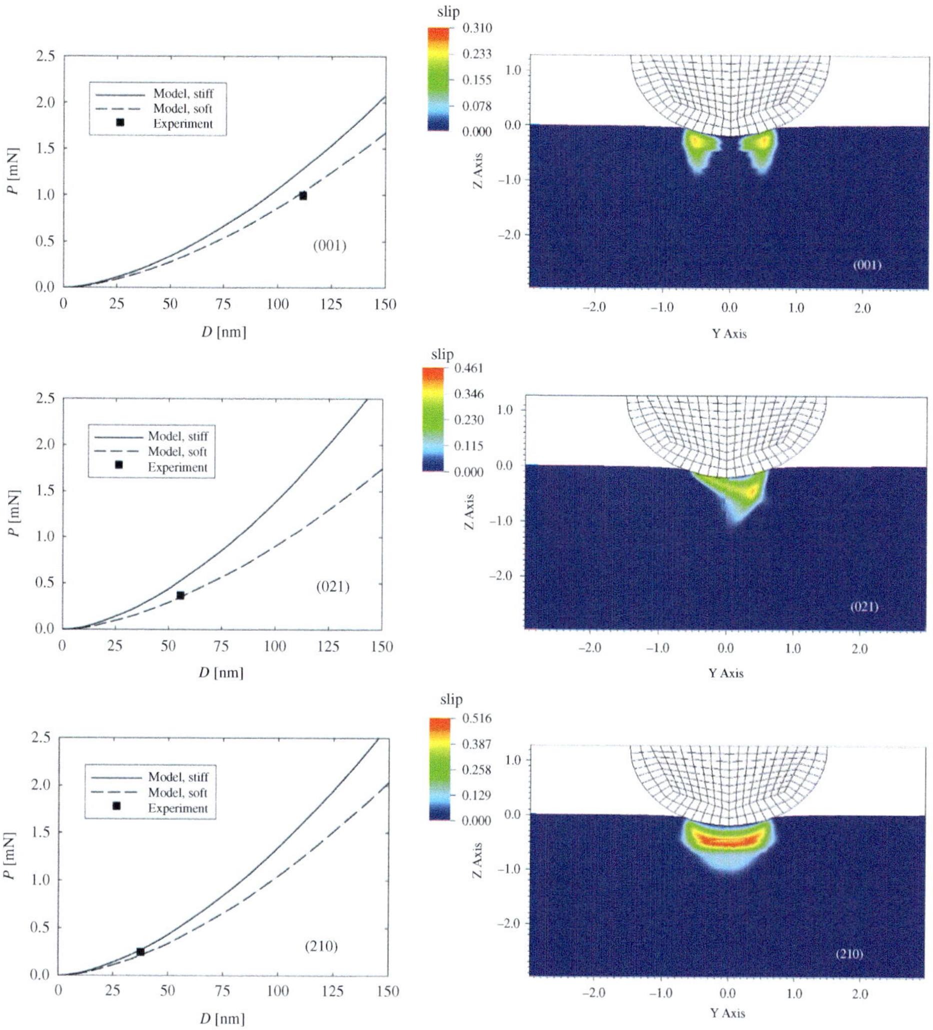

Figure 15.2 Predicted force-displacement curves [17] obtained using soft [35] and stiff [36] elastic constants compared with experimental load excursion data [29] (left) and total slip for indentation depth of 200 nm (right).

those that would be obtained using similar, but not identical, values from more recent experiments [37–39], though this has not been confirmed by additional simulations.

Compared in Figure 15.4 are the residual surface topography from experiment [29] and residual total cumulative plastic slip from the finite element simulation on the (021) crystal surface after unloading. Qualitative correlation is evident. However, the model does not account for the cleavage crack that is highly visible in the experiment. Such cracking would presumably lessen the indentation force P, resulting in closer agreement with experimental data in the inelastic regime. In the future, a phase field fracture representation for anisotropic cleavage [59, 60] or a continuum damage mechanics

Table 15.1 Isothermal elastic constants [GPa] of RDX at 295 K.

Constant	Experiment, soft [35]	Experiment, stiff [36]	Atomic [34]
C_{11}	24.56	36.48	25.0
C_{22}	18.85	24.49	23.8
C_{33}	17.33	20.78	23.4
C_{12}	7.61	0.90	10.6
C_{13}	5.30	1.26	7.6
C_{23}	5.24	8.16	8.8
C_{44}	5.15	11.99	3.1
C_{55}	4.06	2.72	7.7
C_{66}	6.90	7.68	5.2
B'_0 a)	6.95	6.95	9.2

a) Dimensionless; the same experimental value from [58] is used in calculations with both soft and stiff constants.

representation of material separation/failure [61] might be used to augment the constitutive model of Section 15.2.2 implemented in the present numerical simulations.

Compared in Table 15.2 are shear strengths for resistance to dislocation glide/slip among experiment [29, 30], atomic simulation [32], and inferred by the present continuum crystal plasticity model fit to indentation data. Note that experiment and crystal plasticity representations do not distinguish among edge and screw components; in the real crystal, it is likely that dislocation loops with both characteristics are abundant [52]. Values for continuum theory are realistically lower than the upper bounds from experiment and are of similar order of magnitude to those of the atomic calculations. In the finite element simulations, variation of strengths g^k among families of systems by factors of 0.5–2.0 did not usually produce very different results [17]. Values calculated directly from large-scale atomic simulations shown in Table 15.2 [32] are within 25% of values obtained using a Peierls-Nabarro model [62] with stacking fault energy surfaces computed using smaller-scale atomic simulations [31].

In summary, the model developed here and in [16, 17], when used with elastic constants obtained from resonant ultrasonic methods [35] or atomic simulation [34], is thought to provide an accurate representation of the nonlinear anisotropic response of RDX single crystals in the elastic regime of indentation. This model is also thought to provide a qualitatively realistic depiction of activity of different slip systems when a uniform and constant shear strength on the order of $G_0/20$ is prescribed. However, refinements of the model are needed to address any reduction in stiffness associated with discrete or highly localized slip events and cleavage fractures observed at larger indentation depths.

15.4 Application: Indentation of Calcite Single Crystals

Twinning in calcite ($CaCO_3$) was extensively characterized via a number of experiments in the mid-20th century [5, 40, 63–67]. At room temperature, dislocation glide does not

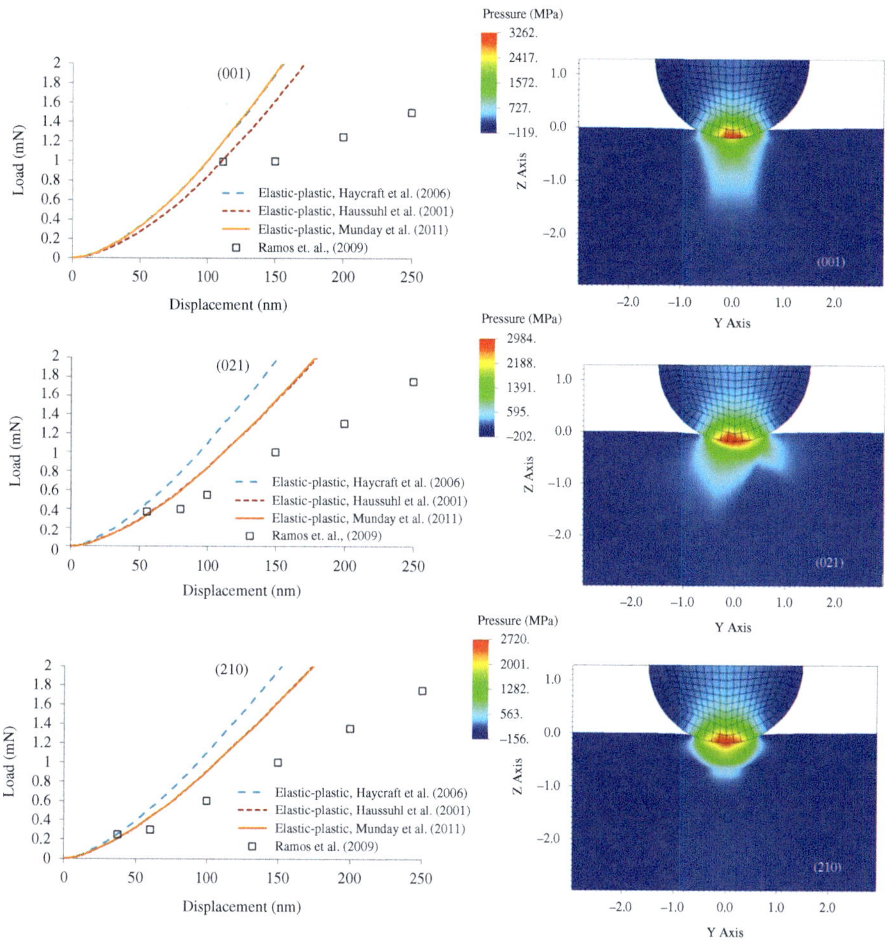

Figure 15.3 Force-displacement predictions from various elastic constants [34–36] compared to experiment [29] (left) and predicted pressures for indentation depth of 200 nm (right) onto (001), (021), and (210) faces of RDX.

occur readily in calcite, but twinning can occur profusely [40, 41]. Thus, calcite is an ideal material for validation of models of deformation twinning such as those described in Section 2.3 and in [42, 47] since twin morphologies can be observed using optical measurements (a benefit of calcite's transparency) and since dislocation-mediated plasticity need not be addressed.

A number of experimental studies of twinning in calcite considered indentation loading, either with knife-edge indenters [63, 65, 67] or with spherical indenters [43, 64, 66]. Calcite crystals were oriented such that the direction of twinning shear was parallel to the loading direction and the habit plane normal was perpendicular to the loading direction. Twinning under such conditions was often reported to be elastic, that is, twins under the indenter disappeared fully or partially upon load removal [5, 40, 43, 64]. Twins originated at the contact surface and were reported to be thin with a sharp tip or apex, with the length of the twin increasing with increasing load. Twins could be maintained in a stable position (i.e., held at a constant length) when

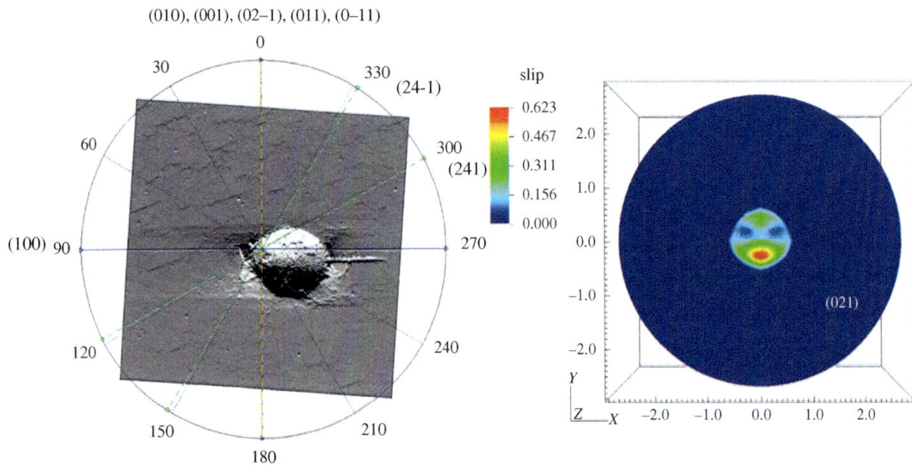

Figure 15.4 Comparison of AFM deflection image from experiment [29] (left) and predicted residual total cumulative slip (right) upon unloading of (021) face of RDX crystal. Left figure reprinted from Figure 7 of reference [29]. *Source*: Ramos 2009 [29]. Reproduced with permission from Taylor & Francis Ltd.

Table 15.2 Resolved shear stresses [GPa] for slip in RDX.

Slip systems	Number	Experiment [29][a)]	Atomic, screw [32]	Atomic, edge [32]	Present model (g^k)
(010)[100]	1	0.885	0.280	0.201	0.320
{021}[100]	2	0.885	0.524	0.576	0.320
{011}[100]	2	0.645	0.518	0.583	0.320
(010)[001]	1	0.885	0.831	0.782	0.320

a) Upper bound from indentation excursion data.

the load was held fixed, with measured sizes reported on the order of micrometers to hundreds of mm depending on the magnitude of applied load and type of indenter geometry [40, 43, 63–67]. When the calcite crystals exhibit no surface defects, twins appear only after a threshold load is applied, whereas twins can appear immediately (i.e., upon application of a minimal load) at surface imperfections when the crystals exhibit such defects [66]. Transparency of calcite enables direct (i.e., *in situ*) optical observation of elastic twinning during indentation loading and unloading, as shown in Figure 15.5. In contrast, appearance and disappearance of elastic twins cannot be visually observed within opaque materials such as metals. Although only one true twin exists in the indentation test in Figure 15.5, the birefringent nature of the crystal creates the optical illusion of two twins [43]. Furthermore, the twin appears at the surface of the calcite crystal in contact with the indenter at a load of around 50 N; refraction can cause an illusion of an offset between twin and indenter.

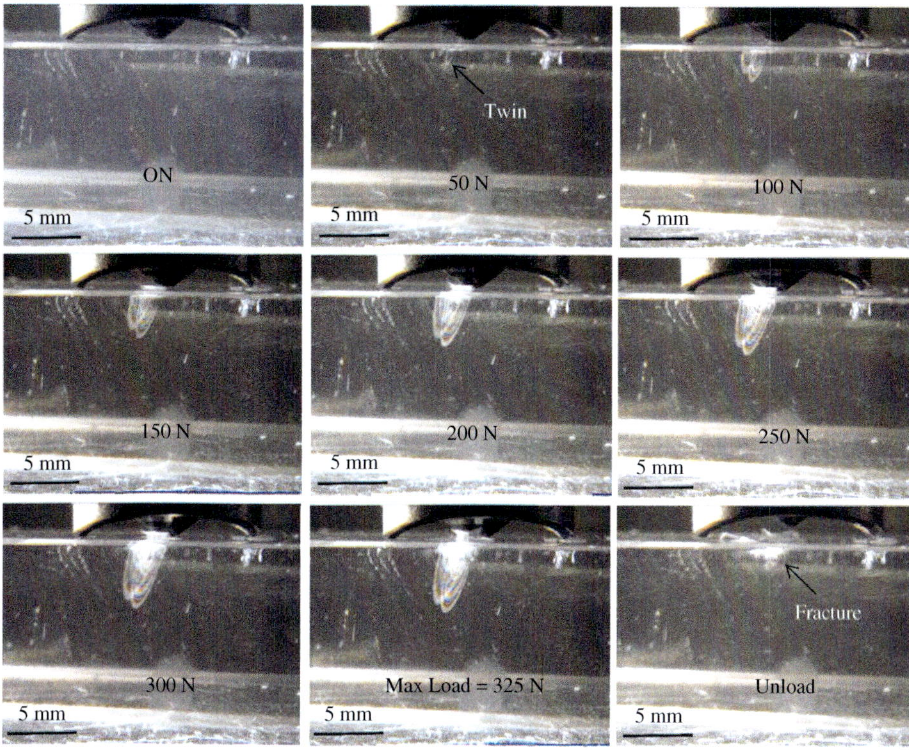

Figure 15.5 Indentation experiments on (100) cleavage plane of calcite: 3 mm diameter diamond indenter [43].

15.4.1 Review of Prior Work

A phase field model for twinning in elastic crystals was developed and exercised for the problem of indentation of calcite single crystals with wedges in [18], following the theory outlined in section 15.2.3. Predictions of nonlinear (i.e., finite strain) theory incorporating a compressible neo-Hookean elastic potential W were compared with those of geometrically linear phase field theory invoking isotropic and anisotropic strain energy potentials. In all cases reported in [18], the polynomial interpolation function in the first of (15.20) was employed.

Phase field predictions of indentation agreed qualitatively with experimental observations: a long thin twin forms asymmetrically beneath one side of the indenter (see left side of Figure 15.6), the tip of the twin is sharp, and the length of the twin increases with increasing force. Normalized, dimensionless variables for indentation depth, indentation force, and twin length were derived. Importantly, physically realistic twin shapes were predicted by a model whose only parameters are the elastic constants, twin boundary surface energy, and twin boundary thickness, all of which can be obtained from independent experiments or quantum/molecular mechanics calculations [68]. Parameters for calcite single crystals with the primary e^+ twin system [18, 19, 68] are listed in Table 15.3 for reference. Qualitatively similar twin shapes were obtained using isotropic

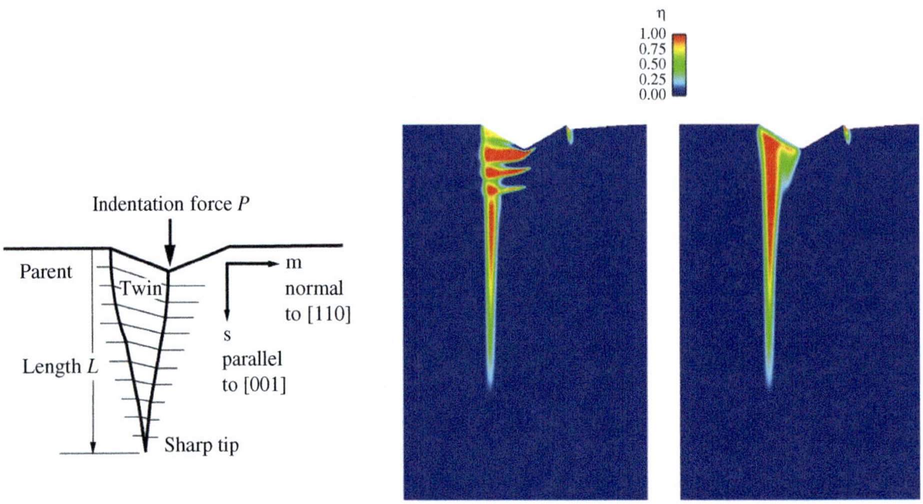

Figure 15.6 Wedge indentation in calcite: geometry (left), nonlinear prediction (center), linear prediction (right). *Source*: Clayton 2011 [18]. Reproduced with permission of IOP.

Table 15.3 Properties of calcite (room temperature) [18, 19].

Constant(s)	Value(s)	Units
$C_{11}, C_{12}, C_{13}, C_{14}, C_{33}, C_{44}$	148.1, 55.8, 54.6, −20.6, 85.6, 32.7	GPa
B_0, G_0	79.1, 36.7	GPa
B'_0	4.0	–
Γ	0.183	J/m^2
l	10^0–10^4	nm
γ^T	0.694	–

and anisotropic elastic constants. Use of anisotropic rather than isotropic surface energy had little effect on twin morphology or indentation force for the particular indentation boundary conditions and ranges of material parameters considered in [18].

A scaling method was developed for modeling behavior of specimens of (mm to cm) sizes on the order of those studied experimentally in traditional indentation, as opposed to nano-indentation. Nearly identical twin morphologies were obtained for nanometer-scale specimens and millimeter (laboratory) scale specimens. Twin length versus indentation force data, when properly normalized to dimensionless form, collapsed to nearly the same monotonically increasing curve for nm- and mm-scale simulations, and for 90° and 120° wedges. Indentation forces were greater in the nonlinear model than the linear model because of the decreasing elastic compliance with increasing pressure in nonlinear model, as manifested by $B'_0 > 0$. Normalized relationships between twin length and indentation force were similar for linear and nonlinear theories at both nanometer and millimeter scales. Twin morphologies were similar for linear and nonlinear theories for indentation with 90° wedges: in each case, a single, continuous primary twin formed under one side of the indenter, and a small secondary twin formed at the free surface at the opposite edge.

Perhaps most interestingly (as shown in the center image of Figure 15.6) in the nonlinear model, indentation of calcite with a 120° wedge produced a lamellar twin structure between the indenter and the long sharp primary twin. The number of twin lamellae increased with increasing indentation depth. This complex, layered microstructure was not predicted by the linear theory, which instead predicted a single continuous primary twin. However, equilibrium lengths L predicted by nonlinear and linear theories were about the same. Previous analysis [69] has suggested that geometrically nonlinear and linear representations of twinning in the context of continuum elasticity theory may predict similar or very different results depending on material, loading, and/or geometry.

15.4.2 New Results and Analysis

Prior work [18, 47] focused on 2D wedge indentation modelling. Here, the focus is on spherical indentation and comparison with experiments [43, 63]. Because of the inherently anisotropic nature of deformation twinning, a 3D model representation is required for correct depiction of spherical indentation of calcite single crystals.

Considered first is a nonlinear anisotropic elastic analysis using the model of section 15.2, in the absence of both plastic slip and twinning (i.e., $\boldsymbol{F} = \boldsymbol{F}^E$). Anisotropic calcite properties are listed in Table 15.3; a spherical diamond indenter of diameter 3 mm is used (diamond material properties in [17]), with frictionless contact. Quasi-static implicit finite element simulations are conducted using the ALE3D multi-physics code under isothermal conditions. Representative results are shown in Figure 15.7 for indentation on a (100) rhombohedral (i.e., natural cleavage) face: specifically, resolved shear stresses on the habit plane in the direction of twinning shear, for loads of 50 N and 1 kN. The former value corresponds to the applied load when twinning is first observed in experiment (Figure 15.5 and [43]), implying a strength of ≈500 MPa for twin resistance if a critical resolved shear stress criterion is valid. Note that because contact is modelled as frictionless, the resolved shear stress at the surface or interface between indenter and substrate (i.e., between diamond and calcite) vanishes in the simulation. Thus, this twin initiation mode corresponds physically to homogeneous nucleation of twinning partial dislocations in the bulk crystal. Asymmetry in the predicted shear stress field is a result of elastic anisotropy.

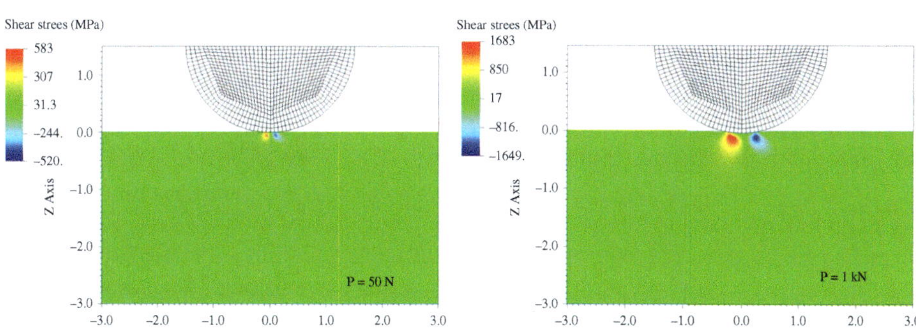

Figure 15.7 Nonlinear anisotropic elastic model predictions of resolved shear stress for twinning in calcite.

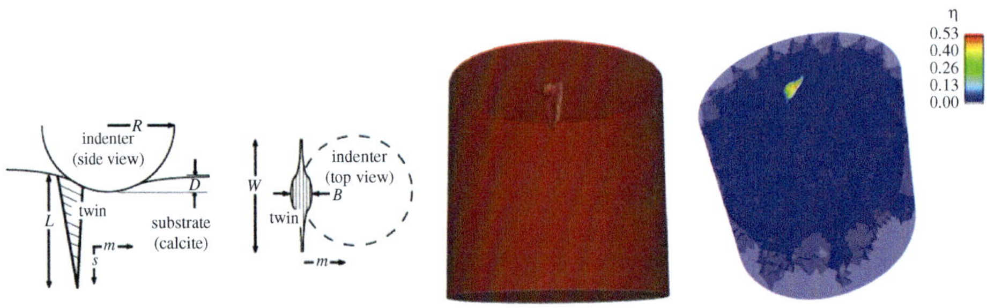

Figure 15.8 Twin morphology: spherical indenter of diameter $2R$, indentation depth D of 0.013 mm (right).

Considered next are new phase field simulations executed using the constitutive model of Section 15.2.3 and [19]. Numerical implementation of the model in a highly scalable parallel finite element code is discussed in [47]. A rigid spherical indenter is used, with frictionless contact. The Fermi-Dirac exponential interpolation function (k=15) in the second of (15.20) was invoked (rather than the cubic polynomial) because it produced a thinner twin more like those observed experimentally [40, 43, 64]. The geometry of the problem necessitated a very large finite mesh, on the order of 10 M tetrahedral elements. Even so, numerical resource constraints imposed that a rather large regularization length l of 10 μm was required (upper limit in Table 15.3), such that twin boundaries could be fully resolved by the finite element discretization.

As is clear from Figure 15.8, the predicted twin morphology is similar to that observed in experiment: asymmetric, long, and thin. However, maximum values of η are less than unity, meaning that the twin is not fully formed (i.e., the twinning eigenshear is incomplete). This discrepancy is attributed primarily to setting the regularization width in the model too large relative to the true width of the twin observed in the experiments [40, 64]: the width of the twin lamella tends towards the lattice spacing (nm dimensions) at the sharp tip or apex. It is speculated that a fully formed twin of similar total energy but thinner and longer shape could be achieved if it were possible to use a very fine mesh (e.g., element sizes as small as the order of nm) and similar small magnitude of resolution length l.

The left side of Figure 15.9 demonstrates that the numerical solution follows the analytical solution [11] until twinning occurs, as it should. Upon load excursion, twinning causes some relaxation of the total force P relative to that obtained from the elastic solution. On the right side of Figure 15.9, twin lengths L are compared from experiments [43, 64] and simulation. The length and force data among experiments and simulations agree reasonably at initiation, but at larger depths, the predicted twin is not as long as observed in experiments. Again, the resolution afforded by the regularization width and finite element mesh may be the source of discrepancy; a finer mesh and much smaller value of l would be expected to resolve a longer, thinner, more fully formed twin at larger values of indentation force. Possible reasons for differences among the sets of experimental data, including indenters and geometry, are discussed in [43].

Reported in Table 15.4 are resolved shear stresses on the twinning plane in the direction of twinning shear for different loading methods. Results are very different for

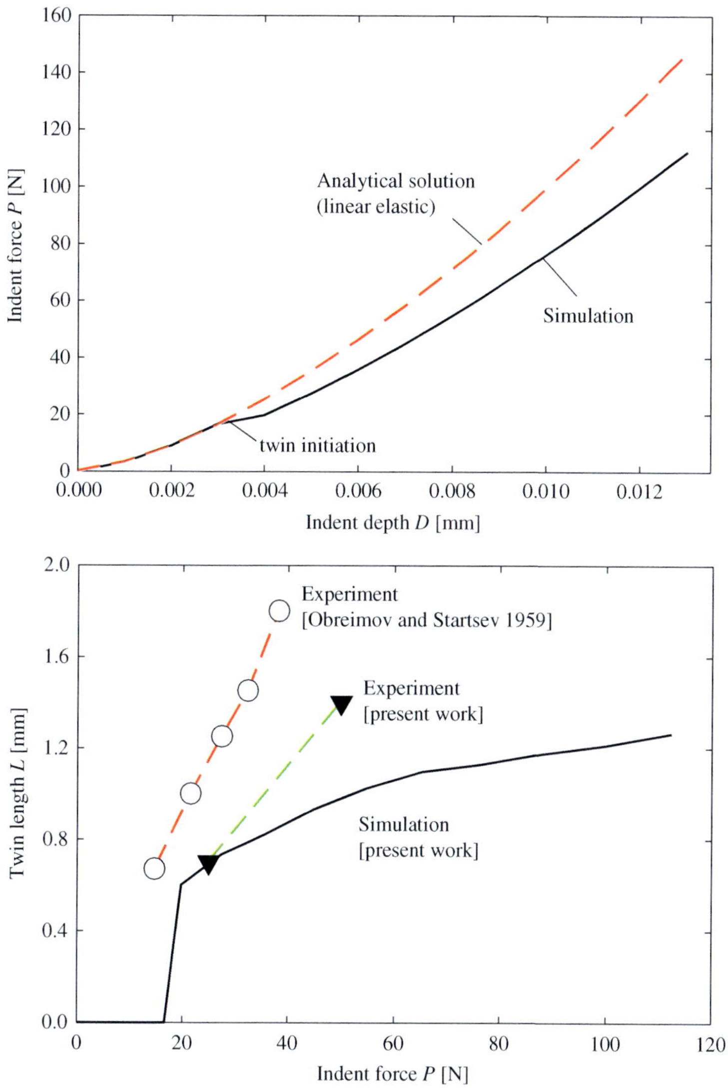

Figure 15.9 Comparison of present phase field simulation results with analytical solution [11] and experiments [43, 64]: force versus depth (left) and twin length versus force (right).

all three cases shown. The largest stress, for spherical indentation, occurs sub-surface (Figure 15.7, nonlinear elastic analysis of the present work), and corresponds to homogeneous nucleation in the bulk of the calcite crystal, i.e., formation and motion of twinning partial dislocations within the otherwise perfect crystal at some finite depth beneath the free surface. However, if twinning instead initiates at the surface, as observed in experiments [40, 43, 64], the actual initiation stress may be much smaller than 500 MPa, and effects of friction, surface flaws, and initial defects enabling heterogeneous nucleation, not addressed in the simulations, become crucial. It is noted

Table 15.4 Resolved shear stresses for twin initiation in calcite.

Loading	Stress value [MPa]	Remarks
Spherical Indentation	500	Maximum shear stress from simulation at load from experiment [43]
Knife-edge indentation	1.0	Calculated using analytical isotropic elasticity with experiment [67]
Direct shear	0.1	Approximated directly from experiment [41]

that the present phase field simulations predict twinning initiation at the surface rather than deep within the bulk, consistent with free surface boundary condition $\nabla\eta \cdot \boldsymbol{n} = 0$ for the order parameter [18, 19], and in agreement with experimental observation. Twinning stress may also increase with pressure/constraint; i.e., there may be a pressure hardening effect on twin resistance not addressed by the model in its present form. For example, dilatancy for various relaxed stacking fault structures was computed atomically in [70] for sapphire, another transparent crystal of trigonal structure. Shearing resistance might increase if such dilatation is impeded by locally high pressures.

15.5 Conclusions

A general framework for modelling finite elastic and inelastic deformations of single crystalline solids has been presented, accounting for nonlinear elasticity, plastic slip, and deformation twinning. Slip is represented in the context of continuum crystal plasticity, and twinning in the context of phase field theory. Specific versions of the theory, corresponding to slip only or twinning only, have been applied to model indentation of oriented single crystals of the energetic molecular solid RDX and the polyatomic mineral calcite. Both materials are transparent. Indenter sizes ranging from nanometer to millimeter scales have been considered in various numerical simulations. Results highlight the importance of consideration of elastic and plastic anisotropy and nonlinear elasticity in some cases. Simulation data, in conjunction with experimental observation, provide new insight into suitability of various sets of elastic constants (RDX) and magnitudes of resolved shear stresses needed to initiate slip in RDX and twinning in calcite. Agreements of model results with experimental data on load excursions in RDX (slip initiation, crystal plasticity theory) and calcite (twin initiation, phase field theory) are reasonably close, but simulation results and experimental results diverge at larger indentation depths for both single crystalline materials. Discrepancies for RDX are attributed to omission of fracture in the constitutive model, which would yield a softer force-displacement response in better agreement with experiment. Discrepancies for calcite are attributed to a minimum regularization width that is too large because of numerical constraints (i.e. mesh size limits) that results in a twin that is too short and thick at larger indentation loads.

Acknowledgements

Technical discussions with Dr. K.J. Ramos (Los Alamos National Laboratory) and Drs. M.A. Grinfeld and J.J. Swab (US Army Research Laboratory) regarding respective behaviours of RDX and calcite single crystals are appreciated.

References

1 R.W. Ogden. Non-Linear Elastic Deformations. Ellis-Horwood, Chichester; 1984.
2 D.C. Wallace. Statistical Physics of Crystals and Liquids: a Guide to Highly Accurate Equations of State. World Scientific, Singapore; 2003.
3 J.D. Clayton. Nonlinear Mechanics of Crystals. Springer, Dordrecht; 2011.
4 S. Nemat-Nasser. Plasticity: A Treatise on the Finite Deformation of Heterogeneous Inelastic Materials. Cambridge University Press, Cambridge; 2004.
5 V.S. Boiko, R.I. Garber, A.M. Kosevich. Reversible Crystal Plasticity. AIP Press, New York; 1994.
6 J.W. Christian, S. Mahajan. Deformation twinning. Progress in Materials Science 39, 1–157; 1995.
7 R. Hill, E.H. Lee, S.J. Tupper. The theory of wedge indentation of ductile materials. Proceedings of the Royal Society A, 188, 273–289; 1947.
8 I.N. Sneddon. Fourier Transforms. McGraw-Hill, New York; 1951.
9 W. Hirst, M. Howse. The indentation of materials by wedges. Proceedings of the Royal Society A, 311, 429–444; 1969.
10 K.L. Johnson. The correlation of indentation experiments. Journal of the Mechanics and Physics of Solids 18,115–126; 1970.
11 S. Timoshenko, J. Goodier. Theory of Elasticity, 3rd edn. McGraw-Hill, New York; 1970.
12 J. Knap, M. Ortiz. Effect of indenter-radius size on Au (001) nanoindentation. Physical Review Letters 90, 226102; 2003.
13 Y-C. Chen, K. Nomura, R.K. Kalia *et al.* Molecular dynamics nanoindentation simulation of an energetic material. Applied Physics Letters 93, 171908; 2008.
14 J. Knap, M. Ortiz. An analysis of the quasicontinuum method. Journal of the Mechanics and Physics of Solids 49, 1899–1923; 2001.
15 J.D. Clayton, P.W. Chung. An atomistic-to-continuum framework for nonlinear crystal mechanics based on asymptotic homogenization. Journal of the Mechanics and Physics of Solids 54, 1604–1639; 2006.
16 J.D. Clayton, R. Becker. Elastic-plastic behavior of cyclotrimethylene trinitramine single crystals under spherical indentation, modeling and simulation. Journal of Applied Physics 111, 063512; 2012.
17 J.D. Clayton, R.C. Becker. Modeling nonlinear elastic-plastic behavior of RDX single crystals during indentation. US Army Research Laboratory Technical Report ARL-TR-5864, Aberdeen Proving Ground, MD. US Department of Defense; 2012.

18 J.D. Clayton, J. Knap. Phase field modeling of twinning in indentation of transparent crystals. Modelling and Simulation in Materials Science and Engineering 19, 085005; 2011.
19 J.D. Clayton, J. Knap. Phase field analysis of fracture induced twinning in single crystals. Acta Materialia 61, 5341–5353; 2013.
20 R. Kondo, Y. Tadano, K. Shizawa. A phase-field model of twinning and detwinning coupled with dislocation-based crystal plasticity for HCP metals. Computation in Materials Science 95, 672–683; 2014.
21 R. Kondo, Y. Tadano, K. Shizawa. Investigation on intragranular stress of Mg including several twin-bands using dislocation-based crystal plasticity and phase-field models. Key Engineering Materials 626, 246–251; 2015.
22 J.J. Gilman. Electronic Basis of the Strength of Materials. Cambridge University Press, Cambridge; 2003.
23 J.D. Clayton. A continuum description of nonlinear elasticity, slip and twinning, with application to sapphire. Proceedings of the Royal Society A 465, 307–334; 2009.
24 M. Koslowski, M. Ortiz. A multi-phase field model of planar dislocation networks. Modelling and Simulation in Materials Science and Engineering 12,1087–1097; 2004.
25 S.Y. Hu, C.H. Henager, L-Q. Chen. Simulations of stress-induced twinning and de-twinning, a phase field model. Acta Materialia 58, 6554–6564; 2010.
26 A. Hunter, I.J. Beyerlein, T.C. Germann, M. Koslowski. Influence of the stacking fault energy surface on partial dislocations in FCC metals with a three-dimensional phase field dislocation dynamics model. Physical Review B, 84,144108; 2011.
27 N. Barton, N. Winter, J. Reaugh. Defect evolution and pore collapse in crystalline energetic materials. Modelling Simul. Materials Science and Engineering 17, 035003; 2009.
28 H. Gallagher, P. Halfpenny, J. Miller, J. Sherwood. Dislocation slip systems in pentaerythritol tetranitrate (PETN) and cyclotrimethylene trinitramine (RDX). Philosophical Transactions of the Royal Society A, 339, 293–303; 1992.
29 K. Ramos, D. Hooks, D. Bahr. Direct observation of plasticity and quantitative hardness measurements in single crystal cyclotrimethylene trinitramine by nanoindentation. Philosophical Magazine 89, 2381–2402; 2009.
30 K. Ramos, D. Bahr, D. Hooks. Defect and surface asperity dependent yield during contact loading of an organic molecular single crystal. Philosophical Magazine 91, 1276–1285; 2011.
31 L. Munday, S. Solares, P.W. Chung. Generalized stacking fault energy surfaces in the molecular crystal RDX. Philosophical Magazine 92, 37–41; 2012.
32 N. Mathew, C.R. Picu, P.W. Chung. Peierls stress of dislocations in molecular crystal cyclotrimethylene trinitramine. Journal of Physical Chemistry A, 117, 5326–5334; 2013.
33 T. Sewell, C. Bennett. Monte Carlo calculations of the elastic moduli and pressure-volume-temperature equation of state for hexahydro-1,3,5-trinitro-1,3,5-triazine. Journal of Applied Physics 88, 88–95; 2000.
34 L. Munday, P. Chung, B. Rice, S. Solares. Simulations of high-pressure phases in RDX. Journal of Physical Chemistry B, 115, 4378–4386; 2011.
35 S. Haussuhl. Elastic and thermoelastic properties of selected organic crystals. Zeitschrift Fur Kristallographie 216, 339–353; 2001.

36 J. Haycraft, L. Stevens, C. Eckhardt. The elastic constants and related properties of the energetic material cyclotrimethylene trinitramine (RDX) determined by Brillouin scattering. Journal of Chemical Physics 124, 024712; 2006.
37 R. Schwarz, D. Hooks, J. Dick *et al.* Resonant ultrasound spectroscopy measurement of the elastic constants of cyclotrimethylene trinintramine. Journal of Applied Physics 98, 056106; 2005.
38 B. Sun, J. Winey, N. Hemmi *et al.* Second-order elastic constants of pentaerythritol tetranitrate and cyclotrimethylene trinitramine using impulsive stimulated thermal scattering. Journal of Applied Physics 104, 073517; 2008.
39 C. A. Bolme, K.J. Ramos. The elastic tensor of single crystal RDX determined by Brillouin spectroscopy. Journal of Applied Physics 116, 183503; 2014.
40 A.M. Kosevich, V.S. Boiko. Dislocation theory of the elastic twinning of crystals. Soviet Physics Uspekhi 14, 286–316; 1971.
41 H. Kaga, J.J. Gilman. Twinning and detwinning in calcite. Journal of Applied Physics 40, 3196–3207; 1969.
42 J.D. Clayton, J. Knap. A phase field model of deformation twinning, nonlinear theory and numerical simulations. Physica D 240, 841–858; 2011.
43 C.D. Hilton, J.H. Yu, J.J. Swab. Elastic twinning and cleavage fracture in calcite crystals, experimental methods and preliminary results. US Army Research Laboratory Technical Report ARL-TR-6144, Aberdeen Proving Ground, MD. US Department of Defense; 2012.
44 J.D. Clayton. Differential Geometry and Kinematics of Continua. World Scientific, Singapore; 2014.
45 R. Becker. Effects of crystal plasticity on materials loaded at high pressures and strain rates. International Journal of Plasticity 20, 1983–2006; 2004.
46 J.D. Clayton. Dynamic plasticity and fracture in high density polycrystals, constitutive modeling and numerical simulation. Journal of the Mechanics and Physics of Solids 53, 261–301; 2005.
47 J. Knap, J.D. Clayton. A computational framework for phase-field modeling. US Army Research Laboratory Technical Report ARL-MR-0760, Aberdeen Proving Ground, MD. US Department of Defense; 2011.
48 R. Bowers, J. Romans, W. Zisman. Mechanisms involved in impact sensitivity and densensitization of RDX. Industrial and Engineering Chemistry Process Design and Development12, 2–13; 1973.
49 M. Kuklja, A. Kunz. Compression-induced effect on the electronic structure of cyclotrimethylene trinitramine containing an edge dislocation. Journal of Applied Physics 87, 2215–2218; 2000.
50 D. Hooks, K. Ramos, A. Martinez. Elastic-plastic shock wave profiles in oriented single crystals of cyclotrimethylene trinitramine (RDX) at 2.25 GPa. Journal of Applied Physics 100, 024908; 2006.
51 W. Connick, F. May. Dislocation etching of cyclotrimethylene trinitramine crystals. Journal of Crystal Growth 5, 65–69; 1969.
52 P. Halfpenny, K. Roberts, J. Sherwood. Dislocation configurations in single crystals of pentaerythritol tetranitrate and cyclotrimethylene trinitramine. Journal of Crystal Growth 65, 524–529 ; 1983.
53 W. Elban, R. Armstrong, K. Yoo *et al.* X-ray reflection topographic study of growth defect and microindentation strain fields in an RDX explosive crystal. Journal of Materials Science 24, 1273–1280; 1989.

54 M. Cawkwell, K. Ramos, D. Hooks, T. Sewell. Homogeneous dislocation nucleation in cyclotrimethylene trinitramine under shock loading. Journal of Applied Physics 107, 063512; 2010.
55 K. Ramos, D. Hooks, T. Sewell, M. Cawkwell. Anomalous hardening under shock compression in (021)-oriented cyclotrimethylene trinitramine single crystals. Journal of Applied Physics 108, 066105; 2010.
56 J. Rimoli, E. Gurses, M. Ortiz. Shock-induced subgrain microstructures as possible homogeneous sources of hot spots and initiation sites in energetic polycrystals. Physical Review B, 81, 014112; 2010.
57 J. Winey, Y. Gupta. Anisotropic material model and wave propagation simulations for shocked pentaerythritol tetranitrate single crystals. Journal of Applied Physics 107, 103505; 2010.
58 B. Olinger, B. Roof, H. Cady. The linear and volume compression of β-HMX and RDX. In: Proceedings of the International Symposium on High Dynamic Pressures. Commissariat a l'Energie Atomique, Paris, pp. 3–8; 1978.
59 J.D. Clayton, J. Knap. A geometrically nonlinear phase field theory of brittle fracture. International Journal of Fracture 189, 139–148; 2014.
60 J.D. Clayton, J. Knap. Phase field modeling of directional fracture in anisotropic polycrystals. Computation in Materials Science 98, 158–169; 2015.
61 J.D. Clayton. Deformation, fracture, and fragmentation in brittle geologic solids. International Journal of Fracture 163, 151–172; 2010.
62 F.R.N. Nabarro. Dislocations in a simple cubic lattice. Proceedings of the Physical Society 59, 256–272; 1947.
63 R.I. Garber. Formation of elastic twins during calcite twinning. Doklady Akademii Nauk SSSR 21, 233; 1938.
64 I.V. Obreimov, V.I. Startsev (1959) Work of formation of an elastic twin in calcite. Soviet Physics Journal of Experimental and Theoretical Physics 35, 743–748.
65 R.E. Cooper. Kinetics of elastic twinning in calcite. Proceedings of the Royal Society A, 270, 525–537; 1962.
66 R.I. Garber, E.I. Stepina. Rules governing the motion of dislocations during deformation twinning. Soviet Physics, Solid State 5, 152–158; 1965.
67 V.S. Boiko, R.I. Garber. Determination of the phenomenological dislocation theory parameters for elastic twins in calcite. Soviet Physics Journal of Experimental and Theoretical Physics 23, 602–605; 1966.
68 M. Bruno, F.R. Massaro, M. Rubbo *et al.* (10.4), (01.8), (01.2), and (00.1) twin laws of calcite $CaCO_3$, equilibrium geometry of the twin boundary interfaces and twinning energy. Crystal Growth and Design 10, 3102–3109; 2010.
69 K. Bhattacharya. Comparison of the geometrically nonlinear and linear theories of martensitic transformation. Continuum Mechanics and Thermodynamics 5, 205–242; 1993.
70 P.R. Kenway. Calculated stacking fault energies in α-Al_2O_3. Philosophical Magazine B, 68, 171–183; 1993.

16

Nanoindentation Modeling: From Finite Element to Atomistic Simulations

Daniel Esqué- de los Ojos[1] and Jordi Sort[2]

[1] *Doctor, Departament de Física, Facultat de Ciències, Universitat Autònoma de Barcelona, Bellaterra, Spain*
[2] *Professor, Institució Catalana de Recerca i Estudis Avançats (ICREA) and Departament de Física, Facultat de Ciències, Universitat Autònoma de Barcelona, Bellaterra, Spain*

16.1 Introduction

Hardness is one of the most common mechanical properties which rise naturally from the classification of objects into *hard* and *soft* ones. The concept of hardness has thus been present since ancient times even though its probing and quantification was only achieved in the past 100 years or so. The first modern technique for measuring the hardness of metals was developed by Brinell, a Swedish engineer who pressed a hard spherical indenter shape against plates of the materials under study. The so-derived Brinell hardness is defined as the load divided by the surface of the indentation. After Brinell's usage of a spherical indenter shape, other geometries have been employed to perform indentation experiments. Generally, these indenters are made of materials (e.g. diamond) such that, upon penetration, their behaviour might be considered as rigid (undeformable). Depending on the shape of the indenter and the experimental set-up, different hardness tests such as the Vickers, Berkovich, Knoop and Rockwell emerged during the 20th century, all of them being nowadays used routinely in materials development and characterization. In these experiments, hardness is defined as the ratio between the maximum applied load and either the actual contact area left by the indenter upon unloading or the projected contact area.

With the advent of instrumented nanoindentation in the 1990s, it has been feasible to systematically measure penetration loads as small as few micro Newtons and depths in the nanometer scale. This technique together with a better knowledge of the indentation mechanics [1–3] and advanced imaging tools allows investigation of the mechanical properties at small length scales.

In recent years, multiscale modelling has become a topic of great interest. Analytical techniques are being constantly applied for describing and understanding material's behaviour through quantum mechanics for a few atoms, molecular dynamics for millions of atoms, dislocation dynamics for interacting dislocations, continuum crystal plasticity at single microstructural units of material, and continuum plasticity theories at macroscopic scales. Due to the ability of instrumented indentation to control the displacement of the indenter, this test has now become a benchmark to study the

Applied Nanoindentation in Advanced Materials, First Edition. Edited by Atul Tiwari and Sridhar Natarajan.

mechanical behaviour of materials across length scales from nano to millimeters. Issues under study essentially involve dislocation nucleation phenomena at surfaces, the development of plastic deformation under the large strain gradients imposed by the indenter tip and the extraction of the uniaxial stress-strain behaviour of the indented material.

In this chapter we will offer a perspective on the usage of modelling at different length scales for the study of nanoindentation experiments. First, we will introduce the concept of scaling and dimensional analysis in the context of indentation measurements along with different methodologies for the assessment of mechanical properties. Secondly, we will put in the context of characterization of new advanced materials the finite element method. Thirdly, recent atomistic simulation results will be presented. Finally, we will draw some conclusions and offer a comprehensive literature on the topic.

16.2 Scaling and Dimensional Analysis Applied to Indentation Modelling

One of the principal, and apparent, drawbacks from indentation modelling is that for each material we will have a single matching simulation. That would turn the modelling of indentation experiments into highly time-consuming parametric studies where mechanical properties would need to be varied until the correct set is found. Luckily, dimensional analysis will help us to identify the parameters of interest during an indentation test and to wisely design our simulation experiments. Also, one should bear in mind that indentation is, basically, the penetration of a tip with a certain geometry (usually taken as non-deformable) into a certain material from which we want to know its properties. It is precisely this tip geometry, along with the material of interest being indented, that can be subjected to scaling analysis.

16.2.1 Geometrical Similarity of Indenter Tips

We can divide indenter tips into two big groups: sharp and spherical. In this division we are not considering flat punches that are more related with compression experiments on, for instance, micropillars. Given two tips, they will be geometrically similar if their length parameters are respectively similar to each other with the same proportionality constant ($a/a' = b/b' = \ldots = z/z' = c =$ constant) and all the respective angles are identical ($\alpha = \alpha', \ \beta = \beta', \ \ldots, \ \omega = \omega'$). The factor c is called the transformation coefficient and a family of geometrically similar objects can be found by changing this factor [4]. In the case of sharp cones and pyramidal (Vickers and Berkovich) indenters, factor c will be the apex angle so, strictly speaking, this indenters do not have a characteristic length. On the contrary, for spherical indenters the parameter describing the tip is its radius, that it is a length. Figure 16.1 shows schematically an object with geometric similarity. This difference between sharp and spherical tips is important because, as shown by Tabor [5], the lack of a characteristic length in sharp indenters is the reason why the mean contact pressure (defined as the applied load divided by the projected contact area left by the indenter) is independent of the penetration depth (h_s) imposed by the indenter. In other words, during sharp indentation (and neglecting effects such as tip rounding, strain gradient plasticity or substrate effects during thin film indentation)

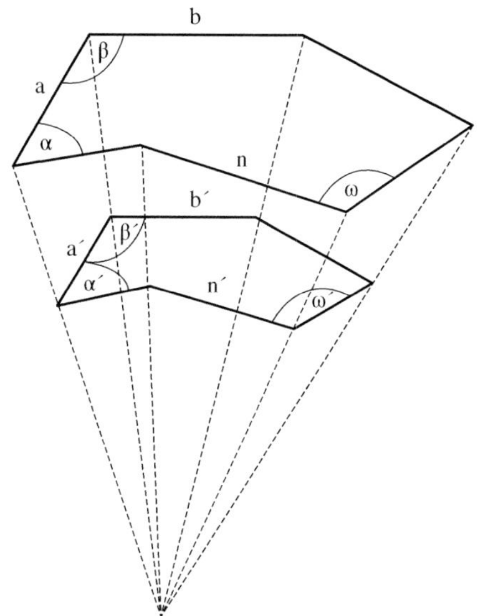

Figure 16.1 Schematic drawing showing geometric similarity. *Source*: Cheng 2004 [4]. Reproduced with permission of Elsevier.

hardness, defined as a mean contact pressure, has a constant value for a given material. On the other hand, during spherical indentation hardness will evolve with a/D, where a is the contact radius between the indenter and the material and D is the diameter of the indenter. As penetration progresses the value of a will increase and, consequently, a/D will also increase. In section 16.3 we will show how dimensional analysis can also be applied to the mechanical properties of the material under consideration depending on the tip geometry used during indentation. But, before, we will introduce some general concepts on dimensional analysis that will help us to particularize it.

16.2.2 Dimensional Analysis

From most fundamental physics we know that the dimensions of a certain physical entity x in the system length/mass/time can be written as the product of units of length, mass, time, each one with a certain exponent, in other words:

$$x = L^{\alpha} M^{\beta} T^{\gamma}, \tag{16.1}$$

where α, β and γ depend on x. In the event that $\alpha = \beta = \gamma = 0$, then x will be a dimensionless quantity. For a given set of physical quantities $x_1, x_2, \ldots, x_i$, they will be said to be dimensionally independent if there is not a set of real numbers group $a_1, a_2, \ldots, a_i$ not equal to zero, such that:

$$[x_1]^{a_1} \cdot [x_2]^{a_2} \cdot \cdots \cdot [x_i]^{a_i} = 1. \tag{16.2a}$$

This implies that from dimensionally independent quantities one cannot generate a dimensionless number. With these two simple concepts shown by Equations (16.1) and (16.2a) we can introduce now what it is known as the Π-theorem.

Given a certain physical relation it is quite evident that we can write it as:

$$z = f(z_1, z_2, \cdots z_i, \cdots, z_n). \tag{16.2b}$$

Let us suppose that $z_1, \ldots, z_i$ are dimensionally independent quantities and that the rest, $z_{i+1}, \ldots, z_n$, are then dimensionally dependent. Given that we can write:

$$[z] = [z_1]^{a_1} \cdots [z_i]^{a_i}. \tag{16.3}$$

$$[z_{i+j}] = [z_1]^{a_{j1}} \cdots [z_j]^{a_{ji}},\ j = 1, \ldots, n-i \tag{16.4}$$

Hence, we can write the dimensionless **Π**-functions:

$$\Pi = \frac{z}{z_1^{a_1} \cdots z_i^{a_i}} \tag{16.5}$$

$$\Pi_j = \frac{z_{i+j}}{z_1^{a_{j1}} \cdots z_i^{a_{ji}}},\ j = 1, \ldots, n-i \tag{16.6}$$

We can then rewrite Equation (16.2) in terms of the **Π** and **Π**$_j$ functions as:

$$\Pi = f(z_1, z_2, \cdots z_i, \Pi_1, \cdots, \Pi_{n-i}) \tag{16.7}$$

From the fact that changing from one system of units to another can make the values of $z_1, \ldots, z_i$ much bigger or smaller and the fact the Π quantities are dimensionless and, hence, independent of the unit system, makes impossible for $z_1, \ldots, z_i$ to appear in Equation (16.7) and so:

$$\Pi = f(\Pi_1, \cdots, \Pi_{n-i}), \tag{16.8}$$

being the number of variables reduced by *i*, all the independent variables, and all the variables in Equation (16.8) will be dimensionless.

16.2.3 Dimensional Analysis Applied to Extraction of Mechanical Properties

It is not the purpose of the present chapter to make a deep analysis on the different methodologies for extraction of mechanical properties that have been proposed during the two last decades. Instead, the reader is encouraged to consult some articles and reviews that cover this topic profusely [2, 6–8]. However, here we will introduce two different methodologies, with their respective dimensional analysis, applied to extract mechanical properties from spherical and Berkovich indentations.

Spherical indentation:

Following dimensional analysis, we can write the Π-function for spherical indentation as [9, 10]:

$$\bar{p}/\sigma_{0.1} = \Pi(a/D; n; E/\sigma_{0.1}), \tag{16.9}$$

where $\bar{p}$ is the mean contact pressure defined as the applied load P exerted by the indenter on the material divided by the projected contact area A between the material and the indenter. $\sigma_{0.1}$ is the uniaxial stress that the material would experience under a uniaxial deformation $\varepsilon = 0.1$. These uniaxial stress and strain appear from considering that the constitutive equation relating stress with strain in this material obeys a piecewise linear-power law relation like:

$$\sigma = \begin{cases} E\varepsilon, & \sigma \le \sigma_{ys} \\ \sigma_o \varepsilon^n, & \sigma \ge \sigma_{ys} \end{cases} \tag{16.10}$$

where σ_o is the uniaxial stress at a unitary deformation, σ_{ys} is the yield stress of the material, E is the Young's modulus and n is the strain hardening coefficient of the material. Tabor's solution [5, 11] has been one of them most used for extraction of mechanical properties and it can be written as:

$$\overline{p}/\sigma_o = 2.8\left(0.4\frac{a}{D}\right)^n \tag{16.11}$$

It is evident that Equation (16.11) follows the same structure as Equation (16.9) given the existing relation betweenσ_o, $\sigma_{0.1}$ and E in Equation (16.10). Works by Biwa and Störakers [12] and Mesarovic and Fleck [13] gave fundamental physical reasoning for Equation (16.11), implying, at the same time, some restrictions for its general application.

To overcome those restrictions, Alcalá and Esqué-de los Ojos [14] came with a more general solution written as:

$$\overline{p}/\sigma_{0.1} = \mathrm{A}(n) - (A(n) - B(n)) \cdot exp[-(K(n)E/\sigma_{0.1})^{D(n)}], \tag{16.12}$$

Being *A(n), B(n), K(n)* and *D(n)* polynomial functions. Determination of the different parameters was achieved by a discrete number of simulations, giving as a result a methodology capable to predict the mechanical properties of a certain material being its Young's modulus known. Details on the simulations and parameters can be found in [14]. Moreover, the methodology allows determination of the active contact deformation regime (elastic, elasto-plastic or fully-plastic) and a possible transition from one to the other depending on the mechanical properties of the material. This transition between contact deformation regimes takes us back to what was said in Section 16.2.1 when talking about the similarity of the indenter tips. The fact that for sharp indenters the characteristic length is the apical angle but that for spherical tips is the diameter of the indenter, not only imposes the constancy of hardness during sharp indentation and its evolution (increase) during spherical indentation. The difference between the characteristic lengths also induces that the contact regime cannot evolve during sharp indentation while it can do so during spherical indentation. This will be translated into the stress contours during indentation as we will see next when comparing spherical and sharp indentation.

Sharp indentation:

In this case, and following the work by Casals and Alcalá [15], the dimensionless function governing the scaled hardness is much more easier than in Equation (16.9), being:

$$\overline{p}/\sigma_{0.1} = \Pi(E/\sigma_{0.1}) \tag{16.13}$$

From Equation (16.13) it can be seen that hardness will not vary with the penetration depth, being a property of the material. In their work, Casals and Alcalá [15] made a detailed analysis on the dimensional analysis of sharp indentation. In particular, for Berkovich indentation, a subsequent work [16] showed that with a set of two-dimensionless equations it is possible to extract mechanical properties from Berkovich indentation, being these equations:

$$\overline{p}/\sigma_{0.1} = \Phi_1(E^*/\sigma_{0.1}) = A + Bexp\left(\frac{-E^*/\sigma_{0.1}}{C}\right) + DBexp\left(\frac{-E^*/\sigma_{0.1}}{E}\right) \tag{16.14}$$

$$\sqrt{\alpha} - 1 = \Phi_2(E^*/\sigma_{0.1}; n) = A(n) + B(n)exp\left(\frac{\sigma_{0.1}/E^*}{C(n)}\right) \tag{16.15}$$

Where $\sqrt{\alpha} - 1 > 0$ denotes pile – up and $\sqrt{\alpha} - 1 < 0$ sinking – in of the free surface around the imprint left by the indenter. E^* is the reduced Young's modulus taking into account the Young's modulus E and Poisson's coefficients ν of both, indenter and material as in:

$$E^* = \left(\frac{1 - \nu_i^2}{E_i} + \frac{1 - \nu_m^2}{E_m}\right)^{-1}, \tag{16.16}$$

corresponding subindexes i and m to indenter and material, respectively. Coefficients A, B, C and D in Equation (16.14) and parameters for functions $A(n)$, $B(n)$ and $C(n)$ can be found in [15].

As we said when discussing extraction of mechanical properties through spherical indentation, for sharp indentation contact properties such as hardness and pile-up/sinking-in will be characteristic of the material and will not evolve with penetration, contrary with what happens during spherical indentation. This also true for the active contact regime, actually both things are connected, so for sharp indentation material will behave as elasto-plastic or fully-plastic irrespectively of the penetration depth while for spherical indentation a contact deformation regime transition can occur depending on the (scaled) mechanical properties of the material. Figure 11 in [14] shows how this transition can occur while indenting the same material with a sharp and a spherical indenter. This drive us to the natural idea that, for a certain *a/D* both spherical and sharp indentation will render the same results, as it was found in [14]

General comments:

Though the introduced methodologies can be highly useful to extract mechanical properties, one should bear in mind that they cannot account for some issues presented during indentation experiments. Strain gradient effects and tip rounding (for sharp indentation) can be overcome by ensuring a sufficient penetration depth when this is possible. Representative mechanical properties can be obtained by the imprint encompassing enough material (for instance 5–7 grains in polycrystalline materials), during spherical indentation that can be achieved by using the appropriate indenter diameter. Also, the tested material has to be homogeneous not having, for instance, a gradient of mechanical properties as can be the case of a nitrited steel, a coating or a multi-layered material. In these cases one should be really careful with the loads and penetrations applied to be sure that the obtained data is not affected by second phases. On the following, we will discuss modelling techniques for advanced materials that can challenge the aforementioned methodologies.

16.3 Finite Element Simulations of Advanced Materials

In this Chapter we will show some modelling examples of advanced materials and the most suitable modelling techniques. Basically, the reader will be introduced to

FE simulations of: (i) nanocrystalline porous materials; (ii) amorphous materials with different dimensions and (iii) thin single crystalline films.

16.3.1 Nanocrystalline Porous Materials and Pressure-sensitive Models

Nanoporous materials are attracting huge interest in the scientific community since they provide new functionalities that bulk fully-dense materials fail to offer. For instance, porous biomaterials have been shown to be of great interest for their high surface area in contact with body fluids (faster regeneration and possibility, if the pores are large enough, to let blood capillaries form and penetrate into the material). However, by inducing porosity the mechanical integrity of these materials can be compromised. Although Ti and its alloys are well-known for their in-vivo behaviour for bone fixation, some routes (e.g. controlled generation of porosity) to reduce their high Young's modulus (~100 GPa) to that of cortical bone's one (~20–30 GPa) are needed. In other words, porosity enhances the biomechanical compatibility. Thus, there is a need to synthesize porous materials with optimum mechanical properties. Previous works studying the mechanical properties of these structures did not take into account the effect of densification, due to pore-size reduction, that occurs during mechanical testing or under the action of an applied load. This is of tremendous importance when considering, for instance, the application of these materials as bone replacement. As the receiver of the implant (i.e., the patient) will more likely be moving, the biomaterial will be under different loading conditions, leading to different compression situations (pore-size reduction). In that case, the mechanical properties of the implant will change depending on the stress configuration, as a result of densification.

Besides porous Ti-based alloys, other metallic porous materials such as nickel (Ni) foams also have a great interest. Ni foam is a low density permeable material that has a very high porosity, with typically 75–95% of the volume consisting of void spaces. Nickel foams have a wide variety of applications in heat exchangers, energy absorption, flow diffusion, lightweight optics and are used as filter materials for gas and liquid. Figure 16.2 shows an example of porous Ni.

For the modelling of the mechanical properties of porous materials, the ideal is to use pressure-sensitive models such as Drucker–Prager (DP) and/or Mohr–Coulomb (MC) for the yielding criteria in the finite element modelling, in order to anticipate the deformation mechanisms active during nanoindentation and how the pore size and their distribution affect the onset of plasticity. Along with common meshing techniques to model the pore distribution in these materials, one can also make use of micro computed tomography to reproduce exactely the actual geometry of the porous material. As an example, Figure 16.3 shows the FE mesh modelling of a material including a uniform distribution of pores. Through this model it is possible to study, for instance, the effect of pore-wall thickness on the resulting mechanical performance.

Figure 16.4 shows the example of a micro-CT image for a porous Ti implant that afterwards can be used for FE simulations.

As has been pointed out previously, while performing a mechanical test on a porous material, indentation or uniaxial compression, it is reasonable to assume that its mechanical properties will change as a result of compaction. In order to study this effect

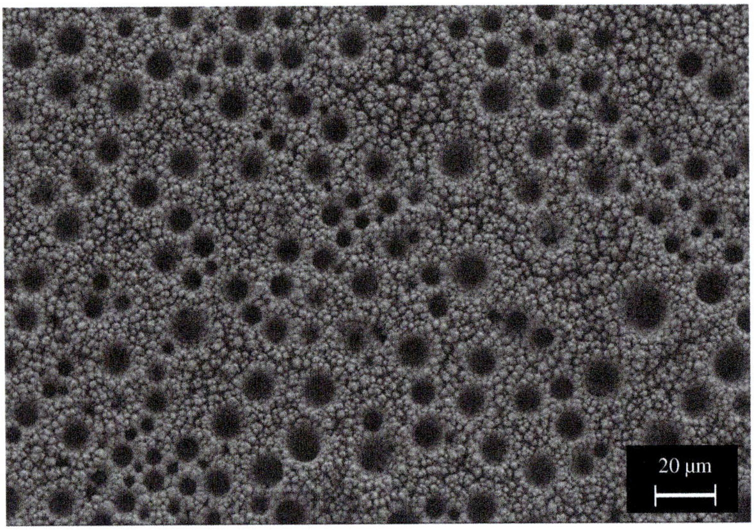

Figure 16.2 Surface SEM image of a Ni-porous sample obtained through electrodeposition. *Source*: Courtesy of Dr Jin Zhang.

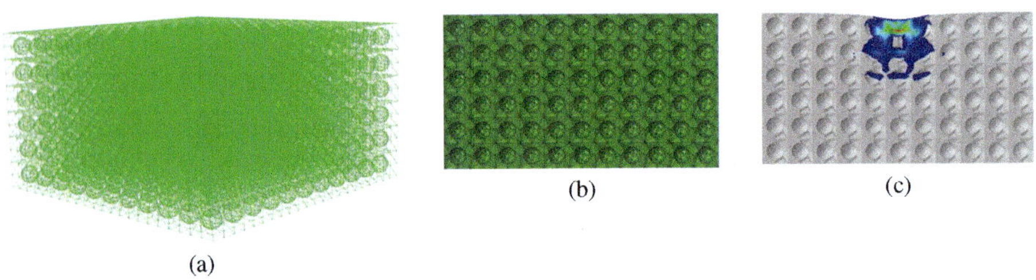

Figure 16.3 Example of a FE mesh for a porous material including pores as spherical cavities on the sketch of the model. Part (a) shows the general perspective with 726 pores, (b) is a cross section showing the inner cavities and (c) shows the same cross-section as in (b) after indenting with a Berkovich indenter, one can observe how the pore (cavity) size is reduced, which will lead to an evolution for the mechanical behaviour of the specimen when compared with its bulk counterpart.

through FE simulations, it is possible to introduce a pressure-sensitivity model which takes into account the evolution of yield stress as a result of the increase of hydrostatic pressure that in turn, will lead to the densification of the material. A first option for a pressure-sensitivity model is the associated linear DP criterion (Equation 16.17) used along with an elastic-perfectly plastic (Equation (16.18)) model to define the uniaxial behaviour of the material, where the Young's modulus E, the frictional angle β and the yield stress σ_{ys} are the required mechanical properties. Basically, what the DP criterion provides is a pressure-sensitivity model in which the von Mises yield surface changes with the hydrostatic pressure according to a frictional angle. In other words, as the material compacts (increase of hydrostatic pressure) the yield surface changes. DP is

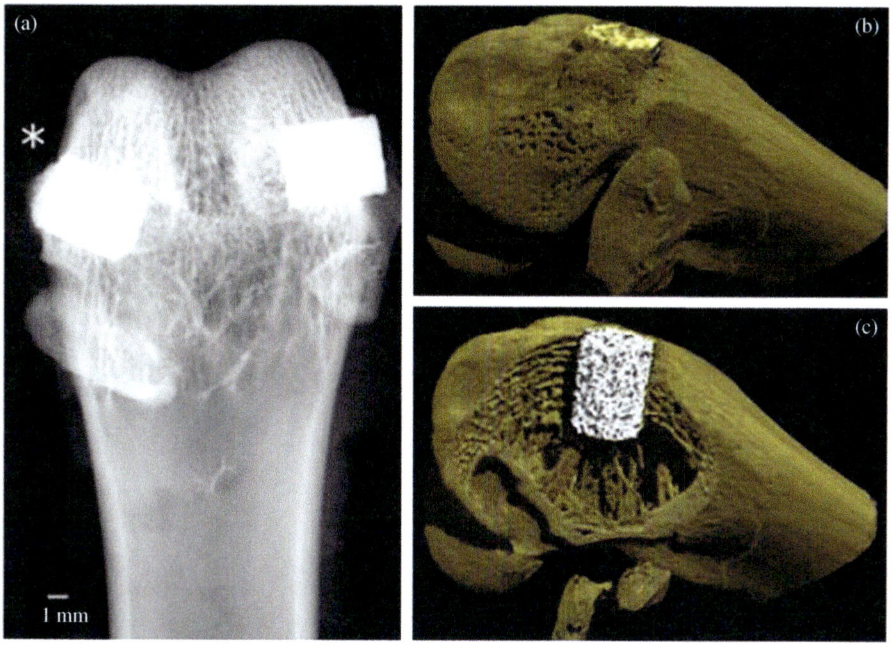

Figure 16.4 (a) High resolution radiographs of the cylindrical porous titanium implants in the condyles of the distal femur after 6 weeks in situ. (b) Micro-CT 3D reconstruction of one of the implants from a in the condyle of the distal femur (asterisked in (a), implant white, bone brown in (b)). (c) 3D visualization of the implant in situ. Part of the bone has been digitally removed to illustrate the trabecular architecture surrounding the implant. Compared to 2D radiographs, 3D reconstructions provide enhanced information regarding implant position and architecture of the surrounding osseous. *Source*: Baril 2011 [17]. Reproduced with permission of Springer.

extensively used during modelling of porous materials [18–21]:

$$F = \frac{1}{2}\sqrt{\frac{3}{2}(\boldsymbol{S} : \boldsymbol{S})} - p\tan\beta - d \tag{16.17}$$

$$\varepsilon = \begin{cases} {}^{\sigma}/_{E} & \mathit{if}\ \sigma < \sigma_{ys} \\ {}^{\sigma}/_{\sigma_{ys}} & \mathit{if}\ \sigma > \sigma_{ys} \end{cases} \tag{16.18}$$

Being F the yield surface, $\boldsymbol{S}$ the deviatoric stress tensor, p the hydrostatic pressure and d the cohesion expressed as:

$$d = \left(1 - \frac{1}{3}\tan\beta\right)\sigma_{ys} \tag{16.19}$$

In the event of hardening, this can be captured also through Equation (16.18) by substituting the branch $\varepsilon = {}^{\sigma}/_{\sigma_{ys}}$ for $\varepsilon = \left({}^{\sigma}/_{\sigma_o}\right)^{1/n}$, where σ_o is a characteristic stress at a unitary deformation and n is the strain hardening coefficient. An alternative to the DP criterion is the MC criterion, which is analogous to the DP, with the only difference that the yield surface is defined by Tresca. Figure 16.5 shows a representation of both pressure – sensitivity criteria.

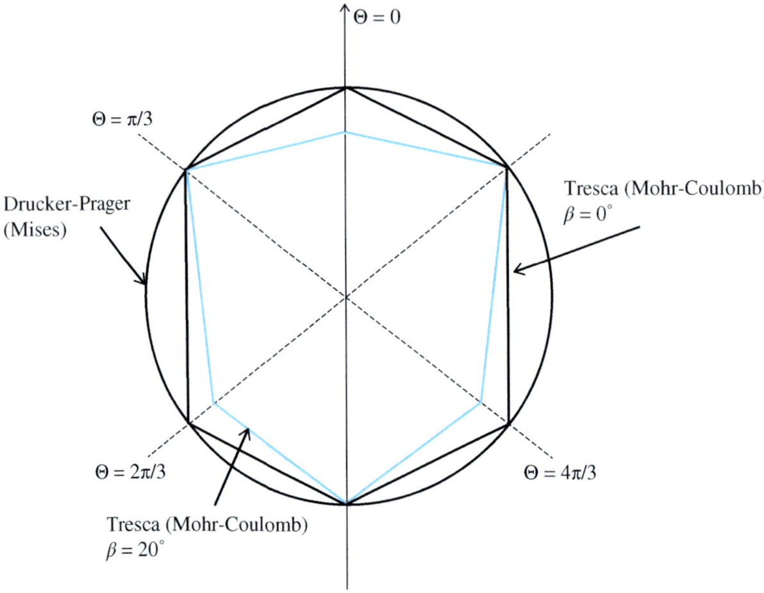

Figure 16.5 Diagram showing the von Mises and Tresca yield surfaces for different friction angles associated to DP and MC criteria respectively. Adapted from [22].

16.3.2 Finite Element Simulations of 1D Structures: Nanowires

One dimensional (1D) materials such as nanotubes, nanorods and nanowires (NWs) are attracting a considerable scientific attention due to their exceptional mechanical, electrical and optical properties. In particular, when studying their mechanical properties, two main problems arise when modelling nanoindentation experiments on these structures: (i) to take into account size – effects on the constitutive equations and (ii) to describe accurately the existing contact between the indenter and the surface of the nanowire. This last point is more important that what could appear at first glance, given the fact that not only the surface of the nanowire will present a certain curvature, but also the tip's indenter is curved due to the shallow penetration depths. Nanowires can have a diameter of ca. 50 nm and the sharpest conventional Berkovich diamond tip may have a tip bluntness of ca. 250 nm. Figure 16.6 shows the implications of not considering this contact between two curved surfaces, especially when taking into account the calculation of hardness as the area involved in its calculation is precisely the contact area between the indenter and the nanowire.

Additionally, the reduced size of these structures will challenge one of the most used relations when relating hardness and yield stress, that is $\overline{p} = C\sigma_{ys}$ where C is the so-called constrain factor taken, usually, equal to three. This factor comes from the effect of the material surrounding the imprint produced by the indenter. In other words, C is related with the hydrostatic pressure exerted by the material around the indent, if the size of the material is reduced, the effect of the hydrostatic pressure is also reduced. Consequently, C will be reduced, eventually to values close to 1.

A survey of the literature shows that the application of FE modelling in the field of NWs is rather scarce and, even more, the few existing works present some fundamental

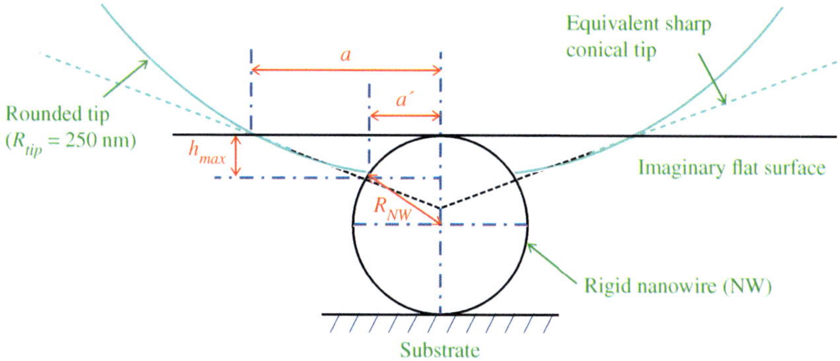

Figure 16.6 Schematic drawing of the contact between the indenter and the nanowire. The 'ideal' Berkovich indenter is depicted here as an equivalent conical tip with an effective cone angle of 70.3° [24, 6]. The 'real' (i.e., blunted) indenter tip, represented with a continuous black line, is more similar to a spherical tip with a radius of 250 nm. The dimensions of the nanowire and the tip are drawn to scale. The contour of the cono-spherical indenter that best approaches the real tip used in our study is drawn in cyan.

problems that give an open opportunity for further study. In Feng *et al.* [24] a two interface contact model was developed for the nanoindentation of a nanowire on a flat surface. That model was used to analyse the nanoindentation data for GaN and ZnO nanowires with radii in the range of 20–50 nm positioned on a silicon substrate and bonded to the substrate with platinum at their ends. The double contact model proposed by Feng *et al.* was used to determine the contact area (A_c) between the indenter (they used a Berkovich tip with a tip roundness of 200 nm) and the nanowire. This goes on the same lines as aforementioned within the framework of Figure 16.6. By considering that the two contacts (indenter/nanowire and nanowire/substrate) are in series, a composite contact can be defined as $1/S = 1/S_1 + 1/S_2$, where S_1 and S_2 are the stiffness of the indenter/nanowire and nanowire/substrate contacts, respectively. With the latter and considering the radii of curvature of the tip indenter and the nanowires, and considering an elastic contact between the spherical indenter tip and the nanowire, it is possible to calculate:

$$S_1 = \lambda_1 2E_{r1}\sqrt{a_1 b_1}, \tag{16.20}$$

where $\lambda_1 = \frac{\pi}{2\sqrt{k_1 K_1'}}$, $K_1' = \int_0^{\pi/2} \frac{1}{\sqrt{1-(1-k_1^2)\sin^2\theta}} d\theta$ and k_1 is the aspect ratio of the ellipse formed during the contact between the indenter and the nanowire. E_{r1} is the reduced Young's modulus between the indenter and the NW, (see Equation 16.16). With this, and knowing the applied load P, one has that hardness can be calculated through:

$$H = \frac{P}{A_{c1}} = \frac{4}{\pi}\frac{P}{S_1^2}E_1^2\lambda_1^2 \tag{16.21}$$

This proves to be an intermediate solution between those provided by the classical Oliver and Pharr solution (lower bound) and the Joslin-Oliver solution (upper bound) [25]. Notice that this contact model is just used for analytical purposes, but has not been implemented and/or compared with FE simulations. This was done a posteriori by Askari and Feng [26]. That study was rather interesting, as it also introduces two limiting cases regarding the movement of the NW with respect of the supporting substrate, namely (i) frictionless contact and freely separable and (ii) bonded along the line initially in contact and frictionless elsewhere. However, the model suffers from an important limitation: all contacts (indenter/NW and NW/substrate) were considered to be elastic, so no permanent deformation was induced and, hence, the extracted hardness is lacking physical significance.

Following the same idea, Chang *et al.* [27] performed FE simulations of conical indentation on Cu wires, where the conical indenter had a tip roundness of 400 nm. On their study, the main purpose was to investigate how the extracted elastic modulus of the wires deviated from the bulk value depending on the ratio of penetrated wire (h/D, where h is the displacement of the indenter inside the wire and D is the diameter of the wire). Notice that this kind of analysis resembles the one generally performed to study the effect of a substrate on the hardness of a thin film. Although this approach proves to be really interesting, there are two main issues that give room for improvement: (i) in their work Chang *et al.* used wires, and not NWs, with 100 μm of diameter. Being the indenter tip radius 0.4 μm ($D = 250 \times R$, where R is the tip radius), a huge influence of the curved contact on the extraction of mechanical properties is not expected. Moreover, in their work $h/D \leq 0.1$, which means that the maximum penetration depth was equal to 10 μm (10% of the wire's diameter), as the tip radius is equal to 400 nm. For that penetration depth the tip is no longer spherical but conical. (ii) Substrate, where the wire is lying, is supposed to be rigid even when they considered it to be made of copper. In the case of a thin structure on top of a substrate made of the same material, the 10% rule, usually assumed for thin film indentation, will be challenged and it could actually be better to consider either shallower penetration depths or a two interface contact model as the one proposed by Feng *et al.* [24].

Finally, Wang *et al.* [28] did an interesting study comparing the lateral in-situ SEM nanoindentation of single crystalline Si NWs, with 250 nm of diameter, with a flat punch with FE simulations considering a purely Hertzian solution. Given the mechanical properties of the Si NWs it was sensible to assume that, without fracture, all deformation would be comprised inside an elastic regime. Comparison between experiments and simulations confirmed that it was possible to extract the elastic modulus of such Si NWs by considering the presented closed-form analytical solution based on a Hertzian contact.

16.3.3 Continuum Crystal Plasticity Finite Element Simulations: Nanoindentation of Thin Solid Films

With the advent of nanoindentation, characterization of materials at different length scales has become possible. In that sense, great advances have been made in the study of coatings [29–37]. These studies are of great importance given the potential industrial impact that the mechanical characterization of MEMS (micro electromechanical systems) or, more recently, NEMS (nano electromechanical systems) will have. Works published in the last 20 years have been mainly focused in establishing at which point a substrate affects the mechanical properties of a coating [29, 32, 33, 35, 36]. For instance, Fabes *et al.* [29] stated that as long as the plastic flow does not extend into the substrate, mechanical properties of the coating will not change as compared with those found in its homogeneous counterpart. In that work, titanium (Ti) coatings with different crystallographic orientations deposited on alumina (Al_2O_3) substrates, also with different orientations, were analysed. Crystallography of both, coating and substrate were chosen to favor epitaxial growth of Ti on Al_2O_3. Although the thickness of the Ti coatings was varied between 75 and 400 nm, the reported finite element simulations, used to compare predictions of the mechanical properties with experimental values, assumed isotropic models for both coatings and substrates, i.e., crystallography was not taken into account. Similar work was done by Gouldstone *et al.* [38] for thin aluminium (Al) polycrystalline films (400, 600 and 1000 nm thick with a grain size in the same order of magnitude as thickness) and thin (111) textured copper (Cu) films deposited on silica (SiO_2) substrates. The purpose of that work was to develop an understanding of the discrete and continuous deformation characteristics of thin metallic films subjected to nanoindentation. Even though the applied load (P)–penetration depth (h_s) curves shown by Gouldstone *et al.* are at penetration depths shallow enough so the indentation imprint could be embedded inside a single grain, their models did not take into account the anisotropic character of the Al and Cu films. Finally, Saha *et al.* [32] studied the behavior of Al and tungsten (W) coatings with thicknesses in the range of 0.5 to 2.16 μm deposited on glass, sapphire and silicon substrates. By analyzing evolutions of hardness in terms of the ratio of indented thin film, and again using isotropic models, they concluded that only in the event that the indenter penetrates the substrate will the hardness of the system be affected. Other works where coating and substrate are supposed to behave as elastic-perfectly plastic materials or where crystallography is not taken into account are [30, 37, 39]. More recently, works by Li *et al.* [40] and Cäer *et al.* [41] introduced the concept of indentation Schmid factors (ISF) to study the effect of crystallographic orientation on nanoindentation pop-in events. Those works contained some drawbacks in the analysis methods: (i) they dealt with bulk spherical indentation and not with thin film sharp indentation; (ii) they supposed purely elastic contacts, i.e. defect-free materials, rather than interaction of preexisting dislocations. Li *et al.* defined the ISF as the ratio of the maximum resolved shear stress to the maximum contact pressure during spherical nanoindentation, being then the critical load for homogeneous dislocation nucleation under Hertzian contact:

$$P_{crt} = \left(\frac{\tau_{th}}{S}\right)^3 \frac{\pi^3 R^2}{6E_r^2} \tag{16.22}$$

τ_{th} is the theoretical strength at which homogeneous dislocation nucleation occurs, equal to the maximum resolved shear stress among all possible slip systems ($\tau_{th} = \tau_{rss}^{max}$). S is the ISF, R the spherical indenter's radius and E_r is the reduced modulus, that depends on the Young's modulus and Poison's coefficient of the indenter and, in the case of anisotropic materials, on the effective indentation modulus E_{eff}. E_{eff} depends, in turn, on elastic anisotropy of the material and indentation direction. In their work, Li *et al.* found through nanoindentation on NiAl single crystals with simple {110}<001> slip systems that: (i) the crystallographic dependence of the pop-in loads agreed with the predictions based on the ISF; (ii) the location of the τ_{rss}^{max} depended on the indenter shape, elastic anisotropy, and active slip systems. Caër *et al.* did a similar study on f.c.c. Ni single crystals using Berkovich indentation. In that work, they were able to explain differences in the pop-in loads through ISF while indenting different crystallographic planes. Caër *et al.* also introduced a modification of the ISF, the indentation Patel-Cohen factor, to explain the different loads at which a martensitic transformation is induced in CuAlBe shape memory alloys depending on the indented plane. Nevertheless, these two works

[40, 41] were limited to the study of anisotropic elastic effects, not accounting for crystal plasticity. Given these facts, it is important to establish some guidelines on how thin single crystalline films could be modelled. In doing so, an archetypical model for f.c.c. single crystals on rigid substrates can be implemented. In particular, we introduce here the basis of continuum crystal plasticity models for Cu single crystals.

While modelling Cu, one can assume anisotropic elasticity and a continuum crystal plasticity model that accounts for plastic deformation through dislocation gliding. Onset of plastic deformation is governed by Schmid's law. Thus dislocation glide can be activated simultaneously in one or more of the 12 slip systems defined by the slip family {111}<110> of the f.c.c. Cu single crystals. This occurs when the resolved shear stress $\tau^{(\alpha)}$ in the slip system becomes equal to a certain critical shear stress i.e., $\tau^{(\alpha)} = \tau_c^{(\alpha)}$ where $\alpha = 1,2 \ldots 12$. Once the applied shear stress approaches the critical value to activate a specific slip system, we will have a shear strain rate in this slip system ($\dot{\gamma}^{(\alpha)}$) written as [42–45]:

$$\dot{\gamma}^{(\alpha)} = \dot{a}\left|\frac{\tau^{(\alpha)}}{\tau_c^{(\alpha)}}\right|^m sign(\tau^{(\alpha)}) \tag{16.23}$$

where $\dot{a}$ and m are constants related with thermally-activated dislocation glide. To take into account the evolution of $\tau_c^{(\alpha)}$ as deformation procedes we use the following strain hardening relation [45]:

$$\dot{\tau}_c^{(\alpha)} = \sum_{\beta} h_{\alpha\beta}|\dot{\gamma}^{(\beta)}| \tag{16.24}$$

where $\dot{\gamma}^{(\beta)}$ is the slip rate in each slip system ($\beta = 1, 2 \ldots 12$ in f.c.c. crystals) and $h_{\alpha\beta}$ is a hardening matrix. The hardening model described by Bassani and Wu (hereafter BW hardening model) in [46] is used here. This model is suitable for single crystalline materials whose uniaxial stress-strain behavior presents stage I deformation (single glide) and transition towards stage II (multiple glide). In this case $h_{\alpha\beta}$ has the following form:

$$h_{\alpha\alpha} = \left\{(h_o - h_I)\left(\text{sech}\left[\frac{(h_o - h_I)|\gamma^{(\alpha)}|}{(\tau_s - \tau_o)}\right]\right)^2 + h_I\right\} \times \left\{1 + \sum_{\beta \neq \alpha} f_{\alpha\beta} \tanh\left(\frac{|\gamma^{(\beta)}|}{\gamma_o}\right)\right\};\ h_{\alpha\beta} = 0 \tag{16.25}$$

where h_o and τ_o are the initial hardening and shear stress after yielding, h_I hardening during stage I, τ_s saturation shear stress at stage I. $f_{\alpha\beta}$ is a dimensionless matrix which measures the interaction strength of dislocations gliding in a slip system α with the forest of dislocation in slip systems β, and γ_o is a phenomenological parameter governing the transition from stage I to stage II deformation. Notice that Equation (16.23) takes into account the sign function of $\tau^{(\alpha)}$, while Equations (16.24) and (16.25) consider the absolute values of $\dot{\gamma}^{(\alpha)}$ and $\gamma^{(\alpha)}$, respectively, in each slip system. The reason for this is that during the kinematic description of the model Equation (16.23) needs to take into account both senses of a given slip direction, while during the description of the increase in the critical resolved shear stress in each slip system ($\dot{\tau}_c^{(\alpha)}$) and the definition of the hardening matrix ($h_{\alpha\alpha}$), only one sense, the one that drives plastic flow beneath the free surface during indentation, is needed.

Looking at Equation (16.24), we can see that the evolution of $\dot{\tau}_c$ in a specific slip system α depends not only on deformation in the primary active slip system α, but also on deformation in the rest of secondary active slip systems, i.e.$\beta \neq \alpha$. In the same spirit, the first part of $h_{\alpha\alpha}$ in Equation (16.25) captures self-hardening during deformation in the primary active slip system, while the second part accounts for hardening due to glide in the rest of slip systems, i.e. secondary active slip systems. In other words, although in Equation (16.25) $h_{\alpha\beta} = 0$, $h_{\alpha\alpha}$ incorporates both self- and latent-hardening. Parameters for the present Cu model can be found in references [47, 48] for a more detailed explanation on the model. One of the limitations of the continuum crystal plasticity model presented here is that it is suitable for bulk Cu single crystals with an starting density of dislocations only because their plasticity is governed by dislocation interactions and not by dislocation nucleation. This means that the stress needed to initiate plastic deformation is smaller within this model than in single crystals with a reduced number of dislocation or, in the extreme case, in defect-free structures where high levels of stress are needed to nucleate the first defects.

Section 16.4 will deal with the nucleation and interaction of dislocations in single crystals during atomistic simulations of indentation experiments.

16.4 Nucleation and Interaction of Dislocations During Single Crystal Nanoindentaion: Atomistic Simulations

With the advent of the twentyfirst century and the increasingly powerful computational clusters, and their reduction in cost, a lot of interesting atomistic simulation works have appeared in the literature dealing with the initiation of plastic deformation in crystalline solids during nanoindentation. Mainly, two different lines have been followed these years: (i) dislocations dynamics (DD) simulations and (ii) molecular dynamics (MD) simulations. While the former has been used to study the interaction and multiplication of dislocations, the latter has proven to be extremely useful to study the formation of stacking faults, shear bands and general dislocation structures during, for instance, indentation simulations. In the following, we will summarize some of the latest developments in DD and MD simulations along with some of its fundamental basis.

16.4.1 Dislocation Dynamics Simulations

During DD simulations, either in two or three dimensions, the objective is to study the movement, interaction, multiplication and annihilation of an initial set of dislocations. The reader is encouraged to read the following references in order to learn the details on how to set – up a DD simulation [49–56]. Benoit Devincre and Ladislas Pierre Kubin (along with their colleagues) have been the main contributors to the fundamental study of dislocation motion and dislocation interactions. Proof of that are the nearly 60 research articles published in the topic. As this is not intended to be an extensive review on dislocation dynamics, here we will just summarize some of their most valuable contributions. Of special interest is the study of strain hardening in single crystals by introducing the concept of the mean free path length [57] which is the distance travelled by a dislocation segment of length l before it is stored by interaction with the microstructure. They applied that formulation to the strain hardening in f.c.c. crystals, given their well – documented, though complicated, stress-strain relation. They concluded that the mean free path length:

$$\frac{1}{L^{(\alpha)}} = \frac{\tau_c^{(\alpha)}}{\mu b K_{hkl}}, \quad \text{with } K_{hkl} = \left[\frac{n(1+\kappa)^{3/2}}{p_0 k_0 \sqrt{\bar{a}}(n-1-\kappa)}\right] \tag{16.26}$$

where $L^{(\alpha)}$ is the mean free path length for each slip system (α), μ is the shear modulus and b is the Burgers vector. $L^{(\alpha)}$ also depends on the Taylor stress $\tau_c^{(\alpha)} = \mu b\sqrt{\sum_\beta a_{\alpha\beta}\rho^{(\beta)}}$, $\rho^{(\beta)}$ being the dislocation density in each slip system and $a_{\alpha\beta}$ is a matrix that accounts for the interaction strength between dislocations (Lomer-Cotrell and Hirth locks and collinear and glissile interactions). Finally, $L^{(\alpha)}$ also depends on K_{hkl} that, in turn, depends on three dimensionless constants (p_0, k_0 and κ) and on the number of active slip systems, n, that changes depending on the orientation [hkl] of the loading axis. For example, $n = 4$, 3 and 2 for loading axis {001}, [111] and [112] respectively. $\bar{a}$ is the average value of the $a_{\alpha\beta}$ matrix. One additional and important aspect of that work is that the authors, by introducing the concept of mean free path length, were able to overcome the limited plastic strains that could be obtained up to that moment due to computational limitations [58–61].

Although it could appear that the topic of dislocation dynamics somehow diverges from the central point of this chapter, this is not the case. As the interest of nanosized structures has been growing, the assessment of their mechanical properties through nanoindentation has become more challenging. This is related to what is known as indentation size effects (ISE), coming not only from the tested sample but also from the indenter tip. For example, when indenting single crystals at low loads, it can happen that the zone affected by the indenter is small enough to be free of dislocations. Hence, in order to accommodate the geometry of the indenter some defects need to be created. At these first stages of deformation it will be interesting to study the movement, multiplication and annihilation of dislocations. The presence of these geometrically necessary dislocations (GNDs) has been exhaustively reported in the literature [62–65]. For instance, in the works by Marc Fivel and colleagues [66, 67] they developed a protocol to study the initial nucleation, movement and interaction of dislocations during spherical indentation of an f.c.c. single crystal through a combination of three dimensional DD and finite element simulations, where the latter were used to calculate the elastic deformation that, subsequently, was incorporated into the DD simulations. The model was checked for Cu single crystals when indenting the (001) crystallographic plane and the dislocation arrangement obtained by the simulations was compared with TEM observations. During the model set-up it was necessary to locate the centre of the nucleated dislocation loops, that will coincide with the point on the indentation axis where the resolved shear stress is the highest, being for the (001) – indented plane at a penetration depth equal to $0.47R_{\text{cont}}$, being R_{cont} the contact radius between the material and the indenter. In this particular case it was observed that each nucleation process was involving four prismatic loops, each gliding in two slip systems, giving a total of eight active slip systems defined by four Burgers vectors: $\boldsymbol{b}_1 = \frac{1}{2}[\bar{1}01]$, $\boldsymbol{b}_2 = \frac{1}{2}[011]$, $\boldsymbol{b}_3 = \frac{1}{2}[01\bar{1}]$, $\boldsymbol{b}_4 = \frac{1}{2}[101]$. The nucleated loops are those necessary to accommodate the imposed plastic displacement. More recent works by Kreuzer and Pipan [68] and Widjaja *et al.* [69, 70] tackled how the material behaviour is influenced by the motion of pre-existing discrete edge dislocations during indentation [68] and the indentation size effects on an initially defect-free material, but with an initial distribution of dislocation sources and obstacles, when indented with different tip geometries [69, 70].

16.4.2 Molecular Dynamics Simulations

While doing a survey on the latest research articles using MD simulations to study nanoindentation experiments, one can find extremely interesting works. For instance, Ma and Yang [71] observed the burst and arrest of stacking faults in nanocrystalline copper (Cu), where indentation in both single grains and grain boundaries were performed. Choi *et al.* [72] studied the mechanisms by which plasticity progresses in aluminium single crystals. Feichtinger *et al.* [73] explored the deformation mechanisms in polycrystalline gold. Zhu and Fang [74] and Lai *et al.* [75] performed MD simulations of nanoindentation experiments on germanium single crystals to study phase transformations as a deformation mechanism. Wu *et al.* [76] studied the response of various face-centered cubic (f.c.c.) and body-centered cubic (b.c.c.) single crystals under indentation and scratching conditions. Verkhovtsev *et al.* [77] focused on the effect of the indenter shape on the nanoindentation curves of single crystalline titanium. Shao *et al.* [78] focused their work on strain hardening effects in nanoscale multilayer metallic composites through atomistic simulations of nanoindentation. Grain size effects in nanocrystalline nickel were studied by Liu *et al.* [79], showing inverse Hall-Petch effects for grain sizes up to 40 nm. Tan *et al.* [80] discussed how to extract the Young's modulus of monolayer graphene through nanoindentation models. Seymour *et al.* [81] investigated the influence of the crystallographic orientation of the indented plane in NiAl and Ni_3Al crystals in terms of load-displacement curves, hardnesses and deformation mechanisms. Goel *et al.* [82] explored the nanomechanial response of iron carbides (Fe_3C and Fe_4C) by means of nanoindentation simulations. Finally, works by Ziegenhain *et al.* [83] and Salehinia *et al.* [84] dealt with initiation of plastic deformation on Cu single crystals while indenting different crystallographic oriented planes. All the referenced works have in common that they do not study, systematically, the effect of the geometry of the indenter and the indentation velocity. Moreover, all the simulations were performed using a number of atoms just slightly above 1 million. Given the highly demanding computational resources of these kinds of simulations, impact velocities (above 4 m/s) are used when modelling the indentation experiment. It is important to take into account the effects of the size of the tip radius as different indenter sizes will induce different pressures inside the material and, hence, it will affect the load and penetration depths at which the pop-in events (nucleation of dislocations) occur. Also, a sufficient number of atoms is needed during simulations to avoid lateral effects. Existing publications deal with tip radii below the 50 nm, that is much smaller that the tip bluntness of Berkovich or conical tips (and slightly closer to the ones shown by AFM tips). This happens because of the extreme large number of atoms needed to consider realistic indenter tips so the imprint left by the indenter is not too close to the borders of the simulation box.

When setting up an MD simulation of an indentation experiment, besides selecting the adequate number of atoms conforming the simulation box, deciding the size of the indenter, the indentation velocity, the test temperature,... the main issue is to select the correct interatomic potential that will describe the interactions between the atoms of the simulation box. Here we will give some details on one of the most widely used kind of potentials, especially for f.c.c. single crystals, that is the embedded atom potential (or method) EAM.

In the embedded-atom method, the total energy of an elemental system is represented by:

$$E_{tot} = \frac{1}{2}\sum_{ij} V(r_{ij}) + \sum_{i} F(\overline{\rho}_i) \quad (16.27)$$

Where $V(r_{ij})$ is a pair potential as a function of distance r_{ij} between atoms i and j, and F is the embedding energy as a a function of the host electron density $\overline{\rho}_i$ induced at site I by all other atoms in the system. The latter is given by:

$$\overline{\rho}_i = \sum_{j \neq i} \rho(r_{ij}) \quad (16.28)$$

where ρ (r) is the electron density function. One can find details on these embedded-atom potentials for different materials such in, for example: Michin *et al.* [85] for Cu single crystals; Liu et al [86] for several f.c.c. metals such us Ni, Cu, Al, Ag, Au, Pd and Pt; Sun *et al.* [87] for Mg single crystals with an h.c.p. crystal structure or for Ta b.c.c. single crystals as in [88]. More recently, Ruestes *et al.* [89] studied the effect of the indenter diameter, penetration velocity and interatomic potential on the defect nucleation and evolution during MD simulations of single crystal Ta nanoindentation. In their work they were able to pose some limitations on the different existing potentials describing Ta single crystals. For instance, they reached the conclusion that the EAM potential designed by Li *et al.* [88] was not working properly during high-pressure and high strain gradients such the ones met during nanoindentation. Additionally, Rodriguez-Nieva *et al.* [90] studied the mechanical properties of porous Au by means of MD simulations of indentation experiments using the embedded atom method. In those simulations, the authors found two different deformation regimes: (i) isolated behaviour of the pores and (ii) interaction of dislocations coming from different pores as the distance between them was reduced. Also, their results allowed them to challenge some of the classical hardening theories used during continuum simulations, such as Taylor hardening and the application of scale-free continuum models for nanostructured materials. Finally, in a recent work from Esqué-de los Ojos et al [91], MD simulations of spherical indentation on porous single crystalline Cu showed the effect of pore size on the determination of the reduced Young's modulus.

References

1 Doerner, M.F., Nix, W.D. A method for interpretating the data from depth-sensing indentation instruments. Journal of Materials Research 7, 601–609; 1986.

2 Oliver, W.C., Pharr, G.M. An improved technique for determining hardness and elastic modulus using load and displacement sensing indentation experiments. Journal of Materials Research 7, 1564–1583; 1992.

3 Casals, O., Alcalá, J. The duality in mechanical property extractions from Vickers and Berkovich instrumented indentation experiments. Acta Materialia 53, 3545–3561; 2005.

4 Cheng, Y-T., Cheng, C-M. Scaling, dimensional analysis, and indentation measurements. Materials Science and Engineering R44, 91–149; 2004.

5 Tabor, D. The Hardness of Metals. Clarendon Press, Oxford; 1951.

6 Oliver, W.C., Pharr, G.M. Measurement of hardness and elastic modulus by instrumented indentation: Advances in understanding and refiniments to methodology. Journal of Materials Research 19, 3–20; 2004.
7 Fischer-Cripps, A.C. Introduction to contact mechanics, 2nd edn. Springer, Australia; 2007.
8 Gouldstone, A., Chollacoop, N., Dao *et al.* Indentation across size scales and disciplines: Recent developments in experimentation and modeling. Acta Materialia 55, 4015–4039; 2007.
9 Cao, Y.P., Lu, J. A new method to extract the plastic properties of metal materials from an instrumented spherical indentation loading curve. Acta Materialia 52, 4023–4032; 2004.
10 Wenyi, Y., Sun, Q.P., Feng, X.Q., Qian, L.M. Analysis of spherical indentation of superelastic shape memory alloys. International Journal of Solids and Structures 44, 1–17; 2007.
11 Hill, R., Störakers, B., Zdunek, A.B. A theorical study of the Brinell hardness test. Proceedings of the Royal Society of London A, 423, 301–330; 1989.
12 Biwa, S., Störakers, B.J. Analysis of fully plastic Brinell indentation. Journal of the Mechanics and Physics of Solids 43, 1303–1334; 1995.
13 Mesarovic, S., Fleck, N. Spherical indentation of elastic-plastic solids. Proceedings of the Royal Society of London A, 455, 2707–2728; 1999.
14 Alcalá, J., Esqué-de los Ojos, D. Reassessing spherical indentation: contact regimes and mechanical property extractions. International Journal of Solids and Structures 47, 2714–2732; 2010.
15 Casals, O., Alcalá, J. The duality in mechanical property extractions from Vickers and Berkovich instrumented indentation experiments. Acta Materialia 53, 3545–3561; 2005.
16 Alcalá, J., Esqué-de los Ojos, D., Rodríguez, S.A. The role of crystalline anisotropy in mechanical property extractions through Berkovich indentation. Journal of Materials Research 24, 1235–1244; 2009.
17 Baril, E., Lefebvre, L.P., Hacking, S.A. Direct visualization and quantification of bone growth into porous titanium implants using micro computed tomography. Journal of Materials Science: Materials in Medicine 22, 1321–1332; 2011.
18 Karakus, M., Fowell, R.J. Effects of different tunnel face advance excavation on the settlement by FEM. Tunneling and Underground Space Technology 18, 513–523; 2003.
19 Guo, T.F., Faleskog, J., Shih, C.F. Continuum modeling of a porous solid with pressure-sensitive dilatant matrix. Journal of the Mechanics and Physics of Solids 56, 2188–2212; 2008.
20 Clément, P., Meille, S., Chevalier, J., Olagnon, C. Mechanical characterization of highly porous inorganic solids materials by instrumented micro-indentation. Acta Materialia 61, 6649–6660; 2013.
21 Domingo-Roca, R., Esqué-de los Ojos, D., Guerrero, M. *et al.* Nanomechanical behavior of 3D porous metal-ceramic nanocomposite Bi/Bi_2O_3 films. Materials Science and Engineering: A, 626, 150–158; 2015.
22 Dassault Systèmes. Abaqus Analysis User's Manual v6.14. Dassault Systèmes, Providence, RI, USA.
23 Fischer-Cripps, A.C. Nanoindentation, 2nd edn. Springer, New York; 2004.

24 Feng, G., Nix, W.D., Yoon, Y., Lee, C.J. A study of the mechanical properties of nanowires using nanoindentation. Journal of Applied Physics 99, 074304; 2006.

25 Joslin, D.L., Oliver, W.C. A new method for analyzing data from continuous depth-sensing microindenter tests. Journal of Materials Research 5, 123–126; 1990.

26 Askari, D., Feng, G. Finite element analysis of nanowire indentation on a flat substrate. Journal of Materials Research 27, 586–591; 2012.

27 Chang, N.-K., Lin, Y.-S., Chen, C.-Y., Chang, S.-H. Size effect of indenter on determining modulus of nanowires using nanoindentation technique. Thin Solid Films 517, 3695–3697; 2009.

28 Wang, Z., Mook, W.M., Niederberger, C. *et al.* Compression of nanowires using a flat indenter: Diametrical elasticity measurement. Nano Letters 12, 2289–2293; 2012.

29 Fabes, B.D., Oliver, W.C., McKee, R., Walker, F.J. The determination of film hardness from the composite response of film and substrate to nanometer scale indentations. Journal of Materials Research 7, 3056–3064; 1992.

30 Leigh, S.-H., Lin, C.-K., Berndt, C.C. Elastic response of thermal spray deposits under indentation tests. Journal of the American Ceramics Society 80, 2093–2099; 1997.

31 Chudoba, T., Schwarzer, N., Richter, F. Determination of elastic properties of thin films by indentation measurements with a spherical indenter. Surface and Coatings Technology 127, 9–17; 2000.

32 Saha, R., Xue, Z., Huang, Y., Nix, W.D. Indentation of a soft metal film on a hard substrate: strain gradient hardening effects. Journal of the Mechanics and Physics of Solids 49, 1997–2014; 2001.

33 Saha, R., Nix, W.D. Effects of the substrate on the determination of thin film mechanical properties by nanoindentation. Acta Materialia 50, 23–38; 2002.

34 Zhu, T., Van Vliet, K.J., Ogata, S. *et al.* Predictive modeling of nanoindentation-induced homogeneous dislocation nucleation in copper. Journal of the Mechanics and Physics of Solids 52, 691–724; 2004.

35 Tsui, T.Y., Pharr G. M. Substrate effects on nanoindentation mechanical property measurement of soft films on hard substrates. Journal of Materials Research 14, 292–301; 1999.

36 Pelletier, H., Krier, J., Mille, P. Characterization of mechanical properties of thin films using nanoindentation test. Mechanics of Materials 38, 1182–1198; 2006.

37 Zhao, M., Xiang, Y., Xu, J. *et al.* Determining mechanical properties of thin films from the loading curve of nanoindentation testing. Thin Solid Films 516, 7571–7580; 2008.

38 Gouldstone, A, Koh, H.J., Zeng, K.Y. *et al.* Discrete and continuous deformation during nanoindentation of thin films. Acta Materialia 48, 2277–2295; 2000.

39 Sakai, M. Substrate-affected indentation contact parameters of elastoplastic coating/substrate composites Journal of Materials Research 24, 831; 2009.

40 Li, T.L., Gao, Y.F., Bei, H., George, E.P. Indentation Schmid factor and orientation dependence of nanoindentation pop-in behavior of NiAl single crystals. Journal of the Mechanics and Physics of Solids 59, 1147–1162; 2011.

41 Cäer, C. Patoor, E., Berbenni, S., Lecomte, J.-S. Stress induced pop-in and pop-out nanoindentation events in CuAlBe shape memory alloys. Materials Science and Engineering: A, 587, 304–312; 2013.

42 Kocks, U.F., Argon, A.S., Ashby, M.F. Thermodynamics and kinetics of slip. Progress in Materials Science 19, 1–281; 1975.
43 Estrin, Y., Mecking, H. A unified phenomenological description of work-hardening and creep based on one-parameter models. Acta Metallurgica 32, 57–70; 1984.
44 Tabourot, L., Fivel, M., Rausch, E. Generalised constitutive laws for f.c.c. single crystals. Materials Science and Engineering: A, 234 639–642; 1997.
45 Asaro, R.J. Micromechanics of crystals and polycrystals. Advanced Applied Mechanics 23, 1–115; 1983.
46 Wu, T.Y., Bassani, J.L., Laird, C. Latent hardening in single-crystals. 1. Theory and experiments. Proceedings of the Royal Society of London: A, 435, 1–19; 1991.
47 Casals, O., Očenášek, J., Alcalá, J. Crystal plasticity finite element simulations of pyramidal indentation in copper single crystals. Acta Materialia 55, 55–68; 2007.
48 Alcala, J., Casals, O., Očenášek, J. Micromechanics of pyramidal indentation in f.c.c. metals: Single crystal plasticity finite element analysis. Journal of the Mechanics and Physics of Solids 56, 3277–3303; 2008.
49 Friedel, J. Dislocations. Pergamon, Oxford; 1967.
50 Hirth, J., Lothe, J. Theory of Dislocations. Krieger, Malabar, Florida, USA; 1992.
51 Devincre, B., Kubin, L., Hoc, T. Physical analyses of crystal plasticity by DD simulations. Scripta Materialia 54, 741–746; 2006.
52 Bulatov, V.V., Hsiung, L.L., Tang, M. *et al.* Dislocation multi-junctions and strain hardening. Nature 440, 1174–1178; 2006.
53 Bulatov, V.V., Abraham, F., Kubin, L. *et al.* Connecting atomistic and mesoscale simulations of crystal plasticity. Nature 391, 669–672; 1998.
54 Madec, R., Devincre, B., Kubin, L. *et al.* The role of collinear interaction in dislocation-induced hardening. Science 301, 1879–1882; 2003.
55 El-Awady, J. A., Biner, S. B., Ghoniem, N.M. A self-consistent boundary element, parametric dislocation dynamics formulation of plastic flow in finite volumes. Journal of the Mechanics and Physics of Solids 56, 2019–2035; 2008.
56 Ghoniem, N.M., Tong, S.-H., Sun, L.Z. Parametric dislocation dynamics: A thermodynamics-based approach to investigations of mesoscopic plastic deformation. Physical Review : B 61, 913–927; 2000.
57 Devincre, B., Hoc., T., Kubin, L. Dislocation mean free paths and strain hardening of crystals. Science 320, 1745–1748; 2008.
58 Arsenlis, A., Cai, W., Tang, M. *et al.* Enabling strain hardening simulations with dislocations dynamics. Modelling and Simulation in Materials Science and Engineering 15, 553–595; 2007.
59 Zbib H, Diaz de la Rubia T, Rhee M and Hirth J. 3D dislocation dynamics; stress–strain behavior and hardening mechanisms in f.c.c. and b.c.c. metals. Journal of Nuclear Materials 276, 154–165; 2000.
60 Devincre B, Kubin L, Lemarchand C and Madec R. Mesoscopic simulations of plastic deformation. Materials Science and Engineering: A, A309–310, 211–219; 2001.
61 Madec R, Devincre B, Kubin L. From dislocation junctions to forest hardening. Physical Review Letters 89, 255508; 2002.
62 Pharr, G.M., Herbert, E.G., Gao, Y. The indentation size effect: A critical examination of experimental observations and mechanistic interpretations. Annual Review of Materials Research 40, 271–292; 2010.

63 Nix, W.D., Gao, H. Indentation size effects in crystalline materials: A law for strain gradient plasticity. Journal of the Mechanics and Physics of Solids 46, 411–425; 1998.
64 Ma, Q., Clarke, D.R. Size-dependent hardness of silver single-crystals. Journal of Materials Researc 10, 853–863; 1995.
65 McElhaney, K.W., Vlassak, J.J., Nix, W.D. Determination of indenter tip geometry and indentation contact area for depth-sensing indentation experiments. Journal of Materials Research 13, 1300–1306; 1998.
66 Fivel, M., Verdier, M., Canoca, G. 3D simulation of a nanoindentation test at a mesoscopic scale. Materials Science and Engineering A, 234–236, 923–926; 1997.
67 Fivel, M.C., Robertson, C.F., Canova, G.R., Boulanger, L. Three-dimensional modeling of indent-induced plastic zone at a mesoscale. Acta Metallurgica 46, 6183–6194; 1998.
68 Kreuzer, H.G.M., Pippan, R. Discrete dislocation simulation of nanoindentation: The effect of statistically distributed dislocations. Materials Science and Engineering A, 400–401, 460–462; 2005.
69 Widjaja, A., van der Giessen, E., Needleman, A. Discrete dislocation modelling of submicron indentation. Materials Science and Engineering A, 400–401, 456–459; 2005.
70 Widjaja, A., van der Giessen, E., Deshpande, V.S., Needleman, A. Contact area and size effects in discrete dislocation modeling of wedge indentation. Journal of Materials Research 22, 655–663; 2007.
71 Ma, X.-L., Yang, W. Molecular dynamics simulation on burst and arrest of stacking faults in nanocrystalline Cu under nanoindentation. Nanotechnology 14, 1208–1215; 2003.
72 Choi, Y., Van Vliet, K.J., Li, J. Size effects on the onset of plastic deformation during nanoindentation of thin films and patterned lines. Journal of Applied Physics 94, 6050–6058; 2003.
73 Feichtinger, D., Derlet, P.M., Van Swygenhoven, H. Atomistic simulations of spherical indentations in nanocrystalline gold. Physical Review B 67, 0244113; 2003.
74 Zhu, P.Z., Fang, F.Z. Molecular dynamics simulations of nanoindentation of monocrystalline germanium. Applied Physics A, 108, 415–421; 2012.
75 Lai, M., Zhang, X., Fang, F. Nanoindentation-induced phase transformation and structural deformation of monocrystalline germanium: a molecular dynamics simulations investigation. Nanoscale Research Letters 8, 353; 2013.
76 Wu, C.-D., Fang, T.-H., Lin, J.-F. Atomistic-scale simulations of material behaviors and tribology properties for F.C.C. and B.C.C. metal films. Materials Letters 80, 59–62; 2012.
77 Verkhovtsev, A.V., Yakubovich, A.V., Sushko, G.B. *et al.* Molecular dynamics simulations of the nanoindentation process of titanium crystal. Computational Materials Science 76, 20–26; 2013.
78 Shao, S., Zbib, H.M., Mastorakos, I., Bahr, D.F. Effect of interfaces in the work hardening of nanoscale multilayer metallic composites during nanoindentation: A molecular dynamics investigation. Journal of Engineering Materials Technology 135, 021001.2013.
79 Liu, X., Yuan, F., Wei, Y. Grain size effect on the hardness of nanocrystal measured by the nanosize indenter. Applied Surface Science 279, 159–166; 2013.

80 Tan, X., Wu, J., Zhang, K. *et al.* Nanoindentation models and Young's modulus of monolayer graphene: A molecular dynamics study. Applied Physics Letters 102, 071908; 2013.
81 Seymour, R., Hemeryck, A., Nomurai, K.-I. *et al.* Nanoindentation of NiAl and Ni_3Al crystals on (100), (110), and (111) surfaces: A molecular dynamics study. Applied Physics Letters 104, 141904; 2014.
82 Goel, S., Joshi, S.S., Abdelal, G., Agrawal, A. Molecular dynamics simulation of nanoindentation of Fe_3C and Fe_4C. Mater. Sci. Eng. A 597, 331–341; 2014.
83 Ziegenhain, G., Urbassek, H.M., Hartmaier, A., Influence of crystal anisotropy on elastic deformation and onset of plasticity in nanoindentation: A simulation study. Materials Science and Engineering 107, 061807; 2010.
84 Salehinia, I., Lawrence, S.K., Bahr, D.F. The effect of crystal orientation on the stochastic behavior of dislocation nucleation and multiplication during nanoindentation. Acta Materialia 61, 1421–1431; 2013.
85 Mishin, Y., Mehl, M.J., Papaconstantopoulos, D.A., *et al.* Structural stability and lattice defects in copper: Ab initio, tight-binding, and embedded-atom calculations. Physical Review B 63, 224106; 2001.
86 Liu, C.L., Cohen, J.M., Adams, J.B., Voter, A.F. EAM study of surface self-diffusion of single adatoms of f.c.c. metals Ni, Cu, Al, Ag, Au, Pd, and Pt. Surface Science 253, 334–344; 1991.
87 Sun, D.Y., Mendelev, M.I., Becker, C.A. *et al.* Crystal melt interfacial free energies in HCP metals : a molecular dynamics study. Physical Review B 73, 024116; 2006.
88 Li, Y., Siegel, D.J. Adams, J.B., Liu, X.-Y. Embedded-atom-method tantalum potential developed by the force-matching method. Physical Review B 67, 125101; 2003.
89 Ruestes, C.J., Stukowski, Tang, Y. *et al.* Atomistic simulation of tantalum nanoindentation: Effects of indenter diameter, penetration velocity, and interatomic potentials on defect mechanisms and evolution. Materials Science and Engineering A 613, 390–403; 2014.
90 Rodriguez-Nieva, J.F., Ruestes, C.J., Tang, Y., Bringa, E.M. Atomistic simulation of the mechanical properties of nanoporous gold. Acta Materialia 80, 67–76; 2014.
91 Esqué-de los Ojos, D., Pellicer, E., Sort, J. The influence of pore size on the indentation behavior of metallic nanoporous materials: A molecular dynamics study. Materials 9, 355; 2016.

17

Nanoindentation *in silico* of Biological Particles

Olga Kononova[1,2], *Kenneth A. Marx*[1] *and Valeri Barsegov*[1,2]

[1] *Department of Chemistry, University of Massachusetts, Lowell, MA, USA*
[2] *Division of Applied Mathematics, Moscow Institute of Physics and Technology, Russia*

17.1 Introduction

In the past few decades, experimental approaches have been developed for nanomaterials characterization. A number of techniques have been designed to characterize the mechanics of materials including: surface forces apparatus, transmission and scanning electron microscopies, optical tweezers, instrumented nanoindentation and scanning probe microscopies. The latter two techniques possess significant advantages in that they allow mechanical response measurements of very small volumes at specific sample locations. Moreover, they provide a force response that can be followed as a function of penetration depth and time. The selection of a specific sample location coupled with a high spatial resolution provides a means to determine localized lateral as well as vertical mechanical response effects in the material being studied. Of particular importance in the study of soft matter (polymers, biomaterials and biological macromolecules) is the ability to measure mechanical responses as a function of time. This leads to an important understanding of these materials' viscoelastic properties.

Both instrumented nanoindentation and scanning probe microscopies (now primarily atomic force microscopy, AFM) were developments of the 1980s. These have grown in instrumental sophistication and capabilities over the ensuing years. However, there are significant advantages to using one technique versus the other and these have been extensively discussed in recent reviews [1, 2]. In summary, with instrumented nanoindentation, there is a larger available dynamic force range and more reliable force measurement. However, AFM possesses the following superior properties: displacement sensitivity (0.05 nm or lower: 20-fold better), load sensitivity (<0.05 nN: 200-fold better) and nanometer scale imaging and resolution sensitivity (table 1 in [1]). These distinct advantages of AFM have made it the main practical method of characterization of individual nanoscale biological particles including biological macromolecules (DNA, RNA, and proteins) and their larger *in vivo* complexes. In this review chapter, focused on the theoretical and computational modelling of the biomechanical properties of biological particles, we only address the use of AFM for modelling force application and a particle's biomechanical response.

The AFM technique was developed in 1986 [3]. It was immediately applied with success to the high resolution characterization of hard surfaces of materials via

Applied Nanoindentation in Advanced Materials, First Edition. Edited by Atul Tiwari and Sridhar Natarajan.

scanning by a sharp cantilever tip. In the initial development of AFM instrumentation, surfaces were scanned in the continuous 'contact mode'. However, despite many attempts to apply AFM to understand the surface structure of soft matter such as biological macromolecules, experiments initially met with little success. The contact mode of analysis inevitably led to significant distortion and damage to the biological molecules being studied. This situation led to the development of the tapping mode of AFM analysis in 1993, where the cantilever tip oscillated, thereby interacting in a significantly reduced way with the sample [4]. The elimination of distortion by tapping mode analysis, combined with the evolution of improved cantilever tip technology, resulted in the productive application of AFM to the study of biological molecules [5, 6]. Currently, state-of-the-art AFM has been applied to study the structure and dynamics of a diverse range of single biological molecules and whole biological particles [7].

Going beyond the use of tapping mode AFM to scan and characterize single molecules, some investigators realized that the cantilever tip could also be used to carefully apply force to a local region of a biological molecule, thereby either pulling to unravel the native structure [8] or to carry out a nanoindentation experiment to deform the structure [9]. In the former experiment, one obtains information about the domain dynamics, energetics and inter-domain interactions in the native structure. In the latter experiment, information is obtained on the deformation dynamics of the structure. However, as advantageous as these force application methods appear, the molecular level interpretation of their experimental results remains problematic for a number of reasons. To begin with, a high resolution secondary-tertiary structure level dynamic view of the systems being studied is not provided by the AFM instrumental output; this must be inferred by the investigator. Second, the lack of high-resolution information makes the exact point of force application unknown. As a result, understanding the complete dynamic, energetic pathway of the resulting structure alterations is not possible from the AFM experiment. Given these limitations to interpreting AFM output from force application, the use of accurate high-resolution computer-based theoretical modelling of the system becomes imperative.

In this Chapter, we describe the use of Molecular Dynamics (MD) simulations of biological systems undergoing nanoindentation by geometrically regular tips. By comparing our detailed *in silico* output to available experimental data, in a number of instances the modelling results agree well with the more limited experimental results. Moreover, as we will show, our high-resolution views of the systems' behaviours provide highly detailed predictive descriptions of the systems' structures, dynamics, thermodynamics and biomechanical properties, including exact control over the point of force application and its consequence for the aforementioned properties. Finally, this high resolution modelling and understanding of biological particles' properties allows one to develop analytically tractable models to describe the biomechanical behaviour of regular geometry biological particles as they undergo mechanical deformation by a regular geometry cantilever tip.

The outline of this chapter is the following. First, we feature our multiscale modeling approach, which we refer to as 'nanoindentation *in silico*'. This method allows one to explore the biomechanics of important biological particles in a computer. The method uniquely combines the all-atom Molecular Dynamics simulations of atomic structural models of biological particles available from the Protein Data Bank (PDB) and the Langevin simulations of their native-topology based coarse-grained reconstructions.

Next, we briefly present the coarse-grained description of a polypeptide chain, which is based on the so-called Self Organized Polymer (SOP) model (we refer an interested reader to the original papers on this subject [10, 11]). Importantly, the SOP model has been extensively used by several research groups by now to accurately characterize the physico-chemical and biomechanical properties of biomolecules [12–20]. We then introduce the reader to our methodology: to what we term 'nanoindentation *in silico*'. We shall show that our unique *in silico* 'experiments' provide a complete and detailed simulation view of the dynamics of compressive force induced viral capsid deformation. The numerical output can then be directly compared with and used to inform the AFM experimental interpretation. These simulation studies can also be used, among other things, to understand capsid fatigue and structural collapse, processes that cannot be easily understood by current experimental methodology. Finally, we demonstrate the utility of our approach to nanoindentation *in silico* by exploring the biomechanical properties of several important biological particles: the capsid of Cowpea Chlorotic Mottle virus (CCMV), and the intact microtubule (MT) and its constituent protofilaments. These specific examples represent biological particles of regular geometry: spherical (CCMV) and cylindrical (MT).

17.2 Computational Methodology of Nanoindentation *in silico*

17.2.1 Molecular Modelling of Biological Particles

The development of novel high-resolution molecular modelling techniques is strongly motivated by the inherent limitations in resolution and interpretability of AFM experiments. A number of theoretical methods have been employed to characterize the deformation dynamics of virus capsids. However, only a brief description of the most successful approaches is provided here. Historically, the first method used to model the mechanical properties of biological assemblies was Finite Element Analysis [21, 22]; specifically - the mechanical deformation of the capsid of Cowpea Chlorotic Mottle virus (CCMV), both native and swollen forms [21], and Herpes Simplex Virus type 1 (HSV1) were simulated [22]. Using this method, a triangulated meshwork is created, where each node is described by a potential energy function; this network is used to represent the protein virus capsid shell. Finite element analysis methodology has a number of drawbacks that include: (i) the technique does not account for the discrete nature of the particle's local structure, and (ii) the stochastic nature of mechanical deformation is not reflected in the particle's dynamics. An alternate approach is based on using atomic structural models of biological particles for direct MD simulations. Using this approach, an all-atom indentation study of Southern Bean Mosaic virus (SBMV) has been performed [23, 24]. However, with this approach the timescale of simulations is limited to a few tens of nanoseconds given the enormously large number of degrees of freedom ($\sim$1 000 000 atoms). Consequently, when performing MD simulations for an all-atom system of such large size, unphysical force-loading rates (i.e. rates of mechanical loading of a particle) must be used; these typically are $\sim 10^6$–10^8-fold faster than the experimental AFM force-loading rates.

More recently, several researchers have proposed using a coarse-grained description of a polypeptide chain, rather than an atomic description, in order to reduce the number of degrees of freedom. Using these methods, groups of atoms are replaced by one bead (unified center), and potential energy functions represent interactions between the beads. The aqueous environment in which a particle is immersed, is represented using the stochastic thermostat for the combined system (i.e. particle plus water bath). Then by propagating forward in time the Langevin equations of motion for each bead, the stochastic dynamics are obtained. As an example, the coarse-grained representation of a virus capsid was created using a method based on Voronoi cells [25, 26]. While this represented a significant step forward, this particular method combined several amino acids into one bead, a significant limitation which lowered the structural resolution. Moreover, the results of experiment *in vitro* and *in silico* cannot be directly compared because the force-loading rates in these simulations are three orders of magnitude faster than their experimental counterparts; this clearly limits the interpretive capacity of the method.

17.2.2 Coarse-graining: Self-organized Polymer (SOP) Model

The C_α-based coarse-grained Self Organized Polymer (SOP) model of a polypeptide chain was used here to describe virus and microtubule polymer mechanical properties. Displayed in Figure 17.1 is an idealized schematic for the coarse-graining procedure. The SOP coarse-grained model for proteins is based on a simplified representation of a polypeptide chain in terms of C_α-atoms (i.e. one bead/amino acid residue). Despite its simplicity, the SOP model has proven to be very accurate at describing the physico-chemical properties of both small biomolecules [10–15, 27] and large biological assemblies [16–19, 28]. Our use of a simplified SOP model is based upon the idea that the unique features embodied in the native topology and local symmetry of capsomer arrangements primarily govern the dynamic behavior of biological particles (viruses and microtubules) under mechanical stress (Figure 17.1), rather than their atomic details. Furthermore, the reduction in system size and the additional computational speedup attained using Graphics Processing Units (GPUs) [27, 28] for simulations, provides a closer match to the experimental conditions of force applications used in dynamic force assays (timescale, force-loading rates). Thus, the SOP-model based description of a biological particle's nanomechanics, accelerated on a GPU, represents a unique tool for computational modeling. In more detail, our multiscale modeling approach of 'nanoindentation *in silico*' utilizes all-atom molecular dynamics simulations of the atomic structure of a biological particle first to resolve the atomic-level details of the intra- and inter-subunit protein-protein interactions. This high resolution information then is used to parameterize the coarse-grained SOP model of the particle. Utilizing this approach, there is no requirement for additional kinetic or thermodynamic information to perform the multiscale modeling of the particle's biomechanics.

The SOP model, as previously described, was originally designed to address the forced unfolding problem of proteins. There is now an impressive track record of success stories involving the application of this model to a variety of biological systems [10–20, 27, 28]. We have taken a step further with the SOP model. The SOP model of the polypeptide chain [10, 11, 27, 28] was adapted by our lab to model entire biological particles (i.e. virus capsid, bacteriophage capsid, microtubule polymer, etc.). Each amino acid residue

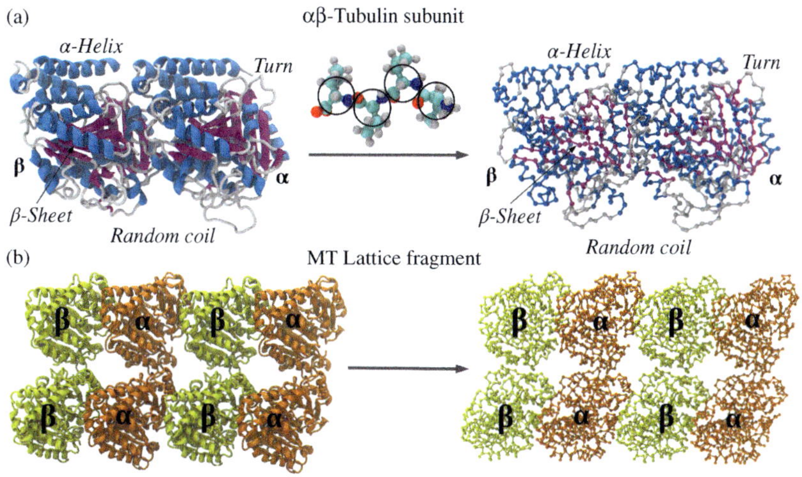

Figure 17.1 Coarse-graining procedure for constructing a self-organized polymer (SOP) model [10, 11] of a polypeptide chain. **Panel A** exemplifies coarse-graining of the atomic structure of the αβ-tubulin dimer – the structural unit of the microtubule cylinder (see Figure 17.3). The amino acid residues are replaced by single interaction centers (spherical beads) with the coordinates of the C_α-atoms (represented by the black circles). Four representative circles are shown to exemplify the coarse-graining process. Consequently, the protein backbone is replaced by a collection of the C_α-C_α covalent bonds with the bond distance of 3.8 Å. The potential energy function (molecular force field) given by Equation (17.1) describes: the binary interactions between amino acids stabilizing the native folded state of the protein, chain collectivity, chain elongation due to stretching, and chain self-avoidance (Equations (2)–(4)). Because the SOP model is based on the native topology of a particle, the coarse-graining procedure preserves the secondary structure of a protein: α-helices (shown in blue), β-strands and sheets (shown in purple), and random coil and turns (shown in gray). **Panel B** depicts the results of coarse-graining of a small fragment of microtubule cylinder. Four identical copies of the tubulin dimer structure, coarse-grained as described in panel **A,** form a C_α-based model of the fragment. These are combined together to form a coarse-grained reconstruction of the full microtubule lattice (see Figure 17.3).

in the native topology-based SOP model is represented by a single interaction center (C_α-atom). A collection of the C_α-C_α covalent bonds with the peptide bond distance of $a = 3.8$ Å (Figure 17.1) is then used to represent the protein backbone. The potential energy function of the protein conformation U_{SOP} specified in terms of the coordinates of the C_α-atoms $\{r_i\} = r_1, r_2, \ldots, r_N$, where N is the total number of amino acid residues, is given by:

$$U_{SOP} = U_{FENE} + U_{NB}^{ATT} + U_{NB}^{REP} \tag{17.1}$$

In Equation (17.1), the first term is the finite extensible nonlinear elastic (FENE) potential:

$$U_{FENE} = -\sum_{i=1}^{N-1} \frac{k}{2} R_0^2 \log\left(1 - \frac{(r_{i,i+1} - r_{i,i+1}^0)^2}{R_0^2}\right) \tag{17.2}$$

where $k = 14$ N/m is the spring constant, and $R_0 = 2$ Å is the tolerance in the change of the covalent bond distance. The backbone chain connectivity is described by the FENE potential. The distance between the next-neighbor residues i and $i+1$, is $r_{i,i+1}$, and $r^0_{\ i,i+1}$

is its value in the native structure. The non-covalent (non-bonded) interactions that stabilize the native folded state of the biomolecule are accounted for by our use of the Lennard-Jones potential:

$$U_{NB}^{ATT} = \sum_{i=1}^{N-3} \sum_{j=i+3}^{N} \varepsilon_h \left[\left(\frac{r_{ij}^0}{r_{ij}} \right)^{12} - 2\left(\frac{r_{ij}^0}{r_{ij}} \right)^{6} \right] \Delta_{ij} \tag{17.3}$$

In Equation (17.3), we assume that if the non-covalently linked residues i and j ($|i\text{-}j| > 2$) are within the cut-off distance of 8 Å in the native state, then $\Delta_{ij} = 1$; $\Delta_{ij} = 0$ otherwise. The strength of the non-bonded interactions are quantified by the value of ε_h. The non-native (non-bonded) interactions are treated as repulsive:

$$U_{NB}^{REP} = \sum_{i=1}^{N-2} \varepsilon_l \left(\frac{\sigma_l}{r_{i,i+1}} \right)^{6} + \sum_{i=1}^{N-3} \sum_{j=i+3}^{N} \varepsilon_l \left(\frac{r_{ij}^0}{r_{ij}} \right)^{6} (1 - \Delta_{ij}) \tag{17.4}$$

In Equation (17.4), an additional constraint is imposed on the bond angle between the triplet of residues i, $i+1$, and $i+2$ by including the repulsive potential with parameters $\varepsilon_l = 1$ kcal/mol and $\sigma_l = 3.8$ Å. These define the strength and the range of the repulsion. The parameter ε_h in the SOP model determines the energy scale. Results of the all-atom MD simulations of the particle at equilibrium are used to estimate this parameter.

17.2.3 Multiscale Modeling Primer: SOP Model Parameterization for Microtubule Polymers

Here we describe the workflow involved in the construction of a simplified (coarse-grained) SOP model of the microtubule polymer, in order to illustrate our multiscale modeling approach. Microtubules are large protein-based polymers that play a critical role in cell division, internal cellular structure and intracellular transport (see section 17.3). The SOP model parameterization, as previously mentioned, is achieved by extracting the atomic-level details available from all-atom MD simulations and exporting these to the SOP model. The parameter ε_h in the SOP model defines the average magnitude of non-covalent residue-residue interactions that stabilize the native state (see potential energy term U_{NB}^{ATT} given by Equation 17.3). Using MD simulations of an atomic structure model of the system being studied at a specific temperature, the values of ε_h need to be calculated directly. For the microtubule polymer, we carried out the following steps. First, for a small four-dimer long microtubule fragment (where *dimer* is a complex comprised of two protein subunits; see Figure 17.1), the atomic-level details that determine the type and number of binary contacts between amino acids and their energies were extracted from the MD simulations. Second, these data were then imported to the microtubule lattice SOP model. More specifically, a total of three 10 ns simulation runs were performed. For each group of contacts, the average non-bonded energy (E_{nb}), given by the sum of the van-der-Waals energy (Lennard-Jones potential) and the electrostatic energy (Coulomb potential) was calculated, and the average number of binary contacts between amino acids (N_{nb}) that stabilize the native microtubule structure (native contacts) was determined. A pair of residues was assumed to form a contact if the distance between their C_α-atoms in the native state did not exceed the cut-off distance R_C. A standard cut-off distance $R_C = 8$ Å value was used.

Table 17.1 Parameterization of the SOP model for the microtubule lattice. Summarized for each group of residue-residue contacts 1–5 are the average energy (E_{nb}), the average number of native contacts (N_{nb}), and the average strength of non-covalent interactions per contact (ε_h).

Contacts type	E_{nb}, kcal/mol	N_{nb}	ε_h, kcal/mol
Intra-monomer contacts (α-tubulin)	2345	1340	1.8
Intra-monomer contacts (β-tubulin)	2320	1240	1.9
Intra-dimer contacts	150	78	1.9
Longitudinal inter-dimer contacts	37.6	38	1.0
Lateral inter-dimer contacts	17.6	20	0.9

Native contacts were all divided into five groups (contact types); these were: (1) intra-monomer contacts in the α-tubulin monomers; (2) intra-monomer contacts in the β-tubulin monomers; (3) intra-dimer contacts stabilizing the dimer's structure; (4) longitudinal inter-dimer contacts between any two dimers along the MT cylinder axis; and (5) lateral inter-dimer contacts between the α-tubulin monomers and between the β-tubulin monomers in adjacent protofilaments. The Solvent Accessible Surface Area (SASA) [29] and Generalized Born (GB) [30] models of implicit solvation, which are based on the CHARMM19 force-field [31] were used to calculate the energy for non-bonded interactions. The output from SASA model based simulations (coordinate and energy files) were used to determine E_{nb} and N_{nb} values for the contact groups 1–3. Electrostatic interactions are known to be important for the formation of longitudinal and lateral tubulin-tubulin bonds. Therefore, we used a more accurate GB model to calculate E_{nb} and N_{nb} for the contact groups 4 and 5. Lastly, by dividing E_{nb} by the corresponding value of N_{nb} for each contact group we obtained the values of ε_h (see Table 17.1 below).

17.2.4 Using Graphics Processing Units as Performance Accelerators

Graphics Processing Units (GPUs) are widely used in state-of-the-art numerical applications, as performance accelerators. Most GPU-based calculations, depending on the numerical algorithm and system size, vary from 4-fold to 200-fold faster than heavily optimized Central Processing Unit (CPU)-based implementations of the same algorithm. In our research group, we have developed a GPU-based computational methodology. It has the following attributes: (1) generation of pseudo-random numbers based on Hybrid-Taus, Lagged Fibonacci, and Mersenne-Twister algorithms, (2) calculation of molecular forces including particle-based and interaction pair-based parallelization approaches, and (3) the numerical propagation of the Langevin equations of motion [27, 28]. Pseudo-random numbers possessing high statistical quality are critical, especially in computational modeling of large-size biological systems in a stochastic thermostat. Force calculations are required to characterize both intra- and inter-chain interactions between different amino acids within and between the protein's polypeptide chains. Numerical solution of the Langevin equations of motion is necessary to generate the dynamics of the system. In order to optimize fully the available computational resources, we also developed: (a) the one-run-per-GPU approach, and (b) the multiple-runs-per-GPU approach that generates several independent trajectories concurrently on a single GPU device [27].

The SOP-GPU software package, for the computational exploration of large-size biological assemblies [27, 28], incorporates the developed numerical algorithms on a GPU, along with the Self Organized Polymer (SOP) model based Langevin Dynamics simulations. Employing the SOP-GPU software package has real advantages for the user. This package enables the user to perform dynamic force measurements *in silico* (see Figure 17.3) within the centisecond timescale of the AFM experiment; therefore, this software uses the experimentally relevant conditions of dynamic force application (pulling speeds or force-loading rates) [15, 17–19, 27, 28]. The computational performance of the SOP-GPU package was profiled in our lab by comparing the computational time of the SOP-GPU program implemented on a single GPU device (GeForce GTX 780) with the computational time of a fully optimized version of the program implemented on a CPU. Figure 17.2 presents the numerical results from these comparative benchmark tests. Considerable performance gains, i.e. ~10–50-fold speedup for a system of $N_{tot} = 10^6$ particles (harmonic oscillators) were observed from our GPU-based implementation. This finding completely justifies our use of high-performance GPU-based computing in the multiscale modeling of biological particles.

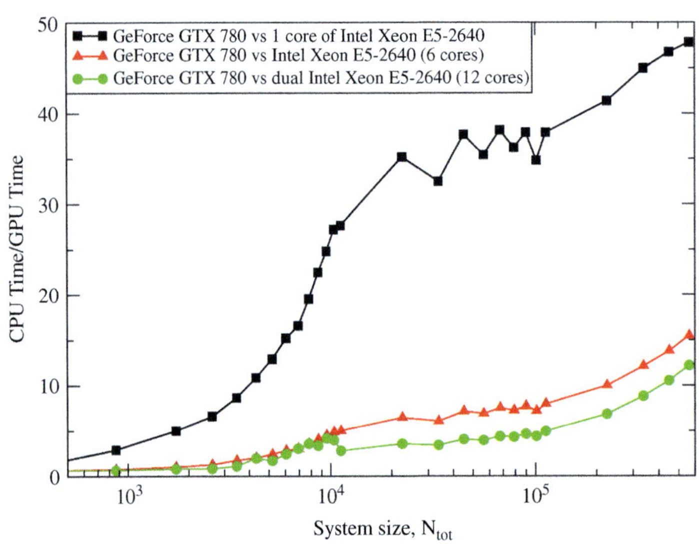

Figure 17.2 Computational acceleration on a GPU for the SOP-GPU software package. Compared is the computational time on a Central Processing Unit (CPU) vs. Graphics Processing Unit (GPU) as a function of the system size (total number of residues) N_{tot} on a logarithmic scale. To perform these comparative benchmark tests, we used a system of N_{tot} independent Brownian oscillators (i.e. uncoupled harmonic oscillators in a stochastic thermostat) at room temperature. We compared the computational performance of the SOP-GPU program on a single GPU device – GeForce GTX 780 (from NVidia) versus 1 CPU core, 6 CPU cores, and 12 CPU cores (Intel Ivy Bridge architecture). We see a substantial computational acceleration when SOP-GPU is used. This acceleration gradually increases with system size, i.e. ~5–40-fold speedup for $N_{tot} \approx 10^5$ particles and ~15–50-fold for $N_{tot} \approx 10^6$ particles, when the GPU device becomes fully loaded with computational tasks.

17.2.5 Virtual AFM Experiment: Forced Indentation *in silico* of Biological Particles

Our *in silico* 'experiment' is critically important, since it can provide a complete and high resolution simulation view of the entire process of particle deformation, relaxation, and structural collapse. As a result, it provides a high resolution detailed mechanistic interpretation and modeling of the low resolution experimental force-indentation (or force-deformation) spectra. In our nanoindentation assays *in silico*, full control over the simulation protocol is maintained by the investigator. Controlled parameters include: the amplitude and direction of applied force, the precise location of contact between the cantilever tip (indenting object) and the biological particle, designating surface-constrained particle residues to prevent the particle from rolling on the substrate surface during the simulation. As in AFM experiments *in vitro*, we can manipulate the particle *in silico*, reverse the direction of applied force, decrease or increase the force amplitude, etc. The investigator's complete control over the system *in silico* during the entire indentation process can be used, e.g. (a) to study the nanoindentation characteristics of the particle at different symmetry points on the particle surface, and (b) to relate energy changes observed in the particle during any point in the simulation to changes in the specific molecular details observed in the particle's structure. Moreover, we are capable of (c) easily switching modes back and forth from the forward indentation to the backward tip retraction [18, 19]. The latter capability can be used to perform simulations of repeated cycles of indentation and retraction, important sources of information in the understanding of viscoelastic properties, mechanical fatigue and structural collapse in soft biological matter.

The biological system's complete dynamics are obtained by numerically propagating forward in time the Langevin equations of motion for each particle position r_i in the overdamped limit (Brownian dynamics):

$$\eta \frac{dr_i}{dt} = -\frac{\partial U_i(r_i)}{\partial r_i} + g_i(t) \tag{17.5}$$

In Equation (17.5), $U_i(r_i)$ is the total potential energy; this accounts for the biomolecular interactions (potential U_{SOP} in Equation 17.1) and interactions of particles with the indenting object – spherical tip (potential U_{tip} in Equation (17.6) below). Also, in Equation (17.5) $g_i(t)$ is the Gaussian distributed zero-average random force, and η is the friction coefficient. To generate the Brownian dynamics, the equations of motion are propagated with the time step $\Delta t = 0.08\tau_H$, where $\tau_H = \zeta\varepsilon_h\tau_L/k_BT$. Here, $\tau_L = (ma^2/\varepsilon_h)^{1/2} = 3$ ps, $\zeta = 50.0$ is the dimensionless friction constant for an amino acid residue in water ($\eta = \zeta m/\tau_L$), $m \approx 3\times10^{-22}$ g is the residue mass, and T is the absolute temperature [20, 32]. To simulate a biological particle's nanoindentation, we set T to room temperature and use the bulk water viscosity, which corresponds to the friction coefficient $\eta = 7.0\times10^5$ pN ps/nm. Numerically, the temperature information is contained in the width of the distribution $g_i(t)$, while information about external pressure is implicit in the water density, which translates into the water viscosity and friction coefficient. Therefore, by fixing the values of the width of distribution $g_i(t)$ for all particles in all three dimensions (x, y, and z) and by using the same value of the water viscosity (friction coefficient η) the conditions of constant external pressure and temperature in the combined system, i.e. particle plus water bath (NPT-ensemble) are maintained.

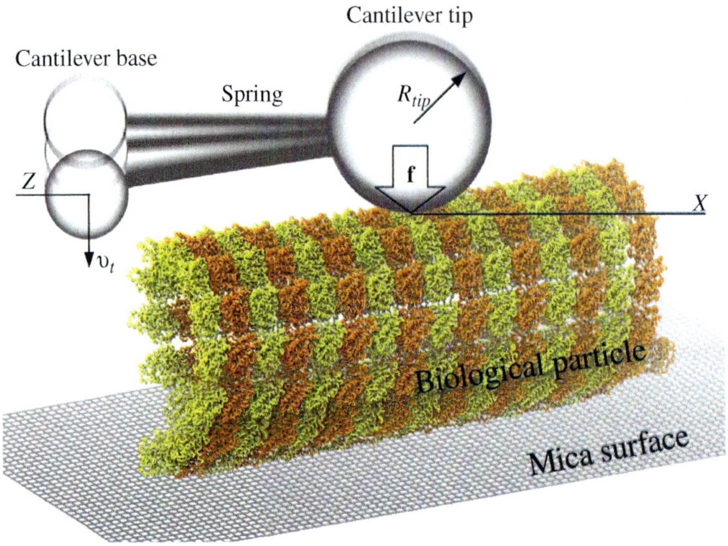

Figure 17.3 Dynamic force experiment *in silico* on a fragment of the microtubule cylinder. The particle is immobilized on the substrate surface. The cantilever base (virtual sphere in simulations or piezo- in AFM) is moving in the direction perpendicular to the particle surface described by the cantilever coordinate Z with constant velocity v_f. This dynamic force-ramp creates a compressive force, which is transmitted to the cantilever tip (sphere of radius R_{tip}) through the harmonic spring with the spring constant κ. The compressive force (with large vertical arrow) ramps up linearly in magnitude with time, $f(t) = r_f t$, with the force-loading rate $r_f = \kappa v_f$. The force loads the particle and produces an indentation in the particle's structure quantified by the coordinate X. The mechanical response, i.e. the restoring force from the particle, is measured by profiling the deformation force (indentation force) F as a function of the cantilever base displacement Z (*FZ* curve) or as a function of the indentation depth X (*FX* curve).

In dynamic force measurements *in silico*, the cantilever base is represented by a virtual sphere, which is connected by a harmonic spring to the cantilever tip (indenter), modeled by a spherical bead of radius R_{tip} (see Figure 17.3). The tip interacts with the biological particles via the repulsive Lennard-Jones potential:

$$U_{\text{tip}} = \sum_{i=1}^{N} \varepsilon_{\text{tip}} \left(\frac{\sigma_{\text{tip}}}{|r_i - r_{\text{tip}}| - R_{\text{tip}}} \right)^6 \tag{17.6}$$

thereby producing an indentation on the particle's outer surface. In Equation (17.6) above, r_i and r_{tip} are coordinates of the i-th particle and the center of the tip, respectively, while $\varepsilon_{\text{tip}} = 1.0$ kcal/mol, and $\sigma_{\text{tip}} = 1.0$ Å are parameters of interaction. Also, in Equation (17.6), the summation is performed over all the particles interacting with the tip.

For the cantilever tip (spherical bead in Figure 17.3), we need to solve numerically (simultaneously with Equation (17.5) for the particle's residues) the following Langevin equation of motion:

$$\eta \frac{dr_{\text{tip}}}{dt} = -\frac{\partial U_{\text{tip}}(r_{\text{tip}})}{\partial r_{\text{tip}}} + \kappa((r^0_{\text{tip}} - v_f t) - r_{\text{tip}}) \tag{17.7}$$

In Equation (17.7), the initial position of spherical tip center is $r_{tip}{}^{0}$ (v_f is the cantilever base velocity; κ is the cantilever spring constant), and the friction coefficient is $\eta = 7.0 \times 10^6$ pN ps/nm. This value corresponds to the dimensionless friction constant $\zeta = 500$ for a spherical tip in water. To generate the dynamics of the biological particle, we solve numerically Equations (17.1–17.5) for the particle and Equations (17.6) and (17.7) for the indenter (tip).

The cantilever base (virtual bead in Figure 17.3) is moving with constant velocity (v_f). Therefore, the time-dependent force $\mathbf{f}(t) = f(t)\mathbf{n}$ is being exerted through the cantilever tip in the direction **n** perpendicular to the particle surface. As a result, the compressive force mechanically loads the particle gradually. As a function of time, the force magnitude $f(t) = r_f t$ ramps up linearly, with the rate $r_f = \kappa v_f$. In the simulations of 'forward indentation', the cantilever base (and spherical tip) is moving towards the particle. In contrast, the direction of motion of the cantilever base is reversed in the simulations of 'force-quenched retraction', resulting in the compressive force decreasing to zero. When performing simulations, both the cantilever base displacement Z and the cantilever tip position X can be controlled. The resisting force F from the particle, which corresponds to the experimentally measured indentation force, can be extracted from the energy output. To prevent the particle from rolling and to mimic adsorption of the particle on the substrate surface, the bottom portion of the particle can be constrained by fixing the relevant C_α-atoms. Owing to the significant computational acceleration observed on a GPU, we can afford to use the experimentally relevant values of cantilever velocity used in AFM ($v_f = 0.1–1.0$ μm/s).

17.3 Biological Particles

17.3.1 Cylindrical Particles: Microtubule Polymers

Microtubules (MTs) are hollow protein cylinders. They are comprised of lateral assemblies of protofilaments: long linear strands of longitudinally-repeating αβ-tubulin dimers oriented along the axis of the cylinder (Figure 17.3) [33]. MTs are fairly rigid (persistence length 3.5 mm [34]). In eukaryotic cells, they serve as important structural and organizing elements. For example, the mechanical properties of MTs help define internal cell architecture and polarity, while the MTs long, rigid and durable linear tracks assist neuronal transport via the operation of molecular motors transporting molecular 'cargo' along its 'tracks' [35]. The dynamic properties of MTs, undergoing stochastic cycles of polymerization and depolymerization, also play a prominent role in many cellular processes [36, 37], most prominently in cell division, where they form a mitotic spindle [38]. During mitosis, the disassembling MTs have been proposed to serve as a primary biological motor creating poleward chromosome motion [39, 40].

Within cells the MT polymer is thought to exist in a structure containing 13 protofilaments. These protofilaments are arranged in a left-hand 3-start helix and they are stabilized by the resulting network of lateral and longitudinal non-covalent bonds (Figure 17.3). This complex multi-protofilament structure makes it difficult to establish a direct correspondence between the observed macroscopic MT properties *in vitro* and the microscopic molecular characteristics of tubulin. Theoretical approaches applied to this system have played an important role in providing such a connection, and to reveal the mechanisms of MT dynamics and force generation [41–43]. These investigations have helped identify high resolution characteristics that exert a significant effect on MT behavior. Three major characteristics have been identified that include: the interfacial thermodynamic characteristics between adjacent tubulins, i.e. the lateral and longitudinal bonds' dissociation energies, as well as the individual tubulin protofilaments' mechanical flexural rigidity. In the past, these properties have often been estimated using theoretical models; this explains the lingering controversy surrounding virtually every aspect of MT thermodynamics.

The lateral tubulin bonds rupture before the longitudinal bonds during MT disassembly. This is evident experimentally at the ends of shortening MT polymers by the appearance of bending 'rams horns' structures [44]. These bending protofilament structures suggest that longitudinal tubulin-tubulin bonds are stronger than the lateral bonds [45], although the absolute values of the interaction energies are unknown. Reported estimates of the dissociation energies of longitudinal and lateral bonds range from 3.6 to 12.0 kcal/mol and from 1.8 to 8.0 kcal/mol, respectively [46–49]. The energy profile distributions as well as the geometry and number of the sites in the MT models remain a subject of debate [42, 47–50]. The published estimates of the value of flexural rigidity of microtubule protofilaments also vary over a considerable range, by an order of magnitude, from 1500 to 13 000 pN nm^2 [51]; this corresponds to energies of 2.2 to 19.2 kcal/mol per dimer for the full straightening of a protofilament. Therefore, accurate determination of protofilament rigidity is important, because defining the mechanism of force generation during the MT depolymerization follows directly from the magnitude of this protofilament parameter.

17.3.2 Spherical Particles: CCMV Shell

The *Bromoviridae* represents an important family of single stranded RNA plant viruses that are distributed worldwide. These viruses infect a range of hosts and some members of the *Bromoviridae* family are the cause of major crop epidemics [52]. This family of plant viruses is comprised of dozens of members, of which CCMV is a well-studied example (Figure 17.4) [53]. Four unique single-stranded viral RNA molecules comprise the entire genome in CCMV. The nucleic acids are packaged within a capsid comprised of 180 identical single subunit capsid proteins. The icosahedral virus capsid is formed by the self-assembly of these 19.5 KDa (190 amino acid sequence) capsid proteins into 20 hexameric and 12 pentameric structural fragments. The CCMV shell (triangulation number $T=3$) exhibits a ~26 nm diameter with average shell thickness of ~2.8 nm [7, 53]. The shell is comprised of 60 trimer structural units. Pentameric symmetry is exhibited at the 12 vertices (pentamer capsomeres) and hexameric symmetry at the 20 faces (hexamer capsomeres) of the icosahedral capsid (Figure 17.4). The CCMV capsid shows a number of interesting physico-chemical characteristics. For example, the CCMV capsid fully self-assembles in the absence of RNA molecules to form an empty shell; or, it can form a fully infectious virion by self-assembling in the presence of its genomic RNAs. The CCMV capsid x-ray structures have been determined for both RNA filled and empty CCMV capsids.

Viral capsids have been shown experimentally to undergo cooperative swelling transitions. Native CCMV is stable in a weakly acidic environment at pH = 5. At higher

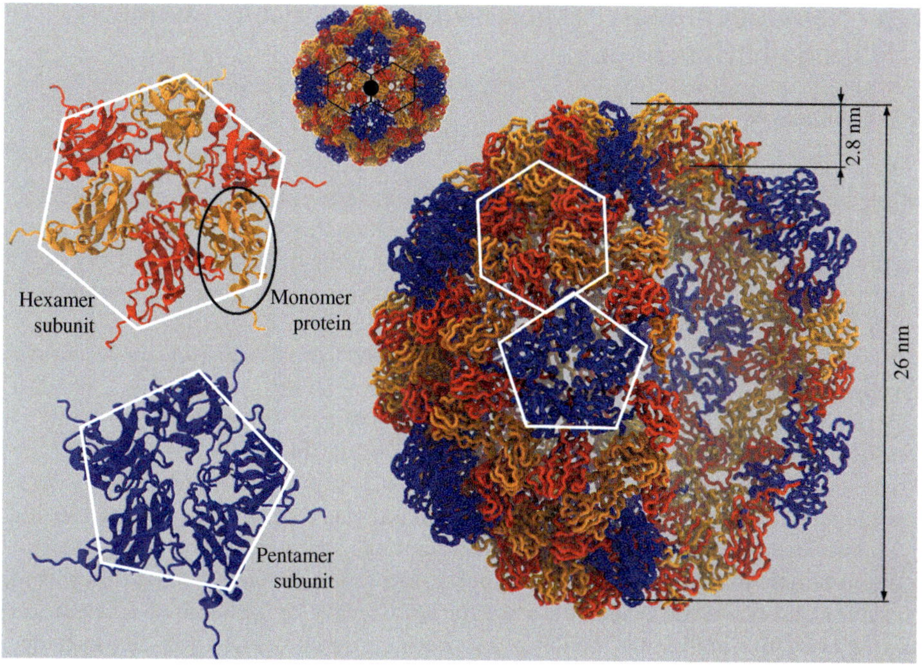

Figure 17.4 The crystal structure of the Cowpea Chlorotic Mottle Virus (CCMV) from VIPERdb (PDB entry: 1CWP). Shown is the side view of CCMV; the protein domains forming the pentamer capsomers are coloured in blue, while the same protein domains in the hexamer capsomers are shown in red and orange. The pentamer and hexamer capsomers composed of five and six copies of the same monomer protein domain (circled in the black ellipse), respectively, are magnified on the right. There are small openings in the centers of the pentamer and hexamer capsomers, which correspond to the 5- and 3-fold symmetry axes. The CCMV capsid is a thick shell. The outer diameter of the CCMV shell is ~26 nm and the shell thickness is ~2.8 nm; hence, the shell thickness is not negligibly small compared to the shell size. The top structure displays the top view of the CCMV particle with the 2-fold symmetry axis at the center of the common edge of adjacent hexamer subunits.

pH = 7.5 and low ionic strength, structural elements forming the CCMV capsid undergo a concerted 'swelling transition'. A new stable form (swollen) results, which is characterized by the capsid's radial expansion of ~10% and the appearance of voids in the 3-fold symmetry vertices [53]. This swelling transition is fully reversible; when the pH decreases back to pH = 5 or the concentration of ions (Ca^{2+} or Mg^{2+}) increases, the capsid contracts in size. Under well-defined conditions of pH, ionic strength, and divalent metal ion concentration, the CCMV capsid self-assembles *in vitro*. Specifically, the CCMV capsid proteins or the capsid proteins and RNA reassemble to form $T = 1$, $T = 3$ particles, swollen $T = 3$ particles, and $T = 7$ icosahedral particles. Also formed from the CCMV capsid proteins are multi-shelled $T = 3$ and $T = 7$ particles, tubes, rosettes, sheets and a variety of laminar structures [53]. More generally, these structural transitions appear to be a generic feature of plant viruses. This property may be directly related to their mechanism of infectivity of plant cells. Therefore, investigating the capsid structures' dynamics, energetics and biomechanical properties of CCMV and other viruses are of great interest.

17.4 Nanoindentation *in silico*: Probing Reversible Changes in Near-equilibrium Regime

17.4.1 Probing Reversible Transitions

Before applying the *in silico* nanoindentation approach to examine the complex deformation dynamics of the two chosen examples of large-size biological particles: the MT polymer and the CCMV capsid, we will first demonstrate the capabilities of the method. Specifically, we focus on two fundamental questions: (a) the (ir)reversibility of mechanical deformation, and (b) the statistical nature of the observed quantities (i.e. equilibrium or non-equilibrium ensemble).

The first question we tested was whether our *in silico* nanoindentation approach was appropriate to probe reversible transitions. This (ir)reversibility question was addressed by performing simulations of compressive force-induced indentation, followed by force-quenched tip retraction, for a small system using the SOP-GPU package. The system we chose to examine was an 8-dimer long single protofilament fragment (PF8/1; see Figure 17.5). In this model system, our thinking was that the entropic barrier for the longitudinal bond formation and protofilament unbending might be low; as a result we might be able to detect longitudinal bond re-formation during tip retraction within the millisecond timescale available via simulations. Owing to GPU-based acceleration and our using the multiscale modeling approach, which combines the biomolecular simulations of atomic-structural models and simplified protein models, this timescale has become available in simulations.

For force-induced forward indentation followed by force-quenched backward tip retraction, the simulation set-up is briefly described in Figure 17.5. The two ends of PF8/1 were constrained and we applied a compressive force at the longitudinal tubulin-tubulin interface connecting the 4-th and 5-th dimer. Then we monitored the dynamics of bending/unbending and longitudinal bond dissociation/association of the protofilament using the structure overlap function χ defined as $\chi(t) = (2N(N-1))^{-1} \sum \Theta(|r_{ij}(t) - r_{ij}^0| - \beta r_{ij}^0)$, where in the Heaviside step function Θ, $r_{ij}(t)$ and $r_{ij}{}^0$ are the inter-particle distances between the i-th and j-th residues in the transient structure and in the native state, respectively, and $\beta = 0.2$ is the tolerance for the inter-particle distance change. In summary, χ is an inter-residue distance-based measure of structural similarity between the chosen reference state and any given conformation. At its extreme values, when the two conformations are very similar (dissimilar), $\chi \approx 1$ ($\chi \approx 0$).

In Figure 17.5, the results we obtained clearly show the following. After the initial longitudinal bond dissociation between the 4-th and 5-th dimers (sharp drop in χ), when the tip motion direction was reversed, both the bond re-formed and the protofilament unbent completely over the 3, 6, 11, and 16 ms timescale (χ approaches a value of unity). This was true for all bond-dissociated conformations, corresponding to $X = 7.0$, 8.0, 8.5, and 9.2 nm forward deformations, respectively. These specific conformations were used as the initial structures in the simulations of backward tip retraction. Therefore, in short protofilament fragments (such as PF8/1) the longitudinal bond dissociation is fully reversible on the millisecond timescale of our simulation experiments. It is worth noting that in the original AFM experiments [54, 55], the authors observed that it would take up to 4 min for the entire MT cylinder lattice (Figure 17.3) to fully self-repair following an

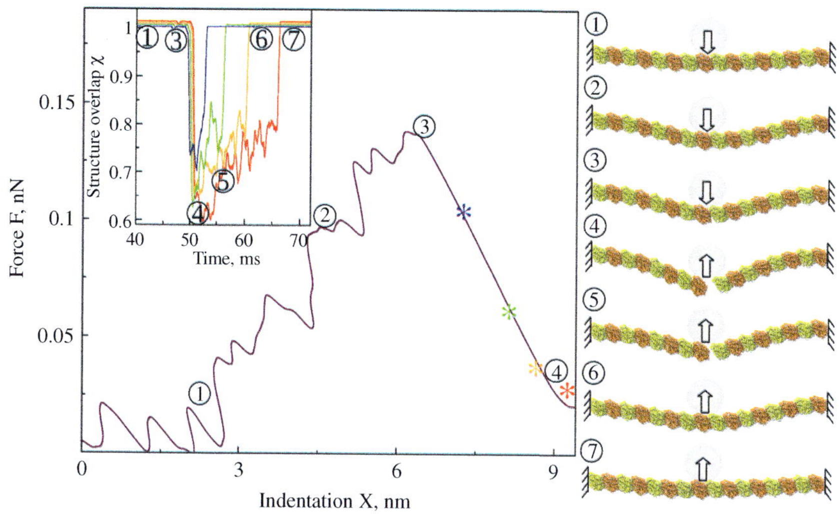

Figure 17.5 Reversibility of bonds' dissociation in MT protofilament PF8/1. The *FX* curve is obtained with $v_f = 0$. μm/s and $R_{tip} = 5$ nm. In the forward indentation (structural snapshots 1–3), the tip deforms the protofilament and disrupts the longitudinal bond. The asterisks mark $X = 7.0$ nm (blue), 8.0 nm (green), 8.5 nm (orange), and 9.2 nm (red) deformations corresponding to the fully dissociated state. These were used as initial structures in the subsequent tip retraction simulations. *The inset* shows the structure overlap χ vs. time, which captures dissociation (drop in χ) and subsequent reformation (increase in χ) of the bond between 4-th and 5-th dimers. Snapshots 4–7 obtained from the initial structure corresponding to 8.5 nm deformation (orange asterisk) show the dynamics of bond reformation and protofilament restructuring.

indentation beyond 10 nm. This finding demonstrates that even following disruption of lateral and longitudinal bonds, the MT lattice remains capable of recovery, making the indentation process fully reversible, but on a significantly longer minute(s) timescale. However, given the current limitations in computing power, numerical exploration of the equilibrium dynamics of large-size biological particles such as the MT cylinder is not yet feasible.

17.4.2 Studying Near-equilibrium Dynamics

Next, we addressed the question of whether the deformation dynamics of the biological particle are detail-balanced locally in time (i.e., each step is detail-balanced). If each step in the dynamics of the system is detail-balanced, then the system is microscopically reversible. This is true even if the system suffers an external perturbation driving it from equilibrium. Here, we utilize the dynamic force-ramp conditions of external compressive force application (mechanical perturbation) in our *in silico* indentations, which mimic the dynamic force measurements. Now, there remains the question of whether one can extract equilibrium information from these non-equilibrium measurements. To address this question, we simulated the forced deformation for a 24-dimer long single protofilament fragment (PF24/1; structure not shown). Here, the simulation setup has the two ends of PF24/1 constrained, and the compressive force was applied in the middle, in order to bend the structure and disrupt the longitudinal tubulin bonds. This is very similar to the set-up used in Figure 17.5.

Since it is impossible to apply infinitely slow force loading (which would represent true equilibrium conditions of mechanical force application) either in simulations or in experiments, investigators apply a sufficiently slow force-ramp (slow force-loading conditions); this translates to using a cantilever velocity $v_f = 0.01\text{–}1.0$ μm/s for systems of different size. Within the context of mechanical nanomanipulation discussed here, these slow force-loading conditions correspond to the rate of force increase being slower than the rate of system re-equilibration. This is true at each point along the deformation (or indentation) reaction path (*FX* trajectory), from the initial state (native state with $F = X = 0$) to the final state (deformed state with nonzero values of *F* and *X*). This would correspond to the detailed balance condition being satisfied under the condition of quasi-equilibrium. To address this question, using the SOP-GPU package, we performed bending deformation simulations for the 24-dimer long single protofilament fragment PF24/1. For a series of these simulations, we varied v_f from 16.0 μm/s to 8.0 μm/s, to 4.0 μm/s, to 2.0 μm/s, to 1.0 μm/s, and, finally, to 0.5 μm/s.

In Figure 17.6 we compare the obtained *FX* curves. It is clear that as v_f decreases the *FX* curves become increasingly similar. As one example, the *FX* curves obtained with $v_f = 1.0$ and 0.5 μm/s look nearly identical. The implication is that the force loading is so slow in this range that the mechanical response (*FX* curve) measured with 1.0 and 0.5 μm/s cantilever velocity no longer differs. Next, we profiled, as a function of cantilever velocity v_f, the quantities: critical force F^*, critical distance X^*, and mechanical work for critical deformation w^*. These quantities help to define the transition of the particle structure to the dissociated state with at least one longitudinal bond fully disrupted.

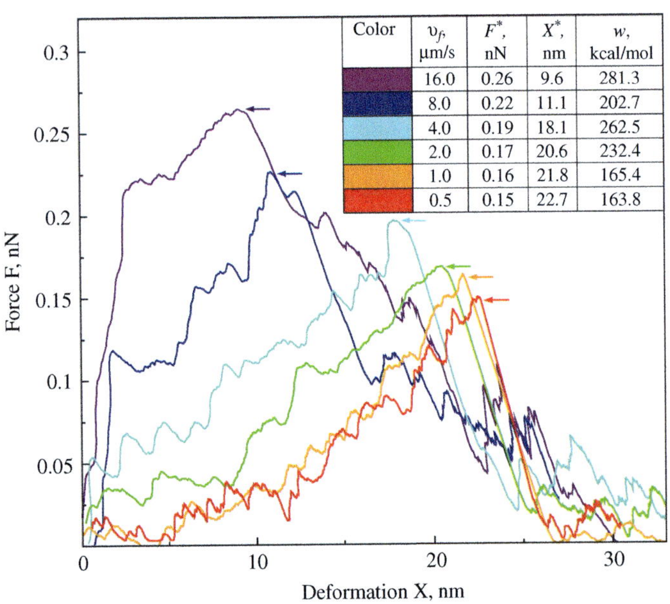

Color	v_f, μm/s	F^*, nN	X^*, nm	w, kcal/mol
	16.0	0.26	9.6	281.3
	8.0	0.22	11.1	202.7
	4.0	0.19	18.1	262.5
	2.0	0.17	20.6	232.4
	1.0	0.16	21.8	165.4
	0.5	0.15	22.7	163.8

Figure 17.6 Dependence of the forced deformation for single protofilament fragment PF24/1 on the cantilever velocity v_f ($R_{tip} = 5$ nm). Shown in different color are the *FX* curves obtained using different values of v_f accumulated in *the table inset*. The horizontal arrows mark the longitudinal bond dissociation transitions which occur at the critical force F^* and critical deformation X^*. *The inset* also lists the obtained values of F^*, X^* and the values of deformation work *w* as a function of velocity v_f.

Here, we calculated the mechanical work for critical deformation w^* to be the area under the FX curve up to the point when $X = X^*$ and $w^* = \int_0^{X^*} F(X)dX$. For these data, a gradual convergence was observed for all three quantities F^*, X^*, and w^* with decreasing cantilever velocity v_f (see the table inset in Figure 17.6) to particular stable values. Clearly these results demonstrate that when slow cantilever velocities of 0.5–1.0 μm/s are used in *in silico* mechanical testing made possible with our SOP-GPU software package, then the simulations can be performed under near-equilibrium conditions of the compressive force application. Therefore, performing our *in silico* mechanical testing of biological particles at slow force loads enables investigators to gather equilibrium-ensemble information (forces, deformations, values of work and Young's modulus, etc.) from the non-equilibrium measurements.

17.5 Application of *in silico* Nanoindentation: Dynamics of Deformation of MT and CCMV

17.5.1 Long Polyprotein – Microtubule Protofilament

We used the ring structure comprised of 13 protofilaments to construct a finite-length fragment of the MT lattice [56]. This structure utilizes atomic coordinates of the αβ-tubulin dimer (PDB code: 1JFF [57]). We obtained the structure of a finite-length 8 dimer long fragment of MT cylinder MT8/13 (Figure 17.3) as described in [19]. Structures of varying length: 8, 16, 24, and 32 dimer long single protofilaments (PF8/1, PF16/1, PF24/1, and PF32/1) were extracted and replicated using the structure of MT8/13. As a representative example, the structure of protofilament PF8/1 is presented in Figures 17.5 and 17.7 (structures of the other protofilaments are not shown). Dynamic force measurements *in silico* were carried out on these single protofilament fragments (Figure 17.3 and 17.5) using the SOP model [10, 11] and simulations on a GPU [27, 28]. Bending deformation of protofilaments were induced using the cantilever velocity $v_f = 0.2$ μm/s.

Summarized in Figure 17.7 are the results we obtained, which shows typical examples of the FX curves for the single protofilament PF8/1. The output from simulation of the bending deformation of PF8/1, and longer protofilaments PF16/1, PF24/1, and PF32/1 was then used in the subsequent determination of flexural rigidity and persistence length values listed in Table 17.2. By carrying out numerical integration, the force-deformation (F vs. X) profiles obtained from bending simulations were transformed into the profiles

Table 17.2 Mechanical bending parameters deduced from *in silico* deformations. Values are averages with standard deviations of the flexural rigidity (*EI*) and persistence length (L_p) obtained from five bending simulation runs for single protofilaments of 32 (PF32/1), 24 (PF24/1), 16 (PF16/1), and 8 (PF8/1) tubulin dimers.

System	PF32/1	PF24/1	PF16/1	PF8/1
$EI \times 10^{-26}$, N m^2	2.6±0.1	2.6±0.1	2.3±0.2	1.8±0.1
L_p, μm	6.3±0.3	6.3±0.2	5.7±0.4	4.5±0.3

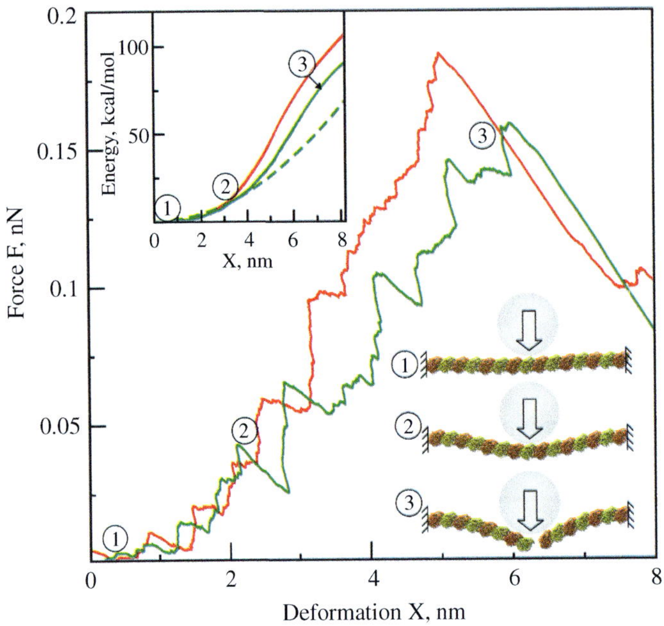

Figure 17.7 Force-deformation spectra for single protofilament fragment of the MT PF8/1. Shown are the examples of the force spectra (*FX* curves) for PF8/1 (green and red solid curves) obtained by using the cantilever velocity $v_f = 0.2$ μm/s and tip radius $R_{tip} = 10$ nm. Structures numbered 1–3, which show the deformation progress, correspond to the green *FX* curve and represent the native state ($X = 0$), weakly bent state ($X = 2$ nm), and dissociated state ($X = 6$ nm), respectively. The tip shown with the vertical arrow deforms the protofilament until the dissociation of the longitudinal bond occurs. *The inset* shows the corresponding profiles of bending energy as a function of deformation for the estimation of the flexural rigidity. The dashed green curve is a fit of the quadratic function to the green curve of the potential energy of deformation *V* vs. *X*, which shows the validity of the harmonic approximation in the calculation of flexural rigidity and persistence length for the protofilament fragments PF8/1, PF16/1, PF24/1, and PF32/1.

of bending energy $V = \int F(X)\mathrm{d}X$ vs. X (*the inset* in Figure 17.7). The flexural rigidity for each protofilament was estimated by calculating the energy of bending of the beam, $V = \frac{EI}{2}\int_L (1/R - 1/R_0)^2 \mathrm{d}l$, where R_0 and R are the radii of the curvature of the protofilament in the initial state, used as a reference structure, and the final state, respectively; dl is the length element; and L is the total length [58]. This is a harmonic approximation valid only for small deformations X. We performed a fit of the quadratic function $V \sim X^2$ to the curves of the bending energy (dashed black line in *the inset* to Figure 17.7) in order to find the range of X values for which this approximation is valid.

The protofilament is straight in the initial state, and, hence, $1/R_0 = 0$. To simplify these calculations, only the protofilament portion of constant curvature was considered, in which case $V = EIL/(2R^2)$. Knowing V and R allows one to calculate $EI = 2VR^2/L$. From the simulation output, we estimated R through the relationship for the radius of the arc $R = X/2 + C^2/(8X)$, where C is the end-to-end distance. For each of the protofilament fragments the persistence length was calculated using the formula $L_p = EI/k_BT$. By utilizing the harmonic approximation we discovered that the flexural rigidity for these protofilament fragments EI are in the (18 000–26 000) pN nm^2 range (Table 17.2).

The values of EI we obtained are somewhat larger than that estimated from the energy of GTP hydrolysis ($EI = 13\ 000$ pN nm^2; [51]). From this comparison, the implication is that almost the entire energy resulting from GTP hydrolysis is stored as mechanical stress in the strained tubulin conformation found in the protofilament structures of the native MT.

17.5.2 Cylindrical Particle – Microtubule Polymer

Dynamic force measurements were carried out *in silico* on fragment MT8/13 (Figure 17.3) using the SOP model [10, 11] and simulations were accelerated on GPUs [27, 28]. In order to account for the absence of long MT length, the positions of the C_α-atoms at the left edge (minus end) and at the right edge (plus end) of the cylindrical fragment MT8/13 were constrained. The dynamic force-ramp measurements *in silico* on MT8/13 (forward indentation and backward tip retraction) were performed as described previously (see section 17.2). We used the cantilever base (piezo) velocity $v_f = 1.0$ μm/s and cantilever spring constant $\kappa = 0.05$ N/m to produce indentations on the outer surface of MT8/13. Different points on the MT surface were nanoindented *in silico* using spherical tips of radii $R_{tip} = 10$ and 15 nm; these are comparable to the 15-nm tip used in AFM experiments [54, 55]. It is important to note that the GPU-based computational acceleration has enabled us to use the experimentally relevant force-loading rate ($v_f = 1.0$ μm/s) and to span the experimental timescale (30–60 ms), making for a clean comparison between the results from simulation and experiment.

In dynamic force experiments, the dependence of the indentation force F from the MT cylinder on the cantilever tip displacement X (indentation depth) and Z (the piezo displacement) was profiled and the results are presented in Figure 17.8. Here, we have compared both the FZ curves and the FX curves. The FX curves show higher sensitivity to the MT deformation dynamics than FZ curves. This is because the FX curves do not contain any contribution from the displacement of the cantilever base, i.e. virtual particle in simulations or piezo in AFM. This fact necessitates that we make the following important observation to experimentalists. Although FZ curves are widely used by experimental groups to profile the dynamics of indentation of biological particles, we find that quantifying the displacement of the cantilever tip via the deformation (indentation) depth X, represents a superior reaction coordinate. Indeed, it is important to realize that the deformation depth X reflects the external compressive force-induced dynamics of the particle deformation. Importantly, the FX curves show some weak nonlinearity, unlike the FZ curves, especially for small values of X, and less sharp force peaks F^* (Figure 17.8). For these reasons, we focused on analyzing the FX curves to resolve the precise mechanism of MT deformation. They were also used in the subsequent determination of thermodynamic quantities and materials properties.

The FX curves (Figure 17.8b) exhibit both single-step transitions, characterized by a single force peak, and multi-step transitions with several force peaks. Although some variability was exhibited by the force spectra, depending on the location of indentation points, each spectrum reveals three distinct regions: (a) the initial regime of continuous elastic deformation ($X < 6$–8 nm); (b) the transition regime, where the MT lattice undergoes discrete structural changes (6–8 nm $< X <$ 11–13 nm); and lastly (c) the post-collapse regime ($X >$ 11–13 nm). The spring constant K_{MT} was estimated from the

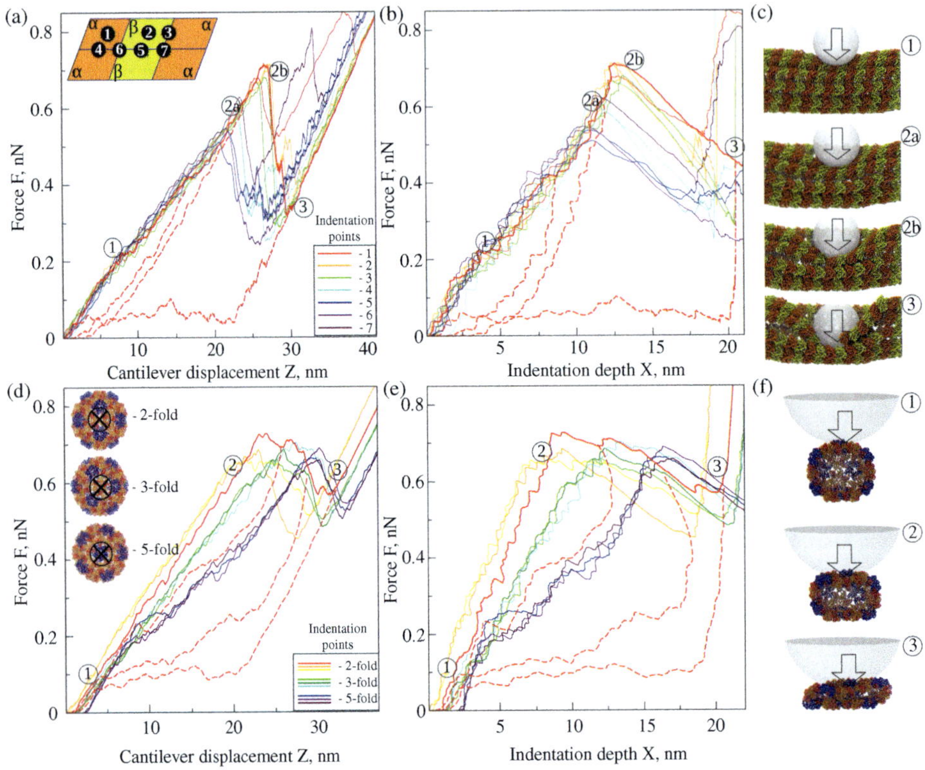

Figure 17.8 Force-deformation spectra for 8-dimer fragment of the MT cylinder MT8/13 and for CCMV capsid. The structures of the MT cylinder particle and CCMV spherical particle are presented, respectively, in Figures 17.3 and 17.4. Panels A-C: The force-deformation curves for 7 indentation points on the outer surface of MT lattice (shown in panel A), each depicted with different color, obtained with $v_f = 1.0$ μm/s and $R_{tip} = 10$ nm. Solid curves and dashed red curves represent the *FZ* profiles (panel A) and *FX* curves (panel B) for the forward deformation and backward tip retraction simulations, which followed the forward indentations (solid red curve) with $Z = 17$, 24, and 35 nm and $X = 7$, 11, 21 nm as initial conditions. Panel C shows the MT structure snapshots 1, 2a, 2b, and 3 illustrating the mechanism of MT deformation and collapse (direction of motion of the cantilever tip is indicated by a large vertical arrow). Structure 1: continuous deformation ($Z < 15$–20 nm, $X < 6$–8 nm; elastic regime). Structures 2a and 2b: disruption of lateral and longitudinal interfaces, respectively (20–25 nm $< Z <$ 25–30 nm, 6–8 nm $< X <$ 11–13 nm; transition regime). Structure 3: post-collapse evolution ($Z >$ 25–30 nm, $X >$ 11–13 nm). These structures correspond to the accordingly numbered regions in the *FZ* and *FX* curves in panels A and B. Panels D–F: The force-deformation curves for the CCMV shell indented along the 2-fold, 3-fold, and 5-fold symmetry axis (shown in panel D), each depicted with different color, obtained using $v_f = 1.0$ μm/s and $R_{tip} = 20$ nm. The force-deformation spectra, i.e. the *FZ* curves (panel D) and *FX* curves (panel E), results for the forward deformation and backward retraction are represented by the solid and dashed red curves, respectively. The retraction simulations are performed using the structures of the deformed CCMV shell occurring at $X = 5$, 11 and 19 nm deformation. The retraction curves show that the 5 nm deformation can be retraced back almost reversibly (small hysteresis), whereas the 19 nm deformation is nearly irreversible (large hysteresis). Panel F shows the CCMV structures numbered 1–3 for the native state of intact CCMV shell (structure 1), for the strongly deformed virus shell right before the transition to the collapsed state occurs (structure 2), and for the globally collapsed state (structure 3).

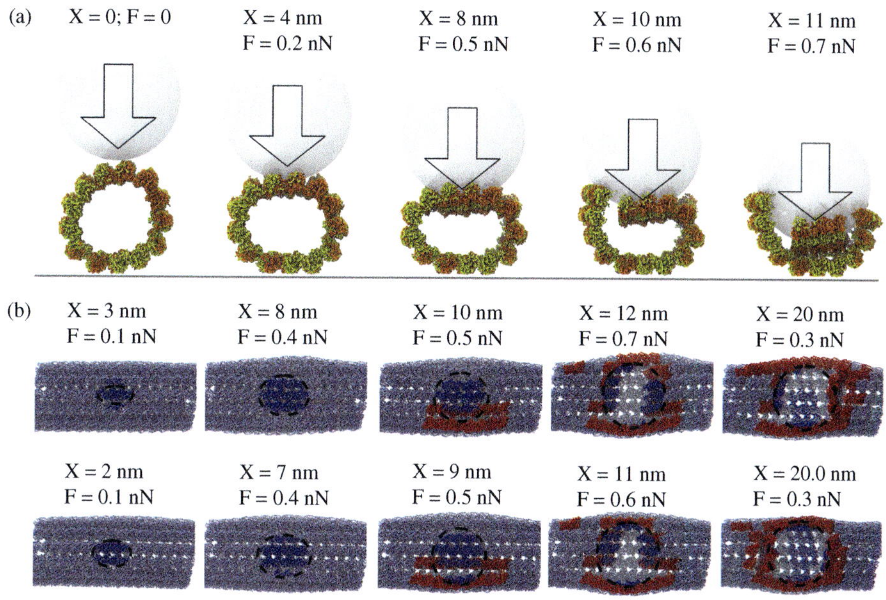

Figure 17.9 Structure changes observed during mechanical compression of MT cylinder MT8/13. Panel A: MT lattice profile viewed along the cylinder axis for different extent of indentation obtained with R_{tip} = 15 nm (indentation point 7; see Figure 17.8a). The cantilever tip is represented by a sphere. Panel B shows top views of the MT lattice for indentation points 3 (upper raw in B) and 6 (lower raw in B). In the course of mechanical compression (indicated by an arrow in panel A), the MT structure (light blue) is deformed which increases the MT-tip contact area (encircled dark blue area). Subsequent force increase results in the dissociation of lateral tubulin-tubulin bonds and then longitudinal bonds. In panel B, the tubulin monomers with disrupted lateral and/or longitudinal interface(s) are shown in dark red.

initial slope of the *FX* curves (linear-like regime), then we extracted the values of critical force F^* (peak force in *FX* curves) and the critical distance Z^*, at which the transition to the collapsed state occurs. All these values agree well with their experimental counterparts from Ref. [54, 55]. Therefore, the SOP model provides an accurate description of the biomechanical properties of the MT lattice.

As our simulations have shown, the MT deformation and collapse occur, as depicted in Figure 17.9, via a conserved mechanism where the MT lattice initially resists deformation; this is seen from the increase in the indentation force, which results in small variations in the local curvature of the MT cylinder under the tip (structure with $X \approx 4$ nm and $F \approx 0.2$ nN in Figure 17.9a). This initial resistance to deformation constitutes the linear-like regime of (continuous) elastic deformation represented in Figure 17.8 by the quasi-linear dependence of *F* on *X*; this behavior persists until $X \approx 8$ nm and $F \approx 0.5$ nN (middle structure in Figure 17.9a). At this point, the compressive force loads an increasingly larger portion of the MT surface. As a result, the MT cylinder flattens (buckles). Beyond $X \approx 6$–8 nm, indentation can no longer be accommodated solely by bending of the MT. Therefore, beyond this point the MT system enters the transition regime, where discrete structural changes occur. In the transition regime, the magnitude of mechanical tension exceeds the strength of lateral and longitudinal bonds; this results in their subsequent dissociation. Lateral interfaces are the first to be disrupted, initially at $X \approx 10$ nm (structure 2a in Figure 17.8c; see also snapshots in Figure 17.9); next the longitudinal bonds dissociate around $X \approx 11$–12 nm (structure 2b in Figure 17.8c; see also snapshots in Figure 17.9). The latter event triggers a sharp force drop, corresponding to force peaks in Figure 17.8a and 17.8b, as the MT lattice transitions to the collapsed state. The crossover from continuous deformation behavior to the multi-step discrete dissociation transitions was observed in all our *in silico* indentation runs. In addition, dissociation of the longitudinal inter-dimer bonds was detected, but not the intra-dimer bonds. Moreover, disruption occurred simultaneously for lateral interfaces between the α-tubulins and between the β-tubulins.

Our *in silico* nanoindentation approach was also utilized to probe MT lattice structural transitions in the experimentally inaccessible sub-nanometer length scale. Under stress the MT filament was observed to behave as a collection of rigid elements – tubulin monomers linked together through elastic lateral and longitudinal bonds. Simulations were performed for seven unique indentation points using different tip sizes. The force spectra are similar comparing the *FX* curves for a 10 nm tip vs. a 15 nm tip. However, the values of F^*, X^* (critical indentation depth) increase slightly with tip size (data not shown). Using a 10 nm tip, we present two examples- forward indentation followed by tip retraction at the surface of a protofilament (indentation points 2 and 3; Figure 17.10a and 17.10b) and between protofilaments (indentation points 6 and 7; Figures 17.10c and 17.10d). Where the compressive force is applied determines the magnitudes of critical force F^* and critical indentation depth X^*: $F^* = 0.65$–0.7 nN and $X^* \approx 12$ nm for compressing a protofilament (Figure 17.10a) are larger than $F^* = 0.5$–0.55 nN and $X^* \approx 10$ nm for compressing the interface between protofilaments (Figure 17.10c).

The slope of the *FX* curve (dF/dX) – a measure of mechanical compliance of the MT particle, was also examined. Here, we found that dF/dX varies significantly with X (Figures 17.10b and 17.10d). Steep increases interrupted by sudden decreases in dF/dX suggest that the MT lattice behaves as a soft material. The dF/dX peak heights denote the limits of deformability of the MT cylinder. Mechanical collapse of the MT is resisted longer when indentation occurs on the protofilament rather than between the protofilaments. The clear indication is that the interfaces between tubulins are softer (more compliant mechanically) than the tubulin monomers. Similar conclusions were reached when we examined results obtained with a 15 nm tip (data not shown). Profiles of the structure overlap function χ (defined in section 17.4) demonstrate that the MT lattice following collapse remains ~80–90% similar to the uncompressed state (see the insets to Figures 17.10b and 17.10d). This interesting result implies that changes in the MT cylinder resulting for the force-induced stress are mainly localized to lateral and longitudinal interfaces rather than being expressed through intramonomer alterations in structural elements (tubulin monomers). Hence, these results demonstrate that the elasticity expressed by the MT lattice tested mechanically originates in the interfacial properties of the network of lateral and longitudinal bonds stabilizing the MT cylinder.

Next, we studied the deformation-based thermodynamics of the MT lattice. The *FX* curves for forced indentation and force-quenched retraction were analyzed to calculate the enthalpy change ΔH, reversible work w_{rev} and free energy change ΔG for the transition of the MT lattice from the native state ($X = 0$) to the collapsed state ($X = 20$ nm). To be more specific, the energy output from simulations (potential energy U_{SOP}; see Equation 17.1) was first analyzed to estimate the enthalpy change of deformation ΔH. Then, by integrating the area under the *FX* curve for the forward indentation,

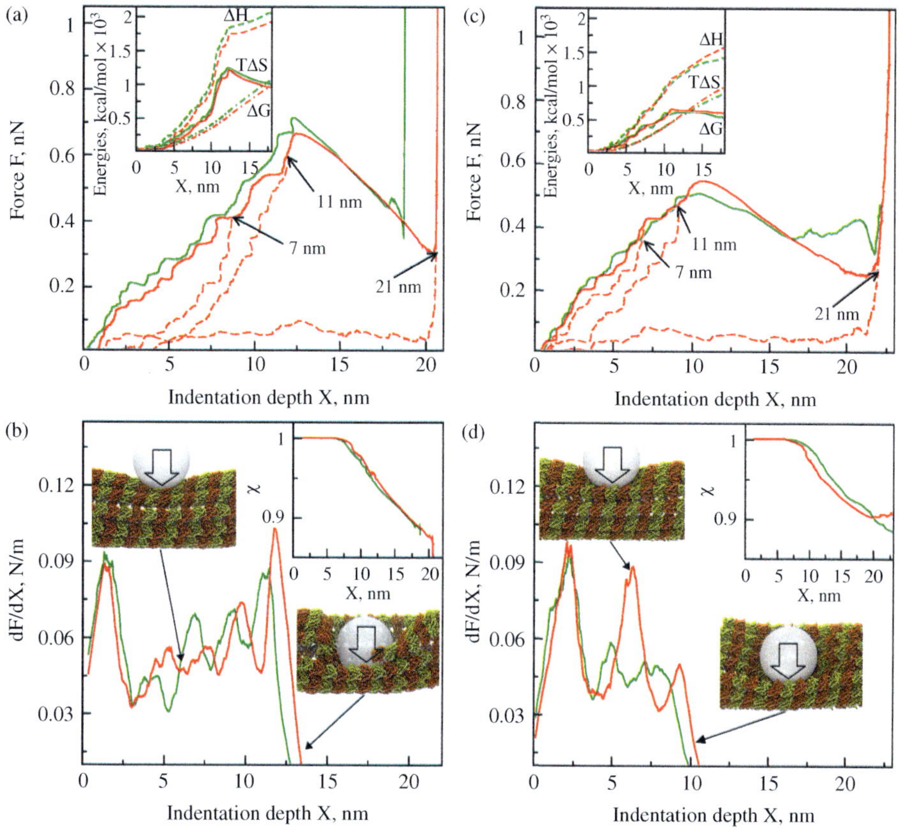

Figure 17.10 Force-indentation spectra for the forward indentation and backward (tip) retraction of the MT lattice MT8/13 *in silico* and corresponding structure alterations (cantilever velocity $v_f = 1.0$ μm/s and tip radius $R_{tip} = 10$ nm). Shown are results for the indentation points 2 (green) and 3 (red) in panels A and B, respectively, and for the indentation points 7 (green) and 6 (red) in panels C and D, respectively. Panels A and C: The *FX* curves for forward indentation (solid green and red curves). Curves for the backward tip retraction (dashed red lines) were generated using the structures with $X = 7$, 11, and 21 nm deformation (indicated on the graphs). *The insets* show the profiles of ΔH (dashed), $T\Delta S$ (solid) and ΔG (dash-dotted) vs. X. Panels B and D: The slope of the *FX* curves - the derivative dF/dX of the curves presented in panels A and C. Snapshots show the side-views of the MT before dissociation of the lateral bonds and after dissociation of the longitudinal bonds. *The insets* show the profiles of χ vs. X demonstrating that the MT in the collapsed state ($X > 20$ nm) is ~80–85% similar to the native undeformed state ($X = 0$).

the total work of deformation w can be obtained. This procedure can be repeated for the retraction curve to determine the reversible part of work $w_{\text{rev}} = \Delta G = \Delta H - T\Delta S$ and the entropic contribution $T\Delta S = \Delta H - w_{rev}$. The enthalpy and free energy changes for disruption of the interfacial contacts at both the lateral (ΔH_{lat} and ΔG_{lat}) and longitudinal (ΔH_{long} and ΔG_{long}) interfaces were next calculated using the following approach. First, two short time intervals within the same trajectory were carefully selected, one showing the dissociation of lateral interfaces and the other showing the dissociation of longitudinal interfaces. Then, for each selected trajectory portion we calculated the number of dissociated lateral interfaces n_{lat} and longitudinal interfaces n_{long}, and estimated the corresponding total reversible work terms w_{lat} and w_{long}. By integrating the corresponding regions in the *FX* curve for the forward indentation (see Section 4.2) and the backward retraction the quantities w_{lat} and w_{long} were determined. An estimate of the amount of dissipated energy contained in the irreversible part of work w_{irrev}, was made using the Crooks theorem (see [59, 60] for more detail). Finally, we calculated the dissociation free energy values of individual lateral and longitudinal bonds by taking the ratio $\Delta G_{lat} = w_{lat}/n_{lat}$ and $\Delta G_{long} = w_{long}/n_{long}$.

By comparing the values of ΔH_{lat}, ΔH_{long}, and ΔG_{lat}, ΔG_{long} it is clear that these state functions show a limited variation with both tip position on the MT surface and tip size (Table 17.3). These data were subsequently used to estimate the average enthalpy change ΔH, free energy change ΔG, and entropy change $T\Delta S = \Delta H - \Delta G$ for the disruption of a single lateral bond and a single longitudinal bond. Values of $\Delta G_{lat} = 6.9 \pm 0.4$ kcal/mol and $\Delta G_{long} = 14.9 \pm 1.5$ kcal/mol were obtained; these establish that the intra-protofilament longitudinal tubulin-tubulin bonds are roughly twice the strength of lateral inter-protofilament tubulin-tubulin bonds. These can be compared to value ranges estimated indirectly by other groups of ΔG_{lat} in the 1.8–8.0 kcal/mol range for the lateral bond, whereas ΔG_{long} is close to the upper limit of the 3.6–12.0 kcal/mol range for the longitudinal bond [46–49]. Interestingly, the corresponding bond difference in $T\Delta S$ is roughly four-fold. The increased flexibility of the newly created protofilament ends can account for the large $T\Delta S$ difference.

We also compared the distance range exhibited by tubulin-tubulin interactions through an examination of the force-induced flattening of the ring of tubulin dimers under the tip, both in the lateral direction ΔY_{lat} as well as the elongation of MT protofilaments in the transverse (longitudinal) direction ΔY_{long} (Figure 17.3). The indentation depth X was then related to the resulting changes in MT-tip contact area along the cylinder axis as well as the transverse direction. When the lateral contacts become disrupted at $X \approx 6$–8 nm indentation, the elongation of the tubulin ring is $\Delta Y_{lat} \approx 3.3$ nm. Through analysis of the structure, the number of lateral interfaces under the tip were found to be $N_{lat} = 3$–4 (Figure 17.9a). These numerical values were then used to estimate the average extension per lateral interface at which dissociation occurs, $\Delta y_{lat} = \Delta Y_{lat}/N_{lat} \approx 0.85$–$1.1$ nm. Next, longitudinal contacts become disrupted at $X \approx 9$–11 nm indentation. The corresponding elongation of a 3–4 dimer-long portion of an MT protofilament under the tip is $\Delta Y_{long} \approx 7.5$ nm. Again, by analysis of structures, the number of longitudinal interfaces involved was observed to be $N_{long} \approx 5$–6. Therefore, for dissociation of the longitudinal interface, the average extension is $\Delta y_{long} = \Delta Y_{long}/N_{long} \approx 1.25$–$1.5$ nm. A comparison of our results indicate that the longitudinal bonds possess a longer interaction range ($\Delta y_{long} \approx 1.25$–$1.5$ nm) relative to the lateral bonds ($\Delta y_{lat} \approx 0.85$–$1.1$ nm). Both of these interaction range values lie within the 1.5–2 nm characteristic of a variety of protein complexes [61, 62].

17.5.3 Spherical Particle – CCMV Protein Shell

Using the SOP model and GPU-accelerated Langevin simulations, we performed dynamic force measurements *in silico* on the CCMV shell, as described in section 17.2. The CCMV virus capsid, devoid of RNA molecules was used (PDB code: 1CWP) [53].

Table 17.3 Comparison of the mechanical and thermodynamic parameters from indentations at different points on the MT lattice. Simulations were carried out using $v_f = 1.0$ μm/s, and $R_{tip} = 10$ and 15 nm. Presented are the values of indentation depth X^*, critical force F^*, and spring constant K_{MT} (average value of the slope dF/dX in the initial portion of the FX curve). Also shown are the free energy change ΔG and the enthalpy change ΔH of dissociation of the lateral and longitudinal tubulin-tubulin bonds.

Indentation Point	X^*, nm		F^*, nN		K_{MT}, pN/nm		ΔG_{long}, kcal/mol		ΔG_{lat}, kcal/mol		ΔH_{long}, kcal/mol		ΔH_{lat}, kcal/mol	
	10 nm	15 nm	10 nm	15 nm	10 nm	15 nm	10 nm	15 nm	10 nm	15 nm	10 nm	15 nm	10 nm	15 nm
1	12.5	12.0	0.67	0.72	51	53	15.0	16.1	6.5	6.6	27.1	29.1	8.4	10.2
2	12.3	12.4	0.71	0.78	48	54	13.4	17.0	6.7	7.2	21.8	23.9	9.3	10.1
3	11.9	12.6	0.67	0.77	48	58	13.9	16.9	6.4	6.8	24.3	27.8	9.5	9.7
4	11.3	10.9	0.60	0.73	55	61	12.5	13.6	7.1	7.6	25.0	27.7	9.4	9.8
5	10.5	11.0	0.54	0.80	53	57	12.5	16.7	6.6	7.2	26.1	26.3	7.9	8.6
6	10.3	10.8	0.54	0.71	54	66	17.8	16.2	6.3	7.1	21.7	24.0	8.2	8.8
7	10.8	10.9	0.63	0.74	55	65	14.3	16.2	7.1	7.3	26.7	27.8	10.4	10.5

To compress the CCMV capsid along the 2-, 3-, and 5-fold symmetry axes a spherical tip of radius $R_{\text{tip}} = 20$ nm was used. The CCMV particle is capable of rolling on the substrate surface during the simulations; to prevent this we constrained, by fixing in place, five C_α-atoms along the bottom of the CCMV perimeter. Then, the cantilever tip exerted the time-dependent force $\mathbf{f}(t) = f(t)\mathbf{n}$, in the direction $\mathbf{n}$ perpendicular to the surface of CCMV shell, in order to test the particle mechanically. In this simulation, the reader is reminded that the force magnitude $f(t) = r_f t$ increased linearly in time t (force-ramp mode) with the corresponding force-loading rate being $r_f = \kappa v_f$ (v_f is the cantilever base velocity and κ is the cantilever spring constant). Our use of the experimentally relevant cantilever velocity $v_f = 1.0$ μm/s and achieving the experimental 0.1–0.2 s timescale was made possible in these simulations through our use of the significant computational acceleration obtained on a GPU (see Figure 17.2).

During 'forward indentation' simulations, the cantilever base moved towards the capsid at a velocity $v_f = 1.0$ μm/s. In the case of force-quenched retraction, we reversed the direction of tip motion. The piezo displacement Z (cantilever base), and the cantilever tip position X are both controlled in our simulations. In the simulations of both forward indentation and backward (tip) retraction, the cantilever base (virtual particle) moved with constant velocity (v_f). With this setup we could increase or decrease force $f(t)$ applied to the capsid through the cantilever tip, by varying the force-loading rate r_f. Using the energy output, the indentation force F from the capsid could then be calculated. Almost quantitative agreement was obtained between the results of dynamic force measurements *in vitro* and *in silico* (data not shown) [18]. As a result of this quantitative agreement, we conclude that our multiscale modelling of the CCMV shell provides an accurate description of the capsid mechanical properties. This validates our theoretical approach and provides a testimony to its importance in both explaining experimental nanoindentation results and predicting properties of systems prior to their study experimentally.

We indented the capsid at the following different points on its surface: at the center of symmetry axes of the hexamer capsomeres (3-fold symmetry), as well as the pentamer capsomeres (5-fold symmetry), and at the interface between two hexamers (2-fold symmetry). These results are presented in Figure 17.8 (panels d–f). As was the case for the MT cylinder, we observed a profile of F versus indentation depth X (FX curve) to be a more sensitive measure of the capsid's mechanical properties, since it reveals finer features of the force spectrum (Figure 17.8e). Because the FZ curves describe the combined response of the 'capsid-plus-tip' system (Figure 17.8d), it appears that the CCMV particle exhibits a constant elasticity. However, this is not the case and we discuss in greater detail the results obtained from simulations of nanoindentation of the CCMV particle along the 2-fold symmetry axis. It is clear that the FX profiles are essentially nonlinear curves of varying slope, and for that reason the notion of 'spring constant' should be used with great caution. Mechanical nanoindentation is a stochastic process due to thermal fluctuations. This fact is reflected in the observation that, even for the same geometry, repeated simulations exhibit FX curves showing variability (Figure 17.8d and 17.8e). The capsid's mechanical reaction is elastic up to $X \approx 3$–5 nm (linear-like regime) and appears quasi-elastic up to $X \approx 8$–11 nm. If the initial portion of the FX curves ($X <$ 3 nm) are fitted to a straight line, this yields the 'spring constant' (average slope) values of $k_{cap} \approx 0.11$ N/m, 0.10 N/m, and 0.12 N/m for 2-, 3-, and 5-fold symmetry, respectively (Table 17.4).

Table 17.4 Thermodynamic properties of the CCMV capsid from *in silico* indentation along the 2-fold, 3-fold, and 5-fold symmetry axes: indentation depth X^*, critical force F^*, spring constant k_{cap}, entropy change $T\Delta S$, enthalpy change ΔH and Gibbs energy change ΔG. These were obtained by averaging the results of 3 nanoindentation runs generated using $R_{tip} = 20$ nm and $v_f = 1.0$ μm/s. The values of $T\Delta S_{ind}$, ΔH_{ind} and ΔG_{ind} correspond to the total change in these quantities observed at $X =$ 20 nm deformation. The range of variation of k_{cap} (from Figure 17.11b) is shown in parentheses.

Symmetry	X^*, nm	F^*, nN	k_{cap}, N/m	$T\Delta S_{ind}$, kcal/mol × 10^3	ΔH_{ind}, kcal/mol × 10^3	ΔG_{ind}, kcal/mol × 10^3
2-fold	9.1±1.0	0.71±0.02	0.11 (0.06–0.14)	1.67	2.75	1.08
3-fold	11.9±0.5	0.68±0.02	0.10 (0.05–0.10)	1.58	2.80	1.22
5-fold	14.2±0.5	0.69±0.02	0.12 (0.04–0.12)	2.01	2.99	0.98

The profile of k_{cap} vs. X obtained by calculating the derivative $k_{cap} = dF/dX$ over the entire range of X is a sensitive measure of CCMV particle deformation since, in the initial deformation regime, k_{cap} varied considerably from 0.06–0.14 N/m, 0.05–0.10 N/m, and 0.04–0.12 N/m for the respective 2-, 3-, and 5-fold symmetry axes. Fluctuations in k_{cap} exhibit systematic differences for the different icosahedral symmetry axes. Along the 2- and 5-fold axes, the k_{cap} versus X curves show two maxima: the first is at $X \approx 2$–3 nm (for 2- and 5-fold symmetry), and the second is at $X \approx 5$–6 nm (2-fold symmetry) and 11–12 nm (5-fold symmetry). One broad skewed peak centered at $X \approx 5$ nm (data not shown) is shown by k_{cap} for the 3-fold symmetry case. The contact surface area between the tip and the CCMV shell changes with the indentation depth (Figure 17.11a); this means that cooperative behaviour is exhibited by different numbers of protein chains forming capsomers to resist the mechanical stress.

The collapse transition of the CCMV capsid occurs in the range of 11–15 nm (Figure 17.11a and b). During the transition, the capsid undergoes a spontaneous change in shape from a roughly spherical state to a collapsed state; the shape transition is reflected in the abrupt decrease in force and decline of k_{cap} to zero (Figure 17.11b). We monitored the structure overlap χ in order to quantitate the extent of the structural collapse. Within the transition region, χ decreased from χ=1 (native state) to χ=0.65 (collapsed state) for simulations of all symmetry types (see inset to Figure 17.11a). Thus, the capsid structure retained 65% similarity to the native state. Next, we carried out simulations reversing the direction of tip motion. This was performed using, as initial conditions, the following four structures for the collapsed state obtained for $X = 5$, 11, 15, and 19 nm deformation. In full agreement with AFM data [18, 63], the CCMV capsid's mechanical compression was observed to be fully reversible in the elastic regime for $X = 5$ nm (no hysteresis), almost reversible in the quasi-elastic regime for $X = 11$ nm (small hysteresis), but irreversible post-transition for $X = 15$ nm (Figure 17.11a).

Next, the total work of indentation w was evaluated as described in section 17.4. Then we repeated this procedure for the retraction curves to calculate the reversible work w_{rev}. The relative difference parameter $(w - w_{rev})/w$ in the elastic and quasi-elastic regime ($X < 11$ nm) is ~0.12; this means that ~12% of the energy of w was dissipated. This fact is in good agreement with the experimental result, namely that the reversible energy exhibited for the retraction process is ~90% [63]. By contrast, for the retraction curves,

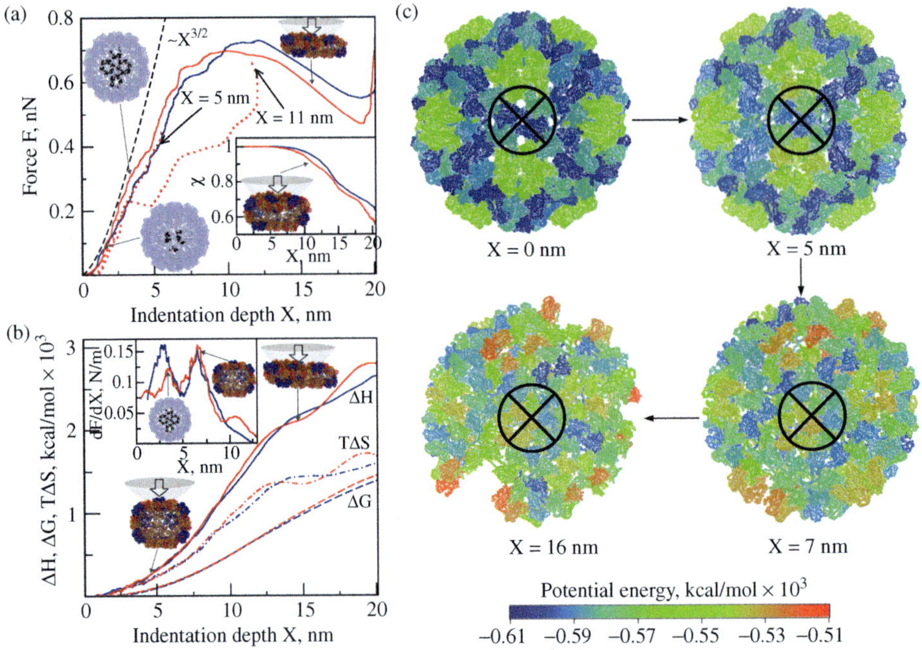

Figure 17.11 Panels A and B: Nanoindentation *in silico* of the CCMV particle. Show in red and blue color are two representative trajectories obtained for 2-fold symmetry axis ($v_f = 1.0$ µm/s and $R_{tip} =$ 20 nm). Panel A: The force-deformation spectra (*FX* curves); results for the forward deformation and backward retraction are represented by the solid and dotted red curves, respectively. The retraction simulations are performed using the structures of the deformed CCMV shell with $X = 5$ and 11 nm. The dashed line represents the fit to the *FX* curve in the elastic regime ($X < 3$–5 nm) by the nonlinear function $\sim X^{3/2}$. Structures on the left show the increasing tip-capsid surface contact area (blackened); the structure on the right shows the CCMV profile in the collapsed state. *The inset* is the profile of the structure overlap $\chi(X)$, which decreases with indentation *X*. The structure in *the inset* is the CCMV particle before the transition to the collapsed state ($X \approx 10$ nm indentation). Panel B: The profiles of ΔH, ΔG, and $T\Delta S$. Also shown are the profile views of CCMV in the elastic deformation regime, where $\Delta H \approx T\Delta S$, and in the plastic regime, where $\Delta H > T\Delta S$. *The inset* shows the profile of the slope of the *FX* curve, i.e. d*F*/d*X* versus *X*, with two peaks which correspond to the mechanically activated (transition) states for the two types of transitions: (i) local curvature change in the tip-capsid surface contact area (first peak at $X \approx 3$ nm and the corresponding top view of CCMV); and (ii) bending deformation of the side portions of CCMV shell (second peak at $X \approx 7$ nm and the corresponding side view of CCMV). Panel C: Surface map of the potential energy (color scale for U_{SOP} is in the graph) for four representative structures of the CCMV shell (top view) observed at $X = 0$, 5 nm, 7 nm, and 16 nm deformation. The direction of motion of the tip is perpendicular to the CCMV surface as indicated by the black cross. The map shows a gradual increase in the potential energy of proteins in pentamers and hexamers as global changes to the structure occur.

a large hysteresis, $(w - w_{rev})/w \approx 75\%$ was exhibited in the transition range (11 nm $\leq X \leq$ 15 nm). Because $w_{rev} = \Delta G = \Delta H - T\Delta S$ and since ΔH can be determined using the simulation's energy output, we were able to estimate ΔH and $T\Delta S$. Results for the ΔG, ΔH and $T\Delta S$ for *in silico* indentation along the axis of 2-fold symmetry are presented in Figure 17.11b. In the initial linear-like regime ($X <$ 10–11 nm), ΔH and $T\Delta S$ describe a parabolic dependence on X and $\Delta H \approx T\Delta S$. However, $\Delta H > T\Delta S$ for the (11–15 nm) transition range and is within the post-collapse regime ($X >$ 15 nm). At $X = 20$ nm the curves of ΔG, ΔH, and $T\Delta S$ achieve constant values ΔG_{ind}, ΔH_{ind}, and $T\Delta S_{\text{ind}}$, which correspond to the Gibbs energy, enthalpy, and entropy of indentation, respectively. For *in silico* indentation along the 2-, 3- and 5-fold symmetry axes, the equilibrium values of ΔG_{ind}, ΔH_{ind}, and $T\Delta S_{\text{ind}}$ are provided in Table 17.4. These thermodynamic quantities indicate that the CCMV capsid's mechanical compression is a costly process energetically, and that the thermodynamic state functions ΔG_{ind}, ΔH_{ind}, and $T\Delta S_{\text{ind}}$ all vary with the type of local capsid symmetry located under the tip.

17.6 Concluding Remarks

The dynamic force spectroscopy *in silico* approach was developed in our group to perform numerical simulations of nanoindentation experiments on biological particles. Our computer-based approach fully mimics AFM based dynamic force *in vitro* experiments in terms of the conditions of force application and general experimental setup. A hallmark of our approach is the full system control we have during the entire process of forced deformation. This includes: the point of tip contact, the direction and magnitude of applied force, choice of constrained residues, and size and shape of the indenting object (Figure 17.3). Structural resolution at the nanometer-scale, which allows us to distinguish intact versus disrupted interfaces, represents a huge advantage; it can be used by investigators to probe e.g. protein-protein interactions between capsomers in virus particles and tubulin monomers in the microtubule lattice. With our nanoindentation *in silico* approach one can also directly correlate energy changes with the corresponding structure alterations at the level of individual amino acids as a simulation progresses. We would like to emphasize the following point. Structural resolution at the residue-level and corresponding information about the thermodynamic state functions, e.g. the enthalpic and entropic contributions to the free energy change are directly available from our nanoindentation measurements *in silico*. However, they are not directly accessible from traditional nanoindentation experiments *in vitro*. This fact makes our approach to *in silico* nanoindentation a powerful new methodology for the theoretical exploration of large biological assemblies.

Our approach involves testing a biological particle mechanically by following stochastic Langevin dynamics. For this reason, little effort was needed to port the numerical algorithm to the Graphics Processing Unit, the result of which provides investigators with a powerful platform to carry out high-performance computations. In our SOP-GPU software package, all computational features are fully implemented on the GPU device. These include all the numerical algorithms: generation of (pseudo)-random numbers, the calculation of forces, generating lists of amino acid residues, and the numerical integration of the Langevin equations of motion. The payback of using this approach is significant and involves the following attributes: we are capable of attaining a 40–50-fold computational acceleration on a single GPU (Figure 17.2). This speedup enables us to model the process of deformation dynamics for biological particles of $N_{tot} = 10^5$–10^6 residues (such as CCMV; see Figure 17.4) in the very long (experimental) $\sim 10^1$–10^2 millisecond timescale. Importantly, we achieve these simulation results using the conditions of compressive force application applied experimentally (cantilever velocity of ~ 0.1–1.0 μm/s). It is worth emphasizing that the use of experimental or biologically relevant conditions of mechanical force load is crucial. This is due to the soft nature of biological matter, which has the properties that both the amplitude (i.e. forces and deformations) and microscopic mechanism underlying the mechanical response of individual biological particles, are force-loading rate dependent (see Figure 17.6).

As we have demonstrated, when a sufficiently slow force loading is employed, the deformation dynamics are microscopically reversible (Figure 17.5). Concerning the application of compressive force, this regime of slow mechanical loading offers a number of advantages. We demonstrated that when the rate of structural re-equilibration of a biological particle exceeds the rate of force increase, then quasi-equilibrium conditions exist at each and every point along the pathway of the deformation process. For example, this can be observed from a comparison of the *FX* curves generated using varying cantilever velocities applied to the 24-dimer long protofilament fragment PF24/1 (Figure 17.6). In particular, the *FX* curves become less different as v_f decreases, implying that increasingly similar mechanical responses are being exhibited. Therefore, the results we have obtained clearly demonstrate that *in silico* nanoindentation experiments reported here are performed under near-equilibrium conditions of force application. Stated differently, our *in silico* nanoindentation approach enables investigators to obtain equilibrium information from non-equilibrium computer-based measurements when the computations are performed under the slow force-loading regime (cantilever velocity). Therefore, the relevant materials properties (critical deformation, force threshold and Young's modulus) and thermodynamic state functions (changes in Gibbs free energy, enthalpy and entropy), that can readily be determined from analysis of the force-deformation spectra (*FX* curves), describe equilibrium properties of the biological particle being studied. In addition, the deformation mechanism of the biological particle, which can be discovered through analysis of the simulation output (coordinate files and structural snapshots), describes the properties of the particle in its equilibrium (native) state.

We have previously applied the dynamic force spectroscopy *in silico* approach we described here to examine the forced unfolding of fibrin polymers [15, 64]. In this chapter, we exemplified the use of this *in silico* approach to map the free energy landscape underlying the dynamics of deformation of three different systems, representing specific examples of large size biological particles- microtubule protofilaments of varying length, the eight-dimer long microtubule (MT) fragment, and the intact Cowpea Chlorotic Mottle Virus (CCMV) capsid, devoid of its packaged genome. This approach is really made possible through a combination of the following advantageous features: the atomic-level description and C_α-based coarse-grained modelling (Figure 17.1) are combined with precise *in silico* nanomanipulation of these biological particles. Then, by taking advantage of the huge GPU-based computational acceleration, we performed a long timescale (~50 ms) detailed exploration of MT dynamics and biomechanics and examined the thermodynamics of lateral and longitudinal non-covalent bonds that stabilize the MT lattice. Moreover, this was performed using the experimentally appropriate conditions of force application (Figure 17.8a–c). Taking this approach, atomic-level details underlying the lateral and longitudinal tubulin-tubulin interactions are observed, this detail being implicit in the MT cylinder SOP model. Next-neighbour interactions that stabilize the MT structure and the lattice confinement for individual dimers are explicitly described as well (Figures 17.1 and 17.3). Accurate measurements of forced nanoindentation *in silico* were also performed for the CCMV capsid

(Figure 17.4). The objectives were to understand in secondary structure molecular detail the nature and extent of all structural transitions, and to illuminate the mechanism(s) of mechanical deformation and transition to the collapsed state in virus capsids (Figure 17.8d–f). The rationale behind our computational approach employed here is that, rather than atomic details, the unique features associated with the particle's geometry, as well as with its native topology and symmetry of capsomer arrangement are what govern the physico-chemical properties of virus capsids.

Force-generating and mechanical properties of MTs play important roles in many cellular processes. Despite this importance to the functioning of the cell, little is known with certainty about the thermodynamics of tubulin-tubulin interactions as well as the biomechanics of single protofilaments that comprise the MT lattice. Lateral and longitudinal tubulin-tubulin bonds' dissociation energies were previously estimated using a variety of molecular-mechanical models, where the smallest unit was the tubulin monomer or dimer [46, 49, 65]. There is a major drawback to this approach. It is that the dynamic parameters of MT assembly and disassembly are used to derive the tubulin bond energies; unfortunately, this process reports only indirectly on the thermodynamics of tubulin-tubulin interactions. By contrast, the *in silico* nanoindentation approach described here provides a definite advantage, because the protofilaments' deformation and tubulin-tubulin bonds' dissociation are recorded in these simulation experiments with high spatial and temporal resolution [19]. Indeed, due to the MT structural complexity (Figure 17.9), the modelling of experimental force-indentation spectra at the level of tubulin-tubulin bonds and the accompanying molecular interpretation is rather difficult. However, the area under the *FX* curve is the total work carried out by the tip on the MT lattice (Figure 17.10), and the reversible portion of that work can be converted to the Gibbs free energy change. Hence, by analysing *in silico* experiments in which these bonds are directly manipulated, the average dissociation free energies of the lateral bond 6.9±0.4 kcal/mol and longitudinal bond 14.9±1.5 kcal/mol (Table 17.3) are obtained. These values, coupled with the large measured flexural rigidity of tubulin protofilaments (18 000–26 000 pN nm^2; see Table 17.2), support the concept that the microtubule is capable of generating a force sufficiently large to move chromosomes during cell division.

Hierarchical supramolecular systems play fundamental roles in biology. Their dynamic properties of spontaneous assembly, disassembly, and self-repair have evolved to be key determinants of their biological functions. Plant and animal viruses and bacteriophages are prime examples of hierarchical supramolecular assemblies. It remains a challenge to understand the structural origins of their unique physico-chemical properties and to resolve the precise mechanisms of their dynamic response to a wide variety of external mechanical factors and biochemical species. Single-molecule techniques such as AFM have become available within the past 15–20 years to explore the physical and materials properties of biological assemblies. AFM experiments, as one specific example, yield valuable information on the capsid particle spring constant and the forces required to distort capsid structure. Nevertheless, experimental results are difficult to interpret, due to the high complexity of viruses ($\sim 10^4$–10^5 amino acid residues; see Figure 17.4), without first having some input from theoretical modelling. To create new applications of virus particles in biotechnology, it is crucial to maintain a proper design control over their physical properties. However, in many cases knowledge of these properties from either experiments or modelling is clearly lacking.

Although biomolecular simulations have now become available for the theoretical exploration of virus particles [23, 26, 66], the very long (ms–s) timescales needed to observe important dynamic features of the structures limits the current theoretical capabilities of the standard methodology. We have overcome this problem through the use of our multiscale modelling approach and GPU-based computational acceleration. This has enabled our high resolution view that shows the capsid's physical properties are dynamic and local characteristics of the structure. Furthermore, they change with the tip indentation depth and depend on the magnitude and geometry of mechanical input. Interestingly, transitioning to the collapsed state by the CCMV capsid occurs without substantial local structural alterations (Figure 17.11); this behaviour implies that the formation of the collapsed state is a shape changing transition, rather than a structural one. The average enthalpy change associated with populating the collapsed state $\Delta H_{ind} = 2.75–3.06$ kcal/mol $\times 10^3$ contributes to the bending of the capsid; in contrast, the average entropy change $T\Delta S_{ind} = 1.22–1.39$ kcal/mol $\times 10^3$ mediates the capsid stiffening (Table 17.4). During CCMV capsid indentation, the extent of (ir)reversibility is directly correlated with the capsid's (in)elastic mechanical response to the compressive force (Figure 17.11).

In summary, our computational modelling approach offers a comprehensive quantitative platform to link the dynamic mechanical behavior of biological particles at the macroscopic level with their molecular characteristics at the nanoscale. The *in silico* nanoindentation method we have developed provides a powerful new characterization tool. It provides a methodology for exploration of biomechanical properties and thermodynamic characteristics, and for the discovery of the mechanism(s) underlying the compressive force-induced deformation of large biomolecular assemblies. Through detailed understanding of the mechanisms of MT deformation and collapse, unique insights into the mechano-chemistry of the MT polymers in living cells will be achieved. The CCMV capsid results we have obtained illuminate how unique physico-chemical properties of protein nanoshells help define their dynamic structure and overall morphology. Although the nanoindentation *in silico* modelling tools we have developed have been used here to describe only a few examples of biological particles, these tools have broad applicability. They can be applied to study a broad range of complex biological assemblies, whose physico-chemical characteristics cannot be resolved using only experimental approaches.

References

1 Cohen, S.R., Kalfon-Cohen, E. Dynamic nanoindentation by instrumented nanoindentation and force microscopy: a comparative review. Beilstein Journal of Nanotechnology (2013) 4: 815–833.
2 Feng, Z.Q., He, Q.C. Zeng, Q., Joli, P. Theory of nanoindentation. In: Handbook of Nanophysics, K. Sattler (ed.). Taylor and Francis; 2010.
3 Binnig, G., Quate, C.F., Gerber, C. Atomic force microscope. Physical Review Letters (1986) 56:930.
4 Zhong, Q., Inniss, D., Kjoller, K., Elings, V.B. Fractured polymer/silica fiber surface studied by tapping mode atomic force microscopy. Surface Science Letters (1993) 290: L688–L692.

5 Zlatanova, J., Lindsey, S.M., Leuba, S.H. Single molecule force spectroscopy in biology using the atomic force microscope. Progress in Biophysics and Molecular Biology (2000) 74: 37–61.
6 Yang, Y., Wang, H., Erie, D.A. Quantitative Characterization of biomolecular assemblies and interactions using atomic force microscopy. Methods (2003) 29: 175–187.
7 Roos, W.H., Bruinsma, R., Wuite, G.J. Physical virology. Nature Physics (2010) 6:733–743.
8 Fernandez, J.M., Li, H. Force-clamp spectroscopy monitors the folding trajectory of a single protein. Science (2003) 303: 1674–1678.
9 Ivanovska, I. L., de Pablo, P.J., Ibarra, B. *et al.* Bacteriophage capsids: tough nanoshells with complex elastic properties. Proceedings of the National Academy of Sciences USA (2004) 101:7600–7605.
10 Hyeon, C., Dima, R. I. Thirumalai, D. Pathway and kinetic barriers in mechanical unfolding and refolding of RNA and proteins. Structure (2006) 14:1633–1645.
11 Mickler, M., Dima, R. I., Dietz *et al.* Revealing the bifurcation in the unfolding pathways of GFP by using single-molecule experiments and simulations. Proceedings of the National Academy of Sciences USA (2007) 104: 20268–20273.
12 Dima, R.I., Joshi, H. Probing the origin of tubulin rigidity with molecular simulations. Proceedings of the National Academy of Sciences USA (2008) 105: 15743–15748.
13 Lin, J., Hyeon C., Thirumalai, D. Relative stability of helices determines the folding landscape of adenine riboswitch aptamers. Journal of the American Chemical Society (2008) 130: 14080–14084.
14 Hyeon, C. Onuchic, J.N. Mechanical control of the directional stepping dynamics of the kinesin motor. Proceedings of National Academy of Sciences USA (2007) 104: 17382–17387.
15 Zhmurov, A., Brown, A.E., Litvinov, R. I. *et al.* Mechanism of fibrin(ogen) forced unfolding. Structure (2011) 19: 1615–1624.
16 Zhang, Z. Thirumalai, D. Dissecting the kinematics of the kinesin step. Structure (2012) 20: 628–640.
17 Theisen, K.E., Zhmurov, A., Newberry, M.E. *et al.* Multiscale modeling of the nanomechanics of Microtubule protofilaments. Journal of Physical Chemistry B (2012) 116: 8545–8555.
18 Kononova, O., Snijder, J., Brasch, M. *et al.* Structural transitions and energy landscape for cowpea chlorotic mottle virus capsid mechanics from nanomanipulation in vitro and in silico. Biophysical Journal (2013), 105: 1893–1903.
19 Kononova, O., Kholodov, Y., Theisen, K.E. *et al.* Tubulin bond energies and microtubule biomechanics determined from nanoindentation in silico. Journal of the American Chemical Society (2014) 136: 17036–17045.
20 Kononova, O., Jones, L. Barsegov, V. Order statistics inference for describing topological coupling and mechanical symmetry breaking in multidomain proteins. Journal of Chemical Physics (2013) 139: 121913–121925.
21 Gibbons M.M. Klug, W.S. Influence of nonuniform geometry on nanoindentation of viral capsids. Biophysical Journal (2008) 95: 3640–3649.
22 Klug, W.S., Roos, W.H. Wuite, G.J. Unlocking internal prestress from protein nanoshells. Physical Review Letters (2012) 109: 168104.

23 Zink, M. Grubmuller, H. Mechanical properties of the icosahedral shell of southern bean mosaic virus: A molecular dynamics study. Biophysical Journal (2009) 96: 1350–1363.
24 Zink, M. Grubmuller, H. Primary changes of the mechanical properties of Southern Bean Mosaic Virus upon calcium removal. Biophysical Journal (2010) 98: 687–695.
25 Arkhipov, A., Freddolino, P.L. Schulten, K. Stability and dynamics of virus capsids described by coarse-grained modeling. Structure (2006) 14: 1767–1777.
26 Arkhipov, A., Roos, W.H., Wuite, G.J. Schulten K. Elucidating the mechanism behind irreversible deformation of viral capsids. Biophysical Journal (2009) 97: 2061–2069.
27 Zhmurov, A., Dima, R.I., Kholodov, Y. Barsegov, V. SOP-GPU: Accelerating biomolecular simulations in the centisecond timescale using graphic processors. Proteins (2010) 78: 2984–2999.
28 Zhmurov, A., Rybnikov, K., Kholodov, Y. Barsegov V. Generation of random numbers on graphics processors: forced indentation in silico of the bacteriophage HK97. Journal of Physical Chemistry B (2011) 115: 5278–5288.
29 Ferrara, P., Apostolakis, J. Caflisch, A. Evaluation of a fast implicit solvent model for molecular dynamics simulations. Proteins (2002) 46:24–33.
30 Hawkins, G.D., Cramer, C.J. Truhlar, D.G. Parametrized models of aqueous free energies of solvation based on pairwise descreening of solute atomic charges from a dielectric medium. Journal of Physical Chemistry (1996) 100: 19824–19839.
31 MacKerell, A.D., Bashford, B., Bellott, M. *et al.* All-atom empirical potential for molecular modeling and dynamics studies of proteins. Journal of Physical Chemistry B (1998) 102: 3586–3616.
32 Barsegov, V., Klimov, D., Thirumalai, D. Mapping the energy landscape of biomolecules using single molecule force correlation spectroscopy (FCS): Theory and applications. Biophysical Journal (2006) 90: 3827–3841.
33 Nogales, E., Whittaker, M., Milligan, R.A. Downing, K.H. High-resolution model of the microtubule. Cell (1999) 96: 79–88.
34 Valdman, D., Atzberger, P. J., Yu, D. *et al.* Spectral analysis methods for the robust measurement of the flexural rigidity of biopolymers. Biophysical Journal (2012) 102:1144–1153.
35 Dogterom, M. Surrey, T. Microtubule organization in vitro. Current Opinions in Cell Biology (2013) 25:23–29.
36 de Forges, H., Bouissoua, A. Pereza, F. Interplay between microtubule dynamics and intracellular organization. The International Journal of Biochemistry: Cell Biology (2012) 44: 266–274.
37 Desai, A. Mitchison, T.J. Microtubule polymerization dynamics. Annual Review of Cell Development and Biology (1997) 13: 83–117.
38 Walczak, C.E., Cai, S. Khodjakov, A. Mechanisms of chromosome behavior during mitosis. National Review of Molecular Cell Biology (2010) 11: 91–102.
39 McIntosh, J.R., Volkov, V.A., Ataullakhanov, F.I. Grishchuk, E.L. Tubulin depolymerization may be an ancient biological motor. Journal of Cell Science (2010) 123: 3425–3434.
40 Joglekar, A.P., Bloom, K.S. Salmon, E.D. Mechanisms of force generation by end-on kinetochore-microtubule attachments. Current Opinion in Cell Biology (2010) 22: 57–67.

41 Grishchuk, E.L., McIntosh, J.R., Molodtsov, M.I. Ataullakhanov, F.I. Force generation by dynamic microtubule polymers. In Comprehensive Biophysical, Vol. 4, Molecular Motors and Motility. Oxford: Academic Press, pp. 93–117; 2011.
42 VanBuren, V., Cassimeris, L. Odde, D.J. A mechano-chemical model for microtubule structure and self-assembly kinetics. Biophysical Journal (2005) 89: 2911–2926.
43 Bowne-Anderson, H., Zanic, M., Kauer, M. Howard, J. Microtubule dynamic instability: a new model with coupled GTP hydrolysis and multistep catastrophe. Bioessays (2013) 35: 452–61.
44 Mandelkow, E.M., Mandelkow, E. Milligan, R.A. Microtubule dynamics and microtubule caps: a time-resolved cryo-electron microscopy study. Journal of Cell Biology (1991) 114: 977–991.
45 Sanchez, S.A., Brunet, J.E., Jameson, D.M. *et al.* Tubulin equilibrium unfolding followed by time-resolved fluorescence correlation spectroscopy. Protein Science (2004) 13: 81–88.
46 van Buren, V., Odde, D.J. Cassimeris, L. Estimates of lateral and longitudinal bond energies within the microtubule lattice. Proceedings of the National Academy of Sciences USA (2002) 99: 6035–6040.
47 Wu, Z., Nogales, E. Xing, J. Comparative studies of microtubule mechanics with two competing models suggest functional roles of alternative tubulin lateral interactions. Biophysical Journal (2012) 102: 2687–2696.
48 Wu, Z., Wang, H.-W., Mu, W. *et al.* Simulations of tubulin sheet polymers as possible structural intermediates in microtubule assembly. PLoS One (2009) 4: e7291.
49 Efremov, A., Grishchuk, E.L., McIntosh, J.R. Ataullakhanov, F.I. In search of an optimal ring to couple microtubule depolymerization to processive chromosome motions. Proceedings of the National Academy of Sciences USA (2007) 104:19017–19022.
50 Molodtsov, M.I., Ermakova, E.A., Shnol, E.E. *et al.* (2005) A molecular-mechanical model of the microtubule. Biophysical Journal 88: 3167–3179.
51 Asbury, C.L., Tien, J.F. Davis, T.N. Kinetochores' gripping feat: conformational wave or biased diffusion? Trends in Cell Biology (2011) 21:38–46.
52 Scott, S.W. Bromoviridae and allies. In: Encyclopedia of Life Sciences. John Wiley and Sons, Chichester; 2006.
53 Speir, J. A., Munshi, S., Wang, G. *et al.* Structure of the native and swollen forms of cowpea chlorotic mottle virus determined by X-ray crystallography and cryo-electron microscopy. Structure (1995) 3: 63–78.
54 de Pablo, P.J., Schaap, I.A., MacKintosh, F.C. Schmidt, C.F. Deformation and collapse of microtubules on the nanometer scale. Physical Review Letters (2003) 91: 098101–098104.
55 Schaap, I. A. T., Carrasco, C., de Pablo, P. J. *et al.* (2006) Elastic response, buckling, and instability of microtubules under radial indentation. Biophysical Journal 91: 1521–1531.
56 Wells, D.B. Aksimentiev, A. Mechanical properties of a complete microtubule revealed through molecular dynamics simulation. Biophysical Journal (2010) 99: 629–637.
57 Lowe, J., Li, H., Downing, K.H. Nogales, E. Refined structure of α-β-tubulin at 3.5 Å resolution. Journal of Molecular Biology (2001) 313: 1045–1057.

58 Landau, L.D. Lifshitz, E.M. Theory of Elasticity, 3rd edn, vol. 7 Theoretical Physics. Elsevier; 1986.
59 Crooks, G.E. Entropy production fluctuation theorem and the non-equilibrium work relation for free energy differences. Physical Review E. (1999) 60: 2721–2726.
60 Collin, D., Ritort, F., Jarzynski, C. *et al.* Bustamante, C. Verification of the Crooks fluctuation theorem and recovery of RNA folding free energies. Nature (2005) 437: 231–234.
61 Hoefling, M. Gottschalk, K.E. Barnase-Barstar: from first encounter to final complex. Journal of Structural Biology (2010) 171: 52–63.
62 Selzer, T. Schreiber, G. New insights into the mechanism of protein-protein association. Proteins (2001) 45: 190–198.
63 Michel, J.P., Ivanovska, I.L., Gibbons, M.M. *et al.* Nanoindentation studies of full and empty viral capsids and the effects of capsid protein mutations on elasticity and strength. Proceedings of the National Academy of Sciences USA (2006) 103: 6184–6189.
64 Zhmurov, A., Kononova. O., Litvinov. R.I. *et al.* Mechanical transition from α-helical coiled coils to β-sheets in fibrin(ogen). Journal of the American Chemical Society (2012) 134: 20396–402.
65 Molodtsov, M.I., Grishchuk, E.L., Efremov, A.K. *et al.* Force production by depolymerizing microtubules: A theoretical study. Proceedings of the National Academy of Sciences USA (2005) 102: 4353–4358.
66 Phelps, D.K., Speelman, B. Post, C.B. Theoretical studies of viral capsid proteins. Current Opinions in Structural Biology (2000) 10:170–173.

18

Modeling and Simulations in Nanoindentation

Yi Sun and Fanlin Zeng

[1] *Professor, Harbin Institute of Technology, People's Republic of China*

18.1 Introduction

Indentation is one of the principal techniques for the mechanical characterization of materials. The method monitors the penetration of an indenter into the material surface during the application and release of a load. Usually, this technique is also referred to as nanoindentation when penetration depths of a few microns are involved. Nanoindentation is a variety of indentation hardness tests applied to very small volumes and hardly brings harm to the samples tested. This property makes it very useful in characterizing some key mechanical properties of nano structural materials [1, 2]. The nanoindentation technique was first used in characterization the properties of metals [3, 4] and a lot of simulation results on metals or crystals has been reported [5–8]. The application of nanoindentaton to polymer nanocomposites has received increasing interest in recent years. The technique has been used to research the filler content, filler dispersion, as well as to the interfacial nanofiller-matrix adhesion [9–12]. Information on heterogeneities of the composite material, either across the thickness or along the surface arising as a consequence of changes in the matrix morphology or uneven distribution of the filler, can be readily detected by means of nanoindentation [13, 14]. However, the simulations of the nanoindentation on polymers have not often been reported.

Although more and more types of indenters, such as Vickers, Berkovich, Knoop, Rockwell, conical, spherical, flat-ended, and so on, and corresponding test methods have been developed to characterize different materials or different properties [15], the behaviors and the mechanism of deformations below the indenter tip is far from clear. Although there are some computer simulation researches in this field [16], as a matter of fact, it is very hard to find applicable theories to explain the simulation results because most current theoretical systems derived from the classic continuum mechanics are always no longer valid in a nano scale, where the model is usually described as discrete particles or atoms. In another aspect, present nanoscale mechanics theories are imperfect and far from meeting the requirements of common problems. Simulations aimed at specific problems are probably the most effective approaches to clarify these issues satisfactorily.

Applied Nanoindentation in Advanced Materials, First Edition. Edited by Atul Tiwari and Sridhar Natarajan.

18.2 Simulations of Nanoindention on Polymers

The nanoindentation simulations of polymers are always more complicated than that on crystals due to the amorphous molecular structures of polymers, which makes the structure analysis become difficult [17, 18]. In this section, we introduce some recent simulation work on the nanoindentation of polyethylene (PE) and its polyhedral oligomeric silsesquioxane (POSS) composites (POSS-PE) [19, 20]. The aim is focused on how to compute the hardness and Young's modulus of polymers under nanoindentation and how to understand the different mechanical behaviours of polymers with different molecular structures.

18.2.1 Models and Simulation Methods

We consider two kinds of polymers, the PE homopolymer and POSS-PE. The polymerization degree of PE is 5000, which means there are 498 CH_2 and 2 CH_3 units in one molecular chain. POSS is a unique class of materials that can be depicted by the formula (RSiO1.5)n (where n is an even number and R = H, Cl or a variety of organic groups). The POSS-PE molecules are built by incorporating 3 POSS into the PE ones, as shown in Figure 18.1. Then 32 PE and POSS-PE molecular chains were packaged into two cubic cells and formed the amorphous PE and POSS-PE models, respectively. The total atom number of PE model is 48064. The initial density is 0.904 g/cm^3 and the cell sizes $a = 69.5410$ Å, $b = 72.2955$ Å, $c = 84.5242$ Å; $\alpha = 86.5212$, $\beta = 88.8148$, $\gamma = 103.920$. The total atom number of POSS-PE is 56896 and the initial density is 0.97 g/cm^3, the cell sizes are $a = 85.9529$ Å, $b = 71.6145$ Å, $c = 91.2180$ Å; $\alpha = 99.5949$, $\beta = 84.9095$, $\gamma = 72.7355$. Three kinds of perfect rigid indenters are considered here. They are cube-cornered, cylindrical with spherical tip and cylindrical with flat tip ones. The cube-cornered indenter consists of 3469 atoms. The diameter of the spherical one is 30 Å and it contains 3889 atoms. The cylindrical one with a flat tip includes 4439 atoms and the diameter is 30 Å too. There are totally six nanoindentation models constructed here from the three kinds of indenters and the PE and POSS-PE cells respectively, which are shown in Figure 18.2(a) and (b). The periodical conditions are performed for each model.

During the simulations, the COMPASS [21] force field is employed to compute the interactions among different atoms. First, we optimized the energies of each model to achieve the energy-minimized structures. The atoms in the indenter and the bottom surface of the sample were then fixed. After that, the indenter was driven into the upper surface of all the samples step by step (0.25 Å at each step) and a full relaxation of the structures was followed at each step. When the strain loading gradually rise up

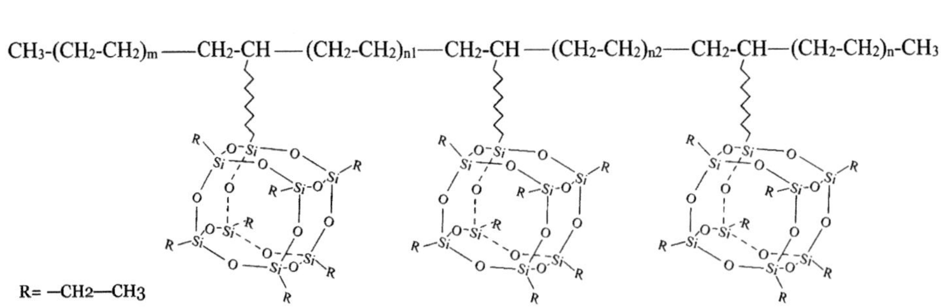

Figure 18.1 The structure of a POSS-PE molecule.

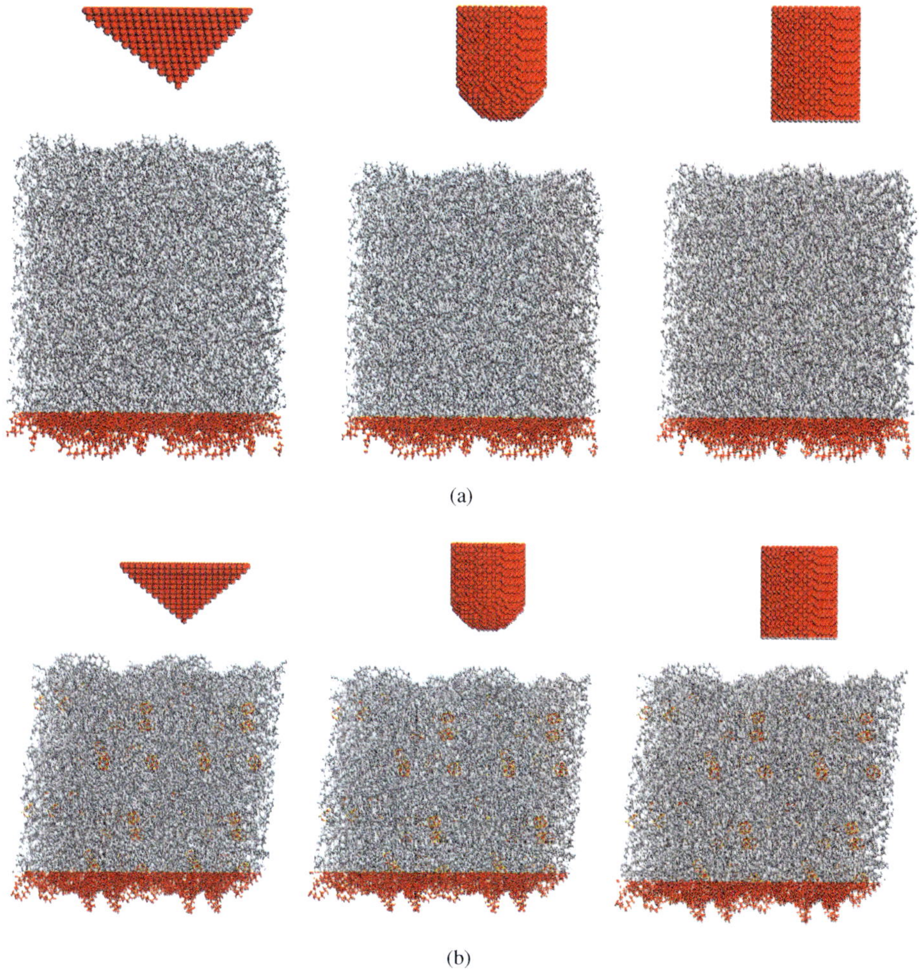

Figure 18.2 (a) The simulation models for the nanoindentation of PE with indenter tips of cube-corner, sphere and flat. (b) The simulation models for the nanoindentation of POSS-PE with indenter tips of cube-corner, sphere and column.

to 20 Å it was unloaded step by step. Then the atomistic motion trajectories beneath the indenters in loading and unloading procedures can be obtained. We consider the load-displacement relationship and the mechanical properties in each case.

18.2.2 Load-displacement Responses

From the simulations of the nanoindentation, the deformation processes of the two models are visible. With the aids of the indent loading and its displacement, the loading-depth curves of PE and POSS-PE under three different indenters are depicted in Figure 18.2a and Figure 18.2b, respectively. Figure 18.3 reveals the forces of the POSS-PE model are much greater than that of PE, which shows the deformation resistibility of POSS-PE model is stronger than that of PE. In other words, the resistance of PE is enhanced by incorporating with POSS. In addition, several loading

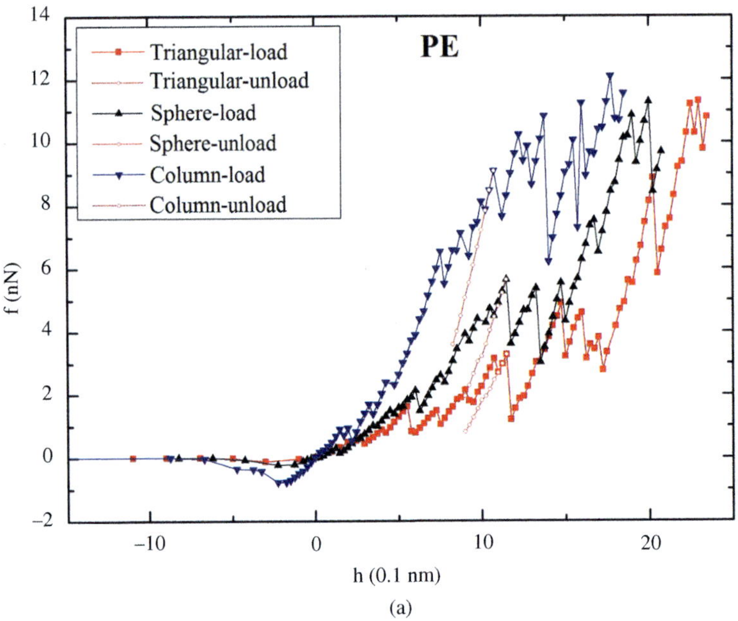

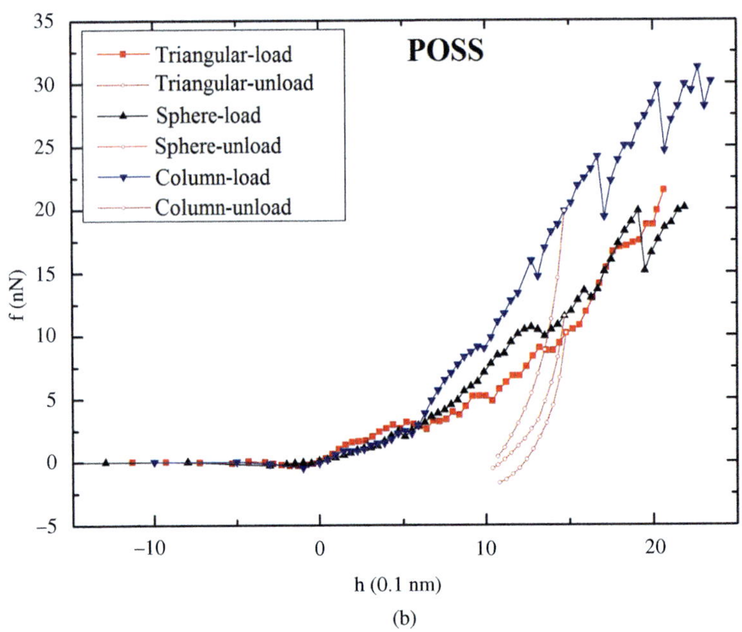

Figure 18.3 The loading and unloading load-displacement curves for (a) PE and (b) POSS-PE under different indenters.

sudden drops (or strain bursts) are observed clearly from the three loading curves in Figure 18.3a. For metal crystal, this phenomenon is believed that the main source is collective nucleation or dislocation avalanches [22]. However, the loading drop reduces obviously in Figure 18.3b, compared with Figure 18.3a. The reasons for this difference will be analyzed in detail later.

18.2.3 Hardness and Young's Modulus

Figure 18.3 shows that the contact force of indenter is nearly zero when the indenter is far from substrate. Then, the force changes from attraction to repulsive, as the distance becomes shorten. And in this process, there is a critical value (the force equal to zero), which corresponds to the initial zero position. Therefore, the contact depth is confirmed. Since the shape of the indenter is fixed, the contact area and the projected area could be calculated approximately through the contact depth. And the contact loading equals to the resultant force of the indenter. In general, two kinds of hardness are defined on the basis of the two areas (contact area A_C and projected area A_P). One is the projected hardness [23, 24], which can be computed from:

$$H = \frac{P}{A_p} \tag{18.1}$$

The other one is the Martens hardness [25], which can be similarly obtained by:

$$H = \frac{P}{A_c} \tag{18.2}$$

Where P is the loading.

Figure 18.4 shows the projected hardness of PE and POSS-PE under different indenters as computed from Equation (18.1). It can be found that only the hardness under the cube-corner indenter converges. While the other two hardness curves show a rising trend, and the values are obviously larger than the value obtained by cube-corner indenter. The reason is that the projected area of the flat indenter is a constant, and the sphere indenter tends to get a constant area. Thus their projected hardness keeps rising as the load increases.

Figure 18.5 shows the Martens hardness of PE and POSS-PE under different indenters as computed from Equation (18.2). First, all the curves of Martens hardness converge. Nevertheless, the projected hardness doesn't converge. In fact, all the curves of PE and POSS-PE under three indenters almost converge to the same values respectively. From this point, the Martens hardness maintains a good consistency when different indenters are used. Actually, the three indenter tips have almost the same contact area under a same penetration depth. Furthermore, we find the hardness of PE computed by this means is in good agreement with the value in reference [26]. This proves that our method and models in this simulation are quite reasonable. As a result, the Martens hardness is more proper to represent the hardness of polymers, such as PE and POSS-PE here. In addition, Figures 18.4 and 18.5 show a same trend that both the projected and the Martens hardness values of POSS-PE are much greater than those of PE for each indenter. This result exhibits clearly that incorporating POSS into PE will greatly improve the hardness of the latter. We obtained the Martens hardness values of PE and POSS-PE via fitting the

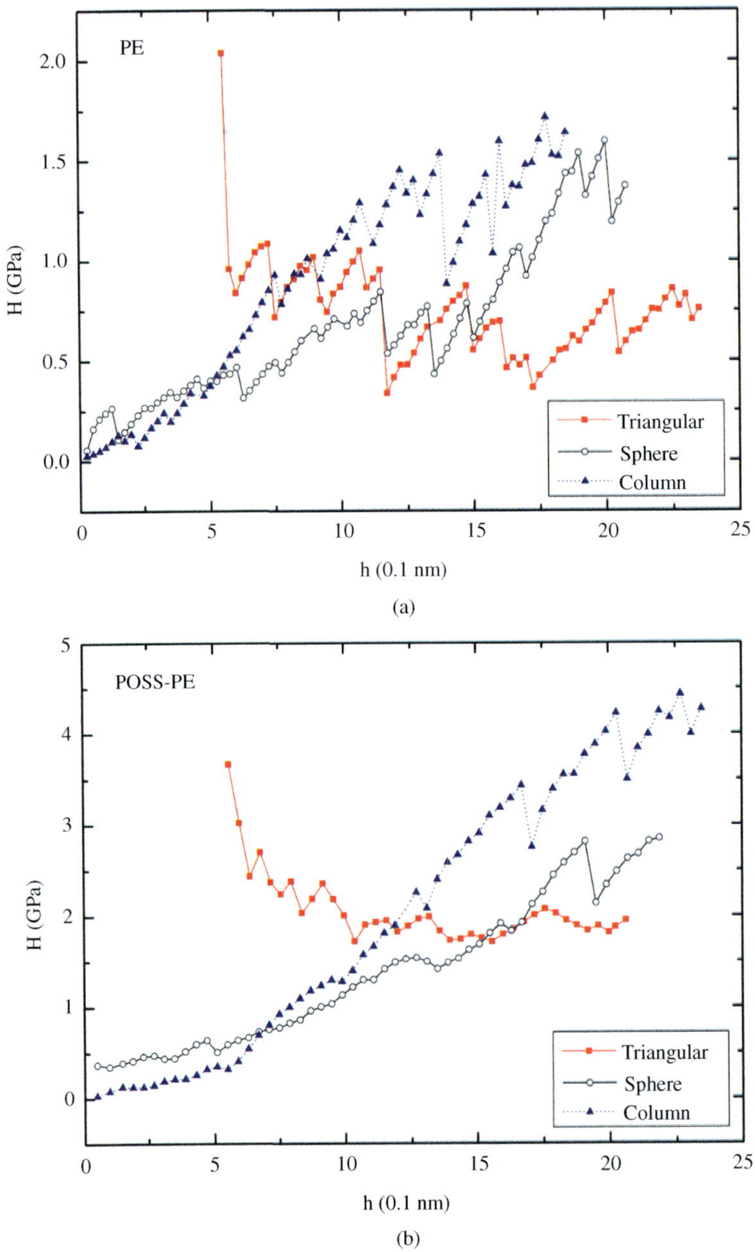

Figure 18.4 The projected hardness for (a) PE and (b) POSS-PE in different indentation simulations.

hardness-penetration depth curves in Figure 18.5 under different indenters and listed them in Table 18.1.

We consider the projected hardness here corresponds to the hardness measured by the depth without considering the pile-up effect in the Oliver and Pharr method [23], while the Martens hardness here is similar to the hardness by considering the pile-up

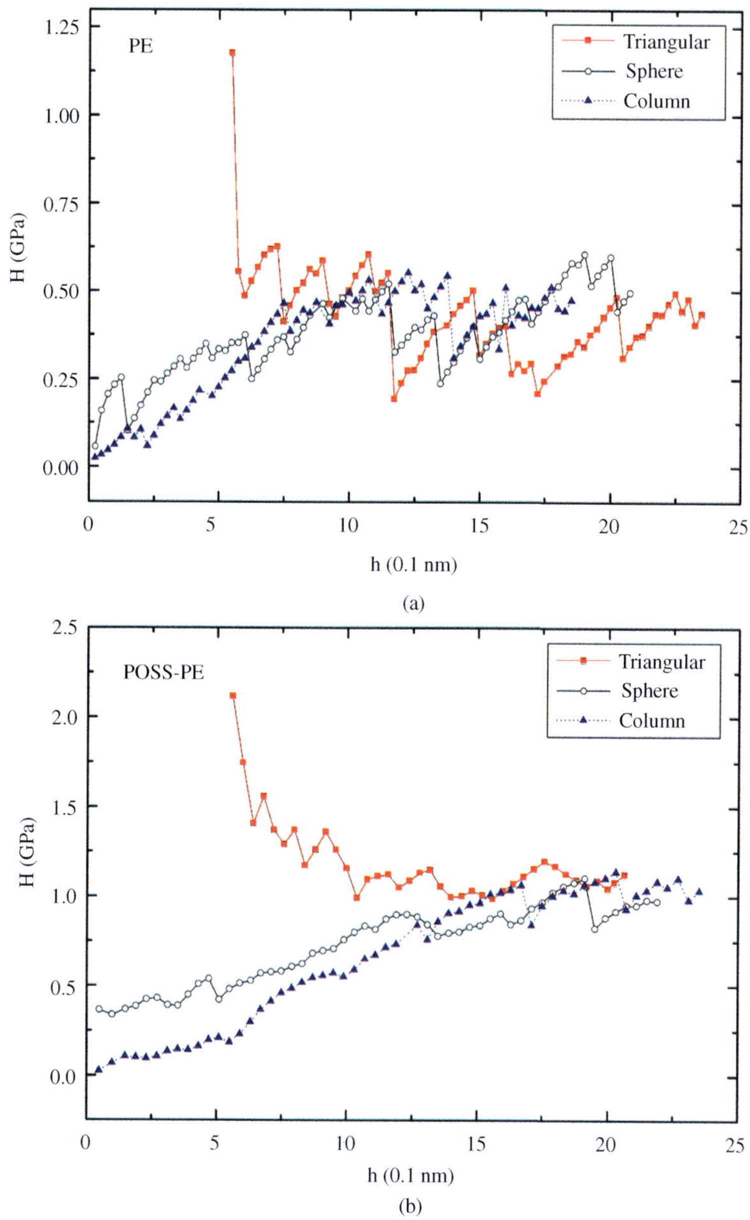

Figure 18.5 The Martens hardness for (a) PE and (b) POSS-PE in different indentation simulations.

correction. Since the contact area is not measured directly but computed from the indentation depth, how to compute the real depth is very important. In the Oliver and Pharr method [23], for an ideal sharp Berkovich indenter, the contact area A can be linked to the contact indentation depth h_c by an area function:

$$A(h_c) = C_0h_c^2 + C_1h_c + C_2h_c^{1/2} + C_3h_c^{1/4} + C_4h_c^{1/8} + \cdots \tag{18.3}$$

Table 18.1 The Martens hardness and Young's modulus of PE and POSS-PE.

	Martens hardness (GPa)			Young's modulus(GPa)		
Model	**Cube-corner**	**Sphere**	**Flat**	**Cube-corner**	**Sphere**	**Flat**
PE	0.42 ± 0.05	0.44 ± 0.09	0.45 ± 0.05	4.22	5.02	7.37
POSS-PE	1.08 ± 0.06	0.98 ± 0.08	1.02 ± 0.07	24.12	21.76	33.20

where C_0, C_1, C_2, C_3 and C_4 are fitting coefficients. According to Oliver and Pharr [23], the indentation contact depth writes:

$$h_c = h_{\max} - \varepsilon \frac{P_{\max}}{S} \tag{18.4}$$

where $h_{\max}$ is the current maximum indentation depth, S is the contact stiffness, corresponding to the slope of the load–displacement curve at the initial stages of unloading curve (will be discussed below), and ε is a geometric constant of 0.75 for Berkovich indenter [23]. It is generally known that the contact area A at a particular depth of indentation depends not only on the shape of the indenter but also on the elastic-plastic response of the material being indented. In some case, the pile-up of material around the indent tends to underestimate the actual contact depth and consequently underestimate the contact area A between the indenter and the sample, resulting in an overestimation of the hardness, sometimes by as much as 50% [23].

In our previous work on the indentation size effect of hardness of metallic glasses [24], we found that the size effect vanished if the pile-up effect had been corrected in the experiments. We use the pile-up height, Δh, to correct the $h_{\max}$ as $h'_{\max} = h_{\max} + \Delta h$, prior to calculating the contact indentation depth following the standard Oliver–Pharr scheme. We found the hardness values after the pile-up effect correction remain fairly constant for metallic glasses, regardless of the indentation depth [27]. We believe the situation is quite similar here. The projected hardness is computed just from the area corresponding to the loading depth, while the Martens hardness is deduced from the real contact area. That is equivalent to the pile-up correction in the latter case. Thus the Martens hardness values converge in each case. More importantly, they converge to almost the same value, which means the shape of an indenter is not a crucial factor when the hardness is measured by this means. Comparing Figures 18.4 and 18.5, we can also find that the initial parts of the hardness-depth curves in two figures are almost the same, this is because the pile-up is not oblivious in the initial stage of the indentation, and the correction effect can be neglected. This result confirms that, from another perspective, the consistency of the Martens hardness in different situations derives from the pile-up correction when we consider the real contact area.

The Young's modulus E of PE and POSS-PE in different simulations are computed from the unloading curves in Figure 18.3 by means of the Oliver and Pharr method [24]:

$$\frac{1}{E_r} = \frac{1 - \nu_1^2}{E} + \frac{1 - \nu_2^2}{E_i} \tag{18.5}$$

where E_r is the effective elastic modulus, which takes into account the fact that elastic displacements occur in both the specimen, with Young's modulus E and Poisson's ratio ν, and the indenter, with elastic constants E_i and v_i. For a diamond indenter, E_i is 1114 GPa and v_i is 0.07. E_r can be described as:

$$E_r = \frac{\sqrt{\pi}}{2\beta}\frac{S}{\sqrt{A}} \tag{18.6}$$

where β is the dimensional parameter related to the indenter shape, A is the projected area. S is the unloading stiffness which is given by:

$$S = \left(\frac{dP}{dh}\right)_{h=h_{max}} = Bm(h_{max} - h_f)^{m-1} \tag{18.7}$$

where P is the loading, h the indentation depth, h_{max} the maximum depth and h_f the final depth after the indenter is fully unloaded. B and m are power law fitting constants, which can be obtained by fitting the data of the upper portion of the unloading curve corresponding to the initial stage of unloading:

$$P = B(h - h_f)^m \tag{18.8}$$

We computed the Young's moduli of PE and POSS-PE for each indenter by this means and listed them in Table 18.1, too. It can be found that the Young's moduli of PE for different indenters are consistent with reference [28], which proves again that our simulations are reasonable.

It should be noted that both the Martens hardness and the Young's modulus of POSS-PE are much higher than those of PE in each case and the enhancement effect is quite similar to the result of POSS on polypropylene as reported in reference [29]. All these results indicate that the mechanical properties of PE have been improve obviously by POSS. In addition, owing to the different shapes of indenters, the influence on the Martens hardness is much less, while it is much greater on the Young's modulus. The reasons that why POSS brings this effect on the hardness and Young's modulus of PE will be discussed below.

18.2.4 The Mechanism of Mechanical Behaviours and Properties

In order to explain the mechanical behaviours of PVDF and POSS-PVDF in the nanoindentation simulations, as well the effects of POSS on the mechanical properties of PVDF, we investigated the deformations beneath the tip of the cube-corner indenter. Figure 18.6 shows this area and all the atoms in this hemisphere with radius r are considered. We computed the absolute displacements for each atom (the 'movement' of the atom) in this area and compared the average values of all these atoms (named 'average displacement') in PE and POSS-PE models. Figure 18.7 shows the average displacements of the two models at different radius. The results indicate that firstly the average deformation of POSS-PE is much less than that of PE, which means that the atomic movements in POSS-PE are much more difficult than that in PE when the same indentation loading is added; then the deformation decreases when r increases, which means that the atoms far from the indenter tip are less influenced than those near the tip. We also observed the stress distribution in the PE model at different depths (shown in Figure 18.8). It can be found that the stress near the indenter tip is greater, and the stress mainly changes near the tip. In general, the stress increases with increasing loadings. Figure 18.8b and 18.8c are two snapshots corresponding to the two states of

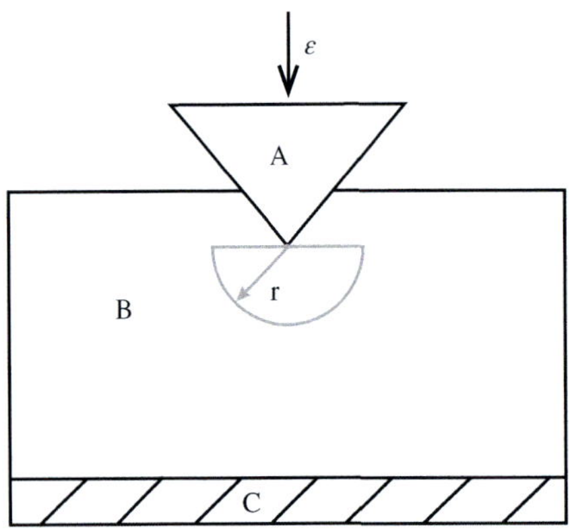

Figure 18.6 The hemisphere area beneath the indenter tip.

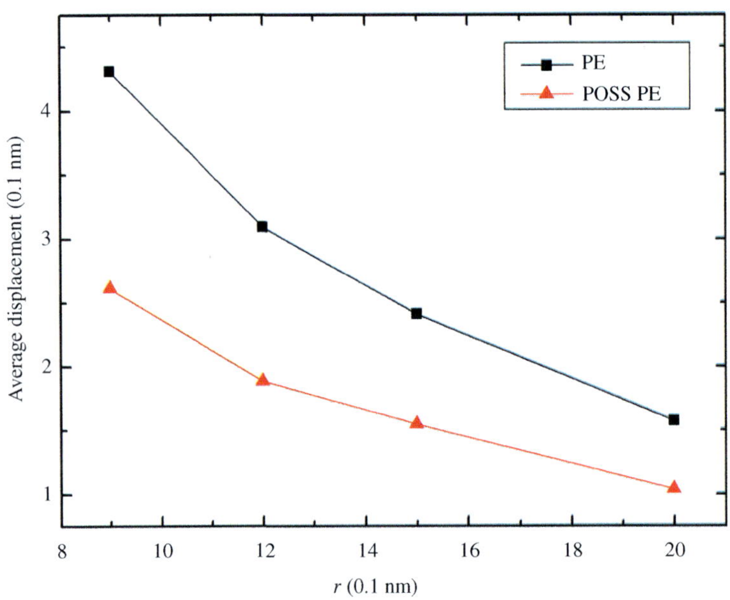

Figure 18.7 The average displacements for PE and POSS-PE at different hemisphere radius.

before and after the loading drop in the load-displacement curve of PE in Figure 18.3. They show that the stress concentration near the indenter tip has been greatly relaxed when the loading drops.

We found the loading drop is mainly derived from the slip among different molecular parts beneath the indenter [19, 20]. The degree of the drop greatly depends on how much the slip happens. We consider the slipping work and slipping energy to simply quantify how much the slip happens between molecular chains via simple simulations. We performed two simulations in which two single PE and POSS-PE molecular chains were drawn out from a regular PE cluster (with an external force F) respectively. When the slip does not happen, i.e., there is no any work done by friction force, all the work W_e done by the external force is consequently transformed into potential energy increment ΔU, Thus $W_e = \Delta U$; While when the slip happens, the friction force will generate the slipping work (negative value), then the potential energy increment is composed of both

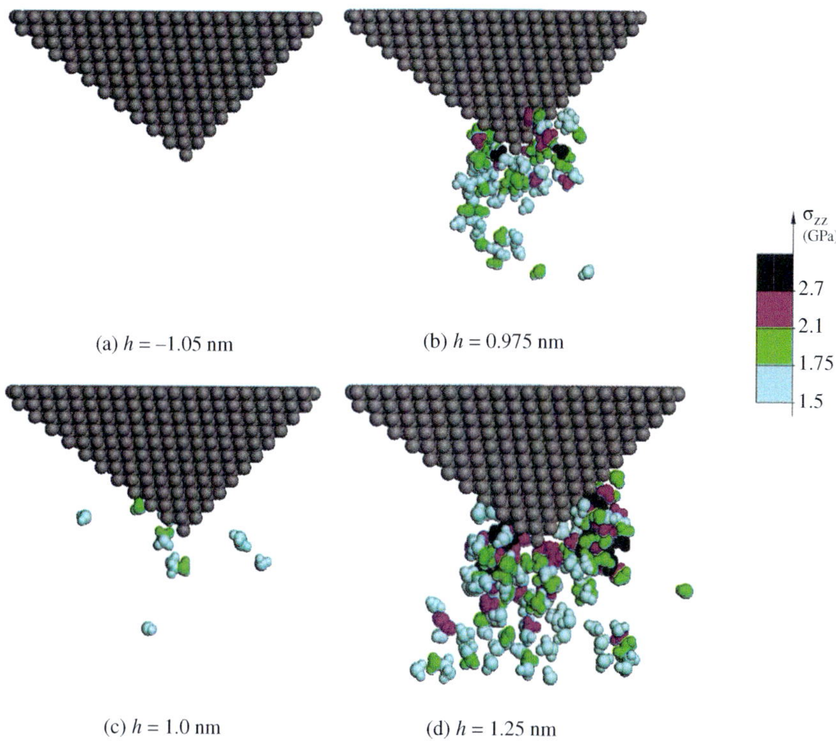

Figure 18.8 Stress distribution in PE at different loadings (only the atoms with stress over 1.5 GPa are displayed).

two works, the external work W_e and the slipping work W_s, thus $\Delta U = W_e + W_s$. ΔU and W_e can be evaluated directly from the simulations. Therefore, the slipping energy (W_s) can be calculated from the external work and the increment of the potential energy straightforwardly from $W_s = \Delta U\text{-}W_e$. Figure 18.9 shows the slipping energy and the indentation load for PE and POSS-PE at each step. It can be found clearly that each sudden drop in the slipping energy curve exactly corresponds to a indentation load sudden drop, and the absolute value of the change of the slipping energy is in amazing agreement with the degree of the loading drop. This relationship undoubtedly indicates that main reason responsible for the loading drop is the interchain slip between different molecular chain parts.

Although the slipping energy shows how much the interchain slip happens, it doesn't reflect the slipping resistance. As mentioned in reference [30], how difficultly the displacement bursts depends on the stacking fault energy (SFE), and higher SFE leads to easier bursts. To some extent, SFE is a kind of energy density. Thus, here we introduce the slipping energy density (SED) as the criterion, like SFE. We believe that the SED could reflect the resistance. Since deformation mainly occurs near the indenter tip, interchain slip should arise in local area, where the stress is concentrated. We distinguish the region approximately by judging whether the stresses of the atoms are one order of magnitude higher than others. To compare the resistance of slip, both slipping energy and the volume of the local (or the local atoms number) should be considered together. Hence the

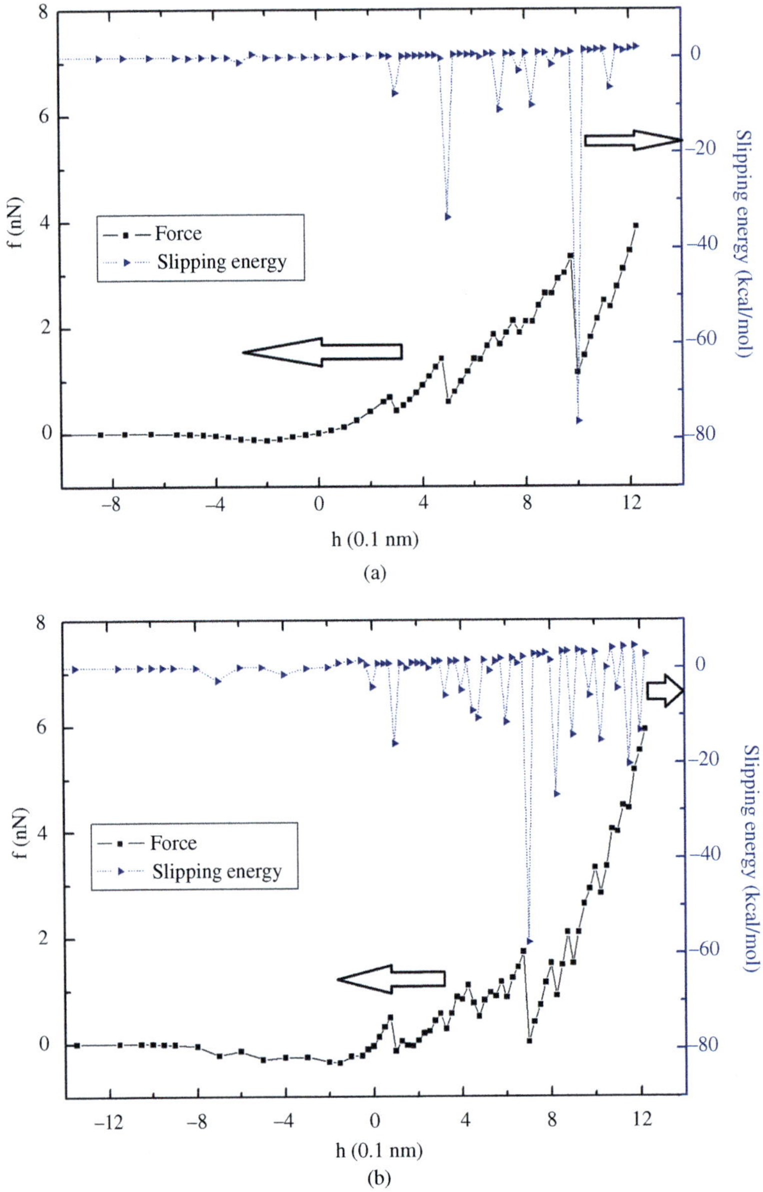

Figure 18.9 The slipping energy and the indention load for (a) PE and (b) POSS-PE under a cube-corner indenter at each step.

SED is derived via dividing the slipping energy by the local volume (or the local atoms number). The average slipping energy of the two models at different depths can be computed (as shown in Figure 18.10). In general, the variation degree of the average slipping energy of POSS-PE is obviously lower than that of PE, which means that the slipping resistance in POSS-PE is much stronger. In other words, the interchain slip in the PE

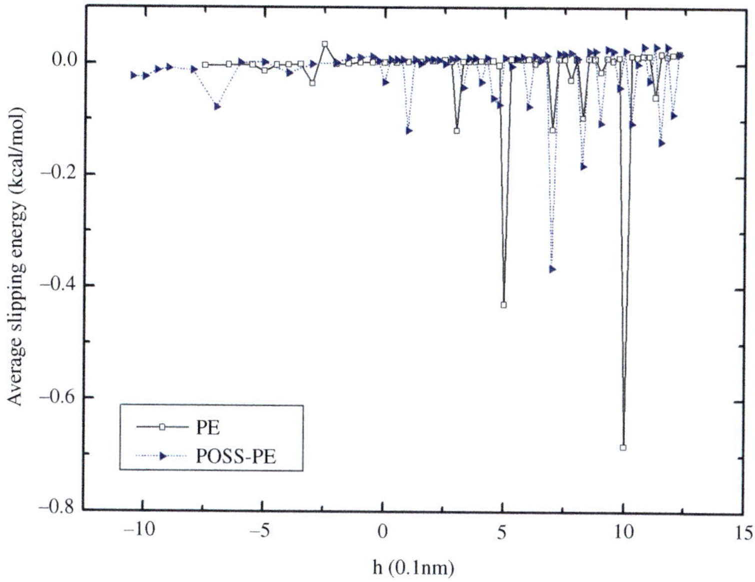

Figure 18.10 The average slipping energy for PE and POSS-PE in the cube-corner nanoindentation simulations.

system can be effectively suppressed by POSS. This is the main reason to explain the effects of POSS on the mechanical properties of PE.

From the above analysis, we can make a conclusion that the interchain slip should be responsible for the sudden drops in the load-displacement curves. POSS will obviously improve the interchain friction and enlarge the resistance force for interchain slip, which should be chiefly responsible for its influences on the mechanical behaviors of PE in the nanoindentation simulations.

18.3 Simulations of Nanoindention on Crystals

In theory, the nanoindentation simulations of crystals are easier to handle compared with polymers since the atomistic structures of crystals are always regular. However, just as mentioned in introduce section; it is always difficult to describe the atomistic problems properly in current theoretical frameworks. Fortunately, dislocation theories probably are appropriate approaches to explain some specific atomic simulation results, especially in some crystal materials, because both the models in these two situations are similarly non-continuous. In fact, some specific deformation mechanism of the face centered cubic (f.c.c.) metals have been analyzed by this means, for example, aluminum [31], nickel [32–34], gold [32, 35], copper [36–38] and so on. It appears from current investigations that different analytical methods were used in different cases. It is difficult to find a set of unified theory and technique with universal validity in the analyzing of different deformation behaviors and their micro mechanism in nanoindentation.

In this section, we introduce some recent quasi-continuum (QC) simulation work on the nanoindentation of nickel [33, 34]. We research the incipient plasticity deformations of nickel beneath different shapes of plane strain indenters. The main work is focused on the dislocation nucleation and emission. The analysis method is derived from the energetic criterion, the elastic contact theories and the Peierls-Nabarro (PN) dislocation model. And the aim is to clarify when and where the dislocation will be emitted under different shapes of indenters.

18.3.1 Models and Simulation Methods

The models of the nanoindentation in this work is shown in Figure 18.11. The flat-ended, tilted flat-ended (with 5° ~ 30° inclined angles, 5° interval) and wedged (with 120° ~ 170° vertex angles, 10° interval) indenters were driven into a thin nickel film (0.15 um thick and 0.3 um wide) resting on a substrate. The substrate was modeled as a rigid surface allowing no displacements, which only provided a displacement constraint along [−110] direction to the bottom surface of the film. The upper surface of the nickel film was kept free and the film was oriented so that the slip direction {−110} was parallel to the indentation direction. In the meantime, the indenters were placed as the edge was parallel to the slip plane (111) and were driven into the surface of (−110) to facilitate dislocation nucleation. The nanoindentation was carried out by moving the atoms on the interface as the shape of the imaginary indenters. We don't give the width of the indenters because we only plan to simulate a tilted punch with receding contact, which means the width is infinite thus the contact is singular at the corner of the punch but bounded at the other end, a point which lies on the punch face. A perfect-stick condition was applied to characterize the contact between indenters and the film surface. The simulation procedure was similarly conducted as follows:

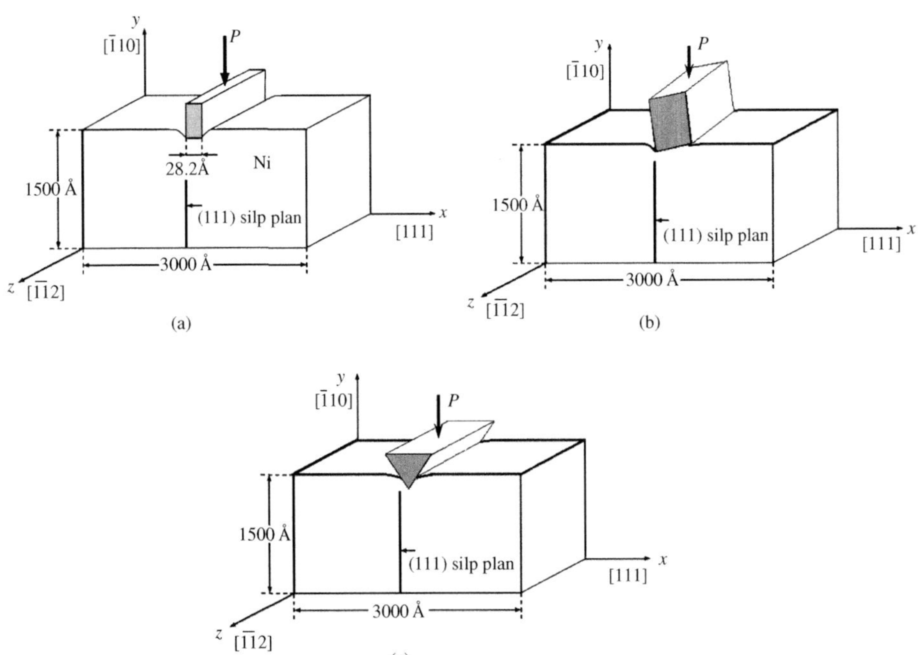

Figure 18.11 The models of the nanoindentation on nickel with (a) flat-ended, (b) tilted flat-ended, and (c) wedged indenters.

(1) indenters were pushed a small distance of 0.1 Å into the crystal at every load step; (2) atoms inside the indenters were relocated onto the interface and constrained to remain there, which means the perfect sticking conditions are used here; (3) atoms already in contact with the interface were moved down with the indenters; (4) the new configurations were optimized according to the energy minimum principle. We performed different loading steps to different models, depending on the convergence situations.

The QC method is an energy-based method which mixes the continuum mechanics and the atomistic mechanics [39–44]. In QC, any node or atom that is retained in the model is identified as a 'representative atom' or simply 'repatom'. These repatoms are distinguished by the terminology 'non-local repatom' and 'local repatom', the former represents the repatoms (all the atoms) in the atomistic region where the atomic bonding is non-local and the latter depicts the repatoms (all the nodes) in the continuum region where a local constitutive relations is applied. The idea is that the total energy of a fully atomistic system is computed from a good approximation. The total energy is a function of the positions of all the atoms in the body and can be written as a sum over the energy of each atom:

$$E^{\text{tot}}(\boldsymbol{x}) = \sum_{i=1}^{N} E_i(\boldsymbol{x}) = \sum_{i=1}^{N} E_i(\{\boldsymbol{x}_1, \boldsymbol{x}_2, \cdots, \boldsymbol{x}_N\}) \tag{18.9}$$

where $\boldsymbol{x}_i$ is the position of the ith atom in the deformed configuration, and E_i is the energy associated with the ith atom, which depends on the positions of the other atoms, and N is the number of total atoms in the solid. If the possibility of external forces applied to the atoms is considered, the total energy of the system can be written as:

$$\Phi(\boldsymbol{x}) = E^{\text{tot}}(\boldsymbol{x}) - \sum_{i=1}^{N} \boldsymbol{f}_i \boldsymbol{u}_i \tag{18.10}$$

where $\boldsymbol{f}_i\boldsymbol{u}_i$ is the work done by the external load $\boldsymbol{f}_i$ on atom i. $\boldsymbol{u}_i = \boldsymbol{x}_i$-$\boldsymbol{X}_i$, is the displacement of atom i, where $\boldsymbol{X}_i$ is the position of the ith atom in the reference configuration.

In order to reduce the number of degrees of freedom (DOF) in the system, an efficient approach is to constrain the motion of most of the atoms to follow the motion of the representative atoms. The density of repatoms varies according to the expected severity of the deformation. In some critical regions (as around the indenter in this article, also called the non-local region) all atoms are selected as repatoms, while very few are selected in regions that are more linear elastic (local regions). In the local regions, a finite element mesh with constant-strain triangles (in 2D) or constant-strain tetrahedra (in 3D) is defined by the repatoms, and the repatoms act as the finite element nodes. The displacements of all other atoms can then be determined from the displacement of these node/repatoms through interpolation. Then the potential energy of all the atoms can be described as a function of only the node/repatom displacement vector, $\boldsymbol{x}^R$:

$$E^{\text{tot}}(\mathbf{x}^R) = \sum_{i=1}^{N} E_i[\mathbf{x}(\mathbf{x}^R)] = \sum_{i=1}^{N} E_i(\mathbf{x}_1^R, \mathbf{x}_2^R, \cdots, \mathbf{x}_n^R) \tag{18.11}$$

Where n is the number of total repatoms in the solid, $n << N$. By this means, the number of DOF in the system is reduced effectively. However, this has not lowered the computational burden associated with computing the energy of the system because one still need to compute the energy of every atom. So it is necessary to divide the body into the atomistic and continuum sub-bodies, and the energy can then be rewritten as:

$$E^{tot}(\mathbf{x}^R) = \sum_{i \in R^A} E_i[\mathbf{x}(\mathbf{x}^R)] + \sum_{j \in R^C} E_j[\mathbf{x}(\mathbf{x}^R)] \tag{18.12}$$

where R^A, R^C represent the atomistic non-local region and the continuum local region, respectively. The first sum in this equation is computed exactly as the method in a fully atomistic model. The second sum, on the other hand, is replaced by finite elements and the Cauchy–Born constitutive rule. For a repatom in R^C, it is seemed as in a region of approximately uniform deformation, which can be characterized by the deformation gradient tensor $\mathbf{G}$, so that:

$$E_j \approx \Omega_0 W(\mathbf{G}) \tag{18.13}$$

where Ω_0 is the Wigner–Seitz volume of a single atom in the reference configuration, W is the strain energy density at the deformation gradient $\mathbf{G}$. Then the total energy can be rewritten by replacing the second sum over atoms in Equation (18.12) with a sum over elements as:

$$E^{tot}(\mathbf{x}^R) = \sum_{i \in R^A} E_i[\mathbf{x}(\mathbf{x}^R)] + \sum_{k \in R^C} \Omega_k W(\mathbf{G}^k) \tag{18.14}$$

where Ω_k is the volume of the element k and $\mathbf{G}^k$ is the deformation gradient there. Then the equilibrium configurations of the system can be identified by minimizing the total energy with respect to the repatom positions. The minimization is carried out by a quasi-Newton solver with a conjugate gradient backup when the initial guess is outside the basin of attraction of the Newton solver. The EAM potential with parameters from Voter [45] was used to compute the energy of the atomistic non-local regions here.

18.3.2 The Load-displacement Responses

Figure 18.12 shows the load-displacement curves for different simulation models. The loads are calculated in Newtons per meter length of the indenter in the out of-plane direction. Figure 18.12a shows that the curve is linear at the initial load stage (A to B). This implies that the crystal is elastic as predicted by elasticity theory. The linearity stops at the point C shows that the crystal is not elastic at this point and the inner structure has changed. After a slight fluctuation the curve reaches to point D, and at the DE part the curve is linear, too. At the F point the linearity stops again. The plasticity deformation has taken place at the point C. The hardness at the point B corresponds to the hardness of nickel. The corresponding displacement and load are 6.6 Å and 16.72 N/m respectively, and the computational hardness is 5.9 GPa. Figure 18.12b reveals that all curves are quite similar and linear at the initial load stage, especially when the depth is less than 1.2 Å. The whole variation trend for different curves, generally speaking, is that the loads grow with increasing depth, and the rate of increase drops with increased inclined angles. It seems that different curves are quite similar, that is, the linear portions of the curves terminate abruptly at some specific points, then linear stages appear again, and then another sudden drops follow. The sudden drops on the loading curves are closely related to the emission of dislocations from beneath the indenter. The dislocations nucleate in a single minimization step, going from incipient slip distributions at the indenter tips to fully edge dislocations beneath the surface. Similarly, Figure 18.12c indicates that all curves are similar and linear at the initial load stage (the depth is less than 1.8 Å). This is because the loading procedures are quite similar at the initial stage, in which the wedged indenters only bear on one column of atoms. After that, curves are quite different. Loads increase with increased indention depth, and the increasing rate, increases with increased vertex angles. It can be found also that the curves are not always linear, that is, loads vary

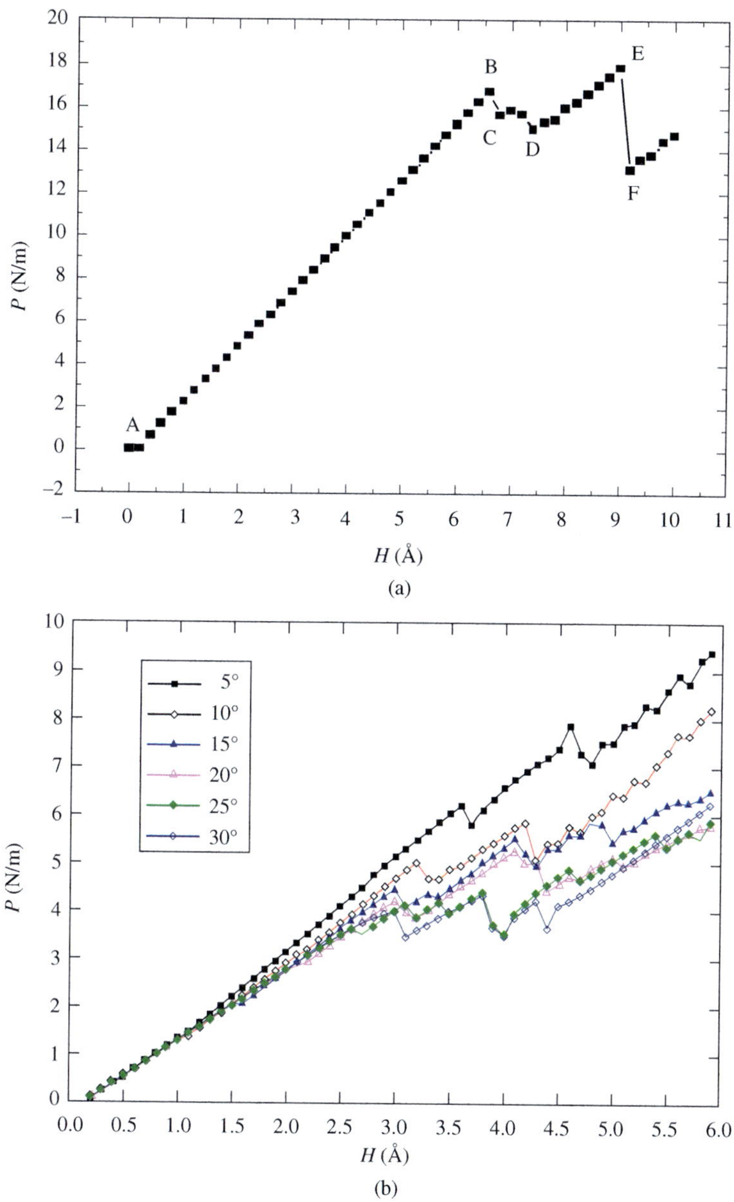

Figure 18.12 The load-displacement curves for the nanoindentation into the (−110) plane of nickel with different indenters. (a) flat-ended, (b) tilted flat-ended, and (c) wedged.

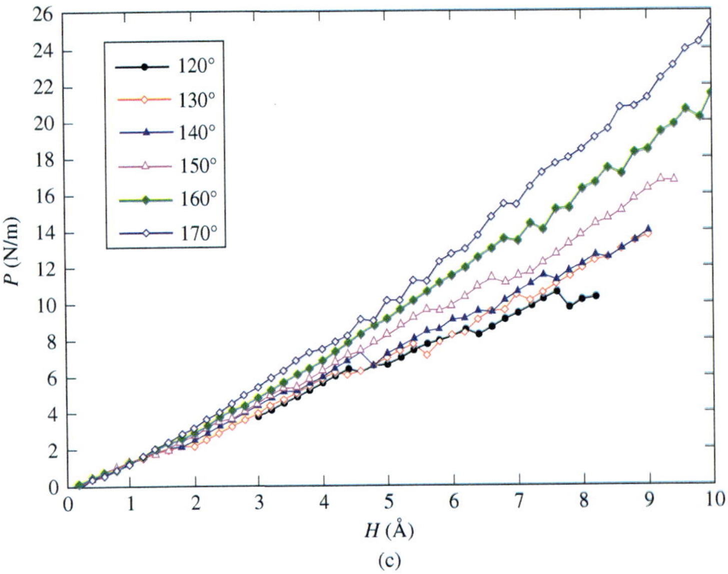

Figure 18.12 *(Continued)*

at some specific indentation depths. This is quite different in different curves. For the indenters with 160° and 170° vertex angles, load curves vary irregularly after the initial linear stage. For other curves, loads drop suddenly after initial linear stages, then linear stages appear again, and then other sudden drops follow. Different load variations, however, indicate different dislocation nucleation and emission situations inner nickel.

18.3.3 Dislocation Nucleation

The linear portion of the load-displacement curve in Figure 18.12a terminates abruptly at *C* (the 34th load step) with the emission of dislocations from beneath the indenter. The inner structure and the atom distribution of the nickel film beneath the indenter are given in Figure 18.13a. Two edge dislocations have been emitted at two sides of the local area beneath the indenter. During the simulation the symmetry was broken due to numerical noise. This is acceptable because it is expected that symmetry will also be broken in a real nanoindentation because of the imperfections of the specimen. The load drops suddenly at point *F* (the 46th load step), and the corresponding inner structure of the crystal is shown in Figure 18.12b and c. Eight edge dislocations (four pairs) have been emitted at this point. There are four at left side and another four at the right side. The asymmetry still exists. The four pairs of nucleated dislocations adopt a staggered configuration so that their extra half-planes line up.

Since there are too many results for the tilted flat-ended indentation simulations, it is not possible to list them all. We just select some typical results at three specific inclined angles of 5°, 15° and 30° and display them in Figure 18.14. The linear portion of the *P-H* curve of the inclined angle of 5° in Figure 18.12 terminates abruptly at depth of 3.7 Å (the 37th load step). The corresponding atomistic structure shows that one edge dislocation has been emitted beneath the sharp corner of the indenter. And this *P-H* curve drops again at the depth of 4.7 Å, Figure 18.14a proves that one more edge dislocation has been

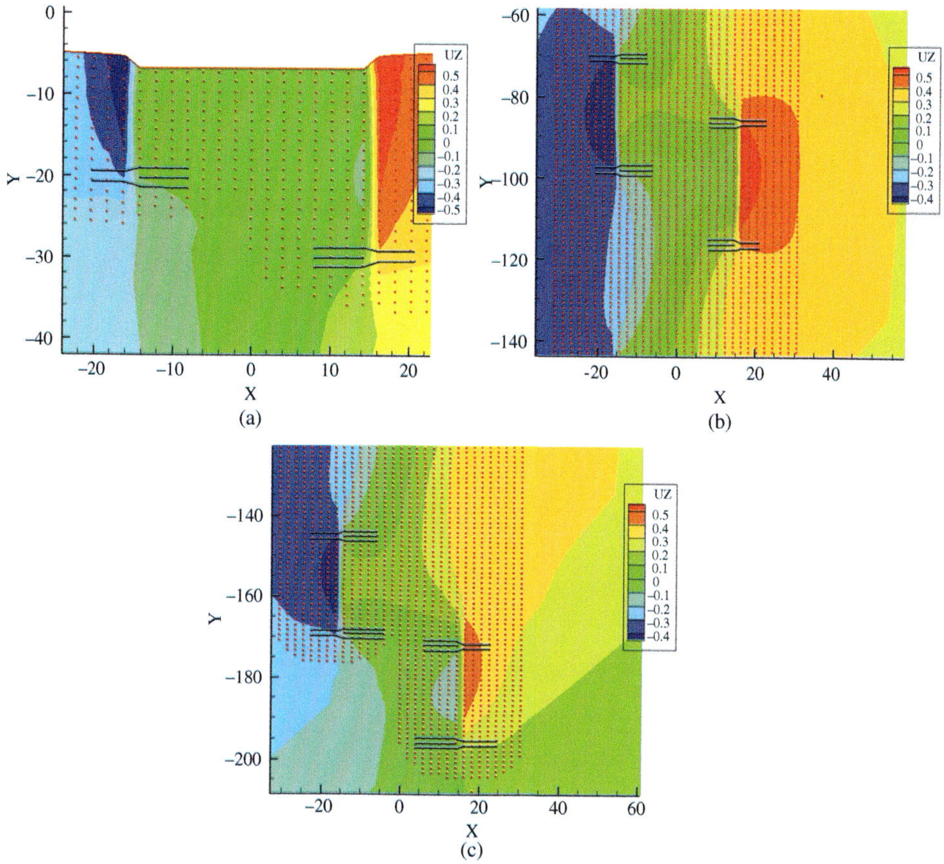

Figure 18.13 The inner structure of nickel beneath the flat-ended indenter at the load depth of (a) 6.8 Å, and (b)(c) 9.2 Å (the dotted lines locate the positions of slip planes, and dimensions and displacements in Å, the same hereafter).

emitted just beneath the former one. It is predictable that the dislocation emission will continue when the indentation depth increases. However, it is difficult to predict where the dislocation will be emitted, or in other words, we don't know if the dislocation will be emitted elsewhere besides beneath the sharp edge of the indenter because there is no indication that the slip plane will appear at other places. Figure 18.14b shows the dislocation emission situation at the depth of 5.0 Å with the inclined angle of 15°. One dislocation has been emitted similarly beneath the sharp edge of the indenter, and a new dislocation has been emitted beneath the lower surface of the indenter. This should be caused by the obvious slip among the atom columns under the lower surface of the indenter derived from a larger inclined angle, which certainly needs to be verified from more other results. Figure 18.14c presents the dislocation emissions of the indenter of 30° at the load depth of 4.4 Å. Edge dislocations have been emitted not only beneath the sharp edge of the indenter but also beneath the lower surface. And compared to the results of 15° there is one more dislocation which appears beneath the lower surface. Therefore, a bigger inclined angle is beneficial to emit dislocations beneath the indenter lower surface.

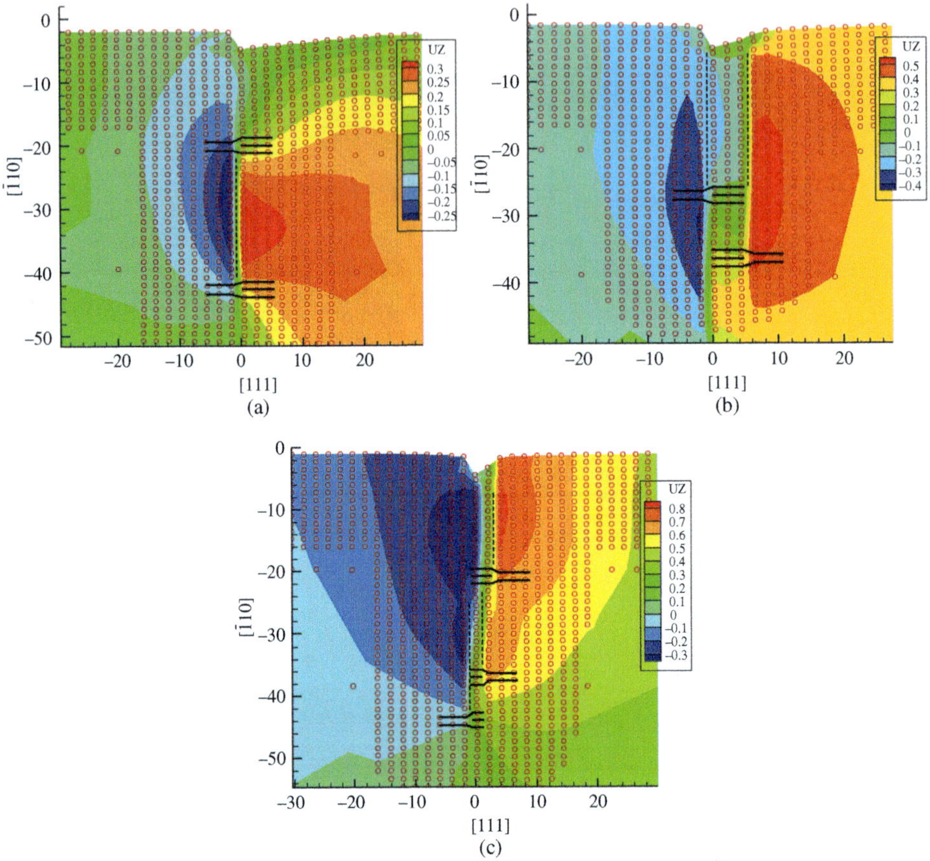

Figure 18.14 Dislocated structures beneath the tilted flat-ended indenter with the inclined angle of (a) 5° (at the load depth of 4.7 Å), (b) 15° (at the load depth of 5.0 Å), and (c) 30° (at the load depth of 4.4 Å).

Similarly to the case of the tilted flat-ended indenter, we only present the results at three vertex angles of 120°, 140° and 160° in the wedged indentation simulations. The linear portion of the load-displacement curve of the indenter of 120° in Figure 18.12c terminates abruptly at depth of 4.6 Å (the 23rd load step) with the emission of dislocations from the ridge of the indenter. Two edge dislocations have been emitted at two sides of the local area beneath the indenter. Then at depth of 6.4 Å, the curve drops again, indicates edge dislocations have been emitted once more beneath the indenter. It is predictable that the dislocation emission will be continued when the indentation depth increases. Figure 18.15a shows the corresponding emitted dislocations. Similarly, the *P-H* curve of the indenter of 140° in Figure 18.12c drops suddenly at depth of 5.0 Å, and an edge dislocation dipole (a pair of dislocations) is found beneath the shoulders of the indenter. When the depth reaches to 7.8 Å, another drop appears on the curve. And the corresponding atomistic structure shows the previous dislocation dipole has travelled into the bulk and a new pair of nucleated dislocations are found above the previous ones (Figure 18.15b). We are quite sure that upper dislocation dipole is the new pair because their slip planes are located at different positions. No linear stages appear

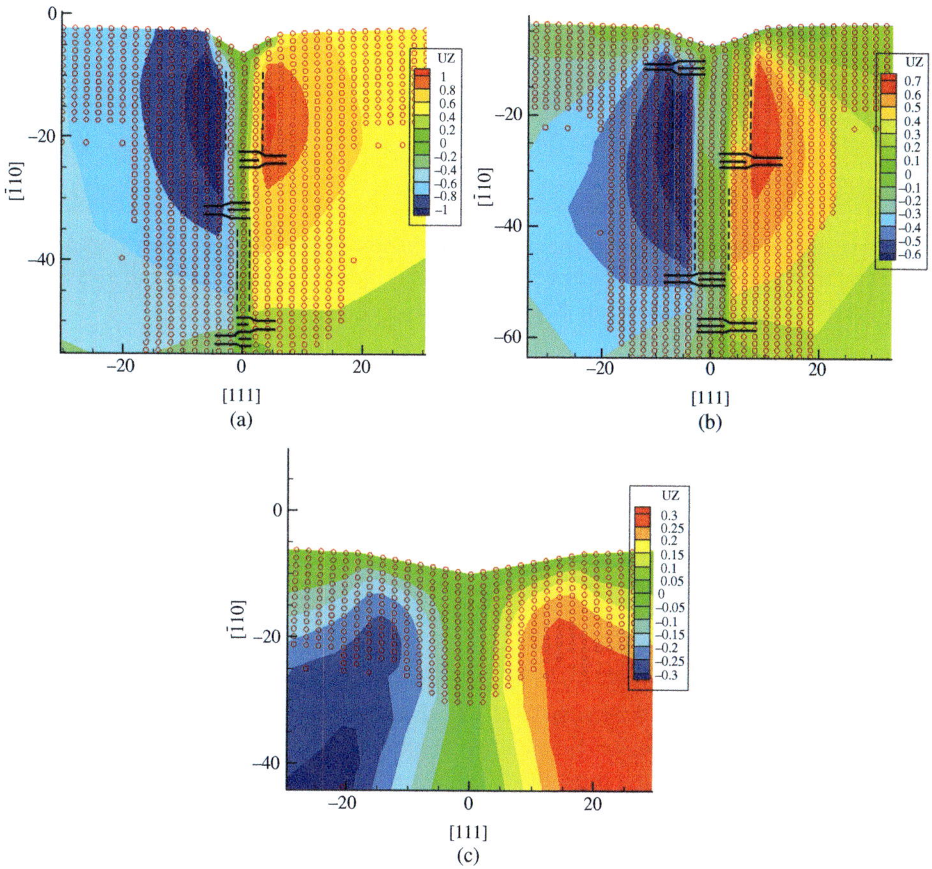

Figure 18.15 Atomistic structures beneath the wedged indenter with vertex angle of (a) 120° (at the load depth of 6.4 Å), (b) 140° (at the load depth of 7.8 Å) and (c) 160° (at the load depth of 10 Å).

after the variations on the *P-H* curves for indenters with 160° and 170° vertex angles, it is unlikely to find dislocations from the atomistic structures of them. Figure 18.15c shows the inner structures beneath the indenter with vertex angle of 160° at the maximum 50th load step. It proves that no any dislocations have been emitted even at a relative high depth. And the two structures are quite symmetric thus the accidental factors in the calculations can be overlooked.

18.3.4 Mechanism of Dislocation Emission

We use the Peierls-Nabarro (PN) dislocation model [32, 46] and the energy method to compute the critical loads at the dislocation emission and predict where and when the dislocations will be emitted. In the PN framework, a slip distribution is imagined to occur along certain anticipated slip planes that are chosen on the basis of the relative orientations of the crystal and the loading axes [32]. With increasing load, the incipient slip distribution becomes unstable and ultimately leads to a fully formed dislocation. For the indentation problems of interest here, we assume that slip occurs along the (111) planes at $x = a_0$. A periodic relation between shear stress and atomic shear displacement, derived using atomistic potentials is used in the calculations. Suppose the slip distance is $\delta(y)$, then the total energy of the system is:

$$U[\delta(y)] = U_0 + 2\int_0^{\infty} \Phi[\delta(y)]\mathrm{d}y + 2\int_0^{\infty} \tau_{xy}^{I}(a_0, y)\delta(y)\mathrm{d}y + 2\int_0^{\infty} \tau_{xy}^{S}(a_0, y)\delta(y)\mathrm{d}y \tag{18.15}$$

where U_0 is the elastic energy of the unslipped system; Φ is the interplanar potential of the crystal; τ^{I}_{xy} is the stress filed produced by the indenter; and τ^{S}_{xy} is the stress field produced by the slip distribution. Minimizing $U[\delta(y)]$, we get:

$$-\frac{\mathrm{d}\Phi}{\mathrm{d}\delta(y)} = \tau_{xy}^{I}(a_0, y) + \tau_{xy}^{S}(a_0, y) \tag{18.16}$$

It is obvious that $\tau^{S}_{xy} = 0$ at $y = 0$. At this time, the interplanar shear stress, $\tau_{xy}(\delta(0)) = \tau^{I}_{xy}(a_0, 0)$. If the shear stress between the slip planes reaches to the maximum shear stress $\tau_{\max}$, that is, when $\tau^{I}_{xy}(a_0, 0) > \tau_{\max}$, the incipient slip distribution becomes unstable, leading to the formation of a full dislocation. This is the key to determining the critical stress. The PN model predicts that dislocations are nucleated when the below inequality is satisfied [32]:

$$p(a_0)/\pi + q(a_0)/2 > \tau_{\max} \tag{18.17}$$

Figure 18.16 shows the elastic models of the three problems in this section. Where h is the indentation depth, a is the contact half-width and P is the load to hold the indenter at this depth. In Figure 18.16b, ε is the inclined angle of the indenter and M is the moment to keep the indenter rotating the angle ε. In Figure 18.16c, α is the half vertex angle of the indenter.

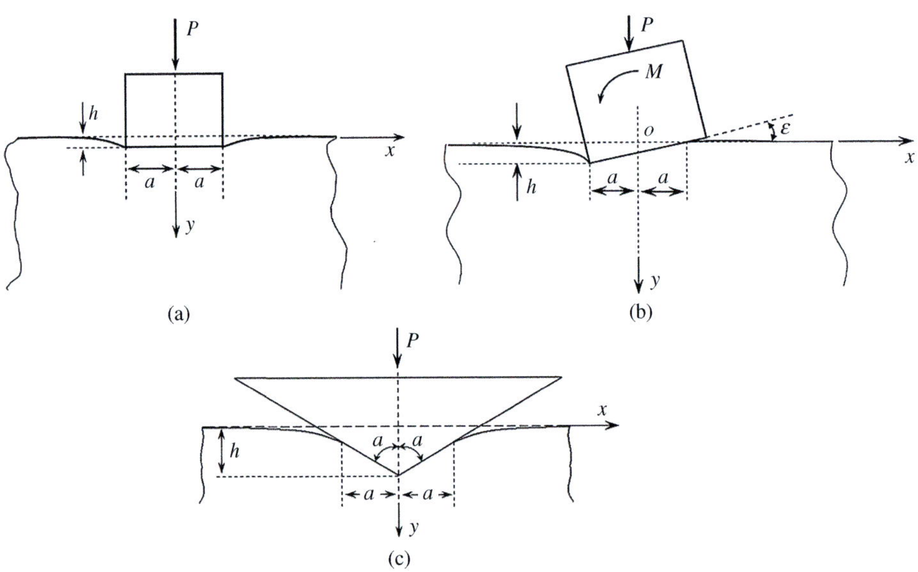

Figure 18.16 The elastic models of the three indentation simulations with (a) flat-ended, (b) tilted flat-ended, and (c) wedged indenters.

For the model of the flat-ended indentation, the $\tau^I{}_{xy}$ can be calculated by the normal and tangential forces at the boundaries from Equation (18.18) [47]:

$$\tau^I_{xy}(a,0) = p(a)/\pi b + q(a)/2b \tag{18.18}$$

To a friction-free indentation, $q(x) = 0$. The $p(x)$ can be determined by the eq. below [47] if it is unsingular at $x = a$:

$$p(x) = p_0/(\pi\sqrt{a^2 - x^2}) \tag{18.19}$$

where P_0 is the critical load added on the indenter when the dislocation is emitted. Using Equation (18.19) we cannot compute the load $p(a)$ at the two boundaries $x = \pm a$ directly. Of course the $p(x)$ cannot be obtained, either. But we can get all the $p(x)$ when $|x| < a$. Because the whole load is resisted by the atoms at the boundaries and elsewhere, the boundary load $p(a)$ can be obtained by subtracting the $p(x)$ ($|x| < a$) from the whole load. We get $p(a) = 0.62487P_0$. From the $p(a)$, the shear intensity $\tau_{max} = u/2\pi$ under the pure shear condition of nickel and Equation (18.18), we get $P_0 = 21.4$ N/m. Compare with the load of 16.72 N/m at the point B in the real simulation of Figure 18.12a, this value is higher. A conceivable reason is the neglecting the influence of the tangential forces. The tangential forces exist in the real simulation, and this can be proved by the different UZ under the indenter in Figure 18.13.

For the model of the tilted flat-ended indentation, the contact theory gives the normal pressure $p(x)$ on the elastic half space as [48]:

$$p(x) = \frac{4\mu\varepsilon}{(\kappa+1)\sqrt{a^2 - x^2}}(a - x) = \frac{4\mu\varepsilon}{\kappa+1}\sqrt{\frac{a-x}{a+x}} \qquad (-a < x \le a) \tag{18.20}$$

where $\kappa = 3–4\nu$ is the Kolosov's constant, and μ is the shear modulus. Although $p(x)$ is singular at $x = -a$, $x > -a$ is the region that we are interested in. Consider the inequality (18.17) in PN model, which provides a criterion to predict where a dislocation will be nucleated from the normal and tangential tractions of distributed loads. We only consider the perfect sticking conditions, the tangential traction of distributed loads, $q(x)$ is not zero, which can be obtained from the function $q(x)/p(x) = \cot\varepsilon$. Since a and x are both variable during the indentation process, it is convenient to treat a/x as a single variable. Then Equation (18.20) can be rewritten as:

$$p(x) = \frac{4\mu\varepsilon}{\kappa+1}\sqrt{\frac{1 - a/x}{1 + a/x}} \qquad (-a < x \le a) \tag{18.21}$$

Subsequently, we can obtain the final expression as follow by substituting $p(x)$ (Equation (18.21)) and $q(x)$ into inequality (18.17):

$$\frac{4\mu\varepsilon}{\kappa+1}\sqrt{\frac{1 - a/a_0}{1 + a/a_0}}\left(\frac{1}{\pi} + \frac{\cot\varepsilon}{2}\right) > \tau_{max} \tag{18.22}$$

If we set the left part of inequality (18.22) as τ_{xy}, then the values of τ_{xy} varied with a/a_0 at different inclined angles ε can be obtained. Figure 18.17 shows the shear stresses at different inclined angles as determined from the left part of inequality (18.22). Taking $\tau_{max} = \mu/2\pi = 12.1$ GPa as the reference value, it is obvious that dislocations can only be nucleated in some specific areas, or rather the region above the dotted line of τ_{max} in

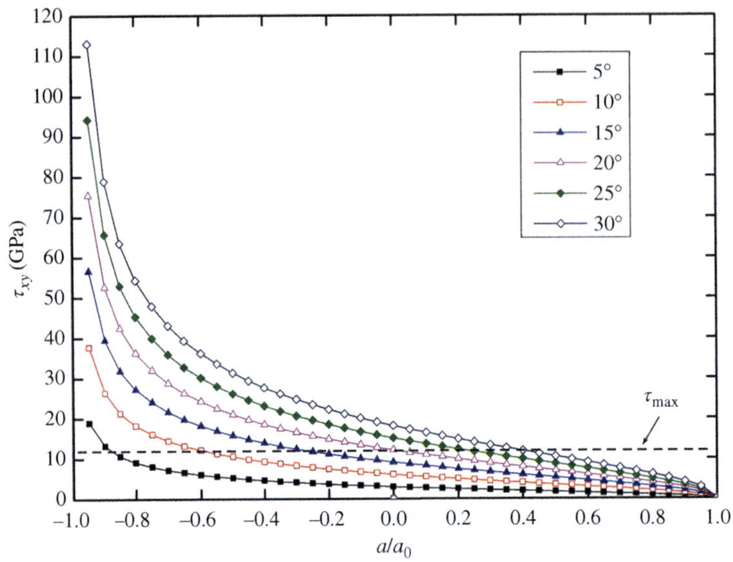

Figure 18.17 The shear stresses at different inclined angles as determined from a/a_0 in the tilted flat-ended model. The dotted line locates the position of τ_{max}.

Figure 18.17. The dislocation emissions in simulations as analyzed in section 1.3.2 can be understood easily. Theoretically, each indenter generates the infinite shear stress at $a_0 = -a$ $(a/a_0 = -1)$, thus dislocations are undoubtedly emitted beneath the edge of the indenters in each case. To the indenters with inclined angles of 5° and 10°, τ_{xy} can only obtain the values above τ_{max} when a/a_0 is very close to −1, which implies dislocations are unlikely nucleated elsewhere unless a_0 is very close to the corner of the indenter. Taking into account that a and a_0 are both discrete (integral multiple of the lattice constant in (111) direction), dislocations have been found only beneath the sharp edge of the indenters. On the contrary, to the indenters with inclined angles of 15° to 30°, τ_{xy} can reach the values above τ_{max} in a large range of a/a_0 (close to or even higher than 0), which means that dislocations could be emitted beneath the middle or the right part of the indenters. Thus if the width of the indentation in these cases contains two or more complete lattices in simulations, it is highly possible that dislocations will be emitted elsewhere. Figure 18.17 also shows the threshold values of a/a_0 to emit dislocations beneath the lower surfaces of different indenters. Only when a/a_0 is less than some specific values can dislocations probably be emitted in different cases. This method provides the analytical approaches to predict when and where the dislocations will be emitted beneath the lower surface of an inclined punch.

For the model of the wedged indentation, the distributed loads on the indenter surface $p(x)$ (which is a function of coordinate x) can be described from the Hertz contact theory as [47]:

$$p(x) = \frac{E^* \cot\alpha}{2\pi}\left(\frac{2a}{\sqrt{a^2-x^2}} - \ln\frac{a+\sqrt{a^2-x^2}}{a-\sqrt{a^2-x^2}}\right) + \frac{P}{\pi\sqrt{a^2-x^2}} \tag{18.23}$$

Where E^* is the reduced Young's modulus when two kinds of materials contact, which is defined as [47]:

$$\frac{1}{E^*} = \frac{1-\nu_1^2}{E_1} + \frac{1-\nu_2^2}{E_2} \tag{18.24}$$

where E_1 and ν_1 are the Young's modulus and Poisson's ratio for the test material, and E_2 and ν_2 are the elastic modulus and Poisson's ratio, respectively, of the indenter. For the indenter used here, since it is perfect rigid so $E_2 \to \infty$. E_1, the Young's modulus of nickel along the [−110] orientation, is not reported specifically. But the Young's modulus values of nickel material obtained from nanoindentation experiments as reported in reference literature are in the range of 164 ~ 218 GPa [36], so we think E_1 should fall within this range. As an anisotropic material, the Poisson's ratio of nickel along [−110] orientation is not specified either. Here the value of 0.31 is adopted for ν_1 as in our previous work [33]. It may seem counterintuitive that one must know the very accurate Poisson's ratio of the material in order to compute its modulus, even a rough estimate. But in fact when $\nu_1 = 0.25 \pm 0.1$, produces only about a 5% uncertainty in the calculated value of the final result [49].

Since $p(x)$ must be zero when $|x| > a$, and taking it into account that $P = aE^* \cot\alpha$, the distributed loads are:

$$p(x) = \frac{E^* \cot\alpha}{2\pi} \ln \frac{a+\sqrt{a^2-x^2}}{a-\sqrt{a^2-x^2}} = \frac{E^* \cot\alpha}{\pi} \operatorname{arcosh}\left(\frac{a}{x}\right) \tag{18.25}$$

Although $p(x)$ is singular at $x = 0$ since the slope of the contact profile is not continuous here, the principal shear stress on x-y plane τ_{xy} is finite and can be computed from $p(x)$ and the final form is:

$$\tau_{xy} = (E^* a/\pi)\cot\alpha/\sqrt{a^2+y^2} \tag{18.26}$$

It gets a finite maximum value $(\tau_{xy})_{\max} = E * \cot\alpha/\pi$ beneath the ridge of the indenter. When $\tau_{xy} > \tau_{\max}$, which is the maximum shear stress the crystal can bear, the incipient slip distribution becomes unstable, leading to the nucleation of a full dislocation. From the shear intensity $\tau_{\max} = u/2\pi$ under the pure shear condition of nickel (where u is the shear modulus, 76 GPa used here) and E^* from Equation (18.24), we obtained the range of the critical vertex angle is 156° ~ 162°, which means when the vertex angle is less than this range then the maximum shear stress can make dislocations be emitted. Since it is relatively softer along [−110] orientation for nickel and the Young's modulus should be lower than the maximum value from experiments (218 GPa), the critical vertex angle range should be a little less than 162°. This result is in surprisingly close agreement with our simulation ones, where dislocations have been emitted at vertex angle of 150° and have not in cases of 160° and 170°.

Moreover, we can find from section 18.3.2 that dislocations have been emitted sometimes beneath the ridge of the indenter and sometimes beneath the shoulders when the indentation depth reaches to some specific values. It seems that the dislocation emission can take place anywhere and there is no any rule to follow. This is quite different from the cases of a rectangular indenter [31–33] (where dislocations are emitted at $x = \pm a$) and a cylindrical indenter [31] (where dislocations are expected to be emitted at $x = \pm\sqrt{3}a/2$). Taking it into consideration that the loads to hold the indenter are perpendicular to the indenter's sloping sides and the perfect sticking conditions used here, they can be resolved into two components, the normal force $p(x)$ and the tangential force $q(x)$. It is quite simple to get $q(x)/p(x) = \cot\alpha$. Consider the PN inequality (18.17), which can be rewritten as:

$$\frac{E^* \cot\alpha}{\pi} \operatorname{arcosh}\left(\frac{a}{a_0}\right)\left(\frac{1}{\pi} + \frac{\cot\alpha}{2}\right) > \tau_{max} \tag{18.27}$$

If we set the left part of the above inequality as τ_{xy}, then the values of τ_{xy} varied with a/a_0 at different half vertex angles α can be obtained easily and showed as Figure 18.18. Taking $\tau_{max} = u/2\pi = 12.1$ GPa as reference value, it is obvious that the half vertex angle of 80° (the vertex angle of 160°) is a critical value. When the vertex angle is higher than 160° it is very hard to get a τ_{xy} higher than τ_{max}, even at a high value of a/a_0. Taking it into account that a/a_0 is not continuous in a crystal; dislocations cannot be emitted even at a relative high indentation depth when the vertex angle is higher than 160°. This result is in good accordance with both the previous analysis in this section and the simulation results in section 18.3.2. When the half vertex angle is less than 80° it is very easy to get a τ_{xy} higher than τ_{max}, which means dislocations will be emitted if a/a_0 reaches to some specific values in these cases.

The threshold values of a/a_0 obtained from inequality (18.25) and the simulated ones are listed in Table 18.2. It can be found that the analytical model we proposed achieves quantitative agreement with our atomistic results. Almost all the values of a/a_0 in real simulations are just over the corresponding threshold values. From this analytical model we know that dislocations are probably not emitted anywhere beneath a wedged indenter, only at the positions where a_0 is small enough compared with a that dislocation emissions are likely to take place. This result also shows that a dislocation is unlikely to appear very close to the boundary of the local region since the value of a/a_0 there is very

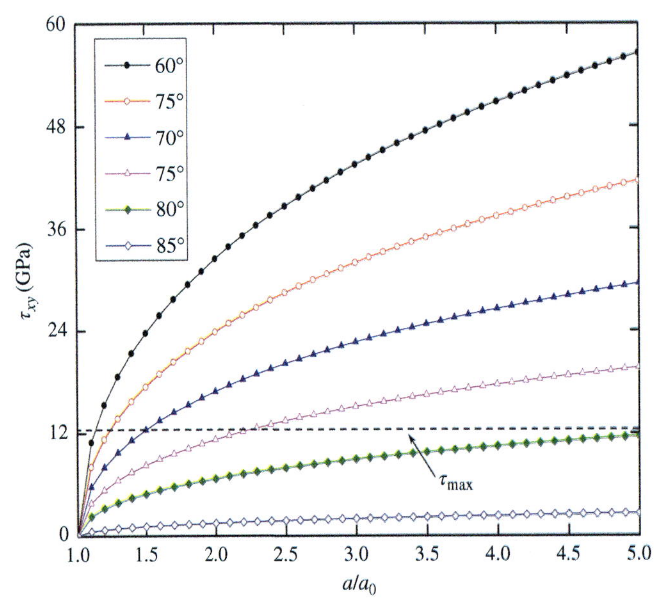

Figure 18.18 The shear stresses at different half vertex angles as determined from a/a_0 in the wedged model. The dotted line locates the position of τ_{max}.

Table 18.2 The threshold and real values of a/a_0 obtained from inequality (18.8) and simulations for dislocation emission at different vertex angles.

Vertex angle	Dislocations emitted?	Values of a when dislocations are emitted	Values of a_0 at dislocation emissions	Real values of a/a_0	Threshold values of a/a_0
120°	Yes	$2b$, $3b$	$0.5b$, $1.5b$	4, 2	1.2
130°	Yes	$2b$, $3b$	$1.5b$	1.33, 2	1.3
140°	Yes	$3b$, $5b$	$1.5b$, $2.5b$	2	1.5
150°	Yes	$5b$, $6b$	$2.5b$, $3.5b$	2, 2.4, 1.71	2.2
160°	No	—	—	—	5.6
170°	No	—	—	—	—

Note: '–' represents this value is not existing or involved in this work and b is the lattice constant.

possibly close to 1 and below the threshold. Although nearly all real simulation results are accordance with the analytical model, an exception is still found in case of 150°, in which some real values of a/a_0 are a little below the threshold. We think this is probably derived from the breakup of the structural symmetry during the simulations.

18.4 Conclusions

The authors have introduced some recent modeling and simulation work of the nanoindentation on some polymers and nickel. The results are quite different in two cases. For the polymers, our interest is focused on how to compute the common mechanical properties (hardness and Young's modulus) and how the organic-inorganic nano filler POSS influences the mechanical properties and behaviors in the nanoindentation. We found both the projected and the Martens hardness are able to describe the hardness of polymers. However the shape of the indenter has a big impact on the projected hardness, but only a little on the Martens hardness. From this perspective, the Martens hardness more properly represents the hardness of polymers. POSS-PE shows an undoubted enhanced effect on the hardness and Young's modulus of PE. A detailed analysis on the atomic structures and movements beneath the cube-corner indenter reveals that the main reason causing the loading drop is the interchain slip between different molecular chain parts. POSS can improve the interchain friction and enlarge the resistance force for interchain slip and leads to the enhanced influences on the mechanical behaviors of PE in the nanoindentation simulations. For the crystal of nickel, we performed three kinds of indentation, and the main aim is to find a connection between the simulation results and present theoretical knowledge. There are also several loading drops in the *P-H* curves of nickel, which are chiefly caused by the dislocation emission. In the case of a flat-ended indenter, the dislocations are always emitted at the positions of $x = \pm 2a$, and the critical load for the first dislocation emission can be predicted from the PN dislocation model; In the case of a tilted flat-ended indenter, not the indentation depth but the ratio of the contact half-width to the position of the slip plane, a/a_0, is the key parameter to determine whether dislocations will be emitted under the lower surface of the indenter. Dislocations could only be emitted beneath the edge when the inclined angle is small.

While in cases of greater inclined angles, the threshold values of a/a_0 growing to 0 or even higher predicts that dislocations probably be nucleated and emitted beneath the middle or the right part of the indenters. The values of a/a_0 also imply that dislocations will never be emitted beneath the right boundary of the contact in any case; similarly in the case of the wedged indenter, there are different threshold values of a/a_0 for indenters with different vertex angles to determine whether the dislocations will be emitted. Dislocations are not able to be emitted anywhere beneath a wedged indenter, only when the real value of a/a_0 is over the corresponding threshold value that dislocations can be emitted. We hope the work presented in this section will be beneficial to those who want to perform some simulations in nanoindentation, both on polymers and crystals.

Acknowledgments

The authors would like to thank the National Natural Science Foundation of China (11272108, 11102053) and the China Scholarship Council (CSC) for the financial support of this research.

References

1 Poon B, Rittel D, Ravichandran G. An analysis of nanoindentation in linearly elastic solids. International Journal of Solids and Structures, 2008, 45(24): 6018–6033.
2 Oliver WC, Pharr GM. Measurement of hardness and elastic modulus by instrumented indentation: Advances in understanding and refinements to methodology. Journal of Materials Research, 2004, 19: 3–20.
3 Pethica JB. Microhardness tests with penetration depths less than ion implanted layer thickness. In: Ashworth V, Grant W, Proctor R, editors. Ion implantation into metals. Oxford: Pergamon Press, pp. 147–157; 1982.
4 Fischer-Cripps AC. Nanoindentation, 2nd edn. New York: Springer; 2004.
5 Jeng Y, Tan C. Investigation into the nanoindentation size effect using static atomistic simulations. Applied Physics Letters, 2006, 89(25): 251901.
6 Fang T, Wu H. Molecular dynamics simulations on nanoindentation mechanisms of multilayered films. Computational Materials Science, 2008, 43(4): 785–790.
7 Jin J, Shevlin S, Guo Z. Multiscale simulation of onset plasticity during nanoindentation of Al (001) surface. Acta Materialia, 2008, 56(16): 4358–4368.
8 Mordehai D, Kazakevich M, Srolovitz DJ, Rabkin E. Nanoindentation size effect in single-crystal nanoparticles and thin films: A comparative experimental and simulation study. Acta Materialia, 2010, 59(6): 2309–2321.
9 Liu T, Phang IY, Shen L, Chow SY, Zhang WD. Morphology and mechanical properties of multiwalled carbon nanotubes reinforced nylon-6 composites. Macromolecules, 2004, 37: 7214–22.
10 Das B, Prasad KE, Ramamurty U, Rao CNR. Nano-indentation studies on polymer matrix composites reinforced by few layer graphene. Nanotechnology, 2009, 20: 125705–125710.
11 Flores A, Naffakh M, Díez-Pascual AM, Ania F, Gómez-Fatou MA. Evaluating the reinforcement of inorganic fullerene-like nanoparticles in thermoplastic matrices by depth-sensing indentation. Journal of Physical Chemistry C, 2013, 117: 20936–20943.

12 Zeng F, Liu Y, Sun Y, Hu E, Zhou Y. Nanoindentation, nanoscratch, and nanotensile testing of poly(vinylidene fluoride)-polyhedral oligomeric silsesquioxane nanocomposites. Journal of Polymer Science Part B: Polymer Physics, 2012, 50(23): 1597–1611.
13 Shen L, Phang IY, Chen L, Liu T, Zeng K. Nanoindentation and morphological studies on nylon-66 nanocomposites. Polymer, 2004, 45: 3341–3349.
14 Dutta AK, Penumadu D, Files B. Nanoindentation testing for evaluating modulus and hardness of single-walled carbon nanotube–reinforced epoxy composites. Journal of Materials Research, 2004, 19: 158–164.
15 Fischer-Cripps AC. Review of analysis and interpretation of nanoindentation test data. Surface and Coatings Technology, 2006, 200(14-15): 4153–4165.
16 Gouldstone A, van Vliet K J, Suresh S. Nanoindentation Simulation of defect nucleation in a crystal. Nature, 2001, 411(6838): 656.
17 Yiapanis G, Henry DJ, Evans E, Yarovsky I. Simulations of nanoindentation of polymer surfaces: Effects of surface cross-linking on adhesion and hardness. Journal of Physical Chemistry C, 2010, 114(1): 478–486.
18 Díez-Pascual AM, Gómez-Fatou MA. Aniab F, Flores A. Nanoindentation in polymer nanocomposites. Progress in Materials Science, 2015, 67: 1–94.
19 Hu EL., Sun Y, Zeng FL, Qu JM. Nanoindentation simulation of PE/POSS under different shapes of indenters. Acta Mechanica Solida Sinica, 2011, 24(4): 365–372.
20 Hu EL, Sun Y, Zeng FL. The enhancement mechanism and deformation analysis of polyethylene incorporated with POSS by nanoindentation simulation. Reviews on Advanced Materials Science, 2013, 33: 85–91.
21 Sun H. COMPASS: An ab initio force-field optimized for condensed-phase applications-overview with details on alkane and benzene compound. Journal of Physics Chemistry B, 1998, 102: 7338–7364.
22 Csikor FF, Motz C, Weygand D, *et al.* Dislocation avalanches, strain bursts, and the problem of plastic forming at the micrometer scale. Science, 2007, 312(5848): 251–254.
23 Oliver WC, Pharr GM. Measurement of hardness and elastic modulus by instrumented indentation: Advances in understanding and refinements to methodology. Journal of Materials Research, 2004, 19(1): 3–20.
24 Oliver WC, Pharr GM. An improved technique for determining hardness and elastic-modulus using load and displacement sensing indentation experiments. Journal of Materials Research, 1992, 7(6): 1564–1583.
25 Shahdad SA, McCabe JF *et al.* Hardness measured with traditional Vickers and Martens hardness methods. Dental Materials, 2007, 23: 1079–1085.
26 Yashiro K, Furuta A, Tomita Y. Nanoindentation on crystal/amorphous polyethylene: Molecular dynamics study. Computational Materials Science, 2006, 38(1): 136–143.
27 Huang YJ, Shen J, Sun Y, Sun JF. Indentation size effect of hardness of metallic glasses. Materials and Design, 2010, 31: 1563–1566.
28 Bischel MS, Vanlandingham MR *et al.* On the use of nanoscale indentation with the AFM in the identification of phases in blends of linear low density polyethylene and high density polyethylene. Journal of Materials Science, 2000, 35(1): 221–228.
29 Misra R, Fu BX, Morgan S E. Surface energetics, dispersion, and nanotribomechanical behavior of POSS/PP hybrid nanocomposites. Journal of Polymer Science Part B-Polymer Physics. 2007, 45(17): 2441–2455.

30 Shibutani Y, Tsuru T, Koyama A. Nanoplastic deformation of nanoindentation: Crystallographic dependence of displacement bursts. Acta Materialia, 2007, 55(5): 1813–1822.
31 Tadmor EB, Ortiz M, Phillips R. Nanoindentation and incipient plasticity. Journal of Materials Research, 1999, 14(6): 2234–2250.
32 Shenoy VB, Phillips R, Tadmor EB. Nucleation of dislocations beneath a plane strain indenter. Journal of the Mechanics and Physics of Solids, 2000, 48: 649–673.
33 Zeng FL, Sun Y. Quasicontinuum simulation of nanoindentation of nickel film. Acta Mechanica Solida Sinica, 2006, 19(4): 283–288.
34 Zeng FL, Sun Y, Liu YZ, Zhou Y. Multiscale simulations of wedged nanoindentation on nickel. Computational Materials Science, 2012, 62: 47–54.
35 Li JH, Zhao X, Wang SQ, Zhang CB. Multiscale simulation of the dislocation emissions of single ni crystal in nanoindentation. Journal of Wuhan University of Technology: Materials Science edition, 2010, 25: 423–428.
36 Wang CT, Jian SR, Jang JSC *et al.* Multiscale simulation of nanoindentation on Ni (100) thin film. Applied Surface Science, 2008, 255(5): 3240–3250.
37 Ju SP, Wang CT, Chien CH, *et al.* The nanoindentation responses of nickel surfaces with different crystal orientations. Molecular Simulation, 2007, 33(11): 905–917.
38 Dupont V, Sansoz F. Molecular dynamics study of crystal plasticity during nanoindentation in Ni nanowires. Journal of Materials Research, 2009, 24(3): 948–956.
39 Tadmor EB, Ortiz M, Phillips R. Quasicontinuum analysis of defects in solids. Philosophical Magazine, 1996, A73: 1529–1563.
40 Tadmor EB, Ortiz M, Phillips R. Mixed atomistic and continuum models of deformation in solids. Langmuir, 1996, 12: 4529–4534.
41 Shenoy VB, Miller R, Tadmor EB, *et al.* Quasicontinuum models of interfacial structure and deformation. Physical Review Letters, 1998, 80: 742–745.
42 Shenoy VB, Miller R, Tadmor EB, *et al.* An adaptive finite element approach to atomic-scale mechanics: The quasicontinuum method. Journal of the Mechanics and Physics of Solids, 1999, 47: 611–642.
43 Miller R, Tadmor EB, Phillips R, *et al.* Quasicontinuum simulation of fracture at the atomic scale. Modelling and Simulation in Materials Science and Engineering, 1998, 6:607–638.
44 Miller RE, Tadmor EB. A unified framework and performance benchmark of fourteen multiscale atomistic/continuum coupling methods. Modelling and Simulation in Materials Science and Engineering, 2009, 17: 053001.
45 Voter AF, Chen SP. Accurate interatomic potentials for Ni, Al and Ni3Al. Materials Research Society. Symposium Proceedings, 1987, 82: 175–180.
46 Rice JR. Dislocation nucleation from a crack tip: an analysis based on the Peierls concept. Journal of the Mechanics and Physics of Solids, 1992, 40 (2): 239–271.
47 Johnson K L. Contact Mechanics. Cambridge University Press, Cambridge; 1985.
48 Sackfield A, Truman CE, Hills DA. The tilted punch under normal and shear load (with application to fretting tests). International Journal of Mechanical Sciences, 2001, 43(8): 1881–1892.
49 Hay JL, Pharr GM. Instrumented Indentation Testing. Materials Park, OH, USA: ASM International, 2000, 232–243.

19

Nanoindentation of Advanced Ceramics: Applications to ZrO_2 Materials

Joan Josep Roa Rovira[1,2], *Emilio Jiménez Piqué*[1,2] *and Marc J. Anglada Gomila*[1,2]

[1] *Universitat Politècnica de Catalunya, CIEFMA, Campus Diagonal Besòs - Edif. DBI, Av. d'Eduard Maristany, 10–14, 08019 Barcelona, Spain*
[2] *Universitat Politècnica de Catalunya, Research Center in Multiscale Science and Engineering, Campus Diagonal Besòs - Edif. DBC, Av. d'Eduard Maristany, 10–14, 08019 Barcelona, Spain*

19.1 Introduction

Advanced ceramic materials are being increasingly used in many engineering fields, where their excellent properties, either mechanical [1–3], thermal, tribological [4–8], chemical or biological, are desired.

However, ceramic materials present a low fracture toughness that may translate into low mechanical reliability [9–17]. In order to overcome the risk of catastrophic failure, the use of ceramic materials is often limited to small parts the components, in which they can fulfil their function without affecting the piece reliability, or to components that work under compression loads. This is especially true for those applications in which the required mechanical properties are related to the surface, such as the tribological or biomedical ones.

In this sense, there is an increasing use of ceramics as reinforcements in composites, or in coatings, where they can strongly affect the mechanical behaviour [17–29]. In this case, the dimensions involved are normally small, making mechanical testing by standard procedures cumbersome.

Testing material specimens with dimensions in the micron and sub-micron range also provides basic understanding of the deformation and fracture mechanisms, and quantifies micromechanical parameters which can be used for micromechanics-based constitutive models and simulations [27–30]. Also, the knowledge of the mechanical response is also significant to optimize the design, fabrication, and material selection for different devices.

During the last 20 years, several techniques have become popular for measuring the mechanical integrity of very small volumes. Nanoindentation is probably the most frequently used (see for example review papers [31–33] and books [34–37]), due to the relative easiness in testing. In fact, nanoindentation has the ability to induce high stresses locally, of the order of magnitude of the material hardness, in very small volumes confined by the hydrostatic pressure caused by the same surrounding material. Thus, nanoindentation allows deforming plastically ceramic materials at room temperature, while classic mechanical testing only leads to fracture [38, 39]. Furthermore, it

Applied Nanoindentation in Advanced Materials, First Edition. Edited by Atul Tiwari and Sridhar Natarajan.

can be applied to measure the hardness and the stiffness of inclusions, coatings and the reinforcements of composite materials among others systems [29, 40–44].

The present chapter seeks to highlight a brief overview of the main points for nanoindentation of ceramics, which are hard, stiff and brittle as well as to give some basic guidelines to ensure proper interpretation of the results. For this purpose we will first briefly present the mechanics of nanoindentation, the main contact modes, and some issues in relation to the surface of the material investigated. We will then present a case study on zirconia under different conditions, in order to show how the technique is applied to ceramic materials.

19.2 Indentation Mechanics

In indentation tests (macro- and/or microindentation), a hard tip whose mechanical properties are well known (frequently made of a hard material like diamond) is pressed on the surface of a sample whose properties are unknown. The load on the specimen is increased as the tip penetrates into the specimen until reaching user-defined values. At this point, the applied load should be held constant for a short period of time in order to accommodate the plastic field induced during the indentation process and other time-dependant processes. Then, the area of the residual imprint is measured and the hardness is directly calculated.

When loads are large, the projected area may be measured directly by means of optical microscopy. However, when imprints are small (on the micron range) the contact area is difficult to measure directly. To overcome this difficulty, the instrumented indentation technique was developed. This technique, also known as nanoindentation because of the high resolution, measures both load (P) and penetration (h) continuously during indentation. From the load-displacement (P-h) curve obtained during loading and unloading, and knowing the exact shape of the indenter, the hardness (H) and the elastic modulus (E) of the material can be extracted. Nanoindentation is a powerful tool which measures mechanical properties such as contact pressure versus deformation, yield strength, fracture resistance, creep, viscoelatic and viscoplastic properties. In this sense, it can be seen as a complete micromechanical toolbox instrument, and not just as a low load indenter.

Nanoindentation presents several advantages over standard and conventional methods when it comes to the quantification of the mechanical integrity of advanced ceramics:

- measurements are local in the micron range
- mechanical properties are depth-sensing so it allows understanding of the mechanical evolution of a material at different penetration depths
- observation of the residual imprints by optical techniques is not required.

However, it has to be taken into account that the stress fields under indentation are complex, and many deformation mechanisms may be active at the same time, complicating the understanding and modelling of the mechanical behaviour, compared to the simpler case of remote tension.

19.2.1 Deformation Mechanics

The main deformation modes of materials subjected to nanoindentation takes place via two different regimes: elastic and elasto/plastic. Since elastic deformation is a reversible process, it exhibits virtually no size dependence unless a large population of pre-existing defects are involved [45]. The plastic deformation response, which takes place as a result of the generation, annihilation, and motion of defects such as dislocations, displays marked size effects when those defect dimensions are in the range of the indentation size. It has to be taken into account that ceramic materials also display dislocation-driven plasticity under nanoindentation, even at room temperature, because the hydrostatic field under the nanoindentation precludes the propagation of cracks under mode I, allowing the attainment of stresses high enough to move dislocations. Other deformation mechanisms may be also present, such as twinning, microcracking, grain boundary sliding or densification (especially in glass).

Since the early work done by Hertz on the contact elastic field, considerable research efforts have been devoted to study the mechanics of contact at the macro-, micro- and nanometric scale [46]. For instance, Sneddon [47] derived a theoretical framework for the contact response between an elastic solid and axisymmetric punches, i.e., disc flat-ended punch, sphere, cone, etc, while the characterization of the elasto/plastic response was undertaken by Johnson [46], providing important guidelines to the analysis of indentation experiments in the elasto/plastic field as we will comment in the following sections.

19.2.2 Elastic Contact

The elastic contact can be studied by means of the Hertz model, which can also be used to study the elasto/plastic transition [48].

It is necessary to highlight that in order to investigate the elastic field the dimension of the contact area must be small compared with the dimensions of the specimen, and with the relative radii of curvature of the surfaces. Moreover, the surfaces just outside the contact region should roughly approximate to a half-space plane surface, and the strains in the contact region should be sufficiently small to lie within the scope of the linear theory of elasticity.

By using an elastic indentation test, valuable information on elastic properties of materials can be obtained, especially when these materials present a poor structural integrity and are susceptible to damage by a sharp indenter, as it make take place in foams or porous materials [49]. In addition, an approximate value of the yield stress can be also obtained by measuring the pressure for initiation of inelastic deformation.

The procedure involved in the analysis of the elastic curves can be described in the following steps:

- **Plot load-displacement curve.** Plot the initial loading part of the curves, which can be satisfactorily fitted with the Hertz equation [33, 50, 51]. Deviations from Hertz curve are attributed to the elasto-plastic transition.
- **Determine the contact point.** Several methods for the correct determination of the contact point have been widely discussed in the literature [52–56]. However, the method most widely used was proposed by Moseson [55] following the Sneddon analysis [50]. In this step the contact radius (a) can be determined using the following equation:

$$a = \sqrt{2Rh_c - h_c^2} \tag{19.1}$$

where R is the tip radius and h_c is the penetration contact depth.

- **Indentation stress-strain curve.** Plot the indentation stress-strain curve by using Equation 19.2, where the left-hand side of this equation is the Meyer's hardness, represented as indentation stress and the term in parenthesis, on the right-hand side, stands for indentation strain [50]:

$$\frac{P}{\pi a^2} = \frac{4}{3\pi} E_{eff} \left(\frac{a}{R} \right) \tag{19.2}$$

where E_{eff} is the effective elastic modulus.

More information about the elastic behaviour as well as determining the indentation stress-indentation curve can be found in many research papers [44, 51, 53, 57, 58] and books [35, 50].

19.2.3 Elasto/plastic Contact

The most well-known criterion for plastic yielding is that of Hertz [59], who postulated that hardness was the smallest pressure beneath a spherical indenter necessary to produce a permanent deformation at the center of the area of contact. Later treatments by Auerbach [60], Meyer [61] and Hoyt [62] were all directed to removing some of the practical difficulties in Hertz's original proposal.

As can be extensively found in the papers commented on in the previous paragraphs, the static indentation hardness test usually involves the application of load to a spherical or pyramidal indenter. Then, the value of the mean contact pressure (p_m), which does not increase by increasing the indenter load, is shown by experiment to be related to the Meyer's hardness. Therefore, the mean contact pressure or the Meyer's hardness beneath the indenter is of particular interest and can be treated by rigid-plastic theory. More information in this regard can be extracted from Refs [50, 63].

The elasto/plastic transition can be investigated by the analysis of the loading (P) and the displacement into surface (h) based on the work done by Doener and Nix [64] and later by the work developed by Oliver and Pharr [44]. As it can be appreciated in these two references, the main mechanical properties determined by means of sharp indentation are the hardness and elastic modulus [44]. Further details of the Oliver and Pharr analysis can be found in Refs. [33, 44] and elsewhere in this book.

This method usually works well when applied to ceramics, because these materials are hard and stiff and therefore do not produce a large amount of pile-up and do not present relevant time-dependant deformation. However, some overestimation of Young's modulus (E) may be produced if the value of the calibration factor (β) between E and contact stiffness is not chosen properly (see Pharr *et al.* for details [65]).

19.3 Fracture Toughness

Because of the inherent brittleness of ceramics, it is important to measure the fracture toughness (K_{Ic}) of these materials. While correct determination of K_{Ic} is cumbersome and requires well-controlled stress states and crack lengths, an estimation of this value can be done by indentation micro-fracture methods (*IM*).

When indenting a brittle material with a sharp indenter, cracking of the material can occur. If radial cracks emanating from the corners of the imprints are produced, the crack length together with the applied load can be related with K_{Ic}, through different equations proposed in the literature. The most commonly used sharp indenter is the Vickers indenter. However, at very small loads, typical of nanoindentation tests, Vickers or Berkovich do not generate cracks. In that case, a cube-corner indenter, with a more acute angle is used, and cracks are more easily generated. For cube corner indenter, Pharr [66] proposed an adjustment of the formula of Anstis *et al.* [67] for Vickers indenters for measuring K_{Ic}:

$$K_{Ic} = \xi_R \left(\frac{E}{H}\right)^{\frac{1}{2}} \frac{P}{c^{\frac{3}{2}}} \tag{19.3}$$

Where c is the length of the generated cracks measured from the center of the indenter and $\xi_R = 0.016$. This formula is valid for semi-circular cracks. However, for materials with high values of fracture toughness, Palmqvist-type cracks may be generated instead, and the Laugier [68] formulation may yield better results:

$$K_c = \chi^p \left(\frac{l}{a}\right)^{\frac{-1}{2}} \left(\frac{E}{H}\right)^{\frac{2}{3}} \frac{P}{c^{\frac{3}{2}}} \tag{19.4}$$

Where l is the crack length measured from the indentation corners, a is the half-diagonal of the indentation imprint and $\chi^p = 0.057$ for cube-corner indenters [69].

It is important to remark that IM methods, though easy to perform, may give inaccurate measurements of K_{Ic}, Specially because several crack systems may be activated at the same time and crack geometries may deviate from the one assumed in the model. In this sense, it is advised to perform indentations at the lowest load possible that generate long enough cracks (with lengths larger than three times the imprint dimensions).

Moreover, it is important that the tested material is fully dense. Residual porosity hampers the generation of cracks because they accommodate the deformation, and can give rise to incorrect values of high toughness (see, e.g. [70]).

19.4 Coatings

When a nanoindentation is performed on a ceramic coating or on specimens superficially modified (i.e. sandblasted, chemically etched, etc.), the hardness and elastic modulus depend on the depth of penetration, as the tested surface layer is constrained by the 'substrate', and the mechanical response strictly depends on the relative indentation depth [35]. Therefore, values should be corrected in order to extract the real value of hardness and elastic modulus of the surface.

19.4.1 Coating Hardness

For relative hard systems, it is generally accepted that a nanoindentation with a maximum displacement into the surface of less than 10% of the coating thickness will measure the coating hardness without any substrate influence [32]. When the coating is very thin it may be not possible to measure directly the properties of the coating without the measurement being affected by the substrate.

In this case, there are several models for fitting the composite hardness response of the coating-substrate as a function of the displacement into surface, in order to isolate the coating hardness, that is without any substrate interaction, as can be seen in: Bückle [71], Jonsson and Hogmark [72], Burnett and Rickerby [73], Tsui, Ross and Pharr [74], Bhattacharya and Nix [75], etc.

19.4.2 Coating Elastic Modulus

Compared to hardness, the contact stiffness is more strongly affected by the substrate, which manifests itself even at $h < 0.1t_{\text{coating}}$ [76]. The measurement of the elastic modulus in coatings presents the same problems as was commented before for hardness. In this case, to accurately characterize the response of the coating, even smaller penetrations depths are required. Then, a careful calibration of the indenter tip is needed.The popular 10% rule previously explained does not strictly apply for the elastic modulus measurements of coatings; in this situation the 1% rule should be used. However, at these small depths the values extracted are not realistic and it is not recommended for determining the elastic modulus since huge errors can be introduced. For this reason, several analytical and parametric models can be found in the literature for determining the elastic modulus for the coating system without any interference of the substrate: King [77], Doener and Nix [64], Gao, Chiu and Lee [78], Bec [79], etc. From our point of view, all these models provide realistic values; however, the simplest model is the Bec's model [79].

19.5 Issues for Reproducible Results

Most of the methods for analyzing nanoindentation *P-h* curves rely on an accurate determination of the initial contact point between the indenter and the specimen surface and on a flat surface. Furthermore, several corrections should be done to assure the reproducibility of the measurements (i.e. compliance of the loading column, shape geometry of the indenter, evaluation of the indentation size effect, among others).

To ensure reproducible results it is necessary to have in mind the following considerations:

- **Tip shape calibration:** this is the most important issue to take into account, since the hardness and the elastic modulus values are strongly related to the contact area and indenter shape. In the literature it is possible to find methodologies to improve the tip shape calibration [33, 80–85]. In these references, several methods to determine the indenter shape are proposed, like using atomic force microscopy [80–83] or indenting reference materials, such as fused silica [80, 84, 85], which is widely used.
- **Choosing an appropriate indenter:** depending on the property to be evaluated it will be necessary to choose between sharp (Berkovich to study the hardness and elastic modulus, cube-corner to study the damage induce under indentation, etc.) or blunt indenter (spherical to study the elasto/plastic transition and determine the yield stress).
- **Environmental control:** uncertainties in the measured displacement arise from two separate environmental sources: vibration and temperature. To reduce vibration, the nanoindenter should be located on a quite solid foundation and mounted on a vibration-isolation system. Thermal stability can be provided by closing the

apparatus in a thermally buffered cabin and controlling the temperature (constant stable temperature throughout the year).

- **Surface preparation:** one of the most important factors is the surface roughness of the specimens of interest in order to assure a correct contact point between the indenter and the surface of the specimen. The main problems are found when the characteristic wavelength of the roughness is comparable to the contact diameter; in this case the mechanical properties will present a huge scatter, masking in some situations the real property. For these reasons, it is necessary to accurately prepare the surface to be mechanically investigated in order to obtain a flat surface. Furthermore, prior the indentation process, it is recommended to ultrasonically clean the specimen with organic solvents and dry with pure air.
- **Testing procedure:** in order to avoid any overlapping effect from previous tests; two successive indentations should be separated at least 20–30 times the maximum displacement into surface when a sharp indenter is used and seven to ten times the maximum contact radius when a blunt indenter is used.

In addition to these general procedures, it is necessary to highlight that each material presents some particular issues to take into account. For example, for coated or superficially modified materials (see section 19.4):

- **$t_{coating}$ > 5 μm thick:** If the roughness of the specimen is higher than 400 nm it is necessary to polish the surface to be tested for a couple of minutes, removing the initial roughness and obtain an appropriate surface to be tested by the nanoindentation technique. The procedure is meaningful as long as the coating has a homogeneous microstructure in the depth direction.
- **$t_{coating} \leq 5$ μm thick:** For these specimens it is not possible to reduce the initial roughness and the indentation test should be done in the as receive surface. In this case, a very large number of imprints should be done in order to obtain statistical signification and the 10% and 1% rule can be used to determine the hardness and the elastic modulus, respectively.
- **$t_{coating} \leq 1$ μm thick:** In this case, the surface should not be modified and only the analytical or parametric models can be employed to obtain the mechanical properties.

Finally, for porous materials the spherical indenter is the most common tip used to determine the Meyer's hardness, mean contact pressure as well as the stress-strain curve. More information can be found elsewhere [53, 86, 87]. For these materials it is not recommended use sharp indenters, due to the presence of porosity that will produce a considerable dropping of their mechanical properties due to the plastic and elastic field induced by the indenter which will interact with the different voids distributed along the specimen.

19.6 Applications of Nanoindentation to Zirconia

In this section, we will present several examples of zirconia based ceramic materials because they are widely employed for technological and biomedical applications. Tetragonal polycrystalline zirconia doped with 3 mol% Yttria (3Y-ZrO_2 or 3Y-TZP) has high mechanical properties such as fracture resistance (~1 GPa), toughness (4–8

MPa $m^{1/2}$), Vicker's hardness (~13 GPa) and elastic modulus (~210 GPa); it is also biocompatible, which, in principle, makes this zirconia an excellent candidate for structural applications such as implants and dental replacements [88–90].

The relative high toughness of zirconia is due to the stabilization of the tetragonal phase at room temperature by the addition of dopants (Y_2O_3, but also CeO_2, CaO, MgO, among others). In the presence of stress, particularly the one generated at the tip of a crack, the tetragonal phase transforms into the stable monoclinic phase, which is accompanied by a ~4% increase in volume [91, 92]. This produces in turn a crack closure translating into a relatively high toughness.

During recent years, there has been a considerable interest in adding carbon nanotubes (CNT) to ceramic materials to improve their mechanical properties, mainly the fracture toughness [93–98].

19.6.1 Hardness and Elastic Modulus

The mechanical integrity of the surface of tetragonal and monoclinic phase has been widely investigated at micro- and nanometric length scale during the last decade [14, 99–110].

With reference to elastic modulus of non-aged 3Y-TZP, several papers have reported an elastic modulus slightly greater than the generally accepted value for the 3Y-TZP, 210 GPa [110, 111]. Gaillard *et al.* [109] suggested that this discrepancy could be attributed to the fact that the β parameter proposed by Oliver and Pharr's method (see equation 19.3) is based on Sneddon's analysis of the elastic contact, so that it does not take into account the lateral stresses generated during the indentation process, which leads to an overestimation of this mechanical parameter [33, 112]:

$$E_{eff} = \frac{1}{\beta}\frac{\sqrt{\pi}}{2}\frac{S}{\sqrt{A(h_c)}} \tag{19.5}$$

The elastic modulus is strongly dependent on two different parameters: grain size as well as sintering temperature. Chintipalli *et al.* [113] produced 3Y-TZP by spark plasma sintering at different temperatures and for sintering below 1175°C reported a grain size in the range between 100 and 200 nm and a considerable drop of the elastic modulus, which is related to the presence of large porosity at these low sintering temperatures. For fully dense materials, the 3Y-TZP sintered by spark plasma sintering (SPS) achieves values around 225 GPa, while for porous 3Y-TZP sintered at 1110°C, this parameter is lower than 125 GPa.

Currently, the addition of CNTs into the 3Y-TZP matrix can produce an increase of their indentation fracture toughness; however, up to date, the behaviour of the addition of CNTs is not fully understood due the complex microstructure as well as possible CNT agglomeration effects at grain boundaries. Chintipalli *et al.* [113] observed that increasing the fraction of CNTs reduces the hardness slightly (16.1 GPa for 3Y-TZP and 15.3 for 3Y-YZP with 2 vol.% CNTs), while the elastic modulus keeps constant to 220–230 GPa. Some studies about the incorporation of CNTs to 3Y-TZP matrix reported an increase in indentation fracture toughness using different techniques for the processing of 3Y-TZP/CNT composites [114, 115]. Regarding the influence of CNT on the hardness of 3Y-TZP composites, most of the studies have concluded that with the addition of CNT, the hardness decreases. Regarding the fracture toughness, the reason of the

Figure 19.1 Effect of pore shape and porosity content on elastic modulus. *Source*: Chintapalli 2012 [113]. Reproduced with permission of Elsevier.

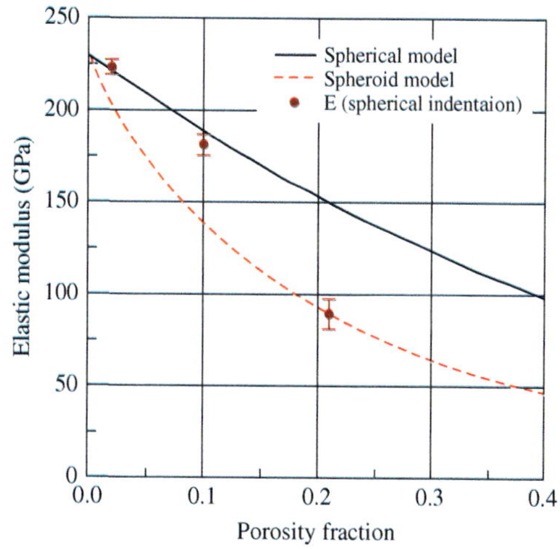

disparity of results in the literature is probably related to several factors, like decohesion effects between the 3Y-TZP and the CNT interface, agglomeration of CNTs as well as to the indentation method itself.

Finally, the hardness (not shown here) and elastic modulus of porous ceramics materials mainly depend on factors like total porosity, pore shape, etc, but no universal model exists to describe the elastic modulus [116] as a function of the pores shape. A suitable simple empirical equation to describe the influence of these parameters in the elastic modulus of 3Y-TZP was proposed by Luo and Steven [117] and it is given by:

$$E = E_0 \frac{1 - P}{1 + \alpha P} \quad (19.6)$$

where E_0 corresponds to the elastic modulus of 3Y-TZP free of pores and P is the porosity volume fraction. The parameter α has been related to the shape of the pores: 1 for spherical pores and increases as the spherical pores change to oblate spheroids. Figure 19.1 represents the effect of pore shape and porosity content on elastic modulus for 3Y-TZP [113]. For porous materials, the reduction in their mechanical integrity in terms of hardness and yield strength follows a similar trend as presented here.

To sum up, it is necessary to highlight that the values of the elastic modulus determined by spherical indentation are more representative of the actual porosity than those measured by using a Berkovich indenter. The reason may lie in that by using spherical indenters the elastic modulus is fitted to the elastic part of the stress-strain curve; while in sharp indentation, the plastic deformation is produced form the initial contact which results in closure of pores and densification of the material.

19.6.2 Stress–strain Curve and Phase Transformation

Using the spherical tip indenter, it is possible to study the elasto/plastic transition [50, 55, 58] as well as to determine the minimum contact pressure to induce phase transformation [118].

Indentation Stress–strain Luo [100] found that the critical load for the elasto/plastic transition for the 8Y-TZP specimens was about 0.15 mN. Furthermore, the critical shear stresses at the elasto-plastic transition position in 8Y-TSZ were obtained by the Hertzian contact theory yielding a value of around 6.4 GPa.

Chintipalli *et al.* [113] determined the indentation stress-indentation strain curve for 3Y-TZP sintered at different temperatures using SPS. He found that the repeatability of the material behaviour under the residual imprint was not affected by the porosity due to the indenter radius and/or the contact radius is bigger than the pore size. Also, it can be observed that the yield strength determined though spherical indentation ranged between 2.1 up to 9.4 GPa for specimens sintered at 1100°C and 1175°C, respectively. In other words, this parameter decreases by increasing the porosity due to lower sintering temperatures.

By using spherical indenters, the elastic modulus can be determined using the Hertz equations. However, scarce information exists in the literature, because it is difficult to determine the correct contact point between the surface and the indenter. Several works [113, 119] investigated and determined the elastic modulus using the Hertzian equations and reported values ranged between 220–241 GPa, which are in concordance with the value obtained through sharp indentation as commented previously.

Phase Transformation The minimum critical pressure to induce phase transformation ranges between 11.2 and 16.7 GPa according to Gaillard *et al.* [118]. Bradby *et al.* [120] predicted that this transformation mainly occurs just under the contact zone. The phase transformation is martensitic, so it is mainly governed by a shear process [121].

19.6.3 Plastic Deformation Mechanisms

Nanoindentation has the ability to induce very high stresses locally, of the order of magnitude of the material hardness, in very small volumes confined by the hydrostatic pressure caused by the same surrounding material. Thus, indentation allows plastic deformation of ceramic materials at room temperature, while conventional techniques only lead to fracture [38, 39].

In this section, we will present the main plasticity events activated under indentation and observed by atomic force microscopy in order to clearly distinguish: dislocations [122–124], twinning [125], domain switching [126], phase transformation [127], and microcracking as a function of crystallographic structure (cubic, tetragonal and monoclinic).

Figure 19.2 presents the AFM images of indentations performed in a cubic grain (Figure 19.2a) and in a grain boundary with tetragonal grains (Figure 19.2b) extracted from Gaillard *et al.* [118]. The main deformation mechanism for the cubic grain is that only pile-up is appreciated around the residual imprint, this phenomenon is associated with the dislocation slip lines as found Gaillard using cube-corner and spherical indenters (images not shown here, see ref. [118]). However, for tetragonal grains (Figure 19.2b) presents surface uplifts that can be associated with two different factors: domain switching [128] as well as the volumetric expansion during the phase transformation produced because of the high stress developed around the imprint.

19.6.4 Mechanical Properties of Damaged Surfaces

Nanoindentation is a suitable tool to study the mechanical integrity in terms of hardness and elastic modulus for damaged surfaces. However, it may be necessary to check

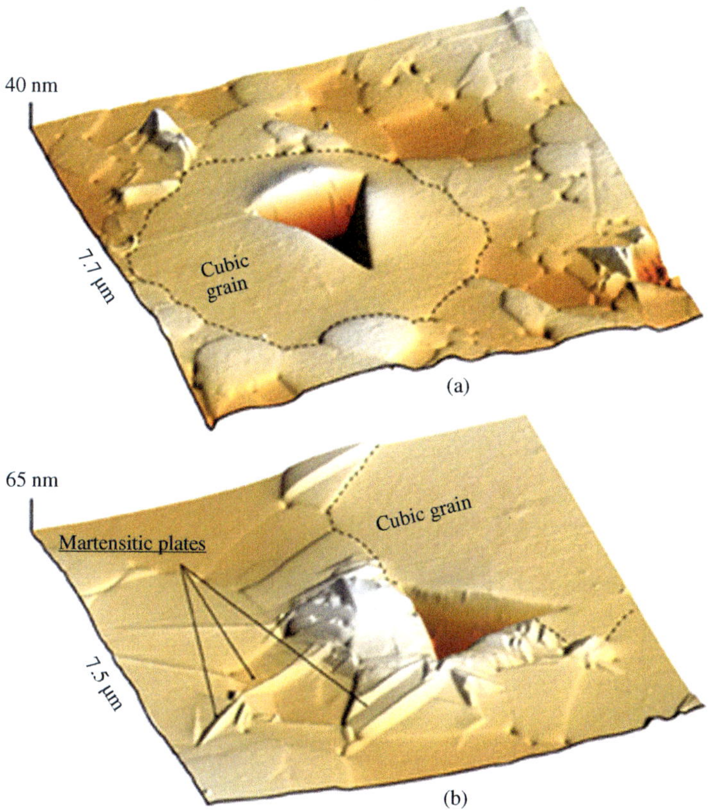

Figure 19.2 AFM topographic images (3D view) of residual Nanoindentation imprint performed until 300 nm in duplex samples. (a) cubic grain, and (b) boundary between a cubic and several tetragonal grains. Reproduced with permissions from reference. *Source*: Gaillard 2009 [118]. Reproduced with permission of Cambridge University Press.

whether the surface to be indented after sandblasting or hydrothermal ageing is smooth enough for nanoindentation. Very often roughness is too large for appropriate nanoindentation.

Knowledge about the mechanical properties near the surface zone is crucial for applications such as dental crowns and implants. In this section, the superficial damage induced during hydrothermal ageing and/or sandblasting of zirconia which alters the performance and reliability of the region near surface will be presented and discussed in detail, as an example of how the nanoindentation technique is able to measure the mechanical properties of modified surfaces.

Hydrothermal Ageing or Degradation: Tetragonal zirconia can spontaneously transform to the monoclinic phase in aqueous environments, by means of a martensitic phase transformation with the corresponding increase in volume mentioned before [85, 86]. This phenomenon, known as hydrothermal ageing or degradation, produces the formation of monoclinic phase at the surface and the generation of microcracks which may reduce the long term performance of components made of 3Y-TZP [93].

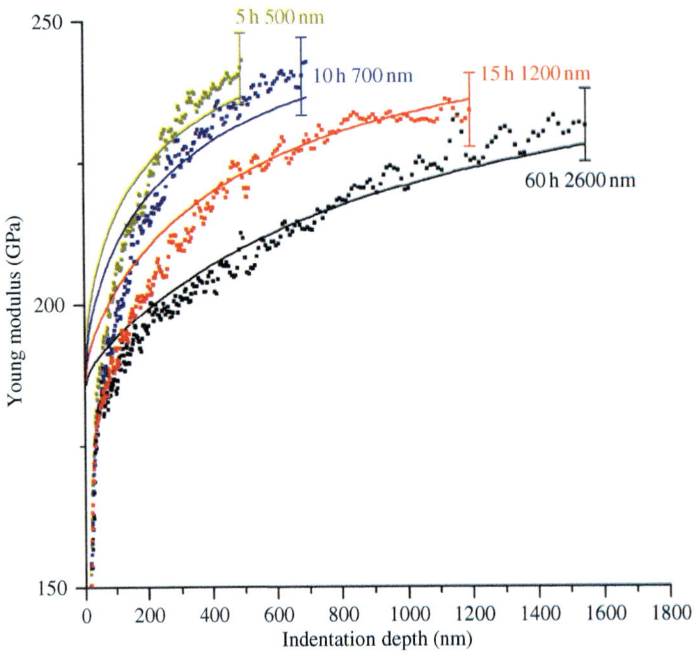

Figure 19.3 Young's modulus against the indentation depth for 3Y-TZP specimens degraded for 5, 10 and 15 h at 134°C. The solid line represents the fitting of the experimental data using the Bec model [79] in order to determine the degraded layer. *Source*: Gaillard 2008 [109]. Reproduced with permission of Elsevier.

After hydrothermal degradation for long periods of time, there is a decrease in hardness measured by indentation reaching values of about 12 GPa which is about 70% the hardness of the initial tetragonal phase ($H_{tetragonal} \approx$ 17 GPa) [103, 107–109]. This dropping on hardness may be attributed to either a lower hardness of monoclinic zirconia or/and to the array of microcracks roughly parallel to the surface that are generated by the volumetric expansion during *t-m* transformation as has been observed on near surface cross-sections prepared by focused ion beam [103, 109, 110]. Gaillard *et al.* [109] correlated the hardness and the elastic modulus with the displacement of the indenter into the surface of degraded tetragonal zirconia after different degradation times and quantified the thickness of the degraded layer using a thin-film model proposed by Bec [79] and briefly described in section 19.4. In this sense, Gaillard *et al.* [109] suggested the intergranular microcrack network [109, 129] generated during the phase transformation as the responsible for the lower values of elastic modulus due to the decrease in contact stiffness.

Furthermore, the thickness of the degraded layer could be determined by the application of one of the models proposed in section 19.4.2 as can be observed in Figure 19.3.

Jiménez-Piqué *et al.* [110] reported a value of elastic modulus for the degraded layer equal to 185 ± 15 GPa. This drop can be also measured directly when the degraded layer is thick enough and there is not significant interaction with the substrate.

Recently, Roa *et al.* [103] studied the annealing effect of hydrothermal aged 3Y-TZP materials by analysing the changes in microstructure and surface mechanical integrity in terms of the annealing temperature. It was shown that recovering the tetragonal structure after ageing by annealing at 850°C does not necessarily lead to recover the deformation and fracture behaviour of the original tetragonal phase at the surface. This observation suggested that this effect is associated with the microcracked grain boundaries induced during the hydrothermal degradation which does not regain their intergranular original strength during the annealing treatment.

On the other hand, the mechanical integrity of specimens with fine grain size (lower than 120 nm) does not change by hydrothermal degradation as can be seen in Ref. [113]. Furthermore, it is necessary to highlight that porosity increases degradation process as has been reported by Masaki [130]. However, for nanometric grain size specimens, the presence of porosity has no effect on degradation of the surface, as may be expected from the results on dense nanometric grain size zirconia, which has a strong resistance to hydrothermal degradation.

Sandblasting: The damage induced during sandblasting of zirconia can also alter the performance and reliability of the near surface of Y-TZP specimens, since it induces phase transformation as well as damage. Thus, concentration of monoclinic phase is found on sandblasted specimens at depths which depend on the conditions of sandblasting [109]. Surface examinations carried out by Kosmac *et al.* [131] in sandblasted specimens revealed the presence of lateral cracks.

Gaillard *et al.* [109] found a considerable reduction in hardness down to 12 GPa in the surface of specimens subjected to sandblasting which was related to the presence of microcracks attributed to the expansion volume associated with the phase transformation as has been previously explained. Nanoindentation is an excellent tool to study these surface variations in mechanical properties and damage.

19.6.5 Relation Between Microstructure and Local Mechanical Properties by Massive Nanoindentation Cartography

In this section, we present the use of massive nanoindentation cartography to characterize the possible differences in the mechanical properties due to microcracking at the tip of a machined shallow single edge V notch. The notch was produced on the surface of rectangular bending bars of 3Y-TZP by ultra-short pulsed laser ablation [104]. The cartography method provides a general overview of the mechanical evolution as well as the superficial damage into the region of interest, when the damage produced is big enough to affect the hardness parameter.

Hardness cartography (Figure 19.4) directly extracted from Turón-Vinas *et al.* [104] was thus used to map out the local surface hardness. A colour chart has been associated with the hardness values, and a bicubic interpolation has been applied to build the map from the discrete dataset; more information about this process can be found elsewhere [132].

As seen in Figure 19.4, nanoindentation carried out in the region in front of the notch, where a microcracked region can be clearly distinguished, reveals a drop in hardness of up to 50% as compared to points outside this region.

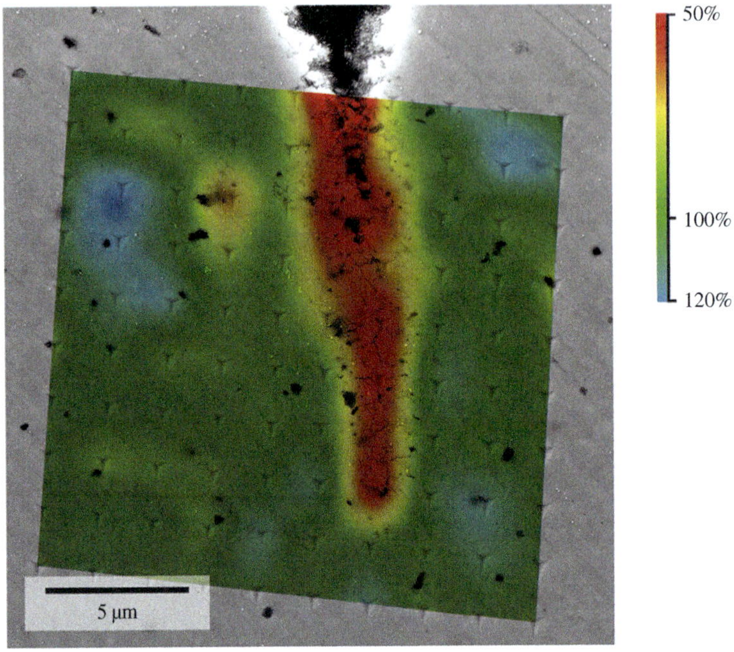

Figure 19.4 Relative hardness cartography superimposed to the scanning electron image showing the indentation array in front of the 3Y-TZP notch. *Source*: Turon-Vinas 2014 [104]. Reproduced with permission of Elsevier.

19.7 Conclusions

The present chapter seeks to present and discuss some observations concerning the mechanical integrity at micro- and nanometric length scale of zirconia based ceramic materials in order to provide an overview of ceramic micro- and nanomechanics to researchers entering the field as well as a guide for experimental advice. Besides, the theoretical framework as well as references necessary to interpret the results concerning the different superficial qualities are also reviewed.

Acknowledgements

The current study was supported by the Spanish Ministerio de Economía y Competitividad (Grant MAT2014-60720-R). The financial support from Direcció General de Recerca del Comissionat per a Universitats i Recerca de la Generalitat de Catalunya is also acknowledged (2014-SGR-130). Dr. J. J. Roa would like to thank the Juan de la Cierva programme (grant number JCI-2012-14454) for its financial support.

References

1 Li X, Bhushan B. Micro/Nanomechanical characterization of ceramic films for microdevices. Thin Solid Films (1999) 34: 210–217.
2 Davidge RW. Mechanical Behaviour of Ceramics. Cambridge University Press, London; 1979.
3 Krell A, Blank P. Grain size dependence of hardness in dense submicrometer alumina. Journal of the American Ceramic Society (1993) 78: 1118–1120.
4 Mishra M, Tangpatjaroen C, Szlufarska I. Plasticity-controlled friction and wear in nanocrystalline SiC. Journal of the American Ceramic Society (2014) 97: 1194–1201.
5 Cain M, Morrel M. Nanostructured Ceramics: A review in their potential. Applied Organometric Chemistry (2001) 15: 321–330.
6 Krell A. Improved hardness and hierarchic influences on wear in submicron sintered alumina. Materials Science and Engineering (1996) 209: 156–163.
7 Wang X, Padture NP, Tanaka H *et al.* Wear-resistant ultra fine grain ceramics. Acta Materialia (2005) 53: 271–277.
8 Kear BH, Colaizzi J, Mayo WE *et al.* On the processing of nanocrystalline and nanocomposite ceramics. Scripta Materialia (2001) 4: 2065–2068.
9 Kuntz JD, Zhan G-D, Mukherice AK. Nanocrystalline-matrix ceramic composites for improved fracture toughness. Materials Research Society Bulletin (2004) 29: 22–27.
10 Ovid'ko IA. Deformation and diffusion modes in nanocrystalline materials. International Materials Review (2005) 50: 65–82.
11 Wolf D, Yamakov V, Phillpot SR *et al.* Deformation of nanocrystalline materials by molecular dynamics simulation: Relationship to experiments? Acta Materialia (2005) 53: 1–40.
12 Dao M, Lu L, Asaro RJ, *et al.* (2007) Toward a quantitative understanding of mechanical behavior of nanocrystalline metals. Acta Materialia 55: 4041–4065.
13 Mukhopadhyay A, Basu B. Consolidation-microstructure-property relationships in bulk nanoceramics and ceramic composites: A review. International Materials Review (2007) 52: 257–288.
14 Afrasiabi A, Kobayashi A. Hot corrosion control in plasma sprayed YSZ coating by alumina layer with evaluation of microstructure and nanoindentation data (H,E). Vacuum (2013) 88:103–107.
15 Pande CS, Cooper KP. Nanomechanics of Hall-Petch relationship in nanocrystalline materials. Progress in Materials Science (2009) 54: 689–706.
16 Padilla HA, Boyce BL. A review of fatigue behavior in nanocrystalline metals. Experimental Mechanics (2010) 50: 5–23.
17 Koch CC, Ovid'ko IA, Seal S, *et al.* Structural nanocrystalline materials: fundamentals and applications. Cambridge: Cambridge University Press; 2007.
18 Cheng S, Ma E, Wang YM, *et al.* Tensile properties of in situ consolidated nanocrystalline Cu. Acta Materialia (2005) 53: 1521–1533.

19 Youssef KM, Scattergood RO, Murty KL, *et al.* Ultrahigh strength and high ductility of bulk nanocrystalline copper. Applied Physics Letters (2005) 87: 091904/1–091904/3.

20 Youssef KM, Scattergood RO, Murty KL, *et al.* Nanocrystalline Al-Mg alloy with ultrahigh strength and good ductility. Scripta Materialia (2006) 54: 251–256.

21 Sergueeva AV, Mara NA, Mukherjee AK. Grain boundary sliding in nanomaterials at elevated temperatures. Journal of Materials Science (2007) 42: 1433–1438.

22 Bhaduri S, Bhaduri SB. Enhanced low temperature toughness of Al_2O_3-ZrO_2 nano/nano composites. Nanostructure of Materials (1997) 8: 755–763.

23 Mirshams RA, Liao CH, Wang SH, *et al.* R-Curve characterization of the fracture toughness of nanocrystalline nickel thin sheets. Materials Science and Engineering A (2001) 315: 21–27.

24 Zhao Y, Qian J, Daemen LL, *et al.* Enhancement of fracture toughness in nanostructured diamond-SiC composites. Applied Physics Letters (2004) 84: 1356–1358.

25 Pei YT, Galvan D, De Hosson JT. Nanostructure and properties of TiC/a-C :H composite coatings. Acta Materialia (2005) 53: 4505–4521.

26 Volinsky AA, Moody NR, Gerberich WW. Interfacial toughness measurements for thin films on substrates. Acta Materialia (2002) 50: 441–466.

27 Kang U, Lee T, Kim Y-H. Pt/Ti thin film adhesion on SiNx/Si substrate. Japanese Journal of Applied Physics (1999) 38: 4147–4151.

28 Lee A, Litteken CS, Dauskardt RH, *et al.* Comparison of the telephone cord delamination and interfacial fracture in thin-film structures. Acta Materialia (2005) 53: 609–616.

29 Cordill MJ, Bahr DF, Moody NR, *et al.* Recent developments in thin film adhesion measurement. IEEE Transactions Device and Materials Reliability (2004) 4: 163–168.

30 Kriese MD, Boismier DA, Moody NR, *et al.* Nanomechanical fracture-testing of thin films. Engineering and Fracture Mechanics (1998) 61: 1–20.

31 Fischer-Cripps AC. A review of analysis methods for sub-micron indentation testing. Vacuum (2000) 58: 569–585.

32 Fischer-Cripps AC. Critical review of analysis and interpretation of nanoindentation test data. Surface and Coatings Technology (2006) 200: 4153–4165.

33 Oliver WC, Pharr GM. Measurement of hardness and elastic modulus by instrumented indentation: advances in understanding and refinaments to methodology. Journal of Materials Research (2004) 19: 3–20.

34 Tromas C, Verdier M, Fivel M. *et al.* Mechanical and nanomechanical properties. Chapter 8 in Bréchignac C, Houdy P, Lahmani M (eds) Nanomaterials and Nanochemistry. Springer, Berlin; 2007.

35 Fischer-Cripps AC. Nanoindentation, 2nd edn. Springer-Verlag, New York; 2004.

36 Gogotsi Y, Dominich V (eds). High Pressure Surface Science and Engineering. Institute of Physics, Bristol, UK; 2004.

37 Hay JL, Pharr GM. Materials Testing and Evaluation. ASM Handbook (2000) 8: 232.

38 Page TF, Oliver WC, McHargue CJ. Deformation behavior of ceramic crystals subjected to very low load (nano) indentations. Journal of Materials Research (1992) 7: 450–473.

39 Tromas C, Girard JC. Study by atomic force microscopy of elementary deformation mechanisms involved in low load indentations in MgO crystals. Philosophical Magazine (2000) A 80: 2325–2335.

40 Lee H, Ko S, Han J, Park H, *et al.* Novel analysis for Nanoindentation size effect using strain gradient plasticity. Scripta Materialia (2005) 53: 1135–1139.

41 Kriese MD, Moody NR, Gerberich WW. Effects of annealing and interlayers on the adhesion energy of copper thin films to SiO_2/Si substrates. Acta Materialia (1998) 46: 6623–6630.

42 Toparli M, Sasaki S. Evaluation of the adhesion of TiN films using nano-indentation and scratch testing. Philosophical Magazine A (2002) 82: 2191–2197.

43 Ollendorf H, Schneider D. A comparative study of adhesion test methods for hard coatings. Surface and Coatings Technology (1999) 113: 86–102.

44 Oliver WC, Pharr GM. An improved technique for determining hardness and elastic modulus using load and displacement sensing indentation experiments. Journal of Materials Research (1992) 7: 1564–1583.

45 Choi Y, Van Vliet KJ, Li J, *et al.* Size effects on the onset of plastic deformation during nanoindentation of thin films and patterned lines. Journal of Applied Physics (2003) 94: 6050–6058.

46 Johnson KL. Contact Mechanics. Cambridge University Press, London; 1985.

47 Sneddon IN. The relation between load and penetration in the axisymmetric Boussinesq problem for a punch of arbitrary profile. International Journal of Engineering Science (1965) 3: 47–57.

48 Herbert EG, Pharr GM, Oliver WC, *et al.* On the measurement of stress-strain curves by spherical indentation. Thin Solid Films (2001) 331: 398–399.

49 Lawn BR. Fracture of Brittle Solids, 2nd edn. Cambridge University press, Cambridge; 1993.

50 Tabor D. Hardness of Metals. Clatendon Press, Oxford; 1951.

51 Field JS, Swain MV. Simple predictive model for spherical indentation. Journal of Materials Research (1993) 8: 297–306.

52 Linss V, Schwarzer N, Chudoba T, *et al.* Mechanical properties of a graded B-C-N Sputtered coating with varying Young's modulus: deposition, theoretical modelling and nanoindentation. Surface and Coatings Technology (2005) 195: 287–297.

53 Basu S, Moseson A, Barsoum MW. On the determination of spherical nanoindentation stress-strain curves. Journal of Materials Research (2006) 21: 2628–2637.

54 Liang Y-H, Aria Y, Ozasa K, *et al.* Simultaneous measurement of nanoprobe indentation force and photoluminescence of InGaAs/GaAs quantum dots and its simulation. Physica E (2007) 36: 1–11.

55 Moseson AJ, Basu S, Barsoum MW. Determination of the effective zero point of contact for spherical nanoindentation. Journal of Materials Research (2008) 23: 204–209.

56 Kalidindi SR, Pathak S. Determination of the effective zero-point and the extraction of spherical nanoindentation stress-strain curves. Acta Materialia (2008) 56: 3523–3532.

57 Field JS, Swain MV. Determining the mechanical properties of small volumes of material from submicrometer spherical indentations. Journal of Materials Research (1995) 10: 101–112.

58 Roa JJ, Jiménez-Piqué E, Capdevila XG, *et al.* Nanoindentation with spherical tips of single crystals of YBCO textured by the Bridgman technique: Determination of indentation stress-strain curves. Journal of the European Ceramics Society (2010) 30: 1477–1482.
59 Hertz H. On the contact of elastic solids. Journal für die Reine und Angewandte Mathematik (1881) 92: 156–171. Translated and reprinted in English as chapter 5: On the contact of elastic solids, in: Hertz's Miscellaneous Papers, Macmillan, London, 1896, p. 146.
60 Auberbach F. Absolute hardness. Annual of Physical Chemistry (Leipzig) (1891) 43: 61–100. Translated by Barus C, Annual Report of the Board of Regents of the Smithsonian Institution, July 1, 1890–June 30 1891. Reproduced in: Miscellaneous documents of the House of Representatives for the First Session of the Fifty-Second Congress, Government Printing Office, Washington, DC 43, 1891–1892: 207–236.
61 Meyer E. Untersuchungen über Harteprunfung und Harte. Physikalische Zeitschrift (1908) 9: 66–74.
62 Hoyt SL. The ball indentation hardness test. Transactions of the American Society of Steel Treatment (1924) 6: 396–420.
63 Hubert MT. Contact of solid elastic bodies. Annaler der Physik (1904) 14: 153–163.
64 Doerner MF, Nix WD. A method for interpreting data from depth-sensing indentation instruments. Journal of Materials Research (1986) 1: 601–609.
65 Pharr GM, Bolshakov A. Understanding Nanoindentation unloading curves. Journal of Materials Research (2002) 17: 2660–2671.
66 Pharr GM. Measurement of mechanical properties by ultra-low load indentation. Materials Science and Engineering A (1998) 253: 151–159.
67 Anstis GR, Chantikul P, Lawn BR, *et al.* A critical evaluation of indentation techniques for measuring fracture toughness I: Direct crack measurements. Journal of the American Ceramic Society (1981) 64: 533–538.
68 Laugier MT. New formula for indentation toughness in ceramics. Journal of Materials Science Letters (1987) 6: 355–358.
69 Cuadrado N, Casellas D, Anglada M, *et al.* Evaluation of fracture toughness of small volumes by means of cube-corner Nanoindentation. Scripta Materialia (2012) 66: 670–673.
70 Wang XT, Padture NP, Tanaka H. Contact-damage-resistant ceramic/single-wall carbon nanotubes and ceramic/graphite composites. Nature Materials (2004) 3: 539–544.
71 Bückle H, Westbrook JW, Conrad H (ed). The Science of Hardness Testing and its Applications. American Society for Metals, Metals Park, OH, 1973, 453.
72 Jonsson B, Hogmark S. Hardness measurements of thin films. Thin Solid Films (1984) 114: 257–269.
73 Burnett PJ, Rickerby DS. The mechanical properties of wear-resistance coatings I: Modelling of hardness behavior. Thin Solid Films (1987) 148: 41–50.
74 Tsui TY, Ross CA, Pharr GM. A method for making substrate-independent hardness measurements of soft metallic films on hard substrates by nanoindentation. Journal of Materials Research (2003) 18: 1383–1391.
75 Bhattacharya AK, Nix WD. Analysis of elastic and plastic deformation associated with indentation testing of thin films on substrates. International Journal of Solids and Structures (1988) 24: 1287–1298.

76 Sakai M, Zhang J, Matsuda A. Elastic deformation of coating/substrate composites in axisymmetric indentation. Journal of Materials Research (2005) 20: 2173–2183.
77 King RB. Elastic analysis of some punch problems for a layered medium. International Journal of Solids and Structures (1987) 23: 1657–1664.
78 Gao H, Chiu CH, Lee J. Elastic contact versus indentation modeling of multi-layered materials. International Journal of Solids and Structures (1992) 29: 2471–2492.
79 Bec S, Tonck A, Loubet JL. A simple guide to determine elastic properties of films on substrate from nanoindentation experiments. Philosophical Magazine (2006) 86: 5347–5358.
80 Jennett NM, Meneve J. Depth sensing indentation of thin hard films: a study of modulus measurement sensitivity to indentation parameters. In: Moody NR, Gerberich WW, Burnham N, Baker SP, editors. Fundamentals of Nanoindentation and Nanotribology. Pittsburg, PA: Materials Research Society; 1998.
81 Meneve JL, Smith JF, Jennet NM, *et al.* Surface mechanical property testing by depth sensing indentation. Applied Surface Science (1998) 100/101: 64–68.
82 VanLandingham MR, Villarrubia JS, Meyers GF. Recent progress in nanoscale indentation of polymers using the AFM. In: Proceedings of the SEM IX International congress on experimental mechanics. Betherl, CT: Society for Engineering Mechanics, Inc.; 2000, pp. 912–915.
83 McElhancey KW, Vlassak JJ, Nix WD. Determination of indenter tip geometry and indentation contact area for depth-sensing indentation experiments. Journal of Materials Research (2002) 13: 1300–1306.
84 Thurn J, Cook RF. Simplified area function for sharp indenter tips in depth-sensing indentation. Journal of Materials Research (2002) 17: 1143–1146.
85 Sun Y, Zheng S, Bell T, *et al.* Indenter tip radius and load frame compliance calibration using nanoindentation loading curves. Philosophical Magazine Letters (1999) 79: 649–658.
86 Roa JJ, Capdevila XG, Segarra M. Mechanical characterization at nanometric scale of ceramic superconductor composites. In: (eds) Tovar H, Jonathon F. Superconducting Magnets and Superconductivity: Research, Technology and Applications. Nova Science Publishers; 2009, p. 117.
87 Chintapalli RV, Marro FG, Jiménez-Piqué E, *et al.* Phase transformation and subsurface damage in 3Y-TZP after sandblasting. Dental Materials (2013) 29: 566–572.
88 Thompson VP, Rekow DE. Dental ceramics and the molar crown testing ground. Journal of Applied Oral Science (2004) 12: 26–36.
89 Piconi C, Maccauro G. Zirconia as a ceramic biomaterial. Biomaterials (1999) 20: 1–25.
90 Meyenberd KH, Lüthy H, Schärer P. Zirconia posts: a new all-ceramic concept for nonvital abutment teeth. Journal of Esthetic Dentistry (1995) 7: 73–80.
91 Hannink RHJ, Kelly PM, Muddle BC. Transformation toughening in zirconia-containing ceramics. Journal of the American Ceramic Society (2000) 83: 461–487.
92 Cutler RA, Reynolds JR, Jones A. Sintering and Characterization of Polycrystalline Monoclininc, Tetragonal and Cubic Zirconia. Journal of the American Ceramic Society (1992) 75: 2173–2183.

93 Deville S, Chevalier J, Fantozzi G, *et al.* Low-temperature ageing of zirconia-toughened alumina ceramics and its implication in biomedical implants. Journal of the European Ceramic Society (2003) 23: 2975–2982.

94 Peigney A. Tougher Ceramics with Nanotubes. Nature Materials (2002) 2: 15–16.

95 Zhan GD, Mukherjee AK. Carbon nanotube reinforced alumina-based ceramics with novel mechanical, electrical and thermal properties. International Journal of Applied Ceramics Technology (2004) 1: 161–171.

96 Shi SL, Liang J. Effect of multiwall carbon nanotubes on electrical and dielectric properties of yttria-stabilized zirconia ceramic. Journal of the American Ceramic Society (2006) 89: 3533–3535.

97 Dusza J, Blungan G, Morgiel J, *et al.* Hot pressed and spark plasma sintered zirconia/carbon nanofiber composites. Journal of the European Ceramic Society (2009) 29: 3177–3184.

98 Datye A, Wu K-H, Gomes G, *et al.* Synthesis, microstructure and mechanical properties of yttria stabilized zirconia (3YTZP)-multi-walled nanotube (MWNTs) nanocomposite by direct in-situ growth of MWNTs on zirconia particles. Composite Science and Technology (2010) 70: 2086–2092.

99 Inam F, Yan H, Peijs T, *et al.* The sintering and grain growth behaviour of ceramic-carbon nanotube nanocomposites. Composite Science and Technology (2010) 70: 947–952.

100 Carpio P, Rayón E, Pawlowski L, *et al.* Microstructure and indentation mechanical properties of YSZ nanostructured coatings obtained by suspension plasma spraying. Surface and Coatings Technology (2013) 220: 237–243.

101 Maerten A, Zaslansky P, Mochales C, *et al.* (2013) Characterizing the transformation near indents and cracks in clinically used dental yttria-stabilized zirconium oxide constructs. Dental Materials (2013) 29: 241–251.

102 Gong J, Peng Z, Jiang D. Nanoindentation characterization of tetragonal zirconia polycrystalline implanted by titanium ions with MEVVA sources. Materials Chemistry and Physics (2014) 147: 268–272.

103 Roa JJ, Anglada M. Annealing aged zirconia: study of surface mechanical properties at the micrometric length scale. Journal of the European Ceramic Society (2015) 35: 1031–1039.

104 Turon-Vinas M, Anglada M. Fracture toughness of zirconia from a shallow notch produced by ultra-short pulsed laser ablation. Journal of the European Ceramic Society (2014) 34: 3865–3870.

105 Alao AR, Yin L. Nano-scale mechanical properties and behaviour of pre-sintered zirconia. Journal of the Mechanical Behavior of Biomedical Materials (2014) 36: 21–31.

106 Luo JM, Dai CY, Shen YG, *et al.* Elasto-plastic characteristics and mechanical properties of as-sprayed 8 mol% yttria-stabilized zirconia coating under nano-scales measured by nanoindentation. Applied Surface Science (2014) 309: 271–277.

107 Catledge SA, Cook M, Vohra YK, *et al.* Surface crystalline phase and nanoindentation hardness of explanted zirconia femoral heads. Journal of Materials Science: Materials in Medicine (2004) 14: 863–867.

108 Guicciardi S, Shimozono T, Pezzotti G. Ageing effects on the nanoindentation response of sub-micrometric 3Y-TZP ceramics. Journal of Materials Science (2007) 42: 718–722.

109 Gaillard Y, Jiménez-Piqué E, Soldera F, *et al.* Quantification of hydrothermal degradation in zirconia by nanoindentation. Acta Materialia (2008) 56: 4206–4216.
110 Jiménez-Piqué E. Ramos A, Muñoz-Tabares JA, *et al.* Focused ion beam tomography of zirconia degraded under hydrothermal conditions. Journal of the European Ceramic Society (2012) 32: 2129–2136.
111 Chan SK, Fang Y, Grimsditch M, *et al.* Temperature dependence of the elastic moduli of monoclininc zirconia. Journal of the American Ceramic Society (1991) 74: 1742–1744.
112 Hay JC, Bolshakov A, Pharr GM. Critical examination of the fundamental relations used in the analysis of nanoindentation data. Journal of Materials Research (1999) 14: 2296–2305.
113 Chintapalli, RK. Influence of sandblasting on zirconia in restorative dentistry. Phd Thesis at the Department of Materials Science and Metallurgy Engineering. Universitat Politècnica de Catalunya. Barcelona, Spain; 2012.
114 Melk L, Roa JJ, García-Marro F, *et al.* Nanoindentation and fracture toughness of nanostructured zirconia/multiwalled carbon nanotube composites. Ceramics International (2015) 41: 2453–2461.
115 Duszová A, Dusza J, Tomásek K, *et al.* Zirconia/carbon nanofiber composite. Scripta Materialia (2008) 58: 520–523.
116 Rice RW. Porosity of Ceramics. New York: Marcel Dekker; 1998.
117 Luo J, Steven R. Porosity-dependence of elastic moduli and hardness of 3Y-TZP ceramics. Ceramics International (1999) 25: 281–286.
118 Gaillard Y, Anglada M, Jiménez-Piqué E. Nanoindentation of yttria-doped zirconia: Effect of crystallographic structure on deformation mechanisms. Journal of Materials Research (2009) 24: 719–727.
119 Trejo-Arroyo D, Zárate-Medina J, Alvarado-Orozco JM, *et al.* Microstructure and mechanical properties of Al_2O_3-YSZ spherical polycrystalline composites. Journal of the European Ceramic Society (2013) 33: 1907–1916.
120 Bradby JE, Williams JS, Wong-Leung J, *et al.* Transmission electron microscopy observation of deformation microstructure under spherical indentation in silicon. Applied Physics Letters (2000) 77: 3749–3751.
121 Green DJ, Hannink RHJ, Swain MV. Transformation Toughening Ceramics. CRC Press, Boca Raton, FL; 1989.
122 Messerschmidt U, Baither D, Baufeld B, *et al.* Plastic deformation of zirconia single crystals: A review. Materials Science and Engineering (1997) 233: 61–74.
123 Martínez-Fernández J, Jiménez-Melendo M, Domínguez-Rodríguez A, *et al.* High Temperature precipitation hardening in Y_2O_3 partially tabilized ZrO_2 (Y-PSZ) single crystals. III. Effect of solute composition and orientation on the hardening. Acta Metallurgica et Materialia (1995) 43: 2469–2484.
124 Martínez-Fernández J, Jiménez-Melendo M, Domínguez-Rodríguez A. Microstructural evolution and stability of tetragonal precipitates in Y_2O_3 partially-stabilized ZrO_2 single crystals. Acta Metallurgica et Materialia (1995) 43: 593-601.
125 Martínez-Fernández J, Jiménez-Melendo M, Domínguez-Rodríguez A, *et al.* An unusual twin structure in transformed precipitates in Y-PSZ single-crystals. Journal of the American Ceramic Society (1994) 77: 57–64.

126 Srinivasan GV, Jue JF, Kuo SY, Virkar AV. Ferroelastic domain switching in polydomain tetragonal zirconia single crystals. Journal of the American Ceramic Society (1989) 72: 2098–2103.

127 Garvie RC, Hannink RHJ, Pascoe RT. Ceramic steel? Nature (1975) 258: 703–704.

128 Chien FR, Ubic FJ, Prakash V, *et al.* Stress-induced martensitic transformation and ferroelastic deformation adjacent microhardness indents in tetragonal zirconia single crystals. Acta Materialia (1998) 46: 2151–2171.

129 Muñoz-Tabares JA, Anglada M. Hydrothermal degradation of ground 3Y-TZP. Journal of the European Ceramic Society (2012) 32: 325–333.

130 Masaki T Mechanical properties of Y_2O_3-stabilized tetragonal ZrO_2 polycrystals after aging at high temperature. Journal of the American Ceramic Society (1986) 69: 519–522.

131 Kosmac T, Oblak C, Jevnikar P, *et al.* Strength and reliability of surface treated Y-TZP dental ceramics. Journal of Biomedical Materials Research (2000) 53: 304–313.

132 Tromas C, Stinville JC, Templier C, *et al.* Hardness and elastic modulus gradients in plasma-nitrided 316L polycrystalline stainless steel investigated by nanoindentation tomography. Acta Materialia (2012) 60: 1965–1973.

20

FEM Simulation of Nanoindentation

F. Pöhl[1], W. Theisen[2] and S. Huth[3]

[1] Dr.-Ing., Ruhr-Universität Bochum, Germany
[2] Prof. Dr.-Ing., Ruhr-Universität Bochum, Germany
[3] Dr.-Ing., Hilti Corporation, Schaan, Liechtenstein

20.1 Introduction

Nanoindentation is widely used for the local mechanical characterization of materials and their phases. It is a relatively simple and quick test without extensive sample preparation. However, the results of indentation tests provide numerous important mechanical parameters and give insight into the complex mechanical behaviour of materials. Especially, nanoindentation allows the mechanical characterization of small phases in the order of microns and nanometers or brittle phases that cannot be tested with standard test methods such as the tensile test. The fields of applications range from characterization of metals, brittle phases (hard phases), coatings, porous structures, shape memory materials, polymers, graded materials to tissues, membranes, bones, teeth, and cells, among many other applications [1–18].

The rapid technical development of computer technology and computational methods led to a widespread application of these methods in materials science and also in nanoindentation within the last decades. In particular, the Finite-Element-Method (FEM) is widely used for simulation of nanoindentation and gives in addition to experiments deep insights into the complex deformation behaviour of materials during indentation. The FEM simulation of nanoindentation covers a broad spectrum: e.g. single-phase isotropic materials, coatings, graded materials, multiphase materials, porous structures, inverse parameter determination, among others [19–31].

This chapter gives an overview of the deformation behaviour of isotropic materials (focused on metallic materials) and presents selected areas of application of FEM simulations in the field of nanoindentation. Firstly, the mechanical behaviour of an isotropic material during indentation is analysed based on a continuum mechanics point of view. An analytical review in combination with FEM simulations gives insights into the deformation behaviour during indentation and shows the relationships between the results of indentation tests (P-h curve and pile-up/sink-in characteristic) and the mechanical properties of materials (Young's modulus, yield stress, and hardening behaviour). The analysis is focused on metallic materials which usually show hardening behaviour. Secondly, different examples of FEM simulations in the field of nanoindentation are

Applied Nanoindentation in Advanced Materials, First Edition. Edited by Atul Tiwari and Sridhar Natarajan.

exemplarily shown. The numerical simulation of the indentation of a film-substrate system and a hard phase embedded in a softer metallic matrix are shown.

20.2 Indentation of Isotropic Materials

The following investigation is an energy-based analysis (continuum mechanics) of the indentation into a homogeneous material with a self-similar indenter according to [32]. The study shows relationships between the indentation results usually obtained from the P-h curve and the material parameters of Ludwik power law materials. Here, it is focussed on the important indentation parameters loading curvature C and indentation energies W_{el}, W_{pl}, and W_{tot}. Those parameters can directly be derived from the P-h curve and are usually used for inverse material parameter determination [31]. Figure 20.1 schematically shows the indentation (indentation depth h) of an isotropic, elasto-plastic material by a rigid, self-similar indenter (conical indenter with an included half-apex angle θ). The sharp indenter induces a plastic zone directly under the tip and at a larger distance an elastic zone. The plastic zone has mean plastic strain $\bar{\varepsilon}_{pl}$ and volume V_{pl}. The mean plastic strain is according to Branch *et al.* defined as the average volumetric equivalent plastic strain of the entire plastic zone (Equation 20.1) [33, 34]. It is not the representative strain according to Tabor or several other studies [31, 35–38]. The elastically deformed zone has analogous to the plastically deformed zone mean elastic strain $\bar{\varepsilon}_{el}$ and volume V_{el}.:

$$\bar{\varepsilon}_{pl} = \frac{\sum \varepsilon_{pl,j} V_{pl,j}}{\sum V_{pl,j}} \quad \text{and} \quad \bar{\varepsilon}_{el} = \frac{\sum \varepsilon_{el,j} V_{el,j}}{\sum V_{el,j}} \tag{20.1}$$

The work done for the deformation of the material can be divided into the strain energy of the plastic zone and the strain energy of the elastic zone. Since the plastic zone is deformed elastically and plastically it contributes two components to the total strain energy (work). Firstly, the elastic strain energy of the plastic zone (W_{el1}) and, secondly, the plastic strain energy of the plastic zone (W_{pl}). Additionally, the elastic zone contributes the elastic strain energy of the elastic zone (W_{el2}). Thus, total indentation work W_{tot} is the sum of the three components and given by Equation (20.2):

$$W_{tot} = W_{el1} + W_{pl} + W_{el2} \tag{20.2}$$

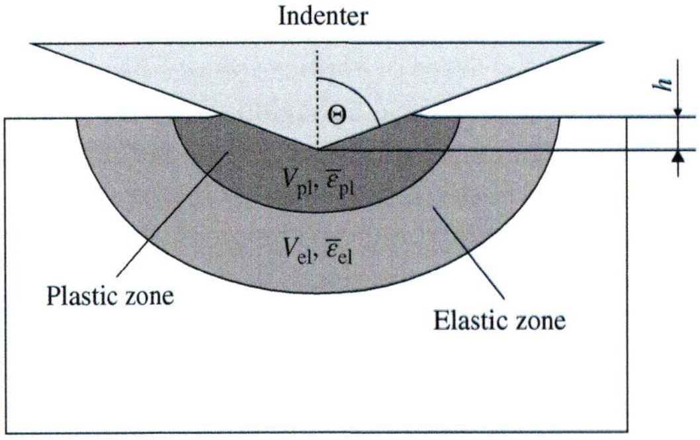

Figure 20.1 Schematic illustration of indentation in an isotropic elasto-plastic material by a self-similar rigid indenter.

The mechanical behaviour of many metallic materials can be described by the Ludwik power law (Equation (20.3)), which is often used [25, 31, 32, 37]. The elastic material behaviour follows Hooke's law with Young's modulus E and plasticity is described by a power-law function with two independent material parameters such as strain hardening coefficient K and strain hardening exponent n. The yield stress is a depending parameter given by Equation (20.4):

$$\begin{aligned} \sigma &= E\varepsilon_{\mathrm{el}} \qquad \sigma \le \sigma_{\mathrm{y}} \\ \sigma &= K\varepsilon_{\mathrm{pl}}^{n} \qquad \sigma \ge \sigma_{\mathrm{y}} \end{aligned} \tag{20.3}$$

$$\sigma_{\mathrm{y}} = \left(\frac{E^n}{K}\right)^{\frac{1}{n-1}} \tag{20.4}$$

In the case of a self-similar indenter geometry and a constant Poisson's ratio ν and with respect to the Ludwik power law, Equation (20.2) can transformed to Equation (20.5). Thus, the total indentation work is a function of the material parameters, the deformed volumes and mean strains:

$$W_{\mathrm{tot}} = \frac{K^2\bar{\varepsilon}_{\mathrm{pl}}^{2}}{2E} \cdot V_{\mathrm{pl}} + \frac{K}{n+1} \cdot \bar{\varepsilon}_{\mathrm{pl}}^{n+1} \cdot V_{\mathrm{pl}} + \frac{E}{2}\bar{\varepsilon}_{\mathrm{el}}^{2} \cdot V_{\mathrm{el}} \tag{20.5}$$

$$\frac{W_{\mathrm{tot}}}{h^3} = \frac{K^2\bar{\varepsilon}_{\mathrm{pl}}^{2}}{2E} \cdot \frac{V_{\mathrm{pl}}}{h^3} + \frac{K}{n+1} \cdot \bar{\varepsilon}_{\mathrm{pl}}^{n+1} \cdot \frac{V_{\mathrm{pl}}}{h^3} + \frac{E}{2}\bar{\varepsilon}_{\mathrm{el}}^{2} \cdot \frac{V_{\mathrm{el}}}{h^3} \tag{20.6}$$

Following observations can be made in case of a perfect self-similar indenter and in absence of indentation size effect (ISE):

- Mean elastic and plastic strains ($\bar{\varepsilon}_{\mathrm{el}}$ and $\bar{\varepsilon}_{\mathrm{pl}}$) remain constant during indentation of an isotropic, elasto-plastic material. There is no depth dependency.
- With dimensional analysis it can be shown that the elastically and plastically deformed zone (V_{el} and V_{pl}) increase with cubed indentation depth. Vice versa, both volumes normalized with cubed indentation depth remain constant.

Dividing both sides of Equation (20.9) by h^3 leads to Equation (20.5). From above observations, $\frac{V_{\mathrm{el}}}{h^3}$, $\frac{V_{\mathrm{pl}}}{h^3}$, $\bar{\varepsilon}_{\mathrm{el}}$, and $\bar{\varepsilon}_{\mathrm{pl}}$ in Equation (20.6) are constant for a given material and self-similar indenter geometry (given K, E and n). Thus, $\frac{W_{\mathrm{tot}}}{h^3}$ is also constant and can be seen as a material-dependent specific indentation work. Rearranging Equation (20.6) leads to Equation (20.7), which is well known as Kick's law and the natural outcome of the energy-based approach [39]. Kick's law is also the result of dimensional analysis as it was shown by Cheng and Cheng [40]:

$$\int_0^h Ch^2 \mathrm{d}h = \frac{K^2\bar{\varepsilon}_{\mathrm{pl}}^{2}}{2E} \cdot V_{\mathrm{pl}} + \frac{K}{n+1} \cdot \bar{\varepsilon}_{\mathrm{pl}}^{n+1} \cdot V_{\mathrm{pl}} + \frac{E}{2}\bar{\varepsilon}_{\mathrm{el}}^{2} \cdot V_{\mathrm{el}} = W_{\mathrm{tot}} \tag{20.7}$$

Equation 20.7 illustrates the general relationship between the loading curvature C and the material parameters K, n, and E and is schematically illustrated in Figure 20.2. FEM simulations show that the deformed volumes and mean strains are highly dependent on the indenter geometry (e.g. half-apex angle θ of a conical indenter) but are also dependent on the material parameters. Furthermore, under certain circumstances (particular combinations of the material parameters), the deformed volumes normalized with h^3 and the mean strains remain constant. These combinations of material parameters

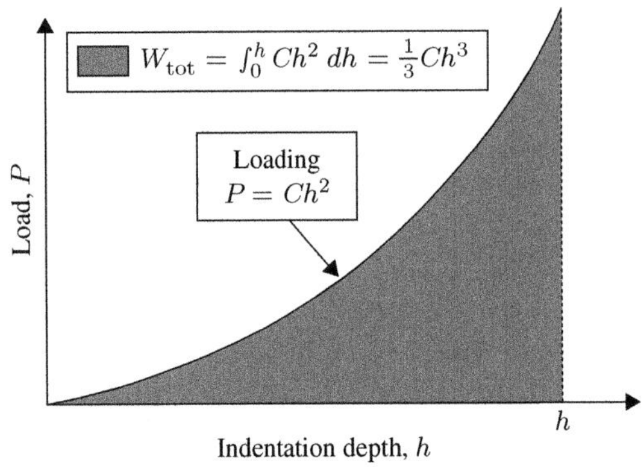

Figure 20.2 Schematic illustration of the loading curve for a self-similar indenter and its analytical description by Kick's law.

are found to exhibit a constant $\frac{K}{E}$ ratio and constant strain hardening exponent n. The relationship found between materials with constant $\frac{K}{E}$ ratio and n and the normalized volumes and mean strains was studied for four different materials by means of FEM simulations.

Figure 20.3 shows the stress and strain field during indentation ($\theta = 70.3°$, $h = 3.5\,\mu m$) of two materials (materials 1 and 2). They obey an equal constant $\frac{K}{E}$ ratio and n. Figure 20.3b) shows that the plastic, equivalent strain field is equal in size, shape, and quantity. Thus, the plastically deformed zone V_{pl} and the mean plastic strain $\overline{\varepsilon}_{pl}$ are equal for both materials. Furthermore, Figure 20.3a) illustrates that the shape and size of the von Mises stress field are equal for materials 1 and 2. Since the plastic zone is constant, both materials exhibit a constant size of the elastically deformed zone V_{el}. The only difference between materials 1 and 2 is the quantity of the stress values. The strains in Figure 20.4 prove that the elastic strains of materials 1 and 2 are equal. Therefore, because V_{el} is constant, the mean elastic strain $\overline{\varepsilon}_{el}$ is also constant for both materials (see Equation 20.1).

Materials 3 and 4 exhibit equal $\frac{K}{E}$ ratio and n, although $\frac{K}{E}$ differs from that of the first pair (materials 1 and 2). Figure 20.3 and 20.4 illustrate that (analogously to the first pair) both have equal V_{el}, V_{pl}, $\overline{\varepsilon}_{el}$, and $\overline{\varepsilon}_{pl}$. The differences between the first (materials 1 and 2) and second pair (materials 3 and 4) are the quantitative results of deformed volumes and mean strains, which is the consequence of the changed $\frac{K}{E}$ ratio.

Figure 20.5 shows the surface profiles of all four investigated materials at maximum load. It can be seen that when the $\frac{K}{E}$ ratio and n are constant, the surface profile of the indented material is equal, thus leading to a constant contact depth h_c. The contact depth of materials 1 and 2 is $h_c = 3.9\,\mu m$ (pile-up), whereas that of materials 3 and 4 is $h_c = 2.8\,\mu m$ (sink-in). This clearly demonstrates that the pile-up and sink-in behavior is not only related to the strain hardening behavior as it was often reduced to [41, 42]. It is rather a function of the strain hardening exponent n and the $\frac{K}{E}$ ratio. Thus, the normalized contact depth $\frac{h_c}{h}$ can be described by Equation (20.8). Similar results were obtained by Matá *et al.* as well as Taljat and Pharr [28, 43]:

$$\frac{h_c}{h_{max}} = f\left(\frac{K}{E}, n\right) \tag{20.8}$$

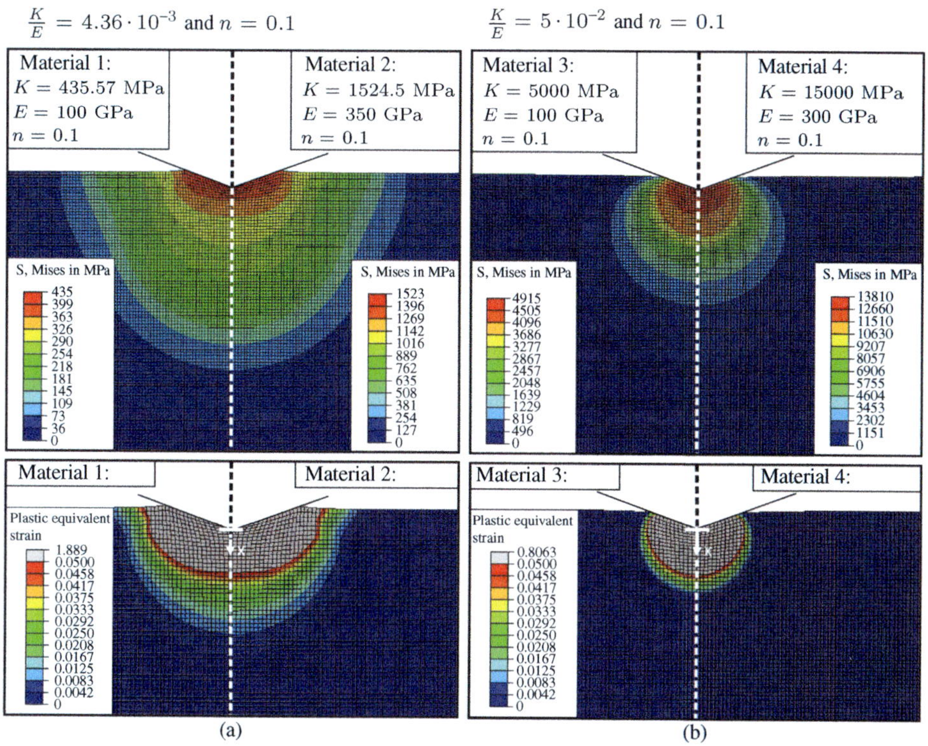

Figure 20.3 Stress and strain field of materials 1–4. a) The von Mises stress field σ_v is equal for materials 1 and 2 as well as materials 3 and 4, respectively; b) The plastic equivalent strain field is equal in size, shape, and quantity for materials 1 and 2 as well as materials 3 and 4, respectively.

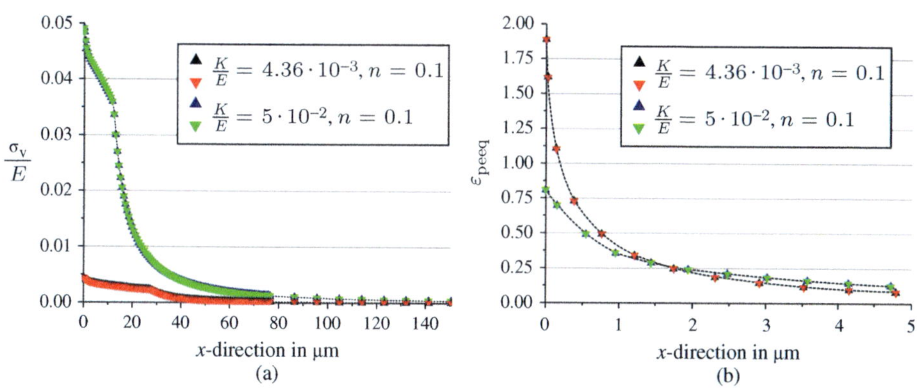

Figure 20.4 Strain profile of materials 1-4 along the path (x-direction, see Figure 20.3): a) Profile of the ratio between von Mises stress σ_v and Young's modulus E. b) Profile of plastic equivalent strain ε_{peeq}.

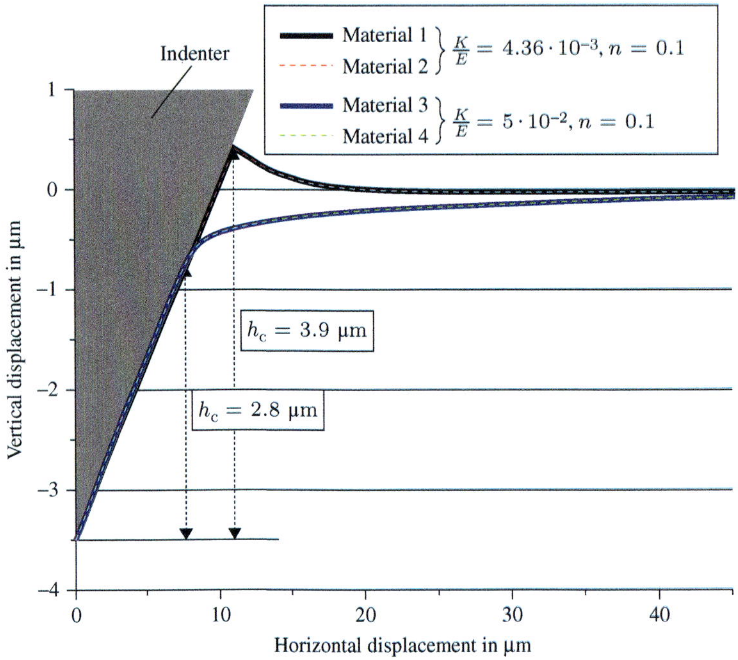

Figure 20.5 Surface profile at maximum indentation depth (3.5 μm). Materials 1 and 2 show pile-up and materials 3 and 4 sink-in behaviour. Materials with constant $\frac{K}{E}$ ratio show identical surface profiles, including equal contact depth h_c.

The result that materials exhibiting a constant $\frac{K}{E}$ ratio and a constant strain hardening exponent n obey equally deformed volumes normalized with h^3 and equal mean strains is useful information. It can be employed to investigate the relationship between the indentation results (C and $\frac{W_{el}}{W_{tot}}$) and the material parameters (K, n, and E). In case of constant V_{el}, V_{pl}, $\bar{\varepsilon}_{el}$, $\bar{\varepsilon}_{pl}$ following relationship between two materials a (K_a, n_a, E_a, and C_a) and b (K_b, n_b, E_b, and C_b) can be deduced:

$$\frac{K_a}{E_a} = \frac{K_b}{E_b} \quad \text{and} \quad n_a = n_b$$

$$C_b = C_a \cdot \frac{K_b}{K_a} = C_a \cdot \frac{E_b}{E_a} \tag{20.9}$$

According to Equation (20.9), a single solution for a given K, n and E should automatically generate all solutions of C for all combinations of materials that obey the same $\frac{K}{E}$ ratio and n. FEM simulations show that Equation (20.9) can be verified [32]. The relationship between C and the material parameters can thus be described by Equation (20.10). If Equation (20.10) is known for a constant Young's modulus E_{fix} it can be transposed (due to the validity of Equation 20.9) to random Young's moduli E, thus leading to Equation (20.12), which can be used with the known function $\frac{K}{E} = f(C \cdot \delta, n)$ (for any chosen E) and the factor δ to calculate C for given material parameters. The factor δ acts as a transposing factor and is the result of Equation (20.9). It transposes the solution of C for a known, constant Young's modulus E_{fix} to random Young's moduli E:

$$C = f\left(\frac{K}{E}, n\right) \tag{20.10}$$

$$C = f\left(\frac{K}{E}, n\right) \cdot \delta^{-1} \quad \text{with} \quad \delta = \frac{E_{\text{fix}}}{E} \tag{20.11}$$

$$\frac{K}{E} = f(C \cdot \delta, n) \tag{20.12}$$

Equation (20.11) formulates the forward problem (calculation of C for given material parameters). It can be seen that given material parameters lead to a unique solution of C, the forward problem is unique. This equation can also be rearranged to formulate the inverse problem (calculation of the material parameters for a known C). This leads to Equation (20.12). In contrast to the forward equation, the solution of the inverse equation is not unique for a known C. There are infinite possible solutions that are mathematically equivalent and indistinguishable. There is one equation for the three unknown material parameters. This problem still occurs even though another material parameter is known (there are still two variable material parameters in one equation). The inverse determination of material parameters is large field of application of nanoindentation. Inverse methods can suffer from a lack of uniqueness as discussed in [44–47]. The results of FEM simulations, as presented in this study, can give insights into the complex relationships between indentation results and material parameters and help to understand the forward and inverse indentation problem.

Another important parameter of the P-h curve is the $\frac{W_{\text{el}}}{W_{\text{tot}}}$ ratio. Analogous to the loading curvature C, it is possible to investigate the relationship between two materials a and b with $(\frac{W_{\text{el}}}{W_{\text{tot}}})_a$ and $(\frac{W_{\text{el}}}{W_{\text{tot}}})_b$ and equal $\frac{K}{E}$ ratio and equal n. It can be shown that $\frac{W_{\text{el}}}{W_{\text{tot}}}$ is constant for a constant $\frac{K}{E}$ and constant n (Equation 20.13). FEM simulations confirm the validity of this relationship [32]:

$$\left(\frac{W_{\text{el}}}{W_{\text{tot}}}\right)_a = \left(\frac{W_{\text{el}}}{W_{\text{tot}}}\right)_b \tag{20.13}$$

Tho *et al.* found a similar relationship between the ratio of indentation work $\frac{W_{\text{tot}}-W_{\text{el}}}{W_{\text{tot}}}$ and material parameters based on the cavity expansion analogy for sharp indentation [44].

The $\frac{W_{\text{el}}}{W_{\text{tot}}}$ ratio can be expressed by a function of $\frac{K}{E}$ and n (Equation 20.14) without a transposing factor δ. If this function for any constant Young's modulus is known, it is generally valid for random Young's moduli.

Equation (20.14) formulates the forward problem. For any given material parameters there exists a unique solution for $\frac{W_{\text{el}}}{W_{\text{tot}}}$. The inverse problem can be formulated according to Equation (20.15). Analogous to the inverse equation of C, the solution for a known $\frac{W_{\text{el}}}{W_{\text{tot}}}$ ratio is not unique (even though another material parameter is known). It is only possible to derive a set of mathematically equivalent inverse solutions that all result in equal $\frac{W_{el}}{W_{tot}}$:

$$\frac{W_{el}}{W_{tot}} = f\left(\frac{K}{E}, n\right) \tag{20.14}$$

$$\frac{K}{E} = f\left(\frac{W_{el}}{W_{tot}}, n\right) \tag{20.15}$$

Every single independent indentation parameter leads to a set of possible inverse solutions (e.g. C or $\frac{W_{el}}{W_{tot}}$). In order to solve the inverse problem uniquely for K, E, and n, at least three independent parameters are necessary as input. If, for example, the Young's modulus is known a priori, a single P-h curve contains theoretically sufficient information for a unique determination of K and n. The parameters C and $\frac{W_{el}}{W_{tot}}$ should lead to independent sets of possible inverse solutions that exhibit a single intersection.

Some researchers state that the P-h curve of a self-similar indenter leads, at most, to two independent indentation parameters [44, 47–51]. As a consequence, derived inverse solutions for three material parameters from a single self-similar P-h curve might not be unique. The use of additional indenter geometries is an expedient method of introducing further independent indentation parameters. The mean strains and normalized volumes are effectively changed by varying the indenter geometry. Thus, derived sets of inverse solutions for different indenter geometries are independent of each other and should lead to unique inverse solutions. For practical use of the forward and inverse equations of this study, knowledge of following functions is required:

$$C = f\left(\frac{K}{E_{fix}}, n\right) \tag{20.16}$$

$$\frac{W_{el}}{W_{tot}} = f\left(\frac{K}{E}, n\right) \tag{20.17}$$

FEM is a powerful tool for solving the forward indentation problem and can be used to determine these functions. The calculation effort is low due to the fact that only two variable parameters need to be taken into account: $\frac{K}{E}$ and n. The initial three-dimensional problem (K, E, and n) can be reduced to two dimensions ($\frac{K}{E}$ and n), thus simplifying the problem and its calculation.

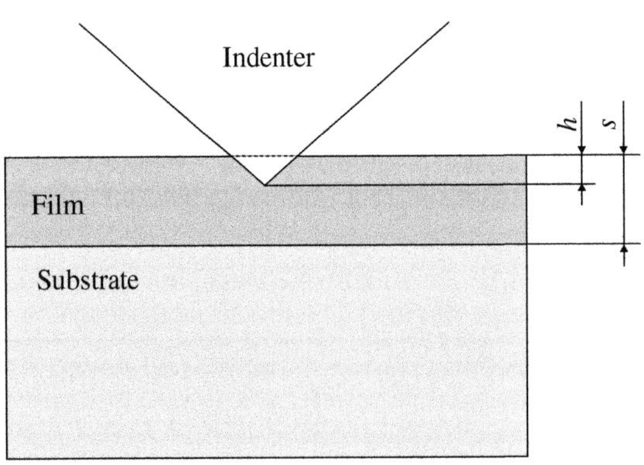

Figure 20.6 Schematic illustration of the indentation of a thin film on a substrate.

Table 20.1 Material parameters used in the FEM simulations.

Material	K (MPa)	n	E (GPa)	ν
Hard	8000	0	300	0.2
Soft	1000	0.15	210	0.3

20.3 Indentation of Thin Films

Nanoindentation is often used for the mechanical characterization thin films (Figure 20.6). Due to small film thicknesses and high indentation depths, respectively, the indentation results can significantly influenced by the substrate [52–56]. For given material parameters of film and substrate the indentation depth to film thickness ratio should not exceed a critical value when indented with a sharp indenter. In practise the indentation depth is often restricted to 10 % of the film thickness, as a rule of thumb [57, 58]. The 10 % restriction cannot be applied to all film-substrate combinations and has no physical basis [59].

In order to investigate the deformation behaviour of the film-substrate compound many numerical studies have been conducted (e.g. [52, 53]). In this respect, FEM simulations can give insight into the deformation behaviour and allow systematic investigations of the influence of film thickness, indentation depth, material parameters of film and substrate on the indentation results. Most studies focus on the accurate determination of mechanical properties such as hardness or Young's modulus of the film.

This chapter gives examples of FEM simulations of indentation (self-similar indenter) of a thin film with different mechanical properties of film and substrate. Firstly, it is established a soft coating on a hard substrate and, secondly, a hard coating on a soft substrate. The material is assumed to follow Ludwik's power law (see section 20.2, Equation (20.3)). The material parameters of films and substrates are summarized in Table 20.1.

The first investigated case is a hard film on a soft substrate. The film thickness is $s = 10\,\mu m$ and the indentation depth $h = 3.5\,\mu m$. Figure 20.7a) illustrates the plastic strain distribution at maximum indentation depth of the compound. Plastic deformation is not limited to the film, both film and substrate are plastically deformed. As can be seen in Figure 20.8, the P-h curve follows Kick's law at low indentation depths and is in agreement with the simulated P-h curve for the pure hard material (reference). With increasing indentation depth, at approx. 1 µm, the curve is progressively deviating from the reference. Since the substrate is softer the force is shifted to lower forces. As a consequence strength or rather hardness is progressively underestimated. The hardness calculated by maximum force divided by projected contact area is 19.2 GPa for the hard material (reference). The hardness for the hard film on soft substrate is significantly underestimated (5.4 GPa). Further parameters such as Young's modulus, indentation energies etc. are also influenced by the substrate. The higher the indentation depth and the smaller the film thickness the larger is the influence of the substrate. The important quantity is the indentation depth to film thickness ratio.

In case of a soft film ($s = 10\,\mu m$) on a hard substrate the substrate influence is considerably lower. Figure 20.7b) shows that at an indentation depth of 3.5 µm the plastic deformation is limited to the film. Also, the P-h curve is almost not influenced by the substrate. The hardness of the compound is calculated to 1.91 GPa which is in good agreement to the reference simulation of the pure soft material (1.87 GPa) and only slightly overestimated. Compared to the hard film on a soft substrate the indentation results of the soft film on a hard substrate is less critical.

20.4 Indentation of a Hard Phase Embedded in Matrix

The mechanical characterization of small phases as features in multi-phase materials is a common application of nanoindentation. In analogy to the indentation of thin films the surrounding matrix can significantly influence the indentation results such as hardness and Young's modulus of a small embedded hard phase. The following study deals with the evaluation of matrix influence during indentation of a spherical hard phase and based on [19].

The indentation of the spherical hard phase embedded in a steel matrix by a self-similar sharp indenter is illustrated in Figure 20.9. In this study the matrix consists of the cold work tool steel X210Cr12 with an embedded M_7C_3 carbide. The indentation depth is h and the spherical hard phase has the diameter d. For FEM simulation the material parameters in Tables 20.2 and 20.3 were used. The Young's modulus of the matrix was 210 GPa and the Poisson's ratio 0.3 [60].

Figure 20.10 shows the equivalent plastic strain distribution during the indentation of the M_7C_3 with diameter $d = 10$ µm in the X210Cr12 soft-annealed matrix at maximum indentation depth ($h = 2$ µm). The hard phase as well as the matrix plastically deform. Even though the quenched and tempered matrix condition possesses a higher matrix

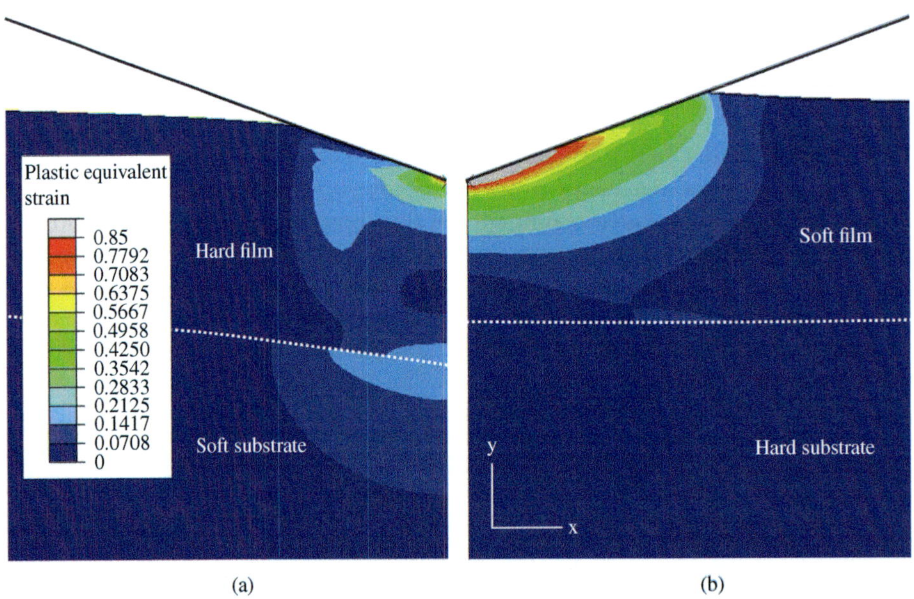

Figure 20.7 Plastic strain distribution of the film-substrate system: a) hard film on soft substrate. b) soft film on hard substrate.

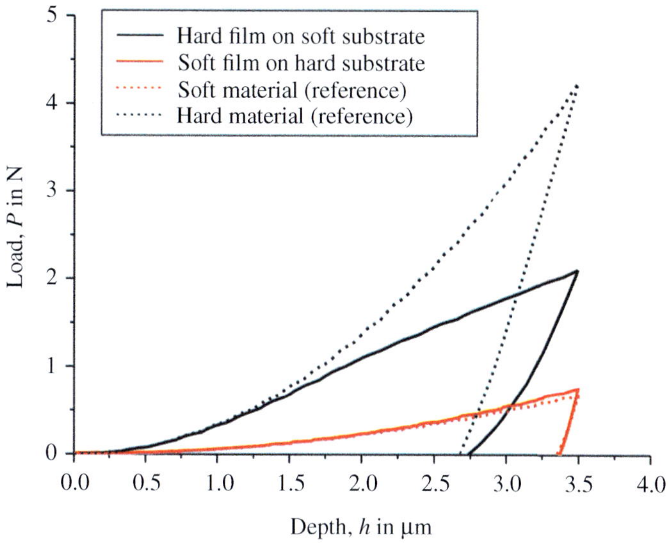

Figure 20.8 P-h curves of hard film on soft substrate, soft film on hard substrate as well as of the pure soft and hard materials.

Figure 20.9 Schematic illustration of the indentation of a hard phase embedded in a softer matrix.

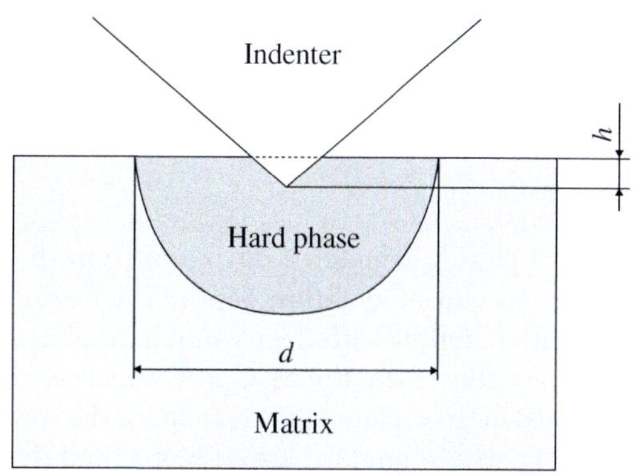

strength, it is still intensely plastically deformed. The strain distribution in Figure 20.10 a) shows a typical Hertz-like shape of spherical indentation beneath the hard phase and reveals that the hard phase itself acts as an indenter which penetrates the matrix. During the indentation of that compound system, the hard phase as well as the matrix undergo elastic and plastic deformation. Thus, the measured load-displacement curve and the calculated mechanical properties are both affected by the matrix influence.

Figure 20.11 shows the simulated load-displacement curves for the compound system with different M_7C_3 carbide diameters d as well as the bulk reference curves. The diameter of the carbide is varied from 10 µm to 100 µm, while the indentation depth remains constant at 2 µm.

The maximum load along each of the curves diminishes with decreasing hard phase diameter for both matrix heat-treatment conditions (grey dashed lines). Depending on

Table 20.2 Plastic behaviour of soft-annealed and hardened + tempered X210Cr12 matrix.

Soft-annealed X210Cr12 matrix		Hardened + tempered X210Cr12 matrix	
True stress σ in MPa	True plastic strain ε	True stress σ in MPa	True plastic strain ε
280	0	550	0
310	0.0015	1550	0.0074
380	0.0058	1620	0.0079
520	0.0153	1730	0.0091
560	0.025	1890	0.0132
670	0.05	2180	0.0253
870	0.11	3270	0.1

Source: Broeckmann 1995 [60].

Table 20.3 Mechanical properties of M_7C_3 carbide used in the FEM simulations [60].

Young's modulus (GPa)	Poisson's ratio	Yield stress (MPa)
280	0.3	8000

the hard phase diameter, a deviation from the M_7C_3 reference curve becomes apparent at a certain indentation depth. The beginning of the deviation is shifted to lower indentation depths with decreasing hard phase diameter.

The deviation from the M_7C_3 reference curve is less significant for the quenched and tempered matrix (Figure 20.11) than for the soft-annealed matrix of equal carbide diameters. The maximum load also is higher and the onset of deviation from reference takes place at higher indentation depths. This indicates a less pronounced matrix influence on the indentation results for the quenched and tempered matrix.

As presented before, the surrounding matrix can highly influence the load-displacement curve. Thus, the mechanical properties such as hardness and Young's modulus derived from that curves are also affected. Figure 20.12 a) shows a plot of apparent hardness vs. the hard phase diameter for a constant indentation depth of 2 μm. The hardness H_i was converted to a Vickers hardness number according to DIN EN ISO 14577-1 [61]. For large diameters, the hardness number converges to that of the M_7C_3 bulk reference value (negligible matrix influence). As previously illustrated, however, with decreasing hard phase diameter the apparent hardness decreases and the influence of the matrix becomes significant. For a 10 μm M_7C_3 in the soft-annealed X210Cr12 matrix, the hardness is almost equivalent to that of the soft-annealed matrix. As previously noted for the load-displacement curves, the quenched and tempered matrix has a weaker influence on the results for the same hard phase diameters.

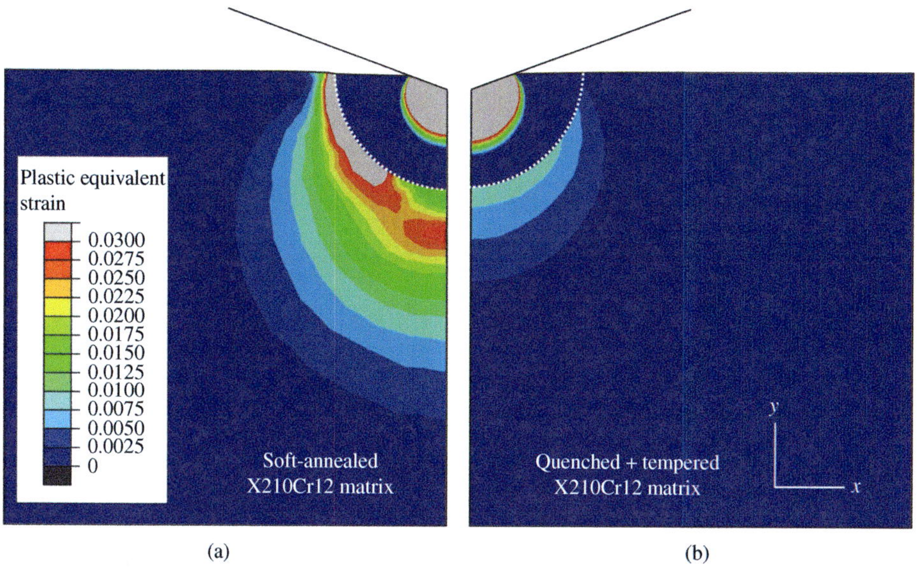

Figure 20.10 Plastic strain distribution of the compound system consisting of M_7C_3 carbide ($d = 10$ mm) and surrounding X210Cr12 matrix in two different heat-treatment conditions at maximum indentation depth (2 μm): a) M_7C_3 in soft-annealed X210Cr12 matrix. b) M_7C_3 in quenched + tempered X210Cr12 matric.

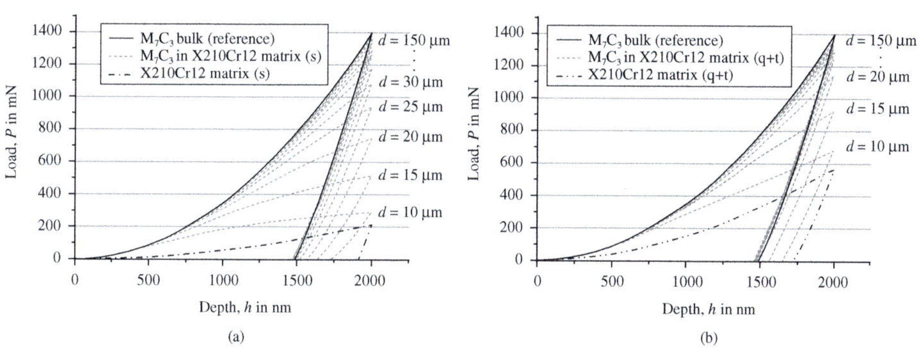

Figure 20.11 P-h curves of M_7C_3 in X210Cr12 matrix for different carbide diameters *d* and bulk reference curves of matrix and carbide: a) M_7C_3 in soft-annealed matrix (s). b) M_7C_3 in quenched + tempered matrix (q+t).

A similar effect can be noticed for the calculated Young's modulus (Figure 20.12b); the loading-unloading part of the load-displacement curve is affected by the matrix influence. The slope of the unloading curve, and in turn the Young's modulus, is increasingly affected as the hard phase diameter decreases. For large diameters the Young's modulus is in good agreement to that of the bulk M_7C_3. With decreasing hard phase diameter, the Young's modulus shows a significant drop.

The determination of hardness and Young's modulus is highly dependent on the correct calculation of the contact depth. The contact depth is not measured directly, but

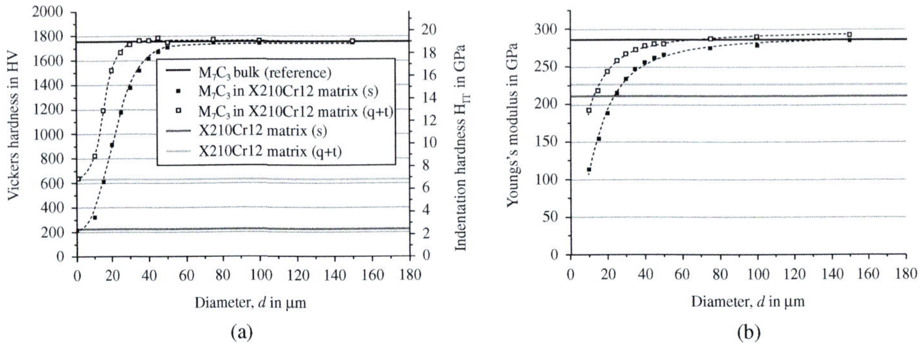

Figure 20.12 Dependence of determined material parameters on the diameter of the M_7C_3 carbide in X210Cr12 matrix in case of two different heat-treatment conditions: soft-annealed matrix (s) and quenched + tempered matrix (q+t). a) Hardness as a function of hard phase diameter d (constant indentation depth of 2 µm). b) Young's modulus as a function of hard phase diameter d (constant indentation depth of 2 µm).

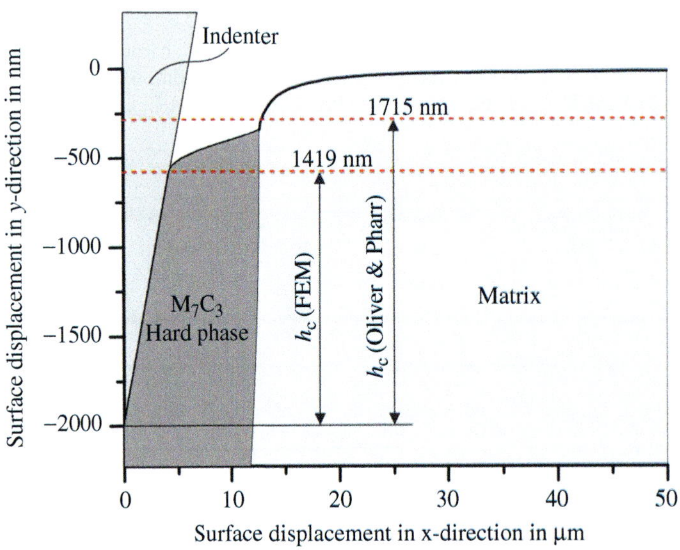

Figure 20.13 Surface profile of the indented M_7C_3 with diameter $d = 25$ µm in the X210Cr12 soft-annealed matrix at maximum indentation depth $h = 2$ µm and declaration of the contact depth h_c calculated with the Oliver and Pharr method and the contact depth evaluated from FEM simulation.

is calculated from the load-displacement curve. It is well known from the pile-up phenomenon that errors in the determination of the contact depth can significantly affect the determination of hardness and Young's modulus. Figure 20.13 shows the surface profile at maximum load of an indented M_7C_3 carbide in the X210Cr12 soft-annealed matrix. The contact depth as calculated by the Oliver and Pharr method [58] is marked in addition to the real contact depth from the simulation. The Oliver and Pharr method predicts a contact depth h_c of 1715 nm, while the simulation shows a considerably lower true contact depth of 1419 nm. The low strength of the soft-annealed matrix leads to a distinctive plastic matrix deformation and thus to a push-in event of the carbide into the matrix. The hard phase indents the matrix at high indentation loads and small hard phase diameters. The push-in event is the reason for the incorrect determination of the contact depth. An overestimated projected contact area leads to lower values of hardness and Young's modulus. The error of the Young's modulus is higher compared to the one of the hardness, although the hardness depends quadratically on the contact depth, while the Young's modulus depends linearly on it. The Young's modulus is calculated with the contact depth (projected contact area) and the contact stiffness S, which is dependent on the maximum force and the slope of the unloading curve. As Figure 20.11 shows, the maximum force as well as the slope of the unloading curve decrease with decreasing hard phase diameter. The decrease of the slope is related to the elastic deformation of the hard phase and the matrix, which is influenced by the push-in event of the hard phase. The reduction of the contact stiffness with decreasing hard phase diameter leads together with the overestimated projected contact area to a distinct underestimation of the Young's modulus.

Similar to the indentation of thin films the matrix can significantly influence the indentation results of an embedded hard phase. Since the hard phase has a higher strength compared to the matrix important indentation parameters such as hardness and Young's modulus are increasingly influenced with increasing indentation depth. Analogous to the indentation of thin films (critical h/s ratio) a critical ratio of indentation depth to hard phase diameter h/d (in case of spherical particles) exists, though, the ratio is strongly dependent on the mechanical properties of both hard phase and matrix.

References

1 Franke O, Durst K, Maier V, *et al.* Mechanical properties of hyaline and repair cartilage studied by nanoindentation. Acta Biomaterialia (2007) 3: 873–881.
2 Ebenstein D, Kuo A, Rodrigo J, *et al.* A nanoindentation technique for functional evaluation of cartilage repair tissue. Journal of Materials Research (2004) 19: 273–281.
3 Oyen M, Ferguson V, Bembey A, *et al.* Composite bounds on the elastic modulus of bone. Jounrnal of Biomechanics (2008) 41: 2585–2588.
4 Rho J, Zioupos P, Currey J, Pharr G.-M. Microstructural elasticity and regional heterogeneity in human femoral bone of various ages examined by nano-indentation. Journal of Biomechanics (2002) 35: 189–198.
5 Rho J, Tsui T, Pharr G.-M. Elastic properties of human cortical and trabecular lamellar bone measured by nanoindentation. Biomaterials (1997) 18(20): 1325–1330.
6 Fan Z, Swadener J, Rho J, Roy M, Pharr G. Anisotropic properties of human tibial corical bone as measured by nanoindentation. Journal of Orthopaedic Research (2002) 20: 806–810.
7 Lee G, Kang S, Kwon D. Characterization of elastic modulus and work of adhesion in elastomeric polymers using microinstrumented indentation technique. Materials Science and Engineering A, (2008) 496: 494–500.

8 Oliveira G, Costa C, Treixeira S, Costa M. The use of nano- and micro-intrumented indentation tests to evaluate viscoelastic behavior of poly(vinylidene fluoride) (PVDF). Polymer Testing (2014) 34: 10–16.
9 Giannakopoulos A. Indentation of graded substrates. Thin Solid Films (1998) 332: 172–179.
10 Choi I, Detor A, Schwaiger R, Dao M, *et al.* Mechanics of indentation of plastically graded materials - II: Experiments on nanochrystalline alloys with grain size gradients. Journal of the Mechanics and Physics of Solids (2008) 56: 172–183.
11 Choi I, Dao M, Suresh S Mechanics of indentation of plastically graded materials - I: Analysis. Journal of the Mechanics and Physics of Solids (2008) 56:157–171.
12 Leite M, Ferland F. Determination of unconfined compressive strength and Young's modulus of porous materials by indentation tests. Engineering Geology (2001) 59: 267–280.
13 Fleck N, Otoyo H, Needleman A. Indentation of porous solids. International Journal of Solids and Structures (1992) 29(13): 1613–1635.
14 Clément P, Meille S, Chevalier J, Olagnon C. Mechanical characterization of highly porous inorganic solids materials by instrumented micro-indentation. Acta Materialia (2013) 61: 6649–6660.
15 Muir Wood A, Clyne T. Measurement and modelling of the nanoindentation response of shape memory alloys. Acta Materialia (2006) 54: 5607–5615.
16 Crone W, Shaw G, Stone D, *et al.* Shape recovery after nanoindentation of NiTi thin films. Society for Experimental Mechanics Annual Conference Proceedings, 2003, pp. 1–6.
17 Pfetzing-Micklich J, Wagner M, Zarnetta E, *et al.* Nanoindentation of a pseudoelastic NiTiFe shape memory alloy. Advanced Engineering Materials (2010) 12(1–2): 13–19.
18 Cheng Y, Page T, Pharr G.-M., Swain M, Wahl K. Fundamentals and application of instrumented indentation in multidisciplinary research. Journal of Materials Research (2004) 19(1): 1–2.
19 Pöhl F, Huth S, Theisen W. Finite element method-assisted acquisition of the matrix influence on the indentation results of an embedded hard phase. Materials Science and Engineering A, (2013) 559: 822–828.
20 Gamonpilas C, Busso E. Characterization of elastoplastic properties based on inverse analysis and finite element modeling of two seperate indenters. Journal of Engineering Materials and Technology (2007) 129: 603–608.
21 Gu Y, Nakamura T, Prchlik L, *et al.* Micro-indentation and inverse analysis to characterize elastic-plastic graded materials. Materials Science and Engineering (2003) 345: 223–233.
22 Nakamura T, Wang T, Sampath S. Determination of properties of graded materials by inverse analysis and instrumented indentation. Acta Materialia (2000) 48: 4293–4306.
23 Alcalá J, Barone A, Anglada M. The influence of plastic hardening on surface deformation modes around Vickers and spherical indents. Acta Materialia (2000) 48: 3451–3464.
24 Daily B, Elson E, Zahalak G. Cell poking. Determination of the elastic area compressibility modulus of the erythrocyte membrane. Biophysical Journal (1984), 671.

25 Kopernik M, Stanislawczyk A. Classical inverse and metamodel approach in identification of nanomaterials properties. Archives of Civil and Mechanical Engineering (2009) 9(3): 77–96.
26 Kopernik M, Pietrzyk M. Rheological model and mechanical properties of hard nanocoatings in numerical nanoindentation test. Proceedings of the 10th ESAFORM Conference on Material Forming, European Scientific Association for material Forming AIP Conference Proceedings (2007) 907: 659–664.
27 Zambaldi C, Roters F, Raabe D, Glatzel U. Modeling and experiments on the indentation deformation and recrystallization of a single-crystal nickel-base superalloy. Materials Science and Engineering A (2007) 454–455: 433–440.
28 Taljat B, Pharr G.-M. Development of pile-up during spherical indentation of elastic-plastic solids. Journal of Solids and Structures (2004) 41(14): 3891–3904.
29 Bouzakis K, Michailidis N, Hadjiyiannis S, Skordaris G, Erkens G. Continuous FEM simulation of the nanoindentation: Actual indenter tip geometries, material elastoplastic deformation laws and universal hardness. Zeitschrift für Metallkunde (2002) (93): 862–869.
30 Bolshakov A, Pharr G.-M. Influences of pileup on the measurement of mechanical properties by load and depth sensing indentation tequniques. Journal of Materials Research (1998) 13(4): 1049–1058.
31 Dao M, Chollacoop N, van Fliet K, Venkatesh T, Suresh S. Computational modeling of the forward and reverse problems in instrumented sharp indentation. Acta Materialia (2001) 49: 3899–3918.
32 Pöhl F, Huth S, Theisen W. Indentation of self-similar indenters: An FEM-assisted energy-based analysis. Journal of Mechanics and Physics of Solids (2014) 66: 32–41.
33 Prasad K, Chollacoop N, Ramamurty U. Role of indenter angle on the plastic deformation underneath a sharp indenter and on representative strains: An experimental and numerical study. Acta Materialia (2011) 59: 4343–4355.
34 Branch N, Subhash G, Arakere N, Klecka M. Material-dependent representative plastic strain for the prediction of indentation hardness. Acta Materialia (2010) 58: 6487–6494.
35 Lee J, Lee C, Kim B. Reverse analysis of nano-indentation using different representative strains and residual indentation profiles. Materials and Design (2009) 30: 3395–3404.
36 Cao Y, Qian X, Huber H. Spherical indentation into elastoplastic materials: Indentation-response based definition of the representative strain. Materials Science and Engineering A (2007) 454–455: 1–13.
37 Ogasawara N, Chiba N, Chen X. Representative strain of of indentation analysis. Journal of Materials Research (2005) 20(8): 2225–2234.
38 Tabor D. The Hardness of Metals. Oxford University Press, Oxford; 2000.
39 Kick F. Das Gesetz der proportionalen Widerstände und seine Anwendungen: Nebst Versuchen über das Verhalten verschiedener Materialien bei gleichen Formänderungen sowohl unter der Presse als dem Schlagwerk. A. Felix, Leipzig; 1885.
40 Cheng Y, Cheng C. Scaling approach to conical indentation in elastic-plastic solids with work hardening. Journal of Applied Physics (1998) 84: 1284–1291.
41 Matthews J. Indentation hardness and hot-pressing. Acta Metallurgica (1980) 28(3): 311–318.

42 Francis H. Phenomenological analysis of plastic spherical indentation. Journal of Engineering Materials and Technology (1976) 98(3): 272–281.
43 Mata M, Anglada M, Alcalá J. Contact deformations regimes around sharp indentations and the concept of characteristic strain. Journal of Materials Research (2002) 17(5): 964–976.
44 Tho K, Swaddiwudhipong S, Liu Z, Zeng K, Hua J. Uniqueness of reverse analysis from conical indentation tests. Journal of Materials Research (2008) 19: 2498–2502.
45 Chen X, Ogasawara N, Zhao M, Chiba N. On the uniqueness of measuring elastoplastic properties from indentation: The indistinguishable mystical materials. Journal of the Mechanics and Physics of Solids (2007) 55: 1618–1660.
46 Liu Y, Ngan A. Depth dependence of hardness in copper single crystals measured by nanoindentation. Scripta Materialia (2001) 44: 237–241.
47 Cheng Y, Cheng C. Can stress-strain relationships be obtained from indentation curves using conical and pyramidal indenters? Journal of Materials Research (1999) 14(9): 3493–3496.
48 Le M. Materials characterisation by dual sharp indenters. International Journal of Solids and Structures (2009) 46: 2988–2998.
49 Le M. A computational study on the instrumented sharp indentations with dual indenters. International Journal of Solids and Structures (2008) 45: 2818–2835.
50 Alcorta J, Martinez-Esnaola J, Sevillano J. Absence of one-to-one correspondence between elastoplastic properties and sharp-indentation load-penetration data. Journal of Materials Research (2005) 20: 432–437.
51 Loubet J, Georges J, Marchesini O, Meille G. Vickers indentation curves of magnesium oxide (MgO). Journal of Tribology (1984) 106(1): 43–48.
52 Tricoteaux A, Duarte G, Chicot D, *et al.* Depth-sensing indentation modeling for determination of elastic modulus of thin films. Mechanics of Materials (2010) 42: 166–174.
53 Cleymand F, Ferry O, Kouitat R, *et al.* Influence of indentation depth on the determination of the apparent Young's modulus of bi-layer material: Experiments and numerical simulation. Surface and Coatings Technology (2005) 200: 890–893.
54 Richter F, Herrmann M, Molnar F, *et al.* Substrate influence in Young's modulus determination of thin films by indentation methods: Cubic boron nitride as an example. Surface and Coatings Technology (2006) 201: 3577–3587.
55 Sun Y, Bell T, Zheng S. Finite element analysis of the critical ratio of coating thickness to indentation depth for coating property measurements by nanoindentation. Thin Solid Films (1995) 258: 198–204.
56 Chollacoop N, Li L, Gouldstone A. Errors in resolved modulus during nano-indentation of hard films on soft substrates: A computational study. Materials Science and Engineering A, (2006) 423: 36–40.
57 Persch G, Born C, Utesh B. Nano-hardness investigations of thin films by an atomic force microscope. Microelectronics Engineering (1994) 24: 113–121.
58 Oliver W, Pharr G. An improved technique for determination hardness and elastic-modulus using load and displacement sensing indentation experiments. Journal of Materials Research (1992) 7(2): 1564–1583.
59 Chudoba T, Schwarzer N, Richter F. Steps towards a mechanical modeling of layered systems. Surface and Coatings Technology (2002) 154(2–3): 140–151.

60 Broeckmann C. Bruch karbidreicher Stähle: Experiment und FEM-Simulation unter Berücksichtigung des Gefüges: Dissertation: Dissertation, Ruhr-Universität Bochum. VDI-Verlag, Düsseldorf; 1995.
61 DIN EN ISO 14577 Instrumentierte Eindringprüfung zur Bestimmung der Härte und anderer Werkstoffparameter (Instrumentation indentation testing for the determination of hardness and further material parameters); 2007.

21

Investigations Regarding Plastic Flow Behaviour and Failure Analysis on CrAlN Thin Hard Coatings

Jan Perne

RWTH Aachen University, Templergraben, Germany

21.1 Introduction

Hard coatings are often used for tools to protect them against mechanical or chemical load like wear, particle sliding or adhesion. A decisive method to mechanically characterize the coating is nanoindentation. Especially for complex loading cases like combinations of cavitation erosion, particle sliding, counterbody impact or melt adhesion in plastic manufacturing the coating layout is more and more supported by Finite Element Method (FEM) Simulations of the loading cases. Therefore a detailed understanding of the coatings mechanical behavior is necessary regarding the plastic deformation at different strain rates and temperatures as well as criterions for failure initiation. Furthermore knowledge of the correlation of these mechanical properties with changes in chemical composition or grain structure is important for improvements in coating design.

The method introduced here is capable of determining a material's flow curves in dependences of strain rate and even temperature, if suitable devices especially for temperature control are available. The method has been applied to fully characterize the (Cr, Al)N coating system regarding its mechanical properties in dependence of chemical composition, grain structure and nanolamination. Also the influence of grain size on strain rate dependency has been investigated. In a last step the responsible stress-strain constellations leading to fracture initiation in the nanolaminated coating have been studied and critical values of size and magnitude of these stresses have been determined.

21.2 Description of the Method

The method presented here to determine the plastic flow curve of thin coatings was developed in a collaborative research center[1] to investigate dependencies of plasma processes and coating properties. It was mainly inspired by the work of Bouzakis *et al.* and

1 Collaborative Research Center 'SFB-TR 87 - Pulsed high power plasmas for the synthesis of nano-structured functional layers', project A4 'Modeling of elastic–plastic behavior of nanostructured coatings'

Applied Nanoindentation in Advanced Materials, First Edition. Edited by Atul Tiwari and Sridhar Natarajan.

combines nanoindentation measurements with FEM simulation. The method was calibrated with different materials of known mechanical properties, furthermore possible influences of different substrate materials as well as statistical calculation methods on the accuracy of the determined flow curves have been investigated [1].

In a second step, nanoscratch investigations combined with FEM simulations on a nanolaminated coating have been used to determine failure criteria and to validate the determined flow curves.

21.2.1 Flow Curve Determination

The plastic flow curve determined by this method is described by the Johnson-Cook equation without the strain rate and temperature depending terms:

$$\sigma(\varepsilon) = A + B \cdot \varepsilon^n \tag{21.1}$$

A represents the materials yield strength, B the (theoretical) strain hardening at 100 % plastic strain and ε the plastic strain. n is a factor to describe the progression of the plastic flow curve between both stress values. The stress value $\sigma(\varepsilon)$ corresponds to the von Mises stress. While the coatings elastic behavior is directly determined from nanoindentation measurement utilizing the Oliver–Pharr method [2], the plastic flow curve parameters are calculated from the force-displacement-curves in several analyzing steps.

21.2.1.1 Nanoindentation Step

In a first step the coating is analyzed by nanoindentation measurements with a spherical indenter of approximately 10 µm radius. In contrary to conical indenter with a very small rounded area at the tip of usually around 50 nm, spherical indenter cause a continually changing stress-strain field in the coating material with increasing indentation depth, providing more information for the following analyzes steps. This is an advantage especially for rough coatings or those with inhomogeneous grain structure because of the greater material volume contributing to the force-displacement response. For very thin coatings, the greater indentation depth needed for spherical indenter with this radius may be a disadvantage because of an increasing substrate influence on the force-displacement curve.

For the measurements a loading rate is defined corresponding to a constant strain rate of the indented material. The loading rate depends on the indenter form function, for conical indenter this results in a quadratic load rate. The maximum indentation depth has to be chosen regarding coating thickness and the ratio of coating and substrate hardness. Due to my experience a comparable hardness of coating and substrate allows significant higher indentation depth than 10% of the coating thickness without measurable effects on the force-displacement curves [1]. From the measured force-displacement curves a mean curve is calculated. This mean curve, as well as the resulting elastic moduli values are used in the following steps to determine the materials plastic properties.

21.2.1.2 Yield Strength Determination

There are several methods published to determine the yield strength value from nanoindentation measurements. Most use data of spherical indentation to determine the yield stress, either of one loading step with continuously raising force like Juliano *et al.* [3] or cyclic measurements with increasing maximum force values [4]. Basu *et al.* [5] found a correlation of material yield strength with Vickers hardness, so for fast investigations a simple Vickers measurement can be used for estimation.

In the investigations presented here mainly the Juliano method is used due to the fact that only one indentation measurement type has to be performed for the whole analysis. The Juliano method uses the force-displacement curves from nanoindentation and calculates stress-strain curves considering the indenter form. The calculated stress-strain curves are not identical with the materials flow curves, but indicate the yield strength at the end of an elastic linear rise of the curve. The Juliano yield strength showed good agreements with those determined from tensile tests.

21.2.1.3 Flow Curve Determination by Iterative Simulation

The determined elastic properties of the coating material and the yield strength are used as input parameters in a FEM simulation model of the indentation process. The model contains two parts, the indenter with its geometry and material properties and the sample containing the substrate with its material properties and the coating. The coating's material properties are set with the known elastic properties and yield strength, while the flow curve parameters B and n are set with estimated values. Then the flow curve parameters are determined iteratively by simulating the indentation process and comparing the simulated and the measured force-displacement curve using an algorithm. The algorithm suggests changes on parameters B and n and the next iteration cycle is started until the deviation of measured and simulated force-displacement curve fall below 1 ‰.

21.2.1.4 Determination of Strain Rate and Temperature Dependency

As mentioned in section 21.2.1.1, the determined flow curve is valid for a specific strain rate induced in the coating material under the indenter. This strain rate can be set by the force growth rate. By determining flow curves of measurements with different indentation strain rates, a strain rate dependency can be measured [6]. The same approach can be used for determining the temperature dependence as shown by Bouzakis *et al.* [7]. For temperature dependency measurements the change of the indenter and substrate material properties has to be taken into account as well as changes in the stiffness of the device. Furthermore significant heat flow has to be avoided to prevent high thermal drift rates.

21.2.2 Failure Criterion Determination with Nano-scratch Analysis

Based on the flow curve determination method described in section 21.2.1, a method to determine failure criterions of thin coatings has been developed based on nano-scratches. Therefore scratches are performed on a material with known plastic forming behavior, ideally strain rate dependent. The scratches can be varied regarding their normal force and their scratch velocity. In the next step pictures of cross-sections of the scratches are made preferentially with scanning electron microscope (SEM). The scratches are simulated with a model of the sample and indenter and considering the different force and scratch velocity values. The simulation results are used to determine the stress-strain values at each point in the cross-sectioned scratch area. If a multi- or nanolayered coating is used for scratching, the remaining plastic deformation normal to the surface can be determined by measuring the layer thickness at any point in the scratch area and comparing it with an undeformed area. This plastic deformation map can be compared with the corresponding values of the simulation to validate the model.

The cross-sectional images of scratches with different forces and velocities are compared and occurrence, size, number and location of delaminations, cracks and fractures are noted. Then the simulation results can be used to study the stress-strain history of these failure points during indenter overrun. The magnitude and size of the stress fields necessary to cause a failure can be determined by comparing the simulated stress fields of scratches with occurring failure and those without.

21.3 Investigations into the CrAlN Coating System

The methods described in Chapter 2 have been evaluated on the CrAlN coating system and were used to widely characterize the influence of chemical composition, deposition process, microstructure and layer-design on the mechanical coating properties. With changes of the target composition the coating's chemical composition is controlled, while deposition parameters like bias voltage, DC and pulsed sputtering mode affect the energy input and therefore the temperature on the substrate. The surface temperature strongly effects density and grain size of the growing coating. Therefore knowledge of the correlations of chemical composition and grain size on the mechanical properties is of great interest for the coating process adaption.

21.3.1 Flow curve dependency on chemical composition and microstructure

The CrAlN coating is a ternary material system. The chemical variations investigated here have only been varied regarding the Cr/Al ratio with similar nitrogen contents of about 0.5. For this nitrogen content two material phases, a chromium rich cubic and an aluminium rich hexagonal phase are detected depending on the Cr/Al ratio (Figure 21.1). Both phases, the cubic pure CrN and hexagonal pure AlN are able to substitute elements of the respectively other metal.

To determine the influence of the chemical composition on the plastic flow behavior, coatings from pure CrN to pure AlN have been deposited and investigated. Furthermore, these investigations have been performed for both Direct Current- (DC) and High Power Impuls Magnetron Sputtering (HIPIMS).

The determined flow curves are presented in Figure 21.2. For the coating variants deposited with HIPIMS we see a very similar flow behavior of pure CrN and pure AlN with slightly higher stresses of the AlN coating. With increasing intermixture of both elements, the flow curves increase especially for higher strains. A maximum can be seen for the $Cr_{0.5}Al_{0.5}N$ coating system, which outperforms both pure coatings over the whole strain rate significantly. For all DC sputtered coatings the plastic flow curves are significantly reduced compared to the corresponding HIPIMS coatings. Here we also find the relatively highest flow curves for the $Cr_{0.42}Al_{0.58}N$ coating that almost reaches the values of the pure HIPIMS coatings, while the flow curves decrease for purer coatings. Especially the most aluminium-rich DC coating shows a significant reduction of plastic flow stresses.

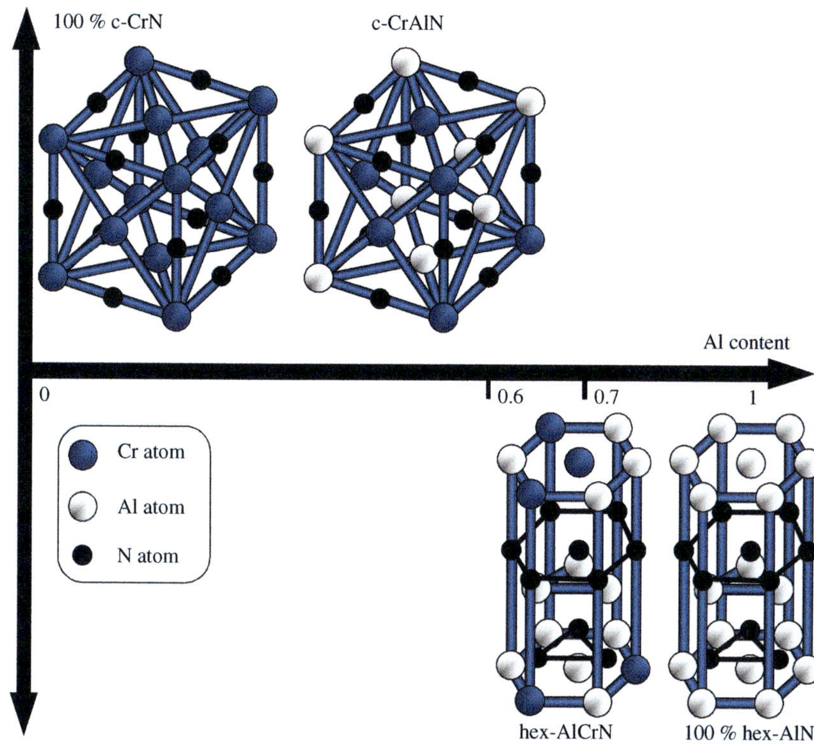

Figure 21.1 Illustration of the $Cr_xAl_{x-1}N$ microstructure in dependence of the Cr/Al ratio for constant nitrogen content of approximately 0.5.

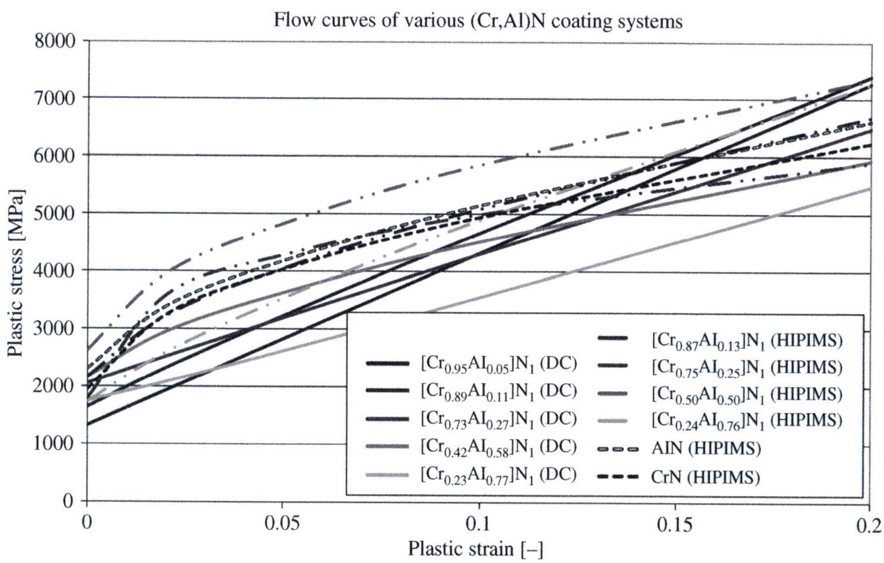

Figure 21.2 Plastic flow curves of $Cr_xAl_{x-1}N$ coatings depending on Cr/Al ratio and deposition method. *Source*: Bobzin 2013 [1]. Reproduced with permission of Elsevier.

The reason for increasing plastic flow curves with higher substitutions of foreign atoms can be explained by deformations of the material phase's lattice structure, leading to higher resistance for the dislocations moving through the grains. The plastic flow curve raise of HIPIMS sputtered coatings is caused by the higher energy input on the coated surface from the plasma, leading to higher diffusivity and therefore finer grained structures with fewer pores [8].

21.3.2 Strain Rate Dependency of Different CrN-AlN Coating Systems

The influence of strain rate on the coatings flow curves has been investigated for both pure CrN and AlN coatings and a CrN/AlN nanolaminate. The purpose was to investigate the influence of grain size on the strain rate hardening behavior of both materials. Due to the single layer thicknesses in the CrN/AlN nanolayer of 14 to 20 nm the grain sizes are limited to these dimensions, in contrary to more than 100 nm for the pure coatings.

The mean strain rate under the indenter has been adjusted by different loading rates during the indentation process. The strain rates have been varied in three steps of 0.05 s^{-1}, 0.25 s^{-1} and 1.25 s^{-1}. For all three coatings, a reduction of mean indentation depth at same load can be observed (Figure 21.3, left column). The resulting flow curves of these loading curves are displayed at the right column of the figure. For all three coatings the flow curve increases with increased strain rate. Nevertheless, significant differences in strain rate dependence can be observed. The nanolaminated coating shows for all three strain rates a continuous flow stress increase for increasing plastic strain values. In contrary, both pure coatings already show for lower strain rates of 0.05 and 0.25 s^{-1} a beginning saturation of flow stress at higher plastic strains. For 1.25 s^{-1} these coatings show very high flow stress values completely missing any strain hardening at all. Furthermore, while for lower strain rates of 0.05 and 0.25 s^{-1} the flow stress values of the nanolaminated sample exhibits the ones for the pure coatings for the same strain rate, for 1.25 s^{-1} the pure coatings flow curves exhibit the nanolaminated ones significantly.

The reasons for these different behaviors can be found in the microstructure. For low strain rates the dislocations moving through the grains and accumulating at barriers like grain boundaries are reduced by time dependent mechanisms like shifting of the gliding plain or annihilation with other dislocations. These mechanisms seem to work more efficiently for coarser grain sizes with their increased volume space for dislocation pileups, leading to a reduced strain hardening for both pure coatings. The reason for the decisive increase of flow stress for high strain rates for both pure coatings is not completely proofed until now. Also the absence of strain hardening at these high stresses is not fully understood. A possible explanation might be the activation of stress induced mechanisms to reduce the dislocation pile up at barriers, for example stress induced shifting of the gliding plain. In contrary for the finer grained nanolaminated coating at higher strain rates an activation of alternative forming mechanisms like grain boundary sliding is assumed, limiting the stresses necessary to keep up the forming velocity. Another possible mechanism reducing the forming stresses for the nanolaminated coating is the absorption of dislocations at the grain boundary, which is assumed to becoming more dominant with finer grain structures. These observations made for the influence of grain size on plastic deformation behavior coincidence quite well with measurements on macroscopic ceramic materials [6].

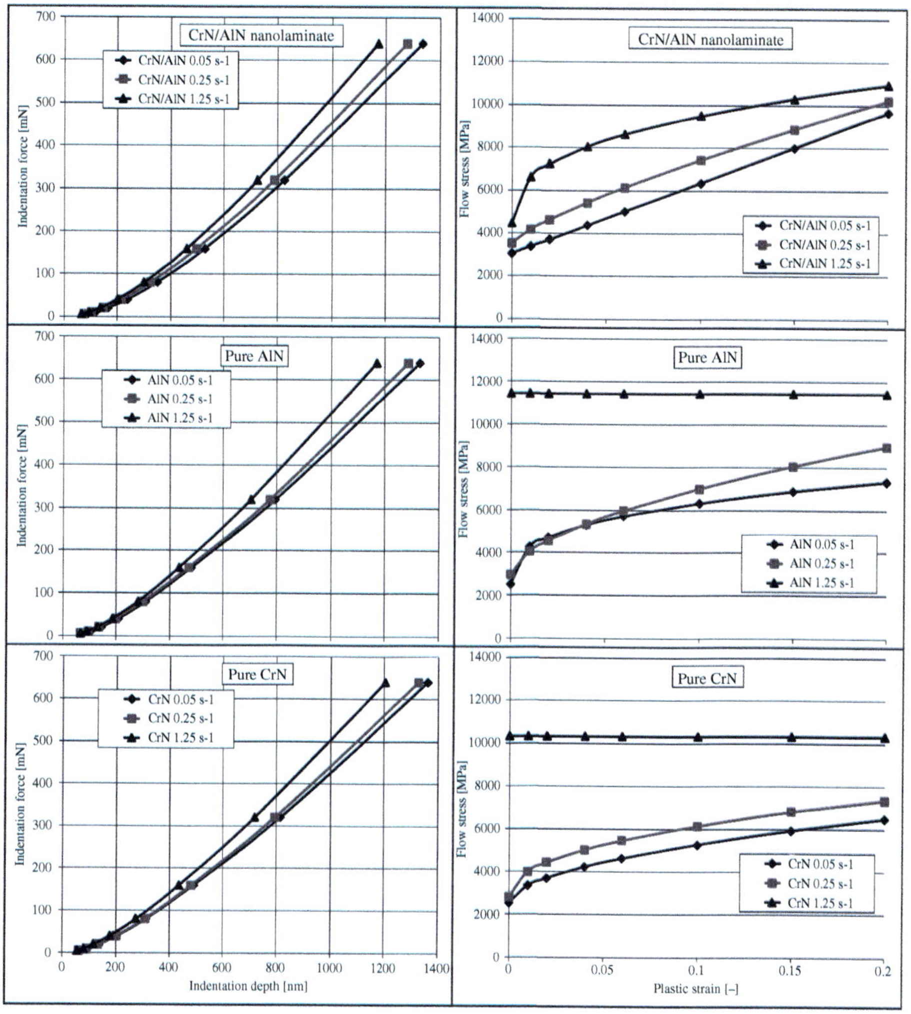

Figure 21.3 Nanoindentation loading curves (left side) and determined plastic flow curves (right side) of a CrN/AlN nanolaminate, a pure AlN and a pure CrN coating for three different loading velocities respectively strain rates. *Source*: Perne 2014 [9]. Reproduced with permission of Elsevier.

21.3.3 Failure criterion determination on a CrN/AlN nanolaminate

To validate the determined strain rate dependent flow curves and to correlate stress-strain occurrence with coating's failures nanoscratches have been performed on the nanolaminated coating. Afterwards the deformation behavior and occurring failures were analyzed on cross-fractured SEM images and have been compared with mechanical simulations of the scratch process. The nanoscratches have been performed with a Berkovich indenter and vertical loads of 10, 40, 160 and 640 mN and scratch velocities of 0.5 and 5 μm/s [9].

Due to the nanolayer structure of the coating it was possible to determine the vertical strain fields with high local resolution by measuring the CrN/AlN bilayer period at any

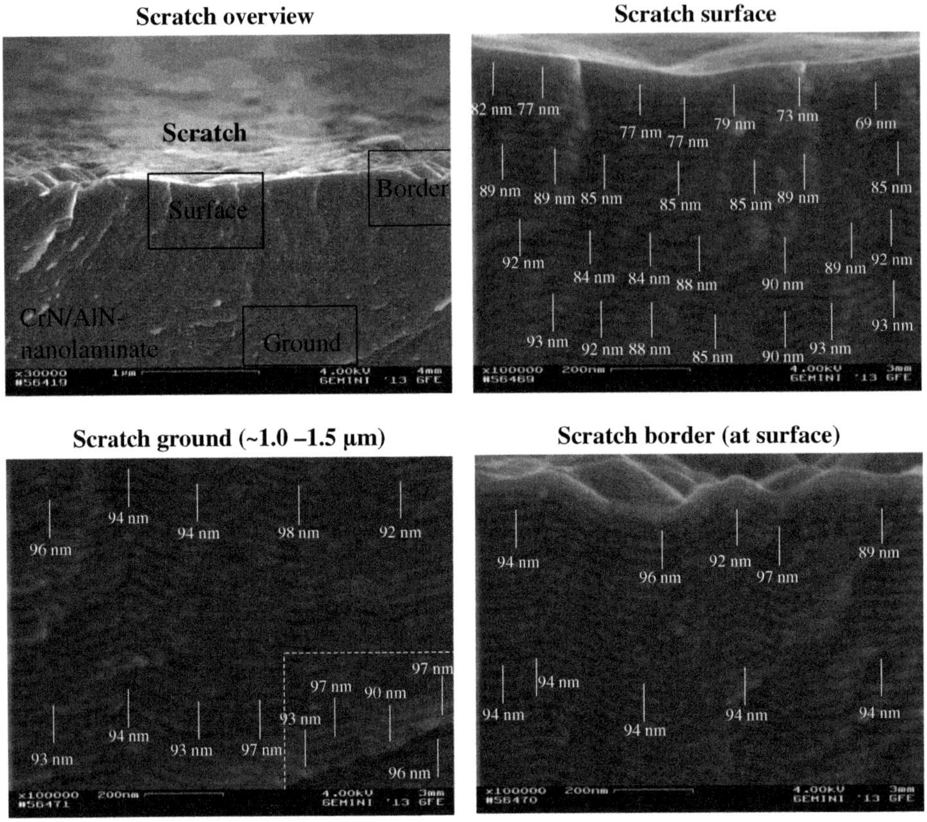

Figure 21.4 SEM picture of a cross-fractured CrN/AlN nanolaminate coating under a 40 mN scratch, applied with 0.5 μm/s (upper left). Due to the mean CrN-AlN single layer length, the remaining vertical strain was determined for an area directly under the scratch surface (upper right), at the scratch ground (lower left side) and the scratch border (lower right side). *Source*: Perne 6 [9]. Reproduced with permission of Elsevier.

location around the scratch and comparing them with the period length at undeformed locations (Figure 21.4). Therefore the remaining vertical strain at any point under the scratch surface could be calculated. These measured strain fields have been compared with those gained from FEM simulations using the determined strain rate dependent flow curves (Figure 21.5). The comparison showed high agreements of the vertical strain fields in the coating under the scratch up to 27% compressive strain. Only on the scratch borders qualitative divergences between experiment and simulation could be found.

To investigate the correlation of stress-strain fields and coating failures like fractures or delamination for all nanoscratches the locations of appearing fractures were compared with the calculated stress fields of FEM simulations (Figure 21.6). From the SEM cross-fractured images it could be found that failures appeared with greater size

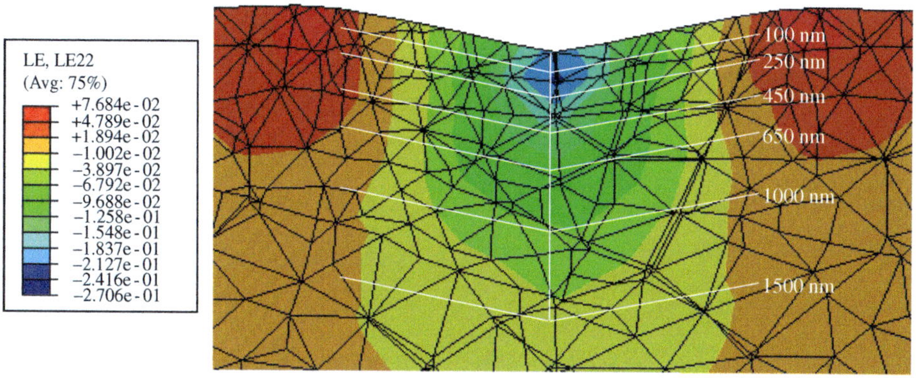

Figure 21.5 Remaining vertical plastic strains from FEM simulation in the nanolaminate coating after a 40 mN scratch, applied with 0.5 μm/s. *Source*: Perne 6 [9]. Reproduced with permission of Elsevier.

and higher number both for increased scratch load and increased scratch velocity. Furthermore the fractures appeared mostly in an area 1 to 2 μm under the scratch surface. In the FEM simulations in these areas high shear stresses could be observed, which changed during the indenter overrun from high negative to high positive values. The maximum shear stress values reached from negative 2–3 GPa and positive 3–4 GPa for the 0.5 μm/s scratches to negative 4 GPa and more than positive 7 GPa for 5 μm/s scratches. The scratch loading only influenced the size of the shear stress area. The analysis of the failure occurrence in the cross fractured SEM figures indicates that both, area size and magnitude of the shear stresses induce crack and fracture initiation.

21.4 Concluding Remarks

The investigations described in this chapter offer the possibility to determine the plastic flow behavior of a material, especially for thin coatings. Thus the influence of different deposition parameters, chemical and microstructure properties can be determined. The combination of this method with experimental and simulative nanoscratch studies further offers the possibility to determine failure criterions.

These investigations have exemplarily been performed on the (Cr,Al)N coating system, which is assumed to behave representatively for other ternary hard coating systems like TiAlN. The results show the great differences in the plastic forming behavior of micro- and nanograined materials, especially regarding the strain rate dependency, as well as the noteworthy influence of the Cr/Al ratio in coating composition. The nanoscratch investigations revealed the decisive influence of shear stresses on the coatings failure initiation.

Altogether, the investigations described here show a small extract of the fascinating opportunities of nanoindentation technology in coating characterization when combined with additional analyzing methods like SEM imaging and FEM simulation.

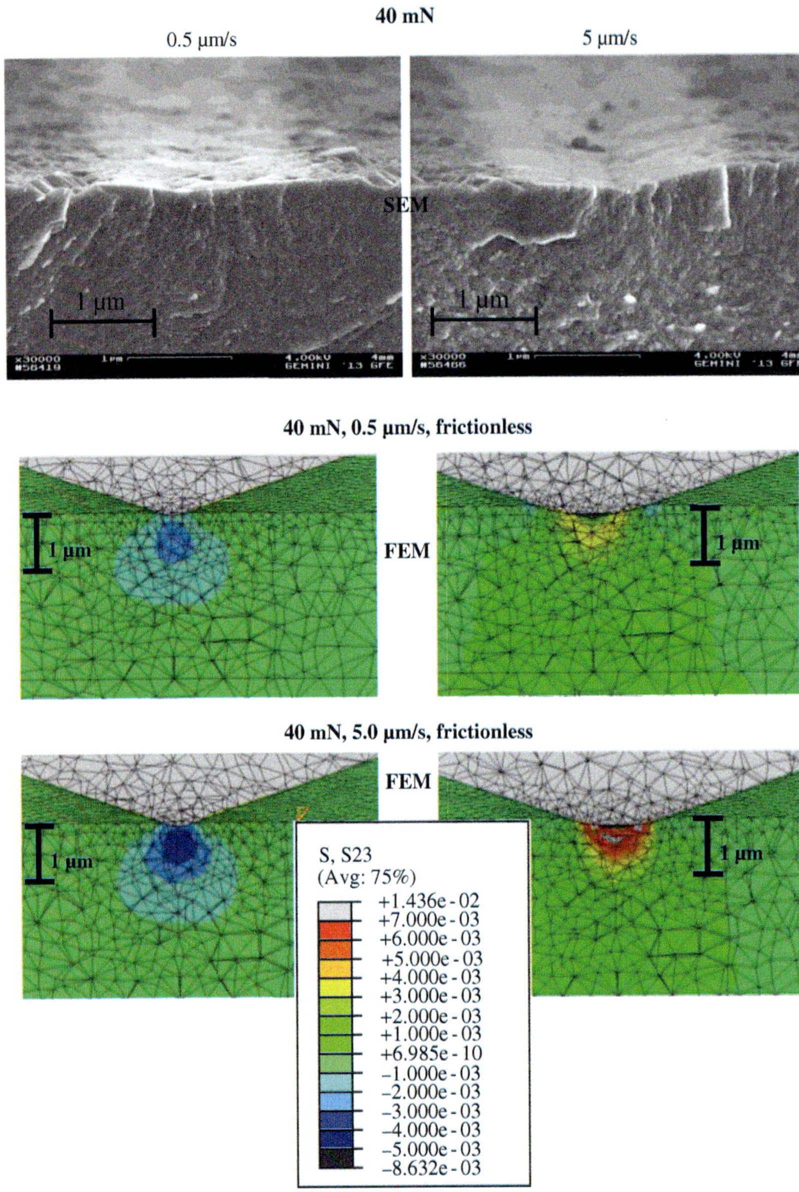

Figure 21.6 Comparison of fracture appearance in SEM image of the cross-fractured nanolaminated coating under 40 mN scratch (upper part) with 0.5 µm/s (left side) and 5.0 µm/s (right side) with appearing shear stresses during indenter overrun in FEM simulation. *Source*: Perne 6 [9]. Reproduced with permission of Elsevier.

References

1 K. Bobzin, N. Bagcivan, R. Brugnara. Flow curve determination of thin films by improved finite element models and different nanoindenter geometries. Thin Solid Films (2013) 549: 313.
2 W.C. Oliver, G.M. Pharr. An improved technique for determining hardness and elastic modulus using load and displacement sensing indentation experiments. Journal of Materials Research (1992) 7: 1564.
3 T.F. Juliano, V.M.R. van Landingham, T. Weerasooriya, P. Moy. Extracting Stress-strain and Compressive Yield Stress Information From Spherical Indentation. Army Research Laboratory Report ARL-TR-4229; 2007, p. 1.
4 K.D. Bouzakis, N. Michailidis. Coating elastic–plastic properties determined by means of nanoindentations and FEM-supported evaluation algorithms. Thin Solid Films (2004) 469: 227.
5 S. Basu, A. Moseson, M.W. Barsoum. On the determination of spherical nanoindentation stress-strain curves. Journal of Materials Research (2006) 21: 2628.
6 J. Perne. Plastic flow behaviour of (Cr, Al)N hard coatings in dependence of strain rate and nanostructure. Thin Solid Films (2014) 556: 390.
7 K.D. Bouzakis, M. Pappa, S. Gerardis, *et al.* PVD coatings strength properties at various temperatures by nanoindentations and FEM calculations determined. Tribololgy in Industry (2012) 34: 29.
8 N. Bagcivan, K. Bobzin, S. Theib. (Cr1 − xAlx)N: A comparison of direct current, middle frequency pulsed and high power pulsed magnetron sputtering for injection molding components. Thin Solid Films (2013) 528: 180.
9 J. Perne. Experimental and simulative strain field investigation of nano- and microscratches on nanolaminated (Cr, Al)N coating. Thin Solid Films (2014) 573: 33.

22

Scale Invariant Mechanical Surface Optimization

Norbert Schwarzer

Saxonian Institute of Surface Mechanics SIO, Ummanz/Rügen, Germany

22.1 Introduction

In order to achieve the goal set, namely, the optimization and parameter identification for layered surfaces in a sufficiently generic manner, it is necessary to combine a variety of scientific fields and/or concepts of material science. Therefore, this introduction needs to cover the following issues:

- First principle based, effective interatomic potential description of mechanical material behavior [1] and its extension into the time domain [2];
- The effective indenter concept [3–6] made time-dependent;
- The extension of the oliver and pharr method to analyze nanoindentation data to layered materials and time dependent mechanical behavior [7–10];
- The incorporation of intrinsic or residual stresses into the model description;
- The physical scratch and/or tribological test and its analysis [11–14].

22.1.1 Interatomic Potential Description of Mechanical Material Behavior

It was shown in [15] that based on an effective potential function like the Morse potential given as:

$$V_{Morse} = \varepsilon[e^{-2p(r-r_0)} - 2e^{-p(r-r_0)}] \tag{22.1}$$

a contact problem can be evaluated using the mechanical parameters derived from such a potential. Here p, ε, σ are material parameters and r_0 usually denotes the equilibrium bond length. In such a case, the potential would define the pair interaction. Here however, as in [1, 15], we will apply the potential as an effective one with r_0 denoting the lattice constant (see also [16]). With respect to molecular dynamic simulation, such an effective potential could be the basis for the extraction of the necessary pair and higher order interactions as demonstrated in [16]. For our study however, we will not need this, because we are only interested in the mechanical constants, especially the Young's modulus and the decomposition strength. Having this, one can apply the method described in [15] to simulate a mechanical contact problem, thereby even taking nonlinear effects like the pressure, shear and temperature dependency of the Young's modulus and other mechanical constants into account [2]. We will show that especially within tribological

Applied Nanoindentation in Advanced Materials, First Edition. Edited by Atul Tiwari and Sridhar Natarajan.

applications temperature fields can play an important role in influencing the mechanical properties of the material.

22.1.2 The Effective Indenter Concept and Its Extension to Layered Materials

For many physical effects, it can be observed that surfaces usually show a different behavior than bulk of the same material. Subsequently, the phenomenological properties of surfaces can be dramatically different from results obtained by applying bulk concepts to surface problems. This holds especially for the mechanical behavior of surfaces, like their reaction to contact situations with very surface-located, surface-dominated stress and strain fields. In tribo and wear problems this very often compromises our ability to simulate and understand the physical processes taking place in certain tests [e.g. 17–21]. In order to improve this situation, we therefore resort to a stringent application of a layered material model. Here, one considers and models the mechanical surface always as (at least potentially) having a property profile starting from the mechanical bulk values in depth, usually well known for a certain material, to sometimes completely different properties on the top-most surface layer. In the case of coated materials, this model extension can easily be justified by the explicit coating structure. However, we should stress the point of also carefully and critically considering apparent 'homogeneous surfaces', which are often of gradient or otherwise layered character. In addition, tribological applications often come with temperature fields due to friction. Be it the internal friction caused by shear or the external one coming from the multiple tribological contact situations, the effect on the top most surface layers can be significant. As shown in [2], such temperature can feed back into the mechanical properties influencing the tribological contact conditions.

The effective indenter concept itself can be described mathematically or understood as a quasi-conform coordinate transformation transforming the difficult problem of a curved surface contacted by a well-defined indenter (like a cone) into a flat surface loaded with a complexly formed indenter. Already in 1995 Bolshakov, Oliver and Pharr [3] introduced the concept of the effective indenter and refined it in a series of wonderful publications until in 2002 the paper on *Understanding of nanoindentation unloading curves* [4] was published by Pharr and Bolshakov.

The extension of this concept is simply performed by substituting the homogenous half space model describing the loaded sample body by a layered half-space model [22].

22.1.3 About Extensions of the Oliver and Pharr Method

As it is a well-established fact, that the classical Oliver and Pharr method [7], as an approach based upon the homogenous half space model, cannot directly be applied to layered materials and small structures, the author here refers to the literature [e.g. 23]. Soon after the publication of the Oliver and Pharr method in 1992 it became clear that there is a physical concept this method can be based on, which was later called the concept of the effective indenter[3, 4]. Later on, a theory has been worked out not only solving the problem for the mechanical contact of an indenter with general shape of symmetry of revolution completely analytically, but also to extend this solution to layered structures [24]. The new method was used in a variety of applications together with Pharr, Chudoba, Richter and others [5, 6, 8, 9, 26, 27]. In order to accelerate and simplify the use of the new and relatively complex mathematical tools, the whole approach was brought into a software package named FilmDoctor® [28].

22.1.3.1 Making the Classical Oliver and Pharr Method Fit for Time Dependent Mechanical Behavior

In an indentation test, we always find complex three-dimensional stress states with all stress components usually being non-zero. Thus, '*the nature of the loading is a complex mixture of hydrostatic compression, tension, and shear*' [29]. In the case of viscous behavior, complex mixtures of stress and strain rates still add to these difficulties. This however, usually influences the time dependent mechanical parameters [30]. Thus, subsequently at different positions we find different stress and strain states and rates leading to different mechanical parameters (time-dependent) at different positions within the material. This automatically makes the system of linear partial differential equations of linear elasticity non-linear in the moment viscosity (even elastic viscosity) comes into play. In order to keep things simple however, here we will not go for a non-linear basic solution, but try to find effective, phenomenological descriptions for visco-elastic, visco-plastic contact problems based on the concept of a time dependent effective indenter. As an additional difficulty coming into play especially in connection with tribological applications we again have to consider temperature fields possibly influencing the mechanical parameters. It was shown in [2] that even smallest temperature gradients could have a dramatic influence on the mechanical parameters. Now, as tribological contacts usually and mainly inevitably produce heat, this effect cannot be ignored when trying to physically simulate tribo experiments.

22.1.4 Introduction to the Physical Scratch and/or Tribological Test and its Analysis

The standard scratch test is a widely used method to test the mechanical stability of coatings on different types of substrates and has become a sensitive technique to control the reliability of the manufacturing process. It is based on various standards [31, 32].

Unfortunately, these standards do not allow us to design and/or perform the tests in a truly physical way, meaning to extract generic physical parameters, which are later applicable in lifetime prediction and optimization. For this, a more sophisticated approach is required which assures physical parameter identification at each and every step towards physical tribological tests. The procedure is outlined in Figure 22.1.

Following the flowchart in Figure 22.1, we have started our surface characterization process with nanoindentation. The section above covered the calculation of true mechanical layered surface properties using these nanoindentation tests. Thus, now all mechanical values are known which are required to properly dimension (fine-tune) a scratch or tribology test for specific surface structures. In addition, these values are also of need for a physical analysis of the scratch and tribo tests. Again the flow chart (Figure 22.1) might give an illustrative understanding into such a test procedure.

22.1.5 Illustrative Hypothetical Example for Optimization Against Dust Impact

We assume an impact situation for an aircraft turbine blade with a 400 μm dust particle hitting the surface at about 200 m/s straight normal. No matter what kind of material the coating consists of and what its yield strength value would be, assuming a rim hardened substrate with a yield strength around 5 GPa or lower, such a system would not survive the impact without suffering severe plastic flow in the substrate (Figure 22.2). In the example shown here an arbitrary hard coating was assumed having a Young's modulus E of 400 GPa and a yield strength Y of 20 GPa. The situation becomes even

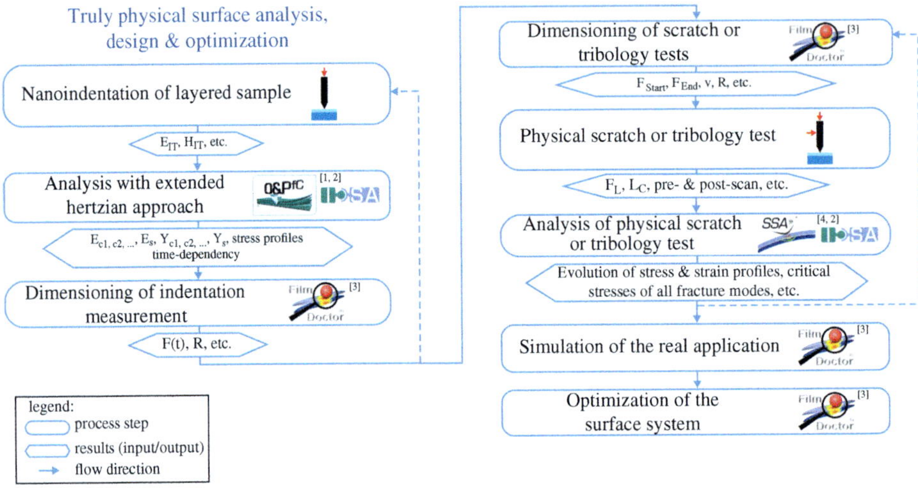

Figure 22.1 A flow chart of the procedure of mechanical characterization and optimization of arbitrary structured surfaces with respect to wear and fretting.

clearer when the stress colors are chosen such that red denotes areas where the yield strength is reached or exceeded (Figure 22.3). Now an interlayer is been searched which will mitigate the situation and helps to protect the substrate under the same impact conditions. A solution is found with a 100 μm interlayer having a yield strength $Y = 13$ GPa (Figures 22.4 and 22.5). The Young's modulus was chosen to be typical for C-rich tungsten carbide films with $E = 450$ GPa. The material properties for the impacting particle were taken to be typical average values for the so-called 'Arizona dust'.

The reader will find more illustrative examples and a very comprehensive elaboration of the method elsewhere [11–14].

22.1.6 About the Influence of Intrinsic Stresses

For many coating systems it is almost impossible to create them without producing residual or intrinsic stresses. As we have seen however, the presence of stresses does not only have an influence on the mechanical parameters like the Young's modulus, but also on the material stability [1, 2], meaning that in general the mechanical parameters like Young's or bulk modulus, Poisson's ratio, yield strength or fracture toughness are no constants. Intrinsic or residual stresses contribute to these effects and therefore, it is important to take them into account.

This becomes even more pronounced whenever time dependent material behavior (e.g. at elevated temperatures or for soft materials like polymers and tissues) plays a role. Then the mechanical parameters themselves depend on time and the actual stress field, especially the hydrostatic stress and the shear stress field.

This leads to some severe conclusions with respect to theoretical strength and stability in the case of contact mechanical situations in comparison with simple tensile or shear test conditions being extremely and what is more, non-linearly dependent on intrinsic stresses. That is why the correct modeling of intrinsic stresses typical for surface treatment and coating production is of special interest here and needs to be discussed in more detail.

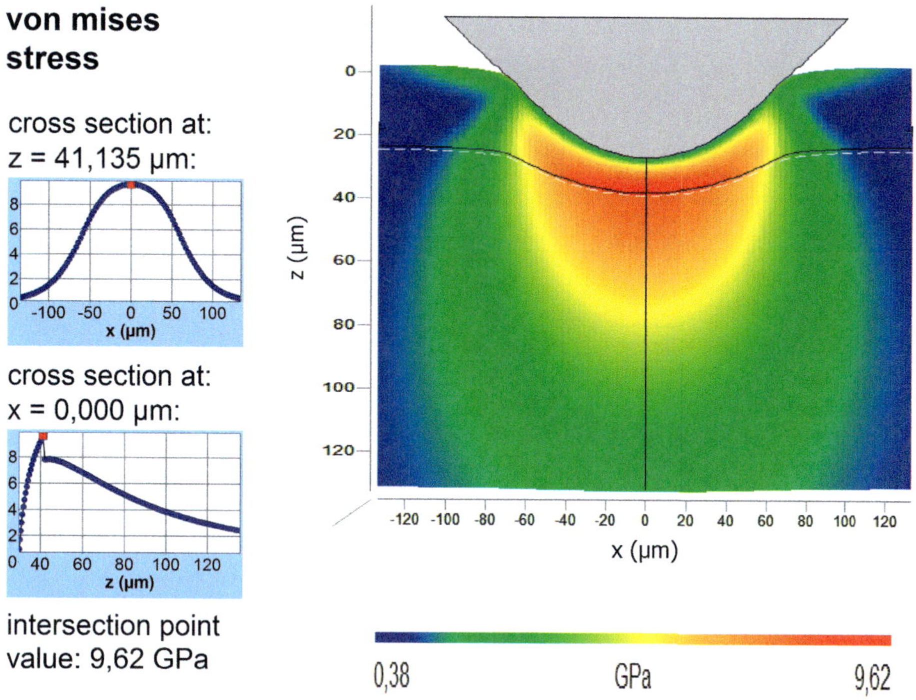

Figure 22.2 FilmDoctor® [28] simulation of a particle impact. Please see text for impact boundaries. The figure shows the von Mises stress distribution with stress values clearly exceeding the yield strength of the substrate. The white dashed line marks the interface, while the black lines are marking the lines along which the diagrams on the left hand side are evaluated. *Source*: Reproduced with permission of FilmDoctor®.

22.2 Theory

22.2.1 First Principle Based Interatomic Potential Description of Mechanical Material Behavior

As shown in [1], the pressure P and bulk modulus B can be derived from an effective potential. Thereby it is convenient to express P and B in units of B_0 (B at $P = 0$) and by substituting the lattice distance r by $r = c \cdot r_0$ one results in the relations:

$$P/B_0 = \frac{e^{r_0 p(1-c)}}{r_0 p\, c^2}[1 - e^{r_0 p(1-c)}]; \quad B/B_0 = \frac{e^{r_0 p(1-c)}}{r_0 p\, c^2}[2(1 + r_0 p\, c)e^{r_0 p(1-c)} - 2 - r_0 p\, c]. \tag{22.2}$$

In order to find an estimate for the critical r leading to decomposition (c.f. [1]), we numerically solve the following expression ($r_{00} = p \cdot r_0$):

$$6(1 + c_m r_{00}) - 3e^{r_{00}(c_m - 1)}(2 + c_m r_{00}) = 0 \tag{22.3}$$

with respect to the critical c-value c_m for maximum $P(c)$.

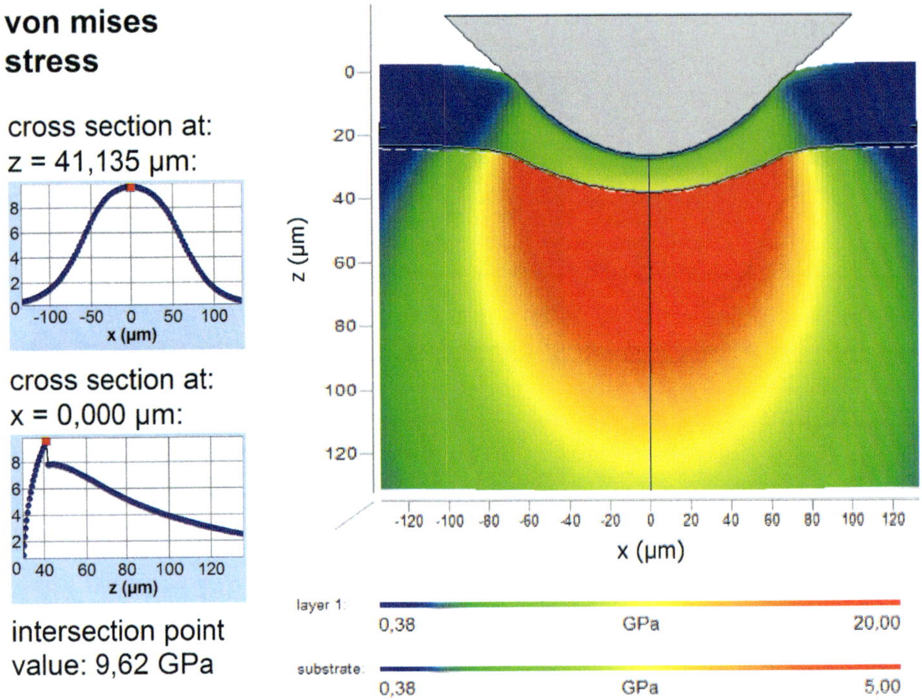

Figure 22.3 FilmDoctor® [28] simulation of a particle impact. Please see text for impact boundaries. The figure shows the von Mises stress distribution with stress values clearly exceeding the yield strength of the substrate. In contrast to Figure 22.2, the color coding is now scaled such that red denotes stresses equal to the yield strength or having it exceeded. The white dashed line marks the interface, while the black lines are marking the lines along which the diagrams on the left hand side are evaluated. *Source*: Reproduced with permission of FilmDoctor®.

As a purely mathematically based measure for the critical bond length (or in our case of an effective potential the lattice distance), the inflexion point for $c_{\mathrm{ifp}} > c_{\mathrm{m}}$ could be used. This can numerically be obtained for the Morse potential via solving the following expression with respect to c_{ifp}:

$$6 + c_{ifp} r_{00}(4 + c_{ifp} r_{00}) - 2e^{r_{00}(1-c_{ifp})}(3 + 2c_{ifp} r_{00}(2 + c_{ifp} r_{00})) = 0. \tag{22.4}$$

As shown in [2], the presence of deviatoric or shear stress components reduces the theoretical strength towards tension by up to 10%. This is of special importance for multiaxial contact situations where the stress tensor always is of mixed form and fully filled.

One also needs to check for temperature gradients caused by external and internal friction due to tribological shear. It was shown [2] that caused by the inhomogeneity of the temperature $T(t, r)$ and subsequently the relaxation $\tau(t, r)$ field, an originally simple approach for the bond interaction like e.g. the standard linear solid (SLS) becomes rather

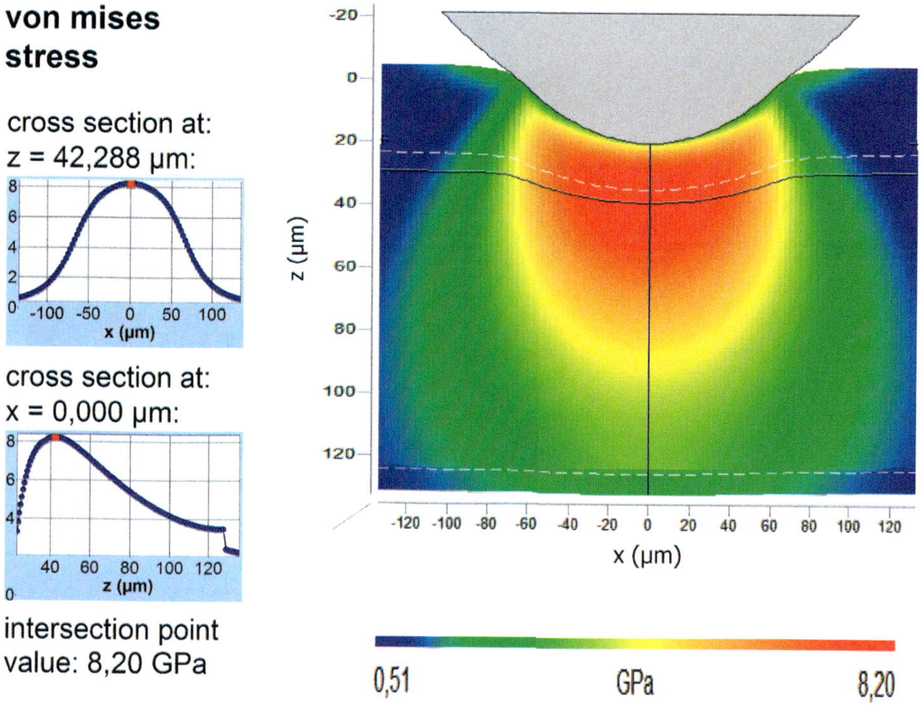

Figure 22.4 FilmDoctor® [28] simulation of a particle impact of an optimized coating structure. Please see text for impact boundaries. The figure shows the von Mises stress distribution with stress values clearly exceeding the yield strength of the substrate. The white dashed line marks the interface, while the black lines are marking the lines along which the diagrams on the left hand side are evaluated. *Source*: Reproduced with permission of FilmDoctor®.

complicated regarding its bulk modulus behavior. Here t and r are denoting time and space coordinates, respectively. Instead of a phenomenological SLS with the well-known bulk modulus of the kind:

$$B_0 = \frac{c}{18N} f^2 \left\{ \left[\sum_{\forall j, j \neq i} \left. \frac{d^2 V_j(\tilde{r}_{ij})}{d\tilde{r}_{ij}^2} \right|_{r_{ij}=r_{0ij}} \right] \frac{1}{r_{0ij}} \right\}$$

$$= \frac{c}{18N} \left(E_0 + E_1 e^{\frac{-t}{\tau}} \right) \left\{ \left[\sum_{\forall j, j \neq i} \left. \frac{d^2 V_j(\tilde{r}_{ij})}{d\tilde{r}_{ij}^2} \right|_{r_{ij}=r_{0ij}} \right] \frac{1}{r_{0ij}} \right\} \tag{22.5}$$

(N gives the number of particles and c defines a parameter connected with the volume of an individual particle) and a subsequent 'classical' Young's modulus time dependency such as $E_0 + E_1 e^{\frac{-t}{\tau}}$ we have a resulting phenomenological Young's modulus structure of

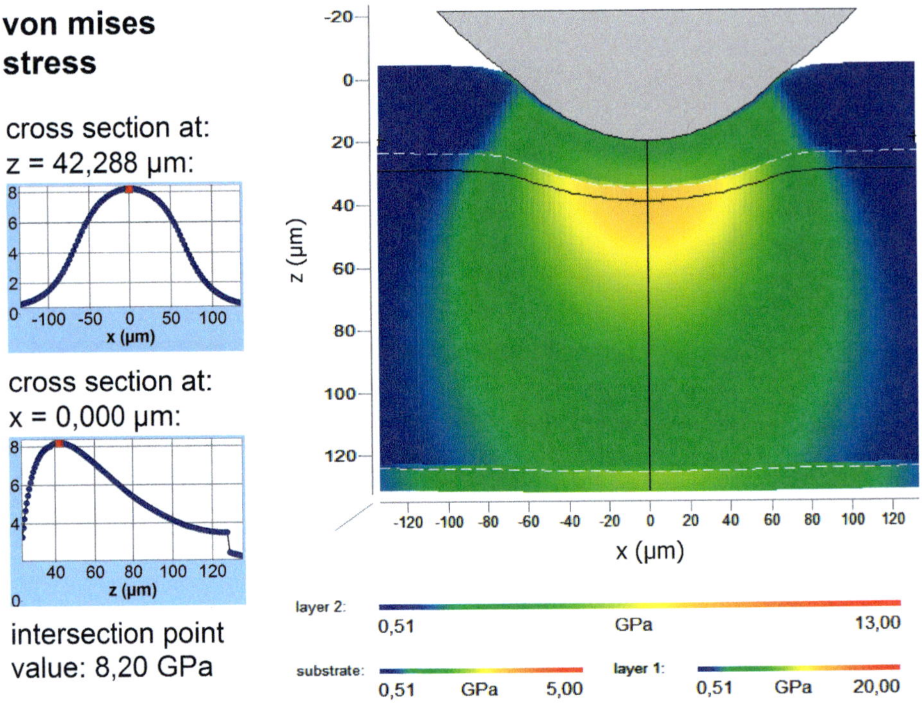

Figure 22.5 FilmDoctor® [28] simulation of a particle impact of an optimized coating structure. Please see text for impact boundaries. The figure shows the von Mises stress distribution with stress values clearly exceeding the yield strength of the substrate. In contrast to Figure 22.4, the color coding is now scaled such that red denotes stresses equal to the yield strength. We see that there are no such areas in the coating system underneath the impacting body. The white dashed line marks the interface, while the black lines are marking the lines along which the diagrams on the left hand side are evaluated. *Source*: Reproduced with permission of FilmDoctor®.

the form:

$$E(t) = E_0 + E_1 e^{\frac{-t}{\tau(T_j(rz),\mu)}} + \vec{r} \cdot E_1 e^{\frac{-t}{\tau(T_j(rz),\mu)}} \frac{t}{\tau^2} \frac{d\tau}{d\vec{r}} + \vec{r} \cdot \vec{r} \frac{\left(E_1 e^{\frac{-t}{\tau(T_j(rz),\mu)}} \frac{t}{\tau^2} \frac{d\tau}{d\vec{r}} \right)^2}{4 \left(E_0 + E_1 e^{\frac{-t}{\tau(T_j(rz),\mu)}} \right)}. \tag{22.6}$$

With a known principle structure for the temperature distribution function deduced from the shear stress distribution (adiabatic approach) such an approach seems to be manageable by the means of iterations. With a typical dependency for the temperature dependent relaxation time like [33, 34]:

$$t = h/E_1 = \frac{h_0 e^{W_A/(k_B T)}}{E_1} \tag{22.7}$$

(with a constant W_A and the $T = T_0$-viscosity η_0 and the Boltzmann-constant k_B) one can evaluate the mechanical contact situations applying the techniques elaborated in [2] (theory part II) under the assumption of completely adiabatic temperature fields resulting from shear and completely following the second invariant distribution of the deviatoric stress tensor. Please note the positive sign in the exponential function for the relaxation time assuring decreasing viscosity η with increasing temperature.

22.2.2 The Effective Indenter Concept

In order to have a sufficiently great variability for the definition of differently shaped effective indenters, we apply the extended Hertzian approach as shown in [35]. With this approach normal and even tangential load distributions of the form [24]:

$$\sigma_{zz0}(r,\varphi) = \sum_{n=0}^{N} c_{\sigma n} r^n \sqrt{a^2 - r^2} \tag{22.8}$$

$$\begin{aligned} \tau_{rz0}(r,\varphi) &= \sum_{n=0}^{N} c_{\tau rn} r^n \sqrt{a^2 - r^2} \\ \tau_{xz0}(r,\varphi) &= \sum_{n=0}^{N} c_{\tau xn} r^n \sqrt{a^2 - r^2} \\ \tau_{yz0}(r,\varphi) &= \sum_{n=0}^{N} c_{\tau yn} r^n \sqrt{a^2 - r^2} \end{aligned} \tag{22.9}$$

with $n = 0$, 2, 4, 6 and arbitrary constants c can be solved completely analytically. By following the instructions of the mathematical procedures for obtaining the complete potential functions as given in [24] also higher numbers of N can be applied.

Together with lateral loads (occurring in all scratch- and tribo-testers or the next generation of nanoindenters and their applications, see e.g. [35–38]) one often faces tilting moments leading to a normal surface stress distribution of the form:

$$\sigma_{zz0}(r,\varphi) = \sum_{n=0}^{N} c_{\sigma n} r^{n+1} \cos(\varphi) \sqrt{a^2 - r^2}. \tag{22.10}$$

These stresses can occur when the indenter shaft is dragged over the surface. Because the shaft itself is elastic and thus would be bent during the lateral loading, an unavoidable tilting moment results and acts on the contacted surface. Furthermore, curved surfaces (e.g. due to roughness) can lead to such tilting moments, too.

In order to simulate the internal complex material structure of porous or composite materials, certain defect fields must be developed and combined with the external loads. Such fields can be extracted from the mathematical apparatus originally developed for the effective indenter concept by a simple generalisation procedure. By developing such a defect model one obtains a very comprehensive tool for the construction of general intrinsic stress distributions caused by internal inhomogeneities. Circular disc-like inclusions could for example be simulated by the use of plane defects within the layered half space. So, introducing circular defects of radii a_i of the loading type:

$$\begin{aligned} \tau_{rz0}\left(r_i = \sqrt{(x-x_i)^2 + (y-y_i)^2}, z_i + 0\right) &= \sum_{n=0}^{N} c_{\tau i,n} r_i^n \sqrt{a_i^2 - r_i^2} \\ \tau_{rz0}\left(r_i = \sqrt{(x-x_i)^2 + (y-y_i)^2}, z_i - 0\right) &= -\sum_{n=0}^{N} c_{\tau i,n} r_i^n \sqrt{a_i^2 - r_i^2} \end{aligned} \tag{22.11}$$

$$\sigma_{zz0}(r_i = \sqrt{(x-x_i)^2+(y-y_i)^2}, z_i+0) = \sum_{n=0}^{N} c_{\sigma i,n} r_i^n \sqrt{a_i^2 - r_i^2}$$

$$\sigma_{zz0}(r_i = \sqrt{(x-x_i)^2+(y-y_i)^2}, z_i-0) = -\sum_{n=0}^{N} c_{\sigma i,n} r_i^n \sqrt{a_i^2 - r_i^2}. \tag{22.12}$$

(with x_i, y_i, z_i denoting the centre of the defect and $n = 0, 2, 4, 6$) directly allows us the application of the extended Hertzian approach [24] that provides a complete solution of the elastic field of the defect loading given above. By superposing a multitude of such 'defect dots', one could model (simulate) a very great variety of material inhomogeneities and intrinsic stress distributions. The evaluation of the complete elastic field is straight forward. It only requires the evaluation of certain derivatives of the potential functions given in [24].

Particularly interesting can be such defects if being placed near or directly on interfaces, because this would simulate weakened, pre-strained or even particle or stress-strengthened interface conditions.

Finally, we need to take into account the curvature of the surfaces in order to its effect on the resulting contact pressure distribution [36]. As the theoretical approach would not find enough space in this short note, the author will publish the necessary details elsewhere. However, the interested reader may derive the results presented here by comparing the solutions of the Laplace equation in cartesian and paraboloidal coordinates.

22.2.3 An Oliver and Pharr Method for Time Dependent Layered Materials

Oliver and Pharr [7] have shown that the force removal curve from elasto-plastic deformations with a Berkovich indenter can be described by a power law function:

$$F = C \cdot (h - h_0)^m \tag{22.13}$$

Applying now the concept of the effective indenter as shown in Figure 22.6, one can easily deduce that such an unloading curve can be connected with an indenter of the shape:

$$Z(r) = B \cdot r^n, \tag{22.14}$$

with:

$$\begin{aligned} m &= 1 + \frac{1}{n} \\ C &= 2 \cdot E_r \cdot \left[\frac{n}{n+1}\right] \cdot \left(\frac{1}{B} \cdot \left[1 - \varepsilon \cdot \frac{n}{n+1}\right]\right)^{\frac{1}{n}} \\ \frac{1}{E_r} &= \frac{1-\nu_i^2}{E_i} + \frac{1-\nu_s^2}{E_s} \\ \varepsilon &= m \cdot \left[1 - \frac{1}{\sqrt{\pi}} \frac{\Gamma\left(\frac{m}{2\cdot m-2}\right)}{\Gamma\left(\frac{2\cdot m-1}{2\cdot m-2}\right)}\right] \end{aligned} \tag{22.15}$$

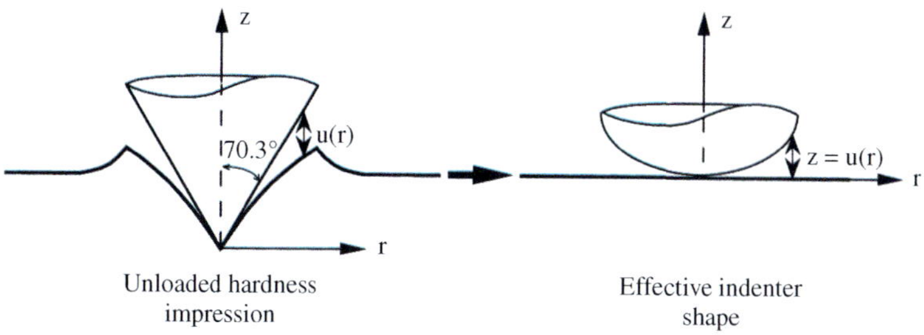

Figure 22.6 The effective indenter concept [4] transferring the theoretically difficult problem of a well defined sharp indenter on an elasto-plastically deformed surface with complex shape (left hand side) to an effective indenter on a flat surface (right hand side). *Source*: Pharr 2009. Reproduced with permission of Cambridge University Press.

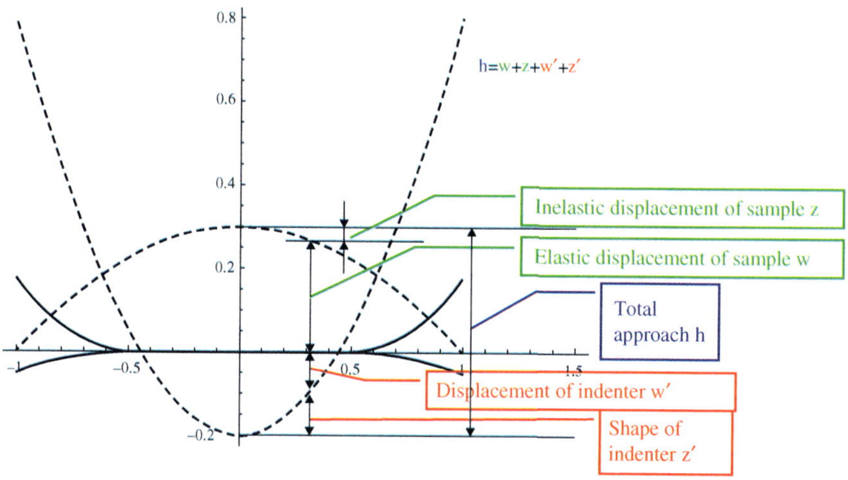

Figure 22.7 Formulating the contact equation $h = w+z+w'+z'$ with respect to partial plastic deformation z of the sample surface.

The indices i and s are standing for the indenter and sample, respectively. The evaluation above is based on the well-known results of Sneddon [39], which require a linear elastic model.

For this paper we need to reformulate the basic contact equation incorporating the plastically deformed surface of the sample with the locally defined shape function (Figure 22.7):

$$z(r) = B_S \cdot r^n. \tag{22.16}$$

Together with the following approach for the indenter part being in contact with the sample surface of the kind:

$$z'(r) = B_I \cdot r^n, \tag{22.17}$$

we obtain:

$$w(r) + w'(r) = h - (B_I + B_S) \cdot r^n, \tag{22.18}$$

as the new governing contact equation with the parameters $B = B_I + B_S$ and n defining the effective indenter shape $Z(r)$, as given above.

Similarly, also more general effective indenter concepts can be derived, like this one:

$$\begin{aligned} w_S(r) + w_I(r) &= h - \frac{r^2}{d_0} - \frac{r^4}{d_2} - \frac{r^6}{d_4} - \frac{r^8}{d_6} \\ &= h - r^2\left(\frac{1}{d_0^{\text{Indenter}}} - \frac{1}{d_0^{\text{plastic shape}}}\right) - r^4\left(\frac{1}{d_2^{\text{Indenter}}} - \frac{1}{d_2^{\text{plastic shape}}}\right), \\ &\quad - r^6\left(\frac{1}{d_4^{\text{Indenter}}} - \frac{1}{d_4^{\text{plastic shape}}}\right) - r^8\left(\frac{1}{d_6^{\text{Indenter}}} - \frac{1}{d_6^{\text{plastic shape}}}\right) \\ &= h - r^2(c_0' - c_0) - r^4(c_2' - c_2) - r^6(c_4' - c_4) - r^8(c_6' - c_6) \end{aligned} \tag{22.19}$$

based on a locally paraboloid surface with r-terms r^0, r^2, r^4, r^6 and r^8. The reader might easily recognize the extended Hertzian character [26] of the basic contact equation given above. Subsequently, the simpler Hertzian contact would read:

$$\begin{aligned} w_S(r) + w_I(r) &= h - \frac{r^2}{d_0} \\ &= h - r^2\left(\frac{1}{d_0^{\text{Indenter}}} - \frac{1}{d_0^{\text{plastic shape}}}\right), \\ &= h - r^2(c_0' - c_0) \end{aligned} \tag{22.20}$$

Some of the following examples and discussions will be based on these more general approaches. However, due to the wide use of the power law fit given above we will explicitly concentrate on practical examples being performed using this power law approach.

In [24] and [26] it is shown how these general contact approaches can be applied and how the complete elastic fields have to be evaluated while the extension to the case of layered materials is been elaborated in [22]. Extension to tilting and lateral loads is given in [24] and [26].

The next thing of need now is a time dependent analysis method for ordinary quasi-static nanoindentation tests (e.g. [2, 40] with application [41]). Our approach is as follows [29]:

$$F = C \cdot (h - h_0)^m \quad \rightarrow \quad F = C(t) \cdot [h - h_0(t)]^{m(t)} \tag{22.21}$$

with t denoting the time.

The next step is the introduction of a time dependent material model. Here we resort to the well-known three parameter approach given by a Young's modulus of the following kind $E(t) = E_0 + E_1\,\text{Exp}[-t/\tau]$, where τ… relaxation time; for the reason of simplicity we here only consider relaxation and retardation separately. In the case of contact experiments usually also retardation has to be taken into account, which leads to additional terms of the form: $C(t) = \frac{1}{E(t)} = \frac{1}{E_0} + \left(\frac{1}{E_0+E_1} - \frac{1}{E_0}\right) e^{-\frac{E_0}{E_0+E_1}\frac{t}{\tau_c}}$ with usually: $\tau_c \neq \tau$.

Now we substitute the time dependent Young's modulus into the function $C(t)$ of Equation (22.21) and obtain:

$$F = 2 \cdot \left(\frac{1-\nu_i^2}{E_i} + \frac{1-\nu_s^2}{\displaystyle\int_V \left\{ E_{s0} + E_{s1} e^{\frac{-t}{\tau(T_j(r,z),\mu)}} + \tilde{r} \cdot E_{s1} e^{\frac{-t}{\tau(T_j(r,z),\mu)}} \frac{t}{\tau^2} \frac{d\tau}{d\vec{r}} + \vec{r}^2 \frac{\left(E_{s1} e^{\frac{-t}{\tau(T_j(r,z),\mu)}} \frac{t}{\tau^2} \frac{d\tau}{d\vec{r}} \right)^2}{4\left(E_{s0} + E_{s1} e^{\frac{-t}{\tau(T_j(r,z),\mu)}} \right)} \right\} \frac{dv}{V}} \right)^{-1}$$

$$\cdot \left[\frac{n}{n+1} \right] \cdot \left(\frac{1}{B} \cdot \left[1 - \varepsilon(n) \cdot \frac{n}{n+1} \right] \right)^{\frac{1}{n}} \cdot (h - h_0)^{\left(1+\frac{1}{n}\right)} \tag{22.22}$$

Please note that in cases of substantial preloading situations integration with respect to time t is required.

This makes the classical fit of the three constants h_0, m and C a time dependent six-parameter fit of h_0, n, B and the material constants E_0, E_1 and τ. It is clear that so many parameters in just one curve will automatically lead to numerical difficulties and instabilities. Thus, even in the pure visco-plastic case it is strongly suggested to use a variety of curves (three might be a good number) resulting from different unloading speeds or at different maximum loads with similar unloading times.

In cases of time dependent inelastic behavior (like visco-plasticity) or more complex constitutive laws also n, B and h_0 have to be taken as $n(t)$, $B(t)$, $h_0(t)$ making the fit even more complex. The same holds in the visco-elastic case for just h_0 when there are great differences between unloading time and τ(comparable strain rates). The parameters n and B on the other hand are then only geometrical parameters and do not explicitly depend on time.

A more detailed derivation and discussion of this extension is to be found in [2].

In recent years more and more experimental concepts have been introduced trying to generalize the classical normal indentation process into a multiaxial, combined tilted or twisted contact test [42–46]. In such cases, also asymmetric and tilted contact situations need to be manageable. We also need to address the problem of how to handle gradient structures, because especially with respect to tribo-protective polymer coatings gradient layers are often of use. However, as closed form solutions of this kind are unwieldy and difficult to handle numerically (c.f. [48, 49] regarding elliptical contact situation or [50–52] in the case of a certain gradient structure) or simply do not exist in sufficiently general manner, we prefer the use of load dots as introduced in [53] or [54]. With this approach a solution of the problem given above can be constructed by combining the superposition of various elastic or visco-elastic spaces as given in [53] with the introduction of interconnected load dots [53, 54]. In our case the new elastic spaces and loads to be superposed would be equivalent to differently loaded elastic spaces at suitably chosen time frames with adapted viscose properties along the virtual time-axis. While each frame satisfies a homogeneous elastic problem, the superposition does not necessarily do so, at least not if the superposition is done in a non-linear way.

This way an approximated solution even for the nonlinear problem given above could be constructed. The method should be demonstrated by the means of interconnected Hertzian load dots with different contact radii a_i [54] being made time dependent as follows:

$$w = \frac{3}{4}\sum_{i=1}^{N} H_i \frac{\lambda_i(t)}{a_i^3}\left\{(2a_i^2 + 2z^2 - r_i^2)\sin^{-1}\frac{a_i}{l_{2i}} + (l_{2i}^2 - a_i^2)^{1/2}\left(\frac{3l_{1i}^2 - 2a_i^2}{a_i}\right)\right\} \tag{22.23}$$

$$u^c = u + i\cdot v = -\frac{3\alpha}{2}\sum_{i=1}^{N} H_i \frac{\lambda_i(t) r_i e^{i\varphi}}{a_i^3}$$

$$\times\left\{(a_i^2 - l_{1i}^2)^{1/2}\left(1 - \frac{l_{1i}^2 + 2a_i^2}{3r_i^2}\right) + \frac{2a_i^3}{3r_i^2} - z\cdot\sin^{-1}\frac{a_i}{l_{2i}}\right\} \tag{22.24}$$

$$l_{1i} = \frac{1}{2}\left(\sqrt{(r_i + a_i)^2 + z^2} - \sqrt{(r_i - a_i)^2 + z^2}\right)$$

$$l_{2i} = \frac{1}{2}\left(\sqrt{(r_i + a_i)^2 + z^2} + \sqrt{(r_i - a_i)^2 + z^2}\right)$$

where we have used the complex presentation of the lateral displacements with $u^c = u + i^* v$ (with $i = \sqrt{-1}$, u displacement in x- and v in the y-direction, while w gives the displacement in the z-direction). The coordinates r_i have to be understood as $r_i = \sqrt{(x - x_i)^2 + (y - y_i)^2}$ with x_i and y_i defining the positions of the various load dots. The material functions respectively constants H_i and α are defined through:

$$H_i = \frac{1 - \nu^2}{\pi E_i(t)}; \quad \alpha = \frac{1 - 2\nu}{2(1 - \nu)}. \tag{22.25}$$

the parameters λ_i have to be fitted to the integral contact equations. They are considered to be only time-dependent, meaning $\lambda_i = \lambda_i(t)$. The reader should note that this approach is just a superposition of Hertzian loads with different contact radii a_i at various positions possibly acting on viscose-elastic spaces being characterized by $H_i = H_i(t)$. From these displacements the strain tensor can be evaluated in the usual manner by using:

$$u_{jk} = \frac{(u_{k,j} + u_{j,k})}{2}; \quad u_k = \begin{pmatrix} u \\ v \\ w \end{pmatrix}. \tag{22.26}$$

The stress components in the case of linear elasticity can be found using the following identities:

$$\sigma_{jk} = \frac{E}{1+\nu}\left(u_{jk} + \frac{\nu}{1-2\nu}u_{ll}\delta_{jk}\right) \quad with \quad j,k = x,y,z \equiv 1,2,3$$

$$u_{xk} = \frac{\partial u}{\partial k}; \quad u_{yk} = \frac{\partial v}{\partial k}; \quad u_{zk} = \frac{\partial w}{\partial k} \tag{22.27}$$

In the visco-elastic case applying a three-parameter Standard Linear Solid model given by a Young's modulus of the following kind $E(t) = E_0 + E_1 \mathrm{Exp}[-t/\tau]$ and assuming a strain driven loading situation we have instead of (27):

$$\sigma_{jk}(t) = \frac{1}{1+\nu}\left[E_0\left(u_{jk}(t) + \frac{\nu}{1-2\nu}u_{ll}(t)\delta_{jk}\right) + \int_0^t G_{jklm}(t-s)\dot{u}_{lm}(s)ds\right];$$
$$\dot{u}_{lm} = \dot{u}_{lm,s} = \frac{\partial u_{lm}}{\partial s} \tag{22.28}$$

with the tensor of relaxation G_{jkl} being a function of time in the case of linear viscosity. Now we know, however (e.g. [1, 2] and from what was said above), that the assumption of linearity with respect to real life problems can be severely inappropriate. In addition, it is almost impossible to make proper assumptions or estimates for the complete tensor function G_{jklm}. Especially in tribological contact situations, we will find stress, strain and temperature fields dramatically influencing the tensor function. However, with our inter-atomic based approach from the beginning of the theory section we are able to give a more correct description by the means of our field dependent Young's modulus for the sample:

$$E_s(t) = \int_V \left\{ E_{s0} + \sum_{q=1}^{j} E_{sq} e^{\frac{-t}{\tau_q}} + \vec{r}\cdot\sum_{q=1}^{j} E_{sq} e^{\frac{-t}{\tau_q}} \frac{t}{\tau_q^2}\frac{d\tau_q}{d\vec{r}} + \vec{r}^2 \frac{\left(\sum_{q=1}^{j} E_{sq} e^{\frac{-t}{\tau_q}} \frac{t}{\tau_q^2}\frac{d\tau_q}{d\vec{r}}\right)^2}{4\left(E_{s0} + \sum_{q=1}^{j} E_{sq} e^{\frac{-t}{\tau_q}}\right)} \right\} \frac{dv}{V}. \tag{22.29}$$

This results in

$$\sigma_{jk} = \frac{1}{1+\nu}\left[\begin{array}{l} E_0\left(u_{jk} + \frac{\nu}{1-2\nu}u_{ll}\delta_{jk}\right) + \\ \int_0^t \left(\dot{U}_{jk} + \frac{\nu}{1-2\nu}\dot{U}_{ll}\delta_{jk}\right)\int_V \left(\begin{array}{c} \sum_{q=1}^{j} E_{sq} e^{\frac{s-t}{\tau_q}} + \vec{r}\cdot\sum_{q=1}^{j} E_{sq} e^{\frac{s-t}{\tau_q}} \frac{t-s}{\tau_q^2}\frac{d\tau_q}{d\vec{r}} \\ + \vec{r}^2 \frac{\left(\sum_{q=1}^{j} E_{sq} e^{\frac{s-t}{\tau_q}} \frac{t-s}{\tau_q^2}\frac{d\tau_q}{d\vec{r}}\right)^2}{4\left(E_{s0} + \sum_{q=1}^{j} E_{sq} e^{\frac{s-t}{\tau_q}}\right)} \end{array}\right)\frac{dv}{V}ds \end{array}\right];$$
$$\dot{U}_{jk} = \dot{U}_{jk,s} = \frac{\partial U_{jk}}{\partial s} \tag{22.30}$$

In some cases it is more convenient to use a continuous relaxation function, which results in:

$$\sigma_{jk} = \frac{1}{1+\nu}\left[\begin{array}{l} E_0\left(u_{jk} + \frac{\nu}{1-2\nu}u_{ll}\delta_{jk}\right) \\ +\int_0^t \left(\dot{U}_{jk} + \frac{\nu}{1-2\nu}\dot{U}_{ll}\delta_{jk}\right) \\ \times \int_V \left(\begin{array}{l} \int_0^\infty E_s(\tau) e^{\frac{s-t}{\tau}} d\tau \\ +\vec{r}\cdot \int_0^\infty \left(\frac{dE_s(\tau)}{d\tau} + E_s(\tau)\frac{t-s}{\tau^2}\right)\frac{d\tau}{d\vec{r}} e^{\frac{s-t}{\tau}} d\tau \\ +\vec{r}^2 \frac{\left(\int_0^\infty \left(\frac{dE_s(\tau)}{d\tau} + E_s(\tau)\frac{t-s}{\tau^2}\right)\frac{d\tau}{d\vec{r}} e^{\frac{s-t}{\tau}} d\tau\right)^2}{4\left(E_{s0} + \int_0^\infty E_s(\tau) e^{\frac{s-t}{\tau}} d\tau\right)} \end{array}\right) \frac{dv}{V} ds \end{array}\right];$$

$$\dot{U}_{jk} = \dot{U}_{jk,s} = \frac{\partial U_{jk}}{\partial s} \tag{22.31}$$

With the assumption of adiabatic temperature fields following the shear field, such an approach appears to be manageable, but its solution strongly depends on the actual shear distribution: see [2]. For the reason of simplicity, we will here proceed the evaluation with the phenomenological SLS-approach being applicable in situations without significant temperature fields.

We start as follows:

$$\sigma_{jk} = \frac{1}{1+\nu}\left[E_0\left(u_{jk} + \frac{\nu}{1-2\nu}u_{ll}\delta_{jk}\right) + E_1\int_0^t \left(\dot{U}_{jk} + \frac{\nu}{1-2\nu}\dot{U}_{ll}\delta_{jk}\right) e^{-\frac{t-s}{\tau}} ds\right];$$

$$\dot{U}_{jk} = \dot{U}_{jk,s} = \frac{\partial U_{jk}}{\partial s} \tag{22.32}$$

We see that by performing the following substitution underneath the time integrals:

$$\int_0^t (\ldots)E_1 e^{-\frac{t-s}{\tau}} ds \Rightarrow \int_0^t (\ldots)\int_V \left[\begin{array}{c} \sum_{q=1}^{j} E_{sq} e^{\frac{s-t}{\tau_q}} + \vec{r}\cdot\sum_{q=1}^{j} E_{sq} e^{\frac{s-t}{\tau_q}} \frac{t-s}{\tau_q^2}\frac{d\tau_q}{d\vec{r}} \\ +\vec{r}^2 \frac{\left(\sum_{q=1}^{j} E_{sq} e^{\frac{s-t}{\tau_q}} \frac{t-s}{\tau_q^2}\frac{d\tau_q}{d\vec{r}}\right)^2}{4\left(E_{s0} + \sum_{q=1}^{j} E_{sq} e^{\frac{s-t}{\tau_q}}\right)} \end{array}\right] \frac{dv}{V} ds. \tag{22.33}$$

we are always able to switch back to the more general form.

By introducing the Hertzian load-dots as:

$$w_i = \frac{3}{4}\frac{\begin{pmatrix}\lambda_i(t)\\ \Lambda_i(t)\end{pmatrix}}{a_i^3}\left\{(2a_i^2+2z^2-r_i^2)\sin^{-1}\frac{a_i}{l_{2i}}+(l_{2i}^2-a_i^2)^{1/2}\left(\frac{3l_{1i}^2-2a_i^2}{a_i}\right)\right\}$$
$$\equiv \begin{pmatrix}\lambda_i\\ \Lambda_i\end{pmatrix}\tilde{w}_i;\quad i=1,2\ldots N \tag{22.34}$$

$$u_i^c = -\frac{3\alpha}{2}\frac{\begin{pmatrix}\lambda_i(t)\\ \Lambda_i(t)\end{pmatrix} r_i e^{i\varphi}}{a_i^3}\left\{(a_i^2-l_{1i}^2)^{1/2}\left(1-\frac{l_{1i}^2+2a_i^2}{3r_i^2}\right)+\frac{2a_i^3}{3r_i^2}-z\cdot\sin^{-1}\frac{a_i}{l_{2i}}\right\}$$
$$\equiv \begin{pmatrix}\lambda_i\\ \Lambda_i\end{pmatrix}\tilde{u}_i^c;\quad i=1,2\ldots N \tag{22.35}$$

$$\tilde{u}_{ijk} = \frac{(\tilde{u}_{ik,j}+\tilde{u}_{ij,k})}{2};\quad \tilde{u}_{ik} = \begin{pmatrix}\tilde{u}\\ \tilde{v}\\ \tilde{w}\end{pmatrix}_i;\quad \tilde{u}_i^c = \tilde{u}_i + i\cdot\tilde{v}_i;\quad \begin{matrix} i=1,2,\ldots N\\ j,k=1,2,3\end{matrix} \tag{22.36}$$

we obtain the rather simple expression for the stress tensor with index i:

$$\sigma_{ijk} = \frac{\left(\tilde{u}_{ijk}+\frac{\nu}{1-2\nu}\tilde{u}_{ill}\delta_{jk}\right)}{1+\nu}\left[E_0\lambda_i(t)+E_1\int_0^t\dot{\Lambda}_i(s)e^{-\frac{t-s}{\tau}}ds\right];$$
$$\dot{\Lambda}_i = \dot{\Lambda}_{i,s} = \frac{\partial\Lambda_i}{\partial s};\quad \begin{matrix} i=1,2,\ldots N\\ j,k=1,2,3\end{matrix}. \tag{22.37}$$

Applying an exponential approach for the $\Lambda_i(t) = \lambda_i(t)$ reading $\lambda_i = c_i\cdot e^{-\frac{t}{\tau_i}}$ (which here only is chosen as an example) we obtain the solution:

$$\sigma_{ijk} = \frac{\left(\tilde{u}_{ijk}+\frac{\nu}{1-2\nu}\tilde{u}_{ill}\delta_{jk}\right)}{1+\nu}\left[E_0c_ie^{-\frac{t}{\tau_i}}-c_i\tau E_1\frac{e^{-\frac{t}{\tau}}-e^{-\frac{t}{\tau_i}}}{\tau-\tau_i}\right];\quad \begin{matrix} i=1,2,\ldots N\\ j,k=1,2,3\end{matrix} \tag{22.38}$$

Now we derive the situation for a stress driven deformation state. By applying the three-parameter model together with our simple load-dot approach again, we have:

$$u_{jk} = \frac{1}{E}((1+\nu)\sigma_{jk}-\nu\sigma_{ll}\delta_{jk})\quad\text{with}\quad j,k=1,2,3;\quad u_{xk}=\frac{\partial u}{\partial k};$$
$$u_{yk}=\frac{\partial v}{\partial k};\quad u_{zk}=\frac{\partial w}{\partial k} \tag{22.39}$$

$$u_{ijk} = ((1+\nu)\widehat{\sigma}_{ijk}-\nu\widehat{\sigma}_{ill}\delta_{jk})\left[\frac{\gamma_i(t)}{E_0}+\left(\frac{1}{E_0+E_1}-\frac{1}{E_0}\right)\int_0^t\dot{\Gamma}_i(s)e^{-\frac{E_0}{E_0+E_1}\frac{t-s}{\tau}}ds\right]$$
$$\dot{\Gamma}_i = \dot{\Gamma}_{i,s};\quad \begin{matrix} i=1,2,\ldots N\\ j,k=1,2,3\end{matrix} \tag{22.40}$$

where the stresses $\widehat{\sigma}$ are now to be understood as:

$$\sigma_{ijk} = \begin{pmatrix}\gamma_i(t)\\ \Gamma_i(t)\end{pmatrix}\widehat{\sigma}_{ijk} = \frac{\begin{pmatrix}\gamma_i(t)\\ \Gamma_i(t)\end{pmatrix}}{1+\nu}\left(\widehat{u}_{ijk} + \frac{\nu}{1-2\nu}\widehat{u}_{ill}\delta_{jk}\right); \quad \widehat{u}_{ixk} = \frac{\partial \widehat{u}_i}{\partial k};$$

$$\widehat{u}_{iyk} = \frac{\partial \widehat{v}_i}{\partial k}; \quad \widehat{u}_{izk} = \frac{\partial \widehat{w}_i}{\partial k} \tag{22.41}$$

$$\widehat{w}_i = \frac{3}{4}\frac{1}{a_i^3}\left\{(2a_i^2 + 2z^2 - r_i{}^2)\sin^{-1}\frac{a_i}{l_{2i}} + (l_{2i}^2 - a_i^2)^{1/2}\left(\frac{3l_{1i}^2 - 2a_i^2}{a_i}\right)\right\}$$

$$= \frac{\tilde{w}_i}{\begin{pmatrix}\lambda_i\\ \Lambda_i\end{pmatrix}}; \quad i = 1,2\ldots N \tag{22.42}$$

$$\widehat{u}_i^c = -\frac{3\alpha}{2}\frac{r_i e^{i\varphi}}{a_i^3}\left\{(a_i^2 - l_{1i}^2)^{\frac{1}{2}}\left(1 - \frac{l_{1i}^2 + 2a_i^2}{3r_i{}^2}\right) + \frac{2a_i^3}{3r_i{}^2} - z\cdot\sin^{-1}\frac{a_i}{l_{2i}}\right\}$$

$$= \frac{\tilde{u}_i^c}{\begin{pmatrix}\lambda_i\\ \Lambda_i\end{pmatrix}}; \quad i = 1,2\ldots N \tag{22.43}$$

$$\widehat{u}_{ijk} = \frac{(\widehat{u}_{ik,j} + \widehat{u}_{ij,k})}{2}; \quad \widehat{u}_{ik} = \begin{pmatrix}\widehat{u}\\ \widehat{v}\\ \widehat{w}\end{pmatrix}_i; \quad \widehat{u}_i^c = \widehat{u}_i + i\cdot\widehat{v}_i; \quad \begin{matrix} i = 1,2,\ldots N \\ j,k = 1,2,3\end{matrix}. \tag{22.44}$$

Generalized with respect to field dependent mechanical properties and sticking to a three-parameter potential interaction the Equation (40) would read:

$$C(t) = \frac{1}{E_0} + \left(\frac{1}{E_0+E_1} - \frac{1}{E_0}\right)e^{-\frac{E_0}{E_0+E_1}\frac{t}{\tau}} - \vec{r}\cdot\left[\frac{E_1}{(E_0+E_1)^2}\right]e^{-\frac{E_0}{E_0+E_1}\frac{t}{\tau}}\frac{t}{\tau^2}\frac{d\tau}{d\vec{r}}$$
$$+\vec{r}\cdot\vec{r}\frac{\left(\left(\frac{E_1}{(E_0+E_1)^2}\right)e^{-\frac{E_0}{E_0+E_1}\frac{t}{\tau}}\frac{t}{\tau^2}\frac{d\tau}{d\vec{r}}\right)^2}{4\left[\frac{1}{E_0} + \left(\frac{1}{E_0+E_1} - \frac{1}{E_0}\right)e^{-\frac{E_0}{E_0+E_1}\frac{t}{\tau}}\right]}. \tag{22.45}$$

Extension to discrete or continuous retardation gives:

$$u_{ijk} = [(1+\nu)\widehat{\sigma}_{ijk} - \nu\,\widehat{\sigma}_{ill}\delta_{jk}]\int_V \left[\begin{matrix} \frac{\gamma_i(t)}{E_0} + \int_0^t \sum_{q=1}^n \frac{1}{E_q}\dot{\Gamma}_i(s)\left(1 - e^{-\frac{t-s}{\tau_{Eq}}}\right)\mathrm{d}s \\ +\mathbf{r}\cdot\int_0^t \sum_{q=1}^n \frac{1}{E_q}\dot{\Gamma}_i(s)e^{-\frac{t-s}{\tau_{Eq}}}\frac{t-s}{\tau_{Eq}{}^2}\frac{\mathrm{d}\tau}{\mathrm{d}\mathbf{r}}s \\ +\mathbf{r}\cdot\mathbf{r}\int_0^t \frac{\left(\sum_{q=1}^n \frac{1}{E_q}\dot{\Gamma}_i(s)e^{-\frac{t-s}{\tau_{Eq}}}\frac{t-s}{\tau_{Eq}{}^2}\frac{\mathrm{d}\tau}{\mathrm{d}\vec{r}}\right)^2}{4\left(\frac{1}{E_0} + \sum_{q=1}^n \frac{1}{E_q}\dot{\Gamma}_i(s)\left(1 - e^{-\frac{t-s}{\tau_{Eq}}}\right)\right)}\mathrm{d}s \end{matrix}\right]\frac{\mathrm{d}\nu}{V} \tag{22.46}$$

$$u_{ijk} = \begin{pmatrix} (1+\nu)\widehat{\sigma}_{ijk} - \\ \nu\,\widehat{\sigma}_{ill}\delta_{jk} \end{pmatrix} \int_V \left[\begin{array}{l} \frac{\gamma_i(t)}{E_0} + \int_0^t \int_0^\infty \frac{1}{E(\tau)}\left(1 - e^{-\frac{t-s}{\tau}}\right) d\tau\, \dot{\Gamma}_i(s) ds + \\ \mathbf{r} \cdot \int_0^t \int_0^\infty \frac{1}{E(\tau)} \left(\frac{t-s}{\tau^2} e^{-\frac{t-s}{\tau}} - \frac{\left(1 - e^{-\frac{t-s}{\tau}}\right)}{E(\tau)} \frac{dE(\tau)}{d\tau} \right) \frac{d\tau}{d\mathbf{r}} d\tau\, \dot{\Gamma}_i(s) ds + \mathbf{r} \\ \mathbf{r} \cdot \mathbf{r} \int_0^t \frac{\left(\dot{\Gamma}_i(s) \left(\int_0^\infty \frac{1}{E(\tau)} \left(\frac{t-s}{\tau^2} e^{-\frac{t-s}{\tau}} - \frac{\left(1 - e^{-\frac{t-s}{\tau}}\right)}{E(\tau)} \frac{dE(\tau)}{d\tau} \right) \frac{d\tau}{d\mathbf{r}} d\tau \right) \right)^2}{4\left(\frac{1}{E_0} + \int_0^\infty \frac{1}{E(\tau)} \left(1 - e^{-\frac{t-s}{\tau}}\right) d\tau \right)} ds \end{array} \right] \frac{dv}{V} \tag{22.47}$$

For reasons of simplicity, however, we will here proceed with the gradient free simple SLS-model.

Together with the normalization condition and the total normal load $p = p(t)$ one obtains:

$$\begin{aligned} p(t) &= \int_0^{2\pi} \int_0^{a_{\max}} \sigma_{zz}(r, \varphi, z),|_{z=0} r dr d\varphi \equiv \int_0^{2\pi} \int_0^{a_{\max}} \sigma_{zz0}(r, \varphi) r dr d\varphi \\ &= 2\pi \int_0^{a_{\max}} \sigma_{zz0}(r) r dr = \sum_{i=1}^{N} \left[E_0 c_i e^{-\frac{t}{\tau_i}} - c_i \tau E_1 \frac{e^{-\frac{t}{\tau}} - e^{-\frac{t}{\tau_i}}}{\tau - \tau_i} \right] \end{aligned} \tag{22.48}$$

for the strain driven case, while one derives in the stress driven situation with $\Gamma_i(t) = \gamma_i(t)$:

$$\begin{aligned} p(t) &= \int_0^{2\pi} \int_0^{a_{\max}} \sigma_{zz}(r, \varphi, z),|_{z=0} r dr d\varphi \equiv \int_0^{2\pi} \int_0^{a_{\max}} \sigma_{zz0}(r, \varphi) r dr d\varphi \\ &= 2\pi \int_0^{a_{\max}} \sigma_{zz0}(r) r dr = 2\pi \sum_{i=1}^{N} \left[\int_0^{a_i} \gamma_i(t) \widehat{\sigma}_{izz0}(r) r dr \right] = \sum_{i=1}^{N} \gamma_i(t). \end{aligned} \tag{22.49}$$

Both cases are relatively easy whenever there is no change of the contact area during the evolution of the deformation time being considered. In most practical applications, however, this is not the case. By concentrating on contact experiments, we can often assume that they are load controlled. However, due to the complex mixed stress states underneath the contact zone, there are also strain rate effects involved and it is not clear yet how to take these into account. Therefore Equations (39) to (44) and (49) are applicable but need to be extended with respect to strain driving effects (e.g. [2], appendix). There are several options how to make the contact area evolvable during the considered time slot. One could either change the number of load dots taken into account respectively contributing to the whole contact or make the contact radii a_i time dependent. As the latter way immediately leads to much more complicated integrals, we are going for the first option not by the apparent simple way of changing the total load-dot number

N, but by introducing Heaviside distributions to the $\gamma_i(t)$ (or respectively the $\lambda_i(t)$ in the strain driven case) as follows:

$$\begin{aligned}\gamma_i(t) &= c_i[\Phi(t - t_{ion}) - \Phi(t - t_{ioff})];\\ \dot{\Gamma}_i(t) &= \dot{\gamma}_i(t) = c_i[\delta(t - t_{ion}) - \delta(t - t_{ioff})]; \quad t_{ion} < t_{ioff},\end{aligned} \tag{22.50}$$

With:

$$\Phi(t) = \begin{cases} 0 \ for \ t \leq 0 \\ 1 \ for \ t > 0 \end{cases} \tag{22.51}$$

and δ giving the Dirac delta distribution. With this approach the integration in (40) can be performed easily:

$$\begin{aligned} u_{ijk} &= ((1+\nu)\widehat{\sigma}_{ijk} - \nu\,\widehat{\sigma}_{ill}\delta_{jk}) \\ &\times \left[\frac{c_i[\Phi(t - t_{ion}) - \Phi(t - t_{ioff})]}{E_0} + \left(\frac{1}{E_0+E_1} - \frac{1}{E_0}\right)\int_0^t c_i[\delta(s - s_{ion}) - \delta(s - s_{ioff})]e^{-\frac{E_0}{E_0+E_1}\frac{t-s}{\tau}}ds \right] \\ &= c_i((1+\nu)\widehat{\sigma}_{ijk} - \nu\,\widehat{\sigma}_{ill}\delta_{jk}) \\ &\left[\frac{1}{E_0}\begin{Bmatrix} 0 & t \leq t_{ion} \\ 1 & t_{ion} < t \leq t_{ioff} \\ 0 & t_{ioff} < t \end{Bmatrix} + \left(\frac{1}{E_0+E_1} - \frac{1}{E_0}\right)\begin{Bmatrix} 0 & t \leq t_{ion} \\ e^{-\frac{E_0}{E_0+E_1}\frac{t-t_{ion}}{\tau}} & t_{ion} < t \leq t_{ioff} \\ e^{-\frac{E_0}{E_0+E_1}\frac{t-t_{ion}}{\tau}} - e^{-\frac{E_0}{E_0+E_1}\frac{t-t_{ioff}}{\tau}} & t_{ioff} < t \end{Bmatrix}\right] \end{aligned} \tag{22.52}$$

For many cases in contact applications we are facing mixed stress–strain driven conditions requiring a combination of (37) and (40). Choosing a sufficiently high number of load dots N one can approximate any displacement (time dependent effective indenter shape) of symmetry of revolution. The approach has the advantage, that the evaluation of the complete elastic field for the whole body would be only a question of summing up a series of Hertzian fields. In addition, the solution can easily be extended to layered materials [28]. In order to fit the model even better to the real test or application, one might even make the constants c_i time-dependent. This, of course, requires reevaluation of the integral in (40) but caused by the distribution structure (properties of the Dirac delta function) in the integrand any additional time-dependency would be very easy to handle. In addition, one could go for extended Hertzian load dots instead of Hertzian ones (e.g. [26]), incorporate tilting, twisting and lateral loads, defects and intrinsic stresses as been done in [12, 35, 36]. Then the method is also applicable to asymmetric scratch- or tribo-like contact situations.

22.2.4 Theory for the Physical Scratch and/or Tribological Test

An approach for stress distributions for more general loading conditions as occurring during tests like scratch and pin on disc has been given already in the section 'the effective indenter concept'. The evaluation of the complex stress and strain fields for these tests is elaborated in the references given there. However, as these evaluations are rather cumbersome and lengthy, the reader is also referred to a software package performing such calculations in an automated and quick manner [28].

It should explicitly be pointed out that knowledge of the complete stress and strain field is essential for a proper failure characterization (scratch) or wear mechanism analysis (tribo). An illustrative example is presented in Figures 22.8–22.10 below.

Hence, only such a physical analysis of mechanical contact measurements like instrumented indentations, scratch and tribo-tests enables one to find out why a surface structure fails under certain loading conditions. These results provide indications on how the investigated coating or surface structure can be improved.

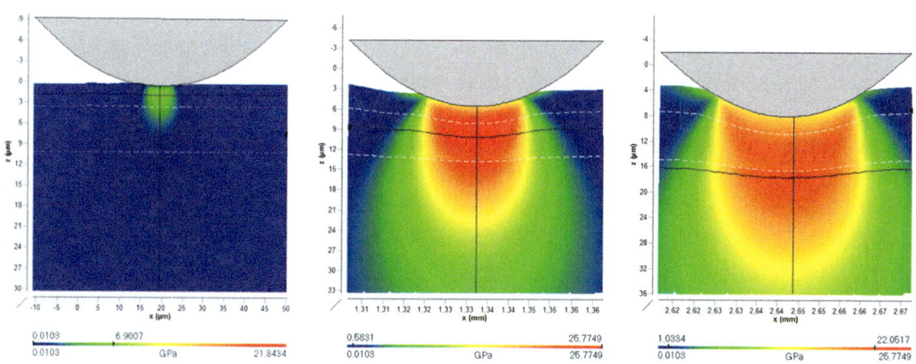

Figure 22.8 The evolution of von Mises stress during the scratch test shown at three measurement points (with a, b, c from left to right): (a) at the beginning of the scratch test, (c) in the moment of L_C failure, and (b) in between. The black crosshairs indicate the location of the maximum.

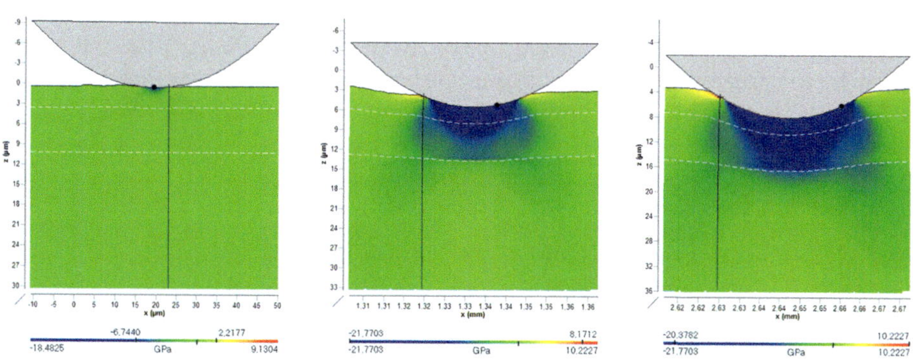

Figure 22.9 The evolution of normal stress in scratch direction illustrated at three measurement points (with a, b, c from left to right): (a) at the beginning of the scratch test, (c) in the moment of L_C failure, and (b) in between. The black crosshairs indicate the location of maximum tensile stress.

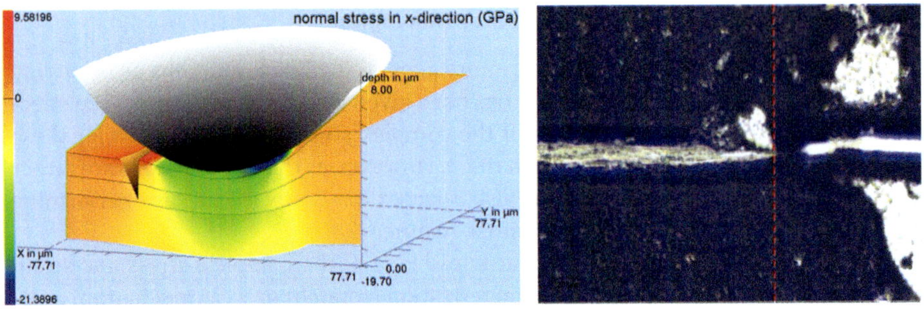

Figure 22.10 Illustrative scheme of the failure mechanism (a) and an optical graph of the post-scratch surface (b) in which the corresponding L_C position is marked by the red dashed line. *Source*: Schwarzer 2011 [12]. Reproduced with permission of Elsevier.

22.2.5 From Quasi-Static Experiments and Parameters to Dynamic Wear, Fretting and Tribological Tests

Now we need to establish the relationship between quasi-static characteristics, like hardness, yield strength, Young's modulus etc. and highly dynamic characteristics like wear, fretting and other tribological processes and effects.

At first it must be understood that within the concept of this approach the process of wear, fretting or general tribological process is to be considered as a multi-physical, multi body (asperity) ensemble of contact situations (called load dots, see [53, 54]). Here each of these contacts has its own geometry, load conditions (including tilting, normal, lateral, twisting loads plus temperature fields caused by internal or external friction) and sometimes also runs – in complex cases with debris – at multiple time scales. This means that various parts of the global tribological contact are running in their own speed. Of course, we will then also have to take temperature effects into account. These are often coupling back into the mechanical properties and time dependent material behavior (c.f. section 'The extension of the Oliver and Pharr method to analyze nanoindentation data to layered materials and time dependent mechanical behavior'). However, the mathematics of partial differential equations does provide an interesting short cut here. It is due to the similarity of the governing differential equations that by having found the solution for the contact problem, the temperature field or any other 'diffusion-like' problem can also be considered solved. So, the layered solution for any diffusion problem can be constructed from the elastic solution by a limiting procedure with respect to the Poisson's ratio, which leads to perfectly non-singular field solutions for the diffusion problem in question.

We are now concentrating on some questions often popping up in connection with tribological tests and observations.

How does hardness influence tribological effects in general?

The connection between non-physical parameters (like hardness) towards wear is not the intention of the paper. In fact, strictly speaking, there is no such connection. Something not generic like hardness and being obtained from a quasi-static test simply cannot – not generally – be extended or applied to a dynamic physical process like wear. Hardness, after all, is a mixed parameter. There are so many things contributing to hardness that – in general – it is not clear what effects affecting the hardness are also influencing – and in what manner – wear, for instance. Thus, yield strength, critical fracture stresses, fracture toughness (energy loss caused by fracture propagation), in short: parameters which could clearly be connected to certain stress fields and stress components, are much better suited to also interpret and discuss tribological effects than hardness.

What is the role of debris particles?
They are just adding up to the complex jumble of contact situations mentioned above.

How are strain-rate or even more general time-dependent effects taken into account?
This is usually been done by applying time-functional dependencies for the mechanical parameters. However, as shown in [2], in the presence of inhomogenous fields, possibly influencing the mechanical parameters, the resulting governing equations become highly nonlinear. This can either be treated by the means of piecewise X-parameter models or by solving the nonlinear governing system of equations with time-stress- and strain-dependent material parameters (see section 22.2.3). First principle derived basics are given in [2].

How can the many faces of wear and tribo effects be incorporated into one theoretical apparatus?
By decomposition limits extracted from first principle approaches. Hereby, the most general way would be to extract decomposition limits from the first principle approaches as described in [2]. These limits have to be compared with the deformation fields obtained in the multiple complex contact model describing our tribo, fretting or wear experiments, where nonlinear effects as described in [1] and [2] should not be neglected. Even though we will demonstrate how this, in principle, can be done within the procedure-plan below, it is eminent that such a procedure is not necessarily fit for daily or even industrial use. Therefore, we also need a phenomenological description of the physics behind tribo-effects in an as general as possible manner.

At first, we explicitly point out, that wear or any other tribological effect cannot be connected by a simple k_d-value or any other scalar value (simple Archard's law) to the deformation or stress field. This is by far too simple and not general enough. Instead, in the simplest (linear) case, the tribo-effect is a tensor, coupling with wear-moduli to every deformation field component in a fully covariant manner. More correctly, any tribo-process could be generalized as:

$$\begin{aligned}
\text{tribo-effect}_{ij} &= k^{\sigma}{}_{ijkl}\,\sigma^{kl} + k^{\varepsilon}{}_{ijkl}\,\varepsilon^{kl} + k^{u}{}_{ijkl}\,u^k u^l + \sum_{\chi=1}^{N} k^{S_\chi}\delta_{ij}S_n \\
&+ k^{\sigma}{}_{ijklmn}\,\sigma^{kl}\sigma^{mn} + k^{\varepsilon}{}_{ijklmn}\,\varepsilon^{kl}\varepsilon^{mn} + k^{u}{}_{ijklmn}\,u^k u^l u^m u^n \\
&+ \sum_{\chi=1}^{N} k2^{S_\chi}\delta_{ij}S_n^2 + k^{\sigma}{}_{ijklmnop}\,\sigma^{kl}\sigma^{mn}\sigma^{op} + k^{\varepsilon}{}_{ijklmnop}\,\varepsilon^{kl}\varepsilon^{mn}\varepsilon^{op} \\
&+ k^{u}{}_{ijklmnop}\,u^k u^l u^m u^n u^o u^p + \sum_{\chi=1}^{N} k3^{S_\chi}\delta_{ij}S_n^3 + \ldots
\end{aligned} \tag{22.53}$$

Here we will concentrate mainly on linear dependencies (first line), where we used the following denotations: $k^{xx}{}_{ijkl}$-tensors are tensors coupling to the various field values or tensors like the stress σ^{kl}, strain ε^{kl}, displacement-vector u^i or scalar values S_n, like free or distortion energy strain work etc. The symbol δ_{ij} is the Kroenecker symbol. In most cases, wear for instance, it should be sufficient to consider only the stresses:

$$\text{tribo-effect}_{ij} \equiv w_{ij} = k_{ijkl}\ \sigma^{kl} \tag{22.54}$$

Where from the scalar wear-depth h_w has to be evaluated via:

$$h_w = w_{ij}\ n^i\ n^j \tag{22.55}$$

with n^i denoting the surface normal unit vector.

We have to point out explicitly, that all the stresses, strains, vectors, energies etc. as being used in (22.53) are field values and not just numbers. So, they are to be understood as functions of the coordinates. However, in the end and in order to obtain a mathematical apparatus one can compare with real experiments, one requires numbers for the tribo-effect or wear instead of complicated functions. Thus, for many applications the components of the k-tensors must even have operational character. For example, if considering the maximum of a certain stress it might be of great importance where exactly that maximum within the material lays. Let us assume a blunt indenter contact situation and a tribo-effect being dominated by fatigue caused by the total shear. In such a case we will always find the von Mises maximum underneath the surface. It seems reasonable in this situation to assume that the amount of destruction respectively the mass of worn off material is not just connected with the absolute value of the von Mises stress maximum σ_{vM}, but also its distance z to the surface respectively the contact centre $|\mathbf{r}|$. The formulation of such a dependency would read:

$$w_{ij} = \delta_{ij}\ (k_{dvM}\sigma_{vM} \cdot e^{-\lambda_k \vec{r}\cdot\vec{r}}) \tag{22.56}$$

with λ_k being yet another parameter in addition to the linear k_{dvM} characterizing the wear performance in dependency of the von Mises stress distribution. We point out that, in using the local load acting on a certain contacted surface point, we have the following local wear depth:

$$h_w = w_{ij}\sigma^{ij}_{\text{surface}} = \delta_{ij}\ (k_{dvM}\sigma_{vM} \cdot e^{-\lambda_k \vec{r}\cdot\vec{r}})\sigma^{ij}_{\text{surface}} = \sigma^{zz}_{\text{surface}}\ (k_{dvM}\sigma_{vM} \cdot e^{-\lambda_k \vec{r}\cdot\vec{r}}) \tag{22.57}$$

Using the quantum mechanical marking to note operators we could use the following generalization for an operational form of (22.53):

$$\text{tribo-effect}_{ij} = \hat{k}^{\sigma}_{ijkl}\ \sigma^{kl} + \hat{k}^{\varepsilon}_{ijkl}\ \varepsilon^{kl} + \hat{k}^{u}_{ijkl}\ u^k u^l + \sum_{n=1}^{N} \hat{k}^{S_n}\delta_{ij}S_n \tag{22.58}$$

For the wear-example given above we can formulate this as:

$$\text{tribo-effect}_{ij} \equiv w_{ij} = \hat{k}_{ijkl}\ \sigma^{kl} \equiv \delta_{ij}\ (k_{dvM}\sigma_{vM} \cdot e^{-\lambda_k \mathbf{r}\cdot\mathbf{r}}). \tag{22.59}$$

As we can see, Archard's law given with a scalar wear coefficient k_d by the simple relation $h_w = k_d \cdot \sigma^{33}$ directly appears out of Equation (22.54). Such a dramatic simplification is always possible wherever either the stress is dominated by its normal component in the surface normal direction (here we named it σ^{33}) or where the coefficient tensor k_{ijkl} is zero except for those components coupling to the normal surface stress σ^{33}, which would then read:

$$h_w = w_{ij}\ n^i\ n^j = k_{ijkl}\ \sigma^{kl} n^i\ n^j = k_{33kl}\ \sigma^{kl};\ \ k_{3333}\sigma^{33} \equiv k_d\sigma^{33} \tag{22.60}$$

We can deduce now that for complex contact conditions, where the stress tensor is fully set and no component is dominant against all the others, one should be rather careful with the assumption of having the wear-tensor being of the most simple, Archard's-law-like kind $k_{33kl} = \begin{pmatrix} 0 & 0 & 0 \\ 0 & 0 & 0 \\ 0 & 0 & k_{3333} \end{pmatrix}$. Usually, one also has to take the other stresses into account and investigate their possible influence regarding the resulting global wear, which, in this case, has to be taken as the sum over all stress-components in connection with the wear-tensor:

$$h_w = k_{33kl}\,\sigma^{kl} = k_{3311}\sigma^{11} + k_{3322}\sigma^{22} + k_{3333}\sigma^{33} + 2(k_{3312}\sigma^{12} + k_{3313}\sigma^{13} + k_{3323}\sigma^{23}) \tag{22.61}$$

Here we have made use of the symmetry of the stress tensor, also requiring a symmetric wear-tensor.

We also point out, that in the general law as given above in Equation (22.54), the hydrostatic (sphere) and deviatoric stress parts are distinguishable. Such a simplified law might read:

$$\text{tribo-effect}_{ij} \equiv w_{ij} = \overbrace{k_{dShear\_ij}\left(\sigma_{ij} - \frac{1}{3}\delta_{ij}\sigma_{ll}\right)}^{\text{shear part}} + \overbrace{\delta_{ij}k_{dSphere}\frac{1}{3}\sigma_{ll}}^{\text{pressure part}} \tag{22.62}$$

Or even simpler and further scalarized:

$$w_{ij} = \delta_{ij}\ \left(k_{dvM}\sigma_{vM} + k_{dH}\sigma_H\right) \tag{22.63}$$

with σ_{vM}, σ_H denoting the von Mises and the hydrostatic stress, respectively.

22.2.6 Including Biaxial Intrinsic Stresses

In order to expand our theoretical developments for layered materials we have to take into account the effects of intrinsic stresses. These stresses in thin films and coatings are usually assumed to be perfectly biaxial. This assumption results in the following expression for pressure in terms of the first invariant stress I_1 for the biaxial stress state:

$$P = \frac{I_1}{3} = \frac{1}{3}\cdot\begin{pmatrix} \sigma_{rr} & 0 & 0 \\ 0 & \sigma_{rr} & 0 \\ 0 & 0 & 0 \end{pmatrix}_{ii} = \frac{2}{3}\cdot\sigma_{rr} \tag{22.64}$$

Even though the perfectly biaxial stress state does not contain any explicit shear components like σ_{ij} (with i and j being different), the resulting total deviatoric stress (shear) is by no means zero. On the contrary, we obtain:

$$\begin{aligned}\sigma_M &= \sqrt{\frac{1}{2}\begin{bmatrix}(\sigma^{11}-\sigma^{22})^2 + (\sigma^{11}-\sigma^{33})^2 \\ +(\sigma^{33}-\sigma^{22})^2 + 6((\sigma^{12})^2 + (\sigma^{13})^2 + (\sigma^{23})^2)\end{bmatrix}} \\ &= \sqrt{\frac{1}{2}}\left[(\sigma_{rr}-0)^2 + (\sigma_{rr}-0)^2 + (\sigma_{rr}-\sigma_{rr})^2 + 6(0^2+0^2+0^2)\right] = \sigma_{rr}\end{aligned} \tag{22.65}$$

meaning, in the perfect biaxial stress state, that the total shear is always as high as the intrinsic stress itself.

Also often ignored is the fact that intrinsic stresses – even though perfectly biaxial – are not leaving the substrate unstressed. In order to end up with a physically correct state of equilibrium, the substrate cannot be stress free, as the following brief evaluation will show.

We start with the following three dimensional, correct solution for the governing equation of elasticity in the isotropic case:

$$\vec{u} = \sum_{i,k=1}^{\infty} c_{ik} \begin{pmatrix} c^2\left((A+Bcz)e^{cz}+(D+Fcz)e^{-cz}\right)\sin[cx]+u^2\left(Ce^{uz}+Ge^{-uz}\right)\cos[uy] \\ u^2\left((\tilde{A}+\tilde{B}cz)e^{uz}+(\tilde{D}+\tilde{F}uz)e^{-uz}\right)\sin[uy]+c^2\left(\tilde{C}e^{cz}+\tilde{G}e^{-cz}\right)\cos[cx] \\ c^2\left((-A-(-3+4\nu+uz)B)e^{cz}+(D+(3-4\nu+uz)F)e^{-uz}\right)\cos[cx] \\ +u^2\left((-\tilde{A}-(-3+4\nu+uz)\tilde{B})e^{uz}+(\tilde{D}+(3-4\nu+uz)\tilde{F})e^{-uz}\right)\cos[uy] \end{pmatrix} \tag{22.66}$$

with $c=\frac{i\pi}{a}, u=\frac{k\pi}{b}$ (a and b denoting the side lengths of the rectangular substrate, the z-axis is parallel to the normal of the substrate) and $i,k = 1, 3, 5, 7\ldots$, which assures the normal stresses σ_{xx} and σ_{yy} being zero at the substrate rim. The constants A, B, C, D, F, G and $\tilde{A}, \tilde{B}, \tilde{C}, \tilde{D}, \tilde{F}, \tilde{G}$ are different for film and substrate so altogether we have 24 constants to determine. The shearing stresses $\sigma_{xz}, \sigma_{yz}, \sigma_{xy}$ should also be zero at the substrate edges in the x and y directions, but this condition cannot be satisfied with the approach suggested in equation (22.66) as easy as the normal stress boundary conditions. However, because here we will only concentrate on the elastic field within or near the centre of the coated sample, the influence of non-zero components $\sigma_{xz}, \sigma_{yz}, \sigma_{xy}$ at the rim can be considered being small. In addition, the condition σ_{rr}^{f} = const over the whole coating area must be satisfied by a proper Fourier series. It can be shown that (22.66) satisfies the equation of equilibrium for an isotropic elastic medium (e.g. [60], but one should notice that our approach is somewhat extended and more general in order to satisfy arbitrary shear stress conditions on z = constant-planes within the coatings-substrate compound). The stress components can be found using the following identities:

$$\sigma_{jk} = \frac{E}{1+\nu}\left(u_{jk}+\frac{\nu}{1-2\nu}u_{ll}\delta_{jk}\right) \quad with \quad j,k = x,y,z;$$
$$u_{xk} = \frac{\partial u}{\partial k};\quad u_{yk} = \frac{\partial v}{\partial k};\quad u_{zk} = \frac{\partial w}{\partial k}. \tag{22.67}$$

The further boundary conditions (h giving the coating thickness):

$$\sigma_{zz}|_{z=0} = \sigma_{xz}|_{z=0} = \sigma_{yz}|_{z=0} = 0,\quad \sigma_{zz}|_{z=h-0} = \sigma_{zz}|_{z=h+0},\quad w|_{z=h-0} = w|_{z=h+0},$$
$$\sigma_{xz}|_{z=h\pm 0} = \sigma_{yz}|_{z=h\pm 0} = \pm f(x,y)$$

give the equations necessary to determine all constants including the coefficients c_{ik}. More interfaces at z = constant-planes can easily be incorporated and satisfied with the approach given above, but here we restrict our considerations to a simple half space with one layer of thickness h. In order to simplify the evaluation we substitute all constants A, B, D, F etc. by products of new constants $A, B, D, F\ldots$ and lateral forces Tx and Ty acting as shearing loads on certain planes $z = \text{constant}$ in the x and y directions, respectively. The new form of (22.66) now reads:

$$u = \sum_{i,k=1}^{\infty} c_{ik}$$

$$\left(\begin{array}{l} T_x c^2 ((A + Bcz) e^{cz} + (D + Fcz) e^{-cz}) \sin [cx] + T_x u^2 (Ce^{uz} + Ge^{-uz}) \cos \left[uy\right] \\ T_y u^2 \left(\left(\tilde{A} + \tilde{B}cz\right) e^{uz} + \left(\tilde{D} + \tilde{F}uz\right) e^{-uz}\right) \sin \left[uy\right] + T_y c^2 \left(\tilde{C}e^{cz} + \tilde{G}e^{-cz}\right) \cos [cx] \\ T_x c^2 ((-A - (-3 + 4\nu + uz) B) e^{cz} + (D + (3 - 4\nu + uz) F) e^{-uz}) \cos [cx] \\ +T_y u^2 \left(\left(-\tilde{A} - (-3 + 4\nu + uz) \tilde{B}\right) e^{uz} + \left(\tilde{D} + (3 - 4\nu + uz) \tilde{F}\right) e^{-uz}\right) \cos \left[uy\right] \end{array}\right) \tag{22.68}$$

Together with (22.67) we can now evaluate the stresses. As this is straightforward and only requires certain derivatives of the displacements given above, one example should suffice:

$$\begin{aligned} \sigma_{xy} &= \frac{-E}{2(1+\nu)} \sum_{i,k=1}^{\infty} c_{ik} (T_x u^3 (C e^{uz} + G e^{-uz}) \sin[uy] + T_y c^3 (\tilde{C} e^{cz} + \tilde{G} e^{-cz}) \sin[cx]) \\ &\equiv \sum_{i,k=1}^{\infty} c_{ik} [f(z) \sin[uy] + \tilde{f}(z) \sin[cx]] \\ &= \sum_{i=1}^{\infty} f(z)\; c_i \sin\left(\frac{i\pi x}{a}\right) + \sum_{k=1}^{\infty} \tilde{f}(z)\; c_k \sin\left(\frac{k\pi y}{b}\right) \end{aligned} \tag{22.69}$$

Here we see, that a relatively simple structure can be obtained for all stress components in Cartesian coordinates. In the same way we extract from (22.68), that the structure of the normal stresses in lateral direction of the film-substrate-compound σ_{xx} and σ_{yy} can be given in the following general form:

$$\begin{aligned} \sigma_{xx} &= \sum_{i=1}^{\infty} fx_x(z)\; c_i \cos\left(\frac{i\pi x}{a}\right) + \sum_{k=1}^{\infty} fx_y(z)\; c_k \cos\left(\frac{k\pi y}{b}\right), \\ \sigma_{yy} &= \sum_{i=1}^{\infty} fy_x(z)\; c_i \cos\left(\frac{i\pi x}{a}\right) + \sum_{k=1}^{\infty} fy_y(z)\; c_k \cos\left(\frac{k\pi y}{b}\right), \end{aligned} \tag{22.70}$$

with the functions fx_y, fy_x, fx_x and fy_y only being abbreviations of the z-dependent terms directly following from (22.67) and (22.68). Now we need to find equations for the determination of the Fourier coefficients c_i and c_k. From the bi-axial stress conditions $\sigma^f_{xx} = \sigma^f_{yy} = \sigma$ at a distinct depth $z = z0$ and together with (22.70) we obtain:

$$\begin{aligned} c_i &= \frac{2}{a} \int_{-a/2}^{a/2} \frac{\sigma(fx_y(z_0) - fy_y(z_0))}{fx_y(z_0)\; fy_x(z_0) - fx_x(z_0)\; fy_y(z_0)} \cos\left(\frac{i\pi x}{a}\right), \\ c_k &= \frac{2}{b} \int_{-b/2}^{b/2} \frac{\sigma(fx_x(z_0) - fy_x(z_0))}{fx_x(z_0)\; fy_y(z_0) - fx_y(z_0)\; fy_x(z_0)} \cos\left(\frac{k\pi y}{b}\right). \end{aligned} \tag{22.71}$$

In the case of thin films and plate-like substrates this new approach should agree with the results given by the equation of Stoney [61], which in fact it does (see [53]) if one does not forget to clearly distinguish between the beam solutions given by Stoney and the necessary adaptations with respect to plates. One can see clearly that in fact none of the stress components is zero, but for geometries with the side lengths a and b being huge compared to the coatings thickness, the stress tensor is mainly dominated by the normal stresses in the x and y directions.

Thus, even though the correct three-dimensional solution of the intrinsic stress problem requires shear stress-components at the interface and also produces normal stresses in the z-direction, the assumption of a stress tensor of the form:

$$\sigma_{ij}^{intrinsic} = \begin{pmatrix} \sigma_{rr} & \varepsilon_{xy} & \varepsilon_{xz} \\ \varepsilon_{xy} & \sigma_{rr} & \varepsilon_{yz} \\ \varepsilon_{xz} & \varepsilon_{yz} & \varepsilon_{zz} \end{pmatrix} \tag{22.72}$$

with $\varepsilon_{ij} << \sigma_{rr}$, meaning negligible values of $\varepsilon_{ij} \approx 0$ almost everywhere in the compound is perfectly plausible.

The important aspect is now coming with the fact that biaxial intrinsic stresses in the coating cannot appear without also producing such stresses in the substrate. Usually the intrinsic stresses within the coating are much smaller than the yield strength of the coating materials (otherwise the coating would not form at all), but this does not necessarily hold for the substrate. Taking the results from above, we find that there is always shear stress at the interface in the substrate which can reach significant values. In [62] (Figure 22.1) an example is shown where compressive intrinsic stresses of about 4 GPa in a TiN-coating of 1 μm thickness already created plastic flow in the interface region of the substrate, which was assumed to be a steel substrate (not rim hardened). Such behavior has also been observed in practice [63].

In the case of metals the plastic flow usually sets in long before any nonlinear behavior would become significant. However, for other substrate materials, as the intrinsic stresses have to be added to the ones coming from any other source (external load, temperature etc.) and contribute to possible dependencies with respect to the mechanical parameters on the stress and strain field, one might be aware of nonlinear behavior as described above. This especially holds for the substrate regions close to the interface simply due to the intrinsic pre-stressing. For this, as shown above, especially the parameter dependency on the first two stress tensor invariants has to be watched. In some cases the pre-stresses simply coming from the test conditions like compression or contact stress fields coming from the sample fixation have to be taken into account (e.g. [64–66]), also.

22.3 The Procedure

Now we need to connect the various models and concepts introduced and elaborated above in order to construct a compact optimization cycle including wear test and simulation.

This will be illustrated by the means of a practical example for cutting tools in Figures 22.11–22.14:

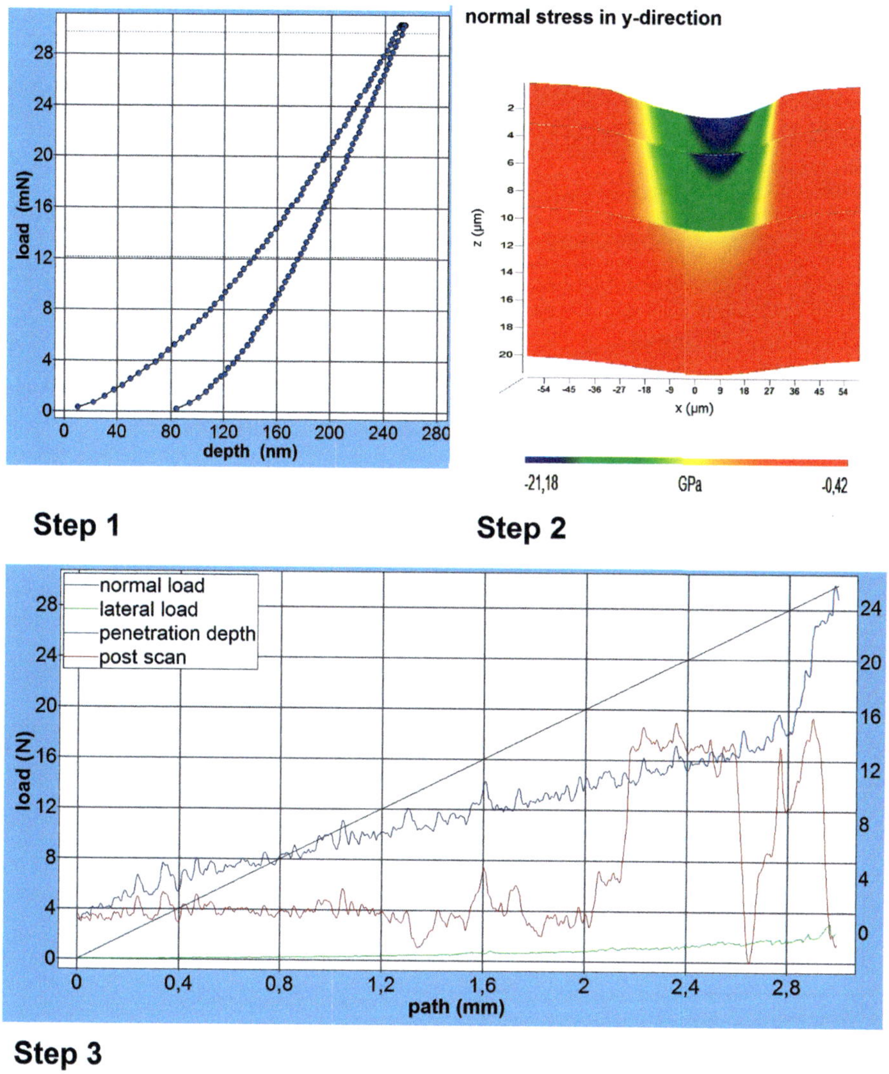

Figure 22.11 Practical example for a mechanical surface optimization cycle, steps 1 to 3 Step 1: Nanoindentation tests are performed to accurately measure the material properties of the substrate and coatings, taking into account possible property profiles [56, 57]. The latter always requires the application of a series of different loads instead of only one load. Step 2: With the mechanical properties previously measured by nanoindentation, a calculation of the stress–strain fields of a simulated scratch allows a perfect dimensioning of the scratch test in accordance with both optimum comparison with worst cases of the later application and maximum sensitivity regarding weak points within the coating-substrate structure [11, 12]. Step 3: Scratch testing with the pre-defined conditions from step 2 is performed.

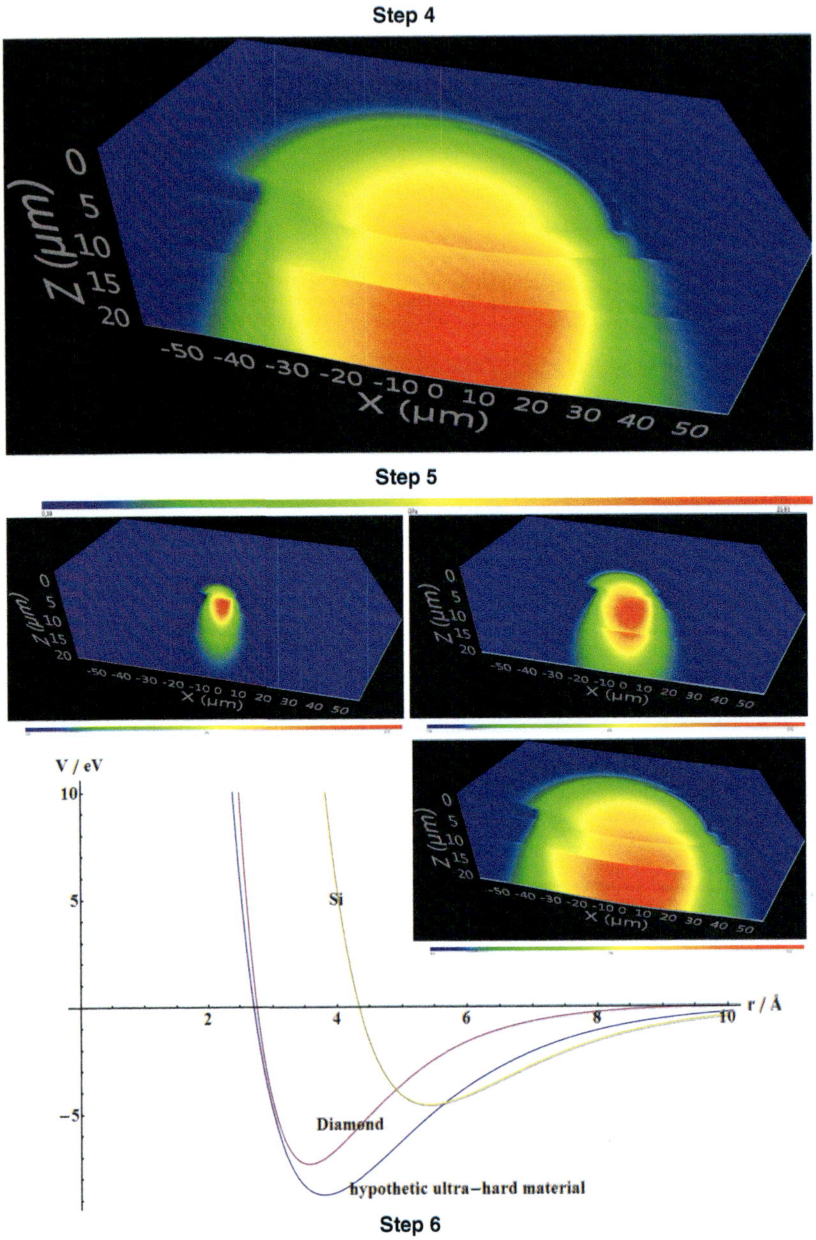

Figure 22.12 Practical example for a mechanical surface optimization cycle, step 4 to 6 Step 4: An advanced analysis with integration of the residual penetration depth, pre-scan surface profile, and topography resolved friction coefficient is been performed. With the help of a simulation software evaluating the complete stress–strain fields at various indenter positions [28], a highly resolved scratch animation can be evaluated. Step 5: For better understanding, an animation of the scratch under various contact situations is created (examples under: www.siomec.de/en/096/Examples). Here the size of the indenter has been varied in order to create von Mises stress maxima at different positions within a 2-layer compound. This was done in order to trigger delamination at either the first or the second interface or (smallest indenter) to provoke plastic flow only in the first layer. Step 6: Using the results of the previous steps, effective interatomic potentials are evaluated and decomposition parameters are derived as shown in [1, 2]. These parameters are used to design a tribo-test optimum for the later application in question.

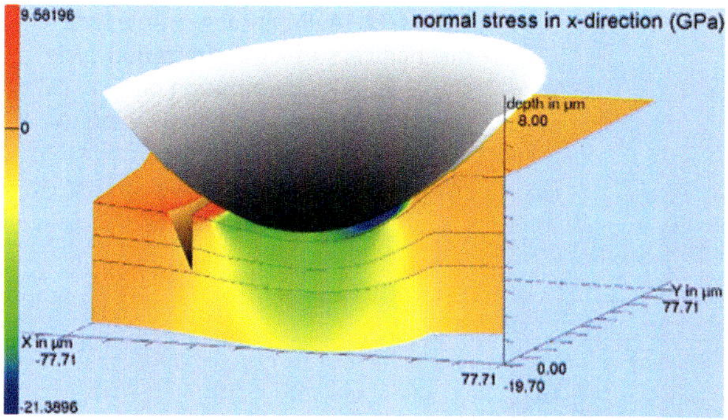

Step 7

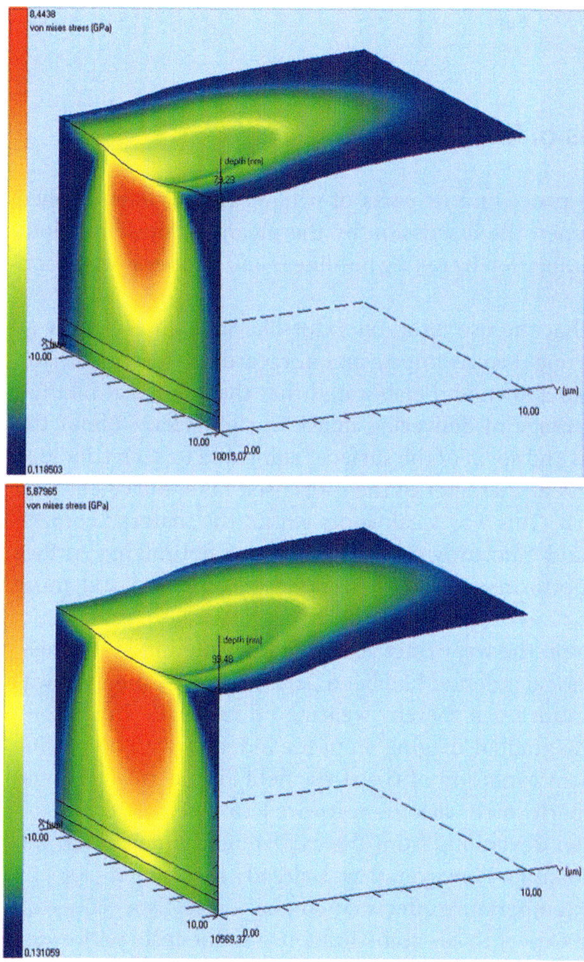

Step 8 (from middle of wear test to its end)

Figure 22.13 Practical example for a mechanical surface optimization cycle, step 7 and 8 Step 7: Performance of tribo-test as designed in step 6 and analysis by inversed global increment wear model [13, 14, 58]. The latter had to be extended by the layered half space model and the extended Hertzian approach as described in the theory section of this paper. Step 8: For better understanding, an animation of the tribological respectively wear-test is created (see Figures 22.15–22.23 below).

Figure 22.14 Practical example for a mechanical surface optimization cycle step 9 Step 9: Optimization of coating-substrate system → Going to next cycle. *Source*: Bemporad, 2008 [55]. Reproduced with permission of Elsevier.

22.4 Discussion by Means of Examples

Within a variety of examples, the procedure or parts of it have already been applied [7–14]. Here we now want to deepen the discussion by the means of these previous examples and try to improve the simulation by taking nonlinear and temperature effects into account.

It was shown in [13] and [14] that the use of an operator-like wear or tribo-law as introduced in the section above brought some improvement regarding the comparison with experimental data. It was also shown and elaborated that the analytical method as outlined here has the big advantage of delivering quasi 'in-sight-views' about the deformation fields, strains, stresses and so on of the surfaces subjected to such tribology tests. Due to the completely analytical character of the tools used here, the calculation usually is extremely fast and robust. This way weak spots within the material composition can be found more easily and efficiently. Subsequently also optimizing surface systems towards a better tribo-performance could be done quicker and still more accurately.

We are demonstrating this here on the wear track example presented in [14], where an oscillating wear test with a spherical indenter had been performed. The indenter was sliding forth and back along a certain track thereby wearing off material. The following Figures 22.15 to 22.23 show the gradual 'digging' in of the indenter with increasing of cycle number and the subsequent evolution of the stress field for a variety of stress components. The evaluation was performed with the software FilmDoctor [28].

Interestingly, there still had been deviations from the experiment which could not been explained in a satisfactory manner. However, it was already mentioned (see [13, 14] and also [2]) that nonlinear or temperature effects could play a role. Especially with respect to temperature one would expect to see some influence as the tests performed were reciprocating wear tests where a spherical indenter is performing an oscillatory sliding movement over the, in some cases coated, sample surface. Here the friction will automatically lead to a temperature increase. With such an increase of temperature, we expect a change of the mechanical parameters, especially Young's modulus and yield strength.

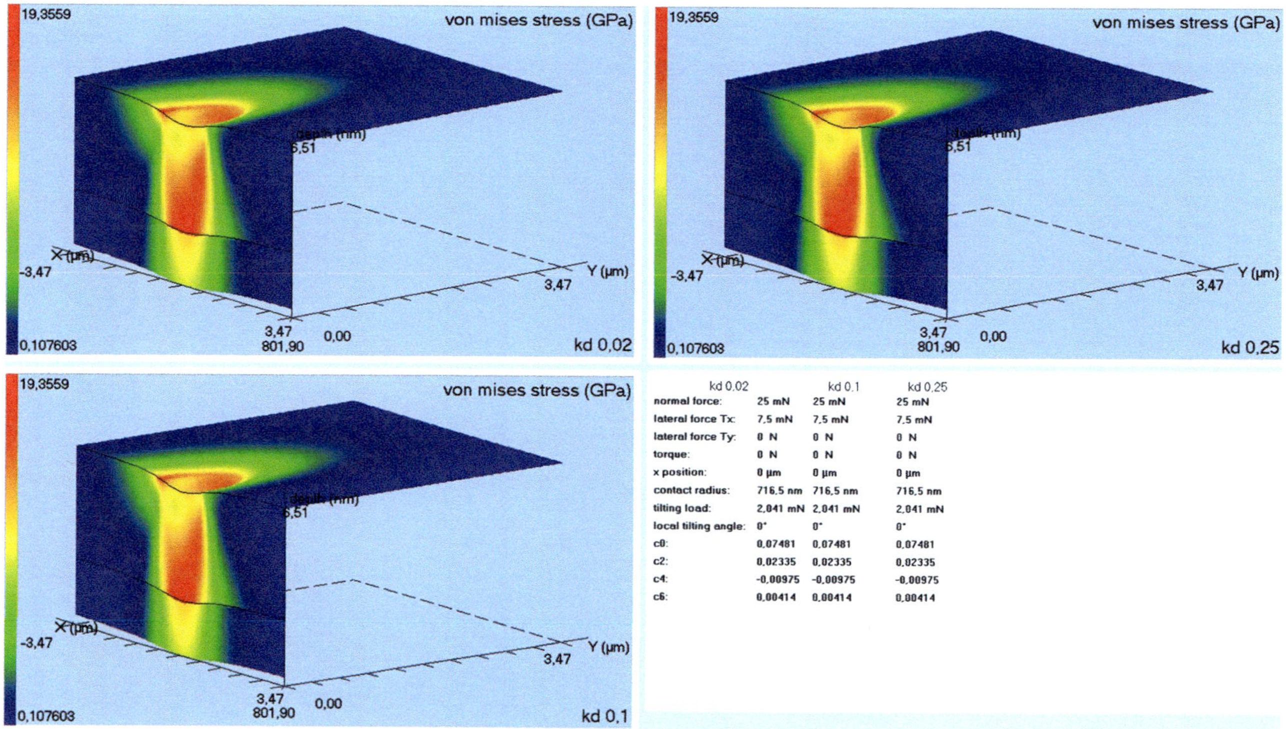

	kd 0,02	kd 0,1	kd 0,25
normal force:	25 mN	25 mN	25 mN
lateral force Tx:	7,5 mN	7,5 mN	7,5 mN
lateral force Ty:	0 N	0 N	0 N
torque:	0 N	0 N	0 N
x position:	0 µm	0 µm	0 µm
contact radius:	716,5 nm	716,5 nm	716,5 nm
tilting load:	2,041 mN	2,041 mN	2,041 mN
local tilting angle:	0°	0°	0°
c0:	0,07481	0,07481	0,07481
c2:	0,02335	0,02335	0,02335
c4:	-0,00975	-0,00975	-0,00975
c6:	0,00414	0,00414	0,00414

Figure 22.15 Wear track evaluation of an oscillating pin (linear wear track) on three different coating materials. Beginning of the wear test. Von Mises stress.

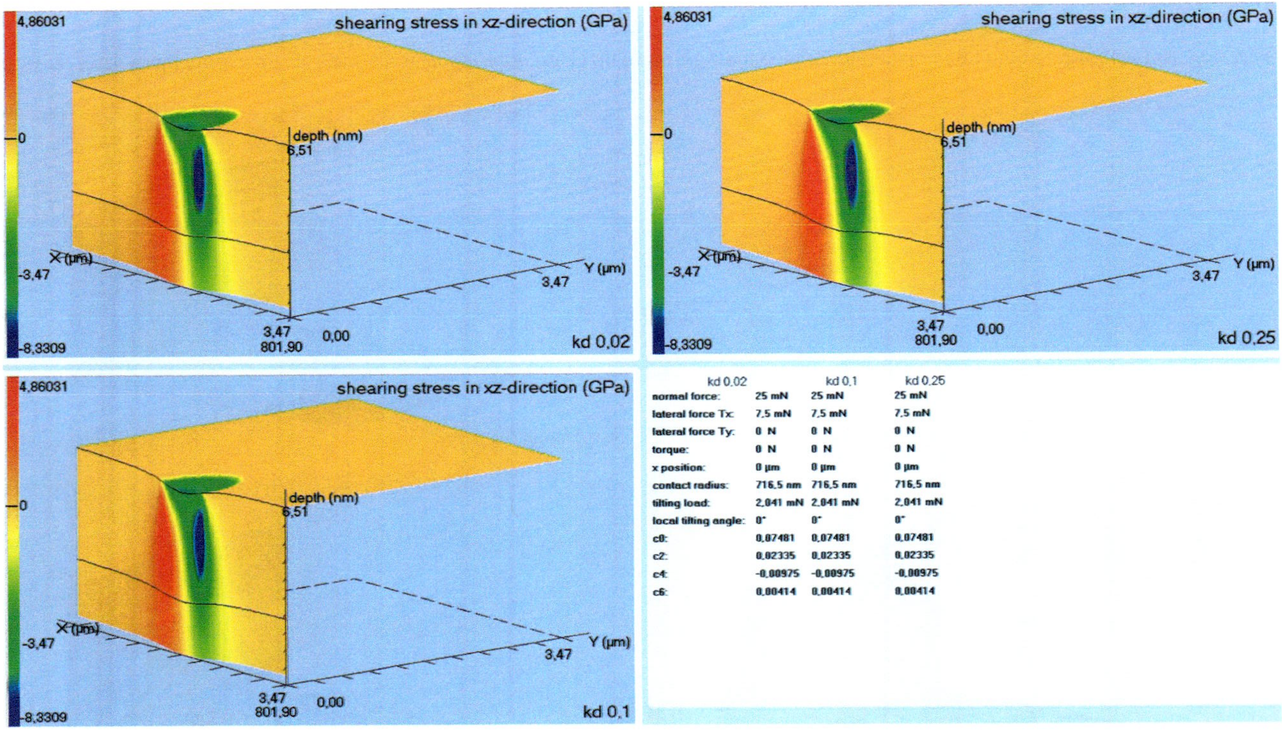

	kd 0,02	kd 0,1	kd 0,25
normal force:	25 mN	25 mN	25 mN
lateral force Tx:	7,5 mN	7,5 mN	7,5 mN
lateral force Ty:	0 N	0 N	0 N
torque:	0 N	0 N	0 N
x position:	0 µm	0 µm	0 µm
contact radius:	716,5 nm	716,5 nm	716,5 nm
tilting load:	2,041 mN	2,041 mN	2,041 mN
local tilting angle:	0°	0°	0°
c0:	0,07481	0,07481	0,07481
c2:	0,02335	0,02335	0,02335
c4:	-0,00975	-0,00975	-0,00975
c6:	0,00414	0,00414	0,00414

Figure 22.16 Wear track evaluation of an oscillating pin (linear wear track) on three different coating materials. Beginning of the wear test. Shear stress in the normal-lateral direction.

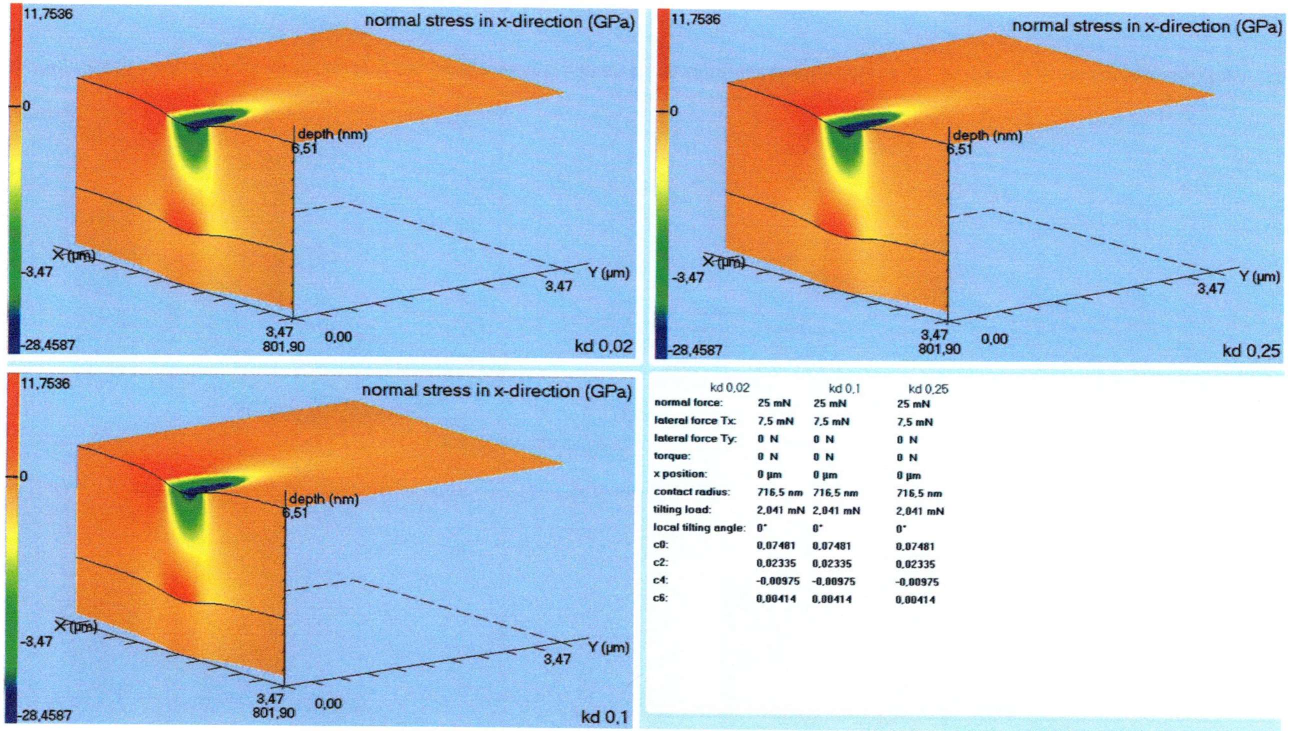

	kd 0,02	kd 0,1	kd 0,25
normal force:	25 mN	25 mN	25 mN
lateral force Tx:	7,5 mN	7,5 mN	7,5 mN
lateral force Ty:	0 N	0 N	0 N
torque:	0 N	0 N	0 N
x position:	0 µm	0 µm	0 µm
contact radius:	716,5 nm	716,5 nm	716,5 nm
tilting load:	2,041 mN	2,041 mN	2,041 mN
local tilting angle:	0°	0°	0°
c0:	0,07481	0,07481	0,07481
c2:	0,02335	0,02335	0,02335
c4:	-0,00975	-0,00975	-0,00975
c6:	0,00414	0,00414	0,00414

Figure 22.17 Wear track evaluation of an oscillating pin (linear wear track) on three different coating materials. Beginning of the wear test. Normal lateral stress in the direction of the moving pin.

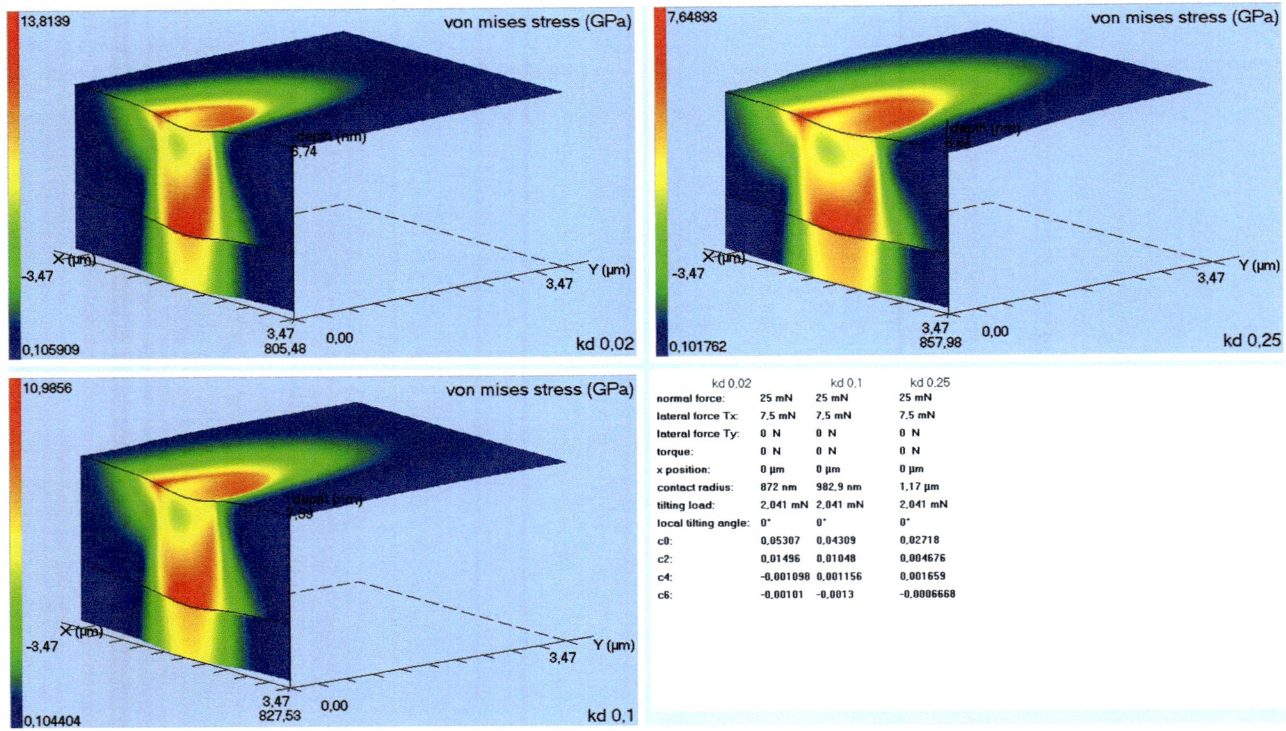

Figure 22.18 Wear track evaluation of an oscillating pin (linear wear track) on three different coating materials. Middle of the wear test. Von Mises stress.

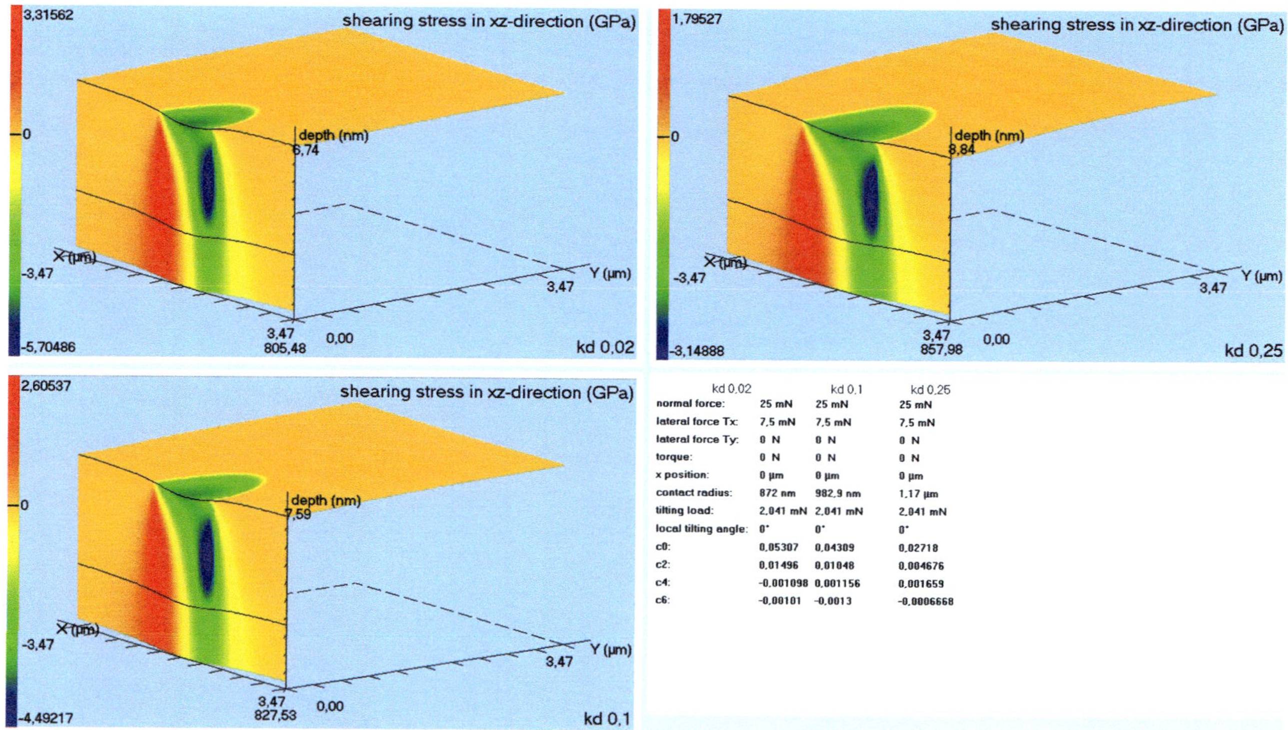

	kd 0,02	kd 0,1	kd 0,25
normal force:	25 mN	25 mN	25 mN
lateral force Tx:	7,5 mN	7,5 mN	7,5 mN
lateral force Ty:	0 N	0 N	0 N
torque:	0 N	0 N	0 N
x position:	0 µm	0 µm	0 µm
contact radius:	872 nm	982,9 nm	1,17 µm
tilting load:	2,041 mN	2,041 mN	2,041 mN
local tilting angle:	0°	0°	0°
c0:	0,05307	0,04309	0,02718
c2:	0,01496	0,01048	0,004676
c4:	-0,001098	0,001156	0,001659
c6:	-0,00101	-0,0013	-0,0006668

Figure 22.19 Wear track evaluation of an oscillating pin (linear wear track) on three different coating materials. Middle of the wear test. Shear stress in the normal-lateral direction.

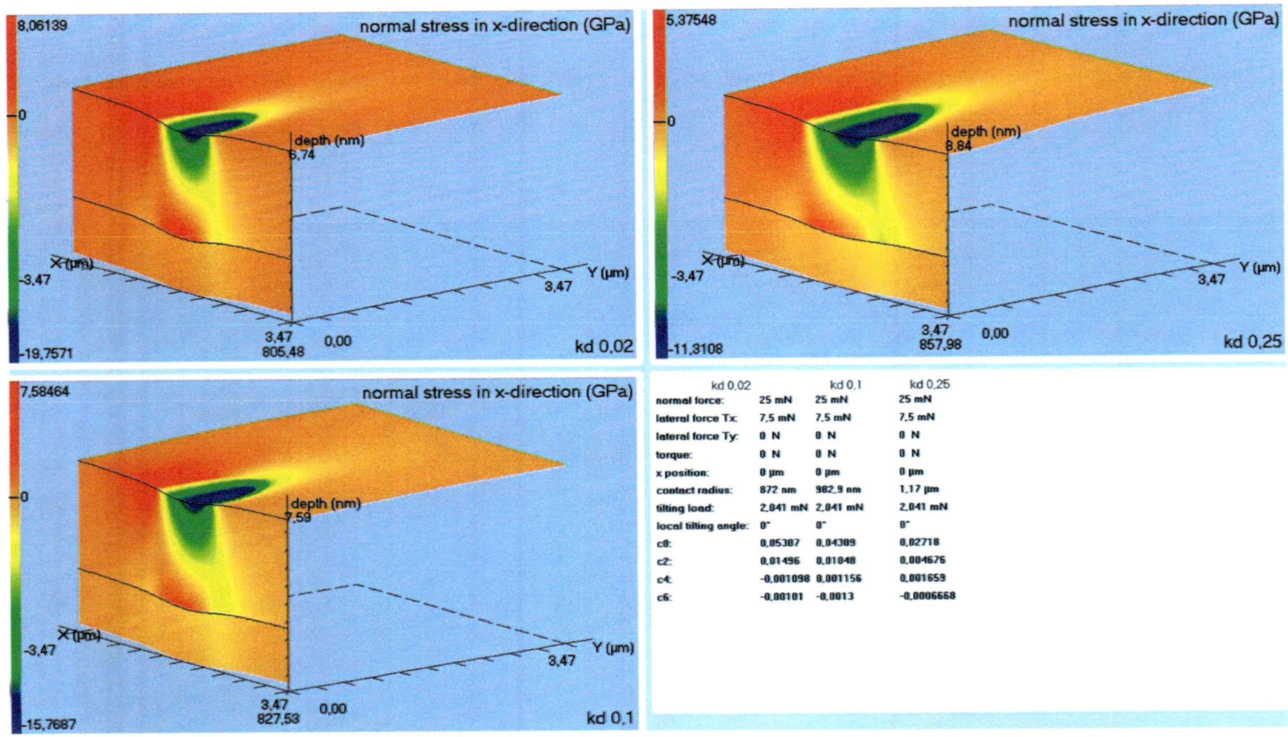

Figure 22.20 Wear track evaluation of an oscillating pin (linear wear track) on three different coating materials. Middle Beginning of the wear test. Normal lateral stress in the direction of the moving pin.

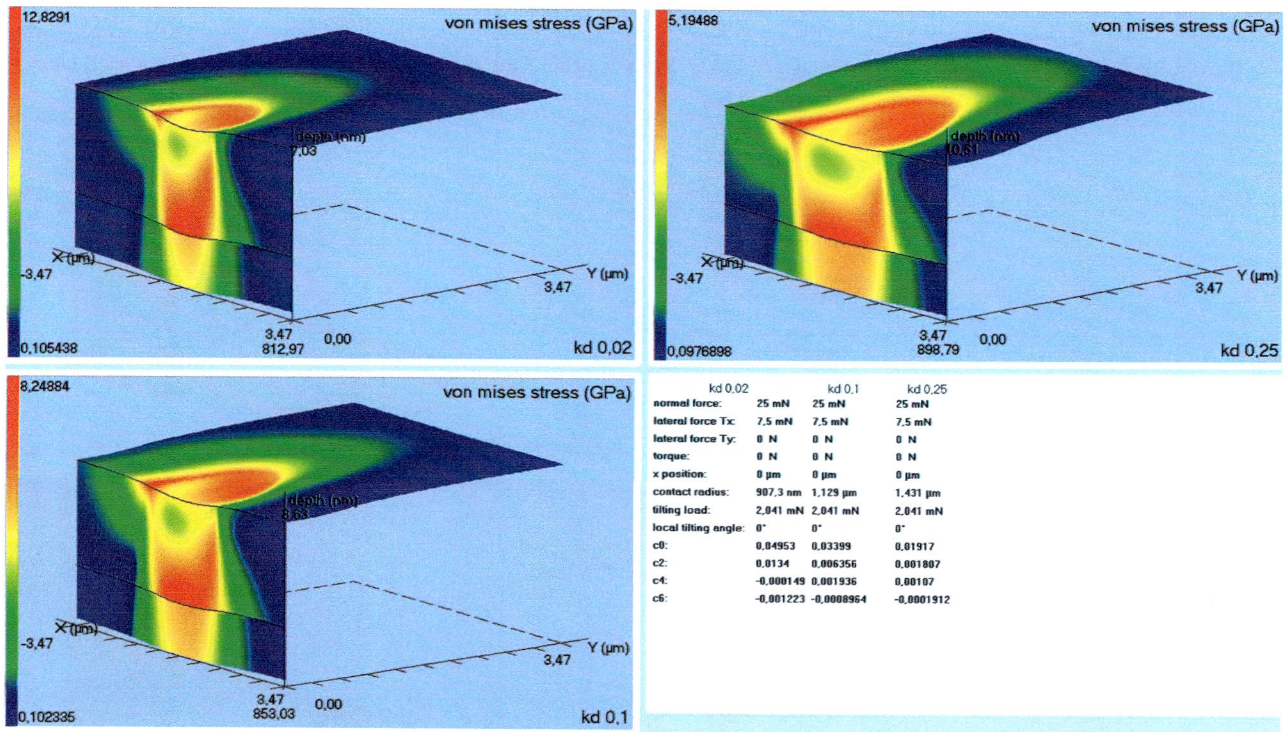

Figure 22.21 Wear track evaluation of an oscillating pin (linear wear track) on three different coating materials. End of the wear test. Von Mises stress.

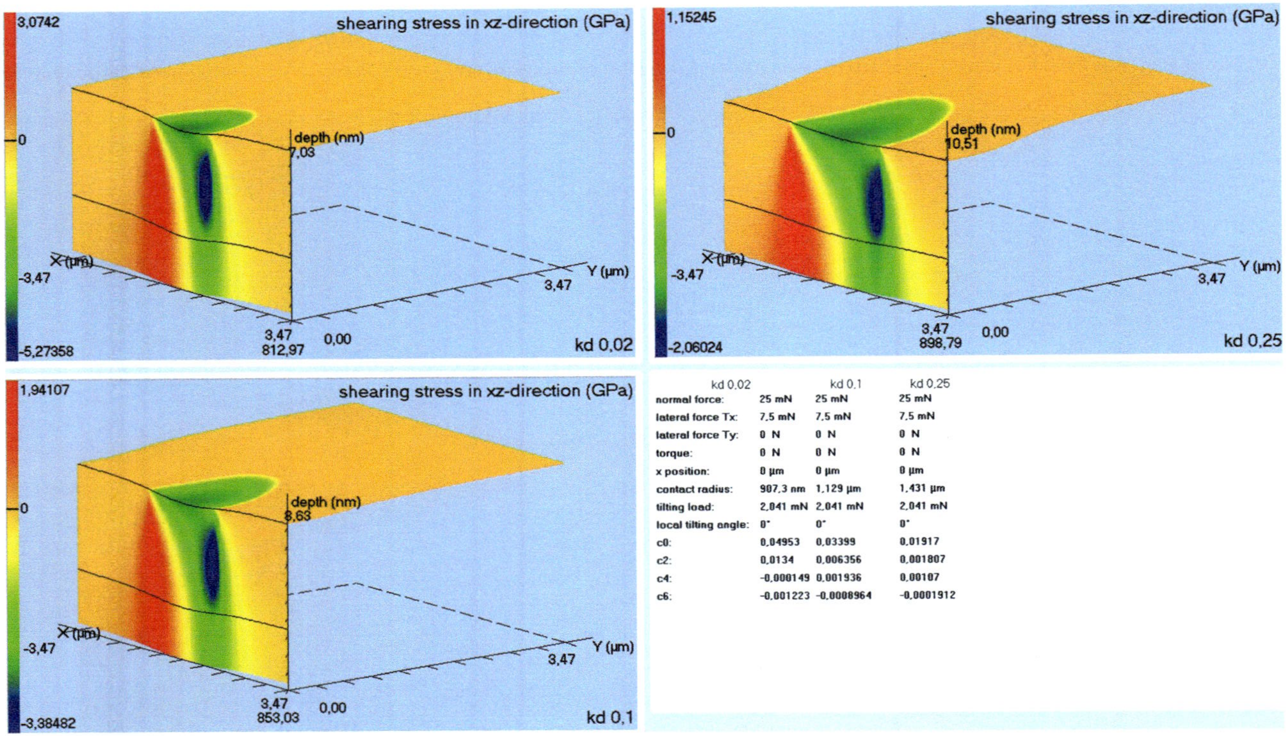

	kd 0,02	kd 0,1	kd 0,25
normal force:	25 mN	25 mN	25 mN
lateral force Tx:	7,5 mN	7,5 mN	7,5 mN
lateral force Ty:	0 N	0 N	0 N
torque:	0 N	0 N	0 N
x position:	0 µm	0 µm	0 µm
contact radius:	907,3 nm	1,129 µm	1,431 µm
tilting load:	2,041 mN	2,041 mN	2,041 mN
local tilting angle:	0°	0°	0°
c0:	0,04953	0,03399	0,01917
c2:	0,0134	0,006356	0,001807
c4:	-0,000149	0,001936	0,00107
c6:	-0,001223	-0,0008964	-0,0001912

Figure 22.22 Wear track evaluation of an oscillating pin (linear wear track) on three different coating materials. End of the wear test. Shear stress in the normal-lateral direction.

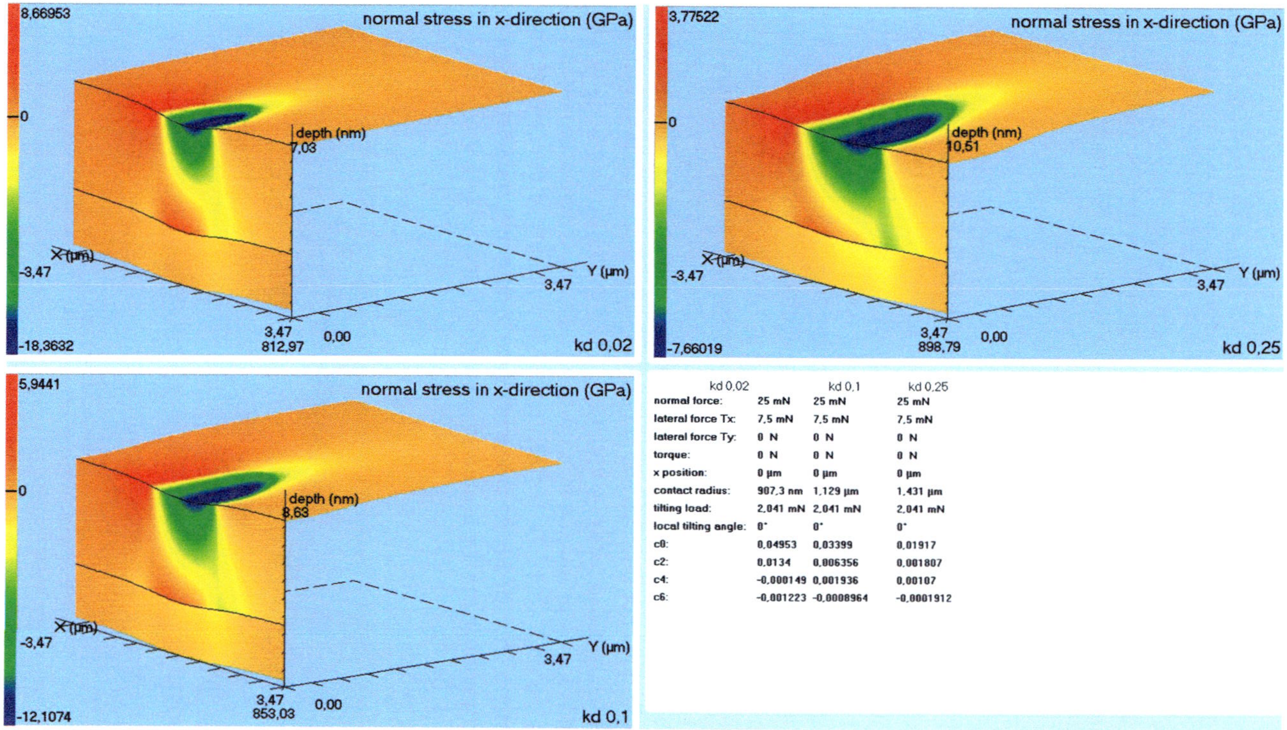

	kd 0,02	kd 0,1	kd 0,25
normal force:	25 mN	25 mN	25 mN
lateral force Tx:	7,5 mN	7,5 mN	7,5 mN
lateral force Ty:	0 N	0 N	0 N
torque:	0 N	0 N	0 N
x position:	0 µm	0 µm	0 µm
contact radius:	907,3 nm	1,129 µm	1,431 µm
tilting load:	2,041 mN	2,041 mN	2,041 mN
local tilting angle:	0°	0°	0°
c0:	0,04953	0,03399	0,01917
c2:	0,0134	0,006356	0,001807
c4:	-0,000149	0,001936	0,00107
c6:	-0,001223	-0,0008964	-0,0001912

Figure 22.23 Wear track evaluation of an oscillating pin (linear wear track) on three different coating materials. End of the wear test. Normal lateral stress in the direction of the moving pin.

Thus, we simply assume such an influence and incorporate it into the wear simulation described in [13] and [14]. Thereby we are using of a Young's modulus function of the kind given in (22.29) assuming a temperature dependency of the relaxation time or viscosity such as:

$$\tau = \eta/E_1 = \frac{\eta_0 e^{W_A/(k_B T)}}{E_1}. \tag{22.73}$$

Applying such an extension, we expect to result into a better fit of our wear or tribo model to the experimental data. Using a similar dependency for the yield strength Y as we did for the Young's modulus and adapting our wear-law (59) such, that the wear should increase with decreasing Y, we set:

$$tribo - effect_{ij} \equiv w_{ij} = \hat{k}_{ijkl}\ \sigma^{kl} \equiv \delta_{ij}\left(k_{dvM}\frac{\sigma_{vM}}{Y(T)}\cdot e^{-\lambda_k \vec{r}\cdot\vec{r}}\right). \tag{22.74}$$

Here we need to pay attention to the fact, that with having a temperature dependent Young's modulus by the means of (29) and (73), we also result in a temperature dependency for the von Mises stress maximum and its position. This makes the evaluation of (74) relatively complicated and we therefore restrict ourselves here to the task of finding a temperature function $T(n)$ with n denoting the number of cycles or oscillations of the indenter which would fit the real wear data significantly better than an assumed constant temperature. For both examples we have at hand [13, 14] we received significantly better fit of the results this way. Interestingly the best fitting function $T(n)$ was of the kind $T(n) = c_0 \cdot \tanh(c_1 \cdot n)$ which simply coincides with a temperature rise at the beginning of the test and reaching a plateau value after only a view cycles. This surprisingly simple result does not only have some beauty with respect to its low number of free parameters, but can also easily be understood by the interpretation of the test evolution. With the indenter producing friction, this leads to a temperature increase until a steady state is reached where the temperature in the vicinity of the contact does not increase any further. This will be the case when heat production and heat diffusion will have reached an equilibrium. The results are shown in Figures 22.24 and 22.25.

More examples, better illustration of the intermediate steps and a description of the experiments are published elsewhere (e.g. [10–14, 55]).

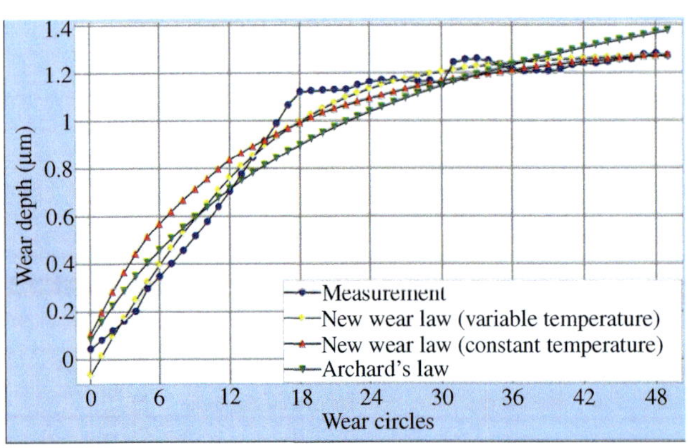

Figure 22.24 Comparison of various wear laws with experimental data (compare with text in section discussion). *Source*: Gies, 2013 [13]. Reproduced with permission of Elsevier.

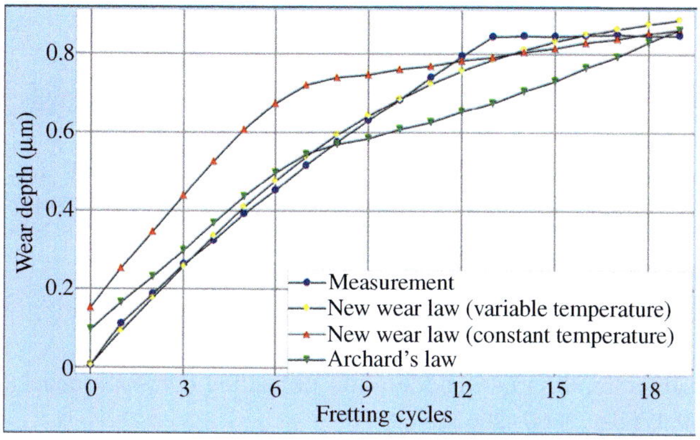

Figure 22.25 Comparison of various wear laws with experimental data (c.f. text in section discussion). *Source*: Liskiewicz 2013 [14]. Reproduced with permission of Elsevier.

22.5 Conclusions

By combining the global increment wear model with the concept of the effective indenter, the extended Hertzian approach and a layered half space solution for contact problems with rather arbitrary combinations of normal, lateral and tilting loads, a general, quick and powerful wear model has been created.

Inversion of the model makes it fit for parameter identification problems.

Adding first principle models, like effective interatomic interaction potentials allows a deeper understanding of those failure mechanisms which are responsible for wear results as observed in practical tests.

Temperature effects had been taken into account by allowing the temperature to couple into the mechanical properties Young's modulus and yield strength.

Possible subsequent time dependent material behavior leads to a relatively complex inhomogeneous Young's modulus distribution which must be taken into account if one aims for a better simulation of wear experiments.

The analytical method as outlined here has the advantage of delivering quasi 'in-sight-views' about the deformation fields, strains, stresses etc. of the surfaces subjected to such tribology tests. Due to the completely analytical character of the tools used, the calculation usually is extremely fast and robust. This way weak spots within the possibly layered material composition can be found more easily and efficiently. Subsequently, also optimizing surface systems towards a better tribo-performance can be done quicker and still more accurately.

Acknowledgements

This work was partially funded through the European Metrology Research Programme (EMRP) Project IND05 MeProVisc. The EMRP is jointly funded by the EMRP participating countries within EURAMET and the European Union. This work has also been supported by Wolfram Research Inc. and Additive GmbH.

Referencess

1 N. Schwarzer. Short note on the effect of pressure induced increase of Young's modulus. Philosophical Magazine, 92(13): 1631–1648.
2 N. Schwarzer. From interatomic interaction potentials via Einstein field equation techniques to time dependent contact mechanics. Materials Research Express (2014) 1: 015042.
3 A. Bolshakov, W.C. Oliver, G.M. Pharr. An explanation for the shape of nanoindentation unloading curves based on finite element simulation. MRS Symposium Proceedings (1995) 356: 675.
4 G.M. Pharr, B. Bolshakov. Understanding nanoindentation unloading curves. Journal of Materials Research (2002) 17(10): 2660–26671.
5 N. Schwarzer, G. M. Pharr. On the evaluation of stresses during nanoindentation with sharp indenters. Thin Solid Films, (2004) 469–470C: 194–200.
6 N. Schwarzer. The extended Hertzian theory and its uses in analysing indentation experiments. Philosophical Magazine, Special Issue: Instrumented Indentation Testing in Materials Research and Development, 86(33–35): 2006, 5153–5767.
7 W.C. Oliver, G.M. Pharr. An improved technique for determining hardness and elastic modulus using load and displacement sensing indentation experiments. Journal of Materials Research (1992) 7: 1564–1583.
8 N. Schwarzer, T. Chudoba, G. M. Pharr. On the evaluation of stresses for coated materials during nanoindentation with sharp indenters. Surface and Coatings Technology (2006) 200/14–15, 4220–4226.
9 N. Schwarzer, T. Chudoba, F. Richter. Investigation of ultra thin coatings using nanoindetation. Surface and Coatings Technology (2006) 200/18–19, 5566–5580.
10 M. Fuchs, G. Favaro. Physical characterization of coated surfaces - Part I: Instrumented Indentation. Available at: www.csm-instruments.com/en/Physical-characterization-of-coated-surfaces-Part-I-%3A-Instrumented-Indentation.
11 M. Fuchs, G. Favaro. Physical characterization of coated surfaces, Part II: Scratch Test. Application Bulletin No. 38 of CSM instruments. Available at: www.csm-instruments.com/en/Physical-characterization-of-coated-surfaces-Part-II-%3A-Scratch-Test.
12 N. Schwarzer, Q.-H. Duong, N. Bierwisch, *et al.* Optimization of the scratch test for specific coating designs. Surface and Coatings Technology, 2011, 206(6): 1327–1335.
13 A. Gies, T. Chudoba, N. Schwarzer, J. Becker. Influence of the coating structure of a-C:H-W coatings on their wear-performance: a theoretical approach and its practical confirmation. Surface and Coatings Technology (2013) 237: 299–304.
14 T. Liskiewicz; B. Beake; N. Schwarzer; M. Davies. Short note on improved integration of mechanical testing in predictive wear models. Surface and Coatings Technology (2013) 237: 212–218.
15 N. Schwarzer. Is ultra-hardness possible: a brief feasibility study. Available at: www.siomec.de/ultra-hard-coatings (accessed November 2016).

16 M. Lie, N. X. Chen. Möbius inversion transform for diamond-type materials and phonon dispersions. Physical Review B, (1995), 52(2): 997–1003.
17 T. Dick, G. Cailletaud. Fretting modelling with a crystal plasticity model of Ti6Al4V. Computational Materials Science (2006) 38, 113–125.
18 J. M. Guilemany, S. Dosta, J. R. Miguel. The enhancement of the properties of WC-Co HVOF coatings through the use of nanostructured and microstructured feedstock powders. Surface and Coatings Technology (2006) 201: 1180–1190.
19 V. Hegadekatte, N. Huber, O. Kraft. Modeling and simulation of wear in a pin on disc tribometer. Tribology Letters (2006) 24: 51–60.
20 R. Holm. Electric Contacts. Almqvist and Wiksells Boktryxkeri AB, Uppsala; 1946.
21 K. Holmberg, A. Matthews. Coatings Tribology, 2nd edn. Springer, Amsterdam; 2009.
22 N. Schwarzer. Arbitrary load distribution on a layered half space. ASME Journal of Tribology, 2000, 122(4): 672–681.
23 T. Chudoba, K. Herrmann. Verfahren zur Ermittlung der realen Spitzenform von Vickers- und Berkovich-eindringkörpern (method for the determination of the real tip shape of Vickers- and Berkovich-tips), Härtereitechnische Mitteilungen (HTM) (2001) 56: 258.
24 N. Schwarzer. Polarization features of optically pumped CdS nanowire lasers. Journal of Physics D: Applied Physics (2004) 37: 2761–2772.
25 N. Schwarzer. Analysing nanoindentation curves using Pharr's concept of tyhe effective indenter shape. Thin Solid Films 494 (2006) 168–172.
26 N. Schwarzer. The extended Hertzian theory and its uses in analysing indentation experiments. Philosophical Magazine Special Issue: Instrumented Indentation Testing in Materials Research and Development, (2006) 86(33–35): 5153–5767.
27 R. Puschmann, N. Schwarzer, F. Richter, *et al.* An applicable concept for the indentation of thin porous films. Proceedings of the NanoMech 5, 7–9 September 2004 in Hückelhoven, Germany; also in Zeitschrift für Metallkunde 96 (2005) 11, 1–6.
28 FilmDoctor®: Software demonstration package, www.siomec.de/FilmDoctor.
29 A.C. Fischer-Cripps. A simple phenomenological approach to nanoindentation creep. Materials Science and Engineering A, (2004) 385: 74–82.
30 D. François, A. Pineau, A. Zaoui. Mechanical Behaviour of Materials. Kluwer Academic Publishers, Netherlands; 1998.
31 ISO 20502 Fine ceramics (advanced ceramics: advanced technical ceramics); Determination of adhesion of ceramic coatings by scratch testing; 2015.
32 ASTM C1624 Standard test method for adhesion strength and mechanical failure modes of ceramic coatings by quantitative single point SCRATCH testing; 2015.
33 J. Ehlert, W. J. Batz. Der Einfluss von Druck und Temperatur auf die effektive Viskosität in geschmierten kontra-formen Kontakten. Rheology Acta (1976), 15: 356–364.
34 R.B. Phillips. Crystals, Defects and Microstructures. Cambridge University Press; 2001.

35 N. Schwarzer. Some basic equations for the next generation of surface testers solving the problem of pile-up, sink-in and making area-function-calibration obsolete, jmr special focus issue on indentation methods in advanced materials research. Journal of Materials Research, (2009) 24(3): 1032–1036.
36 N. Schwarzer. Basic equations for the next generation of surface testers for the case of an elastic indenter and a layered samples. Online archive of the Saxionian Institute of Surface Mechanics, www.siomec.de/pub/2008/001.
37 N. Schwarzer. (2007) Intrinsic stresses: Their influence on the yield strength and their measurement via nanoindentation. Online archives of the Saxonian Institute of Surface Mechanics, www.siomec.de/pub/2007/001(accessed November 2016).
38 N. Schwarzer. Modelling of contact problems of rough surfaces. Online archives of the Saxonian Institute of Surface Mechanics, www.siomec.de/pub/2007/007 (accessed November 2016).
39 I.N. Sneddon. The relation between load and penetration in the axisymmetric Boussinesq problem for a punch of arbitrary profile. International Journal of Engineering Science (1965) 3: 47.
40 N. Schwarzer. Analyse und Simulation der mechanischen Eigenschaften beschichteter Polymere unter Berücksichtigung der meist zeitabhängigen Materialparameter (Analysis and simulation of mechanical properties of coated polymers with time-dependent material parameters), 19 NDVaK, proceedings of the V2011, 19. u. 20.10. 2011 Dresden, Germany, pp. 96 – 101, published by EFDS Dresden, Germany, available at: www.efds.de.
41 J.G. Kohl, N.X. Randall, N. Schwarzer, *et al.* An investigation of scratch testing of silicone elastomer coatings with a thickness gradient. Journal of Applied Polymer Science, (2012), 124: 2978–2986.
42 B.N. Lucas, J.L. Hay, W.C. Oliver. Using multi-dimensional contact mechanics experiments to measure Poisson's ratio of porous low-k films. Journal of Materials Research, (2004) 19(1): 58–65.
43 O. Molnár. Investigations of nanoindentation data obtained by the combination of normal and mixed (normal and lateral) forces. PhD thesis of Technical University of Chemnitz, Germany; 2010, available at: http://archiv.tu-chemnitz.de/pub/2006/0044.
44 M. Karniychuk. Combination of lateral and normal forces for investigation of mechanical properties and tribological behaviour of bulk and coated materials on the micro-scale. PhD thesis of Technical University of Chemnitz, Germany; 2006, available at: http://nbn-resolving.de/urn:nbn:de:swb:ch1-200601120.
45 T. Chudoba, V. Linss, M. Karniychuk, F. Richter. Lateral force–displacement measurements: a new technique for the investigation of mechanical surface properties. Surface & Coatings Technology (2005) 200: 315–320.
46 A. Gies, N. Schwarzer, J. Becker, H. Rudigier. Untersuchung der mechanischen Eigenschaften von DLC-Schichtsystemen mittels Nanoindentation und deren Modellierung Modellierung (Investigation of mechanical properties of DLC-Coatingsystems with nanoindentation and subsequent modelling), Tagungsband der ThGOT (Proceedings of the Thementage Grenz- und Oberflächentechnik) 2010, pp. 104–109.

47 N. Schwarzer. Coating design due to analytical modelling of mechanical contact problems on multilayer systems. Proceedings of the ICMCTF 2000, San Diego, USA, April 2000, 397–402, also in: Surface and Coatings Technology (2000) 133–134: 397–402.
48 V.I. Fabrikant, Several elliptical punches on an elastic half space. Journal of Applied Mechanics (1986) 53(2), 390–394.
49 I.I. Argatov. Asymptotic models of contact interaction among elliptic punches on a semiclassical foundation. International Applied Mechanics, (2006) 42(1): 67–83.
50 G. Ia. Popov. Axisymmetric contact problem for an elastic inhomogeneous half-space in the presence of cohesion. Prikladnaia Mathematika i Mechanika (1973), 37: 1109–1116.
51 V. I. Fabrikant. Application of potential theory in mechanics: A Selection of New Results. Kluwer Academic Publishers, Dordrecht, The Netherlands; 1989.
52 N. A. Rostovtsev. An integral equation encountered in the problem of a rigid foundation bearing on nonhomogeneous soil. Prikladnaia Mathematika i Mechanika, (1961), 25: 164–168.
53 N. Schwarzer. Modelling of the mechanics of thin films using analytical linear elastic approaches. Habilitationsschrift der TU-Chemnitz 2004, FB Physik Fester Körper. Available at: www.siomec.de/pub/2007/013 and http://archiv.tu-chemnitz.de/pub/2004/0077.
54 N. Schwarzer. Effect of lateral displacement on the surface stress distribution for cone and sphere contact. Philosophical Magazine special issue: Instrumented Indentation Testing in Materials Research and Development (2006) 86(33–35): 5231–5237.
55 E. Bemporad, M. Sebastiani, M.H. Staia, E. Puchi Cabrera. Tribological studies on PVD/HVOF duplex coatings on Ti6Al4V substrate. Surface and Coatings Technology (2008) 203: 566–571.
56 V. Linss, N. Schwarzer, T. Chudoba, *et al.* Mechanical properties of a graded bcn sputtered coating with varyingyoung's modulus: deposition, theoretical modelling and nanoindentation. Surface and Coatings Technology (2005) 195: 287–297.
57 T. Chudoba, N. Schwarzer, V. Linss, F. Richter. Determination of Mechanical Properties of Graded Coatings using Nanoindentation. Proceedings of the International Conference on Metal Coatings and Thin Films (ICMCTF) 2004 San Diego, California, USA; also in Thin Solid Films (2004) 469–470C: 239–247.
58 V. Hegadekatte, N. Huber, O. Kraft. , Modeling and simulation of wear in a pin on disc tribometer. Tribology Letters (2006) 24: 51–60.
59 N. Schwarzer, F. Richter. On the determination of film stress from substrate bending: Stoney's formula and its limits. Online archive of the Technical University of Chemnitz: http://archiv.tu-chemnitz.de/pub/2006/0011.
60 D.S. Stone. Elastic rebound between an indenter and a layered specimen. Journal of Materials Research (1998) 13(11): 3207–3213.
61 G. G. Stoney. The tension of metallic films deposited by electrolysis. Proceedings of the Royal Society of London A, (1909) 82: 172.

62 N. Schwarzer. About stress field, time and temperature dependent contact mechanics for thin film structures: especially the role of intrinsic stresses. Proceedings of SVC TechCon 2014, 4 to 8 of May, Chicago, Illinois, USA, paper 58.

63 M. Sobiech, J. Ramm, B. Widrig, N. *et al.* Adhesion of tetrahedral amorphous Carbon (ta-C) Coatings deposited on different substrates: Simulations and experimental verification. Presentation at the ICMCTF 2012 in San Diego, California, USA.

64 S. Nicolle, J.-F. Palierne. On the efficiency of attachment methods of biological soft tissues in shear experiments. Journal of the Mechanical Behavior of Biomedical Materials (2012) 14: 158–162.

65 H See, S Mackenzie, BT Chua. Effect of compression on the response of a magneto-rheological suspension. Korea-Australia Rheology Journal (2006) 18(3): 121–126.

23

Modelling and Simulations of Nanoindentation in Single Crystals

Qiang Liu[1], *Murat Demiral*[2], *Anish Roy*[1] *and Vadim V. Silberschmidt*[1]

[1] *Wolfson School of Mechanical, Electrical and Manufacturing Engineering, Loughborough University, UK*
[2] *Department of Mechanical Engineering, Çankaya University, Ankara, Turkey*

23.1 Introduction

Indentation experiments in a sub-micrometer depth regime are referred to as *nano* or *micro-indentation* depending on the indenter geometry and indentation depth. This technique is widely used to measure resistance of materials to plastic flow in small volumes, e.g. thin films. It is well known that crystalline materials demonstrate a strong size effect, where the measured hardness is observed to decrease with an increase in the indentation depth with conical or pyramidal indenters. This type of size effect is often referred to as *indentation size effect* (ISE). For instance, Figure 23.1 demonstrates a depth dependence of hardness of single-crystal copper and cold worked polycrystalline copper, taken from the work of McElhaney *et al.* [1]. The ISE is more complicated in a case of a spherical indenter, where hardness depends not only on the indentation depth but also on the indenter radius [2]. A size effect was also observed in other experiments at a sub-micron scale; for example, the flow stress increased significantly in micro-torsion experiments in thin copper wires as the wire diameter decreased from 170 to 12 μm [3] and in micro-bending of thin nickel beams with thickness from 50 to 12.5 μm [4].

The prominence of research into the ISE has been partly motivated by large-scale applications of thin films in electronic components and availability of novel methods of probing mechanical properties in very small volumes. Different studies suggested that factors such as: (1) a pile-up or sink-in surface profile [1]; (2) an indenter's loading rate [5]; (3) a grain size [6]; (4) a presence of an oxidation layer or a work-hardened surface layer [2] and (5) a geometric effect of indenter's tip (e.g. its radius) [1] may have an important effect on the strong dependence of indentation hardness on the indentation depth. A number of authors argued that the size dependence of material's mechanical properties resulted from an increase in strain gradients inherent to small localized zones, leading to geometrically necessary dislocations (GNDs), which, in turn, resulted in additional hardening [7–11].

Generally, deformation in metals can be characterized by generation, motion and storage of dislocations. Dislocations can be stored in two ways: they accumulate by trapping each other in a random way, or they are required for compatible deformation between various parts of a material, where they are referred as *statistically stored*

Applied Nanoindentation in Advanced Materials, First Edition. Edited by Atul Tiwari and Sridhar Natarajan.

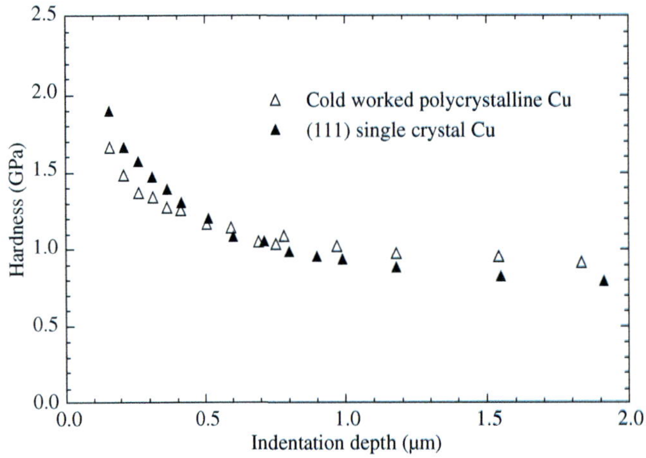

Figure 23.1 Indentation size effect in crystalline materials. Reprinted with permission from [1].

dislocations (SSDs) and *geometrically necessary dislocations*, respectively. SSDs are believed to depend on the effective plastic strain, while the density of GNDs is directly proportional to the gradient of the effective plastic strain. The smaller the deformation scale, the larger the density of GNDs relative to the SSDs and, consequently, the larger the plastic strain gradients compared to average plastic strains. Based on this concept, different strain-gradient plasticity theories were proposed. Nix and Gao [11] modelled the ISE for crystalline materials using the following characteristic form explaining the depth dependence of hardness:

$$\frac{H}{H_0} = \sqrt{1 + \frac{h^*}{h}} \tag{23.1}$$

where H is hardness for a given indentation depth h, H_0 is hardness in the limit of infinite depth and h^* is a characteristic length depending on the indenter shape, shear modulus and H_0. Equation 23.1 predicts a linear relation between H^2 and $1/h$, which agrees well with the micro-indentation hardness data for single-crystal and polycrystalline copper as shown in Figure 23.1, as well as for single-crystal silver [12] (Figure 23.2). However, hardness data obtained in polycrystalline work-hardened and annealed oxygen-free copper [2, 13], where indentation depth was below 100 nm, do not follow Equation (23.1). It was observed that indentation hardness continued to increase as h decreases, but it was significantly lower than the ISE predicted by Nix and Gao [11].

Continuum strain-gradient plasticity theories have been used to demonstrate some success in representing a collective behavior of dislocations and their interactions at the sub-micrometre-scale. It should be noted that these theories are only applicable at a scale much larger than individual dislocation spacing. Gao *et al.* [14] identified this limit to be around 100 nm. Thus, it may be argued that Equation (23.1) is not applicable for indentation depths below 100 nm. An experimental study by Demir *et al.* [15] with a sharp indenter demonstrated that the total GND density below the indents decreased with a decreasing indentation depth. This observation is in sharp contrast to predictions of strain-gradient theories attributing size-dependent material properties to GNDs.

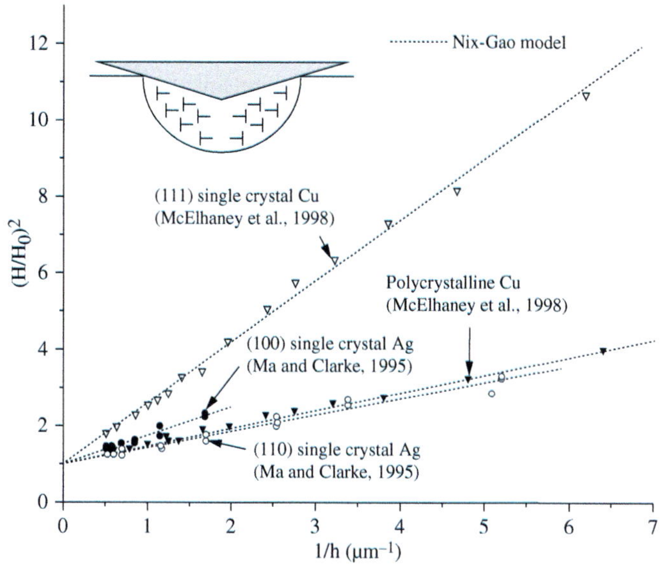

Figure 23.2 Micro-indentation hardness data for single-crystal, polycrystalline copper [1] and for single-crystal silver [12]. The inset shows model of geometrically necessary dislocations [11] underneath a sharp indenter. Reprinted with permission from [54].

In gradient-plasticity theories, an internal constitutive length-scale parameter l is used to scale the effects of strain gradients, which relate to storage of GNDs. The strain-gradient effects become important when the characteristic length associated with deformation becomes comparable to l. This was found to be approximately 4 μm for copper from torsion tests [3] and 6 μm for nickel from bending tests [4]. Nix and Gao [11] estimated the material length-scale parameter l from micro-indentation experiments of McElhaney *et al.* [1] to be 12 μm and 5.84 μm for annealed single-crystal copper and for cold-worked polycrystalline copper, respectively. Begley and Hutchinson [16] estimated that the material length-scale associated with the stretch gradients ranged from 1/4 to 1 μm and rotational gradients of 4 μm by fitting micro-indentation hardness data [17].

A mechanism-based strain-gradient (MSG) plasticity theory proposed by Gao *et al.* [18] is based on the Taylor dislocation model, where the intrinsic material length was identified to be $(\mu/\sigma_{\text{ref}})^2 b$, where μ is the shear modulus, σ_{ref} is a reference stress such as yield stress and b is the Burgers vector. The MSG theory is based on a multi-scale, hierarchical framework linking the microscale notions of SSDs and GNDs to the mesoscale notions of plastic strain and strain gradient. At microscale, the Taylor hardening model in dislocation mechanics was used to characterize dislocation interactions and their effect on the flow stress, and, at mesoscale, constitutive equations were constructed by averaging microscale plasticity laws over a representative volume. The MSG approach falls into the higher-order framework, where the order of equilibrium equations is higher than that of the conventional continuum theories thus requiring additional boundary conditions. The theories developed by Fleck *et al.* [3], Fleck and Hutchinson [9], Gao *et al.* [18], Huang *et al.* [19] and Gurtin [20] can be classified in this group. Huang *et al.* [21] established a conventional theory of MSG, with the plastic strain gradient appearing only in the constitutive model so that the equilibrium equations and boundary conditions remain the same as those in the classical continuum theories. In this lower-order framework, the higher-order stresses are not incorporated in the constitutive equations; hence, no additional boundary conditions are required. The strain-gradient theories proposed by Acharya and Bassani [22], Acharya and Beaudoin [23], Bassani [24], Beaudoin and Acharya [25] and Evers *et al.* [26] are examples of this class.

Most of the strain-gradient plasticity theories were formulated for isotropic materials. However, with decreasing length scale, inherent anisotropy of crystal lattices is expected to become increasingly important, and crystal plasticity offers a better description of lattice anisotropy at the size scale of a few micrometers, at which a significant size effects are observed. Strain-gradient formulations of crystal plasticity were proposed by many groups [22, 27–35]. Regueiro *et al.* [32] suggested a relationship between a dislocation density tensor and back stresses. Cermelli and Gurtin [36] developed a general theory of GNDs by characterizing dislocations by means of a tensor measuring the local Burger's vector per unit area. Shizawa and Zbib [33], Menzel and Steinmann [31] and Gurtin [28] used kinematic-hardening formulations, where second-order strain gradients were related to back stresses resulting from dislocation pile-ups at obstacles. However, in the case of problems with constant strain gradients such as plastically bent beam, these formulations predict no size dependence in contrast to experimental observations [4] and discrete-dislocation simulations [37]. Among the lower-order theories, Acharya and Bassani [22] described a dependence of work-hardening moduli on strain gradients with the help of a quadratic norm of a dislocation-density tensor. In the model of Arsenlis and Parks [38], dislocation-density state variables evolve from initial conditions according to equations based on fundamental concepts in dislocation mechanics such as the conservation of Burgers vector in multiplication and annihilation processes. Groma *et al.* [39] developed models using a statistical description of dislocation interactions. Although these models are in good agreement with dislocation simulations [40], the classical hardening laws used in crystal plasticity were abandoned in these models. On the other hand, Busso *et al.* [8] and Han *et al.* [29, 30] attempted to link the evolution of slip-system strength to the presence of GNDs. In the former, the evolution of GNDs were linked to local slip rate gradients, in the latter, a mechanism-based strain-gradient crystal-plasticity (MSGCP) model, GND densities were obtained from spatial gradients of plastic shear strain on slip systems.

23.2 Review of Indentation Modelling

Modelling of micro-/nano-indentation of single crystals has been developed extensively to elucidate experimentally observed features such as the size effect, pile-up phenomenon and lattice rotations. For instance, Liu *et al.* [41] performed a combined crystal-plasticity (CP) finite-element (FE) numerical simulations and micro-indentation experiments to determine mechanical properties of single-crystal copper. Lee and Chen [42] adopted the MSGCP theory to model the size effect in a micro-indentation process of the same material. Wang *et al.* [43] studied an effect of crystallographic orientation on pile-up patterns and micro-textures using a CPFE model for single-crystal copper. Correct numerical predictions of the surface pile-up patterns were achieved;

however, a difference of one order of a magnitude in the load–displacement curve between experiments and simulations was reported. Liu *et al.* [44] performed a similar study using a spherical indenter instead of a conical one, where satisfactory agreements between the numerical and experimental load–displacement curves were demonstrated. However, different magnitudes of the coefficient of friction were used to represent a contact condition between the indenter's tip and the workpiece material for different orientations to match numerically obtained surface profiles with the experimental data. Still, an error of up to 50% was reported for the maximum pile-up. In the study of Demiral *et al.* [45], where both the incipient and evolving strain gradients were considered in the calibration procedure (unlike prior studies), a better agreement in the load–displacement curves and in the maximum pile-up heights was obtained.

Some studies attempted to characterize the texture evolution in nano-indentation, with the use of different experiments such as a non-destructive 3D synchrotron diffraction technique, 3D electron backscattered diffraction (EBSD) and transmission electron microscopy (TEM). A limited number of numerical studies attempted to analyse physical deformation mechanisms leading to lattice rotations. For instance, Wang *et al.* [43] demonstrated lattice rotations for a single crystal of Cu with different orientations using a 3D elastic viscoplastic CP FE technique. Zaafarani *et al.* [46] proposed a physically based CP model based on dislocation-rate formulations to explain potential reasons for deformation-induced patterns consisting of multiple narrow zones with alternating crystalline rotations. However, the model consistently overestimated the extent of lattice rotations in the experiment. Demiral *et al.* [47] developed a three-dimensional finite-element model of nano-indentation incorporating an enhanced model of the strain-gradient crystal plasticity (EMSGCP) [48] to simulate accurately deformation of a body-centred cubic metallic material. It was noted that deformation-induced lattice rotations can be predicted more correctly using the strain-gradient CP theory since the effect of GNDs were accounted for. This study demonstrated that the introduction of strain gradients altered the activity of slip systems and the relative contribution to the overall plastic slip.

23.3 Crystal Plasticity Modelling of Nanoindentation

In this part, the ISE of three types of single crystals, i.e. f.c.c. copper single crystal, b.c.c. ß phase of Ti-6Al-4V (ß-Ti-64) and b.c.c. ß phase of Ti-15V-3Cr-3Al-3Sn (ß-Ti-15-3-3-3), is investigated using finite-element analysis. In indentation modelling, an idealised conical indenter and spherical indenters were chosen to study the effects of indenter's geometry on the mechanical response of the tested material. A half-angle of the conical indenter was 72° and the radius of spherical indenter 2.5 μm; dimensions of the modelled work-piece were $16 \times 16 \times 8\ \mu m^3$. Its top surface was free of constraints and normal displacements of other surfaces were fixed. Friction between the indenter and the indented surface was ignored for simplicity. The maximum indentation depths for all simulations was $h_{max} = 0.5\ \mu m$.

A crystal plasticity model was used to describe the material's behaviour of these single crystals during nanoindentation. It is well known that conventional crystal-plasticity models that do not account for dislocation motion and evolution are unable to characterise accurately the deformation mechanism in the smallest length scales. In our study, a conventional single-crystal plasticity (SCP) model [49] and the MSGCP model [29] were employed to compare and contrast the (de)merits of the two models. Both models have a similar constitutive framework, with the latter accounting for the effect of strain-gradients. The relationship between the shear rate $\dot{\gamma}^{(\alpha)}$ and resolved shear stress $\tau^{(\alpha)}$ on the slip system α is expressed by the power law proposed by Hutchinson [50], as:

$$\dot{\gamma}^{(\alpha)} = \dot{\gamma}_0 \left| \frac{\tau^{(\alpha)}}{g^{(\alpha)}} \right|^n \operatorname{sgn}(\tau^{(\alpha)}) \tag{23.2}$$

where $\dot{\gamma}_0$ is the reference shear rate, $g^{(\alpha)}$ is the slip resistance and n characterizes the rate sensitivity. The evolution of $g^{(\alpha)}$ is given by:

$$\dot{g}^{(\alpha)} = \sum_{\beta=1}^{N} h_{\alpha\beta} |\dot{\gamma}^{(\beta)}| \tag{23.3}$$

where $h_{\alpha\beta}$ is the hardening modulus that can be calculated in the form proposed by Hutchinson [50]:

$$h_{\alpha\alpha} = h_0 \operatorname{sech}^2 \left(\frac{h_0 \gamma}{\tau_S - \tau_0} \right), \quad h_{\alpha\beta} = q h_{\alpha\alpha} (\alpha \neq \beta), \quad \gamma = \sum_{\alpha} \int_0^t |\dot{\gamma}^{(\alpha)}| dt \tag{23.4}$$

Here, h_0 is the initial hardening modulus, q is the latent hardening ratio, τ_0 and τ_S are the shear stresses at the onset of yield and the saturation of hardening, respectively, and γ is the accumulative shear strain over all the slip systems. Generally, τ_0 is equal to the value of initial slip resistance $\tau_0 = g^{(\alpha)}|_{t=0} = g_0$.

The sole difference between SCP and MSGCP is the calculation of the slip resistance $g^{(\alpha)}$. In the SCP model, $g^{(\alpha)}$ is only determined by SSDs. In contrast, the contribution of GNDs is also taken into account in the MSGCP model, and $g^{(\alpha)}$ is redefined as:

$$g_T^{(\alpha)} = \sqrt{(g_{SSD}^{(\alpha)})^2 + (g_{GND}^{(\alpha)})^2}, \tag{23.5}$$

where $g_{SSD}^{(\alpha)}$ and $g_{GND}^{(\alpha)}$ are contributions to the slip resistance caused by SSDs and GNDs, respectively. The evolution of slip resistance $g_{SSD}^{(\alpha)}$ is given by a strain-hardening equation:

$$\dot{g}_{SSD}^{(\alpha)} = \sum_{\beta} h_{\alpha\beta} |\dot{\gamma}^{(\beta)}| \tag{23.6}$$

The slip resistance $g_{GND}^{(\alpha)}$ is determined by the effective density of GNDs $\eta_G^{(\alpha)}$:

$$g_{GND}^{(\alpha)} = \alpha\mu\sqrt{b\eta_G^{(\alpha)}} \tag{23.7}$$

$$\eta_G^{(\alpha)} = \left| \mathbf{m}^{(\alpha)} \times \sum_{\beta} (s^{\alpha\beta} \nabla \gamma^{(\alpha)}) \times \mathbf{m}^{(\beta)} \right| \tag{23.8}$$

In Equation 23.6, b and μ are the Burgers vector and shear modulus, respectively. In Equation 23.7, $\mathbf{m}^{(\alpha)}$ is the normal unit vector of slip plane, $s^{\alpha\beta} = \mathbf{s}^{(\alpha)} \cdot \mathbf{s}^{(\beta)}$ and $\mathbf{s}^{(\alpha)}$ defines the slip direction. Consequently, the MSGCP model will be reduced to the SCP model if $g^{(\alpha)} = g_T^{(\alpha)} = g_{SSD}^{(\alpha)}$.

The two types of crystal plasticity models were implemented in the commercial FE code ABAQUS/Standard by employing the user interface subroutine UMAT. Calculation of a strain gradient in the MSGCP model was realized with the use of C3D8 element available in the FE package [47]. The simulation results are reported in the form of load-displacement and hardness-indentation depth curves (ISE). Hardness H is defined as:

$$H = \frac{F_{max}}{A} \tag{23.9}$$

where F_{max} is the maximum load applied in indentation and A is the projected area of contact between the indenter and the work-piece. The contact area A was determined by accounting for the contact nodes on the surface of the work-piece as outlined in the work of Lee and Chen [42]. To capture an accurate description of the contact area, a finer local mesh was used in regions underneath the indenter tip.

23.3.1 Indentation of F.C.C. Copper Single Crystal

The material parameters for the f.c.c. copper single crystal used [42] are summarized in Table 23.1. The magnitude of Burgers vector for copper is $b = 0.255$ nm, shear modulus $\mu = 42.0$ GPa and the empirical coefficient in the Taylor model is $\alpha = 0.5$. For the investigated f.c.c. copper single crystal, the slip was assumed to occur on the usual 12 {111} <110> slip systems. For both conical and spherical indenters, the indentation simulations were performed on the [100] crystallographic plane of the work-piece.

Figure 23.3 shows the load-indentation depth curves of for a copper single crystal obtained for the two types of indenters, as predicted using the SCP and MSGCP models. For each indenter, the indentation load obtained with the MSGCP model was higher than that obtained with the SCP model for a fixed indentation depth, implying that strain gradients had a significant effect on the load-indentation depth response in nanoindentation of a copper single crystal.

Variations of hardness with indentation depth in nanoindentation with conical and spherical indenters are presented in Figure 23.4. The ISE is clearly observed for the two types of indenter, although the hardness-indentation depth curves for the spherical indenter exhibits lower depth sensitivity. Interestingly, the trends of hardness change with indentation depth are different for conical and spherical indenters: a decreasing one is observed for the conical indenter; however, this trend is reversed for the spherical indenter. Therefore, the indenter geometry not only affects the magnitude of calculated hardness but also determines the feature of ISE. Due to the ISE, the difference in hardness related to indenter geometry is more significant at a lower indentation depth.

Table 23.1 Material parameters of f.c.c. copper single crystal.

Elastic constants (GPa)	SCP model parameters	Additional parameters for MSGCP model
$C_{11} = 51.227$	$g_0 = 60.84$ MPa	$\alpha = 0.5$
$C_{12} = 36.930$	$\dot{\gamma}_0 = 0.001$	$b = 0.255$ nm
$C_{44} = 22.937$	$n = 50$	$\mu = 42.0$ GPa
	$q = 1.2$	
	$h_0 = 541.50$ MPa	
	$\tau_S = 109.51$ MPa	

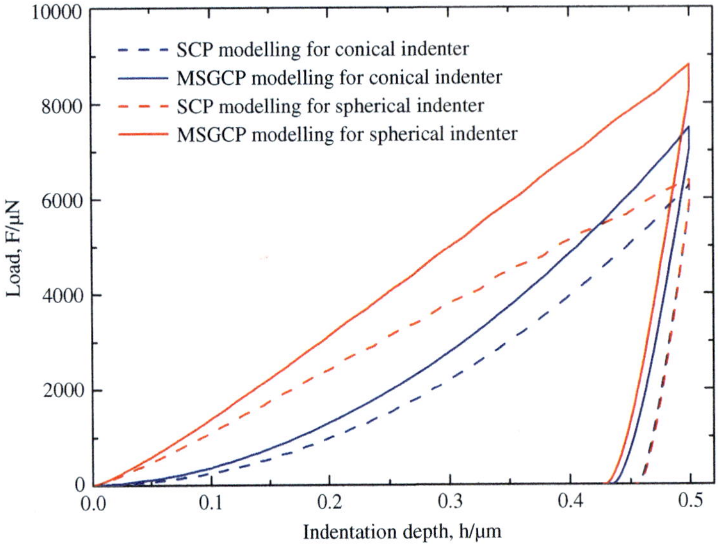

Figure 23.3 Load-indentation depth curves of copper single crystal for conical and spherical indenters.

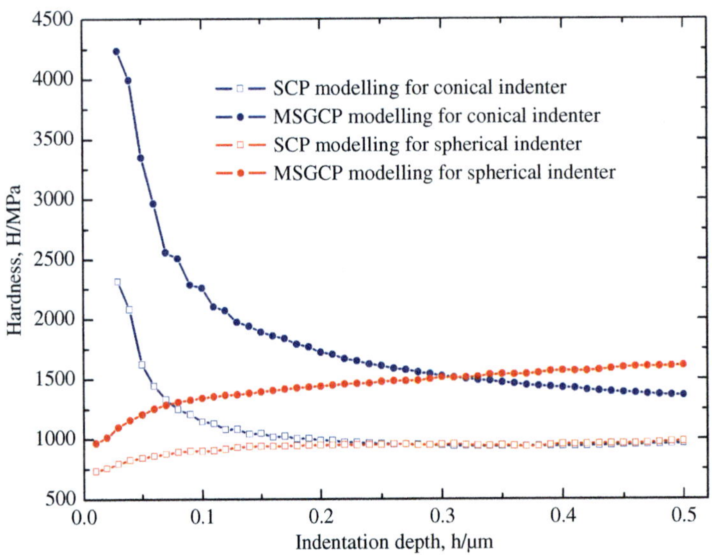

Figure 23.4 Indentation size effect for copper single crystal for conical and spherical indenters.

For the two types of indenters, it is clear that predictions based on the MSGCP model are significantly different from those with the SCP model, as shown in Figure 23.4. This implies that a strain gradient plays a pivotal role in defining hardness. It also implies strain gradient led to a more evident increase in hardness at smaller indentation depths for the conical indenter; however, the tendency is reversed for the spherical one. For the simulation results based on SCP modelling, it was noted that the hardness became insensitive to the depth when the indentation depth in single-crystal copper exceeded 0.15 μm, and the difference between the hardness obtained with the conical and spherical indenters can be neglected for large indentation depths. However, this phenomenon is different for MSGCP modelling. A gradual decrease or increase of hardness with indentation depth was observed for conical indentation or spherical indentation, respectively. Therefore, the effect of indenter geometry on hardness was observed even at a large indentation depth due to the effect of strain gradients. Consequently, the strain gradients caused the hardness-indentation depth curve to exhibit a more significant depth-dependence.

Hence, based on the simulation results in Figures 23.3 and 23.4 it can be concluded that the effect of strain gradient is crucial for nanoindentation of the f.c.c. copper single crystal.

23.3.2 Indentation of B.C.C. Ti-64

Material parameters used for modelling b.c.c. ß-Ti-64 single crystal are available [51, 52]. The magnitude of Burgers vector $b = 0.286$ nm, the shear modulus $\mu = 42.085$ GPa [53] and $\alpha = 0.5$. All relevant material parameters are listed in Table 23.2. Twelve {110}<111> slip systems were considered for b.c.c. Ti-64. For both conical and spherical indenters, the simulations were performed on the [110] crystallographic plane of work-pieces.

The calculated load-indentation depth curves obtained for conical and spherical indenters, using the SCP and MSGCP models, are shown in Figure 23.5. In contrast to the results obtained from indentation of the single crystal of Cu (Figure 23.3), for both conical and spherical indenters, the load-indentation depth curves of Ti-64 are minimally affected by the effect of strain gradients.

In Figure 23.6, the variations of hardness with indentation depth are shown for b.c.c. Ti-64; a significant ISE was observed. For both SCP and MSGCP simulation results, there is an evident discrepancy in hardness caused by the indenter geometry. At a small indentation depth, the hardness obtained for the conical indenter is larger than that for the spherical one; however, a higher hardness was yielded for the spherical indenter at a large indentation depth. It is important to point out that the discrepancy in hardness between two types of indenters existed even if the effect of strain gradients was neglected. Note that Ti-64 has a much higher slip resistance than Cu (see Table 23.1 and Table 23.2). For the range of indentation depths studied, the contact area was a mixture of elastic and plastic zones for the nanoindentation of Ti-64 while in Cu it could

Table 23.2 Material parameters of b.c.c. titanium Ti-64 single crystal.

Elastic constants (GPa)	SCP model parameters	Additional parameters for MSGCP model
$C_{11} = 250.2$	$g_0 = 200.0$ MPa	$\alpha = 0.5$
$C_{12} = 19.0$	$\dot{\gamma}_0 = 0.0023$	$b = 0.286$ nm
$C_{44} = 115.3$	$n = 50$	$\mu = 42.085$ GPa
	$q = 1.0$	
	$h_0 = 1500$ MPa	
	$\tau_S = 500$ MPa	

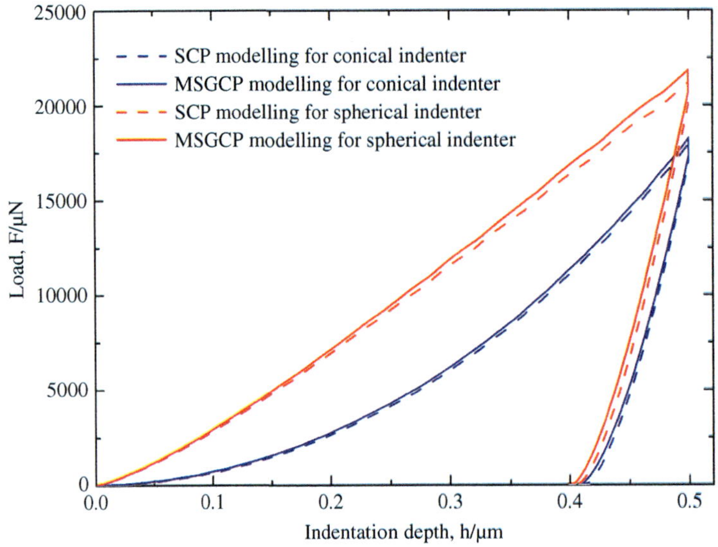

Figure 23.5 Load-indentation depth curves for b.c.c. titanium Ti-64 single crystal for conical and spherical indenters.

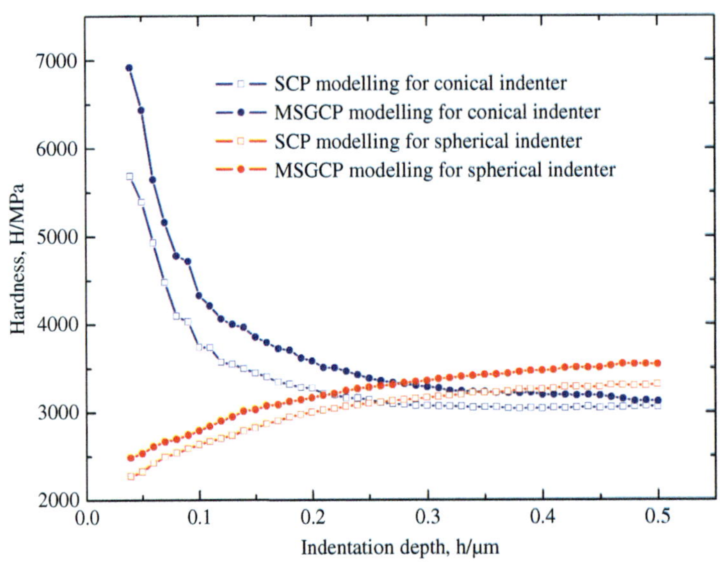

Figure 23.6 Indentation size effects of b.c.c. titanium Ti-64 single crystal for conical and spherical indenters.

be considered fully plastic. Thus, an increase in hardness was observed when using the spherical indenter due to its propensity to induce elastic deformation. In comparison with Figure 23.5, the differences between hardness-indentation depth curves due to strain gradients are more significant than those between load-indentation depth curves even though the strain-gradient effect was small.

In summary, the ISE was significant but the strain-gradient effect diminished in nanoindentation of Ti-64.

23.3.3 Indentation of B.C.C. Ti-15-3-3

Material parameters of a b.c.c. Ti-15-3-3 single crystal were obtained by calibrating experimental results in our previous work [45]; they are summarized in Table 23.3. Here, twelve {112} <111> are the dominant the slip systems. For both conical and spherical indenters, the indentation simulations were performed on the [110] crystallographic plane of work-pieces.

Figure 23.7 demonstrates the load-indentation depth curves obtained for the conical and spherical indenters with both the SCP and MSGCP models used. Similar to the results in Figure 23.5, the load-indentation depth curves obtained with MSGCP modelling are similar to those obtained from SCP-based simulations. A minor deviation between the two types of simulation results is observed at a large indentation depth for the spherical indenter.

The ISE in Ti-15-3-3 is shown in Figure 23.8. The hardness magnitudes obtained for the conical indenter exhibits a significant depth-dependence in comparison to the spherical indenter. For SCP modelling, the effect of indenter geometry on hardness was considerable when the indentation depth was smaller than 0.25 μm but can be neglected at larger depths, similar to the results for the f.c.c. Cu single crystal (Figure 23.4). In contrast, the results from the MSGCP model indicate that a difference in hardness for the two types of indenters was detected, but it is less significant than that for BCC Ti-64 (Figure 23.6).

It is worth noting that strain gradients affect the hardness magnitude (Figure 23.8) more than the imposed indenter load (Figure 23.7) at a fixed indentation depth for both conical and spherical indenters. In order to characterize the effect of strain gradients, a relative error between the results obtained with SCP and MSGCP modelling was defined as:

$$Err = \left| \frac{2(P_{SCP} - P_{MSGCP})}{P_{SCP} + P_{MSGCP}} \right| \times 100\%, \tag{23.10}$$

where P_{SCP} and P_{MSGCP} represent the magnitudes of the respective parameter obtained with SCP and MSGCP modelling, respectively. Figure 23.9(a) shows the comparisons for the indentation load, contact area and hardness obtained with SCP and MSGCP modelling for the spherical indenter. It is clear that the relative error in the contact area is

Table 23.3 Material parameters of b.c.c. titanium single crystal in Ti-15-3-3.

Elastic constants (GPa)	SCP model parameters	Additional parameters for MSGCP model
$C_{11} = 127.740$	$g_0 = 150.0$ MPa	$\alpha = 0.7$
$C_{12} = 118.850$	$\dot{\gamma}_0 = 0.0001$	$b = 0.2803$ nm
$C_{44} = 43.997$	$n = 15$	$\mu = 13.980$ GPa
	$q = 1.0$	
	$h_0 = 10000.0$ MPa	
	$\tau_S = 170.0$ MPa	

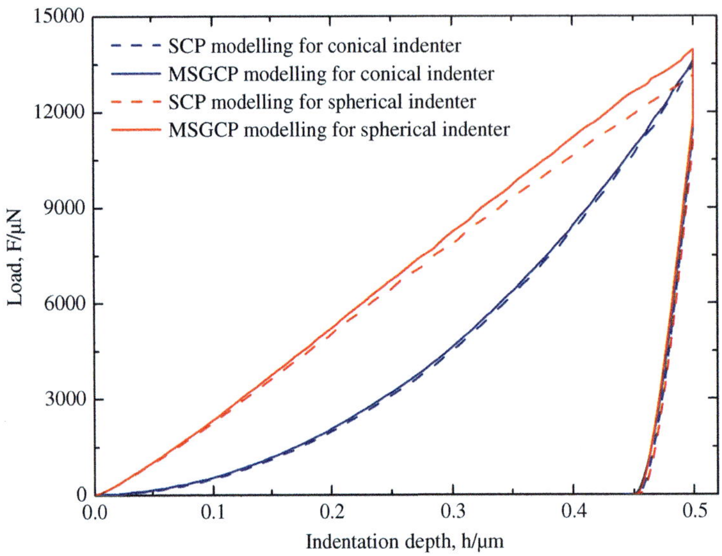

Figure 23.7 Load-indentation depth curves for b.c.c. titanium Ti-15-3-3 single crystal for conical and spherical indenters.

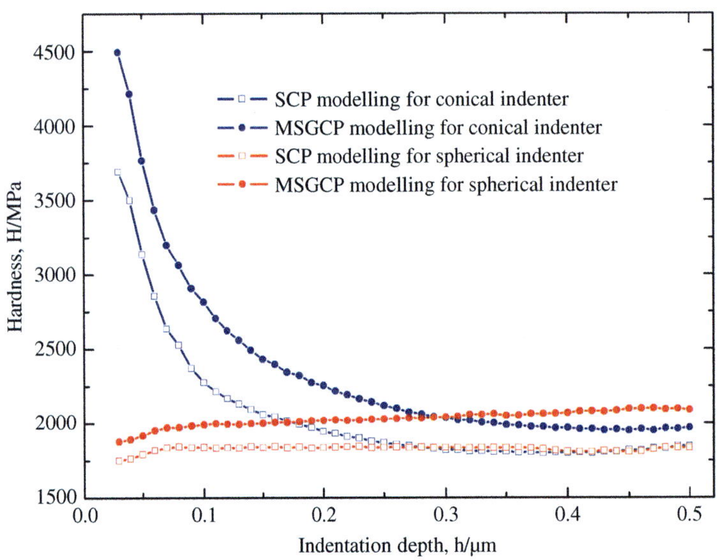

Figure 23.8 Indentation size effects for b.c.c. titanium Ti-15-3-3 single crystal for conical and spherical indenters.

larger than that for the indentation load at each indentation depth. In other words, the decrease of contact area due to strain gradients was more dominant than the increase of the indentation load when the b.c.c. Ti-15-3-3 single crystal was indented. This is the reason that the effect of strain gradients on hardness was more evident than that of the indentation load. Additionally, a direct comparison of contact zones between

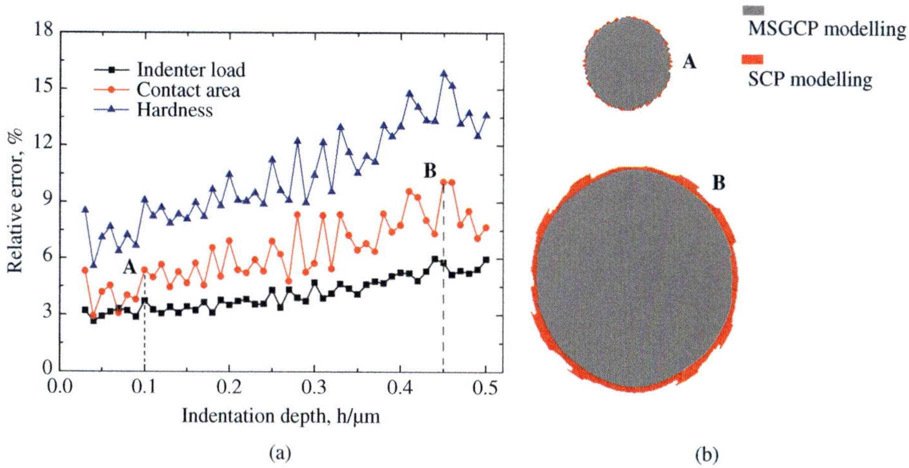

Figure 23.9 Comparisons between SCP and MSGCP modelling for spherical nanoindentation of b.c.c. Ti-15-3-3 single crystal: (a) relative errors of calculated indenter load, contact area and hardness, (b) contact zones corresponding to points A and B in Figure 23.9(a).

the indenter and the work-piece is illustrated in Figure 23.9(b) for SCP and MSGCP modelling. It demonstrates that the difference in the contact area calculated with two modelling schemes is more significant at larger indentation depths for the spherical indenter, consistent with the results shown in Figure 23.9(a).

23.4 Conclusions

The ISE was observed for nanoindentation of the three investigated single crystals with both conical and spherical indenters at small indentation depth. The indenter geometry has a significant effect on the features of ISE: the calculated hardness decreased with the indentation depth for the conical indenter but increased for the spherical one. Therefore, a great discrepancy in hardness for different types of indenters was observed, especially when the indentation depth was relatively small. However, hardness became less sensitive to the depth at larger indentation depths. For all the three types of single crystals studied, the variation of hardness with indentation depth was higher for the conical indenter.

In these monocrystals, strain gradients played different roles in nanoindentation. For f.c.c. Cu, strain gradients led to a significant increase both in the indentation load and the hardness magnitude, as well as the increased ISE. In b.c.c. Ti-64, the effect of strain gradients can be neglected. For b.c.c. Ti-15-3-3, strain gradients greatly affected the calculated level of hardness and the ISE, although their effect on the load-indentation depth curves was small. Generally, strain gradients had a great influence on the contact area in comparison to that of the indentation load. Overall, it was observed that strain gradients played a significant role in nanoindentation of f.c.c. single crystals when compared to b.c.c. ones.

It is emphasized that both the effects of indenter geometry and strain gradients are related to each other. The difference in hardness caused by the indenter geometry can be neglected when the effects of strain gradients are not considered for metal single crystals with low slip resistance (f.c.c. copper and b.c.c. Ti-15-3-3 single crystals). The indenter geometry also influenced strain gradient that, in its turn, affected hardness. The discrepancy in hardness between the results of MSGCP and SCP models decreased with indentation depth for the conical indenter but the trend was opposite for the spherical one. The present simulation results indicate that there are several geometrical and mechanical factors, which affect results obtained with nanoindentation of single crystals. The modelling approached such as those described here help to elucidate the mechanisms that lead to deformation in the small scale.

References

1 McElhaney K, Vlassak J, Nix W. Determination of indenter tip geometry and indentation contact area for depth-sensing indentation experiments. Journal of Materials Research 1998, 13(05): 1300–1306, DOI: 10.1557/JMR.1998.0185.
2 Swadener J, George E, Pharr G. The correlation of the indentation size effect measured with indenters of various shapes. Journal of the Mechanics and Physics of Solids 2002, 50(4), 681–694, DOI: 10.1016/S0022-5096(01)00103-X.
3 Fleck N, Muller G, Ashby M, Hutchinson J. Strain gradient plasticity: theory and experiment. Acta Metallurgica et Materialia 1994, 42(2), 475–487, DOI: 10.1016/0956-7151(94)90502-9.
4 Stölken J, Evans A. A microbend test method for measuring the plasticity length scale. Acta Materialia 1998, 46(14), 5109–5115, DOI: 10.1016/S1359-6454(98)00153-0.
5 Lilleodden E.T. Indentation-induced plasticity of thin metal films. PhD Dissertation. Stanford, USA: Stanford University; 2002.
6 Stelmashenko N, Walls M, Brown L, Milman Y.V. Microindentations on W and Mo oriented single-crystals: an STM study. Acta Metallurgica et Materialia 1993, 41(10), 2855–2865, DOI: 10.1016/0956-7151(93)90100-7.
7 Aifantis E. On the microstructural origin of certain inelastic models. Journal of Engineering Materials and Technology 1984, 106, 326–330, DOI: 10.1115/1.3225725.
8 Buss EP, Meissonnier FT, O'Dowd NP. Gradient-dependent deformation of two-phase single-crystals. Journal of the Mechanics and Physics of Solids 2000, 48(11), 2333–2361, DOI: 10.1016/S0022-5096(00)00006-5.
9 Fleck N, Hutchinson J. Strain gradient plasticity. Advances in Applied Mechanics 1997, 33, 295–361, DOI: 10.1016/S0065-2156(08)70388-0.
10 Gao H, Huang Y. Geometrically necessary dislocation and size-dependent plasticity. Scripta Materialia. 2003, 48(2), 113–118, DOI: 10.1016/S1359-6462(02)00329-9.
11 Nix WD, Gao H. Indentation size effects in crystalline materials. a law for strain gradient plasticity. Journal of the Mechanics and Physics of Solids 1998, 46(3), 411–425, DOI: 10.1016/S0022-5096(97)00086-0.
12 Ma Q, Clarke DR. Size dependent hardness of silver single-crystals. Journal of Materials Research. 1995, 10(04), 853–863, DOI: 10.1557/JMR.1995.0853.

13 Lim YY, Chaudhri MM. The effect of the indenter load on the nanohardness of ductile metals: an experimental study on polycrystalline work-hardened and annealed oxygen-free copper. Philosophical Magazine A. 1999, 79(12), 2979–3000, DOI: 10.1080/01418619908212037.
14 Gao H, Huang Y, Nix W. Modeling plasticity at the micrometer scale. Naturwissenschaften 1999, 86(11), 507–515, DOI: 10.1007/s001140050665.
15 Demir E, Raabe D, Zaafarani N, Zaefferer S. Investigation of the indentation size effect through the measurement of the geometrically necessary dislocations beneath small indents of different depths using EBSD tomography. Acta Materialia 2009, 57(2), 559–569, DOI: 10.1016/j.actamat.2008.09.039.
16 Begley MR, Hutchinson JW. The mechanics of size-dependent indentation. Journal of the Mechanics and Physics of Solids 1998, 46(10), 2049–2068, DOI: 10.1016/S0022-5096(98)00018-0.
17 Abu Al-Rub RK, Voyiadjis GZ. Analytical and experimental determination of the material intrinsic length scale of strain gradient plasticity theory from micro- and nano-indentation experiments. International Journal of Plasticity 2004, 20(6), 1139–1182, DOI: 10.1016/j.ijplas.2003.10.007.
18 Gao H, Huang Y, Nix W, Hutchinson J. Mechanism-based strain gradient plasticity, I: Theory. Journal of the Mechanics and Physics of Solids 1999, 47(6), 1239–1263, DOI: 10.1016/S0022-5096(98)00103-3.
19 Huang Y, Gao H, Nix W, Hutchinson J. Mechanism-based strain gradient plasticity, II: Analysis. Journal of the Mechanics and Physics of Solids 2000, 48(1), 99–128, DOI: 10.1016/S0022-5096(99)00022-8.
20 Gurtin ME. On the plasticity of single-crystals: free energy, microforces, plastic-strain gradients. Journal of the Mechanics and Physics of Solids 2000, 48(5), 989–1036, DOI: 10.1016/S0022-5096(99)00059-9.
21 Huang Y, Qu S, Hwang K, Li M, Gao H. A conventional theory of mechanism-based strain gradient plasticity. International Journal of Plasticity 2004, 20(4), 753–782, DOI: 10.1016/j.ijplas.2003.08.002.
22 Acharya A, Bassani J. Lattice incompatibility and a gradient theory of crystal-plasticity. Journal of the Mechanics and Physics of Solids 2000, 48(8), 1565–1595, DOI: 10.1016/S0022-5096(99)00075-7.
23 Acharya A, Beaudoin A. Grain-size effect in viscoplastic polycrystals at moderate strains. Journal of the Mechanics and Physics of Solids 2000, 48(10), 2213–2230, DOI: 10.1016/S0022-5096(00)00013-2.
24 Bassani J. Incompatibility and a simple gradient theory of plasticity. Journal of the Mechanics and Physics of Solids 2001, 49(9), 1983–1996, DOI: 10.1016/S0022-5096(01)00037-0.
25 Beaudoin A, Acharya A. A model for rate-dependent flow of metal polycrystals based on the slip plane lattice incompatibility. Materials Science and Engineering: A. 2001, 309, 411–415, DOI: 10.1016/S0921-5093(00)01620-8.
26 Evers L, Parks D, Brekelmans W, Geers M. Crystal-plasticity model with enhanced hardening by geometrically necessary dislocation accumulation. Journal of the Mechanics and Physics of Solids 2000, 50(11), 2403–2424, DOI: 10.1016/S0022-5096(02)00032-7.

27 Cheong KS, Busso EP, Arsenlis A. A study of microstructural length scale effects on the behaviour of FCC polycrystals using strain gradient concepts. International Journal of Plasticity 2005, 21(9), 1797–1814, DOI: 10.1016/j.ijplas.2004.11.001.
28 Gurtin, ME. A gradient theory of single-crystal viscoplasticity that accounts for geometrically necessary dislocations. Journal of the Mechanics and Physics of Solids 2002, 50(1), 5–32, DOI: 10.1016/S0022-5096(01)00104-1.
29 Han C, Gao H, Huang Y, Nix WD. Mechanism-based strain gradient crystal-plasticity, I: Theory. Journal of the Mechanics and Physics of Solids 2005, 53(5), 1188–1203, DOI: 10.1016/j.jmps.2004.08.008.
30 Han C, Gao H, Huang Y, Nix WD. Mechanism-based strain gradient crystal-plasticity, II: Analysis. Journal of the Mechanics and Physics of Solids 2005, 53(5), 1204–1222, DOI: 10.1016/j.jmps.2005.01.004.
31 Menzel A, Steinmann, P. On the continuum formulation of higher gradient plasticity for single and polycrystals. Journal of the Mechanics and Physics of Solids 2000, 48(8), 1777, DOI: 10.1016/S0022-5096(99)00024-1.
32 Regueiro RA, Bammann DJ, Marin EB, Garikipati K. A nonlocal phenomenological anisotropic finite deformation plasticity model accounting for dislocation defects. Journal of Engineering Materials and Technology 2002, 124, 380, DOI: 10.1115/1.1480410.
33 Shizawa K, Zbib H. A thermodynamical theory of gradient elastoplasticity with dislocation density tensor. I: Fundamentals. International Journal of Plasticity 1999, 15(9), 899–938, DOI: 10.1016/S0749-6419(99)00018-2.
34 Shu JY, Fleck NA. Strain gradient crystal-plasticity: size-dependentdeformation of bicrystals. Journal of the Mechanics and Physics of Solids 1999, 47(2), 297–324, DOI: 10.1016/S0022-5096(98)00081-7.
35 Smyshlyaev V, Fleck N. The role of strain gradients in the grain size effect for polycrystals. Journal of the Mechanics and Physics of Solids 1996, 44(4), 465–495, DOI: 10.1016/0022-5096(96)00009-9.
36 Cermelli, P, Gurtin, ME. Geometrically necessary dislocations in viscoplastic single crystals and bicrystals undergoing small deformations. International Journal of Solids Structure 2002, 39(26), 6281–6309, DOI: 10.1016/S0020-7683(02)00491-2.
37 Cleveringa H, van der Giessen E, Needleman A. A discrete-dislocation analysis of bending. International Journal of Plasticity 1999, 15(8), 837–868, DOI: 10.1016/S0749-6419(99)00013-3.
38 Arsenlis A, Parks D. Modelling the evolution of crystallographic dislocation density in crystal plasticity. Journal of the Mechanics and Physics of Solids 2002, 50(9), 1979–2009, DOI: 10.1016/S0022-5096(01)00134-X.
39 Groma I, Csikor FF, Zaiser M. Spatial correlations and higher-order gradient terms in a continuum description of dislocation dynamics. Acta Materialia 2003, 51(5), 1271–1281, DOI: 10.1016/S1359-6454(02)00517-7.
40 Yefimov S, Groma I, van der Giessen E. A comparison of a statistical-mechanics based plasticity model with discrete-dislocation plasticity calculations. Journal of the Mechanics and Physics of Solids 2004, 52(2), 279–300, DOI: 10.1016/S0022-5096(03)00094-2.
41 Liu Y, Wang B, Yoshino M, *et al.* Combined numerical simulation and nanoindentation for determining mechanical properties of single-crystal copper at mesoscale.

Journal of the Mechanics and Physics of Solids 2005, 53(12), 2718–2741, DOI: 10.1016/j.jmps.2005.07.003.

42 Lee W, Chen Y. Simulation of micro-indentation hardness of FCC single-crystals by mechanism-based strain gradient crystal-plasticity. International Journal of Plasticity 2010, 26(10), 1527–1540, DOI: 10.1016/j.ijplas.2010.01.011.

43 Wang Y, Raabe D, Klüber C, Roters F. Orientation dependence of nanoindentation pile-up patterns and of nanoindentation microtextures in copper single-crystals. Acta Materialia 2004, 52(8), 2229–2238, DOI: 10.1016/j.actamat.2004.01.016.

44 Liu Y, Varghese S, Ma J, *et al.* Orientation effects in nanoindentation of single-crystal copper. International Journal of Plasticity 2008, 24(11), 1990–2015, DOI: 10.1016/j.ijplas.2008.02.009.

45 Demiral M, Roy A, Silberschmidt VV. Indentation studies in bcc crystals with enhanced model of strain-gradient crystal plasticity. Computational Materials Science 2013, 79, 896–902, DOI: 10.1016/j.commatsci.2013.06.057.

46 Zaafarani N, Raabe D, Roters F, Zaefferer S. On the origin of deformation-induced rotation patterns below nanoindents. Acta Materialia 2008, 56(1), 31–42, DOI: 10.1016/j.actamat.2007.09.001.

47 Demiral M, Roy A, El Sayed T, Silberschmidt VV. Influence of strain gradients on lattice rotation in nano-indentation experiments: A numerical study. Materials Science and Engineering: A, 2014, 608, 73–81, DOI: 10.1016/j.msea.2014.04.033.

48 Demiral M. Enhanced gradient crystal-plasticity study of size effects in bcc metal. PhD thesis. Loughborough University, Loughborough, UK; 2012.

49 Huang Y. A User-material subroutine incroporating single crystal plasticity in the abaqus finite element program. Harvard University, USA; 1991. Available at: http://www.columbia.edu/~jk2079/fem/umat_documentation.pdf

50 Hutchinson J. Bounds and self-consistent estimates for creep of polycrystalline materials. Proceedings of the Royal Society of London A: Mathematical and Physical Sciences 1976, 348(1652), 101–127, DOI: 10.1098/rspa.1976.0027.

51 Lewis AC, Qidwai SM, Geltmacher AB. Slip systems and initiation of plasticity in a body-centered-cubic titanium alloy. Metallurgical and Materials Transactions A, 2010, 41(10), 2522–2531, DOI: 10.1007/s11661-010-0284-5.

52 Thomas J, Groeber M, Ghosh S. Image-based crystal plasticity FE framework for microstructure dependent properties of Ti–6Al–4V alloys. Materials Science and Engineering: A, 2012, 553, 164–175, DOI: 10.1016/j.msea.2012.06.006.

53 Fan X, Yang H. Internal-state-variable based self-consistent constitutive modeling for hot working of two-phase titanium alloys coupling microstructure evolution. International Journal of Plasticity 2011, 27(11), 1833–1852, DOI: 10.1016/j.ijplas.2011.05.008.

54 Huang Y, Zhang F, Hwang KC, *et al.* A model of size effects in nano-indentation. Journal of the Mechanics and Physics of Solids 2006, 54(8), 1668–1686, DOI: 10.1016/j.jmps.2006.02.002.

24

Computer Simulation and Experimental Analysis of Nanoindentation Technique

A. Karimzadeh, M.R. Ayatollahi and A. Rahimi

Fatigue and Fracture Laboratory, Center of Excellence in Experimental Solid Mechanics and Dynamics, School of Mechanical Engineering, Iran University of Science and Technology, Narmak, Tehran

24.1 Introduction

The nanoindentation test has been widely used by many researchers for different types of materials to measure the mechanical properties of both bulk materials and thin coatings [1–7]. A standard method is used in this test in which an indenter tip is immerged into specific sites of the sample material by applying an increasing normal load. Mechanical properties such as Young's modulus, hardness and elastic-plastic deformation are determined using load-displacement data which are recorded during the test by high resolution instruments [1–4]. In the nanoindentation experiment, a complicated stress field is generated in the material below the indenter which makes the interpretation of the results difficult. Computer simulations are powerful methods to study mechanical phenomena which are being improved generally by progress in the computer sciences as well as in solid mechanics. It can visualize the resulting stress fields while providing an important flexibility for different materials and various conditions. Computer simulations for simulating the nanoindentation experiment can be performed at different structural scales. Finite element modeling and molecular dynamic method are two major techniques for simulating the nanoindentation procedure.

Finite element analysis (FEA) is widely used as a powerful computational and modeling tool for simulating different engineering problems.

It uses mesh generation techniques for dividing a complex problem into simple cases within small elements. Although FEM is a strong method for solving continuum mechanics problems, it has certain restrictions for describing material behavior at atomic scales [8].

For investigating the mechanical behavior of structures in the nanoscale, the molecular dynamic (MD) method has been introduced. MD is a computer simulation of physical movements of atoms and molecules in which a view of atomic motions is given by allowing them to interact for a period of time. In this method, for a system of atoms and molecules, the trajectories of particles are specified by numerically solving Newton's equations of motion. For defining the forces between the particles and potential energy, molecular mechanics force fields are used [9]. MD simulation provides detailed information about the materials at the atomic scale, computes the motions of individual

Applied Nanoindentation in Advanced Materials, First Edition. Edited by Atul Tiwari and Sridhar Natarajan.

atoms/molecules and describes how their positions, velocities and orientations change with time. This method is now extensively used to investigate the structure, phase transitions, conductivity and thermodynamic properties of materials, melting and crystallization phenomena, the configuration of crystals, and liquid diffusion at atomic levels [10]. However, the scales of length and time that can be simulated using MD are still restricted [11–13].

In this chapter, both finite element method (FEM) and molecular dynamics (MD) are used to simulate the nanoindentation experiment on aluminum 1100. In order to verify the simulation, the experimental results obtained from the nanoindentation experiment are utilized. A comparison is made between the results of the nanoindentation test and FE simulation to validate the procedure used for finite element modeling of the nanoindentation experiment on aluminum 1100. We also explore how the roundness of the indenter tip can affect the results obtained from the finite element and molecular dynamic simulation of the nanoindentation experiment.

24.2 Finite Element Simulation for Nanoindentation

In computer simulations of the nanoindentation procedure, the stress and strain fields under the indenter tip which are used to determine the basic mechanical properties of materials should be numerically determined. By considering appropriate values for the dimensions of sample material, the geometry of indenter and the indentation depth, finite element modeling can also be used for simulating the nanoindentation procedure [14–21]. Some investigators have compared the results obtained from the FE simulation of nanoindentation and from the experiment to validate finite element modeling of the nanoindentation process, particularly for determination of material hardness [17, 19, 22, 23].

24.3 Finite Element Modeling

In this section, the simulation of the nanoindentation procedure using the finite element method is described.

24.3.1 Geometry

A 2-D axisymmetric model was employed for finite element simulation of the elastic-plastic behavior of material in the nanoindentation process. A conical rigid indenter with a half-angle of 70.3° was used in the axisymmetric model which had the same projected area to depth function as the standard Berkovich indenter [24]. The results presented by [25] show that there is little difference between the finite element results when a 3D simulation of Berkovich indenter is replaced by a 2D axisymmetric indenter if the half angle in the axisymmetric model is considered to be 70.3°. Needless to say that 2D axisymmetric analysis of the nanoindentation process is less complicated and requires less computational effort than a 3D model.

To simulate the nanoindentation process the sample height was considered to be about 50 times larger than the maximum indentation depth. Figure 24.1 shows the schematic representation of indenter and specimen used in the finite element model.

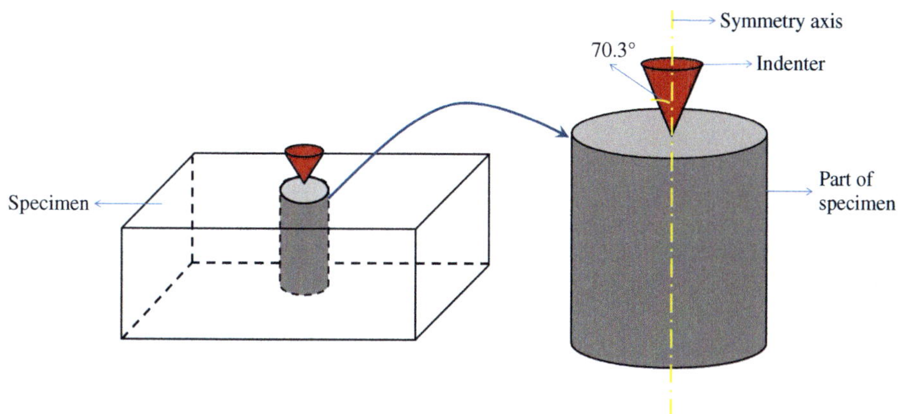

Figure 24.1 The axisymmetric model of indenter and specimen used in finite element simulations.

The indenter tip was initially assumed to be sharp, but in a subsequent analysis, a slight curvature was considered for the indenter tip. This will be described later.

24.3.2 Material Characteristics

Based on ASTM E8/E8M–09 standard [26], a uniaxial tensile test was performed on rectangular specimens to obtain tensile properties of aluminum 1100. A piece-wise linear stress-strain curve obtained from the results of uniaxial tensile test (Figure 24.2)

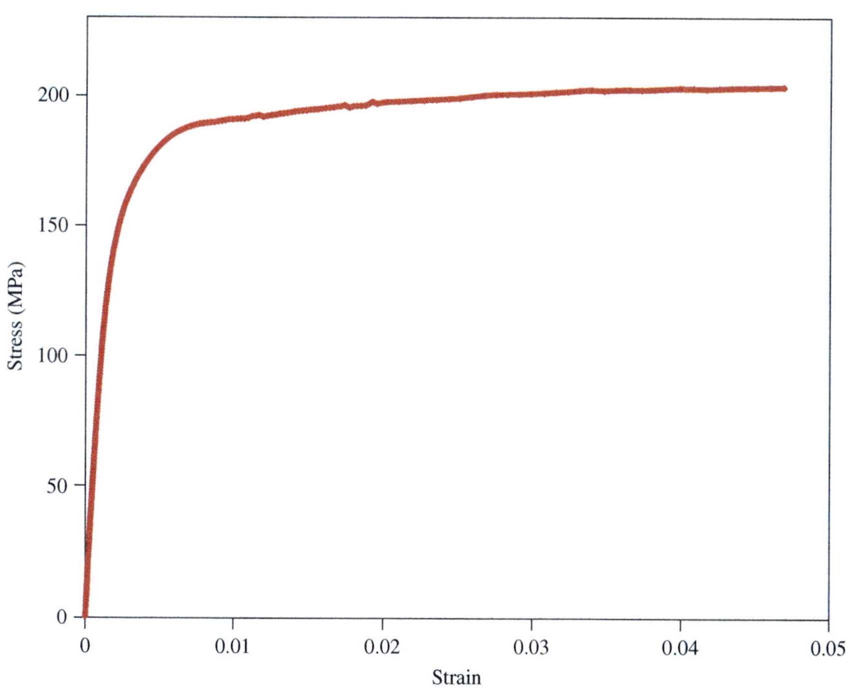

Figure 24.2 Stress-stain curve used for characterization of aluminum 1100.

was used in the finite element analysis to simulate the nanoindentation experiment. As reported in [27], kinematic hardening material behavior can be considered for aluminum 1100.

24.3.3 Boundary Condition

For 2D axisymmetric models, usually the nodes along the axisymmetry axis and across the bottom line of the specimen are fixed, but only normal to the lines [28–30].

In our simulations, the boundary conditions were applied along the centerline and the bottom surface of the specimen by fixing in the horizontal and vertical displacements, respectively.

Both loading and unloading steps of the nanoindentation process were simulated in the finite element model. During the loading stage, the indenter was driven into the specimen surface in the axial direction as deep as 260 nm in 30 seconds with a constant speed. This provides the loading part of the load-displacement curve in the finite element simulation. When the indenter tip reached the preset maximum penetration depth (i.e. 260 nm), the test specimen was unloaded and the indenter tip returned to its initial position with the same speed as that of the loading stage. In this way, the un-loading part of the load-displacement curve is obtained from the finite element simulation.

24.3.4 Interaction

A contact constraint between the indenter and the test sample was defined. The indenter and the test sample were chosen as the master surface and the slave surface, respectively, to simulate the surface to surface contact. The contact direction was then defined from the indenter surface towards the sample due to the fact that only the master surface can penetrate into the slave one. The finite-sliding tracking approach was selected for contact between the sample and the indenter. In order to assess the effect of friction between the two contacting surfaces, a friction coefficient equal to 0.6 was assumed between the indenter tip and the specimen surface [31].

24.3.5 Meshing

The size of mesh in the area beneath the indenter where the material experiences largest deformation is an important parameter in simulating the nanoindentation experiment. This region should be meshed carefully to facilitate the detection of contact and to obtain an accurate distribution of stresses and strains in the material. In order to have an accurate and easy detection of the contact between the indenter and the specimen, the mesh in the area under the indenter must be fine enough.

In this simulation, the specimen was meshed by a four-node quadrilateral bilinear axisymmetric element with reduced integration, as shown in Figure 24.3 A fine mesh was used in the contact area and near the tip of the indenter. In order to reduce the computational time, a coarser mesh was used further away from the indenter tip. A non-linear geometry option was used in the finite element simulation.

A mesh sensitivity study should also be performed to determine the appropriate size and number of elements. Researchers have employed different methods for the mesh sensitivity study. Liu *et al.* [32] investigated the number of elements at the indented area by quantifying the maximum deviations between the load-displacement curves obtained with different element sizes. They found that, for coarse elements the

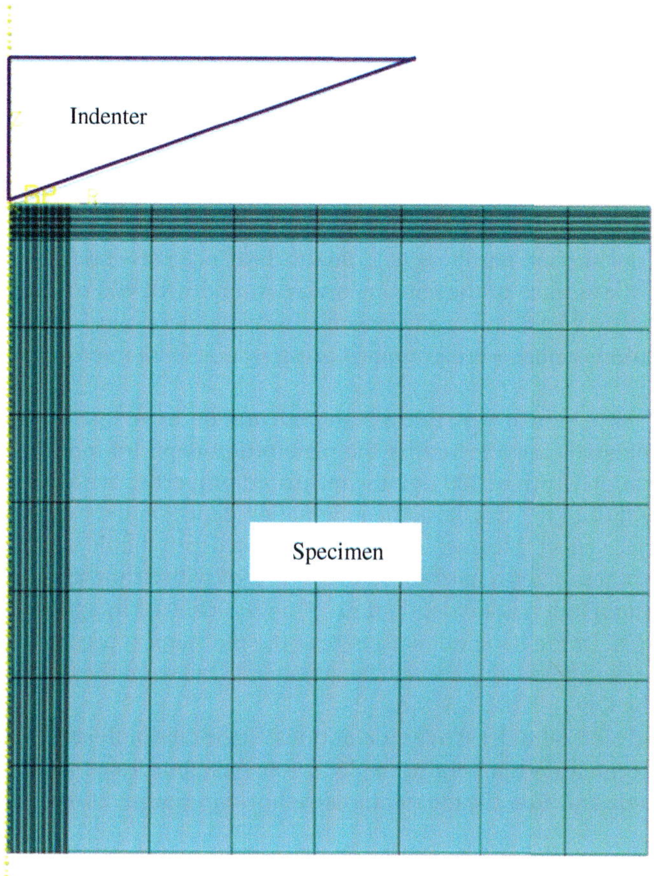

Figure 24.3 Mesh pattern used for axisymmetric finite element simulation of nanoindentation experiment.

load-displacement curve had large oscillations. Bressan *et al.* [33] examined the effect of element sizes in the region under the indenter by comparing the indentation load and global features of the load-displacement curve. Lee *et al.* [34] evaluated their mesh refinement by considering pile-up changes with the increase in the number of elements in the contact area.

In our simulation, mesh sensitivity analysis was carried out by comparing the variation of load-displacement curves for different element sizes. A total number of 805 elements was found to be appropriate for obtaining good agreement between the results of simulation and experiment.

24.4 Verification of Finite Element Simulation

For verifying the results obtained from the finite element simulation, the nanoindentation experiment was performed on Al 1100. Then load-displacement curves and the hardness values obtained from the simulation and from the experiment were compared.

24.4.1 Nanoindentation Experiment on Al 1100

The nanoindentation test was conducted using a Triboscope system (Hysitron Inc., USA) and a Berkovich indenter, which is primarily used for bulk materials. The testing instrument was calibrated according to the ISO 14577 standard [35]. A cubic specimen of thickness 3 mm with the section area of 10×10 mm^2 was prepared from aluminum 1100 material. It is known that the surface finish has a significant influence on the nanoindentation test results. Preparation of the test surface shall be carried out in a way that any alteration of the surface hardness (e.g. due to heat or cold-working) is minimized [36]. Therefore, the surface of the specimen was ground with 400 to 2500 grit sandpapers and then polished by alumina suspension. The roughness values of the tested sample at the indentation regions were examined using atomic force microscopy (AFM).

Previous studies [37, 38] have shown that in the nanoindentation experiment, the mechanical properties of materials are affected by the penetration depth of indenter. In most cases after a penetration depth of 200 nm, the measured properties are almost stable. Indeed, based on the ISO 14577 standard, the indentation depth should be deep enough to minimize the surface effect. Meanwhile, the indentation depth should also be less than 10% of the film thickness when a hard substance in coated with a thin layer. In the current study, the maximum indentation depth of 210 nm was achieved by applying an indentation load of 450 μN. In the nanoindentation test, the mechanical properties of the sample could be calculated by relations which have been derived from the theory of contact mechanics and the Oliver–Pharr method [24].

According to the described method, at least five indentations on the randomly selected sites of the sample were performed at a temperature of 23°C. The AFM images were used for analyzing the surface before and after the indentations as shown in Figures 24.4 a and 24.4 b.

A comparison between the load-displacement curves obtained from five individual nanoindentation tests shows that three tests are repeatable. Therefore, the results of these three tests were considered in our calculations. Figure 24.5 shows the repeatability of three nanoindentation tests.

24.4.2 Comparison Between Simulation and Experimental Results for Al 1100

24.4.2.1 Load-displacement

In order to measure the load versus displacement in the finite element simulation, the displacement of a single node just beneath the indenter tip is depicted according to the reaction force applied to a single node at the indenter tip. These nodes are shown by points A and B in Figure 24.6, respectively.

The displacement and reaction force described above were continuously recorded from the finite element results for both loading and unloading segments of the nanoindentation process.

The load-displacement curve obtained from the finite element simulation of nanoindentation test with a sharp indenter is shown in Figure 24.7. Also shown in this figure is the curve related to the average of experimentally determined results.

A comparison between the curves resulting from the finite element simulation and from the nanoindentation experiment demonstrates that the curve due to the finite element simulation with a sharp indenter tip is significantly different from the experimental

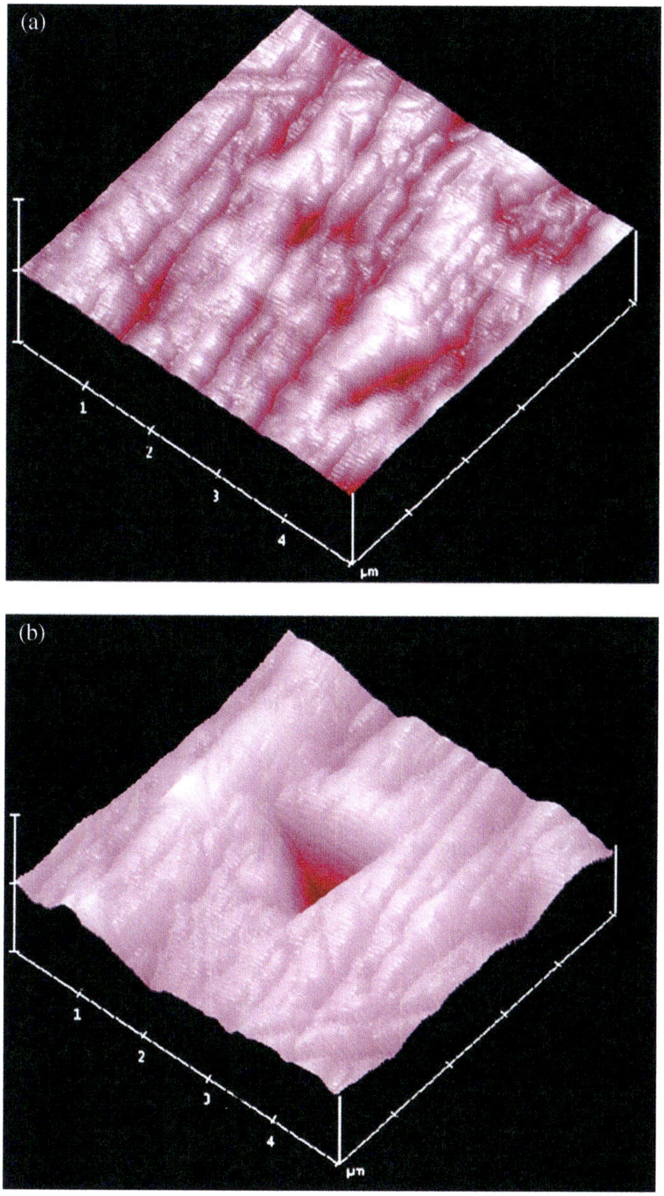

Figure 24.4 AFM images of the specimen surface a) before the nanoindentation test and b) after the nanoindentation test.

one. Then a slight curvature of 200 nm was considered for the indenter tip and the finite element simulation was repeated. It is seen from Figure 24.7 that the finite element results for the rounded tip indenter is in very good agreement with load-displacement curve obtained from the experiment. This may be due to the fact that in practice the indenter tip is not perfectly sharp and based on the ISO 14577 standard, the Berkovich indenter has an average radius of curvature about 150 to 200 nm [35]. Manufacturing

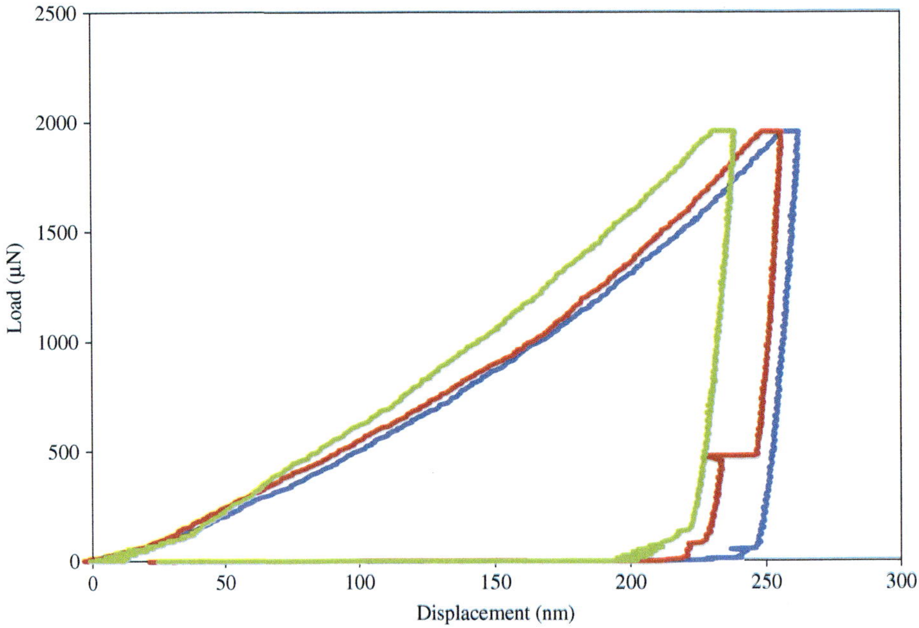

Figure 24.5 The load displacement curves obtained from three repeatable nanoindentation experiments.

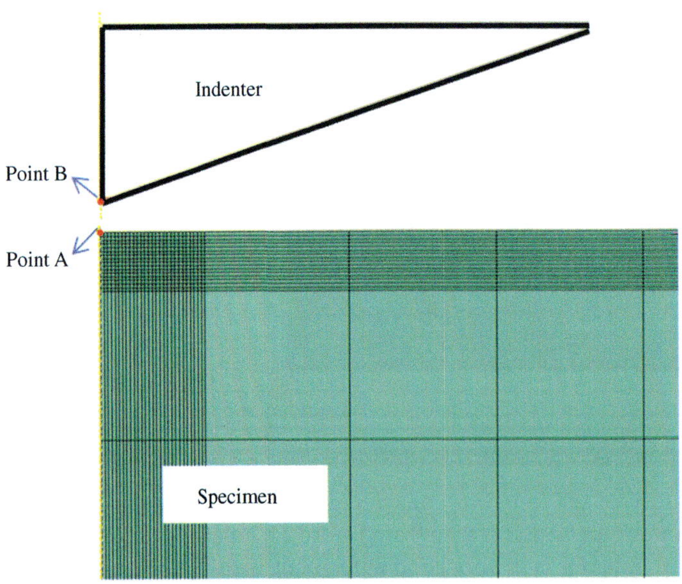

Figure 24.6 Measurement of the indentation displacement and load from the FE model.

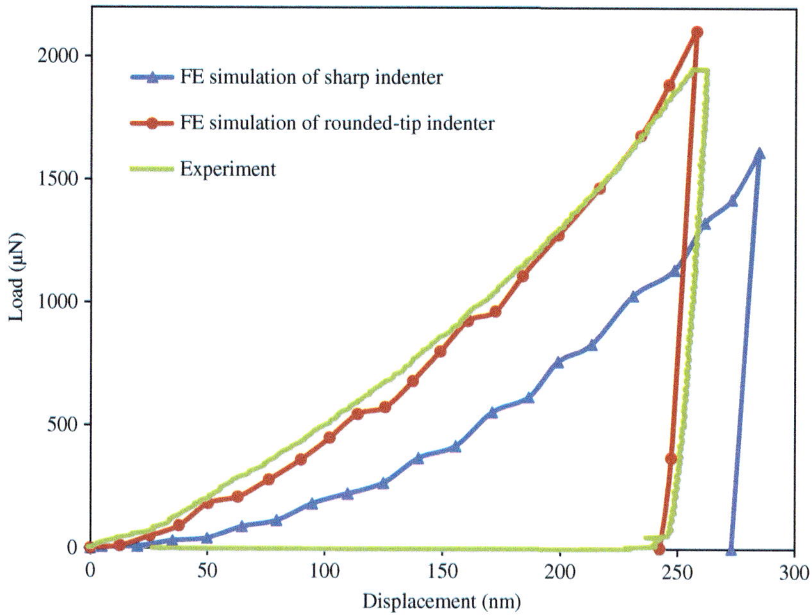

Figure 24.7 Load-displacement curves obtained from finite element simulations of sharp and rounded-tip indenters.

restrictions and structural properties are often two major reasons for reductions in the sharpness of the indenter tip.

As shown in Figure 24.7, there is still a small difference between the finite element results for the round-tip indenter and those of the experiment. This discrepancy can be related to the difference between the value of Young's modulus defined in the finite element model and that obtained from the nanoindentation experiment. One can conclude that, the finite element simulation with a round-tip indenter is more appropriate for simulating the nanoindentation process and for investigating the stress and deformation fields around the indenter.

The nanoindentation of the sample by a friction coefficient equal to zero was also simulated which showed only a small difference. Figure 24.8 shows the load-displacement curves determined from FE simulation for the rounded tip indener by friction coefficients of 0 and 0.6.

As shown in Figure 24.8, using zero friction coefficient in FE simulation decreases the difference between the unloading parts of the load-displacement curves obtained from the FE simulation and from the experiment. However, the loading part of the curve calculated from FE simulation by a friction coefficient of zero has a higher discrepancy with the experimental curve compared to the loading part obtained from FE simulation by a friction coefficient of 0.6.

In the finite element simulation, plastic deformation occurs where the equivalent stress reaches the yield strength of the sample material. The equivalent stress is calculated from:

$$\sigma_{\text{equivalent}} = \sqrt{\frac{(\sigma_1 - \sigma_2)^2 + (\sigma_2 - \sigma_3)^2 + (\sigma_1 - \sigma_3)^2}{2}} \tag{24.1}$$

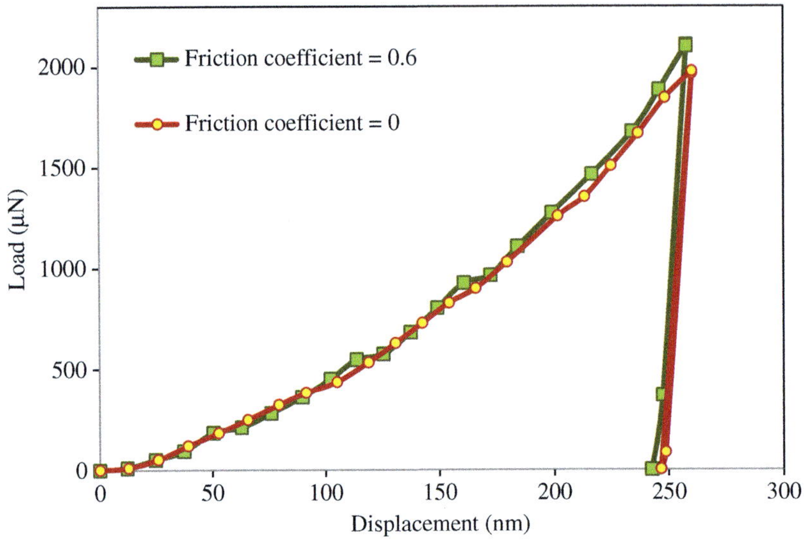

Figure 24.8 Load-displacement curves obtained from FE simulation of rounded tip indenter by friction coefficients of 0 and 0.6.

where σ_1, σ_2 and σ_3 are the three principal stresses generating at each point in the test sample during the finite element simulation of the indentation process. Using the data derived from the direct tensile experiment of aluminum 1100, the value of the yield strength of sample material in the finite element model was taken as 150 MPa. Figure 24.9 shows the stress contour and deformation of aluminum 1100 in the finite element model in different stages of loading.

As shown in Figure 24.9a, at the beginning of the indentation process the maximum equivalent stress occurring beneath the indenter is equal to 203 MPa which is greater than the yield strength of aluminum 1100. Therefore, the plastic deformation takes place around the indenter tip immediately after the indenter touches the sample surface. By increasing the indentation depth, the zone of plastic deformation expands both horizontally and vertically (Figure 24.9b). At the unloading stages (Figures 24.9c and 24.9d) the elastic deformation gradually decreases to zero and an indentation hole is generated due to plastic deformation.

24.4.2.2 Hardness

Hardness is defined by resistance of materials to localized plastic deformation caused by an external load. Hardness tests all make a small (sometimes microscopic) indentation into the surface of a specimen, and then use the force applied and the size of the indentation to calculate the hardness value. Once an indentation test is performed, hardness can bc calculated from [39]:

$$H = \frac{F}{A} \tag{24.2}$$

where F is the maximum load and A is the indentation area. For the nanoindentation test, F is the maximum indentation load and A is the projected area of the contact surface between the specimen and the indenter which is related to the contact depth (h_c) as [24]:

$$A = 3\sqrt{3}h_c^2\tan^2\theta \tag{24.3}$$

where θ is the face angle of the indenter which is equal to 65.27° for the Berkovich indenter. By substituting the value of θ in Equation (24.3) we have:

$$A \approx 24.5h_c^2 \tag{24.4}$$

If h_c is found, the projected area of the contact surface and then the material hardness can be computed from Equations (24.2) and (24.3). The value of h_c can be measured using the load-displacement data obtained from the nanoindentation process according

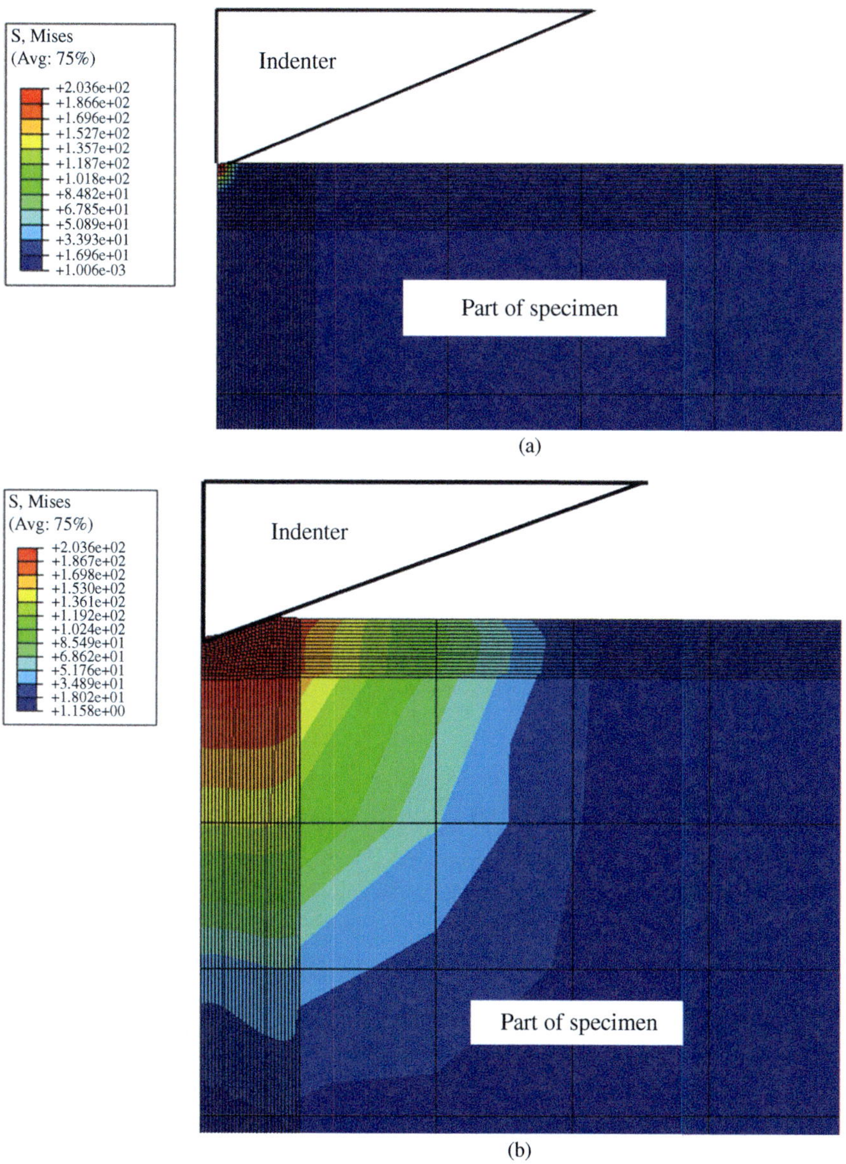

Figure 24.9 Stress contours and deformation of the material in different stages of the finite element model of the nanoindentation experiment.

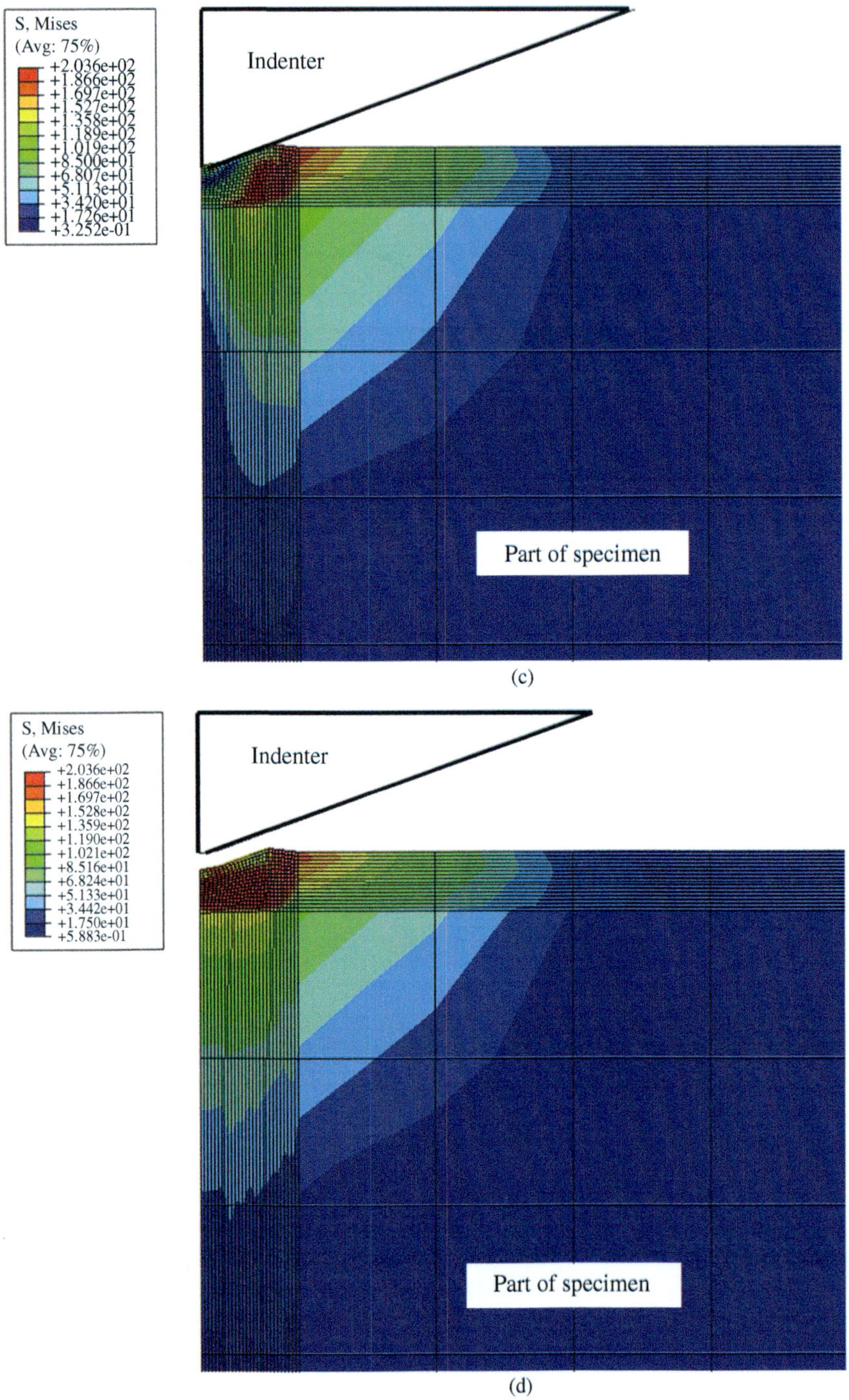

Figure 24.9 *(Continued)*

Table 24.1 Results obtained for material hardness from finite element simulation and from the nanoindentation experiment.

Conditions	Maximum indentation depth (nm)	Hardness (GPa)
Finite element model, Friction coefficient=0	260.28	1.35
Finite element model, Friction coefficient=0.6	257.84	1.34
Experiment	252.46 ± 12.23	1.36 ± 0.13

to the following equation:

$$h_{max} = h_c + h_a \tag{24.5}$$

where h_{max} is the maximum indentation depth and h_a is defined as:

$$h_a = \frac{\pi - 2}{\pi} h_e \tag{24.6}$$

In Equation (24.6), h_e is the elastic displacement during unloading which can be obtained from:

$$h_e = h_{max} - h_r \tag{24.7}$$

where h_r is the depth of the residual impression measured from the nanoindentation test. In the finite element simulation, the values of h_{max} and h_r in Equation (24.7) were obtained from the displacements of point A in Figure 24.8 after the loading stage and unloading stage, respectively. The values of F in Equation (24.2) can also be obtained from the maximum reaction force at point B in Figure 24.7. Finally, the hardness value of aluminum 1100 is calculated from the finite element results by using Equation (24.2). Table 24.1 presents the results obtained for the hardness of aluminum 1100 based on the above description of the finite element model and from the nanoindentation experiment.

It is seen in the table that there is only a small difference between the hardness values obtained from the finite element model and from the experimental results. Therefore, the finite element simulation could be used to successfully estimate the hardness of aluminum 1100 using its stress-strain curve obtained from a uniaxial tensile test.

The slight difference between the hardness values presented in Table 24.1 can be attributed to the variation in the maximum indentation depths in the model and in the nanoindentation experiment which might be reduced by further mesh refinement [40].

24.5 Molecular Dynamic Modeling for Nanoindentation

MD models can be applied to the nanoindentation experiment for investigating the mechanical properties of thin films. For examples Shi and Falk [41] applied MD models to simulate the nanoindentation process for crystalline and metallic thin films to study the structural transformation and atom localizations during the indentation.

Peng *et al.* [42] used the MD simulation to model a three-dimensional nanoindentation test for the aluminium coating on the silicone substrate. Hwang *et al.* [43] simulated the copper cluster deposition on the silicone substrate by using the molecular dynamics approach.

24.5.1 Simulation Procedure

In this section, the MD method was described for simulation of nanoindentation experiment on aluminum 1100. The MD simulation model which was applied in this study consists of two different shapes of Berkovich indenter, including a sharp-tip and a rounded-tip (Figure 24.10). The conical indenter tip with a half-angle of 70.3° and made of diamond atoms of single crystal diamond structure was considered, which has a lattice parameter of 5.43 Å. The indenter tip is supposed to behave as a rigid tip so its shape does not change during the indentation process.

For modeling the interaction between the aluminum 1100 atoms, Morse interatomic potential was employed. This potential can be written as:

$$V(r) = D\left[e^{-2\alpha(r-r_0)} - 2e^{-\alpha(r-r_0)}\right] \tag{24.8}$$

where *V(r)* is the potential energy function, r is the interatomic distance between the atoms i and j, and D, α, and r_0 represent the cohesive energy, elasticity modulus, and equilibrium distance, respectively. In the Morse function, many parameters are needed for different atomic species interactions. These parameters can be calculated by the Lorentz–Berteloth rules [44] as follows:

$$D_{A-B} = (D_A D_B)^{1/2} \tag{24.9}$$

$$\alpha_{A-B} = \frac{1}{2}(\alpha_A + \alpha_B) \tag{24.10}$$

$$r_{0A-B} = (\sigma_A \sigma_B)^{1/2} + \ln\left(\frac{2}{\alpha_{A-B}}\right) \tag{24.11}$$

$$\sigma_{A,B} = r_{0A,B} - \ln\left(\frac{2}{\alpha_{A-B}}\right) \tag{24.12}$$

where $\mathrm{D}_{\mathrm{A-B}}$ represents the cohesive energy of different atomic species, $\alpha_{\mathrm{A-B}}$ is the material parameter that is fitted for two different atoms of A and B, and $\mathrm{r}_{0\mathrm{A-B}}$ is the interaction equilibrium distance between A and B [44].

The parameters required for this MD simulation are shown in Table 24.2, which have been used successfully to study the mechanical behavior of nano-crystalline aluminum [45].

The dimension of the specimen was considered about 500 nm × 400 nm, in the x–y plane and the indenter moved in the y direction. The initial temperature of the work piece was equal to 293 K. The velocity and temperature of atoms through the simulation were re-scaled every 100 time steps, according to the initial temperature. Periodic boundary condition was used in the x and z directions across the lateral planes of the model. In the y direction, the top surface of the specimen was set to be free and its bottom plane was fixed to avoid any movement of the sample. The Verlet velocity algorithm with a time step of one femtosecond (fs) was applied for the time integration of Newton motion equations.

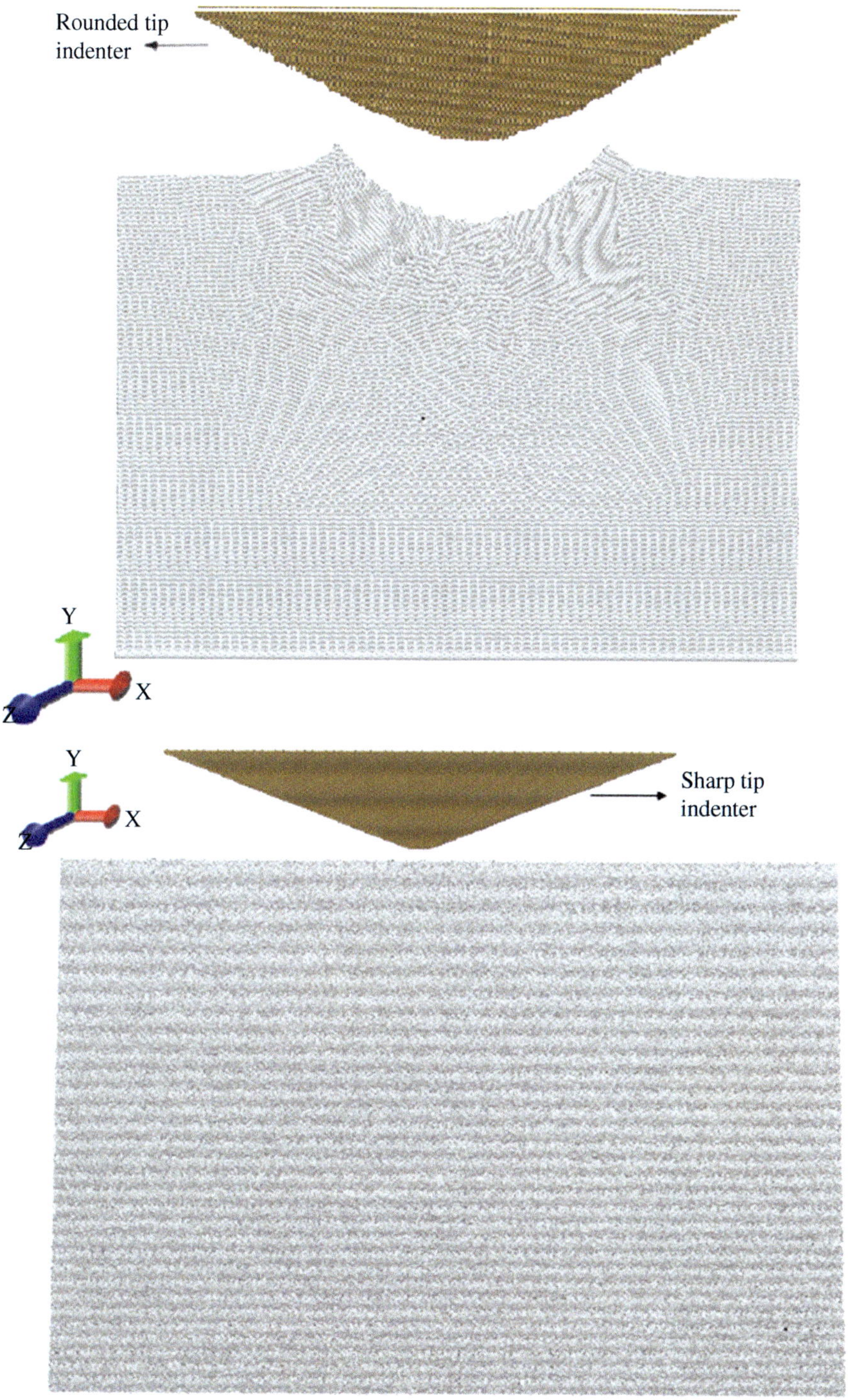

Figure 24.10 MD simulation models.

Table 24.2 Parameters of the Morse function for Al [45].

Material	D (eV)	Compressibility fitted parameter α (1/Å)	Equilibrium radius r_0 (Å)
Al – Al	0.2703	0.011646	32.53

Source: Fang 2008. Reproduced with permission of Elsevier.

At the first step, the simulated sample was left to reach equilibrium at room temperature. For the initial model, the velocities were randomly selected. The velocities of atoms were then such set to represent a total zero momentum of the sample and the ambient temperature. However, these velocities are not constant during the simulation. Therefore, it is necessary to let the model to reach a stable status before indentation. As an important parameter, the value of time step must be determined. Theoretically, smaller time steps provide more accurate results but they need more steps to finish the simulation process leading to higher computational costs. On the other hand, very large time steps will cause numerical instability. Normally, at the nano-level, the time step applied in the molecular dynamics is in the order of femto seconds. In this study, the time step was considered to be one femto second. The frequency of the neighbor list updating and velocity scaling was set to be 50 time steps. Thus, 0.1 picosecond was required to achieve an optimal balance.

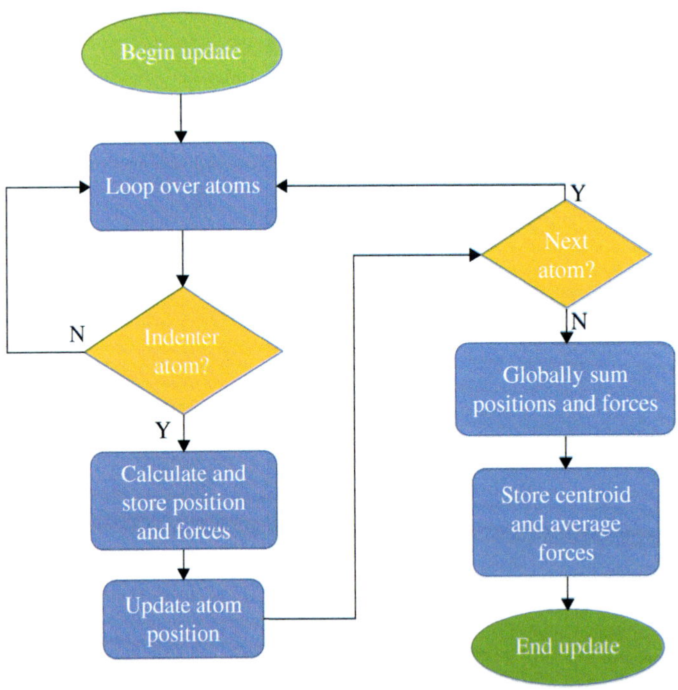

Figure 24.11 The MD simulation logic diagram.

At the start of the simulation, the indenter tip was positioned above the top surface of the specimen at the distance of one inter atomic cutoff radius, so that the indenter would not interact with the specimen before coming down. The indenter was inserted into the free surface of the thin film under displacement control at a constant speed. To maintain the computational time of the simulation within a reasonable range, the indentation speed was considered to be 50 m/s. Although, this speed is relatively high in a general indentation experiment, but it could be used in MD simulation of the nanoindentation procedure, because the indentation speed has no effect on the main features of atomic rearrangements [42]. The indenter moved into the specimen to a depth about 260 nm, which is the maximum indentation depth. At the maximum depth, the indenter was held for about 10 picoseconds in order to allow the model to reach re-equilibrium. Afterwards, the indenter was gradually removed at the same speed of loading until it no longer interacted with the thin film's atoms. The described MD procedure diagram is shown in Figure 24.11.

24.6 Results of Molecular Dynamic Simulation

Sample load-displacement curves obtained from the MD simulation of nanoindentation test are shown in Figure 24.12.

According to Figure 24.12, at the beginning of the indentation process, the load starts with attractive forces which are shown as negative values. Then the repulsive force starts to increase and becomes dominant. After the maximum indentation depth is reached and re-equilibrium takes place, the indenter is removed up from the specimen and the force decreases to zero. The loading parts of the curves are quite fluctuating because the interactions between the indenter and material are based on the discrete atoms. But the unloading parts of the curves are relatively smooth because the attractive force approaches zero in the presence of large interatomic distances.

For extracting the mechanical properties of the specimen, the continuous load-displacement data obtained from a complete cycle of loading and unloading

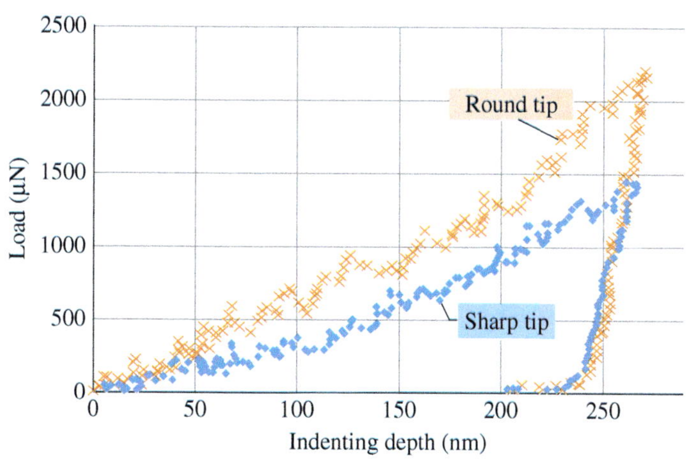

Figure 24.12 Load-displacement curves obtained from the MD simulation.

during the nanoindentation test was used. The effective elasticity modulus can be calculated by using [46]:

$$\frac{1}{E_{eff}} = \frac{1-\nu^2}{E} + \frac{1-\nu_i^2}{E_i} \tag{24.13}$$

where E and ν are the elasticity modulus and Poisson's ratio of samples, E_i and ν_i are elasticity modulus and Poisson's ratio of indenter tip and E_{eff} is the effective elasticity modulus obtained from the nanoindentation test that can be calculated by [47]:

$$E_{eff} = \frac{\sqrt{\pi}}{2}\frac{S}{\sqrt{A_c}} \tag{24.14}$$

A_c is the area of the indented hole and S is the stiffness of the specimen which is the main parameter obtained from the load-displacement curve. This value is the initial slope *S* in the unloading part of the load-displacement curve which is expressed by [47]:

$$S = \left|\frac{dF}{dh}\right|_{F=F_{max}} \tag{24.15}$$

where F_{max} is the maximum force during the nanoindentation test. For calculating the area of the indented hole within the thin film, two parameters should be considered: the indentation depth after the unloading and the geometry of the indenter tip. The procedure described above were used for both sharp-tip and rounded-tip indenters and the corresponding values of elasticity moduli were determined. Table 24.3 shows the calculated elasticity moduli for both indenters.

Equation (24.2) can also be used here for measuring the material hardness H from the MD simulation. The calculated values of hardness for both indenters are shown in the Table 24.4.

As is seen from Tables 24.3 and 24.4, the elastic modulus and hardness values of the aluminum 1100 specimen depend considerably on the type of the indenter tip. The

Table 24.3 Elasticity moduli of Al 1100 obtained from MD simulation.

Type of indenter	Elasticity moduli (GPa)
Sharp tip indenter	78.4
Rounded-tip indenter	85.9

Table 24.4 Hardness values of Al 1100 obtained from MD simulation.

Type of indenter	Hardness (GPa)
Sharp tip indenter	1.86
Rounded-tip indenter	2.02

values obtained from the sharp-tip conical indenter are smaller than the corresponding values of the rounded-tip indenter. This discrepancy can be interpreted by the size of the plastic deformation region at the indented zone. The plastic deformation region in the indentation procedure with the sharp-tip indenter is larger than the plastic deformation region caused by indentation with the rounded-tip indenter. Relatively similar trend was observed in the load-displacement curves obtained from FE simulations of the nanoindentation procedure by the sharp-tip and the rounded-tip indenter. FE simulation is based on the continuum mechanic theory, but the MD method is according to the inter-atomic interactions. Since the bases of FE modeling and MD simulation are completely different, it is not reasonable to compare the results of these two simulations. However, comparing the hardness values obtained from the nanoindentation experiment and from the simulation techniques shows that the FE simulation can estimate the experimental hardness better than the MD method. However, more detailed studies should be performed by these two simulation methods on different materials to find out the advantages and shortcomings of each method for simulating the nanoindentation procedure.

24.7 Conclusions

Finite element and molecular dynamic methods were employed for simulating the nanoindentation experiment on aluminum 1100 as a pilot material. The nanoindentation experiment was performed on the material. The hardness values obtained from the finite element simulation and the molecular dynamic method were compared with the corresponding value obtained from the experiment. Results showed that the finite element method can estimate the experimental hardness better than the molecular dynamic method.

References

1 Pharr GM. Measurement of mechanical properties by ultra-low load indentation. Materials Science and Engineering 1998; A253:151–159.
2 Cuya JL, Mann AB, Livi KJ, *et al.* Nanoindentation mapping of the mechanical properties of human molar tooth enamel. Archives of Oral Biology 2002; 47: 281–291.
3 Ebenstein DM, Pruitt LA. Nanoindentation of biological materials. Nano Today 2006; 1: 26–33.
4 Ayatollahi MR, Alishahi E, Doagou-R S, Shadlou S. Tribological and mechanical properties of low content nanodiamond/epoxy nanocomposites. Composites, Part B 2012; 43: 3425–3430.
5 Ayatollahi MR, Doagou-Rad S, Shadlou S. Nano/micro scale investigation of tribological and mechanical properties of epoxy/MWNT nanocomposites. Macromolecular Materials and Engineering 2012; 297: 689–701.
6 Ayatollahi MR, Karimzadeh A. Determination of fracture toughness of bone cement by nanoindentation test. International Journal of Fracture 2012; 175:193–198.
7 Karimzadeh A, Ayatollahi MR. Investigation of mechanical and tribological properties of bone cement by nanoindentation and nano-scratch experiments. Polymer Testing 2012; 31: 828–833.

8 Park HS, Liu WK. An introduction and tutorial on multiple-scale analysis in solids. Computer Methods in Applied Mechanics and Engineering 2004; 193:1733–72.
9 Marteau J, Bigerelle M, Xia Y, *et al.* Quantification of first contact detection errors on hardness and indentation size effect measurements. Tribology International 2013;59:154–162.
10 Maekawa K, Itoh A. Friction and tool wear in nano-scale machining: a molecular dynamics approach. Wear 1995; 188: 115–122.
11 Jeong-Won K, Ho-Jung H. Mechanical deformation study of copper nanowire using atomistic simulation. Nanotechnology 2001; 12: 295.
12 Diao J, Gall K, L. Dunn M. Atomistic simulation of the structure and elastic properties of gold nanowires. Journal of the Mechanics and Physics of Solids 2004; 52:1935–1962.
13 Rafii-Tabar H. Modelling the nano-scale phenomena in condensed matter physics via computer-based numerical simulations. Physics Reports 2000; 325: 239–310.
14 Sun Y, Bell T, Zheng S. Finite element analysis of the critical ratio of coating thickness to indentation depth for coating property measurement by nanoindentation. Thin Solid Films 1995; 258: 198–204.
15 Bolshakov A, Oliver WC, and Pharr GM. Influences of stress on the measurement of mechanical properties using nanoindentation: Part II Finite element simulations. Journal of Material Research 1996; 11: 760–768.
16 Moy CKS, Bocciarelli M, Ringer SP, Ranzi G. Indentation and imprint mapping for the identification of material properties in multi-layered systems. Computational Materials Science 2011; 50: 1681–1691.
17 Lichinchi M, Lenardi C, Haupt J, Vitali R. Simulation of Berkovich nanoindentation experiments on thin films using finite element method. Thin Solid Films 1998; 333: 278–286.
18 Vlachos DE, Markopoulos YP, Kostopoulos V. 3-D Modeling of Nanoindentation Experiment on a Coating-substrate System. Computational Mechanics 2001; 27: 138–144.
19 Bhattacharya AK, Nix WD. Finite element analysis of cone indentation. International Journal of Solids and Structures 1991; 24: 1047–1058.
20 Knapp JA, Follstaedt DM, Myers SM, *et al.* Finite-element modeling of nanoindentation. Journal of Applied Physics 1999; 85: 1460–1474.
21 Karimpour M, Balint DS, Rzepiejewska-Malyska KA, *et al.* An inverse method for extracting the mechanical properties of the constituent materials of a multilayer from nanoindentation data. Computational Materials Science 2013; 68: 384–390.
22 Stauss S, Schwaller P, Bucaille JL, *et al.* Determining the stress-strain behaviour of small devices by nanoindentation in combination with inverse methods. Microelectronic Engineering 2003; 67–68: 818–825.
23 Zhang J, Niebur GL, Ovaert TC. Mechanical property determination of bone through nano- and micro-indentation testing and finite element simulation. Journal of Biomechanics 2008; 41: 267–275.
24 Fischer-Cripps AC. Nanoindentation. 3rd edn. New York: Springer; 2011.
25 Lichinchi M, Lenardi C, Haupt J, Vitali R. Simulation of Berkovich nanoindentation experiments on thin films using finite element method. Thin Solid Films 1998; 312: 240–248.

26 09 AEEM. Standard test methods for tension testing of metallic materials. ASTM, West Conshohocken, PA,USA; 2009, pp. 1–27.
27 Daly BJ. A Numerical studyof Rayleigh-Taylor instabilityin aluminum and steel plates. Los Alamos National Laboratory, NM, USA; 1998.
28 Khan MK, Hainsworth SV, Fitzpatrick ME, Edwards L. A combined experimental and finite element approach for determining mechanical properties of aluminium alloys by nanoindentation. Computational Materials Science 2010; 49: 751–760.
29 Karthik V, Visweswaran P, Bhushan A, *et al.* Finite element analysis of spherical indentation to study pile-up/sink-in phenomena in steels and experimental validation. International Journal of Mechanical Sciences 2012; 54: 74–83.
30 Marteau J, Bouvier S, Bigerelle M. Review on numerical modeling of instrumented indentation tests for elastoplastic material behavior identification. Archives of Computational Methods in Engineering 2014: 1–17.
31 Lu Y, Kurapati S, Yang F. Finite element analysis of cylindrical indentation for determining plastic properties of materials in small volumes. Journal of Physics D: Applied Physics 2008, 41(11): 115415.
32 Liu Y, Wang B, Yoshino M, *et al.* Combined numerical simulation and nanoindentation for determining mechanical properties of single crystal copper at mesoscale. Journal of the Mechanics and Physics of Solids 2005; 53: 2718–2741.
33 Bressan JD, Tramontin A, Rosa C. Modeling of nanoindentation of bulk and thin film by finite element method. Wear 2005; 258: 115–122.
34 Lee H, Haeng Lee J, Pharr GM. A numerical approach to spherical indentation techniques for material property evaluation. Journal of the Mechanics and Physics of Solids 2005; 53: 2037–2069.
35 ISO-14577. Metallic materials : Instrumented indentation test for hardness and materials parameter. Part 2: Verification and calibration of testing machines; Part 3: Calibration of reference blocks. Geneva, International Standards Organization; 2002.
36 ISO-14577. Metallic Materials: Instrumented indentation test for hardness and materials parameters Part 1: Test Method. Geneva, International Standards Organization; 2002.
37 Hu Y, Shen L, Yang H, *et al.* Nanoindentation studies on Nylon 11/clay nanocomposites. Polymer Testing 2006; 25: 492.
38 Liu TX, Phang IY, Shen L, *et al.* Morphology and mechanical properties of multiwalled carbon nanotubes reinforced nylon-6 composites. Macromolecules 2004; 37: 7214.
39 Oliver WC, Pharr GM. Measurement of hardness and elastic modulus by instrumented indentation: Advances in understanding and refinements to methodology. Journal of Materials Research 2004; 19: 3–20.
40 Bhattacharya AK, Nix WD. Finite element simulation of indentation experiments. International Journal of Solids and Structures 1988; 24: 881–891.
41 Shi YF, Falk ML. Structural transformation and localization during simulated nanoindentation of a noncrystalline metal film. Applied Physics Letters 2005; 86.
42 Peng P, Liao G, Shi T, *et al.* Molecular dynamic simulations of nanoindentation in aluminum thin film on silicon substrate. Applied Surface Science 2010; 256: 6284–6290.
43 Hwang S-F, Li Y-H, Hong Z-H. Molecular dynamic simulation for Cu cluster deposition on Si substrate. Computational Materials Science 2012; 56: 85–94.

44 Ziegenhain G, Hartmaier A, Urbassek HM. Pair vs many-body potentials: Influence on elastic and plastic behavior in nanoindentation of f.c.c. metals. Journal of the Mechanics and Physics of Solids 2009; 57: 1514–1526.
45 Fang T-H, Wu J-H. Molecular dynamics simulations on nanoindentation mechanisms of multilayered films. Computational Materials Science 2008; 43: 785–790.
46 Oliver WC, Pharr GM. An improved technique for determining hardness and elastic modulus using load and displacement sensing indentation experiments. Journal of Materials Research 1992; 7: 1564–1583.
47 Wymysłowski A, Dowhań Ł. Application of nanoindentation technique for investigation of elasto-plastic properties of the selected thin film materials. Microelectronics Reliability 2013; 53: 443–51.

25

Atomistic Simulations of Adhesion, Indentation and Wear at the Nanoscale

Jun Zhong[1], *Donald J. Siegel*[2], *Louis G. Hector, Jr.*[3] *and James B. Adams*[4]

[1] *Associate Professor, School of Materials Engineering, North China Institute of Aerospace Engineering, Langfang, Hebei China*
[2] *Associate Professor, Department of Mechanical Engineering, University of Michigan, Ann Arbor, Michigan USA*
[3] *Senior Research Scientist, Materials and Processes Laboratory, General Motor R&D Center, Warren, Michigan USA*
[4] *President's Professor, School for Engineering of Matter, Transport and Energy, Arizona State University, Tempe, USA*

25.1 Introduction

Nanoscale indentation has been widely applied to identifying mechanical properties and stability of surfaces or thin films during metallic manufactures [1–8]. Throughout experimental and theoretical works, significant efforts have been devoted to exploring nanomechanisms at the interfacial area where two solid surfaces contact. With the aid of some experimental instruments like the scanning tunneling microscope (STM) and the atomic force microscope (AFM), etc., several interesting phenomena at nanoscale were revealed. For examples, the AFM was the first to be used for investigating atomic-scale friction of a tungsten-tip on a graphite surface at low loads, $<10^{-4}$ N, where the lateral frictional force exhibited a stick-slip behavior with the lattice periodicity [9], i.e., when the magnitude of frictional force exceeded a critical value, a sudden tip slip across the surface could take place. In addition, Hector and Schmid [10, 11] developed an AFM technique to study the abrasive wear of metals at nanoscale, with an indentor plowing throughout some surfaces of Al-alloy samples. They observed that the frictional force, the hardness and the plowing track morphology were dependent upon not only alloy-types, but crystallographic directions along which the plowing process occurred with liquid chemical additives onto surfaces [12, 13]. Regarding a relationship between the measured load P on an indentor and the displacement h of the indentor at nanoscale [14–16], the load-displacement (P-h) curve showed that some periodic bursts appeared in penetration depths due to the indentor, which were interpreted as a consequence of dislocation nucleation and the splitting of a perfect dislocation loop [17–19]; While the indentor retraction from the substrate indicated that, the tensile force increased with a constant slope at the beginning. And then it suddenly relaxed: at this point, the contact area exhibited a sequence of structural transitions involving elastic and yielding stages; each curve jump was associated with atomic rearrangement and release of tensile stress.

Applied Nanoindentation in Advanced Materials, First Edition. Edited by Atul Tiwari and Sridhar Natarajan.

Corresponding to experimental studies, during the past two decades, advances in theoretical methodologies like algorithms, interatomic potentials and computational powers etc. have sustained simulations of indentation and wear at nano-dimension and nano-time. Particularly, molecular dynamics (MD) simulation by means of empirically interatomic potentials allowed it for up to millions of atoms and nanoseconds on current massively parallel architectures [20, 21]. As results, for examples, a very top research was a nickel-tip indentation onto a gold-substrate [22, 23], in which people observed the formation of a connective neck at the tip-substrate interface due to strong attractive bonding between two contact materials. In another research, when two Ni(100) crystals approached within a critical distance, an adhesive avalanche occurred whereby atomic layers at contact area fell together across the interface gap between two crystal surfaces [24]. Rafii-Tabar *et al.* [25, 26] ran an Ir-tip motion to contact a Pb-substrate at room temperature. They also observed a phenomenon of jump-to-contact (JC) instability in metallic systems during the approach phase, which was driven by a marked tendency of atoms within an interface gap where tip and substrate formed bonds and thereby reduced their surface areas. They also found that, during a 'pull-out' phase, a substantial neck was formed due to adhesive force. Similarly, Buldum's work [27, 28] also observed the phenomenon of jump-to-contact during the nanoindentation, in which transformations of two consecutive structures occurred periodically but stopped with a layer-wear during a sharp tip sliding. Komanduri *et al.* found that, mainly due to a higher normal force, the frictional coefficient of a friction at nanoscale was rather high due to the breaking and the reforming of atomic bonds, plus a high rake angle [29]. Relatively, the largest simulation of nanoindentation for a Si_3N_4 slab with ten million atoms revealed a significant plastic deformation and a pressure-induced amorphism under a rigid indentor [30]. Besides, Ortiz and Phillips [31–34] developed a so-called 'quasi-continuum method' to investigate an example of nanoindentation, in which a mixed atomistic-continuous (multi-scale) formulation was simulated through a mesh variable that allowed details at atomic-level in the focusing regions for ones at continuum-scale in others. Comprehensively, there have recently appeared several atomistic simulations relevant to nanoindentation, and some very interesting results have been reported, especially several phenomena like the jump-to-contact (JC), the piled-up under an indentor (a neck formation and local melting), the non-monotonic features on the *P-h* curve, and the phase transformation on the indented material substrates (adhesive transfer), etc.

In addition, adhesion and wear also come with indentation at contact areas. For examples, adhesion is critical to successful manufacture for semiconductor circuits: each deposited layer must adhere strongly to that beneath it to survive thermal stresses. Similarly, wear has a tremendous effect on the US industry because it determines the lifetime of many parts, from components of automotive engines to airplane brakes: a particular example is the aluminum-alloys processing which is also relevant to many other metallic alloys, i.e., the Al-alloy products are manufactured through a series of processing steps that begin with casting an aluminum ingot. In order to reduce the ingot thickness, it passes through a series of rolling operations under high loads at speeds up to 22.86 meter per sec (Figure 25.1). In this case, there will take place adhesive wear (aluminum bonds to steel surfaces) and abrasive wear (hard surface asperities in the tool are plowing through soft Al surfaces covered by chemical additives). On the positive side, many machining operations rely upon adhesive wear processes (from oil drilling to polishing an Si-wafer): the processing optimization is critical to commercial success.

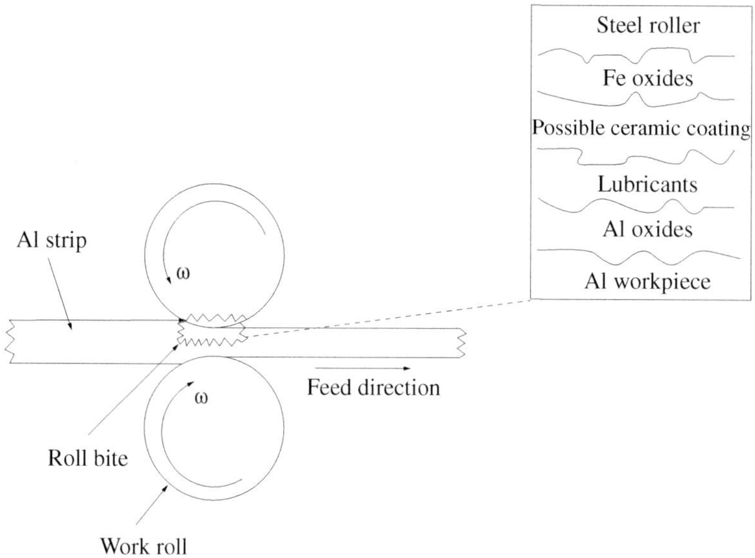

Figure 25.1 The strip rolling process and a realistic toolpiece/workpiece interface.

While on the negative side, there is an increasing interest in dry machining because oils used in the machining are easily evaporated due to frictional heating, and there is a rapidly rising concern about the impact of such vapors on the long-term health of machine operators. Therefore, it is clear that better understandings of mechanisms in adhesion and wear will have positive impacts on many industries and help people fulfill 'greener' manufacturing.

After all, there is still a lack of fundamental knowledge about nanocontact mechanics which links the underlying atomistic mechanisms to nanoscale phenomena, e.g., for the nanoindentation, which in most of the above only investigated an effect of a single variable such as a single temperature, a single indent rate, a single crystal orientation, and a single materials combination, etc. Therefore, it is very important that we carry out a systematic research to investigate, qualitatively and quantitatively, many possible factors that may control nanoindentation behaviors. In the meanwhile, relatively little is understood about atomic-scale mechanisms that occur during adhesion and wear processes because these fields remain dominated by a so-called "yielding short-term solutions." Therefore, we believe that there is a great need for fundamental studies of critical mechanisms involving adhesion and wear at atomic-scale.

In this chapter, three general areas that will be reviewed are adhesion, indentation and wear, with a focus on our group's recent works on aluminum and other alloys: (1) Adhesion, since there is no simple analytical method to predict adhesion behavior between two different contact materials, we will discuss the application of electronic structure to this field between aluminum and various ceramics, to determine factors that control this phenomenon. (2) Indentation, see the part (a) in Figure 25.2, here we focus MD simulation with an empirically interatomic potential on the indentation of a rigid tip into a relatively soft Al-substrate to discuss effects of system (environment) temperature, loading force (stress), tip-substrate bonding and orientation in substrate at nanoscale. Furthermore, another simulation result from *ab-inito* molecular

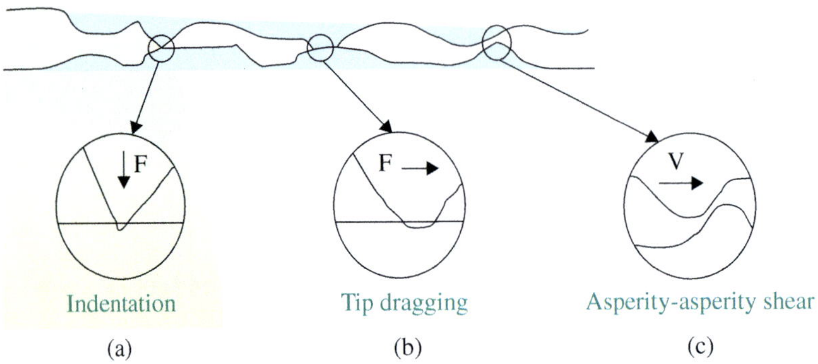

Figure 25.2 Three types of contact processes (left to right): tip-indentation, tip-plowing/dragging and asperity–asperity shear.

dynamics (AIMD) based upon density functional theory (DFT) is presented to extract rich information (e.g., bond breaking and forming and materials transfer etc.) at the reacting interfacial area where two different surfaces of metal slabs compress. (3) Wear, we consider a single crystal Al-asperity to be plowed by a rigid tip at a constant translational velocity (see part (c) in Figure 25.2). Effects of several factors during wear process are investigated, including velocity (V_d), temperature (T), crystal orientation, tip-tip intersection (O_I) and tip-tip bonding (D_{int}). Furthermore, temperature profiles and microdeformed mechanisms of asperity shear are discussed.

In the following sections, although we focus on aluminum samples, without loss of generality, we believe what we discuss is also relevant to many other materials in which adhesion, indentation and wear may occur.

25.2 Methodologies

25.2.1 Density Functional Theory

Materials consist of a huge number of nuclei and electrons with diverse interactions. With the development of quantum mechanics in early 1900s, a law governing those interactions was made clear, i.e., the so-called many-particle Schrödinger equations. In order to solve these equations efficiently, the Born–Oppenheimer (BO) approximation [35] provided a way to separate ionic degrees of the freedom from electronic ones. Even if, an exact solution for such a quantum many-particle problem is still unrealized by means of any combination of wave function because of very limited data storage and very limited computational power in current computational resources. In early 1960s, *density functional theory* (DFT) was constituted to replace *the all-electron wave function* (AE-WF) with the charge density as a fundamental variable. According to this framework, the DFT may drastically reduce degrees of the freedom in a system but still keep its good accuracy and efficiency. Till now DFT has become a more powerful method than others to study most of large molecular clusters and solid systems.

DFT was based upon the theorem developed by Walter Kohn [36, 37], in which it stated that all the ground-state properties of a system were functionals of a solely charge density: $n(\boldsymbol{r})$. This $n(\boldsymbol{r})$, which minimized the energy functional under a given external potential: $V_{\text{ext}}(\boldsymbol{r})$, was named *the ground-state charge density*. Therefore, the total energy of a system can be expressed as a functional of the charge density $n(\boldsymbol{r})$, i.e.:

$$E[n(r)] = \langle \Psi^{el} | \hat{H}^{el} | \Psi^{el} \rangle = \overline{T[n(r)]} + \overline{V_{ee}[n(r)]} + \int V_{ext}(r)\, n(r)\, \mathrm{d}r. \tag{25.1}$$

By using the Kohn-Sham equations [37], one replaced the kinetic term $\overline{T[n(r)]}$ with a kinetic energy of non-interacting electron gas $E_{KE}[n(\boldsymbol{r})]$; the electron-electron interacting potential $\overline{V_{ee}[n(r)]}$ was replaced with a classical Coulombic part: $E_H[n(\boldsymbol{r})]$ – *the Hartree energy*, and an extra non-classical part: $E_{XC}[n(\boldsymbol{r})]$ – *the exchange-correlation* (XC) *energy*. Therefore, Equation (25.1) can be modified as the following formula:

$$E[n(\boldsymbol{r})] = E_{KE}[n(\boldsymbol{r})] + E_H[n(\boldsymbol{r})] + E_{XC}[n(\boldsymbol{r})] + \int V_{ext}(\boldsymbol{r})\, n(\boldsymbol{r})\, \mathrm{d}\boldsymbol{r}. \tag{25.2}$$

The advantage of this modification was that, the first two and the last terms in Equation (25.2) can be dealt with a traditional manner; while the third term, which contributed a small effect of complex quantum on $E[n(\boldsymbol{r})]$, had to be simplified through several approximations.

In order to solve the total energy $E[n(\boldsymbol{r})]$ in Equation (25.2) more accurately, the N single-particle Kohn–Sham equations adopted a set of Lagrange multipliers $\{\varepsilon_i\}$ which corresponded to the orthonormality of the N single-particle orbitals $\{\varphi_i\}$ plus an effective Kohn–Sham Hamiltonian operator H_e^{KS}, so as to have the ground-state energy $E[n(\boldsymbol{r})]$ minimized through the charge density $n(\boldsymbol{r})$, see ref [66]. One advantage of this method was that, as the complexity of a system increased, there was only a linear increase in the number of single-particle equations.

Even if such a Kohn-Sham ansatz is exact in principle for systems at ground-state, it has to be approximated in practice because there is still unknown the exact functional term $E_{XC}[n(\boldsymbol{r})]$ in Equation (25.2). According to this, a precise prediction for the XC energy $E_{XC}[n(\boldsymbol{r})]$ will become crucial for solving material properties. Even though some scientists were devoting themselves to extending the DFT to treat excited states in materials [38], most of current works mainly focus on systems at the ground-state.

25.2.1.1 The Exchange–correlation Functional

Till now several approximate XC methods have been developed. Most commonly used in DFT calculations was so-called *the local density approximation* (LDA) [37, 49], which was based upon a so-called *uniform electron gas*. Theoretically, the LDA was precise for a uniform electron gas and arbitrarily accurate for a system with significant-slowly varying charge density [39]. However, the charge density $n(\boldsymbol{r})$ usually behaved like a rather rapidly varying function in several atomic, molecular clusters and condensed matter systems. Thus in practical calculations, some kinds of corrections to the $n(\boldsymbol{r})$ gradient were required to reduce the LDA error.

One famous extension to the LDA was known as *the generalized-gradient approximation* (GGA) [40, 49], which was developed to restore the sum rule and the non-positive hole conditions [66]. An important GGA functional (PW91) predominantly used in solid state calculations was created by Perdew and Wang [40–42]. While it was recently superseded with a modified formula (PBE) suggested by Perdew, Burke and Ernzerhof [43–45], which differed from the traditional PW91 with a much simplified form of an exchange enhancement factor. Comparing the GGA to the LDA, the former has improved much more accuracy in property calculations for solids and molecular clusters than the latter [40–45].

25.2.1.2 Plane Waves and Supercell

The method of Plane Waves (PW) basis-sets, which was more suitable for calculations of periodic solids in reciprocal space than others on extending its accuracy to any desired convergence, has been widely used for analyzing adhesion properties through calculations of various electronic structures. In general, the representation of an arbitrary orbital in terms of the PW basis-sets would require continuous (infinite) basis-sets. According to Blöch's theorem [46], in a periodic solid one may describe a single-electron wave function as the multiplication of *a cell periodic-function* with *a wavelike factor* as follows:

$$\psi_{n,k}(\boldsymbol{r}) = \varphi_{n,k}(\boldsymbol{r}) \cdot e^{i\boldsymbol{k}\cdot\boldsymbol{r}}, \tag{25.3}$$

where $\boldsymbol{k}$ is the wave vector, n is the band index. And, the cell periodic-function $\varphi_{n,\boldsymbol{k}}(\boldsymbol{r})$ must meet the periodic property: $\varphi_{n,\boldsymbol{k}}(\boldsymbol{r}) = \varphi_{n,\boldsymbol{k}}(\boldsymbol{r}+\boldsymbol{a})$ for any lattice vector $\boldsymbol{a}$ in real solid. According to solid state theory [46], the single-electron wave function can be discretely expanded through a serial number of the PW basis-sets (the Fourier components) whose wave vectors lay in the reciprocal lattice, i.e.:

$$\psi_{n,\boldsymbol{k}}(\boldsymbol{r}) = \frac{1}{\sqrt{\Omega}}\sum_{\boldsymbol{G}} c_{n,\boldsymbol{k}}(\boldsymbol{G}) \cdot e^{i(\boldsymbol{k}+\boldsymbol{G})\cdot\boldsymbol{r}}, \tag{25.4}$$

where Ω is the volume of the primitive cell in real crystal lattice, $\boldsymbol{G}$ is the reciprocal-lattice vector. Therefore, solving the single-electron wave function in Equation (25.4) would involve expressing this wave function with an infinite number of $\boldsymbol{k}$ vectors within the first reduced Brillouin zone of the primitive cell in reciprocal space. To simplify this problem with very accurate approximations, it can be handled by sampling some special k-points in the first reduced Brillouin zone to determine electronic wave function and electric potential through the filled electronic bands. Two widely used methods were Chadi–Cohen [47] and Monkhorst–Pack [48] samplings.

In practical calculation of a large supercell, for the wave vector selection, it was usually sufficient to only consider a particular value-point, the $\boldsymbol{\Gamma}$ point, because, this choice would confer significant savings in data storage and computational cost since this $\boldsymbol{\Gamma}$ point was the origin position in $\boldsymbol{k}$-space.

25.2.2 Pseudopotential Approximation

The Plane Waves (PW) method cannot directly be carried out in the KS theorem since the wave functions of ionic-cores were rapidly varying with many nodes and required enormous PW basis-sets to reach the calculation accuracy. In real practice, the total energy associated with ionic-core orbitals were usually several orders of magnitude larger than those associated with valence orbitals. However, since chemical reactions were solely related to valence electrons while ionic-core electrons were approximately 'frozen' in their original configurations, a *pseudopotential approximation* (PP) would just reproduce the identical curves of valence wave functions in the place sufficiently far from the ionic-core region. In the PP framework, when closing to the ionic-core region, strong oscillations of valence orbitals due to orthogonality in case of all electronic interactions should be smoothed out, but the scattering properties were retained as well. In addition, exchange interactions between ionic-cores and valence electrons must be included in the constituted potential model. To satisfy these, the constructed charge densities outside the ionic-core region by the PP scheme must be identical to the true charge densities, which were called "the norm-conserving condition." So the pseudopotential was usually represented as the following formula [50]:

$$V(\mathrm{r}) = V^{\mathrm{local}}(\mathrm{r}) + \sum_{l=0}^{l_{max}} V_l^{\mathrm{non\text{-}local}}(\mathrm{r})\hat{P}_l, \tag{25.5}$$

where $\hat{\boldsymbol{P}}_l$ is an operator projecting electron wave functions onto eigenfunctions of different angular momentum l.

The biggest merit of pseudopotentials was their formal simplicity. Unfortunately, this simplicity came at a price: the first-row elements, transition elements, and rare-earth elements were computationally demanding to treat with the standard norm-conserving pseudopotentials, i.e., it required a large PW basis-sets. Compromises were made by increasing the core radius significantly beyond the outermost maximum in the AE wave function. But this was usually not a satisfactory solution because the *transferability* was always adversely affected [50]. Therefore, various attempts have been made to generate the *soft* pseudopotentials, most successful of which was the concept of "the ultrasoft-pseudopotential (US-PP)" introduced by Vanderbilt [51]. Blöchl further expanded the US-PP concept by combining ideas from the PP and the LAPW (linearized augmented-plane-wave) methods in a conceptually elegant framework, called "the projector augmented-wave method (PAW)" [52]. And then, Kresse established an exactly formal relationship between the US-PP and the PAW methods [53].

25.2.3 Molecular Dynamics

Molecular dynamics (MD) is a methodology which depicts motions of a many-particle system using Newton's classical equations. It has been over 40 years since the first application of MD to a system of hard sphere by Alder and Wainwright [54, 118]. The MD method is particularly useful for the study of dynamical properties of materials. More important, the study may help ones extract physical-insights from many particle systems.

25.2.3.1 Equations of Motion

In MD simulations, classical trajectories of particles in real space for a finite size system were generated by solving the Lagrange equations numerically. When choosing the Cartesian coordinates, these equations would become Newton's equations:

$$m_i\ddot{\boldsymbol{r}}_i = \boldsymbol{F}_i = \dot{\boldsymbol{P}}_i = -\frac{\partial V_{tot}}{\partial \boldsymbol{r}_i}, \tag{25.6}$$

where V_{tot} is the total potential energy of a system, m_i is the mass of particle i, $\boldsymbol{P}_i$ is the momentum of particle i, and $\boldsymbol{F}_i$ is the total inter-particle force acting on particle i, and the " · " means the first order time derivative; " ·· " means the second order time derivative. The convenience of using Equation (25.6) depends upon particular algorithms adopted in solving them.

25.2.3.2 Algorithms

A standard way of solving Equation (25.6) was the finite-difference method: under given a configuration (positions and velocities of all particles) of a system and other dynamical information at time t, the numerical integration would determine the new configuration of the system at a later time $t + \Delta t$ (Δt is the time step). Commonly used methods in MD were the *Verlet algorithm* [55], the *Leapfrog algorithm* [56], and the *Gear predictor–corrector* (GPC) *algorithm* [57]. An ideal algorithm should be simple, run fast, and require little memory. It should also permit using a long time step Δt to hold the whole trajectory of the system as close as possible, and preserve conservation laws of momentum and energy. According to these, the Verlet algorithm can be regarded as the most widely used one. The leapfrog algorithm was essentially identical to the Verlet algorithm. The GPC method was usually more accurate as well as more complicated than those two algorithms.

25.2.3.3 Statistical Ensembles

The widely used statistical ensembles in MD were usually the microcanonical ensemble (NVE), the canonical ensemble (NVT), the isothermal-isobaric ensemble (NPT) and the grand canonical ensemble (μVT). Thermodynamic variables in parenthesis for each ensemble were fixed in that ensemble. In MD simulation, the (NVE) ensemble was most conveniently realized since all equations of motion conserve the total energy E and particle's number N. The constant volume V can be fixed using free or periodic boundary conditions.

The (NVT) ensemble was very commonly used for practical calculations. Therefore, to control the temperature T, there were a number of ways to simulate it: (1) The simplest one was the velocity scaling method [58], which was simple but crude in MD simulation. (2) The Nosé-Hoover method was more conceptually fundamental [59], but it had an adjustable parameter of thermal inertia in practice.

For the (NPT) ensemble, Anderson [60] provided an 'extended system' method for a constant pressure MD. In this method, an 'external piston' term was added into the Lagrange functional for adjusting a value of external pressure. In another way, Parrinello and Rahman [61] proposed a scheme of 'variable cell constant MD' which allowed the simulation box to change shape as well as size. This method, so far, is the most general constant pressure method in MD simulation.

The (μVT) is very useful for an open system: in this ensemble, the total particle's number N will not be conserved, regarding to this, see ref [101] for details.

25.2.3.4 Interatomic Potentials

Traditionally, the total potential energy of a system V_{tot} in MD simulation was calculated using empirical models. Till now there have constituted many kinds of models like pair-wise potentials: Lennard-Jones or Morse potentials; multi-body interactions [62], etc. In early 1980s, Daw and Baskes [63, 64, 119] proposed a semi-empirical potential model: the *embedded-atom method* (EAM) which reflected a many-body effect in materials. This model was usually written as two terms:

$$E_{\mathrm{tot}} = \sum_{i} F_i(\rho_i) + \frac{1}{2} \sum_{i,j(\neq i)} \phi(r_{ij}), \tag{25.7a}$$

where r_{ij} is the distance between atoms i and j. In Equation (25.7a), on the right-hand side, the first term is many-body in nature, which represents the embedding behavior of ionic-cores in valence electron gas, so the $F_i(\rho_i)$ is named 'the embedding function'; the second term represents the screening Coulombic repulsion among ionic-cores, which is a sum of pair-wise interactions. The total atomic-charge density ρ_i at position $\boldsymbol{r}_i$ can be expressed as:

$$\rho_i(r_i) = \sum_{j(\neq i)} f_j(r_{ij}), \tag{25.7b}$$

where $\rho_i\,(\boldsymbol{r}_i)$ was approximate to a sum of individual atomic-charge density $f_j\,(r_{ij})$ around neighbors of site i as opposed to the true charge distributions that resulted from the self-consistent *ab initio* calculations.

There have been several successful fitting techniques for the EAM models, among of which a particularly reliable method was developed by fitting to a large database of density functional calculated forces and experimental data – the *force-matching method*, see ref [65].

25.2.3.5 *Ab initio* Molecular Dynamics

In general, MD simulation based upon the force calculations through empirical potential is called *classical* molecular dynamics simulation, while that based upon the force calculations using density function theory is called *ab initio* molecular dynamics (AIMD) simulation. Comparing to them, the former can approximately simulate a large system (up to tens of millions of atoms) for long time steps (nanoseconds); while the latter can more accurately simulate a small system of hundreds of atoms, and is better at describing reactions involving the change of atomic bonding types.

In principle, *ab initio* molecular dynamics (AIMD) was an extension of *classical* molecular dynamics, which included the *first principle* derived potentials for the calculation of inter-particle forces and the total energy of a system, see refs [66, 67]. Moreover, in most of other theories on calculations of electronic structure, more powerful computers and wider acceptances of dynamical methods within quantum chemistry were used as well.

Born–Oppenheimer Molecular Dynamics In the AIMD, the total potential energy $V_{tot}(\boldsymbol{R}^N)$ of a system has the same physical meaning as the Kohn–Sham energy (E^{KS}) within the Born–Oppenheimer (BO) approximation. Since the Kohn–Sham energy depended only upon the nuclear positions and the defined hyper-surface for nuclear motions, the Lagrange functional of the Born–Oppenheimer molecular dynamics (BOMD) had the following formula:

$$L_{BO}(\boldsymbol{R}^N, \dot{\boldsymbol{R}}^N) = \sum_{I=1}^{N_I} \frac{1}{2} M_I \dot{\boldsymbol{R}}_I^2 - \min_{\{\varphi_i\}} E^{KS}[\{\varphi_i\}; \boldsymbol{R}^N] + \sum_{i,j} \Lambda_{ij}[\langle \varphi_i | \varphi_j \rangle - \delta_{ij}], \tag{25.8}$$

where $\{\Lambda_{ij}\}$ are the Lagrange-multipliers. So the minimization of Kohn–Sham energy, E^{KS}, was constraint to the orthogonal basis-sets $\{\varphi_i\}$. And then, Equation (25.6) for many-particles can be modified by means of Equation (25.8) as follows:

$$M_I \ddot{\boldsymbol{R}}_I^2 = -\frac{\partial}{\partial \boldsymbol{R}_I}[\min_{\{\varphi_i\}} E^{KS}[\{\varphi_i\};\ \boldsymbol{R}^N]] + \sum_{i,j} \Lambda_{ij} \frac{\partial}{\partial \boldsymbol{R}_I}[\langle \varphi_i | \varphi_j \rangle], \tag{25.9a}$$

And:

$$H_e^{KS} \varphi_i = \sum_j \Lambda_{ij} \varphi_j, \tag{25.9b}$$

where H_e^{KS} denotes an effective Kohn-Sham Hamiltonian operator for a single-electron. Based upon Eqs.(25.9a-b), a *classical* molecular dynamics program can be turned into the BOMD program by replacing the energy and the force routines with the corresponding ones from a quantum chemistry program. The accuracy of the force calculations used in the BOMD depended linearly upon the accuracy of minimizing the Kohn–Sham energy E^{KS}. To realize this, an algorithm of *the conjugate gradient* (CG) minimization was effectively applied to such calculations, see refs [67, 68].

Car–Parrinello Molecular Dynamics The basic idea of the Car–Parrinello molecular dynamics (CPMD) can be regarded as exploring a time-scale separation of fast-electronic and slow-nuclear motions by transforming it into a dynamical framework of "classical-mechanical and adiabatic energies [69]." In classical mechanics, forces on nuclei were obtained through the derivative of a Lagrange functional with respect to the nuclear positions. This suggested that, a functional derivative with respect to the orbitals which were interpreted as classical fields, yield the forces on those suitable orbitals given by the Lagrange functional. Thus the CPMD scheme postulated the following Lagrange functional:

$$L_{CP}[\dot{\boldsymbol{R}}^N, \boldsymbol{R}^N;\ \{\dot{\varphi}_i\}, \{\varphi_j\}] = \sum_{I=1}^{N_I} \tfrac{1}{2} M_I \dot{\boldsymbol{R}}_I^2 + \sum_{i=1}^{occ} \tfrac{1}{2}\mu\langle\dot{\varphi}_i|\dot{\varphi}_i\rangle - E^{KS}[\{\varphi_i\};\ \boldsymbol{R}^N] + \sum_{i,j}\Lambda_{ij}[\langle\varphi_i|\varphi_j\rangle - \delta_{ij}]. \tag{25.10}$$

Like in classical mechanics, both nuclear positions, electronic orbitals, and the Car–Parrinello's equations of motion were found to be the following formulae corresponding to the Newton's equation:

$$M_I\ddot{\boldsymbol{R}}_I^2 = -\frac{\partial}{\partial \boldsymbol{R}_I}\left[E^{KS}[\{\varphi_i\};\boldsymbol{R}^N]\right] + \sum_{i,j}\Lambda_{ij}\frac{\partial}{\partial \boldsymbol{R}_I}[\langle\varphi_i|\varphi_j\rangle], \tag{25.11a}$$

and:

$$\mu\ddot{\varphi}_i = -H_e^{KS}\varphi_i + \sum_j \Lambda_{ij}\varphi_j. \tag{25.11b}$$

where μ is the "fictitious mass" or "inertia parameter" assigned to orbital degrees of the freedom. The unit of this mass parameter μ is the energy multiplies the time square for reasons of dimensionality. Note that constraints within E^{KS} would lead to "constraint forces" in Equation (25.11a). In general, these constraints would depend upon both the Kohn–Sham orbitals and the nuclear positions through an overlap matrix of basis functions – the simulated annealing to search for coefficients of basis-sets that minimize electronic energy. In this terminology, "low electronic temperature" or "cold electrons" in Equation (25.11b) was close to its instantaneous minimum energy: $\min_{\{\varphi_i\}} E^{KS}[\{\varphi_i\};\ \boldsymbol{R}^N]$, i.e., in this case it is close to the exact Born–Oppenheimer hyper-surface. Therefore, a ground-state wave function optimized for the initial configuration of nuclei would stay close to its ground-state during time evolution if the system was kept at a sufficiently low temperature. The SHAKE algorithm was used to impose constraints on the system to ensure that the orbitals remained orthonormal. In practice, it was found that, the Car-Parrinello scheme was unnecessary for electronic configurations to be minimized in coefficient space at each MD time step even though this manner may give errors in force calculations on nuclei. However, it was rather strange to indicate that, errors on nuclear force calculations were canceled by the associated errors in depicting electronic motions. According to this, the Car–Parrinello scheme may tolerate force errors on nuclei motions in AIMD simulation.

In addition, the last terms on the right-hand side of Equation (25.9a) and Equation (25.11a) would vanish if the basis functions were independent of nuclear positions $\{\boldsymbol{R}_I\}$. Comparing the BOMD to the CPMD, the BOMD can be viewed as an alternative the CPMD scheme, which may give very accurate coefficients of basis-sets through minimizing the electronic energy at each MD time step.

25.2.4 Some Commercial Software

25.2.4.1 The VASP

Vienna Ab initio Simulation Package (VASP) is a commercial software under the framework of density functional theory to use the plane wave (PW) basis-sets for an expansion of the single-electron Kohn–Sham wave function and the pseudopotential to describe the electron-ion interactions. In the VASP, the ground-state charge density and the total energy of a system are calculated using an algorithm of the predicted conjugate gradient (CG) minimization coupled with a Pulay-like mixing scheme. Sampling the irreducible wedge of the first reduced Brillouin zone is performed with a regular Monkhorst–Park grid for special k-points. Due to numerical instability associated with integrating the step-function of the Fermi–Dirac distribution at 0 K, partial occupancies of the single-electron wave function are introduced with broadening an energy level of 0.1 eV. The ground-state atomic geometries are constructed through minimizing the Hellman–Feynman forces [70, 82] by the aid of an algorithm: either the conjugate gradient or the quasi-Newtonian steep descent methods. Finally, three separate approximations to the exchange-correlation (XC) energy are employed: the traditionally local density approximation (LDA) developed by Perdew and Zunger, the generalized-gradient approximation (GGA) of the PW91 functional, and the PBE functional, see the http://cms.mpi.univie.ac.at/vasp/ website.

In the VASP, several skills such as the analysis of Bond Strength (BS), the analysis of Electron Localization Function (ELF), the analysis of each Bond Twig (BT) vibration, and the analysis of Density of States (DOS) etc. are fulfilled for relevant research topics.

25.2.4.2 The LAMMPS

Large-scale Atomic/Molecular Massively Parallel Simulator (LAMMPS) is a computational program developed from the Sandia National Laboratories for implementing *classical* molecular dynamics simulation, see the http://lammps.sandia.gov/ website. It uses the MPI for parallel communication, and is free software with open-source, distributed under terms of the GNU general public license.

In the LAMMPS, a skill of neighbor lists is used to keep track of nearby particles in computational program only for the running efficiency. The lists are optimized for systems with particles that are repulsive at short distances, so that the local density of particles would never become too dense. Till now, the LAMMPS is the most efficient tool (in a parallel computing sense) for systems whose tens of million particles fill into a 3D rectangular box with an approximately uniform density.

25.3 Density Functional Study of Adhesion at the Metal/Ceramic Interfaces

A fundamental quantity which influences mechanical properties of an interface is named "the work of separation": W_{sep} (also commonly referred to as "the ideal work of adhesion"). It is defined as the energy needed for breaking interfacial bonds and reversibly separating the interface into two free surfaces, but neglecting diffusion and plastic deformation [71]. Experimental observations on this dissipative process indicated that, due to plastic deformation the W_{sep} was usually proportional to the work done to cleave the interface. Thus formally, the W_{sep} was defined in terms of either surface or interface energies relevant to respective bulk materials, or difference in total energy between interface and two isolated surfaces:

$$W_{sep} = \sigma_{1v} + \sigma_{2v} - \gamma_{12} = \frac{1}{A}(E_1^{tot} + E_2^{tot} - E_{12}^{tot}), \tag{25.12}$$

where σ_{iv} is the surface energy of slab i, γ_{12} is the interface energy, E_i^{tot} is the total energy of slab i, E_{12}^{tot} is the total energy of a system with an interface, and A is the whole interface area.

25.3.1 Calculations

Till now, the available analytical models for predicting the W_{sep} are still limited to liquid-metal and oxide interfaces, and have to rely upon some simple empirical correlations that incorporate either the free energy of formation for oxide and liquid-metal or the mixing enthalpies for respective oxide elements in metals [72–77]. Unfortunately, many of these models were not applicable to those with un-oxidized ceramic interface, i.e., these models may only provide qualitative information about trends in adhesion. Furthermore, their applicable range, even within the class of metal/oxide interfaces, was usually questionable because many of them have been applied only to those using the α-Al_2O_3 (Alumina) as the oxide.

According to these, a number of *ab-inito* studies on metal/ceramic adhesion using density functional theory (DFT) have come to a rapid growth during past two decades. Not only were these studies highly accurate, but may reveal invaluable information of atomic and electronic structures at interfacial area. For examples, Siegel and Adams *et al.* employed the DFT as implemented in the VASP to study the interfacial adhesion for several Al-Ceramic (e.g., Al/α-Al_2O_3, Al/WC, Al/VC, Al/VN, Al/CrN and Al/TiN) groups [78–81]. In their research, they found that, although Equation (25.12) defined the W_{sep} in terms of both surface and interface energies, these quantities did not necessarily play an equal role, i.e., since γ generally behaved much smaller in those metal-ceramic systems, the strength of interfacial bonding relevant to the W_{sep} depended to a large extent upon the reactivity of individual surfaces as reflected by their surface energies: see Table 25.1.

If considering atomic relaxations into the calculation, they also found that, at an Al/Alumina interface, both magnitude and rank ordering of adhesion energies for different stacking sequences changed to the unrelaxed results determined by a Universal Binding Energy Relation (UBER) [83], see Figure 25.3. Regarding the optimized interfacial stacking sequences, at the Al/α-Al_2O_3 and the polar Al/WC interfacial areas, metal atoms preferred to sites which continued the natural stacking sequence

Table 25.1 Calculated surface energies σ and slab dimensions (number of layers) used for the W_{sep} calculations.[a]

Surface	No. of layers	$\sigma(J \cdot m^{-2})$
Al(100)	5	0.89
Al(110)	4	1.05
Al(111)	5	0.81
α-$Al_2O_3(0001)^{Al}$	15	1.59
α-$Al_2O_3(0001)^{O}$	13	7.64 (4.45–10.83)
$WC(0001)^{W}$	9	3.66 (3.43–3.88)
$WC(0001)^{C}$	9	5.92 (5.69–6.14)
$WC(11\bar{2}0)$	4	3.88
VN(100)	7	0.95
VC(100)	7	1.28
CrN(100)	7	0.74
TiN(100)	7	1.25

a) For non-stoichiometric slabs both average σ and possible σ range are given. Superscripts give termination of those surfaces cleaved along a polar plane.

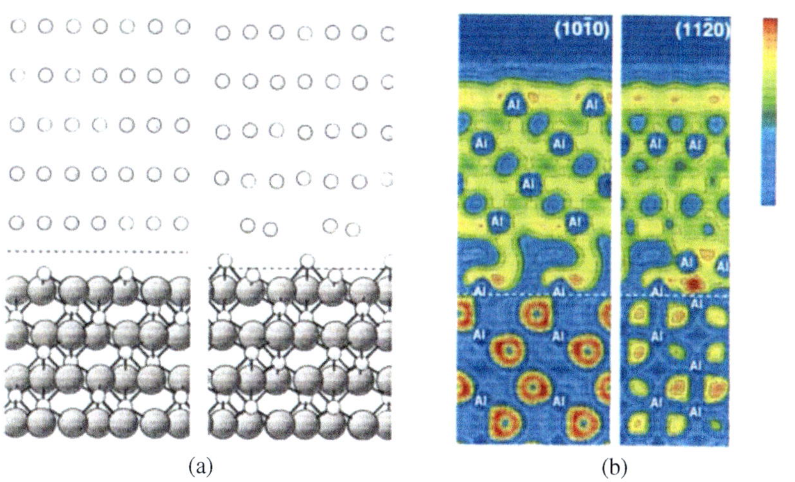

Figure 25.3 (a) (Left) The lowest energy geometry of the fcc-Al/α-Al_2O_3 interface as predicted by the UBER calculations; (Right) the final relaxed structure. Small spheres represent Al atoms; large spheres represent O atoms. The direction of view is along [1$\bar{2}$10] (The lower portion of structure has been omitted for clarity). (b) Two slices through the electron localization function (ELF) for the f.c.c.-Al/α-Al_2O_3 interface taken along (10$\bar{1}$0) and (11$\bar{2}$0) planes, showing four of hexagonal close-packed oxygen layers in oxide (bottom) and all five atomic layers from one of the Al slabs (top).

Table 25.2 Interfacial orientation relationship, polarity (P = polar, NP = nonpolar), strain, interface free energy (γ), and W_{sep}. Terminations of the polar ceramic surfaces are indicated with a superscript. The W_{sep} values correspond to the optimal stacking sequences.

$Al/\alpha\text{-}Al_2O_3{}^{Al}$	$(111)[\bar{1}10]_{Al}\parallel(0001)[10\bar{1}0]_{Al_2O_3}$	NP	4.9	1.34	1.06
$Al/\alpha\text{-}Al_2O_3{}^{O}$	$(111)[\bar{1}10]_{Al}\parallel(0001)[10\bar{1}0]_{Al_2O_3}$	P	4.9	−1.28	9.73
Al/WC^{W}	$(111)[\bar{1}10]_{Al}\parallel(0001)[11\bar{2}0]_{WC}$	P	2.2	0.39	4.08
Al/WC^{C}	$(111)[\bar{1}10]_{Al}\parallel(0001)[11\bar{2}0]_{WC}$	P	2.2	0.72	6.01
Al/WC	$(110)[1\bar{1}0]_{Al}\parallel(11\bar{2}0)[001]_{WC}$	NP	0.4	1.79	3.14
Al/VN	$(100)[001]_{Al}\parallel(100)[001]_{VN}$	NP	2.3	0.11	1.73
Al/VC	$(100)[001]_{Al}\parallel(100)[001]_{VC}$	NP	3.2	0.03	2.14
Al/CrN	$(100)[001]_{Al}\parallel(100)[001]_{CrN}$	NP	0.5	0.18	1.45
Al/TiN	$(100)[001]_{Al}\parallel(100)[001]_{TiN}$	NP	5.3	0.62	1.52

in ceramic bulk across the interface into the Al-slab; while for Al/VC, Al/VN, Al/CrN, Al/TiN and Al/WC($11\bar{2}0$) systems, Al atoms preferred to sit above metalloid atoms (C and N).

The W_{sep} calculations are presented in Table 25.2, where people can find that there are obvious differences in the W_{sep} values between the polar and the non-polar geometries. In particular, three polar surfaces behaved the largest W_{sep} values ranging from 4.08 J m^{-2} for the Al/WC^{W} up to 9.73 J m^{-2} for the Al/α-$Al_2O_3{}^{O}$; for the remaining six (non-polar) interfaces, five had the W_{sep} values of 2.1 J m^{-2} or less. While those for three nitride-ceramics were among the smallest range and looked relatively insensitive to the choice of metallic component, with the W_{sep} values falling within 1.45–1.73 J m^{-2}. On average, the W_{sep} values in polar systems were about 4.6 J m^{-2} larger than in non-polar systems, consistent with their larger surface energies as shown in Table 25.1. It was also noted that, the predicted W_{sep} value of 1.06 J m^{-2} for the Al/α-$Al_2O_3{}^{Al}$ system was in favorable agreement with the experimental value of 1.13 J m^{-2} as scaled to 0 K [83].

25.3.2 Effect of Surface Energies in the W_{sep}

To further verify whether the W_{sep} was mainly determined by surface energies or not, Figure 25.4 plots the W_{sep} *vs.* the ($\sigma_{Al}+\sigma_{ceramic}$); a line with the unit slope is also included to indicate where data points would lie in the idealized case of $\gamma = 0$. In this representation, the horizontal deviation of a given data point from the γ=0-line corresponded to the γ size. For three polar interfaces at top region, data were bracketed by horizontal error bars giving the range of possible σ listed in Table 25.1; while the W_{sep} data in the whole range showed a roughly linear trend with respect to the σ, all but one of data points fell to the line-right. This revealed that the bonding of these interfaces was on average weaker than that of relevant bulk regions. One system falling to the line-left to present a negative γ value, would involve the oxygen-terminated α-Al_2O_3 surface, which had the largest surface energy and the W_{sep}, and could indicate the possibility of fracture within the Al-bulk rather than at the Al-Ceramic junction due to a tensile stress.

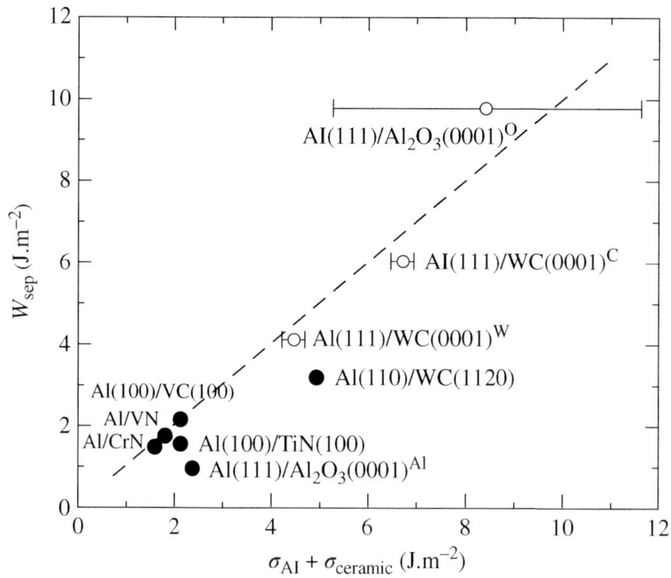

Figure 25.4 The W_{sep} *vs.* $(\sigma_{Al}+\sigma_{ceramic})$. Filled (open) circles correspond to non-polar (polar) interfaces. The dotted line corresponds to the $\gamma = 0$ case.

With the exception of TiN, the rocksalt-structured carbides and nitrides (CrN, VN, and VC) exhibited the largest dependence of the W_{sep} upon the surface energies. These three systems had the smallest interface energies, only 0.03–0.18 J m^{-2}, and their W_{sep} values crowded in a nearly linear fashion (close to the γ=0-line) along the bottom left of Figure 25.4. Presumably, small γ values at these interfaces can be explained by similarities between bulk and interfacial bonding. However, since both bulk components exhibited some extent to metallic bonding, for the Al/TiN system – a singularity with respect to other rocksalt ceramics, the γ value contributed larger, 0.62 J m^{-2}, which could be possibly due to the relatively large interfacial strain of 5.3%.

Figure 25.4 also illustrates that, those W_{sep} values involving α-Al_2O_3 and WC(11$\bar{2}$0) systems were not solely described by their surface energies (i.e., γ effects were more important). For the α-Al_2O_3 interfaces, as with TiN, a possible explanation for this behavior could be the strain effects which were also significant (4.9%) for this system. In addition, the main effect of ionic bonding in bulk α-Al_2O_3 differed substantially from that in Al-bulk. While for the WC system, the strain was very low (0.4%), so differences in respective bulk bonding were less indicated (WC was metallic). Thus this model implemented an incoherent, misfit geometry characterized by irregular bonding across the interface.

25.3.3 Conclusions

In summary, the DFT calculations indicated that surface energies played a dominant role in determining the work of separation (the W_{sep}) for the systems studied in above, while interface energies were generally much smaller in comparison. In most cases, particular for those systems involving cubic carbides and nitrides, the W_{sep} can be described largely in terms of surface energies alone. Exceptions to this rule arose mainly for systems characterized by relatively large interfacial strain or incoherent interfacial bonding. Comprehensively, what Siegel and Adams *et al.* did could serve as a reasonably starting point for the estimation of interfacial strength.

25.4 Molecular Dynamics Simulations of Nanoindentation

Indentation experiments have been performed for more than 100 years to measure the hardness of materials. As we introduced at the beginning, significant findings have been made with the development of the AFM and the STM during last two decades. It is possible to currently determine, with high precision and accuracy, both the load and the displacement of an indentor in each indentation experiment within micro-Newton and nanometer ranges, which are critical to several applications, including tribological coatings, microelectromechanical devices, and adhesion of biomolecules and living cells, etc. Therefore, not only has nanoindentation now become a powerful tool for probing mechanical properties of bulk materials at micro-/nano-meter length scale, but for investigating asperity contact over a wide range of the wear process.

Our understanding of nanoindentation theory has improved in recent years, both on the continuum level and to less extent on the atomic level. For examples, several methods have been proposed to obtain hardness, elastic moduli, initial yield strength and work-hardening exponent from measurements of nanoindentation [2, 84]; the general shape of the loading-unloading curve was understood through the continuum level [85, 86]. However, many questions remained, e.g., the strain rate, the temperature dependence and the deformation mechanisms like dislocation nucleation and motion etc., which were not well described by the continuum-level models. Similarly, even less was known about abrasive wear at nanoscale contact that involved plastic deformation and lubrication when a single tip tool was plowing throughout a material surface adsorbed by additive molecules, see Figure 25.2. Therefore, there is a need for methods to investigate deformation phenomena at the atomic scale, so as to link atomic-scale observations with macroscopic continuum models and experimental results.

25.4.1 Empirical Modeling

Here, without loss of generality, the aluminum (Al) is selected as the modeling sample because of its widespread usage in many industries like aerospace, automotive and microelectronics, etc. Interatomic bonding in the Al-bulk was depicted with a simple EAM potential that was previously developed by fitting to a large DFT and experimental database [65]. This potential has proved to be surprisingly reliable for a wide range of relevant phenomena, including elastic deformation, phonons, thermal expansion, dislocation generation and structure melting, etc., all of which can be involved in the plastic deformation that occurred during nanoindentation and wear processes. Therefore, the modeling results in this section should be generic to most ductile fcc-metals. Rather than constructing a specific indentor material, the indentor (tip) here was chosen as a generic Lennard-Jones interaction (the LJ-tip) with parameters of $\varepsilon = 0.50$ eV and $\sigma = 2.85$ Å to represent it much harder than the Al-substrate because the latter was usually considered as a relatively soft metal easily deformed by many other harder materials. Similarly, the interaction between the Al-substrate and the LJ-tip was also chosen as a simple Lennard-Jones potential because it allowed us to vary the bond strength (the bonding) easily and hence seek the generic effects of a wide range of hard materials: from those that formed passivated surfaces to those that formed strong bonds with the aluminum.

As mentioned in section 25.1, even if several groups have applied molecular dynamics (MD) simulation to the investigation of nanoindentation, and have reported several interesting results, they did not fully report their simulating conditions such as the strain rate or the loading strength, i.e., most of which only included a single factor such as a single temperature, a single indentation rate, a single-crystal orientation and a single material combination, etc. Therefore, it was important to begin a systematic study to investigate, qualitatively and quantitatively, many possible factors that may control nanoindentation and wear behaviors. In this section, we summarize what we did for several factors that determined the extent of nanoindentation, including effects of the (system) temperature, the crystal orientation, the tip shape, the tip-substrate bonding, and the loading force and speed, etc.

25.4.1.1 Modeling Geometry and Simulation Procedures

Adams *et al.* carried out a series of *classical* molecular dynamics simulations with the ParaDyn (a parallel MD code developed by Plimpton *et al.* [87]) for above studies [88, 89], which involved an atomically sharp-indentor (the LJ-tip) moving a normal contact towards a flat Al-substrate (Al-slab). In their research, firstly, both tip and substrate were thermally equilibrated for about 3–10 ps at a desired temperature (e.g., 300 K). A small time step of 0.001 picoseconds (ps) was chosen, which may keep the energy of a system stable during the whole MD simulation. Secondly, the LJ-tip was placed about 6 Å above the substrate, which was beyond the cut-off distance of interatomic potentials to avoid their interactions at the beginning (here, the cut-off distance was determined within the 3rd and the 4th nearest neighbors in a regular crystal); then next, a constant loading force along the z-direction was applied onto top layers of the LJ-tip, accelerating it towards the Al-substrate; simultaneously, an initial velocity along the z-direction, $V_z = -3.00$ Å/ps (300 m/s), was added on the whole tip, which was much higher than that in experiments of nanoindentation (<10 m/s) but was typical of many real processes, including the rolling processes and the vibrating motion of pistons in auto-vehicle engine. Thirdly, three different crystal orientations, (100), (111) and (110), were chosen for constructing the Al-substrate: for the Al(100)-substrate, its dimension was $25a \times 25a \times 9a$, where a was the lattice constant of a regular Al-crystal at a given temperature (e.g., $a = 4.051$ Å at 300 K); their experience proved that $9a$ in height was thick enough for the Al(100)-substrate to withstand the tip impact. Since other two Al-substrates were found to have similar sizes, the Al(100)-substrate was chosen as the standard model in MD simulation. In this standard model, periodic boundary conditions were imposed along the x- and the y-directions, while the z-direction was in free boundary condition. Also, only a small portion in the bottom layer of the substrate was fixed, where dislocations generated due to the tip-indentation were unlikely to reach, see Figure 25.5.

The tip was composed of hard LJ atoms with a same mass unit as an Al atom. It was constructed by a rigid cubic with 15 layers, $9a' \times 9a' \times 9a'$ (e.g., $a' = 4.015$ Å at 300 K), but was shaped as an up-side-down square pyramid with an angle of 90° at its bottom part, also see Figure 25.5. Its normal direction was always [100] even if that of the

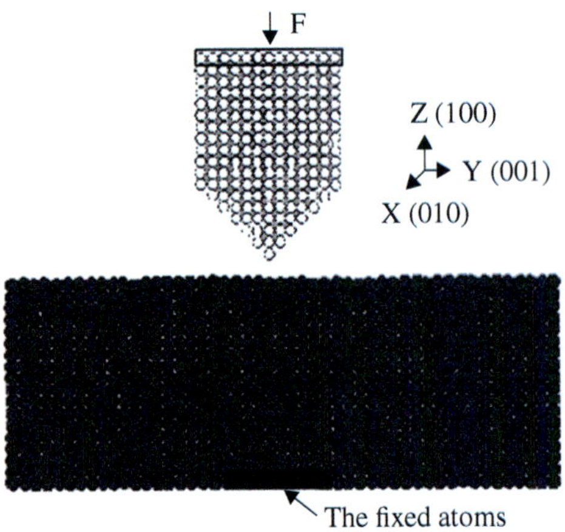

Figure 25.5 Configurations of the LJ-tip and the Al(100)-substrate as the standard model.

Table 25.3 Simulation parameters used for the standard model.

Parameters	Standard Model	Variations
Temperature (K)	300	0, 600, 800
Indent stress (GPa)	2.4	1.2, 4.8, 9.6
Tip-substrate bonding (eV)	0.04	0.10, 0.20
Al-substrate orientation	(100)	(110), (111)

Al-substrate may be varied. And, a loading force was applied onto top layers of the tip along the *z*-direction.

Table 25.3 lists relevant parameters used for the standard model: the temperature $T = 300$ K for system; the orientation (100) for the Al-substrate; the tip-substrate bonding $\varepsilon' = 0.04$ eV for a LJ potential model (still with the same σ); and the loading force $\boldsymbol{F}_z = -2.05\times10^{-18}$ N, which may give an additional loading stress of 2.4 GPa along the *z*-direction, for top layers of the LJ-tip.

25.4.1.2 Results and discussions

Neck Formation, Local Melting, and Dislocation Nucleation and Emission Figure 25.6 shows a partial side-view of the standard system during nanoindentation. In Figure 25.6(a–b), when the tip moved close to the Al(100)-substrate, some Al atoms in top layers of the Al-substrate jumped onto the tip head to form a *neck* to link the tip and the substrate, i.e., as an indentation initially took place, it conformed the tip with some Al atoms being displaced to the periphery of contact area due to plastic deformation and rapid loading (>10 m/s). Note that the temperature at contact area on the Al-substrate rose to 1500 K, which was well above the Al melting point: 933 K. The reason was that, since the rate of heat generation due to the rapid impact of the tip exceeded that of heat diffusion away from contact area into far inner region of the substrate, it would result in neck formation and local melting phenomena, see Figure 25.6(b–c). It should be also noted that, thermal conductivity of the Al-substrate in this simulation was somewhat different from

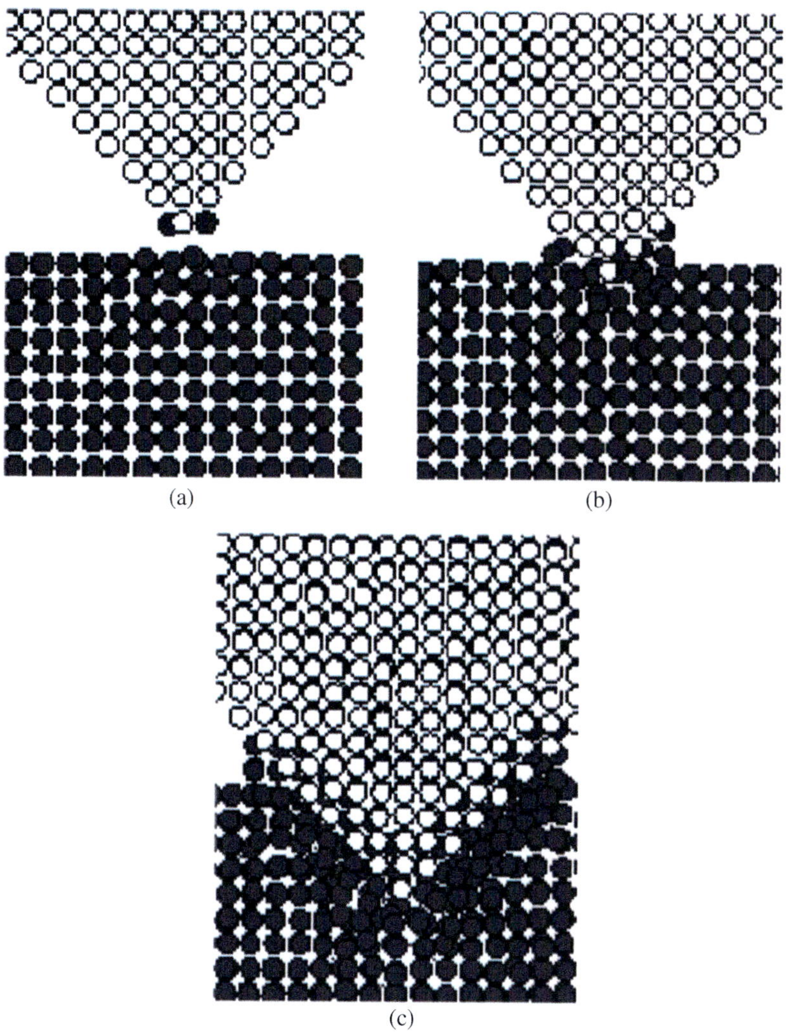

Figure 25.6 Snapshots of partial side-view on the standard system in MD simulation: (a)–(b) Neck formation; (c) Final tip indent depth.

experimental value, because, this simulation just mainly included the nuclear contribution, but not totally the free electronic contribution (in metals, free electrons should play a main role in heat transport). However, due to very large difference in mass between nuclei and electrons, the nuclei heating by rapid indentation would slowly influence profiles of electronic temperature. Therefore, ignoring free electronic contribution to thermal conductivity would behave little effect on simulation results.

By taking advantage of a 'centrosymmetry' method [87], they identified dislocation nucleation in the Al-substrate and found that, dislocation cores were emitted along multiple {111}<110> slip systems and one dislocation loop was approximately formed for every layer indented, which was in agreement with other group's observations [90, 91], see Figure 25.7.

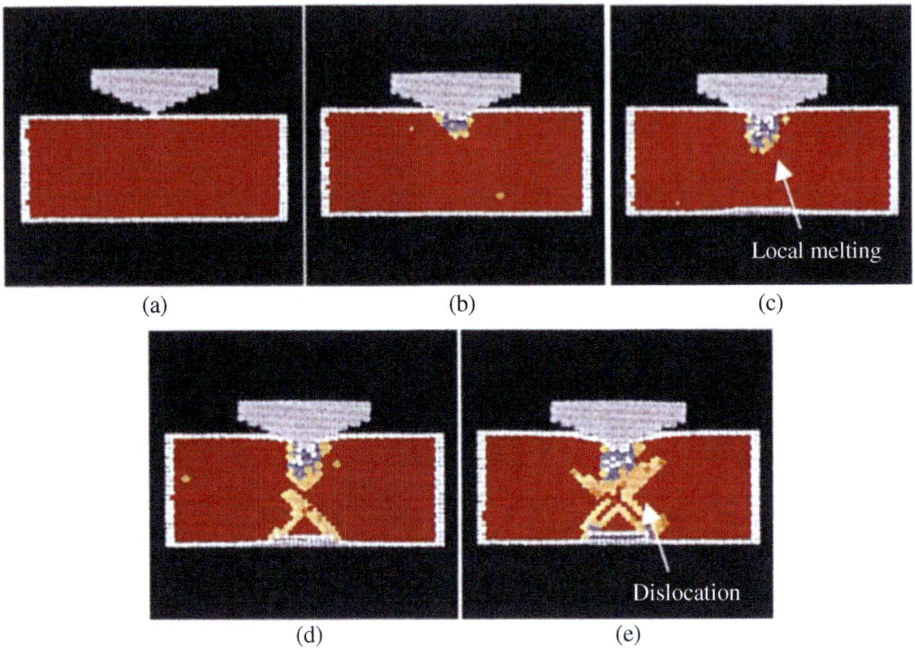

Figure 25.7 (a) neck formation; (b–c) local melting near the LJ-tip at the contact area; (d–e) dislocation nucleation and emission on multiple {111} planes (dislocation cores visualized as orange atoms).

Table 25.4 Results of the loading stress onto the LJ-tip, final indent depth in the Al-substrate, and local stress at contact area.

The loading stress (GPa)	Final indent depth (Å)	Local stress (GPa)
1.2	5.5	6.0
2.4	7.9	5.8
4.8	11.0	6.0

Effect of the Loading Force During each simulation, four different loading forces applying normally onto the top layers of the LJ-tip would convert to extra loading stresses: 1.2, 2.4, 4.8 and 9.6 GPa in that region, respectively. Table 25.4 lists data of the loading stress, final indent depth, and local stress at contact area. Figure 25.8 shows a relationship of indent depth with simulation time under the above four loading strengths.

In Figure 25.8, a higher loading force may result in a deeper indent depth, which was proportional to the square of the final indent depth and the cross-sectional dimension of contact area, so as to maintain the local stress at contact area near the yield value. For examples, under the loading stresses: 1.2, 2.4 and 4.8 GPa, the LJ-tip eventually moved slowly and oscillated at a certain indent depth in the substrate, respectively. The reason was that, when the tip initially reached to the substrate surface, the local stress at contact area became so rapidly high that it may bring about an elastic bond compression to break

Figure 25.8 Relationship between the indent depth and the simulation time under four types of loading.

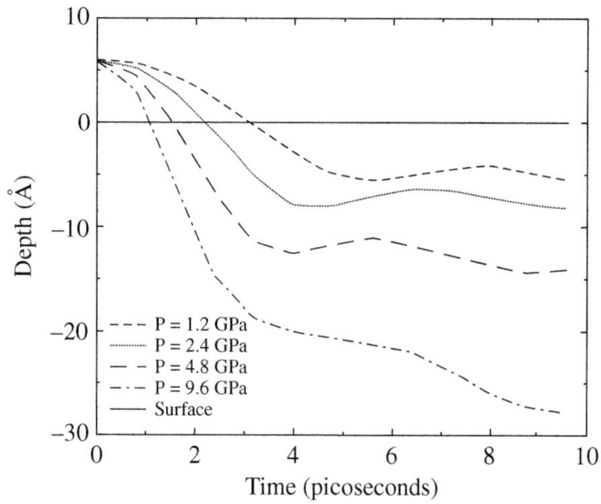

the Al-Al bonds easily. As a result, the tip may continue squeezing into the substrate till it comforted a new equilibrium surrounding in there. Even if a saddle point on these curves was not observed, an extrapolation may predict that a critical value of yield stress was approximately 6.0 GPa for above three indentations. Thus in the last case of the loading stress, 9.6 GPa, the local stress at contact area may have exceeded 6.0 GPa to make the tip moving continuously throughout the whole substrate. In addition, MD simulation results indicated that the tip exact shape behaved little effect on the indentation quality.

Effects of Temperature, Tip-substrate Bonding and Substrate Orientation Such a MD simulation was also conducted at certain environmental temperatures: 0, 300, 600, and 800 K (near the Al melting point: 933 K), respectively. In these cases, at each initial stage, without applying to the loading force, the LJ-tip was directed towards the Al-substrate surface at a same initial speed for iso-thermal tests. It was found that, when the environment temperature increased, dislocation cores became more likely to move. Thus the Al-substrate became softer to have the indent depth enlarged, see ref [89].

Figure 25.9 shows effects of three substrate orientations: (111), (100) and (110) on indentation. It was found that, the Al(110)-substrate was the easiest to indent, which was in qualitative agreement with the plowing experiments done by Hector: in their experiments, after an AFM tip plowed throughout surfaces of the Al-alloys under a 250 mN drag force, the scratch depths were 450, 460 and 540 nm along the (111), (100) and (110) surfaces, respectively. This phenomenon can be explained by the number of atomic dangling bonds onto f.c.c. surfaces: each atom in (111) surface has three dangling bonds, each atom in (100) surface has four dangling bonds, and each atom in (110) surface has five dangling bonds, respectively. Thus the (110) surface should be most reactive among three ones. When the tip was approaching a surface, more dangling bonds in the surface would attract it to indent deeper inside the substrate. Equivalently, as mentioned in above, the tip-substrate bonding ε' in a LJ potential varied from 0.04, 0.10 to 0.20 eV during each MD simulation, respectively. Their outputs also indicated that, a stronger ε' may also result in a larger indent depth, see ref [89]. According to this, the Al(110)-substrate would be the easiest one to deform.

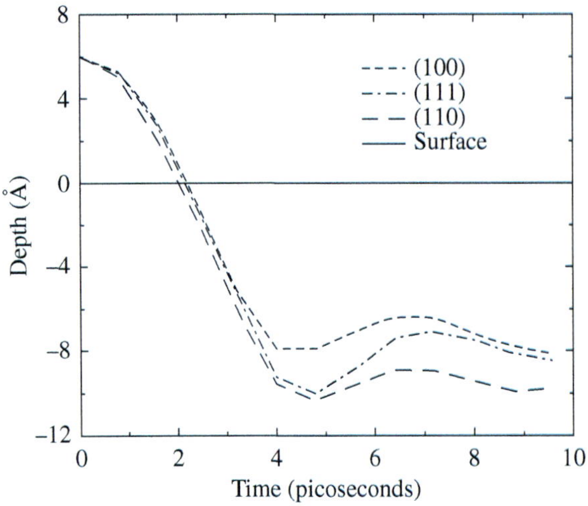

Figure 25.9 Effects of three substrate orientations: (111), (100) and (110) on indentation.

25.4.1.3 Conclusions

In this part, at atomic-scale, a *classical* molecular dynamics simulation is introduced on indentations of a high-speed rigid tip to three Al-substrates. Common phenomena during the MD simulation of nanoindentation such as jump-to-contact (neck formation), local melting, and dislocation nucleation and emission etc. were observed. Effects of several constraint parameters such as the loading force, the system (environment) temperature, the tip-substrate bonding, and the substrate orientation etc. were also discussed. As a result, a higher loading force applied onto the tip usually led to a larger indent depth in the substrate; each loading force was found to be proportional to the square of indent depth and cross-sectional dimension of contact area; critical point of yield stress into the Al-substrate was estimated to be 6.0 GPa; higher environment temperatures also resulted in larger indent depths because much easier dislocation nucleation and emission made the Al-substrate softer in a higher temperature range; the (110) surface was easier to indent than the (100) and the (111) surfaces due to its larger attractiveness of effective dangling bonds to the indentor.

25.4.2 *Ab initio* Modeling

In this section, without any loss of generality, *ab initio* molecular dynamics (AIMD) based upon density functional theory (DFT-GGA) is employed to discuss the indent reaction at an interface between a clean Al(111) and a hematite (α-Fe_2O_3(0001)) slab to obtain invaluable information at contact area. As a result, a compression of these two slabs may form an amorphous alumina at interfacial area through a thermit reaction which welded them together; subsequently, a tensile process resulted in removal of amorphous alumina from the clean Al-slab. The overall reaction was highly exothermic to cause a large temperature increase.

25.4.2.1 Modeling Geometry and Simulation Procedures

Zhong *et al.* [92] carried out the AIMD implemented by the VASP for above study. Simulations for the clean Al(111) surface involved a super-cell geometry with a periodic 4-layer Al(111)-slab (16 ions per layer) in the *x*- and the *y*-directions. This

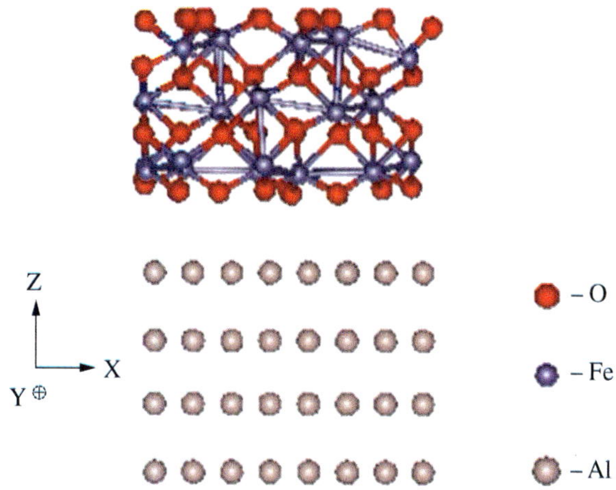

Figure 25.10 Side-view of geometry at the Al(111)/α-Fe_2O_3(0001) interface.

orthorhombic super-cell had three orientated vectors: $a[1\bar{1}0] = 2\sqrt{2} \times a_{Al}$ at the x-axis; $b[11\bar{2}] = \sqrt{6} \times a_{Al}$ at the y-axis; and $c[111] = 40$ Å at the z-axis along with a vacuum distance of 24.0 Å in the z-direction, plus one bottom layer fixed along the z-axis. Here $a_{Al} = 4.053$ Å [93]. Similarly, simulations for a widely studied α-Fe_2O_3(0001) slab with oxygen termination (more reactive than that with iron termination to the Al-surface) were conducted using a periodic 7-layer rectangle slab with two oriented vectors: $\sqrt{5} \times a_{Fe_2O_3}$ at the x-axis and $\frac{\sqrt{15}}{2} \times a_{Fe_2O_3}$ at the y-axis, and a height of 8.40 Å vacuum space with free boundary along the z-direction (total of 36 Fe and 54 O ions). Here, $a_{Fe_2O_3} = 5.031$Å [94]. Comparing to these two slabs, the dimension of the Al-slab was slightly larger than the Fe_2O_3-slab in the x- and y-directions, respectively. Therefore, the super-cell for the Al-slab was chosen as the final frame to contain both the Fe_2O_3 and the Al slabs during the simulation. In this selected super-cell, the mismatch percentages of each of two slab lengths along the x- and y-directions were calculated at very small values: 1.9% and 1.9%, respectively, see Figure 25.10.

The simulating time step was selected as 0.001 ps (i.e. 1ps $=10^{-12}$ s). The whole system was held at constant energy of AIMD simulation (no temperature control) when two slabs contacted. At the beginning of each simulation, top three layers of the α-Fe_2O_3-slab were constrained to move along the z-direction at a constant velocity, $\boldsymbol{V}_z = -4.00$ Å/ps, corresponding to a displacement $\boldsymbol{d}_z = -0.004$ Å per time step. Since this speed, 400 m/s, was typical of real processing conditions, such shock loading may exceed the yield strength of the Al-slab, and the deformation of top layers in the Al-slab can be reasonably observed during contact simulation.

Before carrying out the simulation of compressing two slabs, a favorable interfacial geometry between two slabs was optimized using the energy minimization with DFT. The result indicated that the final spacing between two slabs would be about 1.88 Å to form Al–O bonds at interfacial area, which were similar to those observed in experiments [95], see Figure 25.11. Table 25.5 also lists some relevant bond lengths.

Therefore, in the following AIMD simulation, the initial spacing between two slabs was chosen as 2.50 Å, which was far enough to avoid the significant forming of chemical bonds at the starting stage. During each simulation, as two slabs moved closer, chemical bonds would begin to form in the spacing, so the interfacial reactions would occur. Since

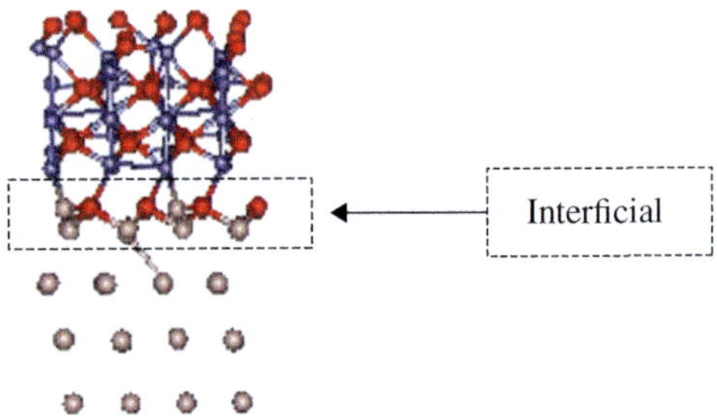

Figure 25.11 The final Al(111)/α-Fe_2O_3(0001) interfacial geometry optimized through DFT.

Table 25.5 Relevant bond lengths at interfacial area [unit: Å].

Bond Pair	Al–O		Fe–O	
Bond length (Å)	Calculation	Experiment[a]	Calculation	Experiment[b]
	1.86 ~ 1.93	1.86 ~ 1.97	1.82 ~ 1.84	1.83 ~ 1.85

a) [116].
b) [117].

top three layers (outside the R–L region) of the α-Fe_2O_3(0001)-slab in Figure 25.12(a) would move at a constant velocity ($V_z = -4.0$ Å/ps) as a rigid block, its final downward displacement was chosen as 0.62 Å, so as to keep the final spacing between two slabs around the optimized value: 1.88 Å.

25.4.2.2 Results and Discussions

Compressive Process Figure 25.12 shows the compressing sequence of the α-Fe_2O_3(0001)/Al(111) interface. A fixed velocity, $V_z = -4.00$ Å/ps, was applied onto top three layers of the α-Fe_2O_3(0001)-slab along the z-direction during the whole simulation.

In this figure, during the first 80 time steps, the α-Fe_2O_3(0001)-slab was approaching the Al(111)-slab without forming Al–O bonds, see Figure 25.12(a); at about 80 time steps, some Al–O bonds (with Al–O bond length about 1.86~1.93 Å) were being formed, see Figure 25.12(b); at about 140 time steps, most of Al–O bonds at interfacial area were formed, see Figure 25.12(c). The local temperature in this area was about 870 K. During this evolution, an amorphous alumina was gradually formed due to further reaction of oxygen-ions with the Al-surface. At 390 time steps, this reaction continued at the uppermost point, see Figure 25.12(d). The local temperature in this area continued rising to 1130 K; by 620 time steps, a new layer made up of amorphous alumina was completely constructed at interfacial area, see Figure 25.12(e). The local temperature in this area increased to 1200 K.

Figure 25.13 indicates the distribution of the stress perpendicular to the interfacial area as two slabs compress together. Initially, the stress showed negative (attractive between two slabs) as Al–O bonds began forming at interfacial area (points a–c); when two slabs was moving closer, the stress tended to positive (repulsive between two slabs) because most of Al and O atoms at interfacial area were forced to come closely together, which resulted in a nearly texture formation of amorphous alumina with thermal

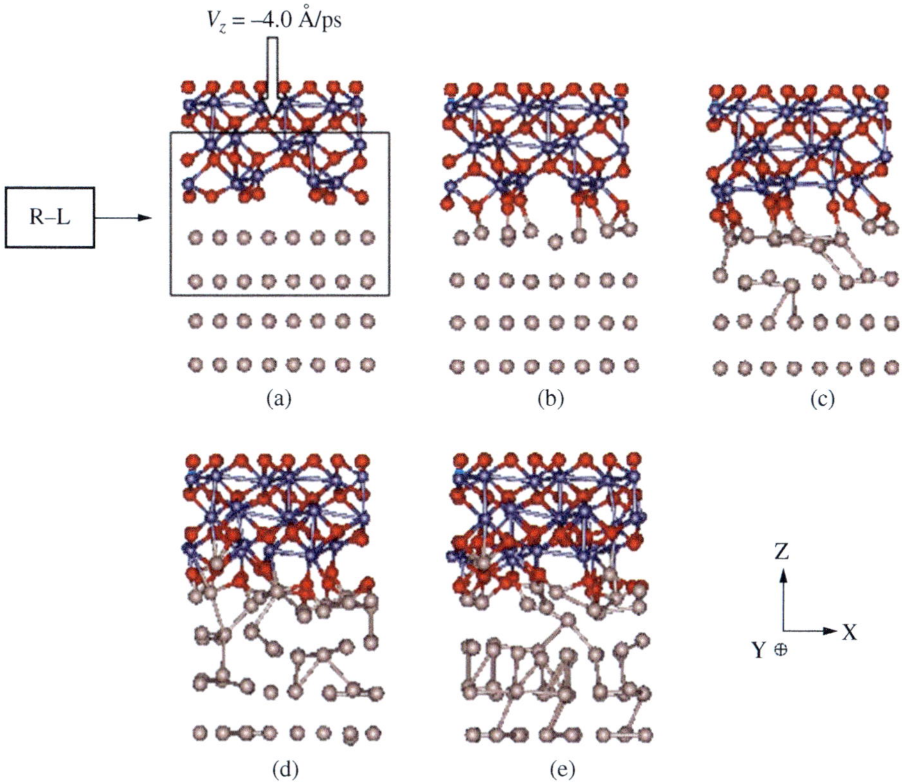

Figure 25.12 Main sequence of compressing the Al(111)/α-Fe_2O_3(0001) interface without lubricants.

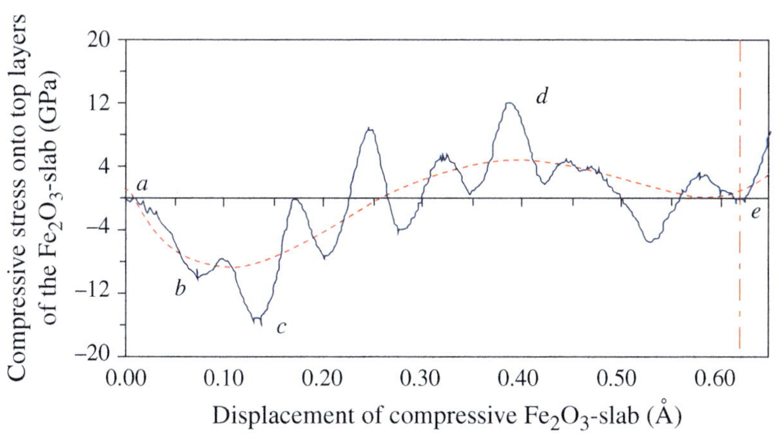

Figure 25.13 The stress perpendicular to the interfacial area as two slabs compress together.

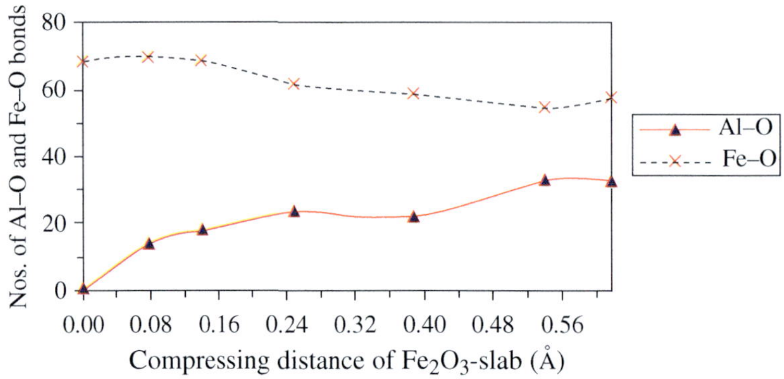

Figure 25.14 Number changes of Al–O and Fe–O bonds in the R–L region during slabs compression.

expansion at there; after two slabs totally reacted, the stress fell down to a near zero point (point *e*) from the highest positive point (point *d*) as an optimal bond distance between two slabs was reached; and then, the stress increased again (after point *e*) when two slabs continued compressing. This figure also represented a polynomial fit to the stress (the dash line), which may regress the stress fluctuation to show its trend clear. Please note that, the stress trend would depict the starting steps of the whole *P-h* curve as stated in refs [17, 91, 96]. However, since two slabs here were too thin and too small in dimension, and at too short time steps throughout the AIMD, it was hardly to rough out the whole *P-h* curve fulfilled by *classical* molecular dynamics on very large size models at very long time steps. Even if, this stress curve may still provide some valuable information at the initial stage of nanoindentation process.

During the slab compression, the compressive stress at point *d* was 13.6 GPa, which was too higher than the bulk modulus of pure Al, 0.76 GPa [97], but was similar to that of nano-crystalline composites (α-Fe_2O_3+fcc-Al), 14.0 GPa, obtained by Tomar *et al.* through *classical* molecular dynamics [98]. This very high stress was largely due to the high rate of deformation between two slabs, resulting in little time for plasticity.

Figure 25.14 shows the number changes of Al–O and Fe–O bonds *vs.* normal distance (time) at interfacial area defined as the R–L region in Figure 25.12(a) during compressive simulation. In this figure, Al–O bonds in the R–L region increased substantially while Fe–O bonds in this region decreased gradually. The reason was that, when two slabs compressed together, the dangling bonds on Al and O atoms at interfacial area would bond together. As the compression continued, more Al–O bonds formed as some Fe–O bonds broke. Nevertheless, only a very porous interface between two slabs was formed during the compression, which was far from the ideally compressed status. Therefore, the thermit reaction at interfacial area only partially occurred, and only a few Fe–O bonds were broken at there.

Tensile Process After compression, an effect of a tensile strain acting on the compressed α-Fe_2O_3(0001)/Al(111) interface was immediately discussed: top three layers of the α-Fe_2O_3(0001)-slab in Figure 25.12(e) moved as a rigid block at a constant velocity, $V_z = +4.00$ Å/ps, reversing the previous compression.

Figure 25.15 shows the stretching sequence of this process: (i) When the top three layers of the α-Fe_2O_3(0001)-slab moved up about 1.30 Å from the starting point in

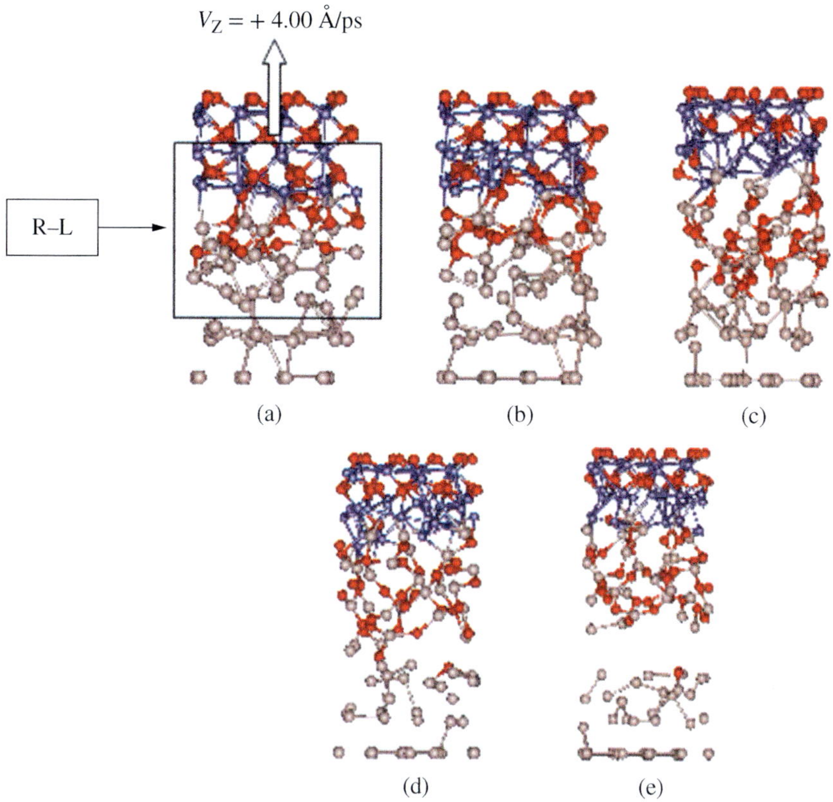

Figure 25.15 The main sequence of stretching the the Al(111)/α-Fe_2O_3(0001) slabs.

Figure 25.12(e), the real tensile process at interfacial area began (point a). (ii) Then next, as two slabs separated, further deformation at interfacial area and thermit reaction would continue, more atomic mixing in this area oxidized the Al-surface (point b). (iii) During transitional stages (points c and d), the whole interfacial area began to break apart, mainly due to the breaking of some more Fe–O and Al–Al bonds. Along with a small increase of Al–O bonds, the amorphous alumina was continuously formed, and adhesive metal transfer took place in this area. (iv) Finally, surface on the Al-slab was broken: adhesive metal transfer had completely occurred. Amorphous alumina was removed from the remnant surface of the Al-slab to adhere onto the bottom of the α-Fe_2O_3(0001)-slab, indicating the final apart step of adhesive indentation or wear. A fresh disordered Al-surface was also exposed on the residual Al-slab after adhesive indentation, see Figure 25.15(e).

Figure 25.16 shows the number of changes of Al–O, Fe–O and Al–Al bonds in a new larger R–L region as shown in Figure 25.15(a) during the tensile process. In this figure, at initial stage, a few more Al–O bonds formed and a few more Fe–O bonds broke, which represented the continuous thermit reaction at interfacial area as two slabs were being pulled apart. Also, there was the breaking of approximately 10~12 Al–Al bonds around top two layers of the Al-slab. At about 8.0 Å, there is a slight increase in Al–O bonds and a slight decrease in Fe–O bonds as the interface was totally separated. By the end of the

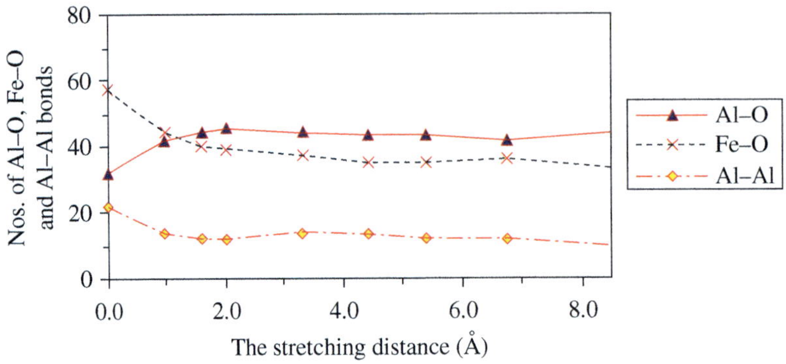

Figure 25.16 Number of changes of Al–O, Fe–O, and Al–Al bonds in the new R–L region during slabs tension.

simulation, the Fe–O coordination was approximately 1:1, while the Al–O coordination was approximately 2:3. Thus the reaction could be approximately $2Al + 3Fe_2O_3 \rightarrow 6FeO + Al_2O_3$, instead of the more favorable thermit reaction, $2Al + 3FeO \rightarrow 2Al_2O_3 + 3Fe$, mainly due to very limited reaction time. The local temperature in the Al-slab was about 810 K, while that in the removed (Fe_2O_3+Al_2O_3) slab was about 3030 K.

Comprehensively, the AIMD simulations revealed that the tensile stress at the Al(111)/α-Fe_2O_3(0001) interface could remove of Al_2O_3 fragments from the Al-slab after a hard iron-oxide covered steel-roller contacted this slab under an external strain load, which would indicate a typical process of adhesive indentation or wear at atomic-scale.

Conclusions The AIMD simulations on adhesive indentation/wear at the Al(111)/α-Fe_2O_3(0001) interface without lubricants were performed for compressive and tensile processes between the Al(111)/α-Fe_2O_3(0001) slabs. The following conclusions can be drawn: (1) Contact of the Fe_2O_3-slab with the Al-slab resulted in a strong exothermic reaction to form a texture of amorphous alumina (Al_2O_3), which welded two slabs together. (2) Separation of two slabs had amorphous alumina removed out of the Al-slab. In addition, this made local temperature at interfacial area increase significantly, leading to the melting of the Al-surface and easy removal of amorphous Al_2O_3. (3) The initial stage of the *P-h* curve can be drawn throughout the AIMD simulation to provide invaluable information which was hardly observed throughout *classical* molecular dynamics simulation. (4) During the AIMD simulation, the whole reaction process at interfacial area between two contact slabs was $2Al + 3Fe_2O_3 \rightarrow 6FeO + Al_2O_3$, i.e., due to the very short simulation time, such a thermit reaction was incomplete.

25.5 Molecular Dynamics Simulations of Adhesive Wear on the Al-substrate

Along with nanoindentation, one important wear mechanism involves the shear of two asperities. Here *classical* molecular dynamics is used to simulate a wear process of an aluminum (Al) asperity impacted by a 'hard' (Lennard-Jones, LJ) tip. This simulation involves an application of a reliable EAM potential based upon *the force-matching method* [65]. Simulations were repeated for a wide range of conditions, including velocities, temperatures, asperity shapes, tip-tip (asperity) intersection, tip-tip (asperity) bonding and crystal orientation to determine their effects on the wear process. The design-of-experiment (DOE) approach [99] was used to analyze the relative importance of these factors and their interactions. Thermal distribution and mechanical deformation in residual Al-substrate during asperity shear were analyzed. Final results indicated that the most significant factors in determining the wear process were the tip-tip bonding (D_{int}) and the tip-tip intersection (O_I), while the starting temperature (T), the translational velocity (V_d), and the crystal orientation would play smaller roles.

25.5.1 Modeling Geometry and Simulation Procedures

Without loss of generality, Zhong *et al.* [100] carried out *classical* molecular dynamics based upon *the force-matching* EAM potential for above research. In his research, an Al-asperity was impacted by a rigid (LJ-)tip moving at a constant translational velocity, see Figure 25.17. Parameters of interatomic potentials in MD simulations were chosen as same as those in Section 25.4.1. For the rigid LJ-tip, a regular hemisphere with fcc structure was created with a diameter of $18a_0^{LJ}$ (e.g., a_0^{LJ} = 4.015Å at 300 K); for the Al-asperity, a sinusoid Al-tip was smoothly connected to a larger rectangle Al-substrate: the diameter of the Al-tip was $12a_0^{Al}$ (e.g., a_0^{Al} = 4.051 Å at 300 K), the dimension of the Al-substrate was $30a_0^{Al} \times 25a_0^{Al} \times 3a_0^{Al}$. A coordinate system was chosen with the x-axis along the $[11\bar{1}]$ direction, parallel to the direction of translational motion of the LJ-tip; the y-axis was along the [112] direction, normal to the surface of the Al-substrate, but directed away from it; the z-axis was along the $[1\bar{1}0]$ direction. All simulations were performed using a periodic boundary condition in x- and z- directions, and a free boundary condition in the y-direction; a spring force was added to the bottom layer of the Al-substrate with a component only in the y-direction — this was added to an approximate effect of a bulk slab onto the substrate. The strength of the spring force was adopted from the elastic constants $\{c_{ij}\}$ for aluminum [97], with a component in the y-direction,

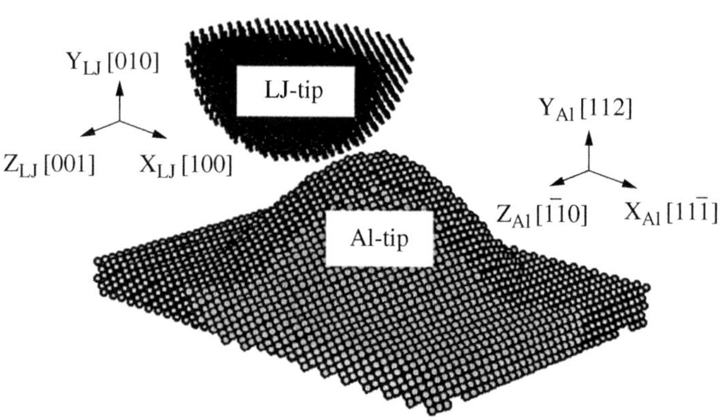

Figure 25.17 Geometries of the LJ-tip and the Al-tip during the MD simulation of asperity shear.

c', being 91.88 GPa, similar to the regular c_{11}. At the beginning of each simulation, the LJ-tip had its thermal velocity plus a translational velocity V_d. During the whole MD simulation, the uppermost layer of the LJ-tip was held at a constant V_d so that this rigid tip could keep moving translationally along the x-direction. The system was equilibrated for approximately 10 000 time steps (1 time step = 0.001 ps) at a desired temperature (e.g. 300 K), which was maintained by rescaling thermal velocities in each time step, see ref [101]. During the shearing simulation, the system was held at constant energy (no temperature control) so that temperature profiles could be well depicted.

25.5.2 Results and Discussions

25.5.2.1 One Common Wear Sequence

Figure 25.18 shows a main sequence of the LJ-tip/the Al-asperity shear starting at T = 300 K, $O_I = 1.50a_0^{Al}$, and V_d = 1.00 Å/ps: at about 20.0 ps, the LJ-tip began plowing the Al-asperity, see Figure 25.18(a); then next, a *neck* formed on the Al-tip at about 40.0 ps, and local heating occurred on this tip followed by the plowing; and then, the deformed Al-neck glided in remnant surface on the Al-substrate along a favorable slip system $(111)[10\bar{1}]$, see Figure 25.18(b–c); finally, adhesive metal transfer of aluminum to the LJ-tip occurred at about 100 ps, the removed aluminum adhered to the bottom of the LJ-tip and was cooled, see Figure 25.18(d).

Such a simulation was also carried out starting at a higher temperature: 600 K. In this case, some layers around the bottom of the Al-substrate became obviously bent, and the Al-tip was almost melting after asperity shear, i.e., the temperature was changing from 600 to above 800 K at top area on residual Al-substrate, but was still lower than the Al melting point: 930 K. In addition, the deformed Al-neck gradually lengthened and broke finally. In this case, the ductility would play a very important role in the fracture of the deformed Al-neck.

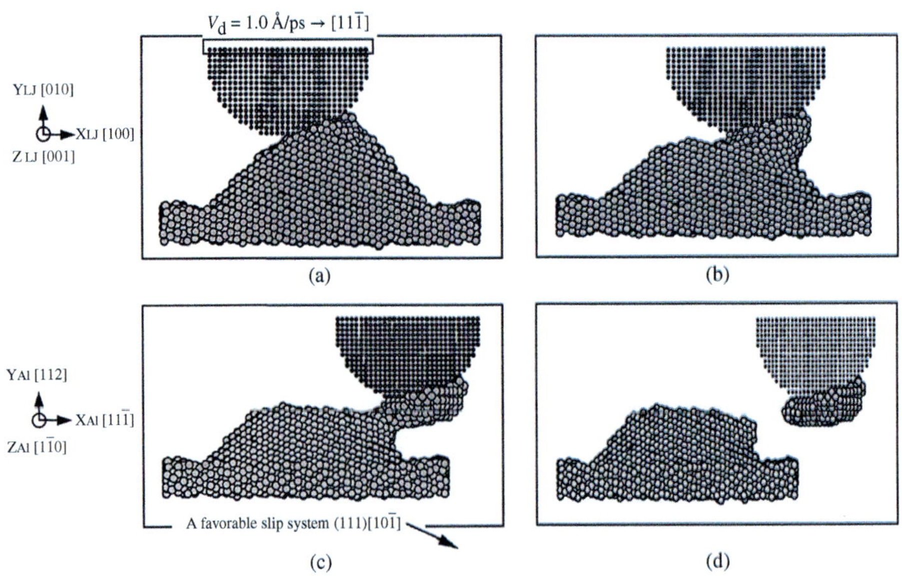

Figure 25.18 The main sequence of the LJ-tip/the Al-asperity shear starting at 300 K.

25.5.2.2 Thermal Analysis for the Wear Sequence

In this section, heat generation and temperature profiles in the Al-substrate are discussed. Similar to section 25.4.1, it should be noted that only nuclear contribution to thermal conductivity was mainly included during discussions, i.e., although free electrons behaved a major component of heat transport in bulk aluminum, the large mass difference between nuclei and electrons slowed the heating of electrons, so the electronic contribution to thermal conductivity would be very small at very short time-scale.

Figure 25.19 shows the main sequence of temperature profiles corresponding to the heat transport during MD simulation as shown in Figure 25.18. In this figure, temperatures of the residual Al-tip and the stationary Al-substrate was estimated by:

$$\frac{3}{2}nk_BT_i = \frac{1}{2}\sum_k m_k v_k^2, \tag{25.13}$$

where k_B is the Boltzmann constant, m_i, T_i and $\mathbf{v}_i$ are mass, temperature, and velocity for atom i, respectively, and n is the number of Al atoms within a sphere of about 32 Å in radius (the so-called *thermal mean path* which may convert atomic temperatures to a median, i.e., a "medium" temperature) at point i considered. For the LJ-tip and the finally adhered aluminum, temperatures were given by:

$$\frac{3}{2}nk_BT_i = \frac{1}{2}\sum_k m_k(\mathbf{v}_k - \mathbf{V}_\mathrm{d})^2. \tag{25.14}$$

Obviously, this may lead to some ambiguities in the definition of temperature at interfacial area between the moving LJ-tip and the stationary Al-asperity. Therefore, that for the moving Al-neck as an arbitrary separation between two tips was also defined by Equation (25.14), see Figure 25.19(a)–(c).

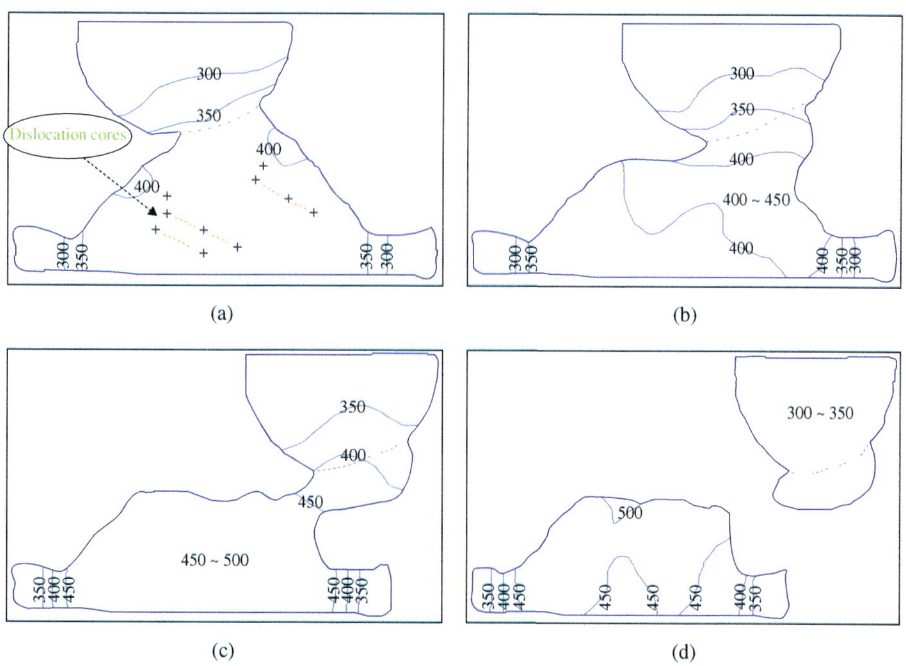

Figure 25.19 Side-view of temperature profiles corresponding to steps in Figure 25.18.

In these figures, temperature became the highest where two tips interacted and then where the deformed Al-neck glided: (i). During the initial contact stage, the tip-tip bonding (D_{int} = 0.20 eV) was strong enough that local heating could occur as bonds formed between two tips, see Figure 25.19(a), so a positive temperature gradient was from interfacial area (dot line) towards its two far sides. Simultaneously, a deformed Al-neck started forming, and dislocation cores (cross signs) identified by a centrosymmetry method [87] were emitted from higher temperature regions along a favorable slip system $(111)[10\bar{1}]$ throughout the Al substrate, see Figure 25.19(a). (ii). During the plowing stage, for the LJ-tip, heat continued diffusing into its bulk inside from interfacial area; while for the Al-asperity, the deformed Al-neck gliding in remnant surface (root area) on the residual Al-substrate, would generate additional heat to increase the temperature there, see Figure 25.19(b–c), so a new positive temperature gradient was from root area towards two far ends of the Al-substrate. (iii). Finally, the deformed Al-neck broke away from remnant surface, and the adhered aluminum part was carried off with the LJ-tip; for the residual Al-substrate, heat continued flowing from its remnant surface towards its two far ends, see Figure 25.19(d).

25.5.2.3 Wear Rate Analyses

The wear rate, W, is defined as the number of atoms removed from the Al-tip during asperity shear simulation, i.e.:

$$W = \text{\# of atoms removed from the Al-tip.} \tag{25.15}$$

In this research, factors influencing the wear rate (the W) included the starting (environment) temperature (T), the translational velocity (V_d), the tip-tip intersection (O_I), the crystal orientation in the Al-asperity, and the tip-tip bonding (D_{int}).

Effects of V_d and T Values on the W At low temperatures, 0 and 300 K, the W changed little because the Al-substrate remained rigid, and deformation mechanism of the substrate involved a relatively brittle fracture of the Al-neck. However at high temperatures, 600 and 800 K, the Al-substrate behaved in a more ductile fashion, leading to more deformation of the Al-neck and higher amounts of wear rate.

Generally, MD simulations were limited to relatively fast velocities, but these velocities were actually typical of some industrial processes. For example, during the rolling of an aluminum sheet, the highest linear speed at an edge of rigid roller contacting the Al-surface was around 60.0–70.0 m/s. According to this, three different V_d values: 50, 100, and 400 m/s (0.50, 1.00, and 4.00 Å/ps) were selected. During the shearing simulation, the W changed less with respect to various V_d values at the low temperatures, 0 and 300 K. While at high temperatures, 600 and 800 K, it looked inversely related to V_d values. The reason was that, there could be less time for ductile deformation of the Al-neck at very high velocities, and thus the W was decreased. Regarding behaviors of these two factors, see ref [100].

Effect of the O_I Value on the W As two tips came into contact, there would generate an initial bonding at interfacial area. In this research, the $N_{r\text{-cut}}$ was defined as the number of aluminum atoms within one r_{cutoff} around the LJ-tip head, where r_{cutoff} was the cutoff distance of interatomic potential between two tips. Four O_I values, $0.0a_0^{Al}$, $1.5a_0^{Al}$, $2.5a_0^{Al}$ and $3.5a_0^{Al}$ were selected. For example, in a case of T = 300 K and V_d = 0.50

Å/ps, the $N_{r\text{-cut}}$ tended to the W, i.e., when the O_I value increased, the $N_{r\text{-cut}}$ would also increase, and same as the W. It was very interesting to find that, if the spacing between two tips lay in their interaction region (within one r_{cutoff}) but the O_I value was still zero, some Al atoms were removed from the LJ-tip as well. This could involve the mechanism of neck formation in nanoindentation. As expected, the $N_{r\text{-cut}}$ looked obviously proportional to the O_I value under various other factors in MD simulation, and so did the W.

Effects of the D_{int} Value and Crystal Orientation on the W During the above simulations, the D_{int} was chosen as a value: 0.20 eV. It seemed that the D_{int} would influence more to the W than other factors. However, if it changed to a very weak value, 0.04 eV, and above simulations were repeated, some aluminum atoms on the Al-tip head were not removed but displaced away by the LJ-tip plowing. And then these atoms subsequently re-adhered onto residual Al-substrate, which could be like a coating type on remnant surface.

For the Al-asperity, if it was reconstructed with a slip plane (111) parallel to the V_d without changing other directions, the moving LJ-tip would behave like an impact stress on the Al-tip. In this case, the Schmid factor $m = \cos(\varphi)\cdot\cos(\lambda)$ could be applied to investigating the influence of crystal orientation on the W (where φ is the angle between the V_d vector and the normal vector of slip plane (111), and λ is that between the V_d vector and the [101] direction). Here $m = 0$, while that in Figure 25.15 was 0.2981. Even so, different crystal orientations were found to have little effect on the W.

Analysis of the Design-of-experiment on the W In Analysis of the design-of-experiment (DOE), factorial designs were widely used in experiments and simulations involving several factors, where it was necessary to study the joint effect of factors on a response [99]. Here, a 2^k *factorial design* with k factors each at two levels is used to analyze effects of those variables (factors) plus their interaction response: the W. Table 25.6 lists factors contributing to the 2^k factorial design for this work. In this table, two "low (–)" and "high (+)" levels were chosen for each factor. Thus $k = 4$, and the total number of combinations was 16. In this analysis, 16 runs were made in random order, which corresponded to 16 contrasts. Table 25.7 shows factorial effects for these contrasts, and the percentage of the response: the W. According to this table, primary effects that emerged from this analysis were main factors A, B, D, and those of AB and BD interactions.

Table 25.6 Factors contributing the 2^4 factorial design.

Factors	Factor levels	
	Low (−1)	High (+1)
$A(T)$	300 K	600 K
$B(O_I)$	$1.5a_0$	$3.5a_0$
$C(V_d)$	0.50 Å/ps	1.00 Å/ps
$D(D_{int})$	0.04 eV	0.20 eV

Table 25.7 Effects on the contrasts and the percentage of the response.

Model term	Effect estimate	Percentage (%)
A	2.2650	1.17
B	7.3450	12.30
AB	0.9375	0.20
C	0.4350	0.04
AC	0.5375	0.06
BC	0.7575	0.13
ABC	0.2350	0.01
D	17.8175	72.20
AD	2.2650	1.17
BD	7.3450	12.30
ABD	0.9375	0.20
CD	0.4350	0.04
ACD	0.5275	0.06
BCD	0.7575	0.13
ABCD	0.2350	0.01

Based upon the final analysis, the most significant factor in determining the *W* was the tip-tip bonding (D_{int}), i.e., when it dropped pretty low (e.g., 0.04 eV), no materials were removed in any of eight simulations; the tip-tip intersection (O_I) was also important; while the starting (environment) temperature (T) and the impact velocity (V_d) seemed to play smaller roles.

Later, Zhong and Adams carried out similar MD simulations with some modifications on this research [102]: they increased the bond strength of the LJ-tip to 1.10 eV, redistricted the D_{int} within (0.20, 1.00) eV, the V_d within (1.00, 10.0) Å/ps, the T within (100, 700) K, and the O_I within $(1.5, 3.5)a_0$, respectively. Besides, four mid-points: 0.65 eV, 5.50 Å/ps, 400 K and $2.5a_0$ were added into the DOE analysis, so the selected technique in design was renamed "The 2^4 full factorial design" with one central point for each factor (the mid-point of the range).

In their comprehensive analysis, any replicates or blocking were not considered because these would not offer additional information due to the nature of computer simulations. So totally, 17 MD simulations were ran for different combinations of four selected factors in above. Since the *W* varied over a large range, i.e., since the ratio of the maximum to the minimum *W* values was 12, the *W* was transformed using a natural log algorithm. Also, it can be more useful to describe the *W* by the normalized variables (i.e., variables were normalized to a scale from −1 to +1, the low and high levels for each variable) because the magnitude of above four factors allowed one to easily determine the relative importance of each variable.

Finally, the modified magnitude of factor coefficients indicated that, the T (= 0.19) term had a small effect, the V_d (= −0.12) term had a minor inverse correlation, while the O_I (= 0.42) and the D_{int} (= 0.37) terms showed major effects on the *W*, i.e., effects of the tip-tip intersection (O_{I}) and the tip-tip bonding (D_{int}) would both influence more to the wear rate than the starting temperature (T) and the impact velocity (V_{d}).

Effect of the LJ-tip Plowing on the Frictional Coefficient The frictional coefficient at atomic-scale, μ_f, was defined as an absolute mean value (in some cases, it was also defined as a value of root-mean-square) of the lateral force over the normal force on the bottom layer of the LJ-tip during the plowing stage, i.e.:

$$\mu_f = \frac{1}{N_f}\sum_i \left|\frac{\text{lateral force}}{\text{normal force}}\right|_i, \tag{25.16}$$

where N_f was the total number of time steps counted in the calculation. Harrison *et al.* [103] investigated an atomic-scale friction of a diamond-surface sliding along the $[11\bar{2}]$ direction. They found that the frictional coefficient was slightly decreased with the temperature increase. In this shearing simulation, no matter how the LJ-tip was plowing along the $[11\bar{1}]$ direction, the frictional coefficient always varied inversely with the temperature, ranging from 0 to 800 K.

In addition, results of the shearing simulations under various constraints (e.g., D_{int} = 0.04, 0.20 eV) indicated that, the frictional coefficient varied by less than 0.5%, which seemed to be independent of these constraints. Similarly, Raff *et al.* ran a series of MD simulations of atomic-scale scratches on a regular surface of single-crystal aluminum [29]. Their calculated frictional coefficient, 0.661±3%, seemed to be independent of the scratch depth, the strength of interatomic bonding, the sliding condition (orientation), and the external load, etc. And, their calculations yielded a value of static frictional coefficient: 0.61 between aluminum–soft iron contact. Calculations here for the sliding frictional coefficient gave 0.435±2.6%, which matched a value of the sliding frictional coefficient, 0.46, for an aluminum sliding on another soft iron [104].

Conclusions In this section, adhesive wear of a fast moving rigid LJ-tip towards an Al-asperity was investigated using molecular dynamics simulation. The simulations were conducted for a dozen different combinations of four main variables: translational velocity of rigid tip, starting (environment) temperature, tip-tip intersection and tip-tip bonding. A method in design-of-experiment (DOE) was adopted in analyzing effects of these variables on the wear process. In addition, tip shapes and crystal orientation in the substrate were added into MD simulations. The goal of this research was to investigate the wear mechanism at atomic-scale.

It was found that (1) the wear process occurred in three stages: the rigid tip contact, the rigid tip plowing, and the necking and the fracture of soft asperity (adhesive metal transfer of aluminum to the rigid LJ-tip). Bond formation and mechanical deformation during asperity shear resulted in a large increase of local temperature in shear region, so the primary deformation (e.g., lengthening and fracture) mechanism of soft asperity was due to its local melting and amorphous plasticity. Generation and emission of dislocation cores along favorable slip systems were observed in the substrate under milder conditions where little melting occurred.

(2) Thermal analysis of MD simulations indicated that, (i) As two tips came into contact, the forming of tip-tip bonds heated the soft asperity; (ii) The gliding of the soft deformed-neck produced new heat in remnant surface; (iii) Finally, the heat diffused into residual soft substrate from remnant surface, which increased temperature there.

(3) Analysis of design-of-experiment indicated that, the most significant factors influencing the wear rate were both tip-tip bonding (D_{int}) and tip-tip intersection (O_I). While others such as the starting (environment) temperature (T), the translational velocity (V_d), and the crystal orientation in soft asperity were less important.

25.6 Summary and Prospect

In this chapter, atomistic simulations of adhesion, indentation and wear at nanoscale are comprehensively discussed, which may illuminate a reasonable way for people's further investigations in this fascinating field. Some main points in the discussion include, without any loss of generality:

1) In adhesion phenomena, the surface energy may play a dominant role in determining the work of separation (W_{sep}) between two contact surfaces as the interface energies were generally much smaller in comparison if ignoring effects of large stain and incoherent bonding at interfacial area.
2) Common phenomena in nanoindentation, including jump-to-contact (neck formation), local melting, adhesive materials transfer (phase transformation of indented materials), non-monotonic features on the *P-h* curve due to dislocation nucleation and emission throughout the substrate, and incomplete bond reactions at interfacial area etc., were simulated through *classical* and *ab initio* molecular dynamics. In addition, effects of several constraint parameters such as the loading force, the system (environment) temperature, the tip-substrate bonding, and the substrate orientation etc. were also discussed. As a result, a higher loading force usually led to a larger indent depth; each loading force value was found to be proportional to the square of indent depth and cross-sectional dimension of contact area, which may help find out the critical point of yield stress of the indented materials; higher temperatures also resulted in larger indent depths because much easier dislocation nucleation and emission in this case would make the substrate softer; more effective dangling bonds in surface of the indented materials would attract the indentor to react more intensively with the substrate. The initial stage of the *P-h* curve can be drawn throughout *ab initio* molecular dynamics simulation to provide invaluable information which was hardly observed throughout *classical* molecular dynamics simulation. Due to the very short simulation time, a thermit reaction at the interfacial area between to contact slabs was incomplete.
3) In wear process, *classical* molecular dynamics was conducted for investigating a dozen different combinations of four main variables: translational velocity of rigid tip, starting (environment) temperature, tip-tip intersection and tip-tip bonding. A method in design-of-experiment was adopted in analyzing effects of these variables on the wear process. Simulation results indicated that, (i) The wear process occurred in three stages: the rigid tip contact, the rigid tip plowing, and the adhesive materials transfer from the soft asperity; generation and emission of dislocation cores along favorable slip systems took place in the substrate under milder conditions where little melting occurred. (ii) Thermal analysis of MD simulations showed that, heat was generated from adhesive reaction area where two tips initially contacted, and then in remnant surface where the deformed materials glided on the

residual substrate. (iii) Analysis of design-of-experiment indicated that, the most significant factors influencing the wear rate were both tip-tip bonding and tip-tip intersection, while other factors seemed to be less important.

Future Tasks Future works in this field would involve lubricant reactions with contact area of two solid surfaces. Some aspects may include: (1) Prediction of chemical reactions of lubricants on pure and oxidized metal surfaces using density functional theories – lubricant formulations used to control indentation, friction and wear processes typically contained mixtures of additive molecules (such as fatty acids, alcohols and esters, etc.) which performed as boundary-layer lubricants in the base oil. They bound to surfaces to form protective thin films several molecular layers thick, which was critical for applications involving high loads because they were able to stay in contact wells after oil was forced away. Previous research has been done on some preliminarily explorations into this question [105–108], various electronic structures were employed to distinguish the binding strengths of adsorbates onto surfaces. As results, mechanisms like the binding sites and the corresponding reaction enthalpies etc. were therefore deemed to be essential characteristics of adsorption/desorption feasibilities, which mainly depended upon crystallography, mutual interactions of surface geometries and additive molecular structures (with long molecular chains and particularly polar functions). Such studies may also help elucidate adhesive mechanisms at interfacial area in composite materials.

(2) Predictions of nanoindentation, abrasive friction/wear properties of two contact solid slabs with dense additives (made up of fatty-acid, alcohol and ester molecules, etc.) adsorbed onto one of their surfaces using molecular dynamics – some previous works have paved a way for this goal. For instance, perfect gold surfaces scratching with perfect adsorbed layers has been described in detail [109, 110]. So, the future works will involve temperature effects on kinetics of adsorbate layers and stress behaviors at contact area, e.g., at low temperatures, additive layers should strongly anchor onto surfaces so that indent and frictional energies can be mostly spent by conformation changes of molecular chains; whereas at higher temperatures, especially close to the stage of spontaneous thermal desorption, indentation and shear could occur by either disruption of additive layers-surface bonding or by diffusion of molecular chains, which may result in synergistic or antagonistic effects, or molecular segregation into clusters. According to these, such studies would separate effects of molecular interactions within additive layers and those between molecules and solid surfaces.

Development of a New Methodology Q. Peng and G. Lu *et al.* were developing a multi-scale approach that was based entirely upon density functional theory [36, 37]. This method was termed *quasi-continuum density functional theory* (QCDFT) [110–114], which extended the idea of coarse-grain in quasi-continuum (QC) approach (a coupling of quantum mechanics/molecular mechanics (QM/MM) to allow quantum simulations at micro-scale and beyond).

As they announced: 'the QCDFT was not a 'brute-force' method for electronic structure, but rather a multi-scale approach that can treat large systems – effectively up to billions of valence electrons. According to this, in the QCDFT some of electronic degrees of the freedom may reduce to continuum degrees of the freedom.' Although the QCDFT utilized the idea of the QM/MM coupling, it did not involve any information from empirical models (e.g., the force-fields), i.e., the energy calculation by the QCDFT was entirely based upon orbital-free DFT. This was an important feature in the QCDFT, which may qualify it as a 'bona-fide' quantum simulation method.

Recently, the QCDFT have been employed to study several questions like nanoindentation of the pure Al [113], random distribution of Mg impurities in the Al thin film [114], effects of solid solution on dislocation nucleation during nanoindentation [112], and the fracture in the Al bulk [111], etc. Regarding its aspects, please see ref [115].

Acknowledgments

This chapter was supported by the National Science Foundation (Grant no.: DMR9619353), USA; the Hebei Province Science and Technology Support Program (Grant no.: 15961006D and Grant no.: 16211031), the Startup Fund of Doctoral Research at North China Institute of Aerospace Engineering (Grant no.: BKY201405), and the Opening Fund of State Key Laboratory of Nonlinear Mechanics at Institute of Mechanics in Chinese Academy of Sciences (Grant no.: LNM201503), China.

References

1 Doerner, MF, Nix WD. A method for interpreting the data from depth-sensing indentation instruments. Journal of Materials Research. 1986, Vol. 1, 601–609, DOI: http://dx.doi.org/10.1557/JMR.1986.0601.
2 Oliver, WC, Pharr GM. An improved technique for determining hardness and elastic modulus using load and displacement sensing indentation experiments. Journal of Materials Research. 1992, Vol. 7, 1564–1583, DOI: http://dx.doi.org/10.1557/JMR.1992.1564.
3 Gerberich, WW, Nelson JC, Lilleodden ET, Anderson E, Wyrobek JT. Indentation induced dislocation nucleation: The initial yield point. Acta Materialia. 1992, Vol. 44, 3585–3598, DOI: 10.1016/1359-6454(96)00010-9.
4 Gouldstone, A, Koh HJ, Zeng, KY. Discrete and continuous deformation during nanoindentation of thin films. Acta Materialia. 2000, Vol. 48, 2277–2295, DOI: 10.1016/S1359-6454(00)00009-4.
5 Page, TF, Oliver WC, McHargue CJ. The deformation behavior of ceramic crystals subjected to very low load (nano) indentations. Journal of Materials Research. 1992, Vol. 7, 450–473, DOI: http://dx.doi.org/10.1557/JMR.1992.0450.
6 Blau, PJ. Friction, Lubrication, and Wear Technology – ASM Metals Handbook. 1992, Vol. 18, 419.
7 Thurn, J, Cook RF. Indentation-induced deformation at ultramicroscopic and macroscopic contacts. Journal of Materials Research. 2004, Vol. 19, 124–130, DOI: http://dx.doi.org/10.1557/jmr.2004.19.1.124.
8 Jungk, JM, Mook WM, Cordill MJ, Chambers MD. Length-scale-based hardening model for ultra-small volumes. Journal of Materials Research. 2004, Vol. 19, 2812–2821, DOI: http://dx.doi.org/10.1557/JMR.2004.0384.
9 Mate, CM, McClland GM, Erlandsson R, Chiang S. Atomic-scale friction of a tungsten tip on a graphite surface. Physical Review Letters. 1987, Vol. 59, 1942–1945, DOI: http://dx.doi.org/10.1103/PhysRevLett.59.1942.

10 Hector, L.G., Jr., Schmid SR. Simulation of asperity plowing in an atomic force microscope Part 1: Experimental and theoretical methods. Wear. 1998, Vol. 215, 247–256, DOI: 10.1016/S0043-1648(97)00245-7.
11 Hector, LG, Jr., Schmid SR. Simulation of asperity plowing in an atomic force microscope. Part II: Plowing of aluminum alloys. Wear. 1998, Vol. 215, 257–266, DOI: 10.1016/S0043-1648(97)00244-5.
12 Hector, L.G., Jr., Opalka SM. Fundamentals of nanoindentation and nanotribology. Materials Research Society Symposium – Proceedings. Platinum, San Francisco, USA, Spring 1998.
13 Opalka, SM, Hector LG Jr., Schimid SR. Boundary Additive Effect on Abrasive Wear During Single Asperity Plowing of a 3004 Aluminum Alloy. Journal of Tribology – Transactions of the ASME. 1999, Vol. 121, 384–393, DOI: 10.1115/1.2833951.
14 Rubio, G, Agrait N, Vieira S. Atomic-Sized Metallic Contacts: Mechanical Properties and Electronic Transport. Physical Review Letters. 1996, Vol. 76, 2302–2305, DOI: http://dx.doi.org/10.1103/PhysRevLett.76.2302.
15 Agrait, N, Rubio G, Vieira S. Plastic Deformation of Nanometer-Scale Gold Connective Necks. Physical Review Letters. 1995, Vol. 74, 3995–3998, DOI: http://dx.doi.org/10.1103/PhysRevLett.74.3995.
16 Cheng, YT, Cheng CM. On the initial unloading slope in indentation of elastic–plastic solids by an indenter with an axisymmetric smooth profile. Applied Physical Letters. 1997, Vol. 71, 2623–2625, DOI: http://dx.doi.org/10.1063/1.120159.
17 Suresh, S, Nieh TG, Choi BW. Nano-indentation of copper thin films on silicon substrates. Scripta Materialia. 1999, Vol. 41, 951–957, DOI: 10.1016/S1359-6462(99)00245-6.
18 Tsui, YY, Oliver WC, Pharr GM. Influences of stress on the measurement of mechanical properties using nanoindentation: Part I. Experimental studies in an aluminum alloy. Journal of Materials Research. 1996, Vol. 11, 752–759, DOI: http://dx.doi.org/10.1557/JMR.1996.0091.
19 Figuera, JD, Gonzalez MA. Epitaxial Growth – Principles and Applications. Materials Research Society Symposium – Proceedings. Platinum, San Francisco, USA, Spring 1997.
20 Lehoucq, RB, Silling SA, Seleson P, Plimpton SJ, Parks ML. Peridynamics with LAMMPS: a user guide. http://www.sandia.gov. 2011. DOI: 10.2172/1031301.
21 Vashishta, P, Bachlechner ME, Campbell T, Walsh P. Multimillion Atom Simulations of Nanostructured Materials on Parallel Computers: Sintering and Consolidation, Fracture, and Oxidation. Progress of Theoretical Physics Supplements. 2000, Vol. 138, 175–190, DOI: 10.1143/PTPS.138.175.
22 Landman, U, Luedtke WD, Burnham NA, Colton RJ. Atomistic Mechanisms and Dynamics of Adhesion, Nanoindentation, and Fracture. Science. 1990, Vol. 248, 454–461, DOI: 10.1126/science.248.4954.454.
23 Gao, J, Luedtke WD, Landman U. Nano-Elastohydrodynamics: Structure, Dynamics, and Flow in Nonuniform Lubricated Junctions. Science. 1995, Vol. 270, 605–608, DOI: 10.1126/science.270.5236.605.
24 Smith, JR, Bozzolo G, Banerjea A, Ferrante J. Avalanche in adhesion. Physical Review Letters. 1989, Vol. 63, 1269–1272, DOI: http://dx.doi.org/10.1103/PhysRevLett.63.1269.

25 Rafii-Tabar, H, Kawazoe, Y. Dynamics of Atomically Thin Layers-Surface Interactions in Tip-Substrate Geometry. Japanese Journal of Applied Physics. 1993, Vol. 32, 1394–1400, DOI: 10.1143/JJAP.32.1394.
26 Rafii-Tarbar, H. Modelling the nano-scale phenomena in condensed matter physics via computer-based numerical simulations. Physics Reports. 2000, Vol. 325, 239–310, DOI: 10.1016/S0370-1573(99)00087-3.
27 Buldum, A, Ciraci S. Atomic-scale study of dry sliding friction. Physical Review B. 1997, Vol. 55, 2606–2611, DOI: http://dx.doi.org/10.1103/PhysRevB.55.2606.
28 Buldum, A, Ciraci S, Bartra IP. Contact, nanoindentation, and sliding friction. Physical Review B. 1998, Vol. 57, 2468–2476, DOI: http://dx.doi.org/10.1103/PhysRevB.57.2468.
29 Komanduri, R, Chandrasekaran N, Raff LM. Molecular dynamics simulation of atomic-scale friction. Physical Review B. 2000, Vol. 61, 14007–14019, DOI: http://dx.doi.org/10.1103/PhysRevB.61.14007.
30 Walsh, P, Kalia RK, Nakano A, Saini S. Amorphization and anisotropic fracture dynamics during nanoindentation of silicon nitride: A multimillion atom molecular dynamics study. Applied Physics Letters. 2000, Vol. 77, 4332–4334, DOI: http://dx.doi.org/10.1063/1.1328371.
31 Phillips, R, Rodney D, Shenoy V, Tadmor E, Ortiz M. Hierarchical models of plasticity: dislocation nucleation and interaction. Modelling and Simulation in Materials Science and Engineering. 1999, Vol. 7, 769–780, DOI: 10.1088/0965-0393/7/5/309.
32 Tadmor, EB, Phillips R, Ortiz M. Hierarchical modeling in the mechanics of materials. International Journal of Solids and Structures. 2000, Vol. 37, 379–389, DOI: 10.1016/S0020-7683(99)00095-5.
33 Shenoy, VB, Miller R, Tadmor EB, Phillips R, Ortiz M. Quasicontinuum Models of Interfacial Structure and Deformation. Physical Review Letters. 1998, Vol. 80, 742–745, DOI: http://dx.doi.org/10.1103/PhysRevLett.80.742.
34 Ortiz, M, Cuitino AM, Knap J, Koslowski M. Mixed Atomistic-Continuum Models of Material Behavior: The Art of Transcending Atomistics and Informing Continua. MRS Bulletin. 2001, Vol. 26, 216–221, DOI: http://dx.doi.org/10.1557/mrs2001.45.
35 Born, M, Oppenheimer R. Zur Quantentheorie der Molekeln (On the Quantum Theory of Molecules). Annalen der Physik (in German). 1927, Vol. 389, 457–484. DOI: 10.1002/andp.19273892002.
36 Hohenberg P, Kohn W. Inhomogeneous Electron Gas. Physical Review. 1964, Vol. 136, B864–871, DOI: 10.1103/PhysRev.136.B864.
37 Kohn, W, Sham LJ. Self-Consistent Equations Including Exchange and Correlation Effects. Physical Review. 1965, Vol. 140, A1133–1138, DOI: 10.1103/PhysRev.140.A1133.
38 Kresse, G, Joubert D. From ultrasoft pseudopotentials to the projector augmented-wave method. Physical Review B. 1999, Vol. 59, 1758–1775, DOI: http://dx.doi.org/10.1103/PhysRevB.59.1758.
39 Jones, RO, Gunnarsson O. The density functional formalism, its applications and prospects. Reviews of Modern Physics. 1989, Vol. 61, 689–746, DOI: http://dx.doi.org/10.1103/RevModPhys.61.689.
40 Perdew, JP. Accurate Density Functional for the Energy: Real-Space Cutoff of the Gradient Expansion for the Exchange Hole. Physical Review Letters. 1985, Vol. 55, 1665–1668, DOI: http://dx.doi.org/10.1103/PhysRevLett.55.1665.

41 Perdew, JP, Burke K, Ernzerhof M. Generalized Gradient Approximation Made Simple. Physical Review Letters. 1996, Vol. 77, 3865–3868, DOI: http://dx.doi.org/10.1103/PhysRevLett.77.3865.
42 Perdew, JP. Electronic Structure of Solids '91. Akademie Verlag, Berlin, Germany, 1991.
43 Perdew, JP, Chevary JA, Fiolhais C. Atoms, molecules, solids, and surfaces: Applications of the generalized gradient approximation for exchange and correlation. Physical Review B. 1992, Vol. 46, 6671–6687, DOI: http://dx.doi.org/10.1103/PhysRevB.46.6671.
44 Perdew, JP, 1991. Generalized gradient approximations for exchange and correlation: A look backward and forward. Physica B. 1991, Vol. 172, 1–6, DOI: 10.1016/0921-4526(91)90409-8.
45 Perdew, JP, Burke K, Ernzerhof M. ERRATA: Generalized Gradient Approximation Made Simple. Physical Review Letters. 1997, Vol. 78, 1396, DOI: http://dx.doi.org/10.1103/PhysRevLett.78.1396.
46 Ashcroft, NW, Mermin ND. Solid State Physics. Saunders, Philadelphia, USA, 1976.
47 Chadi, DL, Cohen ML. Special Points in the Brillouin Zone. Physical Review B. 1973, Vol. 8, 5747–5753, DOI: http://dx.doi.org/10.1103/PhysRevB.8.5747.
48 Monkhorst, HJ, Pack JD. Special points for Brillouin-zone integrations. Physical review B. 1976, Vol. 13, 5188–5192, DOI: http://dx.doi.org/10.1103/PhysRevB.13.5188.
49 Bachelet, GB, Hamann DR, Schlüter M. Pseudopotentials that work: From H to Pu. Physical Review B. 1982, Vol. 26, 4199–4228, DOI: http://dx.doi.org/10.1103/PhysRevB.26.4199.
50 Pasquarello, A, Laasonen K, Car R, Lee C, Vanderbilt D. Ab initio molecular dynamics for d-electron systems: Liquid copper at 1500 K. Physical Review Letters. 1992, Vol. 69, 1982–1985, DOI: http://dx.doi.org/10.1103/PhysRevLett.69.1982.
51 Vanderbilt, D. Soft self-consistent pseudopotentials in a generalized eigenvalue formalism. Physical Review B. 1990, Vol. 41, 7892(R)–7895, DOI: http://dx.doi.org/10.1103/PhysRevB.41.7892.
52 Blöchl, PE. Projector augmented-wave method. Physical Review B. 1994, Vol. 50, 17953–17979, DOI: http://dx.doi.org/10.1103/PhysRevB.50.17953.
53 Kresse, G, Hafner J. Norm-conserving and ultrasoft pseudopotentials for first-row and transition elements. Journal of Physics: Condensed Matter. 1994, Vol. 6, 8245–8258, DOI: 10.1088/0953-8984/6/40/015.
54 Alder, BJ, Wainwright TE. Phase Transition for a Hard Sphere System. Journal of Chemical Physics. 1957, Vol. 27, 1208–1209, DOI: http://dx.doi.org/10.1063/1.1743957. DOI:10.1063/1.1743957;
55 Verlet, L. 1967. Computer "Experiments" on Classical Fluids. I. Thermodynamical Properties of Lennard-Jones Molecules. Physical Review. 1967, Vol. 159, 98–103, DOI: http://dx.doi.org/10.1103/PhysRev.159.98.
56 Beeman, D. Journal of Computational Physics. 1976, Vol. 20, 130–139, DOI: 10.1016/0021-9991(76)90059-0.
57 Gear, GW. Numerical Initial Value Problems in Ordinary Differential Equations. Prentice-Hall Inc., Englewood Cliffs, New Jersey, USA, 1971.
58 Heermann, DW. Computer Simulation Method in Theoretical Physics. Springer Verlag, Berlin, Germany, 1990.

59 Nosé, S. A molecular dynamics method for simulations in the canonical ensemble. Molecular Physics. 1984, Vol. 52, 255–268, DOI: 10.1080/00268978400101201.
60 Anderson, HC. Molecular dynamics simulations at constant pressure and/or temperature. Journal of Chemical Physics. 1980, Vol. 72, 2384–2393, DOI: http://dx.doi.org/10.1063/1.439486.
61 Parrinello, M, Rahman A. Polymorphic transitions in single crystals: A new molecular dynamics method. Journal of Applied Physics. 1981, Vol. 52, 7182–7190, DOI: http://dx.doi.org/10.1063/1.328693.
62 Kaplan, IG. Intermolecular Interactions – Physical Picture, Computational methods and Model Potentials. John Wiley & Sons Ltd., New York, USA, 2006.
63 Daw, MS, Baskes MI. Semiempirical, Quantum Mechanical Calculation of Hydrogen Embrittlement in Metals. Physical Review Letters. 1983, Vol. 50, 1285–1288, DOI: http://dx.doi.org/10.1103/PhysRevLett.50.1285.
64 Daw, MS, Baskes MI. Embedded-atom method: Derivation and application to impurities, surfaces, and other defects in metals. Physical review B. 1984, Vol. 29, 6443–6453, DOI: 10.1103/PhysRevB.29.6443. DOI:10.1103/PhysRevB.29.6443;
65 Ercolessi, F, Adams JB. Interatomic Potentials from First-Principles Calculations: The Force-Matching Method. European Physics Letter. 1994, Vol. 26, 583–588, DOI: 10.1209/0295-5075/26/8/005.
66 Kohn, W. Nobel Lecture: Electronic structure of matter – wave functions and density functionals. Reviews of Modern Physics. 1998, Vol. 71, 1253–1266, DOI: http://dx.doi.org/10.1103/RevModPhys.71.1253.
67 Payne, MC, Allan DC, Arias TA. Joannopoulos JD. Iterative minimization techniques for ab initio total-energy calculations: molecular dynamics and conjugate gradients. Reviews of Modern Physics. 1992, Vol. 64, 1045–1098, DOI: http://dx.doi.org/10.1103/RevModPhys.64.1045.
68 Marx, D, Hütter J. Ab Initio Molecular Dynamics: Basic Theory and Advanced Methods. Cambridge University Press, Cambridge, UK, 2009.
69 Car, R, Parrinello M. Unified Approach for Molecular Dynamics and Density-Functional Theory. Physical Review Letters. 1985, Vol. 55, 2471–2474, DOI: http://dx.doi.org/10.1103/PhysRevLett.55.2471.
70 Feynman, RP. Forces in Molecules. Physical Review. 1939, Vol. 56, 340–342, DOI: http://dx.doi.org/10.1103/PhysRev.56.340.
71 Finnis, MW. The theory of metal – ceramic interfaces. Journal of Physics: Condensed Matter. 1996, Vol. 8, 5811–5836, DOI: 10.1088/0953-8984/8/32/003.
72 Humenik, M, Kingery WD. Metal-Ceramic Interactions: III, Surface Tension and Wettability of Metal-Ceramic Systems. Journal of the American Ceramic Society. 1954, Vol. 37, 18–23, DOI: DOI: 10.1111/j.1151-2916.1954.tb13972.x.
73 Howe, JM. Bonding, structure, and properties of metal/ceramic interfaces: Part 1 Chemical bonding, chemical reaction, and interfacial structure. International Materials Reviews. 1993, Vol. 38, 233–256, DOI: http://www.maneyonline.com/doi/abs/10.1179/imr.1993.38.5.233.
74 McDonald, JE, Eberhart JG. Transactions of the Metallurgical Society of the AIME. 1965, Vol. 233, 512.
75 R.G. Barrera, RG, Duke CB. Dielectric continuum theory of the electronic structure of interfaces. Physical Review B. 1976, Vol. 13, 4477–4489, DOI: http://dx.doi.org/10.1103/PhysRevB.13.4477.

76 Laurent, V, Chatain D, Chatillon C, Eustathopoulos N. Acta Materialia. 1988, Vol. 36, 1797.
77 Eustathopoulos, N, Chatain D, Coudurier L. Wetting and interfacial chemistry in liquid metal-ceramic systems. Materials Science and Engineering A. 1991, Vol. 135, 83–88, DOI: 10.1016/0921-5093(91)90541-T.
78 Siegel, DJ, Hector LG Jr, Adams JB. Adhesion, atomic structure, and bonding at the Al(111)/α–Al2O3(0001) interface: A first principles study. Physical Review B. 2002, Vol. 65, 085415, DOI: http://dx.doi.org/10.1103/PhysRevB.65.085415.
79 Siegel, DJ, Hector LG Jr, Adams JB. Adhesion, stability, and bonding at metal/metal-carbide interfaces: Al/WC. Surface Science. 2002, Vol. 498, 321–336, DOI: 10.1016/S0039-6028(01)01811-8.
80 Siegel, DJ, Hector LG Jr, Adams JB. First-principles study of metal–carbide/nitride adhesion: Al/VC vs. Al/VN. Acta Materialia. 2002, Vol. 50, 619–631, DOI: 10.1016/S1359-6454(01)00361-5.
81 Siegel, DJ, Hector LG Jr, Adams JB. Ab initio study of Al-ceramic interfacial adhesion. Physical Review B. 2003, Vol. 67, 092105, DOI: http://dx.doi.org/10.1103/PhysRevB.67.092105.
82 Hellmann, H. Einfuhrung in die Quantumchemie. Deuticke, Leipzig, Deutschland, 1937.
83 Ackland, GJ, Bacon DJ, Calder AF, Harry T. Computer simulation of point defect properties in dilute Fe-Cu alloy using a many-body interatomic potential. Philosophy Magazine A. Vol. 76, 713–732, DOI: 10.1080/01418619708207198.
84 Cheng, YT, Cheng CM. Relationships between hardness, elastic modulus, and the work of indentation. Applied Physics Letters. 1998, Vol. 73, 614–616, DOI: http://dx.doi.org/10.1063/1.121873.
85 Cheng, YT, Cheng CM. Scaling approach to conical indentation in elastic–plastic solids with work hardening. Journal of Applied Physics. 1998, Vol. 84, 1284–1291, DOI: http://dx.doi.org/10.1063/1.368196.
86 Cheng, YT, Cheng CM. Can stress–strain relationships be determined from indentation curves using conical and pyramidal indenters? Journal of Materials Research. 1999, Vol. 14, 3493–3496, DOI: http://dx.doi.org/10.1557/JMR.1999.0472.
87 Kelchner, CL, Plimpton SJ, Hamilton JC. Dislocation nucleation and defect structure during surface indentation. Physical Review B. 1998, Vol. 58, 11085–11088. DOI: http://dx.doi.org/10.1103/PhysRevB.58.11085.
88 Adams, JB, Hector LG Jr, Siegel DJgg, Yu H, Zhong J. Adhesion, lubrication and wear on the atomic scale. Surface and Interface Analysis. 2001, Vol. 31, 619–626, DOI: 10.1002/sia.1089.
89 Yu, H, Adams JB, Hector LG Jr. Molecular dynamics simulation of high-speed nanoindentation. Modelling and Simulation in Materials Science and Engineering. 2002, Vol. 10, 319–330, DOI: 10.1088/0965-0393/10/3/305.
90 Li, J, Ngan HW, Gumbsch P. Atomistic modeling of mechanical behavior. Acta Materialia. 2003, Vol. 51, 5711–5742, DOI: 10.1016/j.actamat.2003.08.002.
91 Zhu, T, Yip S. Predictive modeling of nanoindentation-induced homogeneous dislocation nucleation in copper. Journal of the Mechanics and Physics of Solids. 2004, Vol. 52, 691–724, DOI: 10.1016/j.jmps.2003.07.006.

92 Zhong, J, Adams JB. Adhesive metal transfer at the Al(1 1 1)/α-Fe_2O_3(0001) interface: a study with ab initio molecular dynamics. Modelling and Simulation in Materials Science and Engineering. 2008, Vol. 16, 085001, DOI: 10.1088/0965-0393/16/8/085001. – This article was selected for the MSMSE's 2008 highlight which was rated in the top 3% out of all publications at Institute Of Physics, UK, 2008.

93 Zhong, J, Adams JB. Adsorption and Decomposition Pathways of Vinyl Phosphonic and Ethanoic Acids on the Al(111) Surface: a Density Functional Analysis. Journal of Physical Chemistry C. 2007, Vol. 111, 7366–7375, DOI: 10.1021/jp0667487.

94 Yamamoto, N. The Shift of the Spin Flip Temperature of α-Fe2O3 Fine Particles. Journal of the Physical Society of Japan. 1968, Vol. 24, 23–28, DOI: http://dx.doi.org/10.1143/JPSJ.24.23.

95 Coast, R, Pikus M, Henriksen NP, Nitowski AG. A vibrational spectroscopic comparison of vinyl-triethoxysilane and vinyl-phosphonic acid adsorbed on oxidized aluminum. Journal of Adhesion Science and Technology. 1996, Vol. 10, 101–121, DOI: 10.1163/156856196X00805.

96 Szlufarska, I. Atomistic simulations of nanoindentation. Materials Today. 2006, Vol. 9, 42–50, DOI: 10.1016/S1369-7021(06)71496-1.

97 Simmons, G, Wang H. Single Crystal Elastic Constants and Calculated Aggregate Properties: A Handbook. 2nd ed, Cambridge, MIT Press, USA, 1971.

98 Tomar, V, Zhou M. Tension-compression strength asymmetry of nanocrystalline α-Fe_2O_3+fcc-Al ceramic-metal composites. Applied Physics Letters. 2006, Vol. 88, 233107, DOI: http://dx.doi.org/10.1063/1.2210797.

99 Montgomery, DC. Design and Analysis of Experiments. 5th ed, Wiley, New York, USA, 2001.

100 Zhong, J, Adams JB, Hector LG Jr., Molecular dynamics simulations of asperity shear in aluminum. Journal of Applied Physics. 2003, Vol. 94, 4306–4314, DOI: http://dx.doi.org/10.1063/1.1558966. – This article was selected for one of the outstanding posters in W5.3 Symposium, the MRS Spring Meeting, USA, 2002.

101 Hoover, WG. Computational Statistical Mechanics. Elsevier, Davis, USA, 1991.

102 Zhong, J, Shakiba R, Adams JB. Molecular dynamics simulation of severe adhesive wear on a rough aluminum substrate. Journal of Physics D: Applied Physics. 2013, Vol. 46, 055307, DOI: 10.1088/0022-3727/46/5/055307.

103 Harrison, JA, White CT, Colton RJ, Brenner DW. Molecular-dynamics simulations of atomic-scale friction of diamond surfaces. Physical Review B. 1992, Vol. 46, 9700–9708, DOI: http://dx.doi.org/10.1103/PhysRevB.46.9700.

104 Tuma, JJ. Handbook of Physical Calculation. McGraw-Hill, New York, USA, 1976.

105 Jiang, S, Yamaguchi S, Goddard WA III. Structures, Vibrations, and Force Fields of Dithiophosphate Wear Inhibitors from ab Initio Quantum Chemistry. Journal of Physical Chemistry. 1996, Vol. 100, 15760–15769, DOI: 10.1021/jp960649j.

106 Jiang, S, Yamaguchi S, Goddard WA III. The SAM Model for Wear Inhibitor Performance of Dithiophosphates on Iron Oxide. Journal of Physical Chemistry B. 1997, Vol. 101, 7702–7709, DOI: 10.1021/jp963835y.

107 Hector, LG Jr, Siegel DJ, Adams JB, Yu H. Investigation of vinyl phosphonic acid/hydroxylated α-Al_2O_3(0001) reaction enthalpies. Surface Science. 2001, Vol. 494, 1–20, DOI: 10.1016/S0039-6028(01)01387-5.

108 Zhong, J, Hector LG Jr, Adams JB. Dynamic decomposition of aliphatic molecules on Al(111) from ab initio molecular dynamics. Physical review B. 2009, Vol. 79, 125419, DOI: http://dx.doi.org/10.1103/PhysRevB.79.125419.

109 Leng, Y, Jiang S. Dynamic Simulations of Adhesion and Friction in Chemical Force Microscopy. Journal of the American Chemical Society. 2002, Vol. 124, 11764–11770, DOI: 10.1021/ja026274z.

110 Zhang, L, Leng Y, Jiang S. Tip-Based Hybrid Simulation Study of Frictional Properties of Self-Assembled Monolayers: Effects of Chain Length, Terminal Group, Scan Direction, and Scan Velocity. Langmuir. 2003, Vol. 19, 9742–9747, DOI: 10.1021/la034007g.

111 Peng, Q, Lu G. A comparative study of fracture in Al: Quantum mechanical vs. empirical atomistic description. Journal of the Mechanics and Physics of Solids. 2011, Vol. 59, 775–786, DOI: 10.1016/j.jmps.2011.01.008.

112 Peng, Q, Zhang X, Carter EA, Lu G. Quantum mechanical study of solid solution effects on dislocation nucleation during nanoindentation. Modelling and Simulation in Materials Science and Engineering. 2010, Vol. 18, 075003, DOI: 10.1088/0965-0393/18/7/075003.

113 Peng, Q, Zhang X, Lu G. Quantum mechanical simulations of nanoindentation of Al thin film. Computational Materials Science. 2010, Vol. 47, 769–774, DOI: 10.1016/j.commatsci.2009.11.002.

114 Peng, Q, Zhang X, Lu G. Quantum simulation of materials at micron scales and beyond. Physical Review B. 2008, Vol. 78, 054118, DOI: http://dx.doi.org/10.1103/PhysRevB.78.054118.

115 Němeček, J. Nanoindentation in Materials Science. Oliver Kurelic Press, Croatia, 2012.

116 Eng, JP, Newville M, Brown EG Jr., Waychunas AG. Structure of the hydrated α-Al_2O_3 (0001) surface. Science. 2000, Vol. 288, 1029–1033, DOI: 10.1126/science.288.5468.1029.

117 Hay, MT, Geib SJ. Tetrabutylammonium isobutyl silsesquioxane monochloroferrate (III). Acta Crystallographica Section E. 2007, Vol. 63, m445–m446, DOI: https://doi.org/10.1107/S1600536807000360.

118 Alder, BJ, Wainwright TE. Studies in Molecular Dynamics. I. General Method. Journal of Chemical Physics. 1959, Vol. 31, 459–466, DOI: http://dx.doi.org/10.1063/1.1730376.

119 Baskes, MI. Modified embedded-atom potentials for cubic materials and impurities. Physical Review B. 1992, Vol. 46, 2727–2742, DOI: http://dx.doi.org/10.1103/PhysRevB.46.2727.

26

Multiscale Model for Nanoindentation in Polymer and Polymer Nanocomposites

Rezwanur Rahman

Department of Petroleum and Geosystems Engineering, University of Texas at Austin, USA

26.1 Introduction

Polymeric materials have become very popular in numerous applications due to their promising mechanical, electrical and thermal properties along with lower weight. It is important to characterize the mechanical properties (i.e. stiffness, strength, fatigue life etc.) of different polymers. Several researchers have been studying such macroscale properties of polymers using various techniques [1–3]. In last few years nano-particles have been incorporated into different polymers in order to enhance its mechanical, electrical or/and thermal properties [4, 5]. The nano-particles could be carbon nanotubes (CNT) [6], graphenes (GNP) [7, 8], nanoclay (NC) [9] etc. The polymeric materials with embedded nano-reinforcements are called *polymer nanocomposites* (PNC).

Graphene is a 2D carbon-based nanomaterial which has drawn a significant attention from the scientific community from all over the world. It possesses almost all the characteristics of carbon nanotubes (CNTs), and it is comparatively affordable. Hence graphene is considered to be one of the most popular nano-reinforcing materials in a wide variety of applications. In the last few years, researchers have been investigating the mechanical, electrical and thermal properties of GNP reinforced nanocomposites. Among several types of polymer resins, thermoset epoxy (such as EPON 862) is known to be very compatible for modern lightweight structures for various automotive, aerospace or civil structural applications. *GNP-epoxy nanocomposites (GPNC)* have a great potential in developing structural components with cheaper manufacturing process [10]. Several researchers have been working on GPNC both theoretically [11–14] and experimentally [15, 16]. The nanoscale information such as GNP distribution in the polymer, GNP-epoxy load transfer mechanism, tendency of agglomeration of GNP sheets etc. are vital information in predicting mechanical properties both qualitatively and quantitatively. In that context classical molecular dynamics (MD) simulation has proven to be a robust tool since the experimental investigation becomes incredibly challenging in terms of dealing all the aforementioned information within few angstroms or nanometers. Despite several difficulties in order to characterize the PNCs at reasonably smaller length scale *nanoindentation* can be very useful. In last decade nanoindentation techniques have been significantly improved in order to characterize materials at nanoscale with force varying in the order of nanonewtons [17–20]. Nanoindentation

Applied Nanoindentation in Advanced Materials, First Edition. Edited by Atul Tiwari and Sridhar Natarajan.

can evaluate local bulk mechanical properties at different locations on the surface. It is evident that the variation in the result can be interpreted in terms of the presence of graphenes, voids, damages etc. The evaluated bulk properties at nanoscale allow us to validate the mechanical properties calculated by MD simulation. The nanoscale information is highly localized. The response in the polymer at small length scale under indentation can be linked with the macroscale through multiscale modeling.

Over the last decade multiscale modeling has become a very interesting topic within the computational materials science and mechanics community. The fundamental idea of multiscale modeling is built by connecting different length and time scales (e.g. coupling between atomistic model with a continuum model) in order to extract the information at atomistic scale while the far field boundary conditions are applied at the continuum scale. One way to link the length scales is done by developing concurrent multiscale model. In past decade several researchers have been working on developing concurrent multiscale modeling scheme by interpolating the degrees of freedom from FEA nodes and atoms, blending forces from different length scales [21–23] or applying deformation gradient through quasi-continuum models [24]. Besides concurrent methods, hierarchical schemes are also able to capture the response of materials at different length scales. The hierarchical multiscale model relies on passing information back and forth among different representative volume elements (RVE) at different length scales by homogenization method [25–28]. Most of the existing multiscale schemes focus on linking the classical continuum model with the nonlocal atomistic model. It is well understood that the locality in the classical continuum formulation starts to become incompatible at the atomistic length scale [29]. This introduces some challenges in multiscale modeling of complex heterogeneous materials (e.g. polymers, bio-molecules etc.). Due to the absence of any local order, the FE or meshfree nodes may not be tied up with the atoms in the handshake regions as it can be done for materials with high space group symmetry. Hence nonlocal continuum formulation (e.g. peridynamics) is introduced in multiscale modeling of complex materials [30] as *it can be used at lowest possible length scales without requiring any coarse-grained potentials.* Peridynamics uses the governing equations in classical continuum mechanics as integro-differential equations [31–33]. Hence the displacement of any PD particle is influenced by the particles in its neighborhood. Due to nonlocality it can be shown that peridynamics (PD) is analogous to upscaling of classical molecular dynamics (MD) formulation [34, 35]. Hence peridynamics based hierarchical multiscale modeling scheme (PFHMM) was proposed in order to pass information between PD RVE(s) and atomistic RVEs [30]. PFHMM possess both top-down and bottom-up information passing schemes. For amorphous polymers nonlocality at nanoscale exists within a wide range of length scales (e.g. few nanometers). The peridynamic nano-indentation model can be developed and linked with an atomistic model using PFHMM which is referred in this article as the *multiscale nanoindentation model.*

26.2 Modeling Scheme

Epoxy resins are an important class of compounds containing two or more epoxy groups which react with different curing agents consisting of chemical groups like amines and anhydrides. The result is cross-linked epoxy systems which exhibits a series of excellent

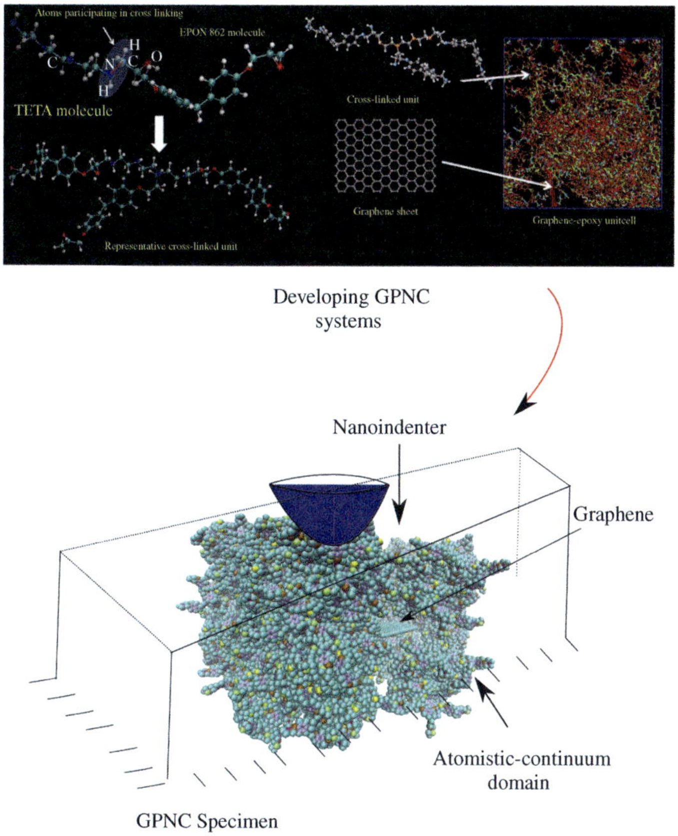

Figure 26.1 Multiscale modeling for nanoindentation in GPNC specimen.

performance, i.e. high modulus and fracture strength, low creep and high-temperature performance. The epoxy resin EPON 862 (diglycidyl ether of bisphenol F) can be mixed with graphene sheets in order to develop GPNC. The variation in weight concentration of graphene gives us the idea on the effect of graphene nano-reinforcement. As shown in Figure 26.1 nano-indentation captures much localized information. Due to the variation in graphene's spatial distribution the mechanical properties may vary within a range. Considering 1%, 3% and 5% weight concentrations the local distribution of graphenes in epoxy matrix are seen in the scanning electron microscopy (SEM) images (Figure 26.2). A nanoindenter with 50–100 nm diameter can determine the local stiffness of the GPNC at different locations.

26.2.1 Details of the MD Simulation

Molecular dynamics (MD) simulations were carried out on GPNC models with different weight concentrations of graphenes. As discussed earlier (Figure 26.1) the MD models can be thought of as representative unitcells for the GPNC. So the mechanical properties determined by the nano-indentation tests will be comparable with the MD results as the length scales are not significantly different. Considering length to width aspect ratio: AR

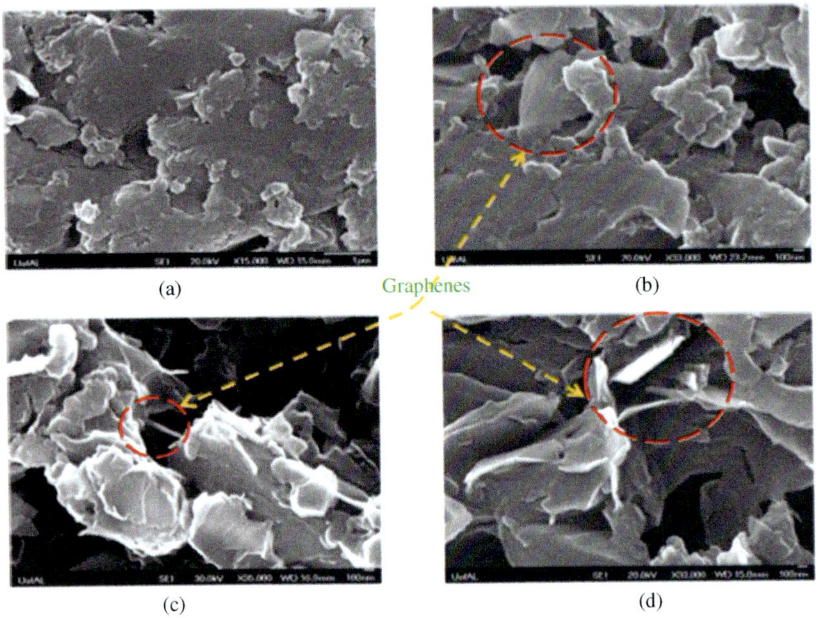

Figure 26.2 SEM images of graphene distribution in epoxy matrix: (a) Neat epoxy: EP (b) GPNC-1 with 1% graphene (c) GPNC-2 with 3% graphene (d) GPNC-3 with 5% graphene.

Table 26.1 GPNC systems for MD simulations.

GPNC systems	Number of atoms	Unitcell dimension (Å)	Number of graphenes
GPNC-1	42289	$L_a = L_b = L_c = 124$Å	1
GPNC-2	42698	$L_a = L_b = L_c = 125$Å	3
GPNC-3	42537	$L_a = L_b = L_c = 130$Å	5

LW = 10 the unitcells for the MD simulations are shown in Table 26.1. A PCFF force field was used in order to define inter- and intra-molecular interactions with non-bonded cutoff distance 8 Å. After energy minimization, MD equilibration was carried out under NVT for 1 000 000 steps at 300 K followed by 1 000 000 steps of NPT at same temperature with external pressure of −0.1 atm. After equilibrating the systems the unitcells were deformed uniaxially with strain rate 0.0001 in every 420 steps for 4200 steps. The length of each time step varied within 0.5–0.7 femtoseconds [35].

26.3 Nanoindentation Test

EPON 862 and EPIKURE 3234 were mixed with 1%, 3% and 5% graphene in order to produce GPNC samples. The graphene sheets were available in the form of powder. Each particle contains several graphene sheets stacked together. In order to disperse the graphene sheets in epoxy solution the graphene and epoxy mixture was sonicated

Table 26.2 Mechanical properties for GPNC.

GPNC systems	MD (GPa)	Nanoindentation test (GPa)	Tensile test (GPa)
GPNC-1	6.16	5.35 +/−1.37	2.17
GPNC-2	4.00	5.08 +/−1.71	2.50
GPNC-3	2.45	4.43 +/−1.03	2.07

approximately for an hour using VXC 750 tip sonicator followed by adding a curing agent. Once the GPNC panel was cured for over 24 hours, specimens for nanoindentation (length = 1.27 cm, width = 0.84 cm and thickness = 0.091 cm) test were prepared. Nanoindentation tests were done using Hysitron TI 900 Triboindenter with Berkovich tip (radius = 100 nm, angle = 142.3°). The quasi-static nanoindentation tests were carried out at sixteen different locations of the GPNC samples in order to see the variation in local properties. In each test the applied load varied from 0 N to 5000 N. Between a loading and unloading regime there is a 5 to 8 seconds *hold regime*. The maximum indentation depth in hold regime was 2190.94 nm for specimen with 5% graphenes and minimum depth was 1181.92 nm in 1% graphene-based specimen. The larger depth indicates evidence of voids in the polymer due to higher weight concentration of graphenes and their agglomeration. The Young's modulus, hardness etc., were calculated by the software incorporated with the nanoindenter. It is well known that the stiffness of the material is evaluated from the slop of the load vs. displacement curve in the unloading regime.

26.4 Theoretically and Experimentally Determined Result

Young's moduli calculated from the MD simulations were compared with the ones obtained from nano-indentation in Table 26.2. As mentioned earlier, the MD models are in the order of few angstroms. Hence, the length scale for nano-indentation is close to the length scales of the unitcells for MD models. Theoretically calculated results have a reasonable agreement with the experimental data. The nanoindentation results are represented with a variance as the indentation was taken at different locations of the samples. It is also very interesting to observe that the macroscale tensile test data are comparatively smaller than the nanoscale data. The combined role of micro-damages seems to affect the overall macroscale data. The atomistic scale models help us further analyze the polymer topology, its deformation while graphene based nano-reinforcement is used [35–37]. Furthermore, a reasonable MD model can be incorporated into the proposed nonlocal multiscale modeling scheme in order to link macro and nanoscales.

26.5 Multiscale of Complex Heterogeneous Materials

The primary goal of this work is to develop a unified multiscale computational scheme for complex materials by considering the nonlocal physics in the material. At smaller

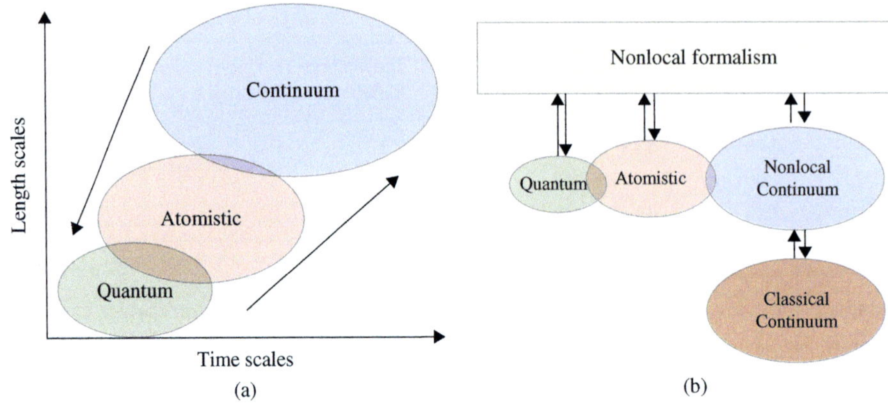

Figure 26.3 A generalized nonlocal multiscale modeling scheme.

length scales the nonlocality becomes more prominent. In common practice a typical multiscale scheme covering a wide range of time and length scales does not address the connection among the length scales through nonlocal interaction. Instead, different scales are modeled independently and information in passed to/from the next length scale. As shown in Figure 26.3, the smallest length scale considering electronic structure is modeled with quantum mechanics. The information is passed to classical molecular dynamics model which is strictly nonlocal. Eventually, MD model is coupled with continuum model (either local or nonlocal). But in actual case there is an overlap regime between the length scales (see Figure 26.3a) due to the nonlocality. In order to develop a seamless link among different length scales this nonlocal physics needs to be addresses. In this work the primary objective is to define a common multiscale platform based on nonlocal formalism. Within the umbrella of this unified scheme, different physical formulation corresponding to either electronic, atomistic or continuum scales will interact with each other through the nonlocal formalism. Due to strong resemblance between peridynamics and classical molecular dynamics, the unified formalism is based on peridynamic theory [31–33]. For electronic structure calculation the first principle based formulation may be linked with equivalent peridynamics model which will connect the classical molecular dynamics model. Between continuum and atomistic scales, extended PFHMM will be used. Eventually, the peridynamics model will be coupled with a local classical mechanics based model. The overall idea is shown in Figure 26.3b. So, the local classical continuum model is shown separately. In the current work the multiscale scheme is limited between classical MD and peridynamics.

26.5.1 Introduction to Peridynamics

The peridynamic theory of solid mechanics [31–33] has been proposed as an alternative to the classical theory, and is offered as a mathematically consistent technique for modeling solid bodies with continuous and discontinuous displacements as well as a method that unifies the mechanics of particles and continuum bodies through the utilization of long-range forces. The balance equation between rate of change of linear momentum and applied force on a deformable body develops the fundamental equation in classical

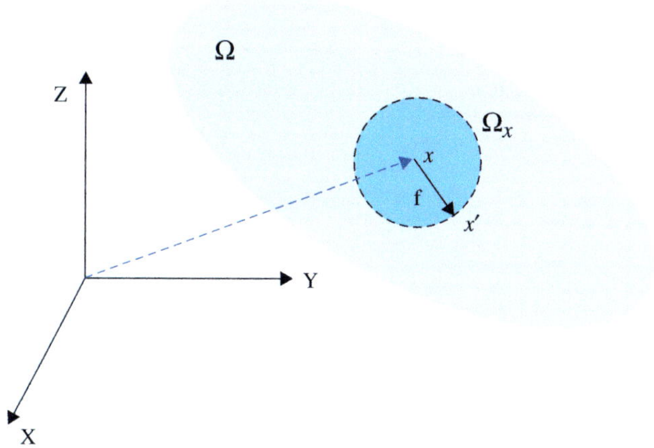

Figure 26.4 Schematic diagrams for peridynamics modeling.

continuum mechanics is written in Equation (26.1)

$$\rho(\mathbf{x})\ddot{\mathbf{u}}(\mathbf{x},t) = \nabla \cdot \boldsymbol{\sigma} + \mathbf{b}(\mathbf{x},t) \tag{26.1}$$

where, $\mathbf{x} \in \Omega$, t is time, ρ is the density, $\ddot{\mathbf{u}}$ is the acceleration vector, $\boldsymbol{\sigma}$ is the stress tensor and $\mathbf{b}$ is the body force. This differential equation is not well defined at the discontinuities. PD formulation introduces integral form of this equation in order to mitigate this issue by calculating the force density on each material point. The new form of equation of motion is

$$\rho(\mathbf{x})\ddot{\mathbf{u}}(\mathbf{x},t) = \int_{\Omega_x} \mathbf{f}(\eta,\xi)\,\mathrm{d}V_{x'} + \mathbf{b}(\mathbf{x},t) \tag{26.2}$$

where, deformable body Ω is represented with respect to an arbitrary frame of reference, $\mathbf{f}$ is the pairwise force applied on particle at x by a neighborhood particle at x' and Ω_x is a spherical region in the neighborhood region of x with radius δ. According to Figure 26.4, this region is referred to as 'horizon'. The relative position and displacement vectors are $\xi = x' - x$ and $\eta = \mathbf{u}(x',t) - \mathbf{u}(x,t)$ respectively. For a certain $\delta > 0$, $\mathbf{f}(\eta,\xi) = 0$ for all η when $\|\xi\| > \delta$. This integral equation can be written more elaborately in terms of bond between x and x' [33]. The peridynamic state-based integral equation and in equation (26.3):

$$\rho(\mathbf{x})\ddot{\mathbf{u}}(\mathbf{x},t) = \int_{\Omega_x} \{\underline{\mathbf{T}}[\mathbf{x},t]\langle\mathbf{x}' - \mathbf{x}\rangle - \underline{\mathbf{T}}[\mathbf{x}',t]\langle\mathbf{x} - \mathbf{x}'\rangle\}\,\mathrm{d}V_{x'} + \mathbf{b}(\mathbf{x},t) \tag{26.3}$$

The force vector state $\underline{\mathbf{T}}[\mathbf{x},t]$ is interpreted in terms of mapping the bond between x and x' to a force density per volume which has a cutoff range. This leads to bond based PD model by defining $\underline{\mathbf{T}}[\mathbf{x},t] = \underline{t}\underline{\mathbf{M}}$. Here, $\underline{t}$ is the force scalar state. This is related to the bond between two particles along the direction $\underline{\mathbf{M}}$:

$$\underline{\mathbf{M}}\langle\xi\rangle = \begin{cases} \dfrac{\xi + \eta}{\|\xi + \eta\|} & \forall\ \|\xi + \eta\| \neq 0, \\ 0\ \forall\ \text{Otherwise.} \end{cases} \tag{26.4}$$

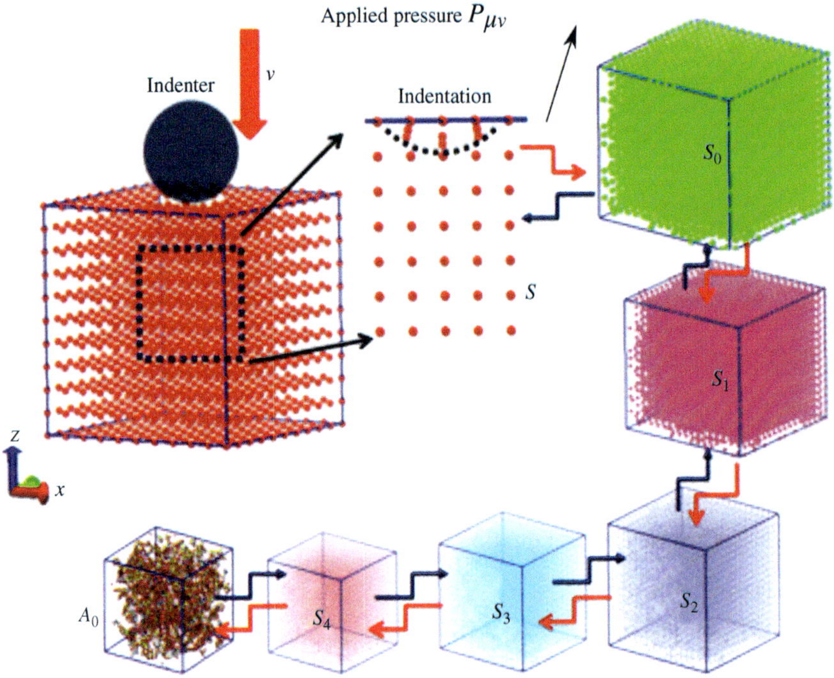

Figure 26.5 Peridynamics and multiscale modeling.

26.5.2 Nonlocal Multiscale Modeling using Peridynamics: Linking Macro- to Nano-scales

Atomistic scale models are nonlocal whereas classical continuum models are not. The nonlocal multiscale scheme eliminates this disparity. Peridynamics has a strong similarity with classical molecular dynamics. This feature is used to capture multi-scale deformation of complex materials. Hence, the peridynamic models at different length scales are subjected to deform based on a common external pressure tensor $P_{\mu v}$ generated at macro scale. As shown in Figure 26.5 the local pressure $P_{\mu v}$ developed at macro scale region is applied on the next peridynamic model S_0 in order to calculate the deformation (i.e. Isobaric-isothermal ensemble) [30, 39]. The average pressure developed in S_0 is applied on next peridynamic model in the hierarchy (S_1) as a volumetric constraint and so forth. In this manner the local region of interest at macro scale can be linked to the atomistic scale model (A_0). During micro to macro upscaling the deformation gradient for (i-1)th peridynamic model: $\mathbf{F}_{S_{(i-1)}}$ is $\mathbf{h}_{S_i} \cdot \mathbf{h}_{S_{i0}}$. Here, $\mathbf{h}_{S_i}$ and $\mathbf{h}_{S_{i0}}$ are the final and initial cell vectors, respectively. For S_4 in Figure 26.5, $\mathbf{F}_{S_4}$ is calculated based on the deformation of atomistic model A_0. Any ith peridynamic model is subjected to the external pressure $p_{\mu v}^{S_i} = \sum_{i \neq j} w_i p_{\mu v}^{S_j}$ preceded by applying the deformation gradient. Here, $p_{\mu v}^{S_i}$ is the weighted summation of internal pressures from the peridynamic models up to a certain length scale and w_i is the weight function. At small scale the deformation of complex materials is highly dependent on the length scale due to the nonlocality [40–42]. For an example, the nonlocality in amorphous polyethylene varies up to 30 nanometers [29].

The strength of nonlocality gradually decreases as length scale increases. So the peridynamic models are deformed in such a way that volumes of the unitcell is constrained by the external pressure obtained from neighboring peridynamics models at comparable length scales. At first, $\mathbf{F}_{S_2}$ will deform S_2 by applying uniform deformation. Afterwards, the external pressure will readjust the unitcell volume under isothermal-isobaric (NPT) condition. For an example, S_2 has the nearest neighboring peridynamic models S_3 and S_1. So the expected external pressure for S_2 would be the weighted summation of these two other models:

$$p_{\mu\nu}^{S_2} = w_3 \cdot {}^{(n+1)}p_{\mu\nu}^{S_3} + w_1 \cdot {}^{n}p_{\mu\nu}^{S_1} \tag{26.5}$$

It is important to observe that the pressure from S_1 is ${}^{n}p_{\mu\nu}^{S_1}$ and the pressure from S_3 is ${}^{(n+1)}p_{\mu\nu}^{S_3}$. The indices $(n + 1)$ and n refer to current and previous time steps, respectively. The pressure in S_1 was calculated during *macro to micro* downscaling or 'top-down PFHMM.' On the other hand, the pressure in S_3 is is recalculated during *micro to macro* upscaling or 'bottom-up PFHMM'. The objective of the bottom-up scheme is to update all the peridynamic models in response to the atomistic scale deformation. If there is bond breakage or damage at the atomistic scale, the calculated pressure will be influenced. Hence, all the peridynamic models in the hierarchy should be updated accordingly i.e. propagating the information from micro to macro through hierarchical length scales. In this hierarchical model the number of peridynamic models depend on the nonlocality among the length scales. If the length scales are strongly nonlocal, there will be larger number of peridynamic models than the case for weakly interacting length scales. By combining macro to micro and micro to macro length scales coupling allows us to develop a nonlocal multiscale modeling scheme based on peridynamics i.e. *PFHMM* [30, 38, 42]. In the typical PFHMM the atoms in A_0 are updated on-the-fly by a classical molecular dynamics scheme. The forcefield is chosen based on the material of interest. The atomistic model A_0 can be further downscaled to quantum scale. At this scale the *Ab-initio molecular dynamics* scheme is applied on updating the positions and velocities of the nuclei considering the force obtained from the electronic potential. The equation of motion in classical molecular dynamics can be upscaled to continuum scale peridynamic equation of motion [34, 35]. The primary objective is to incorporate quantum mechanical information in order to determine critical bond stretch criteria. Peridynamic bond failure criteria involves critical bond stretch. This term depends on parameters like length scale, chemistry of material (e.g. type of bond, topology) etc. In the proposed model the bond breakage is initiated based on the charge distribution between the atoms. As shown in Figure 26.6 the nanoscale continuum model (PD) is linked with atomistic model (MD) through PFHMM at the same length scale. This is further extended to a pair of nanoscale continuum and atomistic models where the continuum model is defined by *ab initio peridynamics (AIPD).*

26.6 Multiscale Modeling for Nanoindentation in Epoxy: EPON 862

PFHMM can be applied to model nanoindentation in GPNC. 3D nanoindentation was carried out within a domain $\tilde{S}$: 1000 Å × 1000Å × 1000Å. The radius of the spherical nanoindenter is 500 Å and the contact stiffness between the indenter and peridynamic

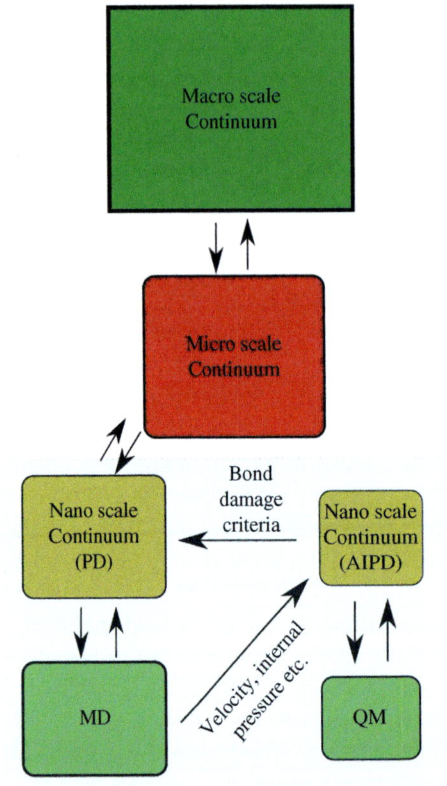

Figure 26.6 Embedding classical and *ab initio* MD in PFHMM.

(PD) surface is $K = 10\frac{\text{kcal/mole} \cdot \text{A}}{\text{A}^2}$. In this demonstration the goal is to capture the effect of nanoindentation within the polymer molecules. The length scale for the indenter is close to the one used in experiments. For 3500 NVE steps ($\Delta t = 0.001$ fs) the contact force was applied on the surface from the indenter. The average pressure $P_{\tilde{S}} = 651{:}974$ atm developed within $\tilde{S}$ is passed to the atomistic model for Epoxy: $A_0\ :\ 41.7\text{Å} \times 41.7\text{Å} \times 41.14\text{Å}$ through a set of peridynamic unitcells of decreasing sizes (Equations 26.6–26.10). Here, the term $\delta_i | i = 0 \cdot 4$ is the cutoff distance or *horizon* for each peridynamic meshfree particle in $S_i | i = 0 \cdot 4$. The cutoff distance for $A_0\ :\ \delta_{A^0} \approx 8\text{Å}$. The elastic modulus for the PD models is used approximately 3.42 GPa and Poission's ratio 0:3. Here, E_{Epoxy} was calculated using MD simulation. The predicted elastic modulus from nanoindentation test is 4.586 ± 1.47 GPa. The schematic diagram for these hierarchical models is shown in Figure 26.5.

Here,

$$S_0\ :\ L = 800\text{Å}, \delta_{S_0} = 144.6\text{Å} \tag{26.6}$$

$$S_1\ :\ L = 480\text{Å}, \delta_{S_0} = 86.75\text{Å} \tag{26.7}$$

$$S_2\ :\ L = 320\text{Å}, \delta_{S_0} = 57.83\text{Å} \tag{26.8}$$

$$S_3\ :\ L = 200\text{Å}, \delta_{S_0} = 36.14\text{Å} \tag{26.9}$$

$$S_4\ :\ L = 120\text{Å}, \delta_{S_0} = 21.07\text{Å} \tag{26.10}$$

According to *top-down PFHMM* the developed pressure in the set of PD models: S_0, S_1, S_2, S_3 and S_4 are 644.41 atm, 666.289 atm, 650.32 atm, 695.53 atm and 752.55 atm, respectively. Pressure developed in S_i is applied as a boundary condition for S_{i+1} considering S_i as a reservoir for S_{i+1}. This concept is very similar to the idea of *canonical ensemble* in classical statistical mechanics. Finally, the average pressure in the atomistic model is 933.66 atm based on the external pressure from S_4. From *micro to macro* upscaling (i.e. *bottom-up PFHMM*) the PD model S_{i+1} is uniformly deformed under the deformation gradient applied from S_i [30, 38, 42]. The unitcell dimensions are changed under no external pressure during bottom-up PFHMM as each model is allowed to strain recovery. This is analogous to the unloading of nanoindentation. After the bottom-up PFHMM, unitcell lengths for S_0, S_1, S_2, S_3 and S_4 are changed to 799.9947 Å, 479.99 Å, 320 Å, 200 Å and 120 Å, respectively. The potential energy in S changes from 9.716 Kcal/mole to 9.7142 Kcal/mole after the bottom-up PFHMM. The displacements of the PD particles or the atoms from the polymer in different PD models at the normalized locations $0.3 \leq \mathbf{r} \leq 0.7$ are shown in Figure 26.7. The displacement gradient decreases with decreasing length scale. This is expected as the macro model is down scaled to micro model. The displacement gradient at the atomistic model can be considered as a weighted summation of the displacement gradients from the PD models at different length scales. Due to periodicity the magnitude of the displacement field is mirrored symmetric with respect to $\mathbf{r} = 0.5$. Using PFHMM we can predict the atomistic scale response to the nanoindentation at larger length scales. The classical Cauchy-Born rule may not be applicable in this case as the polymers or PNCs have complex heterogeneous microstructures [43]. The advantage of the current model is that it is consistent with the scaling in material stiffness as the length scale gets smaller [38]. However, the

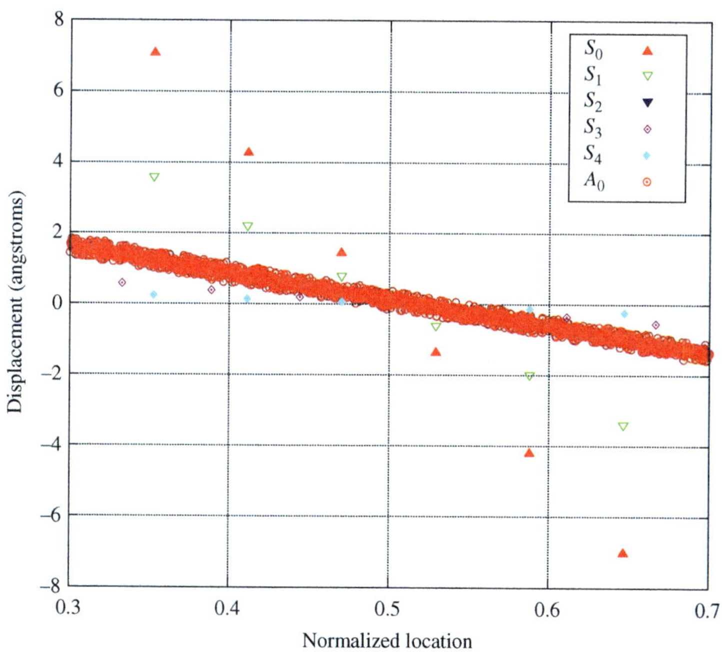

Figure 26.7 Displacements vs. normalized positions from different models in the hierarchy.

elastic properties in the PD models were not scaled with the length scales. Hence, the displacement gradient at A_0 is considered to be a weighted summation from the coarser models. This can be further investigated by incorporating the nonlocality among the length scales and microscale damage/voids in the PD models [42].

26.7 Unified Theory for Multiscale Modeling

The future goal is to develop a common umbrella for nonlocal multiscale modeling framework. According to Figure 26.6, the AIMD model A_0^{AIMD} resides at the smallest length scale communicating with AIPD model S_n^{AIPD}. The MD model A_0^{MD} is larger than the AIMD model because classical MD scheme is computationally less expensive than AIMD. So a reasonably larger MD model for complex heterogeneous models will be used to define the atomistic scale pressure tensor as well as mechanical properties. At every load step S_n^{AIPD} will be updated with the electron density n_e $(\mathbf{r})$. The depletion in the electron density will redefine the bond failure criteria for the peridynamics models. Unlike QM/MM method [44], MD and AIMD models are not directly linked. This will lead us to pass the average damage in the overall system rather than focusing on a specific set of bonds. On the other hand, the local classical continuum region of interest $\tilde{S}^{Local}$ will be linked to the nonlocal peridynamics model at the top of the hierarchy: $S_0{}^{Nonlocal}$. The involvement of fractional decay term in classical continuum model introduces nonlocality [45, 46]. This leads to formulating the classical continuum model with *fractional calculus* (equations 26.11–26.13). This theoretical extension will be implemented for modeling nanoindentation in the upcoming works.

$$\text{Local}: \rho\ddot{u} = (\lambda + \mu)\nabla^2 u + \mu\nabla \cdot \nabla u \tag{26.11}$$

$$\text{Fractional calculus}: \rho\ddot{u} = (\lambda + \mu)\nabla^2 u + \mu\nabla \cdot \nabla u - \frac{C_\alpha \alpha}{\Gamma(1-\alpha)} \int_B \frac{u(x) - u(x')}{(x - x')^{1+\alpha}} \mathrm{d}x \tag{26.12}$$

$$\text{Fractional calculus}: \rho\ddot{u} = \int_B [\underline{\mathbf{T}}[xt]\langle x' - x\rangle - \underline{\mathbf{T}}[x', t]\langle x - x'\rangle] \mathrm{d}V_{x'} \tag{26.13}$$

26.8 Conclusion

Bridging length scales of different types of materials has become a mainstream research topic among computational physicists, material scientists and mechanicians. The application of multiscale modeling is widely spread over different disciplines of science and engineering. The proposed nonlocal multiscale modeling scheme will be a significant addition to this fast growing interdisciplinary field. The peridynamics based multiscale modeling scheme will lead us to have a common platform to study the deformation and failure of complex materials at macroscale, mesoscale and nanoscale by passing information seamlessly back and forth. On the other hand, local and nonlocal continuum models at macroscale can be blended by superimposing the integro-differential equation of peridynamics and fractional governing equation of classical elasticity. This makes *PFHMM* very promising in capturing and analyzing damage, failure, deformation at nanoscale and linking it to the classical continuum mechanics at macroscale through intermediate peridynamic models. PFHMM was able to capture the atomistic scale deformation based on quasistatic nanoindentation on complex heterogeneous material like amorphous epoxy. The displacement field is consistent with the decrease in length scales towards atomistic scale. The similar apporach can be further implemented to capture information from electronic structure of this types of complex materials. In the context of the existing works in the literature the idea for PFHMM is novel. It is strongly understood that this multiscale modeling scheme will be used in wide range of applications for different types of materials in future.

References

1 Nielsen LE. Mechanical properties of polymers. Van Nostrand Reinhold Press; 1962.
2 Landel RF, Nielsen LE. Mechanical properties of polymers and composites. CRC Press; 1993.
3 Swallowe GM. Mechanical properties and testing of polymers: an AZ reference. Vol. 3. Springer; 1999.
4 Koo JH. Polymer nanocomposites. McGraw-Hill Professional; 2006.
5 Krishnamoorti R, Vaia RA. Polymer nanocomposites: synthesis, characterization, and modeling. Washington, DC: American Chemical Society vol. 804; 2002.
6 Moniruzzaman M, Winey KI. Polymer nanocomposites containing carbon nanotubes. Macromolecules 39(16): 5194–5205; 2006.
7 Kim H, Abdala AA, Macosko CW. Graphene/polymer nanocomposites. Macromolecules 43(16): 6515–6530; 2010.
8 Potts JR, Dreyer DR, Bielawski CW, Ruoff RS. Graphene-based polymer nanocomposites. Polymer 52(1): 5–25; 2011.
9 Patel HA, Somani RS, Bajaj HC, Jasra RV. Nanoclays for polymer nanocomposites, paints, inks, greases and cosmetics formulations, drug delivery vehicle and waste water treatment. Bulletin of Materials Science 29(2): 133–145; 2006.
10 Stankovich S, Dikin DA, Dommett GH *et al.* Graphene-based composite materials. Nature 442(7100): 282–286; 2006.
11 Wu C, Xu W. Atomistic molecular modelling of crosslinked epoxy resin. Polymer 47(16): 6004–6009; 2006.
12 Fan HB, Yuen MM. Material properties of the cross-linked epoxy resin compound predicted by molecular dynamics simulation. Polymer 48(7): 2174–2178; 2007.
13 Zhu R, Pan E, Roy A. Molecular dynamics study of the reinforced EPON 862 composites. Materials Science and Engineering A 447(1): 51–57; 2007.
14 Ji XY, Cao YP, Feng XQ. Micromechanics prediction of the effective elastic moduli of graphene sheet-reinforced polymer nanocomposites. Modelling and Simulation in Materials Science and Engineering 18(4): 045 005; 2010.
15 Yasmin A, Daniel IM. Mechanical and thermal properties of graphite platelet/epoxy composites. Polymer 45(24): 8211–8219; 2004.
16 Rafiee MA, Rafiee J, Wang Z, *et al.* Enhanced mechanical properties of nanocomposites at low graphene content. American Chemical Society Nano 3(12): 3884–3890; 2009.

17 Bhushan B. Handbook of micro/nanotribology. CRC Press; 1998.
18 Tabor D. Indentation hardness: fifty years on a personal view. Philosophical Magazine A 74(5): 1207–1212; 1996.
19 Pethicai J, Hutchings R, Oliver W. Hardness measurement at penetration depths as small as 20 nm. Philosophical Magazine A 48(4): 593–606; 1983.
20 Li X, Bhushan B. A review of nanoindentation continuous stiffness measurement technique and its applications. Materials Characterization 48(1): 11–36; 2002.
21 Gumbsch P, Beltz G. On the continuum versus atomistic description of dislocation nucleation and cleavage in nickel. Modeling and Simulation in Materials Science and Engineering 3: 597–613; 1995.
22 Broughton J, Abraham F, Kaxiras E. Concurrent coupling of length scales: methodology and application. Physical Review B 60: 2391–2403; 1999.
23 Xiao S, Belytschko T. A bridging domain method for coupling continua with molecular dynamics. Computer Methods in Applied Mechanics and Engineering 193: 1645–1669: 2004.
24 Miller R, Tadmor E. The quasi-continuum method: overview, applications and current directions. Journal of Computer Aided Materials Design 9: 203–239; 2002.
25 Fish J, Shek K. Multiscale analysis of composite materials and structures. Composites Science and Technology 60: 2547–2556; 2000.
26 Yuan Z, Fish J. Hierarchical model reduction at multiple scales. International Journal for Numerical Methods in Engineering 79: 314–339; 2009.
27 Zohdi TI, Oden JT, Rodin GJ. Hierarchical modeling of heterogeneous bodies. Computational Methods in Applied Mechanical Engineering 138: 273–298; 1996.
28 Oden JT, Vemaganti K, Moes N. Hierarchical modeling of heterogeneous solids. Computational Methods in Applied Mechanical Engineering 172: 3–25; 1999.
29 Maranganti R, Sharma P. Length scales at which classical elasticity breaks down for various materials. Applied Physics Letters 98: 1–14; 2007.
30 Rahman R, Haque A. A peridynamics formulation based hierarchical multiscale modeling approach between continuum scale and atomistic scale. International Journal of Computational Materials Science and Engineering 01: 1–19; 2013.
31 Silling S. Reformulation of elasticity theory for discontinuities and long-range forces. Journal of Mechanics and Physics of Solids 48: 175–209; 2000.
32 Silling S, Zimmermann M, Abeyarante R. Deformation of peridynamic bar. Journal of Elasticity 73: 173–190; 2003.
33 Silling SA, Epton M, Weckner O, Xu J, Askari E. Peridynamic states and constitutive modeling. Journal of Elasticity 88: 151–184; 2007.
34 Seleson P, Parks ML, Gunzburger M, Lehoucq RB. Peridynamics as an upscaling of molecular dynamics. Multiscale Modeling and Simulation 8: 204–227; 2009.
35 Seleson P, Parks ML, Gunzburger M. Peridynamic state-based models and the embedded-atom model. Communications in Computational Physics 159: 179–205; 2014.
36 Rahman R. The role of graphene in enhancing the stiffness of polymeric material: a molecular modeling approach. Journal of Applied Physics 113(24): 243 503; 2013.
37 Rahman R, Foster J. Deformation mechanism of graphene in amorphous polyethylene: a molecular dynamics based study. Computational Materials Science 87: 232–240; 2014.

38 Rahman R, Haque A. Molecular modeling of crosslinked graphene–epoxy nanocomposites for characterization of elastic constants and interfacial properties. Composites Part B: Engineering 54: 353–364; 2013.
39 Rahman R, Foster JT, Haque A. A multiscale modeling scheme based on peridynamic theory. International Journal of Multiscale Computational Engineering 12: 223–248; 2014.
40 Nikolov S, Han C, Raabe D. On the origin of size effects in small-strain elasticity of solid polymers. International Journal of Solids and Structures 44: 1582–1592; 2007.
41 Lam D, Yang F, Chong A, Wanga J, Tong P. Experiments and theory in strain gradient elasticity. Journal of the Mechanics and Physics of Solids 51: 1477–1508; 2003.
42 Challamel N, Wang C. The small length scale effect for a non-local cantilever beam: a paradox solved. Nanotechnology 19: 1–7; 2008.
43 Rahman R, Foster J. Bridging the length scales through nonlocal hierarchical multiscale modeling scheme. Computational Materials Science 92: 401–415: 2014.
44 Shan W, Nackenhorst U. Applying Cauchy-Born rule for converting atomistic model to continuum model. Proceedings in Applied Mathematics and Mechanics 10(1): 425–426; 2010.
45 Warshel A, Levitt M. Theoretical studies of enzymic reactions: dielectric, electrostatic and steric stabilization of the carbonium ion in the reaction of lysozyme. Journal of Molecular Biology 103: 227–249; 1976.
46 Paola MD, Zingales M. Fractional differential calculus for 3D mechanically based non-local elasticity. Journal of Multiscale Computational Engineering 9: 579–597; 2011.

Index

Page references followed by *f* indicate an illustrated figure; followed by *t* indicate a table

Applied Nanoindentation in Advanced Materials, First Edition. Edited by Atul Tiwari and Sridhar Natarajan.

e

q

r

t

u

Printed and bound by CPI Group (UK) Ltd, Croydon, CR0 4YY
16/09/2026
14963474-0005